PRINCIPLES
OF
ELECTRONIC
PACKAGING

COVER

The photo micrograph on the cover is a 45-degree cross section of the High Performance Circuit Board used in IBM's high-end data processing systems. This particular board, measuring 600 by 700 mm by 4.59 mm thick, would carry nine Thermal Conduction Modules (TCMs) containing the active integrated circuit chips that make up the computer's logic and control components.

Buried on six planes within the board is nearly a kilometer of signal wiring only 0.081 mm wide, providing electrical pathways between the TCMs, cable connectors and terminating resistors. Over 25,000 holes drilled through the board and plated internally with copper interconnect the six signal and twelve power distribution wiring layers. Adjacent signal planes are also interconnected with interstitial vias (two are visible in the photo) during board fabrication. In addition, the through-holes provide mounting for the thousands of tiny bifurcated spring connectors used to connect the TCMs and cables to the board.

Photo by Michael D. Dinardo, Geoffrey R. Mariner and John F. Turner, IBM Corp., Endicott, NY.

PRINCIPLES OF ELECTRONIC PACKAGING

Edited By

Donald P. Seraphim

IBM Corporation

Ronald C. Lasky

IBM Corporation

Che-Yu Li

Cornell University

McGraw-Hill, Inc.

New York St. Louis San Francisco Auckland Bogotá
Caracas Lisbon London Madrid Mexico City Milan
Montreal New Delhi San Juan Singapore
Sydney Tokyo Toronto

This book was set in Times Roman by Publication Services.
The editors were Alar E. Elken and John M. Morriss;
the production supervisor was Friederich W. Schulte.
The cover was designed by Albert M. Cetta.
Project supervision was done by Publication Services.

PRINCIPLES OF ELECTRONIC PACKAGING

4 5 6 7 8 9 10 11 12 13 14 BKMBKM 9 9 8 7 6 5 4

ISBN 0-07-056306-3

Library of Congress Cataloging-in-Publication Data

Seraphim, Donald P.
    Principles of electronic packaging.

    (McGraw-Hill series in electrical engineering)
    Includes bibliographies and index.
    1. Electronic packaging. I. Lasky, Ronald.
II. Li, Che-Yu. III. Title. IV. Series.
TK7870.S417 1989   621.381'73   88-13013
ISBN 0-07-056306-3

ABOUT THE EDITORS

Dr. Donald P. Seraphim, IBM Fellow and consultant, holds bachelor and masters degrees in applied science from the University of British Columbia, and a Dr. of Engineering in metallurgy from Yale University. As an IBM research scientist and manager in superconductivity and semiconductors from 1957 to 1965, his group built the first high mobility N channel metal oxide field effect integrated circuits in IBM. He led the design and development of IBM's first bipolar integrated circuits and chip packaging between 1965 and 1969, and was responsible for the IBM General Technology Division's printed circuits programs from 1969 until 1977.

Dr. Seraphim was a member of IBM's Corporate Technology Committee between 1977 and 1979, and was elected an IBM Fellow in 1981. He served as manager of materials science and engineering in IBM's Systems Technology Division from 1981 to 1986. He has written more than 50 articles and has over 30 patents and patent publications.

Dr. Ronald C. Lasky, P.E, has been with IBM since 1967. During this time, he has had numerous assignments in materials engineering in both technical and managerial capacities. He is currently an advisory engineer. He received a B.S. from Cornell University in engineering physics in 1970, an M.S. in applied mathematics from the State University of New York in 1974, and a Ph.D. in materials science from Cornell University in 1986. Dr. Lasky is a member of the Bohmische Physical Society and is a professional engineer.

Dr. Che-Yu Li, Professor of materials science and engineering at Cornell University and a member of the Cornell faculty since 1962, has been engaged in electronic packaging research during the past ten years. He is a specialist on solder interconnections, metallization and TAB, and has made important contributions to the understanding of mechanical reliability and micromechanical testing of components down to submicron sizes. He is presently coordinating electronic packaging research sponsored by the Semiconductor Research Corporation and the U.S. Army Research Office at Durham, N.C., and the electronic packaging course at Cornell University.

CONTENTS

2 Interaction of Materials and Processes in Electronic Packaging Design

3 Electrical Design Concepts in Electronic Packaging

4   Printed Circuit Board Signal Line Electrical Design   104

5   Thermal Management In Electronic Packaging   127

6   Mechanical Design   159

11 Polymers and Polymer-Based Composites for Electronic Applications

12 Lithography In Electronic Packaging

13   Vacuum Metallizing in Packaging

17 Plated Through-Hole Technology

18 Etching of Metals by Wet Processes in Electronic Packaging

21 Bulk Analysis in Electronic Packaging 649

22 Scanning Electron Microscopy in Electronic Packaging 669

26   Case II Diffusion

27   Chemical Bonding and Stress Relaxation at Metal-Polymer Interfaces

FOREWORD

It is to state the obvious that the integrated circuit has virtually revolutionized modern technology. What is not so obvious is the degree to which those tiny chips of silicon have revolutionized the field of circuit packaging. Driven first by hundreds, then thousands and tens of thousands of circuits per chip, the demand for even greater density at the higher levels of packaging is seemingly insatiable. And every leap forward in density at the chip level creates a new set of problems in materials, development, and manufacturing at the packaging level.

Thus a need was created, both in industry and in colleges and universities, for a complete, comprehensive book on the elements of electronic circuit packaging. This book covers virtually every aspect of the subject, from the brief yet dynamic history of circuit packaging, through related materials science and development considerations, culminating with a look into the circuit-producing factory of the future.

This book was written to provide a nucleus for a program to prepare a future generation of scientists and engineers who will continue to further the science of circuit packaging. Armed with the knowledge of their predecessors, the curiosity and creativity of the mind, and the drive of the human spirit to stretch technology to the limits, there is every reason to believe this goal will be achieved.

Patrick A. Toole

PREFACE

A special education program was developed in the early 1980s in conjunction with Cornell University to provide IBM manufacturing and development engineers with courses in scientific disciplines applicable to IBM's printed circuit manufacturing technologies. Related short courses were also conducted by leading experts from several universities. The topics in this book were covered by those courses.

Collaboration in research topics of mutual interest between IBM, Cornell University, and many other universities further identified the long term need for such an education program. Professors at Cornell shared the idea, as they were active in a variety of projects with the SRC Packaging Science Center, NSF Center in Materials Science, and an Industry-University Center for Ceramics. As a result, Cornell and IBM developed in 1984, a course on the Principles of Electronic Packaging. The course consisted of weekly lectures given by university professors and IBM engineers and scientists. It soon became evident that a comprehensive textbook on electronic packaging would be of considerable value to students, engineers, and scientists in the field.

This book was written to meet that need. It is intended primarily for students with at least a basic chemistry, physics, and mathematics background typically found in a junior-year engineering program. However, a significant amount of the material is also applicable to graduate study.

The book has extra benefits as well. In addition to emphasizing the fundamental principles in electronic packaging, and the application of those principles, the book includes current engineering practices related to each topic . . . thus providing additional value to its readers.

ACKNOWLEDGMENTS

Completing a book of this size, scope, and complexity required the talents and skills of many people. Special thanks go to the 75 authors who wrote specific chapters. As experts in their fields, they demonstrated a unique combination of dedication and depth of knowledge during the book's preparation.

Technical editing of the initial drafts and final manuscript fell to seven people in IBM: Jutta Boedecker-Frey, Lillian Brown, Lorraine Coughlin, Nancy Musgrove, Pat Niehoff, Steve Savacool, and Dave Yetter. A special note of appreciation goes to Ms. Musgrove and Mrs. Coughlin, who coordinated the book project from start to finish. We also thank the many others who handled various aspects of manuscript preparation.

Bill Chen, Tom Gall, Bob Vasilow, and Jim Wilcox provided consultation, technical assistance, and helpful comments that added significantly to this book. John Lo provided a total technical review of the book and special help with several chapters. Thanks should also go to Art Ruoff, who provided support throughout the project's duration.

We are further indebted to IBM management, particularly Bob Corrigan and Patrick Toole, for providing the environment and encouragement that ensured the book's completion.

McGraw-Hill and the authors would like to thank the following reviewers for their many helpful comments and suggestions: LaRue R. Durn, Lehigh University; Aiche Elshabini-Riad, Virginia Polytechnic Institute; F. D. Rosi, University of Virginia; and C. A. Steidel, Intel Corporation.

R. C. Lasky
C. Y. Li
D. P. Seraphim

PRINCIPLES
OF
ELECTRONIC
PACKAGING

INTRODUCTION

Enormous strides have been made in microelectronics technology over the past several decades. Computations done on mechanical calculators costing $1,000 each 20 years ago can now be made on electronic calculators worth $4.95. Weather forecasts that require astronomical numbers of computations became a reality only after the advent of the high-speed electronic computer. A major step in this revolution was the development of silicon chip-based integrated circuits.

With continued demand for better performance, the electronics industry has been forcing more and more circuitry onto a silicon chip, but has not put equivalent emphasis on the structures that carry the chips. As the circuit density increases on the chip, the speed of functions it performs increases; however, a chip is not an isolated island. To perform its functions it must communicate with other chips in a system. When computers were rather slow and the circuit density was lower, designers were concerned mainly with how many electrical connections were needed between the chips. But as the density increased, several things happened. The number of electrical connections on a given chip increased from ten or so to millions, and the time it takes for an electrical signal to travel from one chip to another has become an important consideration [1]. Even more important is retention of signal integrity [2].

Other problems have arisen as well. Power requirements of the chip have increased dramatically, and providing that power at a precisely controlled voltage is a major challenge [3]. Because of the large amount of power needed, heat generated by a chip can reach several watts. Dissipating the heat to keep the chip running at its design temperature is an important requirement. And with the

1

widespread use of microelectronic systems, many of them will see service in rather hostile environments such as an automobile [4]. Protecting the silicon chip with its many delicate structures presents additional challenges [5].

The package that holds and protects the chip, provides electrical connections, and removes the excess heat has become increasingly important [6] and is recognized by the industry to be the gating factor in achieving high performance and meeting cost objectives in microelectronics and computer systems in the 1990s and beyond [1,7].

This book is designed to provide the necessary scientific and engineering background to address these electronic packaging issues. It focuses on design, materials, and manufacturing.

The first part of the book is oriented toward architecture and design. In these chapters we attempt to relate some of the design considerations to those that determine or influence the materials and manufacturing processes. Electrical, thermal, and mechanical considerations are treated in these chapters. Reliability and testing is also included. The second part of the book provides overviews of the variety of materials sciences and analytical skills [8] that are applied to electronic packaging. Authors then focus on analytical techniques proven to be useful in solving many of the problems associated with developing and manufacturing packaged circuits. In the next part, engineering aspects of manufacturing are discussed, as well as trends in automation and an overview of financial aspects of development and manufacturing. The book concludes with looks back into the past and ahead into the future of electronic packaging.

FUNCTIONS AND REQUIREMENTS OF AN ELECTRONIC PACKAGE

The design of an electronic package first requires definition of the various functions and requirements or the architectural considerations (Chapter 1), of the package. Consider a personal computer as an example of effective packaging. Today's personal computer has the same or more computing power than a large mainframe computer of the 1960s, yet its cost is a small fraction of that of the mainframe. The circuitry is much more complex, yet it is packaged in a much smaller space.

It all starts with the silicon chip, a small slice of silicon containing electronic circuits that perform basic operations such as addition and subtraction [5]. Transistors on the chips are called *devices* and the connection of a number of devices to perform a function is called a *circuit*. During the processing steps, a single chip is but a portion of a large silicon wafer. Even though the processing steps necessary to make a chip are complicated and expensive, multiple chips made at one time from one wafer reduce the cost of the chip to a very low value.

The circuits in the chip are built using processing steps such as oxide formation, etching, diffusion doping, and thin-film metal deposition. A few years ago a chip might have 10 or 20 circuits; in 1988 it had up to 30,000 on the same size chip. A circuit typically contains about five transistors and resistors. Chips vary greatly in size, but a typical one is a centimeter square.

This book is not concerned with the design and manufacturing of a silicon chip; it focuses on how a chip is packaged efficiently and reliably. A typical microelectronic package is designed to provide the following structures and functions:

Connections for signal lines leading onto and off of the silicon chip

Connections for providing electrical current that powers the circuits on the chip

A means of removing the heat generated by the circuits

A structure to support and protect the chip

A wiring structure for signal and power interconnections within a system and for input/output

The input/output (I/O) interconnections for a silicon chip provide an example of how important mechanical requirements can be. It may appear that interconnections could be made by simply soldering the chip to a substrate (the wiring structure on which the chip resides). However, the thermal coefficient of expansion of the chip and the substrate are usually different, and the resulting temperature change can produce large thermal stresses. The temperature change can be caused by the outside environment, or can result from the action of the circuits themselves. To prevent stresses from causing mechanical failures, chips and substrates can be connected with compliant members. For example, if thin wires are used, temperature-induced relative motions between the chip and substrate will be absorbed in the wires with very little stress being produced. Such arrangements are called stress-relief structures. When thin wires are used, the method of interconnection attachment is called *wire bonding*.

The input/output requirements vary significantly on the system of interest. For example, the I/Os for a hand-held calculator are the keyboard and the display. In a large computer system the I/Os are the tape drives, disk drives, printers, display terminals, consoles, etc. The design of the keyboard is not a simple task. For a keyboard to have good "human factors," the keys should move sufficiently so that the response to the touch can be felt. Current human factors studies also indicate that there should be an audible noise and tactile feel when contact is made. These requirements place stringent demands on the materials, since there will be fairly large movements that require materials with good fatigue resistance.

The term *I/O* usually refers to signal lines leaving or entering a system, whether the system is a hand-held calculator or a mainframe computer. In addition, the term *I/O* is sometimes used to refer to the signals entering or leaving a level of packaging. The chip has signal lines that can be called I/Os even though the interconnections are only going to the next level of packaging—typically the module, which is the chip carrier. This is identified as the first level of packaging.

In addition to the number of I/Os and the human factors of a design, considerable effort is usually placed on satisfying environmental requirements. For example, the system may be exposed to chemicals in the air; these must be removed, or the parts must be designed to tolerate them. Other environmental factors include ambient termperature and humidity. In addition to considering the environment in which the system operates, one must consider the environment during shipping. The system may be run in an air-conditioned computer center where the ambient temperature only varies a few degrees; however, it must also tolerate sitting in a truck in extremely cold or hot weather during shipping.

The performance requirements of large computer systems, as discussed in Chapter 1, demand special considerations. Large systems have many of the same requirements as a desktop personal computer, including an ability to operate for many years with very few problems. However, in order to obtain high speed, larger electrical currents must be used, since the circuit speed depends on the current density as well as several other factors. Also, since the circuits on a chip are very fast today, the time of flight of a signal from one circuit to another is becoming a significant part of the overall delay. If a circuit is built that can switch in one nanosecond (10^{-9} second), it should not take many nanoseconds to transport the signal from one part of the computer to another. Signals propagate at a speed of 30 cm per nanosecond in a vacuum and at about half this speed in a typical electronic package. As circuit speeds go below the nanosecond range as they do today, it becomes very important for the packaging engineer to work to decrease this time of flight by using insulating materials with lower dielectric constants and by packaging the chips closer together. Scientists and engineers with different backgrounds have to work together effectively to meet these performance challenges. Otherwise, the performance bottleneck will remain in the package.

THE TYPICAL ELECTRONIC PACKAGE

This book is concerned primarily with packaging silicon-based integrated circuits in computer systems. However, the principles involved are equally relevant to other applications. The usual packaging hierarchy involved can be illustrated by using the card-on-board (COB) structure for a computer system. Figure I-1 shows a typical COB structure for packaging. Such a structure is three-dimensional, allowing interconnections in all three directions. The chip, containing many circuits, is attached to a ceramic substrate (chip carrier). The substrate is, in turn, attached to the card that is plugged into the board (back panel); both the card and the board contain wiring structures. The levels shown in the figure are referred to as the packaging hierarchy. The first-level package is the chip on the substrate (sometimes called a module), the second level is the substrate on the card, and the third level is the card on the board. At each level of this hierarchy, connections must be made in order to deliver power (enough electrical current at a specified voltage) and signals to the lower level. Wire size used in the interconnections increases as one moves through this hierarchy from the chip

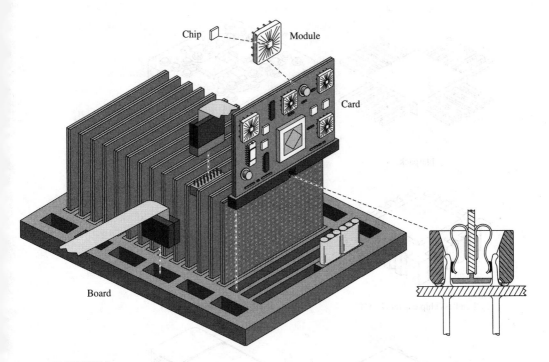

FIGURE I-1
A packaging hierarchy card on board (COB). Frequently used chips are bonded to a chip carrier. Chip carriers are soldered into cards. Cards are plugged into conectors in a board, and cables interconnect between boards.

to the I/O components. On the chip, the devices and the circuit lines are very close together but, at the second level of package and I/O cabling portions of a system, the wires are much larger. While the size of devices on the chip are about a micrometer and the line spacing a few micrometers, the wires leading to an external device might be spaced at only ten per inch—a scaling factor of about three orders of magnitude.

Figure I-2 shows some of the first-level packages in use today. The dual in-line package (DIP, mostly plastic and some ceramic) has been the mainstay in packaging for many years and will remain in high-volume production for many years to come. However, as the number of terminals increases on the chip, it has been necessary to move to different configurations, such as the pin grid array (PGA). The advantage of the PGA is that the I/O pins are spread over the area of the substrate making it possible to have far more I/Os than in a DIP. Peripherally leaded devices such as the flatpack, leadless chip carrier (LCC), and the plastic-leaded chip carrier (PLCC) are used in applications that require more I/Os than a DIP but do not need a PGA. Also shown is the small-outline integrated circuit (SOIC), and the PLCC and Flatpack which uses surface mount technology (SMT) to attach to the card. All of these terms are used generally in the industry and in this book.

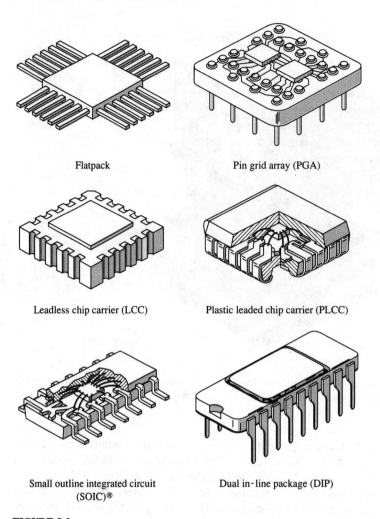

Flatpack

Pin grid array (PGA)

Leadless chip carrier (LCC)

Plastic leaded chip carrier (PLCC)

Small outline integrated circuit
(SOIC)®

Dual in-line package (DIP)

FIGURE I-2
Chip carriers or modules (first level packages). These carriers, diagrams from ref 4, hold semiconductor chips and supply the connections or leads between the chip and printed circuit composites.

Chips are attached to the chip carrier in several ways. The most popular way is wire bonding. In wire bonding, the back (nondevice side) is attached to the substrate and the electrical connections are made by attaching very small wires from the device side of the chip to the appropriate points on the substrate. The wires are attached to the chip by thermal compression bonding. Another method of attachment uses small solder balls to both physically attach the chip and make the required electrical connections. This is sometimes called face down bonding, flip-chip bonding, or controlled-collapse chip connections (C-4).

In addition to the trend toward packages with more I/Os is the trend toward surface mounting of modules instead of using modules with pins soldered into holes. An advantage of surface mounting is that the leads are attached to pads on the surface rather than inserted in relatively large drilled holes in the printed circuit card or board. This provides more space for printed circuit wiring in the inner layers of a card or board.

As the scale of integration increases (which means the number of circuits on the chip increases), the number of interconnections to the next level increases greatly. Figure I-3 shows this increase in the number of I/Os from 1960 to 1985. As the number of I/Os increases, it becomes more and more difficult to make the connections reliably. Improvement in the cost of an electronic system depends on the growth of very-large-scale integration (VLSI), which is occurring at the chip level. To take advantage of this rapid advancement in VLSI, it is necessary for the packaging to improve at a similar pace [9]. In a sense,

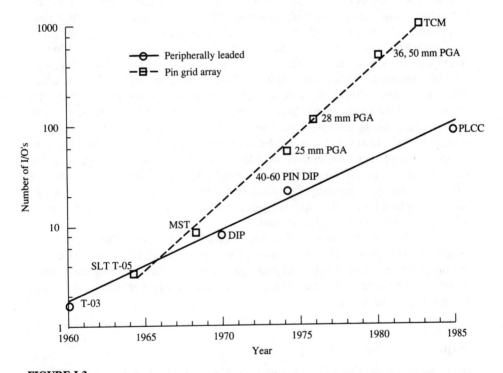

FIGURE I-3
Observations on function growth. The number of I/Os on electronic components have grown by a few orders of magnitude in the last two decades. The pin grid arrays have grown faster since it is possible to place more connections in a given area using an area array than by connecting along the periphery of a component. The 36 mm and 50 mm PGA hold 9 chips while the TCM holds 120 chips. The high I/O capability allows the package to hold more chips than are usually packaged on a card.

those interconnections, previously made by cables and by second-level packaging interconnections, costing from pennies to dollars per interconnection, are now found on the chip at a fraction of a penny in cost. In fact, the price/performance of computing power has decreased by many orders of magnitude since the advent of the digital electronic computer.

Chapter 30 of this book is devoted to the history of electronic packaging and reviews this dramatic increase in performance over the last few decades.

DESIGN

The design of an electronic system is the first step in the development and manufacturing of a package. The system designer must begin by determining the functional specifications. Based on these functional specifications an architecture is devised that will perform the function in the most efficient manner. Many tradeoffs are made during this step. For example, one might be able to meet the performance requirements by making a wide data path (meaning many logic lines are switched at one time) but, by so doing, the target cost of the system might be exceeded because of the associated wiring structure costs. The overall logical structure of the system is finalized based on this type of tradeoff. Next, the logic must be partitioned to identify what functional elements will reside on which chip, which chips will reside on which card, etc. Then it is possible to enumerate the number of interconnections between the functional elements. Many tradeoffs are also made during this process. The designer wants circuits close together to shorten delays. However, there is a limit to how many circuits can be placed on a chip. Some of these limitations are economical and some are physical. If too many circuits are placed on a chip, it may not be possible to manufacture the chip without an excessive number of rejected parts. On the other hand, if not enough circuits are placed on each chip, the system package may not be cost-competitive or its performance may not be adequate.

Once the chip and associated wiring structures have been defined, the development phase begins, including a study of which materials and processes can be best utilized to produce the desired package. Also, the manufacturing capabilities are reviewed to determine what new processes and design compromises, if any, will need to be developed.

Chip

Most electronic circuits today are based on devices created on silicon. Single crystals of silicon are grown from the melt of silicon (1420°C), and a cylinder of silicon is produced with a diameter of about 10 centimeters or even larger. Wafers, about a third of a millimeter thick, are then sliced from the cylinder. Each wafer is ground and polished mechanically and chemically. Then an epitaxial layer of silicon of either p or n type is grown on the surface. This is followed by an oxidation step. The oxide is then etched by photographic processes to form a mask pattern in the oxide, exposing the area where doping is wanted. Exposure to gasous phosphorus compounds, for example, at high temperatures produces a doping of the semiconductor according to the mask pattern, to a depth determined by the diffusion conditions. This is one of the simplest device structures. Metallization and patterning at this point can produce field effect transistors. Metal evaporation and etching for an interconnection pattern follow to produce circuits and interconnections between circuits (integration). Typical integrated circuits today require multiple diffusion steps and several layers of interconnection. In all, there may be several hundred segmented process steps required to manufacture modern, high-performance integrated circuit chips.

Chip Carrier

A silicon chip is attached to a substrate that contains the wiring structure for the power and signal connections. The substrate may also provide a path for heat transfer from the chip.

The "wires" on the substrate are produced by the so-called thin-film or thick-film technology, or they may consist of a metal lead frame, in which case there is no substrate. In thin-film technology the wiring pattern is produced by photolithography, while in thick-film technology the pattern is formed by screening a conductive paste through a metal mask pattern onto the substrate.

The substrate is often made of ceramic. Processes and materials used for ceramic substrates are discussed in Chapter 10. Aluminum oxide is a commonly used ceramic that has adequate thermal conductivity. However, its thermal coefficient of expansion is different from that of silicon and any chip attachment method must incorporate means of absorbing any differential thermal displacement. The back of the chip may be adhesively bonded or metallurgically bonded to a substrate or lead frame to provide mechanical support and a heat transfer path.

As discussed previously, one common method of interconnecting the chip and substrate is through a process called wire bonding. Wire bonding consists of attaching flexible wires from the bonding pads on the top of the chip to the substrate. Gold-plated wires are typically used, and the bonding pads on both the chip and the substrate are also plated with gold. The bond is then made by thermal compression bonding with a heated tool. The bonds are formed as the result of a metal-to-metal diffusion process as discussed in Chapter 25.

In place of wire bonding, IBM has developed the method of C-4 (controlled-collapse chip connections) flip-chip interconnections. This method allows the connections on the chip to be arranged in an area as opposed to being peripherally arranged. The C-4 connections are made by very small solder balls formed during the sequence of process steps for the wafer. Later, when the chip is placed on a substrate, the solder balls contact and melt onto an array of metal pads on the substrate. The surface tension of the liquid solder helps align the chip properly on the substrate. With this solder bonding, the chip is electrically, mechanically, and thermally connected to the substrate. The solder has adequate ductility to endure the relative thermal displacements.

To interconnect many silicon chips effectively in large computers, multi-layered ceramic substrates have been developed on which as many as 100 chips can be attached. Alternative approaches for packaging a large number of chips are covered in Chapter 1. The substrate itself may have many layers of wiring planes. An example of this type of chip carrier is the IBM thermal conduction module (TCM), (Figure I-4), which has 33 layers of wiring [10]. In the TCM, the wiring patterns are screened through a metal mask pattern onto thin layers of green (unfired) ceramic. The layers are placed on top of each other and fired to produce the final substrate with embedded wiring. The chips are attached to the substrate with C-4 solder balls [11].

A relatively new method of attaching chips to cards or substrates is tape-automated bonding (TAB), which uses a thin polymer tape containing metallic circuitry (Figure I-5). The connections from the tape to the chip are called inner lead bonds. These bonds are made by dissolving some of the polymer away so the electrical connections are left as small cantilever beams that are thermal compression bonded or liquid phase bonded to the pads on the chip. Connections

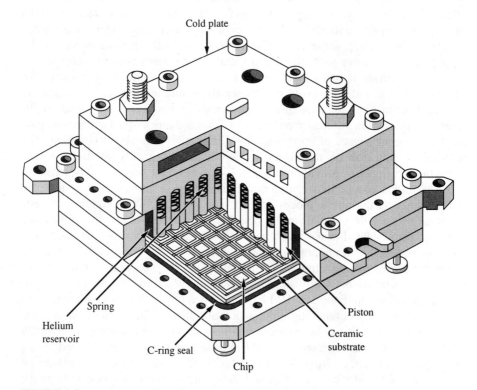

FIGURE I-4
A high performance multichip module that takes the place of both single chip modules and cards and contains up to 100 chips. Special cooling is required due to the high density.

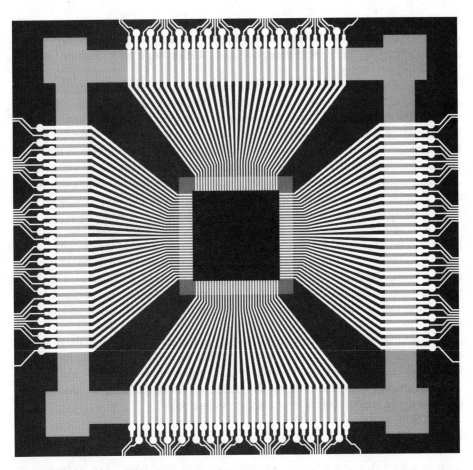

FIGURE I-5
TAB Tape Automated Bond (TAB). Tape Automated Bond parts with inner leads attached to chips and outer leads attached to a test card.

from the tape to the card or substrate are called the outer lead bonds, and are made by soldering or by thermal compression bonding.

Printed Circuit Card and Board

Early in the design of any packaging system, the specifications of the printed circuit cards and boards must be determined. Thickness and width of the printed circuit lines, and the line spacing, are important to both the performance and the manufacturability of these components. Considerable research has been done on predicting wiring capacity requirements [12]. Wiring capacity is the total length of electrical connections that can be placed in a given structure. This field of study is referred to as a wireability and is covered in Chapter 2 and its appendix. It is important to understand that the capacity may be achieved with a combination of layers of interconnection and an interconnection structure on each layer.

Printed circuit cards are typically made of glass cloth impregnated with epoxy that is "B" staged (or partially cured). The epoxy-glass, called "prepreg," serves as an insulating dielectric between the two copper surfaces. A core is made by laminating a layer of copper on both sides of the prepreg in a heated press. Holes are drilled where electrical connections must be made between the front and back surfaces or where components must be soldered to the card. These holes are then plated with copper to make the electrical connections to both sides of the core. Finally, patterns are applied to the core with photolithography and etching processes. This finished core has layers of circuitry on both surfaces.

If a card is to have internal layers of wiring, this process must be repeated for each layer. Some of the copper layers might carry power to the circuits, while others will interconnect the signal lines. Therefore, a typical card or board is made by laminating together several *signal cores* and *power cores*. Great care must be taken to assure alignment of the layers, which are then laminated together. Holes are drilled and plated again to make connections between external and internal layers.

Connections must be provided at the card edge to carry signals from the card to other parts of the system, and to provide power to the card and a ground reference. These connections are usually accomplished mechanically with metal-to-metal contact between noble metals such as gold or palladium. Chapter 7 is devoted to the fundamentals of connector technology.

MATERIAL SCIENCE CONSIDERATIONS

Various material science topics are introduced in this book. For example, in Chapter 18 the chemistry of etching metals is discussed, which is applicable to any level of package.

The fundamental aspects of polymers are reviewed in Chapter 11. Their physical properties, environmental capabilities, and processability are important to their use as well as their manufacturability. Epoxy materials still predominate in the industry for a variety of reasons. They are low in material cost and processing cost, and they have a long history of excellent reliability. When making insulating cores for printed circuits, cards, and boards, the epoxy is impregnated into a glass yarn fabric, usually handled in yard-wide or wider rolls 500 to 1000 feet long, which move through a process at many feet per minute. These materials and new materials, such as polyimide, with improved physical and chemical properties are discussed.

The formation of metal layers and metal interconnection techniques receives substantial attention. Plating reactions are particulary complex but they are very adaptable to a variety of interconnection structures, holes, lines, connectors, and more. Plating may be autocatalytic, driven by chemical potential energy, or driven by electrolytic means external to the solution. The autocatalytic technique gives the freedom to make fine interconnections without externally making contact with them, other than with the chemicals.

For the most advanced structures, thinner, denser layers, and interconnections using thin films are becoming important. Thus, thin-film evaporation and sputtering as well as plasma processing are included in this book.

Ceramics generally have an important role throughout the electronics industry, and are included for that reason. Strength, possible thermal expansion matching to chips, excellent thermal conductivity, and other factors influence their potential use. Ceramic sciences are becoming increasingly sophisticated and good progress is being made in new materials and processes.

Sol-gel science is a precursor role for producing finer and purer materials, and is adaptable to making films and fabrics. Although this technology has not yet taken a lead in product applications, we include it for its strong potential.

Assembly processing is covered, based on the revolutionary changes taking place in this part of the industry. TAB bonding is responsible for new types of metal-to-metal material contacts, and outer lead bonds are increasing in density by an order of magnitude to keep up with the increasing number of I/O connections at the chip level.

INSTRUMENTAL AND ANALYTICAL TECHNIQUES IN PACKAGING SCIENCE

The microelectronics industry relies heavily on standard analytical techniques and also on very specialized instrument methods. The analytical approach is used to characterize products, determine their composition, study the effect of simulated process parameters, assure quality, or determine the most practical conditions to yield good product. In many other cases, the analytical approach must be included as part of failure analysis so that the type of reactions or sequence of events that led to a failure are easily understood. A recent approach known as the miniline method has been successfully used to simulate a process and actually provide better controls than those in conventional process lines [8]. Once the simulated process is completely understood, the solution to a problem may be used to provide better controls in actual manufacturing stages. The need for instrument and spectroscopic analysis in electronic packaging is based on the variety of organic and inorganic components being used today. Due to the complexity of materials, there is virtually no single technique that has the ability and sensitivity to solve a problem completely. Since physical as well as chemical properties must be measured and determined, other techniques that provide precise information on surface topography must be considered. A multitechnique approach is always desirable to fully understand properties of materials.

In a successful analytical approach to any problem, the information to be derived determines whether surface or bulk effects are important and relevant. These effects, in turn, determine what instrument methods to select. Bulk properties are average characteristics of a material, such as density, melting point, chemical composition, molecular weight distribution of a polymer, or tensile strength. Surface properties, on the other hand, are related to interfacial phenomena or those present at discontinuities between two materials. Electronic packaging depends on interfaces to provide chemical or mechanical interactions resulting in enhanced adhesion, corrosion inhibition, and preferential electrochemical plating or etching of organic materials. Maintaining a contaminant-free surface is critical to the quality of the final product. In deposited metallic films of high purity (>99.99 percent) analyzed by spectrochemical methods, for example, it is not surprising to find oxide-rich surfaces and carbonaceous material that differ greatly in composition from the results of the bulk analyses. Because surface interactions are important, the depth of surface layers must be defined. In this book, surface phenomena are restricted to the top 100 angstroms of a film. A transitional near-surface zone (from a depth greater than 100 angstroms to several micrometers) has to be considered to describe profiles resulting from diffusion phenomena such as intermixing or growth on what was initially a surface layer.

The analytical chapters deal with four specific topics: bulk analysis, scanning electron microscopy, ion beam assisted analysis, and electron spectroscopy. In each chapter an introductory section provides a fundamental understanding of the techniques described. These four chapters have been organized to show the transition from bulk- to surface-sensitive techniques. Since all packaging structures are predominately composites with metals, ceramics, and polymers, we introduce the interfacial considerations (adhesion) in Chapter 27. Polymer-to-polymer diffusion and adhesion are important as are metal-to-metal diffusion and adhesion.

MANUFACTURING AND THE FUTURE

The drive toward increased circuit density within electronic packages is expected to continue. This trend will manifest itself in the manufacturing arena in the form of high-precision tools, which can be very expensive. A balance of the tolerances that derive from the tools (Chapter 20 on mechanical processes) and those that derive from the processes must be maintained so that the total tolerances will meet the performance requirements. An engineer must also be able to take advantage of the benefits of automation (Chapter 28 on robotics and automation), keeping in mind the desired tolerances. Automation may provide the capability to handle fragile materials and small components better than conventional approaches, thus providing an additional cost benefit.

The financial aspects of development and manufacturing are discussed in Chapter 29, followed by Chapter 30 on the factory of the future, which covers management of information systems as well as strategies such as just-in-time manufacturing. The book closes with a summary of historical trends and a projection of future trends based on the fundamental limits of electronic packaging.

SUMMARY

As VLSI generates higher and higher levels of circuit integration, the principles and practices delineated in this book and earlier publications [13,14,15] should serve as a general guide toward more reliable and cost-effective electronic packages to meet future challenges.

REFERENCES

1. D. Balderes and M.L. White, "Package Effects on CPU Performance of Large Commercial Processors," *Proceedings of the Technical Program, 20th Electronic Components Conference, May 1985, Washington, D.C.,* pp. 351–355.
2. E. E. Davidson, "Electrical Design of a High Speed Computer Package," *IBM J. Res. Develop.* 26, pp. 349–361 (1982).
3. J. H. Seely and R. C. Chu, *"Heat Transfer in Microelectronics Equipment,"* New York: Marcel Dekker, Inc., (1982).
4. D. P. Seraphim, D. E. Barr, W. T. Chen, and G. P. Schmitt, "Electrical and Electronic Applications," in *Advanced Thermoset Composites,* edited by Margolis, New York: Van Nostrand Reinhold.
5. S. M. Sze, *VLSI Technology,* New York: McGraw-Hill, 1983.
6. J. L. Sloan, *Design and Packaging of Electronics Equipment,* New York: Van Nostrand Reinhold, 1985.
7. J. Balde and G. Messner, "Low Dielectric Constant—The Substrate of the Future," *Proc. Printed Circuits World Convention IV, June 2–5, 1987,* paper 59.
8. D. Barr, W. Chen, R. Rosenberg, D. Seraphim, and P. Toole, "Miniline Research Applied to Manufacturing," *IEEE Trans. on Components, Hybrids and Manufacturing Technology* CHMT-8, 4, p. 410, (1985).
9. G. Messner, "Derivation of Uniform Interconnection Density Analysis," *Proc. Printed Circuit World Convention IV, June 2–5, 1987.*
10. B. T. Clark, and Y. M. Hill, "IBM Multi-Chip Multi-Layer Ceramic Modules for LSI Chips—Design for Performance and Density," *IEEE Trans. on CHMT,* pp. 89–93 (1980).
11. D. P. Seraphim and I. Feinburg, "Electronic Packaging Evolution in *IBM,*" *IBM J. Res. Develop.* 25, pp. 617–629 (1981).
12. W. R. Heller, C. G. Hsi, W. F. Mikhail, "Wirability—Designing Wiring Space for Chips and Chip Packages," *IEEE Design and Test,* August 1984, pp. 43–51.
13. C. L. Harper, editor, *Handbook of Electronic Packaging,* New York: McGraw-Hill, 1969.
14. D. Baker, D. C. Koehler, W. O. Fleckenstein, C. E. Roden, and R. Sabia, *Physical Design of Electronic Systems,* vol. 4, *Design Process,* Englewood Cliffs, N. J. Prentice-Hall 1972.
15. R. Tummala and E. Rymaszewski, *Microelectronics Packaging Handbook,* Van Nostrand Reinhold, to be published 1988.

CHAPTER 1

PACKAGING ARCHITECTURE

DEMETRIOS BALDERES
TEH-SEN JEN
MARLIN L. WHITE

INTRODUCTION

The architecture of electronic packaging involves the practice of both art and science, just like the design and construction of a house, with which the conventional architect deals. In general, the architecture of any product or system is influenced by many factors, some measurable and others intangible. A clear set of objectives and requirements is needed to define the architecture of such products or systems. Some of the objectives and requirements are general in nature, and some are specific to a particular product. They often interact with each other. A unique solution to a given set of objectives and requirements usually does not exist.

The architecture of electronics packaging begins with three basic questions: "what?" "why?" and "how?"

The product spectrum that an electronics packaging engineer may be asked to design is very broad indeed. Table 1.1 shows the diversity of the products but is by no means comprehensive. Each of these products represents a range of performance, environmental, and cost requirements. In order to answer the question "what" it is necessary to discuss these further. In the following, these requirements will be treated separately, and the packaging architecture for large general-purpose computers will be used as an example.

16

TABLE 1.1
Examples of packaged
electronics products

Electronic watch
Hand calculator
Personal computer
Minicomputer
Intermediate computer system
Large general-purpose computer
Supercomputer

OVERVIEW

The considerations involved in the question "what" are design requirements. They are:

Performance (applications, life, human factors)
Environmental considerations (size, weight, power, cooling)
Reliability, availability, serviceability
Manufacturability/cost

Once these requirements are defined the "why" question is considered. The packaging architects have to translate the interests of system designers into parameters that are understood by the component engineers and materials scientists. Making this translation clearly is crucial to motivating the component and material technologists to provide the optimum solution; that is, the answer to the "how" question.

PERFORMANCE

The choice of a package architecture and of how to fulfill the various requirements cannot begin until the factors affecting performance are fully defined. It is then possible to isolate those factors in which the package plays a role. Quantitative analysis is necessary to properly specify the materials, processes, and designs that will best satisfy the performance requirements.

We will take as an example the central processor of a general-purpose computer. We will identify quantifiable parameters, or *metrics*, that are necessary to define and analyze the key elements of the package technology [1].

The performance of the central processing unit (CPU) limits the performance of a computer system. The performance of large commercial processors, measured in MIPS (million instructions per second), is mainly determined by two parameters—the basic cycle time of the system and the average number of system cycles required to process the instruction set. These two parameters in turn are functions of many other parameters, which are related as shown in Figure 1-1 [2].

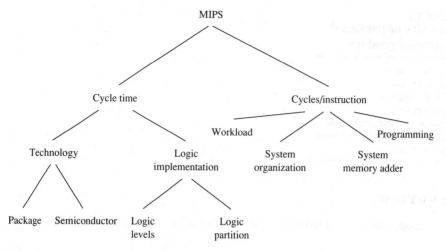

FIGURE 1-1
Internal system performance.

Performance is inversely proportional to cycle time and cycles per instruction, as shown by Equation (1.1):

$$\text{Performance (MIPS)} = \frac{1000}{\text{cycle time} \times \text{cycles per instruction}} \qquad (1.1)$$

where the cycle time is in nanoseconds. It is obvious from the above relationship that minimizing cycle time is of primary importance in striving to increase system performance. The package architecture has the primary function of supporting the circuit technology and system design requirements of the product.

Cycle Time

The cycle time for a central processor is determined by the logical functions that must be performed in each machine cycle. These logical functions historically have used 10 to 20 logical stages, where each stage is a logic gate that performs a boolean arithmetic function. Depending on the semiconductor chip, package, and level of integration, a logical function may require the information to be processed through 5 to 15 different chips and 3 to 5 package entities to complete the function. Signal propagation is discussed in Chapters 3 and 4. As circuit density on a chip increases these requirements are reduced. For instance, when the chip level can contain an entire system, the package entities are reduced to the requirements for input/output.

    The total system cycle time has two main components, as shown in Table 1.2: circuit delay (the time the logical gates require to switch), and package delay (the time for the signal to propagate to all necessary gates). To achieve a shorter

TABLE 1.2
Cycle time contributors and metrics

Circuit delay	
Circuit technology	
Chip power—cooling	W/cm^2
Package delay	
Circuit density	
I/O	$Connections/cm^2$
Wireability	Lines/cm
Interconnection length	
Propagation medium	Circuits/ns

system cycle time, and therefore better performance, these two key areas must be addressed.

Circuit Delay

Circuit delay is a function of the circuit technology chosen and the amount of power, at the correct voltage levels, that can be provided to the circuits by the package. To maintain the circuits at design temperatures, the package must also provide the necessary cooling capabilities. We define a package *metric* of power per unit area (in watts per centimeter squared) at the chip interconnection level to quantify the package capability to support the semiconductor.

The first parameter to be considered in determining the package delay is the circuit density the package can support. This is a very important parameter in that it determines the number of different chips and package entities that need to be part of the cycle time determining logical functions. The chip circuit density a package can support depends on, in addition to cooling and power, I/O capabilities and wireability (see Chapter 2). Input and output pins are required in order to allow access to the logic gates or circuits, and a wireable structure is required to interconnect them.

Other factors affecting package delay include interconnection length and propagation medium (Chapters 3 and 4). They determine the amount of time required to propagate a signal between different chips and package entities [3]. We define additional package metrics as: connections per unit area (I/O cm^2), circuit traces per unit length (lines/cm), and propagation delay per unit length (ns/cm).

The full capabilities of a package can be represented by a final metric that is dependent on all previous metrics. This metric, which we will call circuits per nanosecond (circuits/ns), reflects the total accessible circuits available to the system design per unit time.

As demonstrated in Figures 1-2 and 1-3, the amount of power that can be supplied and dissipated by the package plays a significant role in what circuit

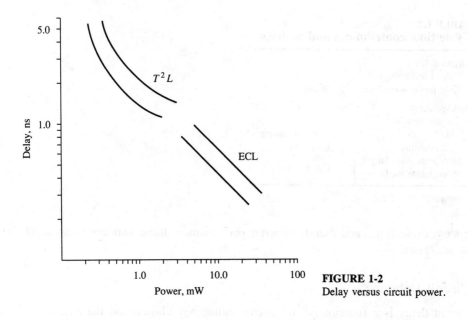

FIGURE 1-2
Delay versus circuit power.

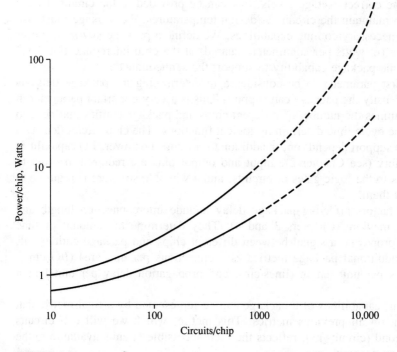

FIGURE 1-3
Power.

delays can be achieved at the chip level. As evident from the speed power curve (nanoseconds versus milliwatts) of Figure 1-2, lower circuit delay requires increased power capability for a given circuit technology. Circuit design and process improvements have historically improved the circuit speed at a given circuit power.

The cooling capability of a package has been a key limiting factor in achieving circuit density at the desired logic gate performance levels. Figure 1-3 shows a projection of chip power versus circuits per chip. It is apparent that the drive for a lower gate or circuit delay and greater circuit density is accelerating the need for chip power and thermal capability.

Package Delay

The type of circuit and circuit density that the package can support affects the package delay because it determines the number of times a package entity is used in any logical path. Figure 1-4 illustrates the number of *entity crossings* in a typical logic path as a function of circuit density. An entity crossing could be from chip to chip on a substrate or from module to module on a board, for example. The plots in Figure 1-4 are the result of analyses on design statistics from existing machines and very large scale integration (VLSI) studies at the chip level. They may be applied to various levels of package beyond the chip. It is evident from Figure 1-4 that the number of entity crossings in a logical path may

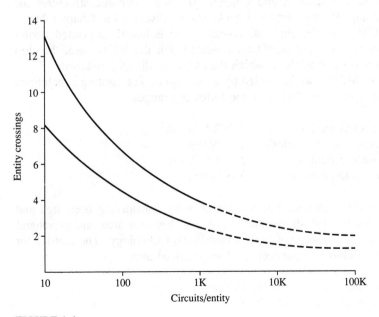

FIGURE 1-4
Crossing effects of circuit density.

be reduced by increasing circuit density, provided the I/O and wireability also increase to support that density [4,5].

The package delays associated with a crossing (e.g., chip to chip, module to module) are a function of the materials (propagation media) and dimensions of the package. The time of flight (TOF) is dependent on the materials and more specifically the dielectric constant of the package material. TOF delays range from that of high speed cable, near 40 ps/cm, through glass epoxy card/board, near 70 ps/cm, to alumina ceramic multichip modules, near 120 ps/cm. These TOF delays can be combined with the associated connector penalties and supportable circuit densities to yield the package metric circuits/ns.

DESIGN ALTERNATIVES

According to the discussion so far, a proper package design must support the chip requirements for power/thermal features, signal I/O, and wireability in addition to minimizing the package delay. Some of the package design choices will be illustrated in more detail. Details of the electrical design considerations are covered in Chapters 3 and 4. The basic package design choices can be considered in two categories; a single chip mounted in a carrier (single-chip module, SCM) or multiple chips within a single carrier (multichip module, MCM). The SCM is used primarily as a means to expand the signal and power connections from the small on-chip grid to the larger grid of the package. An MCM provides for direct mounting of the chip and interconnection of signal and power on the package.

Note that the Introduction and Chapter 10 discuss ceramic structures for first-level packaging. Plastic first-level packages are discussed in Chapter 11.

For the SCM case, the chip interconnect level is usually a printed wiring card or board on which many SCMs are mounted. For the MCM case, the chip interconnect level is the module to which the chips are directly attached.

The SCM or MCM may be cooled by air or liquid. The cooling capabilities for technology available in 1985 have the following ranges:

Single-chip air-cooled modules	0.2–0.5 W/cm^2
Single-chip liquid-cooled modules	1–2 W/cm^2
Multichip air-cooled modules	1.5–3.5 W/cm^2
Multichip liquid-cooled modules	3–8 W/cm^2

A comparison of these choices can be illustrated by considering both area and peripheral I/Os. Figure 1-5 shows the comparison between area and peripheral connections for the dimensions of typical mid 1980s technology. The metrics for this example at the chip interconnect level as practiced are:

Area chip connection	75–250 I/O/cm^2
Peripheral chip connection	13–30 I/O/cm^2

It should be pointed out that these I/O numbers must include the necessary power distribution I/O.

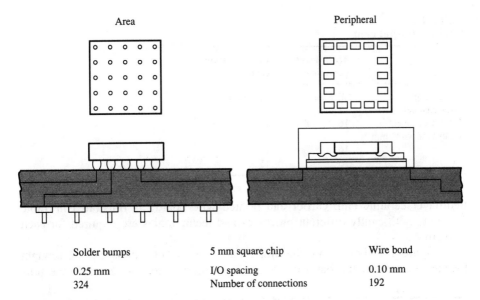

	Area			Peripheral

	Solder bumps	5 mm square chip	Wire bond
	0.25 mm	I/O spacing	0.10 mm
	324	Number of connections	192

FIGURE 1-5
I/O technologies.

The wireability metric depends on the geometrical configuration of the package structure including the number of wiring layers that are available within the package. Again it should be pointed out that the structure must include the necessary power distribution and signal line impedance control layers. The wiring metric can be simply represented by Figure 1-6, which gives a first order indication of the wiring available. Similar wireability can be achieved with

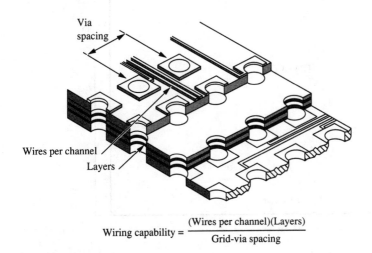

$$\text{Wiring capability} = \frac{(\text{Wires per channel})(\text{Layers})}{\text{Grid-via spacing}}$$

FIGURE 1-6
Wiring capability.

TABLE 1.3
Wiring comparison

	Multi-layer ceramic	Printed circuit card board
Via spacing	0.5 mm	1.2 mm
Wires/channel	1	2
Number of layers	16 (8 x, 8 y)	18 (9 x, 9 y)
Number of wires/mm		
x	16	15
y	16	15

multilayer ceramic chip carrier and printed wire board, as illustrated in Table 1.3, but significantly different processes and technologies are required for each of them.

Figure 1-7 shows how the aforementioned considerations can be brought together to provide the basis for determining usable circuits per unit time (cir-

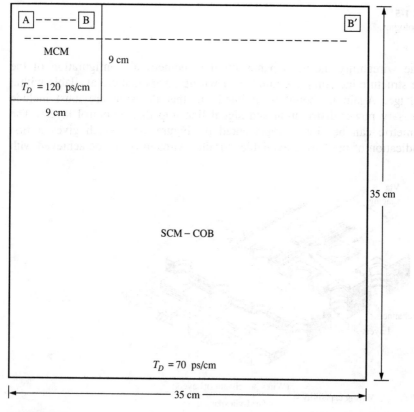

FIGURE 1-7
Propagation delay.

cuits/ns) of the package. Consider the following:

The usable circuits on chip A, B, and B^1 are dependent on:
Power/thermal capability
I/O capability
The time required to propagate a signal from Chip A to Chip B or B^1 is dependent on:
Time of flight/unit length (materials and configuration)
Chip-to-carrier interconnect delays
Chip-to-chip separation distance
Wireability
Power/thermal capability

Figure 1-7 illustrates a typical configuration in which both air cooled SCM on card and water cooled MCM packages would provide the same power/thermal capabilities, I/O, and wireability. With the same 500 usable circuits per chip, the ratio of MCM circuits/ns to SCM-COB (card on board) circuits/ns is the inverse ratio of their TOF $\times$ distance: $(70) \times (35)/(120 \times 9)$, or approximately 2.25:1. This ratio means:

1. The MCM could contain more than twice the circuits for the same allowable delay.

or

2. The delay penalty of the MCM is 1/2 the delay of the SCM-COB for the same circuit function.

With diversity of available technologies and the interrelationship between the package metrics and system performance, it is important that the package architect be able to quickly quantify or rank the technology options with respect to the system performance. For a particular product, a rigorous methodology can be established that will allow this evaluation, without undertaking a full design analysis. The development of an appropriate methodology requires that the package architect fully understand past products and components and their relationships to the product in development.

ENVIRONMENTAL PARAMETERS

Environmental considerations for an electronics system are twofold. First is the environment the system will be operating in, and second are the environmental requirements of the product. The first category is obviously dictated by the product and its intended use. An electronic watch has to fit on a human wrist, should not be too heavy, should contain its own power, and should be somewhat shockproof and in some cases waterproof. An airborne or spaceborne computer would have size and weight limits, as well as power and cooling restrictions, depending on the capabilities of the aircraft or spacecraft. A large general-purpose computer system may be less restricted in these considerations. The second category of environmental requirements is what the product demands, including the amount of power, the type of cooling (air or water), and the electromagnetic environment. Again the same parameters are involved—size, weight, power, and cooling requirements. This is where the product user and the package architect need to fully understand each other.

RELIABILITY, AVAILABILITY, SERVICEABILITY

System reliability is highly dependent on the failure rate of its components (see Chapter 9). The failure rate of a component, which will be treated in depth in Chapter 9, reflects the maturity of the technology involved. It is for this reason that new packaging technologies, especially for large systems, have a long lead time in development.

Availability of the product to the user refers to the time the product is functional, and is dependent on the reliability of the various components and the ability of the system to function after a component failure. In general-purpose computers the latter is accomplished by electronic design for data recovery and error correction. The packaging technologist and the system designer need to work together to optimize failure rates against the overhead of redundant components required to fix the errors.

Serviceability of the product is a demand placed on the pakcaging architect as a result of component failure. Components and subsystems have to be accessible for service. This demand is usually in complete opposition to the performance demand, for which close placement of components is tantamount. If the product is to be repaired or replaced in the field, it will require ease of assembly and removal, and the cost of field replaceable units (FRU) will be important. The higher the value of the field replaceable units, the higher the cost of the field inventories. On the other hand, to lower the cost of FRUs, a larger quantity of smaller and less expensive parts have to be stocked in the field. In the limiting situation, each module or resistor could be an FRU, which would complicate inventory control and system design. Whichever approach is taken to handle FRUs, they represent an associated cost that must be factored into the cost of the entire system.

In some products, the best approach would be to use throwaway parts. In this case, one would choose the architecture that would allow easy removal of parts, fault isolation, and the minimization of the cost of each throwaway part. An example of this type of design may be seen in the consumer electronics area, such as in televisions and radios. Then the question of diagnostic ability arises. Will the system be self-diagnosing, and how large is the cost associated with adding this feature?

Another important aspect is the requirement of field repair. This is especially important in the case of computers used for real-time operation. The system has to operate while the repair or service is taking place, which drives the architecture to allow a "hot-plug" of components that are power isolated from the system. Various other factors such as safety and human factors must be taken into consideration with this requirement.

PACKAGING ARCHITECTURES OF CURRENT LARGE GENERAL-PURPOSE PROCESSORS

The packaging architectures developed by the large general-purpose computer manufacturers of the mid 1980s can be used to illustrate how the various considerations discussed are implemented. These architectures draw on special technology strengths of a particular company, to optimize the architecture. Examples to be discussed include:

1. IBM multichip carrier, multilayer ceramic, planar board, and flip-chip area array connector technologies
2. NEC flip-tab chip carrier, multichip package with fine-line wiring, planar board, and area array connector technologies.
3. Fujitsu single chip carrier, surface mount, double sided board, and edge connector technologies.
4. Hitatchi single chip carrier, surface mount, card on board, and edge connector technologies.

A brief description of each approach will be given with some comments on the unique feature.

The IBM-Approach: Thermal Conduction Module on a Large Board

Since the 308X series computers (1981), IBM has been using the approach of the thermal conduction module (TCM) on a large board (see Figure 1-8). The basic strategy is to package 100-plus chips on a module that provides chip-to-chip interconnections, with six or nine of these modules plugged onto a large board that provides the module-to-module interconnections [6].

The module is made of multilayer ceramic substrate with a size of 90 mm by 90 mm. It contains 100 to 133 chips interconnected and powered through 33 layers of wiring. The total number of signal and power I/Os for each TCM is 1800. Although the ceramic environment with a dielectric constant of nine has relatively high propagation delays for signals, the TCM approach allows chips to be packaged closely together (8.5 mm pitch) and provides a very short interchip

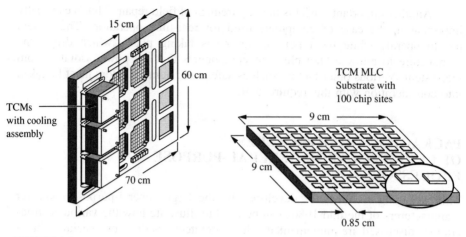

FIGURE 1-8
TCM-TCM board.

distance. In addition, the cooling scheme used has the ability to remove 500 watts of power per TCM. By keeping the chips at a reasonable low temperature, the circuits can be powered to achieve high performance.

Up to nine of these TCMs are then interconnected on a large board, which has dimensions approximately 24 inches by 28 inches. To provide the power and interconnections required by the TCMs, the board has six to eight layers of signal planes, 12 power and ground layers, and supports more than 1300 amperes of current. The board-to-board interconnections are made by high-speed cables. The system is designed in such a way that the cycle time determining path is primarily contained on two TCMs, and the board is essentially the CPU of a system.

To support such a packaging architecture, a multiplicity of technologies has to be developed. These technologies include the multilayer ceramic substrate, the printed circuit board with nearly a mile of interconnection capability in the wiring planes, a high-density zero-insertion-force connector system (for the TCM pins), cooling techniques, and EC (engineering change) rework schemes, at all package levels.

The Liquid-Cooled Module on Large Board of NEC

NEC has used an approach similar to the TCM approach for both their supercomputer and commercial processors [7]. The package architecture is very similar to that of IBM. The liquid-cooled module (LCM), however, is not a single-level design as is IBM's. Instead, the chip is first mounted on a multilayered chip carrier to provide the pin-out and the cooling interface (see Figures 1-9, 1-10).

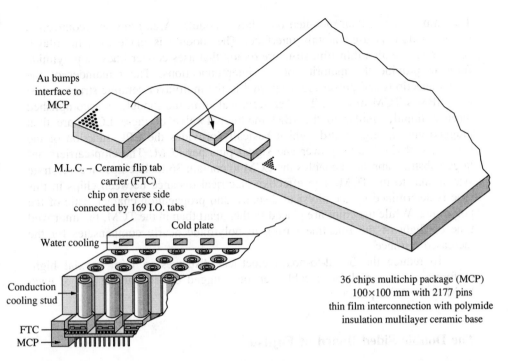

Au bumps
interface to
MCP

M.L.C. – Ceramic flip tab
carrier (FTC)
chip on reverse side
connected by 169 I.O. tabs

Cold plate

Water cooling

Conduction
cooling stud

FTC

MCP

36 chips multichip package (MCP)
100×100 mm with 2177 pins
thin film interconnection with polymide
insulation multilayer ceramic base

FIGURE 1-9
NEC liquid-cooled module.

Liquid-cooled module (LCM)

Board

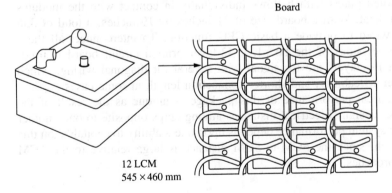

12 LCM
545 × 460 mm

FIGURE 1-10
NEC.

The chip carrier is then mounted on a large module. Area array interconnection to the module is used at this interface. The module is made of a multilayer ceramic base with a thin-film structure on top that uses copper lines on polyimide films to provide the majority of the interconnections. The remainder of the interconnections and power are supported in the multilayer ceramic structure just as in IBM's TCM module. The heat removal from the module is accomplished through liquid, similar to the IBM module. Twelve of these LCMs are then plugged onto a large board, which basically provides the CPU function of the system, with 2177 total power and signal pins per LCM. The chip-carriers-on-large-substrate approach enables the LCM to contain 36 to 42 chips on a substrate size similar to the TCM. The effective electrical distance between chips in this case is determined by the physical distance and propagation characteristic of the polyimide. While the chips are placed farther apart than in the TCM, the improved time of flight of the fine line wiring in polyimide nearly compensates for the increased distance.

To reduce the board-to-board electrical distances, a mini coaxial high-performance (low-dielectric) cable can be plugged directly at the pin of each LCM in the NEC package.

The Double-Sided Board of Fujitsu

A very different architecture is used by Fujitsu in its M780 system [8]. The chip is mounted on a 22.5 mm square single-chip module (SCM) with peripheral leaded connections (see Figure 1-11). One hundred sixty-eight of these modules are then surface mounted on one side of a large board, and another 168 are mounted on the opposite side of the board. This assembled board is then sandwiched between two water-cooled plates with bellows individually in contact with the modules to remove the heat. With a board size of 21 inches by 19 inches, a total of 336 modules are wired to provide a basic CPU function. To interconnect all these modules on both sides of the board, a 42-layer printed circuit board is used. About half of the layers are for signal with several for diagonal wiring, which provides better wireability, and shorter connection length in some cases.

The electrical performance of this package is unique as the result of the "back to back" design. Electrical distances among chips opposite to one another are short. The system performance depends upon one's ability to capitalize on this design approach, since the SCM-to-SCM distance is large relative to the TCM and LCM approaches.

The Card-on-Board Package of Hitachi

The solution of Hitachi to the same problem is more conventional for their 680H computer system [9]. The packaging architecture is a quasi three-dimensional card-on-board approach, with enhanced cooling and connectors (see Figure 1-12).

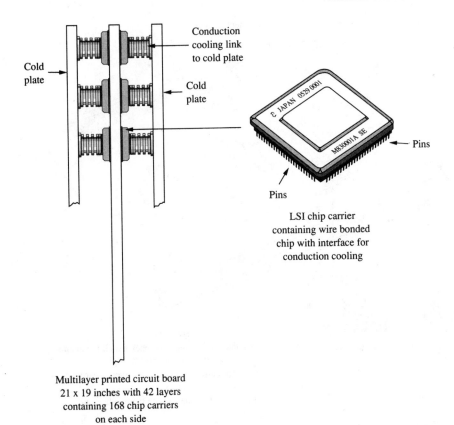

Multilayer printed circuit board
21 x 19 inches with 42 layers
containing 168 chip carriers
on each side

FIGURE 1-11
FUJITSU double-sided board.

The card is 11 inches by 15.75 inches and contains up to 72 single-chip logic modules. Eight signal layers are used to interconnect the modules on this card. The cards are plugged onto a board through high-density card edge connectors with over 1000 contacts per card. Top card connectors are also provided to allow additional connections between cards and for cables. The total number of I/Os per card is similar to the IBM TCM.

A forced-air approach is used to remove the heat from the chip. The chip is first back mounted to silicon carbide as the heat spreader, and a heat sink tower is then connected to the silicon carbide. Both single-chip and multichip modules are used. The multichip module is double-sided and is used to package array chips and support logic to improve density and performance.

The rather large distances between the chips results in a larger component of package delay than in the other approaches, and careful attention to placement is required to take advantage of the relatively short connection length from card to card.

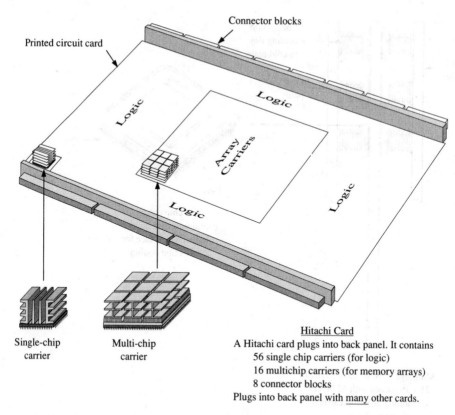

FIGURE 1-12
HITACHI card-on-board.

QUANTIFICATION OF PACKAGE METRICS

The design and performance of package architectures discussed in the previous section can be compared as shown in Table 1.4. The table lists the metric previously defined and reflects the capabilities published by each manufacturer, computed at the chip interconnect level (card, module, or board) for functions of

TABLE 1.4
Package metrics for CPU package approaches

System	Lines/cm*	Connections/cm²*	Watts/cm²*	TOF (ps/cm*)	Circuits/ns**
IBM-TCM	16	250	6	105	28K
NEC-LCM	19.5	75	3	65	24K
FUJITSU-SSC	15	22	1.5	70	29K
HITACHI-COB	6.3	19	0.5	70	10K

* Linear and area dimensions are at the chip interconnection level.

** Normalized to circuit power of 10 mw per circuit.

approximately 100 chips. This comparison illustrates the relative capabilities of the various manufactures in the areas of semiconductor chip technology, circuit density, power, I/O, and cooling.

MANUFACTURABILITY/COST

In the previous section, four types of packaging architecture for large general-purpose processors were compared. It is evident that the question of "how" has been answered differently by each manufacturer. One of the reasons for the diversity in packaging architecture is the variation in the particular strengths of each manufacturer in manufacturing and assembly. Thus, as discussed in the overview section, a packaging architecture reflects its manufacturability.

The surface mount technology of Hitachi and Fujitsu was developed in the 70s. Area chip-to-module array connections (C-4), multilayer ceramic modules, area array module-to-board connections, and large planar boards have a history that also goes back to the late 70s and early 80s. The availability of a mature manufacturing technology is one of the most important considerations in the choice of packaging architecture by the various manufacturers.

Although cost requirements by each manufacturer are not public domain information, every manufacturer strives for the lowest product cost in order to have a competitive price performance product. We have seen package design choices made by different manufacturers based on their manufacturing strengths. These choices are in part determined by cost considerations, i.e., minimizing capital outlay and maximizing quantity production.

MEMORY PACKAGING

In the previous sections, technologies were discussed for CPUs, from logic chip through the total package. Different considerations are made for memory packaging. Memory requirements have changed as CPU speed has increased. A hierarchy of memory levels has been adopted to accomodate the CPU's increased demand for data and still meet the requirements for low cost and high capacity.

High-capacity main and extended memory, at the mega- and gigabyte levels of the hierarchy, are characterized by relatively slow performance and are most often implemented with simple cards, boards, and connectors for low cost. Intermediate capacity, generally with few megabytes of memory and comparatively faster than main memory, may be implemented in card-on-board or more aggressive packages approaching the technology of the CPU. The choice is driven by cost, performance, and capacity requirements. The cycle-by-cycle demand for data by present-day CPUs is satisfied by a local memory buffer (cache). This cache must buffer the slower, high-capacity elements of the memory hierarchy. The requirements for cache (hundreds of kilobytes with access comparable to the

CPU cycle time) is necessarily implemented in the fastest available technology and packaged in the environment of the CPU.

SUPERCOMPUTERS AND FUTURE COMPUTER SYSTEM PACKAGING CHALLENGES

The term *supercomputer* encompasses many different system designs and applications, the majority of which are in the field of scientific computing. The largest part of scientific computing deals with large data arrays, which require high-speed computation, and large internal memories, with very large bandwidths between memory and the processing elements.

The need for high-speed computation, coupled with large amounts of data and ready access to it, presents new challenges to the system, subsystem, and component packaging engineer. Data availability and memory performance is key to the processing speed of the scientific systems. The system main internal memory needs to be fast and physically close, in terms of access and propagation delay, to the processing elements.

The large amounts of parallel data required by vector, matrix, or array data forms demand high signal I/O to main memory, which results in high I/O requirements at all levels of package, including the chip. The requirement for fast access to large amounts of data requires that the volumetric packaging density of memory bits be an order of magnitude higher than general purpose systems. Different chip packages (small outline) and liquid cooling schemes will probably be required in these memory systems.

To provide the high computing speed required, processor parallelism is utilized in these systems. This approach obviously multiplies the packaging demands on all aspects of packaging design. Many more components need to be packaged in the same or smaller volume, which increases the need for better power distribution, better cooling, more I/O in the system control elements, and more wirable structures. There are indications that future general-purpose computers, in order to provide the projected computing performance needs, will incorporate some of the package design approaches of the present supercomputers.

REFERENCES

1. D. Balderes and M. L. White, "Package Effects on CPU Performance of Large Commercial Processors," *Proceedings of the 35th ECC,* pp. 351–355 (1985). Copyright ©1985 IEEE, excerpted with permission.
2. C. A. Collins, "The Shrinking Design Space," presented at the Computer Systems Seminar at the University of Illinois, Urbana, Ill., October 1984.
3. W. R. Heller, C. George Hsi, and W. F. Mikhail, "Wirability—Designing Wiring Space for Chips and Chip Packages," *IEEE Design and Test of Computers* 1, no. 3, pp. 43–50 (1984).
4. B. S. Landman and R. L. Russo, "On Pin versus Block Relationship for Partitions of Logic Graphs," *IEEE Transactions on Computers* C-20, no. 12, pp. 1469–1479 (1971).
5. W. V. Vilkelis, "Lead Reduction Among Combinational Logic Circuits," *IBM Journal of Research and Development* 26, no. 3, pp. 342–348 (1982).

PROBLEMS

1.1. The problems for Chapter 1 and Chapter 2 are combined in Chapter 2.

CHAPTER

2

INTERACTION OF MATERIALS AND PROCESSES IN ELECTRONIC PACKAGING DESIGN

DONALD P. SERAPHIM
ALAN L. JONES

INTRODUCTION

Once the architecture is set, as discussed in Chapter 1, for a given electronic system, one must begin the detailed design of the system components. At each step of the way tradeoffs are made. As with the tradeoffs in architecture [1], these tradeoffs involve performance, cost, size, and so on.

The design of the interface between the chip and the module, the module and the card, the card and the board, and the board and I/O and other boards is a key to a successful package. For performance reasons, one would like to place components as close together as possible. Such a design can place large demands on the interconnections and the wiring density of the printed circuit boards from a materials and process view. The tradeoffs between design, materials, and processes are thus crucial to producing a system which is competitive.

When designing an electronic package, an engineer is continually faced with tradeoffs. For example, it may seem logical to put as many circuits on a chip as possible to increase performance. However, as the number of circuits increases, the number of chip I/Os increase. The consequence may be that one needs a very expensive package because of materials and processing factors as well as developmental costs. If the package is to be low cost, it might be more cost effective to put fewer circuits on a chip and have more chips with fewer I/Os per chip. That is, one has traded off performance against cost.

This chapter discusses the tradeoffs one encounters in design. Knowledge of the trends in system architecture, material usage, and in manufacturing techniques is a necessary part of the input. It is important to note here that the introduction of new processes and materials must be practical in the manufacturing plant.

In this chapter we will discuss the above considerations from semiconductor chip to finished system. The various levels of packaging—chip, chip carrier or module, printed circuit card or planar, back panel or printed circuit board and, finally, cabling will be examined, with special attention given to the design of the interface between these levels.

In all electronic packaging, the semiconductor chip is the driving factor [2,3]. Defect levels in the silicon are so well controlled today that integration levels can be increased to the limits determined by the interconnection capacity on the chip. The logical devices on the chip can be made sufficiently small and close together. The problem is that these devices must be interconnected and, as the number of devices increases, these interconnections become very complicated, requiring several layers of metallic conductors and, thus, many process steps. In a sense, packaging at the chip level is a key to function growth, and what cannot be done there with adequate yield is relegated to the packaging hierarchy outside the chip. But since interconnections on the chip cost only small fractions of a cent, the industry approach has been to take as much advantage of the progress at the chip level as possible and be ready with timely introductions of packaging interface technology (connector capability, heat extraction, etc.) to link effectively with the chip. The link must meet stringent thermal, electrical, and mechanical interfacial requirements for performance that reflect the advances in materials, processes, and designs.

To understand the direction that electronic packaging is taking, it is interesting to consider the demands on the package being made by large computer processors. The performance of large processors is represented in MIPS (millions of instructions per second). The MIPS level of a system is determined by taking the reciprocal of the product of the average number cycles per instruction and the cycle time. The factors that affect each of these contributors to the MIPS level have been discussed by Balderes [1] (see Chapter 1, "Packaging Architecture," Figure 1-4 on page 21). In the figure, an "entity" can be any portion of a system such as a chip, a module, a card, and so on. What the figure shows is that the more circuits one can place on an entity, the fewer times one has to leave the entity to complete a given operation such as a machine instruction. It is costly in terms of time each time an entity boundary is crossed, so it is desirable to package as many circuits as possible within the entity.

The cycles per instruction are an organizational contribution of the system while the cycle time has two main components—circuit delay and package delay. The package can assist in decreasing the circuit delay by providing a method to power the circuits effectively while at the same time extracting the heat effectively. In fact, in order to decrease circuit delay, an increase in power is required. As stated by Balderes, "Key elements of a system's cycle time and therefore performance are dependent on the package technology. Future improvements in cycle time put demands on the package to support faster and denser semiconductors. This, in turn, requires advances in the areas of power, cooling, I/O, wireability, and physical space reduction." We will discuss these factors in detail in later sections. In addition, there is an appendix on wireability to which the reader may want to refer. The key to performance is minimizing the interconnection length while providing a low dielectric propagation medium. It stands to reason that a figure of merit for performance is therefore density (circuits) within a given packaging distance. If we convert this packaging distance to a time of flight (TOF) then this figure of merit could be expressed as circuits per nanosecond. That is, if we take the number of circuits that can be reached in a cycle and divide this number by the length of time of one cycle (expressed in nanoseconds), we will have this figure of merit of circuits per nanosecond, as suggested by Balderes.

TYPES OF PACKAGING HIERARCHY

The packaging hierarchy—semiconductor chips, chip carriers, printed circuit cards or planars, connectors, printed circuit boards, and cables—is often assembled in the arrangement shown in Figure I-1 in the introduction [4,5]. A chip is connected to a chip carrier, often called a module. This assembly may be accomplished in a variety of ways. The predominant method in the industry is wire bonding.

The chip carrier is connected to the printed circuit card through soldered pins or leads. Until recently, the principal method of joining chip carriers to cards or planars was by inserting the pins into plated through-holes in the card and forming a solder joint there. In the *planar* method, two or more panels (cards or planars) are cabled together with the cables plugging into connectors on the card. If many panels must function together they are often arranged with cards plugged directly through connectors into a back panel printed circuit composite or board [5]. The latter arrangement is called a card-on-board package (COB). Thus, the cards are multilevel printed circuit packages holding the modules or chip carriers, and the back panels are also multilevel printed circuit packages holding the cards. Systems are constructed either with a single board containing a few or many cards, or with multiple boards cabled together, each containing many cards.

The simplest systems may contain only one card or planar with a few or many modules. A typical example is the microcomputer used in an automobile which contains engine control circuits, memory retaining instructions, fuel injection control circuits, and a small central processing unit along with many discrete components (capacitors, resistors) [6]. Such one-card systems are called planars.

The technique of reducing the number of levels of a package, as is done in low-cost systems, may also be used in building the largest systems. In this case the chip carrier containing a single chip is replaced by a complex, large multichip module (Figure I-4, Introduction) in which the chips are packaged very close together [7]. Since 100 or more chips are packaged on each module, the need for cards and conventional printed circuit boards is eliminated. The large modules are plugged directly into a back panel planar [9,10]. The planar provides the printed circuit cabling between the modules. Since many chips are placed very close together in this type of planar package, electronic signal transit time is greatly reduced [11]. However, the interface requirements, particularly heat removal, become very difficult to meet because of the increased circuit density. Furthermore there is a demand for a large number of connections between the module and the back panel, driven by the need for communication between the large number of circuits contained on each module. In the case of the IBM 3081 processor, each module contains about 30,000 circuits and requires 1800 connector contacts for power and signals. Thus the connector contact system becomes complex. To achieve the large number of interconnections between each module, the printed circuit back panel (Figure 2-1) becomes very complex. The factors that influence these interfacial considerations will become clear as we discuss each component in this chapter [2,12,13].

ELECTRICAL INTERCONNECTIONS BETWEEN PACKAGING LEVELS

The primary function of the interface is to provide the electrical interconnection between the two levels of packaging. The number of terminals required is related to the communication needed between a given component and the rest of the system. Thus, as the function within a component increases, so do the number of inputs, outputs, and controls, including, in some cases, test points required to guarantee the function. Packaging engineers work toward promoting growth at these interfaces.

As noted earlier, interfacial growth is driven by the functional increase at the semiconductor chip. The number of terminals increases as a power function of the number of circuits. This relationship was discovered in 1960 by IBM engineer E. Rent when he plotted the number of input-output lines against the number of circuits in the cards of the IBM 1401 computer [14]. This relationship can be expressed as

$$I = bC^p$$

where I is the number of I/O lines, b is a constant, C is the number of circuits on a package, and p is a positive exponent.

In Rent's original work, p was found to be approximately equal to $\frac{2}{3}$ while b was about 2. Figure 2-2 shows a relationship similar to Rent's. However, this figure is a plot of the number of I/O connections versus the number of

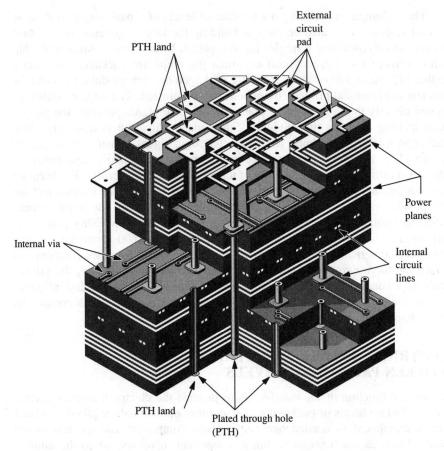

FIGURE 2-1

Cross section of a few grid sites. In a large high-performance printed circuit board, plated through-holes which contain the connector springs for connection to module pins are shown. Also shown are the high-density printed circuit interconnections and plated vias that join pairs of circuit planes.

circuits across many technologies, whereas Rent's work applied to the hierarchy within a given system. That is, his relationship is concerned with the number of I/O connections at each packaging level within a given system. When a plot is made for various components across different technologies as we have done, the exponent is much lower than the $\frac{2}{3}$ found by Rent but the constant b is much higher.

The Rent relation can be derived heuristically based on the arrangement of circuits on a chip. For example, a circuit has several inputs and outputs. Accordingly, the intercept at one circuit should represent the number of signal leads for an average circuit, which is typically in the range of three to five. Let us consider the case of an average of four leads per circuit. Now, consider an

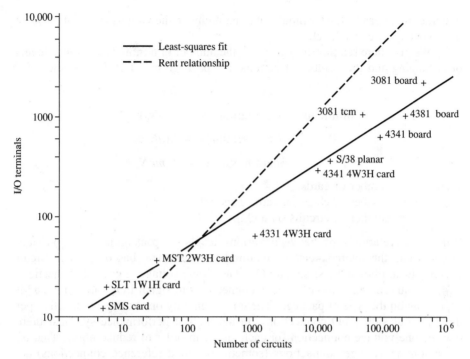

FIGURE 2-2

The Rent trend line relating terminals to circuit function. The trend line appears to be consistent for chips, multichip modules, cards, and boards over six orders of magnitude in circuit function. The least-squares fit is $I = 6.54C^{0.435}$. The steeper line is the Rent line for various levels of packaging within a system.

array of circuits on a chip and cut out a representative square. Along one side of the square let there be N circuits. Then the number of circuits within the square is $C = N^2$ and the number of circuits along the periphery is $4N$. Therefore, the number of I/Os along the periphery is related to the number of circuits within the square as

$$I = bC^{1/2}$$

where b is a constant that depends on the number of I/Os per circuit. This square root rule is very interesting, since it reflects more truly the relationship between circuits and I/Os in today's VLSI world than the $\frac{2}{3}$ power of Rent in the days of unit logic.

The number of terminals or wire bonds connecting a chip to the module has been increasing rapidly, and the number of pins on single-chip modules has been increasing equally rapidly. Furthermore, with the introduction of more than one chip on a module, the number of terminals on the module increases as the product of chips and terminals per chip taken to the power p. Characteristic of integration increases, the terminal count hierarchy on the various levels of package is expanding at every level.

We give here relationships between the number of terminals at various levels of packaging and the number of circuits in a package based on Rent's rule.

$$\text{Chip terminals} = bN_c^p$$

$$\text{Multichip module terminals} = b(nN_c)^p$$

$$\text{Card terminals} = b(nN_c)^p$$

$$\text{Board terminals} = b(mnN_c)^p$$

where m = number of cards
n = number of chips on card or module
N_c = number of circuits on a chip

With these relationships between various levels of packaging, it is possible to determine the interconnections required between them based on their circuit content. In addition to these signal I/O connections, voltage or ground connections will be required. The ratio of voltage connections to signal I/Os varies quite a bit depending on the type of package. There is commonly one voltage connection per three or four signal I/O connections. In the highest performance systems, there may be one voltage connection for each signal in order to reduce noise. That is, the adjoining voltage connections (sometimes called reference connections) act as shields to the signal I/Os.

CHIP CARRIERS OR MODULES

Along with the electrical interconnection discussed in the section above, the joining element between chip carrier and chip must provide the proper mechanical environment. The first semiconductor carriers were ceramic-based, hermetically sealed packages, called TO-5s. The back of the chip was eutectic (AuSi) bonded to the gold-coated ceramic. Thermal compression bonding of thin wires from the chip terminals (evaporated aluminum pads) to glass-sealed Kovar pins held in place by the ceramic glass seal provided a strain-relieved mechanical link. A metal can, also glass sealed, encapsulated the device from environmentally degrading reactions.

A variety of single chip carriers have gradually replaced the hermetically sealed TO-5 can [15]. Predominant among these and a standard for the industry has been the dual in-line package (DIP), constructed primarily from a metal lead frame and plastic. The chip sits on a metal ground strap for good thermal conduction while the interconnections are wire bonded from the aluminum pads on the chip to the frame leads. The entire package is then molded with plastic followed by forming of the leads. Until recently, the leads have been limited to two sides of the chip carrier, but the increased integration at the chip level (Figure 2-3) has required more interconnection. This increased interconnection

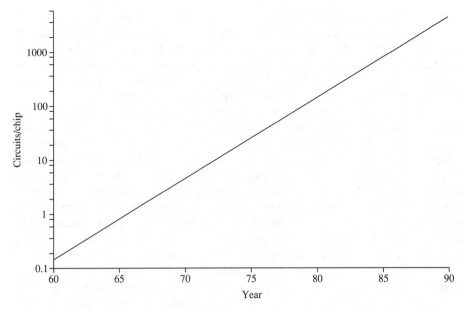

FIGURE 2-3
Chip circuit integration. Increase as a function of time is more than a factor of ten each decade. These are bipolar circuits. CMOS integration leads this by a factor of 5 to 10.

has been accomplished by placing leads on all four sides of the chip carrier and spacing them closer together. Denser spacings are also required to keep pace with the requirements of increasing function at the chip level. The plastic-leaded chip carrier (PLCC) and flatpack are good examples of the progress toward smaller components with high I/O density.

Ceramic chip carriers are also popular in the industry. There are two varieties. One type with peripheral leads is called a CLCC (ceramic-leaded chip carrier) and the other with pins in an area array is called a PGA (pin grid array). They may have a full hermetic seal that is a ceramic cap attached with a glass or solder seal, or they may have a semihermetic seal, with a plastic or metal cap sealed with plastic. PGAs have been made with as many as several hundred leads in an area 36 mm square. Because they can support a large number of I/Os, PGAs have had a predominant role in high-density packaging for many years. They will continue to play an important part in many packaging strategies for years to come by virtue of their area array format. This array format allows more I/Os than peripherally leaded components of the same size. The growth in I/O density with circuit function contained on various packages with edge and area arrays is illustrated in Figure I-3 in the introduction. The virtue of the area array is that it allows the designer to maintain larger spacing between leads, which eases tolerances at the next level of packaging hierarchy. The area array also makes it somewhat easier to accomplish interconnections in multichip arrays. Still another performance attribute is the direct communication that can be achieved by joining the chips to a nearby pin. Since the pin enters the interconnection network vertically, the interconnection distance can be minimized for selected critical paths from chip to chip where necessary. Although ceramic substrates have been favored because of their better thermal communication to the environment, plastic packages are heavily used where low cost is a requirement.

The wire bond from the chip to the module leads provides a strain relief to allow for the difference in thermal expansion between the chip and the module material. Since the wire can be as small as 0.025 mm (0.001 inches, commonly called 1 mil), there is adequate room around the periphery of the chip for all of the interconnections even with spacings of several mils. The bonding is done automatically so the sequential nature of the operation is still quite productive. Human intervention is used primarily for setup of an entire wafer of chips, so that more than one bonder may be run per person. Nevertheless, the growth in I/O density to several hundred per chip appears to be near the limit for this technology even though previous expectations have been exceeded several times already.

A new technology called tape-automated bonding (TAB), very similar to wire bonding, has been in development and was in limited use in 1987. It is expected that this technology will be one of several taking the place of wire bonding for applications where there are more than 100 leads from a single chip.

TAB (Figure I-5, Introduction) is a composite of printed circuit wires held in an area configuration by a thin organic substrate (usually polyimide). The printed circuit wires are in a window and bent down for strain relief where they join the chip. As with wire bonding, the joining is by automatic thermal compression. One of the advantages of TAB is the handling ability in manufacturing. The tape is in roll form similar to 35 mm photographic film with location references punched along the edges. Both wire bonding and TAB configurations are good for thermal conduction through the back of the chip, which is back bonded to a heat sinking material. One drawback is that the relatively long leads can cause degradation of the signal.

A bonding technology called *controlled-collapse chip connections* (C-4) developed by IBM has been receiving increased attention from other chip and package producers due to the ease of handling very large arrays of contacts that the method provides [3]. In this approach all of the wafers in a batch have a sequence of evaporations that produce the C-4 solder metallurgy (PbSn solder) bumps on a chip with a metallurgical barrier between solder and aluminum chip interconnection layers to prevent their interaction. This barrier may be chromium or titanium. The pad size on the chip and module, along with the volume of solder, determine the height of the solder balls. The solder balls are reflowed and connected to a pattern of pads on ceramic substrates. Since the solder is very ductile, it can withstand many thermal cycles even though there is substantial shear on the joints due to differences in thermal expansion between substrate and chip. Many fatigue studies have been completed on this system to determine its reliability.

The C-4 process is an ideal mass production process for several reasons. First, all of the pads on the chip are evaporated at once in large batches of wafers. Second, surface tension of the solder provides self location and centering of the chip on the module during reflow. Third, the reflow is done on a continuous belt furnace. Finally, very large area arrays of contacts on the chip are possible at reasonable tolerances. The electrical performance of these joints is very good, but there are thermal limitations due to a poor conductive thermal path through the insulation on the chip surface and limited metallurgical contact area. Although this thermal limitation has not been critical to date due to decreasing power levels per circuit [16], the rate of increase in integration is gradually pushing up the power per chip. In fact, if one extrapolates the integration levels and power per circuit into the future we may expect to see chips that require 5 to 20 watts or more. Special configurations for cooling are being introduced (see Chapter 5) to handle this level of power [17,18].

An alternative to bipolar semiconductor technology with its associated high power, is the use of complementary metal oxide semiconductor devices (CMOS). In CMOS the power per circuit is lower but it is possible to place more circuits in a given area. Since CMOS densities of 10,000 to 50,000 circuits per chip have already been achieved, the relief appears to be only temporary. A further consideration is the desire to package these chips closely together to minimize the impact of packaging on the circuit performance. Thus, due to high chip density, the power density is increasing dramatically and is becoming a major limitation in packaging at the chip and module level.

INTERCONNECTIONS BY PRINTED CIRCUIT CARDS AND PLANARS

The chip carriers or modules are generally interconnected on a printed circuit as shown in Figure I-1 in the Introduction. The card or planar has pads or holes to accept the terminals from the modules. In the case of surface-mounted components, e.g., PLCCs, the connection points are pads, while in the case of pinned grid arrays (PGA) or dual in-line packages (DIP), the pins interface into plated through-holes in the card. In both cases, they are soldered into place (see Chapter 19, "Joining and Interconnections"). Thus, the mechanical linkage is usually strain relieved to some degree by the flexibility of the leads and the compliance of the card. Nevertheless, the ductility of the solder becomes increasingly important for the larger components, and finally the largest components, such as the thermal conduction module (TCM) are handled by a flexible connector system [17,19].

The amount of printed interconnection per component is determined by the number of terminals on the component and by the spacing of the components on the card or planar or, alternatively, on a multichip module such as the TCM. If there are only a few components adjacent to each other, nearest neighbors so to speak, then the length of each interconnection is approximately the nearest neighbor distance or *pitch P*. If there is a large number of components, then some of the interconnections will be short such as those between nearest neighbor components and others will be long, perhaps extending across the printed circuit card. The average connection length is a function of the number of modules on the card and the distance (or pitch) between them. A great deal of work has appeared in the literature on this aspect of wiring. Some of this work is summarized in an appendix to this chapter.

It is clear that the wiring demand increases as the number of I/O pins increases. It is interesting to plot the wiring capacity of system cards and boards versus the number of I/O pins (N) times the pitch P as in Figure 2-4. The figure shows that wiring capacity versus NP falls on a straight line on a log-log plot. In the plot, some points are above the line, which indicates that the wiring efficiency of some systems is less than that of the other processors. This lower efficiency is due to stringent electrical wiring rules as well as blocking of areas as the wiring paths are filled. On the other hand, some systems utilize the wiring capacity quite well. This was achieved by the system designers who used a great deal of "hand" wiring as opposed to automatic wiring. Final manual imbedding of

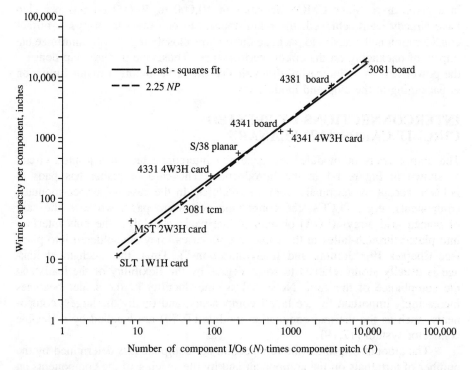

FIGURE 2-4
Printed circuit interconnection capacity. The number of inches of capacity required per module site is plotted against the product of the I/O (N) times the spacing or pitch (P). The solid line represents a best fit approximation to the data ($C = 3.52(NP)^{0.922}$). The dotted line is given by 2.25 times NP. The data are from cards and boards in the IBM systems noted.

overflow wires is often accomplished to save the cost of additional planes of wiring. More discussion of this plot can be found in the appendix on wireability.

It may be noted that the designer has the ability to increase the wiring capacity per component by increasing the pitch. Although this strategy increases the interconnection length required in proportion to P, the total capacity increases by the square of P, or area. Thus the wiring becomes easier to accomplish with more widely spaced components. An alternative strategy is to provide more layers of interconnection. Still another strategy is to increase the interconnection density by increasing the number of lines per channel. Since manufacturing cost is roughly proportional to the surface area of the printed circuit produced and because performance is of prime importance, it is best to build fewer layers by pressing interconnection density toward its practical yield limits.

SUMMARY PROCESS SEQUENCES
FOR MULTILAYER PRINTED CIRCUITS

In their simplest form, single- and double-sided printed circuit cores (Figure 2-5) are made by two conventional methods. One is called *additive* and the other *subtractive*. See Table 2.1.

For two-sided applications, copper is laminated on both sides of a base insulating material, such as epoxy-glass. Holes are made where conductor paths will be positioned by the lithography. The two sides are joined electrically by plating the hole walls and surfaces. For the subtractive process (Figure 2-6) the

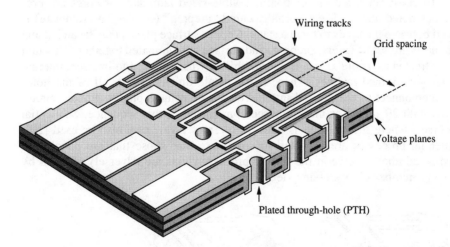

FIGURE 2-5
Composite with two signal-interconnection layers. The plated through-holes carry the signals between the wiring planes and are also used to mount chip carriers. The voltage reference planes carry current to the components and provide shielding and impedance matching. There are two wiring tracks between grid sites (two lines per channel).

TABLE 2.1

Typical circuit core fabrication

Subtractive	Additive
Glass fabric	Same
Silane coupling treatment	Same
Resin coating and curing	Same
Copper lamination (both sides)	Same
Drill holes	Same
	Photo process
Plate holes and surface	Plate holes and pattern
Photo process	Protect pattern
Etch pattern	Strip resist
Strip photoresist	Etch copper and adhesive layer

circuitization step is accomplished with photoresist protecting (tenting) the holes and defining the interconnection lines. In the alternative additive process or pattern plating techniques (Figure 2-7), the resist is applied prior to plating. Channels in the resist define the lines and openings at the holes, which are all plated at the same time. Since, normally, a copper nucleation and adhesive layer is applied everywhere, a final step is required to etch away this layer. During this step the lines and holes may be protected by tin or solder [9].

The subtractive process is somewhat simpler than the additive process but has tolerance limitations due to the chemistry and physical limits of the process.

To construct a multilayer board, double-sided laminates or cores are produced as noted above. Then, partially cured "prepreg" (epoxy-glass laminate) is placed between the cores to form a stack. Copper surface planes may be added and then the entire set of layers, prepreg, and surface are laminated together. This unit is now treated as a core again. Holes are drilled where desired to intersect internal copper pads joined to lines. Then, if the whole unit is plated, all of the holes will interconnect to the desired internal circuit or voltage planes. Printed circuit boards with 20 layers or more may be constructed in this sequence. Sequential processes, where the circuitry is built up layer by layer, may also be used [20]. Sequential processes are expected to cost more than those that are *parallel* as constructed above, since in the former process, yields are adversely affected by the large number of processing steps.

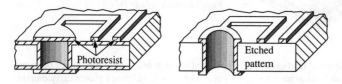

FIGURE 2-6

Circuit line and plated hole fabrication by the subtractive process [4]. See Table 2.1

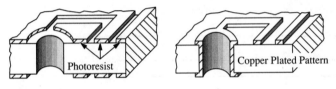

FIGURE 2-7
Circuit line and plated hole fabrication by the additive process [4]. See Table 2.1

DESIGN FACTORS
FOR PRINTED CIRCUITS

Interconnection Lines

As density increases, electrical wiring rules play a highly significant role. (As an example see Figure 2-8.) The spacing between signal lines and the closeness of signal lines to the ground planes are determined by electromagnetic coupling and impedance control [21]. If a higher density is desired, the dimensions must be scaled to preserve the impedance and coupling restrictions. This means the dimensions all decrease by the same factor to meet the same electrical ground rules. See Figure 2-9. That is, if we have designed a circuit board with, say, 80 ohm impedance, and with allowable coupled noise, we could change all dimensions by the same factor and would still have an 80 ohm structure with allowable coupled noise. However, the one electrical parameter that does not scale is the resistance. Therefore, if the scaling factor were one-half, the electrical resistance of the new design would be one-quarter of the original, since resistance is proportional to the inverse of cross-sectional area of the signal line. Thus the packaging density (interconnection density) is limited by electrical factors for high performance systems as well as the ability in photolithographic technology to space the interconnections closely.

The dimensions that control horizontal density are width (W) and spacing (S), along with the hole diameter (d). Typical line widths may be in the range of 0.13 to 0.25 mm (0.005 to 0.010 inch) and the hole diameter may be 0.75 to 1.00 mm (0.030 to 0.040 inch). State-of-the-art dimensions in 1989 are in the neighborhood of 0.5 mm (0.002 inch) and holes are 0.25 to 0.40 mm (0.010 to 0.016 inch). As noted in the above example, the spacing to minimize electrical coupling is quite large, 0.43 mm (0.017 inch), even for the more advanced printed circuits with 0.01-mm wide (0.004 inch) interconnections in an 80-ohm structure.

An approximate horizontal design is illustrated in Figure 2-10. Here the holes are 0.75 mm (0.030 inch) in diameter and the lines, $W = 0.10$ mm (0.004 inch), are spaced at $S = 0.43$ mm (0.017 inch) center to center for minimum coupling. Thus, a clearance of 0.19 mm (0.0075 inch) is left for the lines near the holes. Only four lines per 2.5 mm (0.100 inch) can be achieved. This density is called four lines per channel. It is obviously limited by the line spacing and the rather large holes, i.e., there are both electrical and physical limitations.

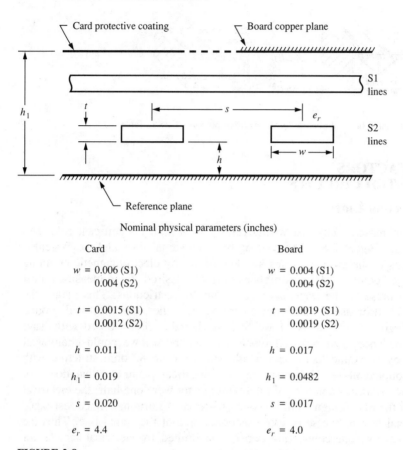

FIGURE 2-8
Card or board printed circuit structure designed for impedance and electromagnetic coupling optimization.[21]

With surface-mount technology the holes are only there for communication from the surface into the circuit planes, and thus the size of the holes can be decreased. The leads are attached to the surface rather than inside the holes. If, for example, the holes could be as small as 0.20 mm (0.008 inch), with a decrease in the spacing between lines and holes to 0.10 mm (0.004 inch), a design could be supported with three lines in 1.25 mm (0.050 inch), or six lines in 2.5 mm (0.100 inch) [4]. Thus the density increase could be about 50 percent in proceeding from pin-in-hole technology to surface-mount technology (six lines versus four lines per channel of 2.5 mm [100 mil]). In the designs, if the spacing between the hole and line is too small, it may be relieved by decreasing S locally in the area near the holes. For an approximate design with $W = 0.05$ mm (0.002 inch) with a closer spacing (0.18 mm) to the ground plane to maintain an 80-ohm impedance, the line spacing decreases to about 0.19 mm while still maintaining low coupling. Thus, the density can increase with these design factors to perhaps

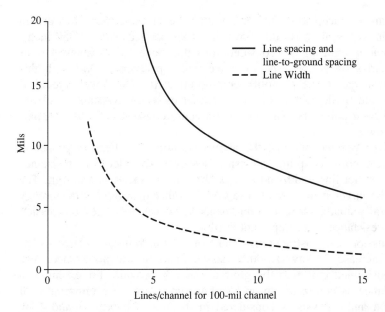

FIGURE 2-9
Interconnection spacing and height from ground plane for controlled impedance and coupling: Triplate Structure, Impedance 80 ohm, Dielectric Constant 4, di/dt same as Figure 2-8.

five lines per 1.25 mm (0.050 inch). When the number of lines per inch increases, there must be incorporated into the design a strategy for programmable vias so that line turning may be accomplished. Programmable vias are discussed more fully in a later section. Although further increases in density would appear to be beyond conventional printed circuit card technology, they are interesting from the point of view of extrapolation to higher chip function growth. As shown in Figure 2-9, 0.025 mm (0.001 inch) lines spaced 0.09 mm (0.0035 inch) to the

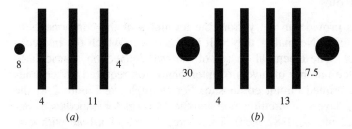

FIGURE 2-10
Approximate design for three and four lines per channel. The three-lines-per-channel design with a grid array of 0.050 inches is for use with surface-mounted devices. The holes for communication to internal planes are 0.008 inches in diameter, the line width is 0.004 inches, and the space between lines is 0.011. The four-lines-per-channel design is for pin grid array elements. The holes are 0.030 inches in diameter, the grid array is 0.100 inches, the line width is 0.004 inches, and the space between lines is 0.013 inches.

ground plane can be spaced about 0.09 mm (0.0035) center to center. Thus design extrapolation through scaling would allow nine lines per 1.25 mm (0.050 inch). These are probably some of the considerations that have led to investigations of packaging chips on silicon, the so-called wafer packaging. Although this would take advantage of the photolithographic art at the chip level, electrical performance would be degraded due to the higher dielectric constant of silicon compared with most polymers. This higher dielectric constant will cause signals to propagate slower.

It should be pointed out that the above example of scaling is proper in keeping electrical noise coupling constant. However, the electrical resistance increases with smaller line width and may be the limiting factor in a design. The resistance can be larger if the circuits use CMOS rather than bipolar technology due to the generally smaller current requirements. However, for large transmission distances, low resistance is an important attribute.

This discussion has illustrated only some of our flexibility to change design parameters to increase density. Another design factor is the impedance. The line may be positioned closer to the ground plane to provide, for example, a 50-ohm impedance as contrasted to an 80-ohm impedance. The importance of electrical design and analysis is considered in detail in Chapters 3 and 4 on electrical design. The increased shielding from the ground plane when the lines are closer to it allows the lines to be spaced closer together for equal width at equal coupling. But since the spacing to the ground plane becomes very small for 50-ohm structures, a new approach to the design of the insulating material may be needed for a guarantee of the tolerances and the integrity of the insulation. Still another design factor is the dielectric constant of the material. The use of lower-dielectric-constant materials would allow proportionally closer spacing of signal lines for the same impedance and equal coupling. No doubt the materials engineer interested in new composite materials will have a large influence in achieving higher-density printed circuit structures. One possibility is the use of Teflon-like[1] materials which have dielectric constants near 2.

Plated Through-Holes

Plated through-holes provide an entry from the terminals of the components to the interconnection lines. Normally, they will have a pad to which the interconnection line is joined. The electrical design of the interconnection, particularly the impedance and the number of layers of interconnection required, determines the thickness of the printed circuit composite. For example, in Figure 2-8, the thickness for just two layers of circuitization and the spacings for impedance control add to $h_1 = 1.22$ mm (0.0482 inch). Therefore, if we drill a hole with $d = 0.20$ mm (0.008 inch) diameter, the aspect ratio would be $h_1/d = 6$. This would require quite a long drill with a small diameter, which would be quite flexible.

---

[1]Teflon is a registered trademark of DuPont Corporation.

Thus, a combination of the tolerances for drill entry, drill flexibility, shrinkage or expansion of the composite material on processing, degradation of the hole wall due to mechanical fractures, and tolerances in registration and insulation specification all will determine how much horizontal space is allowed for the drilled hole. If a four signal-layer composite is required, the drill diameter may have to be increased to decrease flexibility and drill wander. On the other hand, if a lower impedance is used, the composite becomes thinner and thus more practical for small hole drilling. However, with lower impedance, the power per circuit increases so this may not be acceptable for the highest-density circuit packaging where cooling requirements are near the practical limits.

The stability or motion of the composite dimension during processing, discussed in Chapter 11, "Polymers in Electronics," may be controlled in a variety of ways. One of the ways is to control the impact of the processing parameters, such as moisture content. Another is to position the signal lines on either side of a copper ground plane, i.e., to work with a copper core structure, which has a high elastic modulus. In the most precisely designed structures, the dimensional stability is compensated for in the artwork. This may require artwork shifts or expansions of about one part per thousand. In short, what this compensation does is to retain the integrity of the grid positions in the internal structure so that when the drill penetrates it will contact precisely and cut the internal planes at the desired points.

Holes are required at turning points in the interconnection, i.e., where a street meets an avenue or an x track meets a y track. That is, the interconnection must have versatility to move out of one track and into another when the channel becomes blocked by a wire already printed there. Drilled through-holes are one means of achieving the required turning points. Several turning points per connection are required on the average to maintain wiring versatility and efficiency.

Programmable Vias

Turning points may be achieved by using programmable vias. The position of these vias is determined by the automatic wiring programs, hence the name programmable. These holes can be smaller than the through-holes, since they are only equal in depth to the signal core thickness, perhaps 0.006 to 0.010 inch. They may be drilled mechanically or by laser. (See Chapter 20, "Mechanical Processes," Chapter 17, "Plated Through-hole Technology;" and Chapter 15, "Lasers and Laser Applications.") For low-aspect-ratio programmable vias, the laser appears to be an ideal tool [22]. In designing a programmable via, one should keep the diameter small enough so that the via requires no more space than the interconnection line.

Power Distribution
and Impedance Structures

Copper voltage planes buried within the composite establish voltage levels for the circuits and provide an impedance referencing system for the signal lines. It may be required to switch hundreds of amperes at once in high-performance dense printed circuit systems. Hence, there is a desire to provide a substantial conducting system of copper from the power bus on the outside of the composite to the devices. The voltage planes minimize the voltage changes at the semiconductor to below a predetermined tolerance level, perhaps 100 millivolts or less, by supplying the smallest possible resistance through a large area of copper. The voltage cores are often complemented by large decoupling capacitors to smooth out the voltage fluctuations in magnitude and time. Several publications [11,21,23] provide pertinent information on the electrical modeling. These structures will be analyzed further in Chapters 3 and 4 on electronic design.

Voltage structures have clearance holes wherever a signal plated throughhole penetrates the voltage structure. The clearances in the structures are much like those for signal lines discussed earlier. However, the strong copper structures do not have as large a dimensional instability due to processing variables as do the printed circuit structures.

Card Planar Composite Structure

We are now in position to do an approximate design exercise to construct a composite to interconnect a set of components. Let us consider connecting 16 PLCC components, each with 100 I/Os. With 100 I/Os we could expect to support about 1000 circuits per component, or 16,000 circuits total.

Since the PLCCs are surface-mounted components and small holes are required, the six-line-per-channel structure discussed earlier can apply. The length of interconnection, 2.25 NP (Figure 2-4), amounts to 11.25 m (450 inches) of wiring per component if we space them on a 50-mm (2-inch) pitch. However, the area beneath each component only provides for 6.01 m, or 240 inch (20 grid sites × 6 lines/channel × 50 mm or 2 inches), for each signal plane. Therefore, two signal or circuit planes are required along with a minimum of two voltage planes for referencing if a triplate structure is desired. (A triplate structure is one where a reference plane is on either side of each signal line pair.) Since there are 16,000 circuits, the connections to the card or planar, following a Rent-like calculation, should be in the range of $4(16,000)^{0.5} = 500$. This is a large number of contacts to space along one edge of a 200-cm (8-inch) card with a four-by-four array of components with a pitch of 50 mm (2 inches). Contacts on a 0.635-mm (25-mil) grid along the 20-cm (8-inch) edge would only provide 320, which is about half the requirement. Thus, two edges of the card are required for connector contacts until higher contact densities are practical. Historically, these connections have been supplied by top card connectors. More recently, zero insertion force connectors on two sides of the cards have been used to increase the number of contacts.

From this discussion, it appears desirable to increase contact counts. One alternative is to use the entire area for contacts. This is, in fact, a feature of the large ceramic multichip module technology, which replaces both the card planar discussed above and the modules on the card planar. One can, in retrospect, view the 1800 connector contacts per module in an area array on the 3081 printed circuit boards announced in 1981 as still state-of-the-art.

There are a variety of additional attributes of ceramic multichip modules. Perhaps the most valuable of these is the ability to space the chips very closely, approximately 100 in a 10 cm (4 inches) square array, rather than, as above, only 16 for a 20 cm (8 inches) square array on a circuit card using PLCCs. The strength, dimensional stability, and ability to provide an almost unlimited number of layers establishes the ceramic multichip module in performance leadership. The complexity of the planar board package required to interconnect these large and high-I/O multichip modules is discussed in several papers [2,9,10].

THE CARD-ON-BOARD PACKAGE

When many cards must function together they may be plugged directly (Figure I-1, Introduction) through connectors into a back-panel composite or a printed circuit board. One printed circuit board may hold 10 to 20 of these cards or planars and the wiring density is affected in direct proportion to the connector I/O at the card edges and the pitch between cards. That is, wiring considerations discussed earlier are applicable here; just replace a module or chip carrier in your thoughts with a card. Thus, the printed circuit board becomes very complex due to the high I/O count and the rather close spacing when cards are spaced edge to edge. One example of this type of package is the 4300 series of processors announced by IBM in 1979 [5]. Here the card edge contacts numbered 268, not counting the top-card connectors, which often added another 50 to 100 contacts. The reader may do some rough calculations here to be convinced that an eight-signal-layer printed circuit board is required for this application [5]. Since the connector springs are inserted into holes in the board, a 0.45-mm (0.018-inch) hole is required. This hole and its clearance use a significant fraction of the space, leaving only four lines per channel (2.5-mm or 100-mil grid) for wiring.

Air is blown through the space between the cards to cool the components. Thus, the card spacing has a practical limitation determined by the cooling capacity of the air and the total power per card, which must be dissipated by the air. Since the thermal path starts at the chip junction, temperature drops are significant—junction to semiconductor chip, chip to module, thermal spreading on the module, and, finally, module to air and air over the card length. (See Chapter 5, "Thermal Design.") Large volumes of air at up to 300 meters (984 feet) per minute flow must be passed over the cards to maintain semiconductor junction temperatures in the range of 100°C or less.

The capacity for even higher circuit densities is obviously limited in the card or board configuration by a variety of factors. We have discussed the wiring density factor, the connector density factor, and the thermal density factor. Chapter 5 will expand on the progress in achieving higher densities by applying new cooling techniques.

MEMORY
AND MICROPROCESSOR PACKAGING

Memory and microprocessor components use fewer I/Os than conventional random logic components. Since the wiring density is in direct proportion to the I/O, substantially simpler packages are required. Also, since these components are "bus" wired, one finds a higher ratio of wires per I/O, perhaps close to one to one, as compared to three interconnections for four I/Os in the random logic packaging. However, with bus wiring, most of the contacts are nearest neighbors. These two factors are opposite so we expect the interconnection demand requirements to be quite similar for all components. Therefore our previous estimates are also good for interconnection requirements for microprocessor and memory packaging. The principal factor, accounting for a lower interconnection density, is the low number of I/O counts per component.

The growth of memory requires only one additional I/O each time the size of the memory doubles, so memory packaging is easily contained as a subset of logic packaging rather than requiring any special attention. On the other hand, since the wiring is easily available, there is a desire to make the memory first-level packages very small so they can be packaged at low cost and still meet high performance requirements. Microprocessor I/O demands, however, are increasing as wider bit widths are used for the more advanced systems. In the past ten years the bit widths have doubled twice—from 8 to 16 to 32 bits—and continued rapid growth is expected. Thirty-two-bit systems require from 120 to 240 I/Os. Therefore, assuming that these systems will continue to require a processor on a chip and a variety of adapter chips, the card complexity is expected to increase and ultimately look similar to the complex logic packaging we have been experiencing with random logic on the highest-performance systems.

Surface solderable components, i.e., PLCCs and TABs, are removing the previous limitation of small I/O counts compared to those for DIP. Therefore, as these new technologies grow along with the higher demand for wider bit width microprocessors, we expect to see a large increase in the use of complex multilayer printed circuit boards even for small systems.

LONGER-RANGE CONSIDERATIONS

There is no foreseeable end to the drive for performance, which depends on density. The I/O density dependence on function is expected to continue as far as we can project. The limiting density appears to be approached by packages that place chips edge to edge (often called "brick walled") or places the chips on their edge as in single in-line packages (SIP). Whether or not this will be accomplished by strain relief or by matching thermal expansion does not appear to matter. Both are appropriate as long as cost performance requirements are met. Accordingly, there appears to be a lot of room for innovative advances in material composite concepts that will handle these interfaces appropriately. The TCM in the IBM 3081 and 3090 is just one example that has reached the manufacturing stage.

Several examples have been demonstrated in new packaging trends. One of these involves chips that are back bonded and wire bonded directly onto printed circuit boards [24] and another has leadless chip carriers attached to dense circuitry on insulated copper with a laminate having a matched coefficient of expansion [25,26]. References [27] and [28] show some of the nonconventional laminate and foil materials available. Furthermore, new ceramic materials (Chapter 10, "Ceramic Structures for First Level"), including glass ceramics, aluminum nitride, and silicon carbide, that match the chip better in thermal expansion are in the forefront of development thrusts. Aluminum nitride has a much improved thermal conductivity. Thin-film structures for greatly improved interconnection density are receiving increasing attention. These trends in designs and technology are discussed in [28], which has a very large bibliography. There is, in fact, such a large variety of opportunities in applications of new materials and techniques that it is going to be very difficult to predict how many or how few of these will emerge in mass production.

The key will not rest in whether or not such technologies can be developed; rather the issue will be mass production at the lowest system cost. The lowest system cost will depend on the interfacial requirements from package level to level as well as on the productivity of each manufacturing process. To accomplish this, one should apply the following rules:

1. The architecture of the system must be in place prior to development.
2. During development, emphasis must be placed on the manufacturability of the components.

REFERENCES

1. D. Balderes and M. L. White, "Package Effects on CPU Performance of Large Commercial Processors," *Proceedings of the 35th Electronic Components Conference*, IEEE, May 1985, Washington, D.C., pp. 351–356.
2. D. P. Seraphim, "Chip to Module Interfaces," *IEEE Transactions on CHMT 1*, no. 3, pp. 305–309 (1978).
3. D. P. Seraphim and I. Feinberg, "Electronic Packaging Evolution in IBM," *IBM J. Res. Develop.* 25, pp. 617–629 (1981).
4. D. P. Seraphim, D. E. Barr, W. T. Chen, and G. P. Schmitt, "Electrical and Electronic Applications," in *Advanced Thermoset Composites*, ed. James M. Margolis, Van Nostrand Reinhold Co., New York (1986) in Chapter 4, p. 110.
5. G. G. Werbizky, P. E. Winkler, and F. W. Haining, "Making 100,000 Circuits Fit Where at Most 6,000 Fit Before," *Electronics*, 52, no. 16, pp. 109–114 (August 2, 1979).
6. J. Horton and N. Compton, "Technological Trends in Automobiles," *Science* vol. 225, no. 4662, pp. 587–593 (1984).
7. B. T. Clark and Y. M. Hill, "IBM Multi-Chip Multi-Layer Ceramic Modules for LSI Chips-Design for Performance and Density," *IEEE Transaction on CHMT-3*, pp. 89–93 (1980).
8. A. J. Blodgett, "A Multi-Layer Ceramic, Multi-Chip Module," *Proceedings of the Technical Program*, 30th Electronic Components Conference, IEEE, New York, 1980, pp. 283–285.
9. D. P., Seraphim, "A New Set of Printed-Circuit Technologies for the IBM 3081 Processor Unit," *IBM J. Res. Develop.* 26, pp. 37–44 (1982).

10. R. F. Bonner, J. A. Asselta, and F. W. Haining, "Advanced Printed-Circuit Board Design for High-Performance Computer Applications," *IBM J. Res. Develop.* 26, pp. 297–305 (1982).
11. E. E. Davidson, "Electrical Design of a High Speed Computer Package," *IBM J. Res. Develop.* 26, pp. 349–361 (1982).
12. E. S. Landman and R. L. Russo, "On a Pin versus Block Relationship for Partitions of Logic Graphs," *IEEE Trans. Computers* C-20 pp 1469–1479 (1971).
13. W. R. Heller, W. F. Mikhail, and W. E. Donath, "Wireability Analysis for LSI" *J. Design Aut. Fault-Top. Comput.* 2, pp. 117–144 (1978).
14. C. Radke, "A Justification of and an Improvement on a Useful Rule for Predicting Circuit to Pin Ratios," *Proc. 1969 Design Automation Conf.*, IEEE, Computer Society, pp. 257–267.
15. J. W. Balde and D. Brown, "VLSI and the Substrate Connection, the Technological Tradeoffs of the Package Board Interface," a multiclient technical review and market survey, done by D. Brown Associates, Inc., 1982.
16. D. R. Franck and E. Kellerman, "System Performance and Technology Trends," *IEEE Computer Packaging Conference*, Wescon, Los Angeles, May 1985.
17. A. J. Blodgett and D. R. Barbour, "Thermal Conduction Modules: A High-Performance Multi-layer Ceramic Package," *IBM J. Res Develop.* 26, pp. 30–36 (1982).
18. E. A. Wilson, "Integral Liquid-Cooling System Simplifies Design of Densely Packaged Computer," *Electronics*, vol. 57, no. 2, pp. 123–126, (January 26, 1984).
19. G. G. Werbizky and F. W. Haining, "New Bifurcated Spring Connector ZIF-System for the IBM 3081 Processor," *Proceedings of the Technical Program*, Productronica, Munich, West Germany, December 1981.
20. J. R. Bupp, L. N. Chellis, R. E. Ruane, and J. P. Wiley, "High-Density Board Fabrication Techniques," *IBM J. Res. Develop.* 26, pp. 306–317 (1982).
21. N. C. Arvanitakis and J. J. Zara, "Design Considerations of Printed Circuit Transmission Lines for High Performance Circuits," *Proceedings of the Technical Program*, Wescon/81 Professional, San Francisco, Wescon, Los Angeles, September 15–17, 1981.
22. W. R. Wrenner, "Laser Drilling P.C. Substrates," *Circuits Manuf.* 17, no. 5, pp. 28–32 (1977).
23. D. Carlson, "High Performance Gate Array Packaging Design Tradeoffs." *Proceedings of the Technical Program*, IEEE Computer Packaging Spring Workshop, Split Rock, Pa., May 1982.
24. J. W. Beyers, E. R. Zellers, and S. D. Seccombe, "VLSI Technology Packs 32 Bit Computer System into a Small Package," *Hewlett Packard Journal* 34, no. 8, pp. 3–6 (1983).
25. R. A. Reynolds, "Clad-Metal Core PC Boards Enhance Chip-Carrier Viability," *Electronic Daily News*, Vol. 29, no. 17, pp. 211–215 (Aug. 23, 1984).
26. R. W. Wright, "Polymer Metal Substrates for Surface Mounted Devices," *Proceedings of the Technical Program*, National Electronic Packaging and Production Conference, Anaheim, Calif., March 1–3 Cahners Exposition Group, Chicago, 1983, p. 47.
27. G. Messner, "Nonconventional Substrates," *Circuit World*, vol. 11, no. 2, pp. 39–41 (1985).
28. R. Tummala and E. Rymaszewski, "Microelectronics Packaging Handbook," Van Nostrand Reinhold, to be published 1988.

PROBLEMS

2.1. A signal begins from an output of a device on a chip. It then crosses the chip to a wire bond which joins the chip to a ceramic substrate. The signal traverses the substrate and enters a card through a pin, traverses the card to a board, and enters the board through a connector. It traverses the board and enters a cable through a connector termination. Then the path goes through a similar sequence but in reverse. That is, it goes from the board to a card to a substrate to a chip. Distances and dielectric constants are shown as follows.

Element	Distances	Reverse distance	Dielectric constant
Chip	1 cm	0.5 cm	12 (effective)
Ceramic	3 cm	2 cm	9
Card	15 cm	8 cm	4.3
Board	40 cm	10 cm	4.3
Cable	100 cm	—	2.8

(*a*) Roughly how much circuit time would be saved if the modules contained many chips allowing all of the interconnections to be on the module? No cards, boards, or cable would be in this particular circuit. Assume the new module to be able to place these two communicating chips within 4 cm.

(*b*) How many significant electrical discontinuities are seen by this signal and how many are saved by the alternative design?

(*c*) Think about what kind of minor discontinuities could also have an impact due to the fact that there are many of them. Name them and provide a rationale.

(*d*) What is the cycle time, due to transmission delay alone, in the two cases, if there are 12 instructions per cycle and the paths noted are average lengths for each case? Assume a very low level of integration in the first case causing every signal to go off the chip. In the second case assume four of the instructions to take place on the chip while the 4th instruction drives off the chip onto the module to get to the next chip.

(*e*) If all of the communication was on the chip but the chip size and the average distance traversed had to be doubled, what would the cycle time be? How important is the effective dielectric constant of the chip if the circuit switching time (so far neglected) is 100 picoseconds. Discuss several methods for improving this situation.

2.2. Discuss what a "COB" system might look like containing 800,000 circuits. Consider these limitations: (a) maximum chip I/O equals 80, (b) maximum card connector I/O equals 400, (c) 10 cards per board. Let the power connections be one in four of these I/O.

(*a*) How many chips can be packaged per card?

(*b*) How many boards will there be?

(*c*) How many cable connections from each board will there be?

(*d*) Structure a card wiring pattern assuming the card is eight inches square.

APPENDIX
WIREABILITY CONSIDERATIONS

When designing a computer system or a component of such a system, it is very important to be able to predict how much wiring capability is needed. This information is needed very early in the design cycle so that the wiring structure can be set. That is, one usually needs to predict the number of lines per channel and the number of signal planes that will be used.

As the design progresses, actual wiring runs will determine the particular point-to-point wiring. At that time it will be possible to determine if the initial predictions made by wireability studies were successful. The automatic wiring programs in existence usually do not permit the *embedding* of each and every line. There are almost always *overflows*. An overflow is a wire that cannot be embedded by an automatic program. The common practice is to embed as many wires as possible with automatic wiring programs and then do "hand" wiring to embed the overflows. Hand wiring usually consists of a designer using a CAD/CAM terminal to move wires to different locations in an attempt to embed all wires.

A great deal of work has been done on the subject of wireability [1–6]. There are two aspects to wireability. One is *escape* and the other is *global wiring*. Escape is the ability to run lines from interior leads on a module to wiring tracks.

ESCAPE

Escape has to be considered when the module I/Os are arranged in an array— whether surface mount or a pin grid array attached using plated through-holes. Peripherally leaded devices have no escape constraints since the leads are on the periphery of the module and have therefore already escaped.

To determine escape from an array of points interior to the module, one has only to count the number of escape routes at the periphery and take into account that the outer row of pins has already escaped. If the array is in the form of an N-by-N grid, then the number of lines that can escape, N_e, is

$$N_e = 4(N - 1) + 2SL(N - 1) \tag{A.1}$$

where S is the number of signal planes, L is the number of lines per channel at the outer row, and N is the size of grid.

The first term, $4(N - 1)$, is the number of pins on the outer row that are already escaped. The second term, $2SL(N - 1)$, is the number of lines that can escape from interior pins. The 2 represents that on each signal plane, lines can escape in two directions. On x planes, escape can be horizontal in two directions, and on y planes, escape can be vertical in two directions. The above equation can only be used if all escape channels can be used. In practice, a contingency factor of 75 to 85 percent is usually applied. The reason for the contingency factor is to take into account the fact that it is not always best to escape in the shortest direction since it might lead to considerable "wrong way" wiring once the line has escaped. It is best to escape in the direction of the target pin. The other reason for using a contingency factor less than 100 percent is to allow some *pass-through* wiring. Pass-through wiring is wiring connecting two modules on opposite sides of a given module, the shortest path being through the module site.

GLOBAL WIREABILITY

Global wireability is used to predict the number of signal planes and the number of lines per channel on each plane. See the article by Koch, Mikhail, and Heller [7] for a good summary of the global wireability analysis with applications to actual wiring runs.

Four methods of predicting the wiring capacity are mentioned below. Only the first two will be examined in detail, since the third and fourth methods, while requiring more work, are easy to understand.

Simple theory, where the connection length is independent of the number of modules.

Donath, Mikhail, and Feuer theories. In this method, the average connection length is expressed as a function of the number of modules, and this length used to predict the capacity needed.

Measured length. The measured length method depends on having more information than the first two methods require. To use this method, one needs to know how many connections are used between each and every module. The length of each connection is determined, and by counting, the wiring demand is computed. A rule of thumb is that if the wiring demand is less than 40 percent of the track lengths available, the card can probably be wired.

Cuts. The same information is needed as in the length method. However, instead of measuring the lengths, one determines how many connections cross given planes. Generally, a card can be wired if the number of tracks required is not more than 85 percent of the tracks available at the most congested part, as shown by W.G. Thompson [8].

PRELIMINARIES

This wiring capability is based on the number of modules on a structure and the number of input/output (I/O) pins on each module.

One might be interested in the amount of wiring one has to provide on a card based on the number and complexity of chips on the card. Or, in a card-on-board system, one might be interested in determining the wiring needed in the board.

Rent's Rule

In predicting wiring requirements, one often starts with Rent's rule [9], which states that the number of I/O connections is related to the number of circuits in a module as a power function:

$$I = bC^p \tag{A.2}$$

where I is the number of I/O connections, b is the average number of connections per circuit, C is the number of circuits in a package, and p is a positive exponent less than 0.67.

WIRING PREDICTIONS

The ability to predict the wiring requirements of a given configuration is based on three quantities:

Wiring capacity. The wiring capacity is defined as the total length of wire that could be embedded in a component regardless of whether this wiring is doing anything useful. That is, the wiring capacity is determined by computing the length of each wiring channel and multiplying by the number of such channels. It is convenient to normalize this for one module, where it is assumed that the modules are spaced a distance P apart. Therefore, the wiring length in a square of size P centered on a module is

$$W_c = P^2 \frac{SL}{G} \tag{A.3}$$

where P is the pitch, S is the number of signal planes, L is the number of lines per channel, and G is the grid size (channel width).

Rather than describing wiring capacity always in terms of lines per channel and then having to specify the channel, one can define the *wiring factor*, which is the product of the number of signal planes and the number of lines per channel divided by the width of a channel. That is:

$$W_f = \frac{SL}{G} \tag{A.4}$$

This wiring factor has the dimensions of reciprocal length. That is, it could be appropriate to describe W_f in terms of lines per inch or, equivalently, inches per inch squared. That is, it is a measure of the number of inches of wire that can be embedded in one square inch of a structure.

Wiring demand. The wiring demand is a much more complicated quantity. It represents the total length of wire needed to interconnect the components in a system, regardless of the routing of the wires. That is, the demand can be computed by assuming that any time it is desired to route a wire from point A to point B, it can be run with no consideration of other wires blocking the path. The sum of the lengths for each connection represents the wiring demand. However, usually wires can only be run in orthogonal directions. Therefore the distance is measured as the "Manhattan length," which is merely the x distance separating point A from point B added to the y distance separating point A from point B.

It is convenient to define a nondimensional quantity called the average length. This quantity is the average length of all of the interconnections divided by the module pitch. If there were only two modules on a card and one did not consider wires leaving the card, the average length would be 1, since all interconnections would go from one module to the other. Of course, in general the average length would be greater than unity. In the next section we will review some of the expressions that can be used to help predict the average length.

Since the wiring demand represents the total length of all wires, it is obtained by multiplying the number of interconnections times the average length times the module pitch.

Another parameter that has an effect on wireability is the average number of pins per net. A net is defined as an electrically connected set of wires and pins. If all nets just connect two pins, then the average number of pins per net is, of course, two. However, it is more common for some nets to connect several pins. Usually the average number of pins per net is between two and four.

It can be seen that if there are, for example, three pins per net, then there are two interconnections per net, but each net will interconnect three pins. In general, the number of interconnections per net is one less than the number of pins per net. That is, the wiring demand W_d per module can be expressed as

$$W_d = N_t \frac{P_n - 1}{P_n} \overline{R} P \qquad (A.5)$$

where N_t is the number of terminals per module, P_n is the number of pins per net, and $\overline{R}$ is the average, nondimensional connection length.

Wiring efficiency. It is obvious that we cannot fill all of the wiring channels to their capacity. A measure of how much of the capacity we use is the wiring efficiency, which is defined as the wiring demand divided by the wiring capacity. That is,

$$W_d = \epsilon W_c \qquad (A.6)$$

where ϵ is the wiring efficiency. The wiring efficiency is determined empirically, and it is generally in the range of 40 to 50 percent. It depends on many factors, such as how good the automatic routing algorithms are and how many vias are available. Also, an automatic wiring result can usually be improved with some human (hand wiring) intervention.

AVERAGE INTERCONNECTION LENGTH

A large body of literature has grown which addresses this question: What is the average length of a wire in a package? This parameter is important since the wiring demand can be computed by multiplying this number by the number of interconnections, as shown above.

In the theories below, the quantity N stands for the number of modules on a card, if one is trying to predict card wireability, or the number of circuits on a chip, if one is trying to predict chip wireability.

Simple theory. One of the simplest ways to estimate the average interconnection length is to assume that exactly half of the wires go to a module's nearest neighbors (P) and exactly half go to the next nearest neighbor ($2P$). Then the average wire length is

$$\overline{R} = 1.5 \tag{A.7}$$

Donath's theory. Donath [10] used a probabilistic, hierarchical scheme and Rent's rule to develop an equation for average wire length. His expression is an upper bound for the average connection length:

$$\overline{R} = \frac{2}{9}\left(7\frac{N^{p-0.5} - 1}{4^{p-0.5} - 1} - \frac{1 - N^{p-1.5}}{1 - 4^{p-1.5}}\right)\frac{1 - 4^{p-1}}{1 - N^{p-1}} \tag{A.8}$$

where N is the number of modules and p is the Rent exponent.

Feuer's theory. Feuer [11] also used a probabilistic, hierarchical scheme and Rent's rule, but obtained a different result than Donath's:

$$\overline{R} = 2^{0.5}\frac{2p(3 + 2p)N^{p-0.5}}{(1 + 2p)(2 + 2p)(1 + N^{p-1})} \tag{A.9}$$

where N is the number of modules.

Mikhail's formula. Mikhail [1] applied heuristic arguments and some curve fitting to approximate Donath's theory with the somewhat simpler formula:

$$\overline{R} = k(1 + 0.1 \ln N)\overline{N}^{1/6} \tag{A.10}$$

where N is the number of modules and k is a positive constant empirically determined. The quantity k is usually taken to be about 0.8 for cards, whereas for chips it is about 0.6. To develop the above formula, Mikhail used a Rent exponent of 2/3.

It is interesting to compare these different theories graphically, as shown in Figure 2A-1. While the rule-of-thumb prediction is not close to that of the theories for large numbers of modules, note that it can be used with fair accuracy when dealing with small numbers (around 10) of modules.

While Donath and Feuer achieved the same exponential factor [N raised to the ($p - 0.5$)th power, they are not asymptotically the same due to different constant factors. In fact, when $p = 2/3$, Donath's theory gives a number about two times that of Feuer's as N becomes large. When p is less than 2/3, this difference is even larger. However, the difference of a factor of two is not surprising when one considers that Donath only attempted to produce an upper bound. His own investigations let him conclude that he was high by about a factor of two. Mikhail's theory seems to support experimental results, and it is roughly halfway between the theories of Donath and Feuer (at least for $p = 2/3$).

It is interesting to compare the above results with the expression used by Seraphim in the introductory chapter. There he uses the formula

$$W_c = 2.25N_t P \tag{A.11}$$

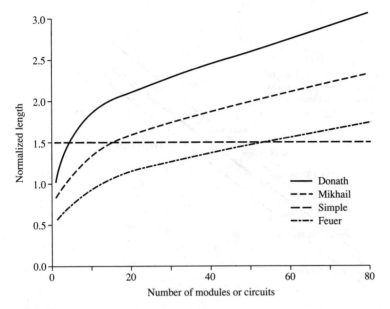

FIGURE 2A-1
Comparison of various theories for predicting average connection length. The theories of Donath, Feuer, Mikhail, and Seraphim are compared.

If we use his figure of an average of 4 pins per net and we use a wiring efficiency of 50 percent, we obtain

$$W_c = 1.5\overline{R}N_t P \qquad (A.12)$$

This shows that Seraphim's figure for average connection length $\overline{R}$ is 1.5, since this value gives 2.25 in Equation (A.11). We now have a simple formula for computing wiring capacity. The wiring capacity in length of wire per module is merely 1.5 times $N_t P$ times $\overline{R}$, where $\overline{R}$ is based on the number of modules used on the structure. When there are about 15 modules, $\overline{R}$ is about 1.5 regardless of which theory is used. As the number of modules increases, $\overline{R}$ should be increased accordingly.

If we use the previous results, we can write the somewhat more general formula:

$$W_c = \frac{\overline{R}(P_n - 1)}{P_n \epsilon} N_t P \qquad (A.13)$$

That is, if the number of pins per net, the average connection length, and the wiring efficiency are all constant, the wiring capacity is merely a linear function of $N_t p$, as shown by Seraphim. The slope is given by

$$\frac{\overline{R}(P_n - 1)}{P_n \epsilon}$$

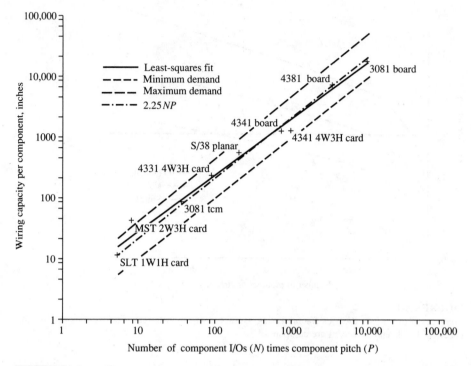

FIGURE 2A-2
Wiring capacity in inches per component versus component I/O × pitch. This figure is the same as Figure 2-7 except that the theory in this appendix has been applied to establish limits. The solid line is a least-squares fit to the data while the center dotted line represents the simple theory where the wiring capacity can be estimated by $W_c = 2.25 N_r P$. The average connection length $\overline{R}$ is computed from the Mikhail formula.

Figure 2A-2 shows the function plotted for some reasonable maximum and minimum values of the slope. The minimum demand is when $\epsilon = 0.5$, $P_n = 2$, and $\overline{R} = 1.08$ (four modules), which gives a slope of 1.08. The maximum demand assumed is with the value $\epsilon = 0.4$, $P_n = 4$, and $\overline{R} = 2.36$ (100 modules), which gives a slope of 4.42. Note that all the examples plotted in the figure fall between these two limits.

EXAMPLE

Let us use the above information in an example to show that it is fairly easy to predict wiring requirements. Recall that the equations for wiring capacity, wiring demand, and wiring efficiency for one module are

$$W_c = P^2 W_f \tag{A.14}$$

$$W_d = N_t \left(\frac{p_n - 1}{P_n} \right) \overline{R} P \tag{A.15}$$

$$W_d = \epsilon W_c \tag{A.16}$$

where W_c = wiring capacity (in./module)
W_d = wiring demand (in./module)
ϵ = wiring efficiency (percent)
$\overline{R}$ = average wiring length (nondimensional)
W_f = wiring factor(in./in.^{2})
P = module pitch (in.)

If Equations (A.14) and (A.15) are substituted into Equation (A.16) and the result solved for P, we obtain an expression for the module pitch in terms of the other parameters:

$$P = N_t \frac{(P_n - 1)\overline{R}}{P_n W_f \epsilon} \tag{A.17}$$

This expression allows us to predict how large a card is needed for a given set of modules.

Let us consider an example of mounting 12 modules on a card. Assume each module has 50 input/output terminals (signal only). If we use the Mikhail formula, this gives us the average connection length of $\overline{R} = 1.42$, where we have taken $k = 0.75$. Also, assume that there are, on the average, three pins per net ($P_n = 3$), the wiring efficiency is 40 percent, there are two lines per 100-mil channel, and there are four signal planes. Therefore $W_f = 80$ in./in.^{2}, $P_n = 3$, and $\epsilon = 0.4$. Then the pitch is $P = 1.84$ inches. That is, to place the 12 modules in a 3-by-4 array would require a card approximately 6 by 8 inches.

By going through examples like this, one can see how the different wiring parameters affect the design of a package. Of course, in the final analysis, we can only find out if our prediction was correct when we do the actual wiring of the card. If the actual wireability of a card is much different than that predicted by the above equations, it could be for any of the following reasons:

The wiring is not random logic. The wireability equations are derived assuming random logic. A memory card, for example, would have bussed logic and may not fit the prediction well.

There may not be enough vias to allow channel switching.

The modules may not be homogeneous. That is, some modules might produce much more wiring demand than others, which can cause congestion in parts of the card.

Escape requirements may require extra wiring space in the vicinity of the modules.

REFERENCES

1. W. R. Heller, C. G. Hsi, and W. F. Mikhail, "Wirability—Designing Wiring Space for Chips and Chip Packages," *IEEE Design Test.*, August 1984, pp. 43–51.
2. D. C. Schmidt, "Circuit Pack Parameter Estimation Using Rent's Rule," *IEEE Trans. Computer-Aided Design Integrated Circuits Systems* CAD-1, no. 4, pp. 186–192 (1982).

3. W. H. Knausenberger and N. A. Teniketges, "High Pinout IC Packaging and the Density Advantage of Surface Mounting," *IEEE Trans. Components, Hybrids, Manufacturing Technology* CHMT- 6, no. 3, pp. 298–304 (1983).

4. G. Messner, "Cost-Density Analysis of Interconnections," *IEEE Trans. Components, Hybrids, Manufacturing Technology* CHMT-10, no. 2, pp. 143–151 (1987).

5. D. S. Landman and R. L. Russo, "On a Pin versus Block Relationship for Partitions of Logic Graphs," *IEEE Trans. Computers* C-20, pp. 1469–1479 (1971).

6. W. R. Heller, W. F. Mikhail, and W. E. Donath, "Wireability Analysis for LSI," *J. Design Aut. Fault-Tol. Comput.* 2, pp. 117–144 (1978).

7. J. H. Koch III, W. F. Mikhail, and W. R. Heller, "Influence on LSI Package Wireability of Via Availability and Wiring Track Accessibility," *IBM Journal of Research and Development* 26, no. 3, pp. 328–341 (1982).

8. Personal communication.

9. C. Radke, "A Justification of and an Improvement on a Useful Rule for Predicting Circuit to Pin Ratios," *Proc. 1969 Design Automation Conf.*, pp. 257–267.

10. W. E. Donath, "Placement and Average Interconnection Lengths of Computer Logic," *IEEE Transactions on Circuits and Systems* CAS-26, no. 4, pp. 272–277 (1979).

11. M. Feuer, "Connectivity of Random Logic," *Workshop on Large Scale Networks and Systems*, IEEE 1980 Symposium on Circuits and Systems (ISCAS 1980), pp. 7–11.

CHAPTER
3

ELECTRICAL DESIGN CONCEPTS IN ELECTRONIC PACKAGING

ROBERT H. KATYL
JOHN G. SIMEK

INTRODUCTION

Electronic packaging is a term that encompasses the group of chip carriers or modules, printed circuit cards and boards, and cabling and connectors that make up an electronic assembly. Packaging provides a variety of mechanical, thermal, and electrical functions. The two primary electrical functions of an electronic package are:

1. Provide a mechanism for delivering power to the semiconductor circuits with a high degree of stability, and free of electrical noise, which can arise when many digital circuits switch simultaneously.
2. Carry electrical signals from one circuit to another with some form of current-carrying element, such as a discrete or printed circuit wire, while maintaining fidelity of the signal even at the highest speeds at which the circuitry can operate.

To maintain signal fidelity, much effort in packaging design for digital computer systems is devoted to minimizing unwanted or parasitic electrical effects, such as delta-I noise and cross-talk between lines, described in detail below. They arise because real electronic components do not have all the properties of idealized circuit elements. For example, real capacitors, resistors, and even simple wiring can have inductive properties, which may cause significant voltage drops in a power distribution system. These can occur during fast switching transients when digital circuitry switches from one state to the next. Also, signals on adjacent lines are not completely isolated electrically from one another, and unwanted signals may couple from one line to another. In this chapter we will discuss the results of some of these parasitic effects in digital computer systems. We start with a review of basic electrical parameters.

BASIC ELECTRICAL PARAMETERS

Resistance. The resistance parameter yields the potential or voltage drop across a circuit element due to dissipative processes. For a current flow of I (amperes, A) into a resistance of R (ohms, Ω), the voltage drop is given by Ohm's law:

$$V = IR$$

Capacitance. The capacitance parameter is a measure of the charge-storing ability of a structure. A capacitance of C (farads, F), charged to a potential V, holds a charge Q (coulombs, C):

$$Q = CV$$

By taking the derivative with respect to time, we find that the current flow into a capacitor ($i = dQ/dt$) is proportional to the derivative of the voltage across it.

$$i = C\frac{dv}{dt}$$

Note the use of a common convention, by which capital (upper-case) letters express DC or steady state voltage or current, and small (lower-case) letters express AC voltage or current.

Mutual capacitance. Mutual capacitance can exist between circuit elements. This occurs because electric fields from charges on one line can attract or repel charges on another. The charge-voltage relation for a system of three capacitively coupled elements held over a common or ground plane is the following:

$$Q_1 = C_{1G}V_1 \qquad + C_{12}(V_1 - V_2) + C_{13}(V_1 - V_3)$$
$$Q_2 = C_{12}(V_2 - V_1) + C_{2G}V_2 \qquad + C_{23}(V_2 - V_3)$$
$$Q_3 = C_{13}(V_3 - V_1) + C_{23}(V_3 - V_2) + C_{3G}V_3$$

where the voltages are measured with respect to the ground plane G. The C matrix $\{C_{ij}\}$ is symmetric, so that $C_{12} = C_{21}$ and so on. Examples of matrix elements are the lumped capacitances shown below on schematic diagrams in Figures 3-5 and 3-13.

Inductance. Inductance is an electrical parameter that determines the strength of induced voltages due to the Faraday effect. The self-inductance of a circuit element, L, (in henries, H) relates the value of the voltage induced on the circuit element caused by magnetic flux set up by the element itself. For a time-dependent current i, this is:

$$v = L \frac{di}{dt}$$

Mutual inductance. Magnetic fields from one conductor can intersect other conductors, producing signal coupling between the two. The coupling is described by a mutual inductance between circuits. For two magnetically coupled conductors, the combination of mutual and self-inductance is described by two equations:

$$v_1 = L_{11} \frac{di_1}{dt} + L_{12} \frac{di_2}{dt}$$

$$v_2 = L_{12} \frac{di_1}{dt} + L_{22} \frac{di_2}{dt}$$

The matrix $\{L_{ij}\}$ is symmetric. Diagonal elements L_{ii} are the self-inductances; the off-diagonal terms are the mutual inductances.

DISTRIBUTION OF DC POWER IN A COMPUTER SYSTEM

In a computer system, DC power is supplied to the electronic circuit chips at a specific set of voltages. The power supply assembly develops these voltages with some form of tolerance control so that correct voltages are maintained over the life of the system, over all the expected variations of operating conditions, such as temperature and AC power voltage.

From the power supply, voltage is brought over to the package, usually by means of heavy bus wire connected to the power input terminals of a printed circuit board. In the larger computer systems, the current required may exceed 500 A, at a voltage of 1.5–3 V. In these computers, the power may be distributed with large rectangular copper *bus* bars instead of wires.

A good distribution system should have all of the following features:

The correct value of voltage at the IC module within a prescribed tolerance

Stability of the voltage under varying conditions of incoming power line disturbance, circuit loading, and switching requirements

Uniformity of the voltage levels to all parts of the electronic system

Ability to supply adequate amounts of current to all circuits in the entire system

Isolation of hazardous voltage levels from the operator

The delivery of DC power is best controlled by designing the path from the power supply to the circuits to have as little resistance as possible. In the following section, a procedure for calculating the resistance of rectangular bus bars and printed circuit traces is developed.

Electrical Resistance of Rectangular Conductors

Many electrical conductors used in electronic packaging are formed from rectangular-shaped sections. Examples include: bus bar wiring for DC power distribution in a large computer, wiring traces on printed circuit boards, thin- or thick-film wiring on IC chip carriers, and thin-film wiring on the chips themselves. An extremely useful concept for calculating the DC resistance of these conductors is the *sheet resistance*.

Consider a section of a long rectangular printed circuit conductor as shown in Figure 3-1. A steady current I applied as shown in the figure will result in a voltage drop V over a section of length d. To derive an expression for the resistance R of the section, we use a generalized form of Ohm's law from field theory:

$$J = \frac{E}{\rho}$$

where J is the current density in A/cm^2, E the electric field in V/cm, and ρ the volume resistivity in $\Omega \cdot$ cm.

From the geometry of the problem, note that $E = V/d$, and $J = I/$(cross-sectional area) $= I/wt$. Using the circuit form of Ohm's law, $R = V/I$, we get the following expression for the line resistance:

$$R = \frac{\rho d}{wt}$$

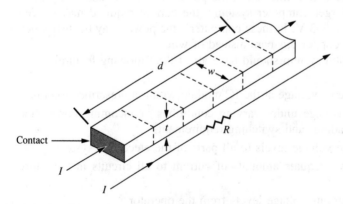

FIGURE 3-1

Section of rectangular resistor. Number of squares along resistor length is d/w. Resistance R is found by multiplying the number of squares by the sheet resistance.

or alternatively,

$$R = \frac{\rho}{t} \cdot \frac{d}{w} = R_{\text{sheet}} \cdot N_{\text{squares}}$$

where R_{sheet} = sheet resistance = ρ/t = resistivity/thickness, commonly described in units of Ω/square (or $\Omega/\square$), and N_{squares} = d/w = number of squares in conductor length.

For example, to find the resistance of a rectangular circuit line that may contain right angle bends, first find the number of squares in the straight sections of its length (length divided by width), and sum up the numbers for all of these sections. The corner squares of right angle bends add 0.559 squares each (see reference [1]). Then simply multiply the total number of squares by the sheet resistance to get the line resistance.

Example 3.1: Resistance calculation. Calculate the resistance of a 200-mm-long printed circuit line of width 0.1 mm, fabricated from 1-oz copper. (Copper sheet thickness is commonly specified by weight. One-ounce copper denotes a sheet that weighs 1 oz/ft^2. Such a sheet has a thickness of 0.0014 inches = 1.4 mils = 35.6 μm = 0.0356 mm.)

Using the conversion factors 25.40 mm/inch and 25.40 μm/mil, we calculate as follows:

$$\text{Conductor length} = 200 \text{ mm} = 200/25.4 = 7.9 \text{ inches}$$

$$\text{Conductor width} = 0.1 \text{ mm} = 0.10 \text{ mm} \cdot 1000\mu\text{m/mm}$$

$$= 100 \ \mu\text{m} = \frac{100\mu\text{m}}{25.4\mu\text{m/mil}} = 3.94 \text{ mils.}$$

$$\begin{array}{l}\text{Number of squares} \\ \text{in conductor length}\end{array} = \frac{200 \text{ mm}}{0.1\text{mm}} = 2000.$$

Taking the resistivity of copper to be $\rho = 1.8\mu\Omega \cdot$ cm,

$$\text{Sheet resistance} = \frac{1.8 \cdot 10^{-6}\Omega \cdot \text{cm}}{0.1\text{cm/mm} \cdot 0.0356 \text{ mm}}$$

$$= 5.06 \cdot 10^{-4}\Omega/\square = 0.506\text{m}\Omega/\square$$

Thus,

$$\text{Line resistance} = 0.506\text{m}\Omega/\square \cdot 2000\square = 1012\text{m}\Omega = 1.01\Omega$$

Example 3.2: Resistance of heavy bus bar. A certain large computer uses copper bars for DC power distribution. The thickness is 6 mm, the width 20 mm, and the length 80 cm.

The sheet resistance for this case is

$$\frac{1.8\mu\Omega \cdot \text{cm}}{0.6 \text{ cm}} = 3\mu\Omega/\square$$

$$\text{Number of squares} = \frac{800 \text{ mm}}{20 \text{ mm}} = 40\square$$

$$\text{Total line resistance} = 3\mu\Omega/\square \cdot 40\square = 120\mu\Omega$$

To find the current for a voltage drop of 10 mV, which may be the maximum allowable for this section of the circuit, we use Ohm's law.

$$I = \frac{V}{R} = \frac{0.01}{120 \cdot 10^{-6}} = 83.33 \text{ A}$$

$$\text{Power dissipated} = IV = 0.83 \text{ W}$$

For a current of 2000 A, the voltage drop would be 0.24 V, and the power dissipated 480 W. Such a computer would use voltage sense and regulation circuits to correct for the voltage drop over the conductor. This bus bar is definitely undersized for a 2000 A current because of the high 480-W power dissipation and the excessive voltage drop. A maximum distribution voltage drop of 1 to 2 percent is a typical specification. For a 3-V supply, this is only 30–60 mV.

DYNAMIC POWER DISTRIBUTION

The dynamic power distribution requirement is probably the most difficult to satisfy. The difficulty arises because of the nature of digital circuitry. The amount of current drawn by a circuit changes with the digital state, and thousands of circuits switch from state to state at exactly the same time with each clock pulse. Thus, the current demand can change by many amps in a few nanoseconds (10^{-9} second), giving rise to a very large time derivative, di/dt. The voltage drop across any small parasitic inductance in the power wiring itself will impose on the supply voltage an unwanted waveform, called power distribution noise, or delta-I noise [2, 3, 4]. Reference [4] discusses a computer model for delta-I noise and provides a useful bibliography.

Power distribution noise (PDN) is an important component of electrical noise in digital computer circuitry. PDN is not a random type of noise, but the result of unwanted waveforms that occur on the power and ground lines to a chip due to the parasitic inductances of the connections and wiring themselves. For most applications a model based on simple $L \, di/dt$ drops is adequate. However, because of the presence of $L, C,$ and R electrical parameters, the waveform may have an oscillatory behavior for circuit technologies such as CMOS that draw little current between clock pulses. This type of behavior is usually only significant in high-performance systems.

The $L \, di/dt$ voltage drop can be large enough to cause circuit failure. If the supply voltage is reduced sufficiently, latches (computer circuitry used to store 1s and 0s) may change state, or internal chip voltages may change in value so that input circuits, which receive incoming signals, do not properly switch. Electrical noise on the ground connections can cause false inputs, or cause inputs to be misinterpreted by the receiving circuitry.

The most significant cause of PDN for all logic circuit families is the large supply/ground current impulses from the output drivers. These circuits are output amplifiers that send signals to other chips via transmission lines in the packaging, modules, printed circuit lines, cables, and connectors. The current demand during the clock transition can be quite large, as each driver must provide current to a fairly low impedance load, typically 50–90 Ω. In addition, the drivers must supply current to charge and discharge capacitances of the external lines. Since all drivers can switch simultaneously, this component of the transient on the power/ground lines can be quite large.

PDN sets a limit on the number of drivers that can be placed on a digital logic chip. That is, it determines the number of output lines that can be simultaneously switched (SS). Thus, in this case, electronic package design directly influences the architecture and chip design of a computer system. Because new computer architectures need to have larger and larger numbers of bits in data and address buses, much effort is presently being applied to develop packaging structures that have low power/ground inductances. The reference by Rainal [3] discusses the calculation of parasitic inductances in chip packages. Two general rules given in [3] for low-inductance chip package design are restated as follows:

1. Apply power and ground over as many lines as possible, distributed symmetrically throughout the wiring pattern.
2. Locate signal lines as close as possible to power and ground lines.

Other guidelines to be followed to minimize PDN noise call for shorter paths, less current, slower switching speeds, and larger copper distribution links. All of these are directed almost opposite to the requirements of a high-speed, high-performance system.

The approach universally applied to minimize PDN is to provide a localized source of current near the chips. This supplemental current source is a small, low-parasitic-inductance capacitor, called a decoupling capacitor. As the circuits switch, current is supplied through the short path from the decoupling capacitors with a minimum of drop. In some designs the decoupling capacitor can be located within the chip package itself. It is most frequently seen mounted directly on a circuit board directly adjacent to the chip carriers. The inner large-area copper planes of a multilayer circuit board provide the low-inductance connection from the decoupling capacitors to the chips.

Most successful designs have decoupling capacitors located on all phases of the system package, from the power supply circuit board all the way down to the silicon chip.

Recent approaches in chip packaging such as TAB (tape-automated bonding) and surface-mount components such as PLCCs (plastic-leaded chip carriers) provide lower inductance than earlier packages such as DIPs (dual in-line packages). Another low-inductance approach is the PGA (pin grid array; see Figure 3-2), which is commonly used in high-performance computers. The TCM (thermal conduction module) [5] provides low-inductance packaging for many chips on one metal ceramic assembly while providing a very flexible wiring capability through many wiring layers within the ceramic itself. The wiring has the proper-

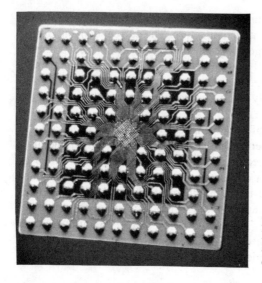

FIGURE 3-2
Photograph of low-inductance pin grid array package with wire traces placed on a dielectric layer over a ground plane.

ties of controlled impedance transmission lines to minimize effects of reflections, discussed later in this chapter.

The laser-patterned interconnect described in [6] provides a package with extremely low parasitic inductance. It uses many extremely short circuit lines deposited directly on the edge of the silicon chip, which connect to wiring on a silicon carrier to which it is mounted with a microscopic bond. Other low-inductance approaches used with gallium arsenide logic chips are discussed in [7].

Example 3.3: Estimation of PDN for chip with 20 SS drivers. Figure 3-3 shows a chip with 20 drivers, each of which must provide 5 V into a $Z_{0i} = 80 \ \Omega$ transmission line, terminated on the far end with $R_{Li} = 80 \ \Omega$. The rise time of the output signal of a driver is 2 ns. The parasitic inductance of the $V_{cc} = 5$ V supply line is $L_{cc} = 500$ pH, and $L_{gnd} = 750$ pH for the ground lines. These inductances include the contributions from wiring, pins, connectors, and so on. The inductance values are relatively small and are typical of pin grid array chip carriers. Decoupling capacitors have been placed as close as possible to the chip to minimize these parasitic inductances.

From Ohm's law, each driver must provide approximately $V_{cc}/R = 5$ V/80 $\Omega =$ 62.5 mA of current when on. Thus, the time derivative of current, assuming a linear ramp, is 31.3 mA/ns per driver, or 625 mA/ns if all 20 switch simultaneously. This converts to a time derivative of current of $6.25 \cdot 10^8$ A/s!

The voltage drop across the supply parasitic inductance is $L \ di/dt = 313$ mV, and is a negative transient when current is forced out of the chip—positively rising output. The chip ground terminal would have a 470 mV positive transient for the same magnitude of current transient. Transient ground current can flow through the chip during the turn-off transition. The actual waveshape and distribution of current over the supply and ground pins depend on the details of the circuit and switching process. Computer simulation models are extensively used to study the details of the switching process.

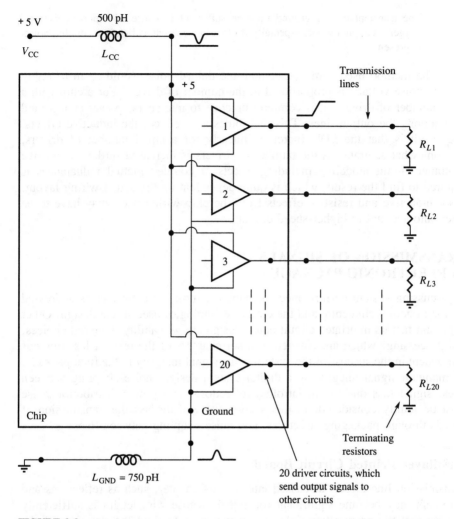

FIGURE 3-3
Illustration of chip with 20 simultaneous-switched drivers connected to transmission lines terminated with 80-Ω resistors. The sudden demand for current when all drivers switch together can cause a sizable transient voltage to appear on the power or ground lines, depending on the current path.

Several problems could arise from these effects:

1. The total chip voltage is reduced during the transient. Depending on all the other conditions, chip tolerances, temperature, other noise contributors, some circuits may change operation sufficiently to fail. Inputs may be missed, and latches (digital circuits used to store 0s and 1s) may change state, for example.
2. Either transient could propagate out through a driver that is not changing state, a "quiet" driver. If sufficiently large, a circuit on a receiving chip may falsely interpret this impulse as data.

3. The transient on the ground may be sufficient to cause an input receiver to trigger on a quiet input, especially if other noise, such as line-line coupled noise is present.

The method used here to calculate the delta-I noise would seem to imply that the noise voltage is proportional to the number of drivers. For a chip with a large number of drivers, it is common practice to intersperse power and ground lines among the output lines. This distributes and reduces the inductive effects. The result is that the PDN increases linearly for a small number of drivers, but somewhat saturates as the number of drivers is increased further. Extensive computer circuit modeling including effects of coupled mutual inductances is required to find the result, which is quite dependent on the actual wiring layout. Also, inductive and resistive effects from the chip wiring itself may have to be taken into account in higher-speed circuits.

TRANSMISSION OF SIGNALS IN ELECTRONIC PACKAGES

In discussing system performance, perhaps the most common factors addressed are the system architecture and the circuit technologies used in the design. Other important factors in printed circuit board design are wireability, material choices, and processing, which are covered in other chapters of the text. A less obvious component in the success of the system is the signal integrity in the final package. To maintain signal integrity in a digital system design, noise coupling between lines, signal loss due to uncontrolled reflections, and power distribution noise must be jointly considered. In this section we will discuss the transmission of signals through packaging structures, and noise coupling between lines.

Multilayer Printed Circuit Boards

Transmission line effects discussed later in this chapter, such as reflections and cross-talk may become significant for signals whose wire length is sufficiently long so that the propagation delay is more than $\frac{1}{3}$ to $\frac{1}{2}$ of the rise time of the signal. The usual means of incorporating transmission lines into a package is by means of traces on printed circuit boards that contain a ground plane or planes. In high-performance digital systems, it may also be necessary to use transmission line design within the chip carrier itself. Figure 3-2 shows an example of a high-performance pin grid array chip carrier in which the signal lines form striplines on a dielectric layer placed over a ground plane [2].

Figure 3-4 shows the cross sections of three different types of multilayer printed circuit boards, where the signal layers are either being imbedded within conductive power/ground planes (Options A and B), or are placed over the planes (Option C). The signal lines of Options A and B are located between power planes; these structures are termed triplate or shielded stripline. The signal lines of Option C have power/ground planes only on one side; this is termed microstrip or open stripline.

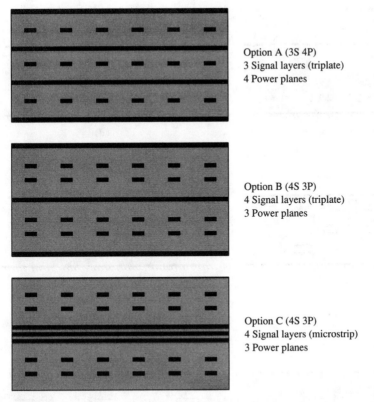

Option A (3S 4P)
3 Signal layers (triplate)
4 Power planes

Option B (4S 3P)
4 Signal layers (triplate)
3 Power planes

Option C (4S 3P)
4 Signal layers (microstrip)
3 Power planes

FIGURE 3-4
Some possible multilayer circuit board cross sections.

Figures 3-5 and 3-6 show the calculated capacitances per millimeter of line length for three different geometries, using computer routines of the type discussed by Weeks [8]. The inductance coefficients are shown in Table 3.1.

Transmission Lines

On a trace of a printed circuit board or chip carrier, the capacitance and inductive effects are distributed quantities. That is, they act not at a particular discrete point in space, but rather are distributed everywhere on the conductor. The net result is best described with a transmission line model. For a detailed discussion, see references [9–11].

We break the trace up into differential sections of length Δx (see Figure 3-7). The circuit problem can be solved by setting up the two loop equations:

$$v_{x+\Delta x} - v_x = -L\Delta x \frac{\partial i_x}{\partial t} \qquad \text{or} \qquad \frac{\partial v}{\partial x} = -L \frac{\partial i}{\partial t}$$

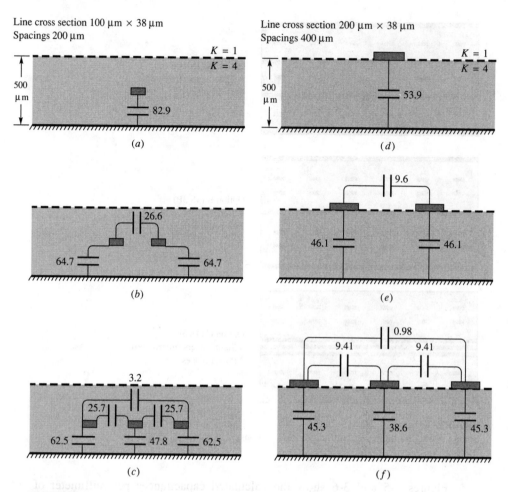

FIGURE 3-5
Examples of microstrip of infinite dielectric width and length showing lumped capacitance calculated numerically. Dielectric constant = 4. The width and spacing of cases (d), (e), and (f) is 2 times the width and spacing of (a), (b), and (c). Capacitances per unit length in units of fF/mm (fF/mm = 10^{-15} F/mm).

$$i_{x+\Delta x} - i_x = -C\Delta x \frac{\partial v_x}{\partial t} \qquad \text{or} \qquad \frac{\partial i}{\partial x} = -C \frac{\partial v}{\partial t} \tag{3.1}$$

where L and C are inductance and capacitance per unit length of the circuit line. These equations can be solved to produce the telegrapher's or wave equation:

$$\frac{\partial^2 v}{\partial x^2} = \frac{1}{c^2} \frac{\partial^2 v}{\partial t^2}$$

This equation has traveling wave solutions $v_+ = F(x - ct)$ and $v_- = F(x + ct)$,

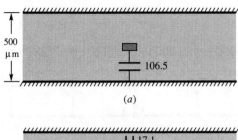

(a)

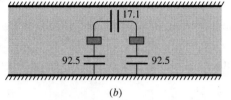

Line cross section
100 μm × 38 μm

Spacings 200 μm

(b)

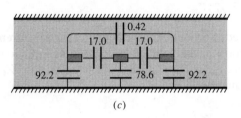

(c)

FIGURE 3-6
Examples of triplate of infinite dielectric width and length showing lumped capacitance calculated numerically. Dielectric constant = 4. Width and spacing is the same as that of Figure 3-5 (a), (b), and (c).

TABLE 3.1
Inductance coefficients for three stripline systems
(Units: pH/mm)

Figure	L_{11} or L_{33}	L_{22}	L_{12} or L_{23}	L_{13}	
3-5(a)	506				
3-5(b)	503		135		Imbedded wires
3-5(c)	503	500	134	55.3	in microstrip
3-5(d)	550				
3-5(e)	548		142		Top lines in
3-5(f)	548	545	141	58.6	microstrip
3-6(a)	417				
3-6(b)	416		64.7		Line in midplane
3-6(c)	415	414	64.4	11.6	of triplate

See Example 3.7 for results from a somewhat different line geometry.

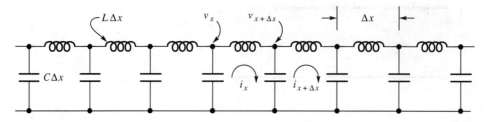

FIGURE 3-7
Differential circuit model of a transmission line used to derive the wave equation.

which represent waves traveling in either the $+x$ or $-x$ direction, respectively. The wave velocity is

$$c = \frac{1}{(LC)^{1/2}}$$

Equations (3.1) can also be solved to produce a wave equation for the current i. Placement of the traveling wave form of the solution into either of Equations (3.1) yields the relation for the current, $i_+ = F(x - ct)/Z_0$, $i_- = -F(x + ct)/Z_0$, where Z_0 is the characteristic impedance

$$\frac{v_+}{i_+} = \frac{v_-}{-i_-} = Z_0 = \left(\frac{L}{C}\right)^{1/2}$$

We see that Z_0 is the ratio of voltage to current at any point when excited by a single wave. Recall that the Z_0 is 377 Ω for a wave in free space, 50–93 Ω for coaxial cables, and 300 Ω for the familiar twin-line used with television reception. The Z_0's of signal lines in printed circuit boards and chip carriers are generally in the range of 25–120 Ω.

An alternative way of stating the velocity is to use the inverse velocity, or time required for a wave to travel a fixed distance, d, of 1 mm. This propagation delay time per unit length is $\tau/d = (LC)^{1/2}$ ps/mm, where the inductance and capacitance per unit length are expressed in H/mm or F/mm. Propagation times for many packaging structures are in the range of 5–8 ps/mm, where 1 ps = 1 picosecond = 10^{-12} second. Table 3.2 lists Z_0 and τ/d for the stripline examples of Figures 3-5 and 3-6.

The fastest wave would be one at the speed of light, c_0. For such a wave, the propagation delay time per millimeter of travel is

$$\frac{d}{c_0} = \frac{1 \text{ mm}}{3 \cdot 10^{11} \text{ mm/s}} = 3.33 \text{ ps}$$

In this approximate treatment, line resistance has been ignored. When losses are included, and for sinusoidal excitation, the wave is a decaying exponential in space. For many printed circuit and chip carrier applications, line loss can be ignored, since distances are small compared to the exponential decay length. However, as cross-sectional dimensions become smaller and computer perfor-

TABLE 3.2
Characteristic impedance and propagation delay per unit length for stripline examples

	Z_0	τ/d
Microstrip imbedded line Figure 3-5(a)	78.1 Ω	6.48 ps/mm
Microstrip line on top Figure 3-5(d)	101 Ω	5.44 ps/mm
Triplate line at midplane Figure 3-6(a)	62.5 Ω	6.66 ps/mm

mance increases, line resistance become important. Also, skin effect increases the resistance at high frequencies, since it causes the current in a conductor to be confined to a thin surface layer. Resistive line loss can degrade the rise time of a wave propagating along a transmission line. It effectively adds another contribution to the propagation delay; see [12] for a Laplace transform solution to the problem.

Traces on boards or chip carriers are frequently placed in close proximity to other traces for considerable distances. The characteristic impedance and propagation velocity relations above can be used for these cases, provided that the capacitance per unit length, C, is the total capacitance to ground, including interline capacitances to adjacent lines that are assumed to be grounded. For example, $C = C_{1G} + C_{12}$ for two lines placed adjacent to one another.

Example 3.4: Propagation delay and computer cycle time. The time that it takes for a signal to propagate across a printed circuit board must be taken into account during the design of a high-speed computer. Other significant effects include the gate delays of the digital circuits themselves, the propagation delays through cables, connectors, and modules, and the tolerances on the clocking signals due to circuit variations and propagation effects. The net effect of all these phenomena determines the ultimate speed or cycle time of a computer system.

For example, for the printed circuit board used in the IBM 3081 mainframe computer system [13], the capacitance and inductance per unit length for a printed circuit wire on the board are

$$C = 85 \text{ fF/mm}$$

$$L = 550 \text{ pH/mm}$$

This results in a characteristic impedance of 80 Ω, and a propagation delay per unit length of 7 ps/mm.

For this board, the average length of a section of printed circuit wire is about 70 mm. Thus, the delay encountered for a signal traversing a network of three such line sections would be:

$$\tau \text{(ckt board)} = 3 \cdot 70 \text{ mm} \cdot 7 \text{ ps/mm} = 1470 \text{ ps} = 1.47 \text{ ns}.$$

For comparison, the gate delay of the basic logic gate in the system is approximately 1.15 ns, which is of the same order of magnitude as a typical board signal propagation time [14]. Other significant propagation delays, such as those on lines in TCM modules and in cables, would be computed in a similar manner.

Reflections at Line Termination

The value of the resistance connected across the end of a transmission line, R_T, determines the current-voltage relation at that point. If the termination value does not equal the characteristic impedance, then a reflected wave is generated [9, 11, 17]. The ratio of the amplitudes is the reflection coefficient, $r = v_-/v_+$, for a line terminated to the right.

Assuming an incident wave in the positive x direction, $v_+ = F(x - ct)$, the reflected wave, $v_- = G(x + ct)$, is similar in shape, but has the following differences:

1. The direction of propagation is reversed.
2. The reflected wave is a mirror image of the incident wave. Its amplitude is determined by the reflection coefficient r and can be either positive, negative, or zero.
3. It is positioned in space so that corresponding points on the incident and reflected waves meet simultaneously at the termination.

The general shape of the reflected wave having these properties is the following:

$$G(x + ct) = rF(-(x + ct) + 2d)$$

for an incident wave $F(x - ct)$. The $+$ sign in the $-(x + ct)$ term provides for propagation in the reverse direction, and the $-$ sign forces the mirror-image shape. The spatial shift by $2d$ causes corresponding points on the two waves to meet at the termination, $x = d$. Substituting $x = d$ into the expressions for v_+ and v_- gives $F(d - ct)$ for both, which demonstrates this property.

Figure 3-8 illustrates the formation of a reflected wave at a short circuit. Note that the reflected wave is inverted in sign and has a mirror-image shape.

To determine the reflection coefficient, we assume the transmission line is terminated with resistance R_T at $x = d$. This requires the ratio of voltage to total current into the termination to be just R_T:

$$R_T = \frac{v_+ + v_-}{i_+ + i_-} = \frac{1 + r}{(1 - r)/Z_0}$$

Solving this equation for the reflection coefficient r, we get an expression for r in terms of R_T/Z_0:

$$r = \frac{(R_T/Z_0) - 1}{(R_T/Z_0) + 1}$$

Three important cases are the following: $R_T \ll Z_0, r \to -1$, short; $R_T = Z_0, r = 0$, matched; $R_T \gg Z_0, r \to +1$, open. Terminations with $R_T < Z_0$

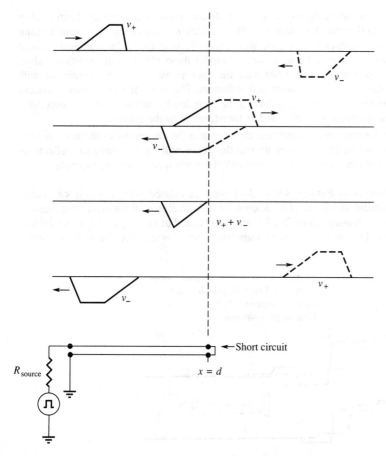

FIGURE 3-8
Snapshot of incident and reflected waves as a function of position on a transmission line terminated with a short circuit. Reflected wave is mirror image of incident wave, and is inverted in sign. Reflection coefficient is $r = -1$.

produce inversions; matched terminations, $R_T = Z_0$, produce no reflections; and terminations with $R_T > Z_0$ produce reflections of the same sign.

> **Example 3.5   Multiple reflections on a printed circuit line.** Reflections of signals are an important consideration in the operation of digital circuitry. We consider here a simple example of a drive circuit on one chip connected by an isolated transmission line to a receiver circuit on another chip. Line-line coupling is ignored here, but is covered in the next section of the text.
>
> For the purposes of illustration, some assumptions are made:
>
> 1. The drive circuit is taken to have a resistive impedance R_D independent of voltage levels, that is, it is assumed to be a linear voltage source. Actual transistor drive circuitry is nonlinear, and real waveforms will differ in detail.

2. Parasitic circuit elements are ignored. In an actual packaging application chip carriers and connectors may contribute additional capacitance to ground along with series inductance, which will affect the waveform. In high-performance applications, it may be necessary to model these effects with additional short sections of transmission lines with the appropriate Z_0 values, combined with additional discrete capacitors and inductors. Coupling of signals from adjacent lines within the module may have to be included by adding coupling capacitors, coupled transmission lines, and/or transformers to the circuit.

3. In a real application, additional circuitry may be connected to the transmission line at various places along its length. These will cause additional reflections and waveform changes. For simplicity these are ignored in this example.

The waveforms of Figures 3-9 to 3-11 were calculated with a circuit simulation program similar to ASTAP [15]. Shown are the waveforms at the driver and receiver end for driver resistances of 20, 78, and 320 Ω, and at varying receiver termination resistances. The input is a 20-ns trapezoidal pulse with 1-ns rise and fall times.

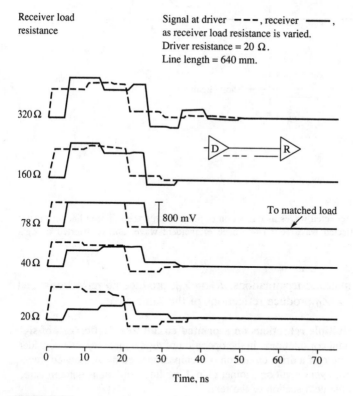

FIGURE 3-9

Multiple reflections on a transmission line. Signal at driver and receiver, driver resistance fixed at 20 Ω. R_T is varied from low to high values. Characteristic impedance Z_0 of line is 78 Ω. Driving waveform is a 20-ns wide trapezoidal pulse with 1 ns rise and fall times and 1 V peak amplitude into an open circuit.

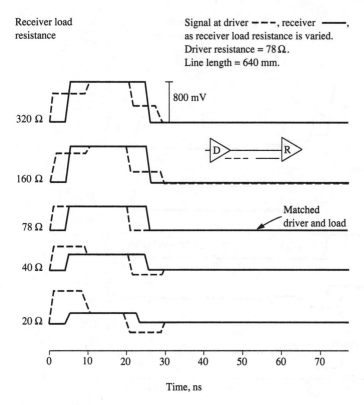

Receiver load
resistance

Signal at driver – – –, receiver ——,
as receiver load resistance is varied.
Driver resistance = 78 Ω.
Line length = 640 mm.

800 mV

320 Ω

160 Ω

78 Ω

40 Ω

20 Ω

Matched
driver and load

Time, ns

FIGURE 3-10
Same as Figure 3-9, but driver resistance is 78 Ω.

The transmission line is assumed to be the imbedded microstrip shown in Figure 3-5(a), which has a characteristic impedance of $Z_0 = 78$ Ω. Its length of $d = 640$ mm results in a propagation delay of $\tau = 4.15$ ns, or about 1/5 of the pulse width. This delay is sufficient to clearly separate multiple reflections in the waveforms.

The value of the receiver waveform at the first step can be shown to be $[Z_0/(Z_0 + R_D)](1 + r)$ for a 1-V peak waveform at the source. Multiple reflections gradually allow the line to charge up to the ultimate dc level predicted by Ohm's law, $R_T/(R_T + R_D)$. For some choices of parameters, the value at the first step is noticeably smaller than the final values reached after several reflections have occurred. For example, note the bottom waveform of Figure 3-9, or the top waveform of Figure 3-11. This effect can cause a circuit to miss data if the first step of the waveform at the receiver is not sufficient to trigger the receiver circuit.

The driver waveform is affected by the reflected wave as it passes back and forth down the line. Large overshoot and undershoot is present at the driver for some receiver load resistance values. In some CMOS technologies, reflections such as these can cause chip failure, if sufficiently large. An impulse with a peak value of 0.6 V or more above the supply voltage or below ground can trigger SCR latch-up of some types of output drivers. This effectively causes the drive transistors to short circuit to ground, which can result in chip burnout.

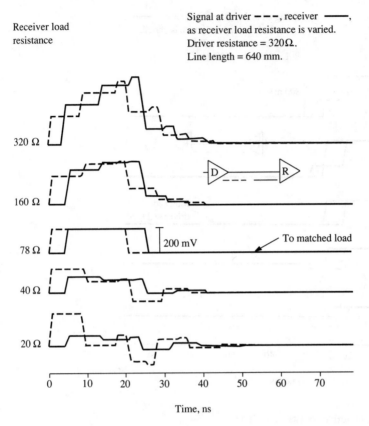

Receiver load resistance

Signal at driver – – –, receiver ——,
as receiver load resistance is varied.
Driver resistance = 320 Ω.
Line length = 640 mm.

FIGURE 3-11
Same as Figure 3-9, but driver resistance is 320 Ω.

SCR refers to a parasitic four-layer pnpn semiconductor device structure that is inherently formed along with the normal gates when a CMOS chip is made; SCR stands for *silicon-controlled rectifier*. SCR devices act as diodes that are blocked from conducting in the forward direction until triggered. Once triggered into latch-up they conduct with very low resistance until the supply voltage is removed. All recent CMOS technologies are designed to minimize SCR effects.

Coupled Noise or Cross-Talk

To achieve maximum wiring density in a circuit board, chip carrier, or connector, one attempts to place signal lines as close to one another as is physically practical. This increases packing density, but also increases electrical noise coupled between lines. This phenomenon is sometimes termed *cross-talk* from its appearance in early telephone and telegraph systems. If not controlled through proper design, it can cause system failure in digital circuitry by the sudden erratic appearance of false data in a circuit. See [16] for a discussion of electrical design issues.

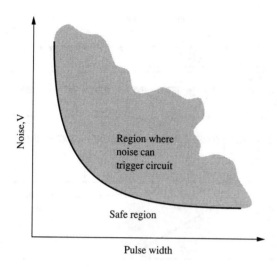

Noise, V

Region where
noise can
trigger circuit

Safe region

Pulse width

FIGURE 3-12
Typical noise tolerance curve for a circuit
that receives an incoming digital signal.

The ability of noise to trigger a circuit is a function of both the peak amplitude of the signal and the noise pulse width. This dependence can be illustrated
with a noise tolerance curve. Figure 3-12 shows a typical noise tolerance curve.
This curve is determined by the circuit design. With it the regions of safe operation of a circuit can be determined.

The coupled noise waveform is best found either experimentally or with
the use of a digital computer circuit simulation program. There are two types of
coupling: capacitive and inductive. The noise waveform usually has no resemblance to the input signal. Reflections on both the main and coupled lines produce
echos which further distort the noise waveform. Key parameters determining the
coupling are the inductance and capacitance matrices, the line length, and the
termination impedances of the lines.

Only coupling from adjacent printed circuit lines is considered here. Line-
line coupling may be significant within a module or chip carrier because the
wiring is usually closely spaced. The ground plane may also be distant from the
wiring plane for some module designs, which tends to increase both capacitive
and inductive coupling. Coupling between module pins may also be significant.

Estimate of Coupled Noise

The calculation of coupled noise between transmission lines is treated in a book
by Matick [17], which assumes sine wave excitation, and also in references [18,
19]. In the appendix an approximate analysis is made of the coupling of a transient
between two loosely coupled identical transmission lines.

Consider the discrete circuit model of two identical transmission lines coupled over a distance d, as pictured in Figure 3-13. Progressing down Line 1
is a trapezoidal wave with rise and fall times both equal to t_{rise}. As the wave
progresses down the line, the coupling capacitances are charged during the rise

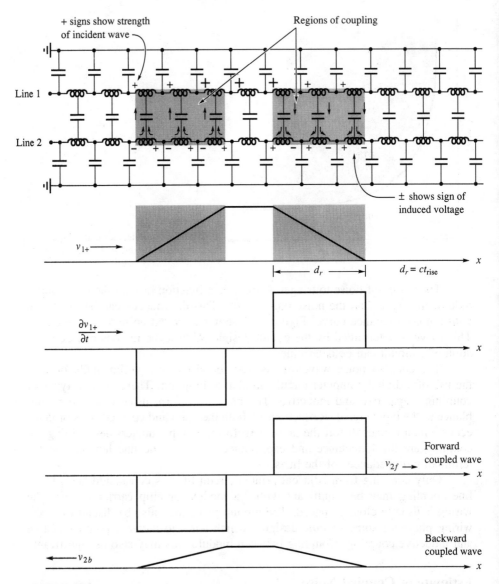

FIGURE 3-13
Circuit element representation of two coupled transmission lines showing coupling by means of mutual capacitances and inductances. Coupling currents flow and inductive voltages are induced only along the rise and fall regions of a trapezoidal impulse.

time, maintain a constant charge during the plateau, and discharge during the fall time. Also, the current through a differential section increases during the rise time, becomes constant through the plateau, and decreases during the fall time.

Thus, within our approximation, current flows through the coupling capacitor of a differential section only during the rise and fall times, when dv/dt is nonzero. Also, voltage is induced across a differential section by the mutual inductance only during the rise and fall times, when di/dt is nonzero.

The analysis in the appendix shows that a signal progressing down one line induces two different noise signals in an adjacent line, a forward coupled signal and a backward coupled signal, which propagates in the opposite direction. The shapes of these waveforms differ considerably from one another, and from the incident applied signal.

FORWARD COUPLED WAVE. The noise coupled in the forward direction, that is, the same direction as the incident wave, consists of two short rectangular impulses of alternate polarity. The pulse widths are equal to the rise and fall times of the trapezoidal wave. The coupled wave amplitude is proportional to the coupled length d, but the pulse width remains constant with changes in length.

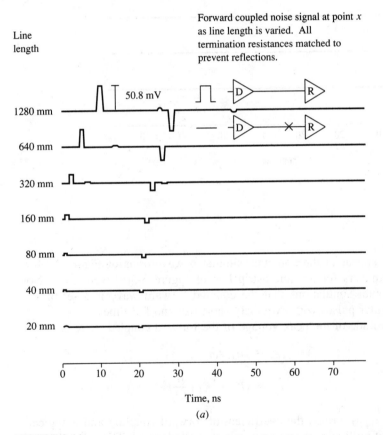

(a)

FIGURE 3-14
Simulated forward coupled signal (a) and backward coupled signal (b) on two coupled transmission lines of varying lengths terminated with $R_T = Z_0$. Microstrip pair of Figure 3-5(b) is used. Driving waveform is a 20-ns-wide trapezoidal pulse with 1 ns rise and fall times, and 0.5 V peak amplitude. The coupling length is the same as the line length. The small secondary impulses on the longer lines are due to higher order effects not discussed here.

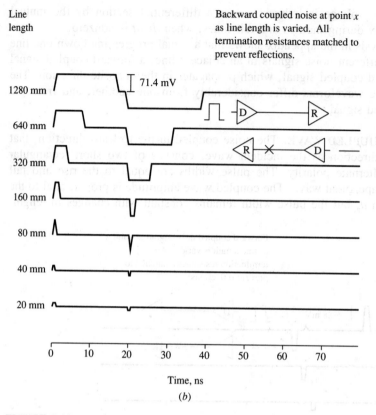

Line length

Backward coupled noise at point x as line length is varied. All termination resistances matched to prevent reflections.

Time, ns

(b)

FIGURE 3-14
(*continued*)

These features appear in the computer-simulated waveforms of Figure 3-14(*a*), where the parameters for the microstrip line of Figure 3-5(*b*) were used. Note from the computer simulations that the coupled forward wave is a set of very narrow rectangular pulses with extremely short rise and fall times.

The magnitude of the peak voltage of the forward coupled wave is

$$v_{2f, peak} = 2a_{12}v_{10}$$

where

$$a_{12} = (K_C - K_L)\frac{d}{4d_r}. \tag{3.2}$$

The quantity $2a_{12}$ is termed the coefficient of forward coupling and is typically stated in units of millivolts of noise per volt of applied signal. The value d_r equals ct_{rise} or ct_{fall}, the distance traveled by the wave during the rise or fall time, and v_{10} is the peak value of the trapezoidal wave on Line 1. The K's are the coupling coefficients for the two identical lines: the capacitive coupling coefficient $K_C = C_{12}/(C_{1G} + C_{12})$, and the inductive coupling coefficient $K_L = L_{12}/L_{11}$.

The forward coupled wave is proportional to the difference between the coupling coefficients. For triplate (Figure 3-6), the coupling coefficients can be shown to be equal because of the homogeneous dielectric cross section. Thus, a triplate line pair terminated in matched impedances has no forward coupled wave. However, forward noise is usually present in real applications due to backward coupling of reflected signals or higher-order coupling in systems of more than two lines (see reference [20]). Also, via structures and crossing orthogonal lines on neighboring layers can cause increased coupling.

BACKWARD COUPLED WAVE. As the trapezoidal wave propagates in the $+x$ direction, a backward coupled wave is excited in the reverse or $-x$ direction on Line 2. Since the region of coupling, the region of rise and fall times, is moving in a direction opposite to the backward coupled wave, the coupling is spread out in space and the coupled wave has a much longer duration than the forward coupled wave. However, the amplitude does not increase indefinitely as the line is made longer. For coupled lengths greater than $1/2$ the rise time length d_r, the amplitude saturates with respect to line length, and remains constant.

The shape of the backward coupled wave is shown in the appendix to be

$$v_{2b}(t) = b_{12}[v_{1+}(t) - v_{1+}(t - 2\tau)]$$

For a trapezoidal wave, the peak or saturated amplitude (for $d > d_r/2$) has magnitude

$$v_{2b, peak} = b_{12}v_{10}$$

where
$$b_{12} = \frac{K_C + K_L}{4} \tag{3.3}$$

The quantity b_{12} is the coefficient of backward coupling and, like the forward coupling coefficient, is usually stated in units of millivolts per volt. The value τ is the one-pass propagation delay of the wave down the line. Note that v_{2b} is proportional to the sum of coupling coefficients. Thus no cancellation occurs, and backward coupling occurs for every line cross section.

Figure 3-14(b) shows an example of a computer-simulated waveform. For all cases the 20-ns pulse width is greater than twice the propagation delay (17.4 ns for the longest length, 1280 mm). The coupled waveform is made up of two trapezoidal waves of alternating polarity and of equal width 2τ.

The backward coupled wave becomes significant in a situation where the signals in two adjacent lines normally have opposite directions of propagation. In this case the backward coupled signal from each line can propagate directly into the receiver of the adjacent line with no propagation delay. Even in the case where adjacent signals have parallel directions, the backward wave can propagate into the receivers after a reflection, if line terminations are not matched to the characteristic impedance.

Figure 3-15 shows the computed coupled waveforms at two nodes for a particular example of line terminations that might be found in a typical design (R(driver) $= 20\ \Omega$, R(receiver) $= 200\ \Omega$). Note the complex nature of the total

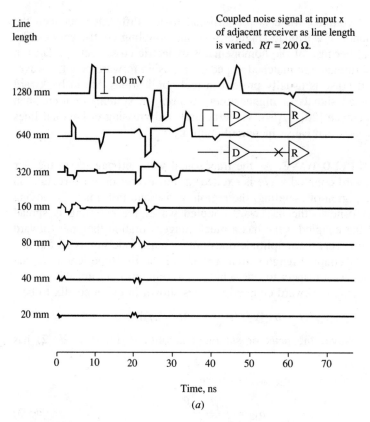

FIGURE 3-15
Simulated coupled noise when termination impedances are not matched to Z_0 producing reflections. Driver resistance = 20 Ω, receiver resistance = 200 Ω. Signal on inactive or quiet line adjacent to an active line, (a) far end, (b) near end. Driving waveform and line geometry is same as Figure 3-14.

coupled wave and its total lack of resemblance to the simple incident trapezoidal wave. The narrow spikes arise from the forward coupling, the broad waveforms from the backward coupling. The waveforms of Figure 3-15(b), which has the receiver at the near end, are larger and wider and reach maximum amplitude at a shorter line length than those of Figure 3-15(a), which has the receiver at the far end.

> **Example 3.6 Estimation of amplitudes of coupled noise.** Table 3.3 compares the results obtained with the coupled noise equations, (3.2) and (3.3), and with a computer circuit simulation (top traces of Figure 3-14). The parameters for the imbedded line pair of Figure 3-5(b) and a line length of 1280 mm were used. The incident wave was a trapezoid with peak value 0.5 V, rise and fall times of 1 ns, and plateau width of 20 ns. The coupling coefficients for this example are $K_C = 0.2913$ and $K_L = 0.2684$.

Line
length

Coupled noise signal at input x
of adjacent receiver as line length
is varied. $RT = 200 \, \Omega$.

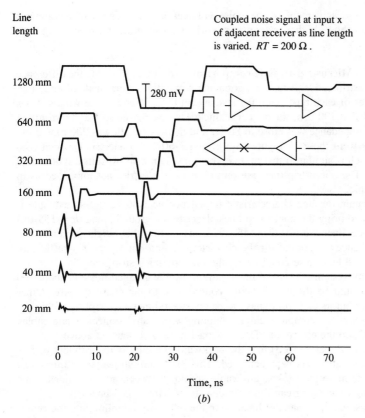

1280 mm

280 mV

640 mm

320 mm

160 mm

80 mm

40 mm

20 mm

0   10   20   30   40   50   60   70

Time, ns

(*b*)

FIGURE 3-15
(*continued*)

TABLE 3.3
Comparison of results from computer circuit simulation and coupled noise equations

	Computer simulation	Coupled noise equation	*a* or *b* coefficient
Forward coupled wave peak voltage v_{2+}	50.8 mV	47.6 mV	$2a_{12} = 95.2$ mV/V
Backward coupled wave peak voltage v_{2-}	71.4 mV	70 mV	$b_{12} = 140$ mV/V

Geometry is microstrip line pair, Figure 3-5(*b*), line length = 1280 mm. Incident wave used is trapezoid of 0.5 volt peak amplitude, rise and fall times of 1 ns, and plateau width of 20 ns.

Reasonable agreement of the noise voltages is seen. The actual computer simulation also shows higher-order effects such as noise on Line 2 recoupling back onto Line 1.

Example 3.7 Microstrip-triplate comparison. To further illustrate the difference between microstrip and triplate lines, computer simulations were made of three line pair geometries of each type, with the spacing set to 2, 3, and 5 line widths. Line width was fixed at 100 μm, the dielectric thickness was chosen so that the resulting characteristic impedance was close to 80 Ω, and the line length was 200 mm. Line cross sections were those of Figures 3-5(e) and 3-6(b). A dielectric constant of 4 was used, which is similar to that of an epoxy-glass laminate. Results are tabulated in Table 3.4. The microstrip lines are on top of the dielectric, not imbedded as in the last example.

To maintain the line characteristic impedance near 80 Ω in this example it was necessary to bring the microstrip lines closer to the ground plane ($h = 155\mu$m) than the triplate lines were ($h/2 = 450$–525μm). While this height produced the correct capacitances to ground for the characteristic impedance, the K coefficients were smaller and less noise coupling resulted for the microstrip line.

For many systems, a typical maximum allowable coupled noise is about 200 mV/V. This would be the circuit board contribution to the coupled noise. Other important contributors that would have to be considered include coupling within the modules that carry the computer chips, coupling within any connectors and cables in the system, and the effects of PDN or delta-I noise as discussed above.

Note from Equation (3.3) that the backward coupling is independent of line length, once saturation has been achieved. This maximum allowable coupling can be used to determine the closest allowable spacing between adjacent lines. This important design parameter can limit the density of wiring on a board.

If we assume three parallel lines, with the outer two coupling into the center one, and consider only the backward coupling, we see that for our geometry the microstrip lines can be placed closer together than the triplate lines. Total coupled noise of 200 mV/V, summing contributions from both outer lines, is obtained at somewhat less than two line widths for microstrip versus three line widths for triplate. A similar example in a 50 Ω system, and the important effects of reducing the dielectric constant, are discussed in Chapter 4.

Whether or not microstrip has lower coupling than triplate is very much geometry dependent, and a general rule cannot be stated since many design factors are involved. The examples discussed here have been somewhat idealized. In many real structures two orthogonal wiring planes are placed adjacent to one another to improve wireability. Vertical via structures are used for interplane connection. The presence of crossing orthogonal lines and vertical vias can cause important changes to the coupling coefficients and may even dominate in some geometries; see Chapter 4 and [20].

Forward coupling may or may not be ignored, depending on the noise tolerance curve for the technology used (see Figure 3-12) and on the line lengths. Recall that portions of the forward coupled noise can be quite narrow in pulse width so that a larger peak value may be tolerated by the receiving circuits.

TABLE 3.4
Comparison of properties of three microstrip and triplate line pairs designed for 80 Ω characteristic impedance

Line spacing, μm	Dielectric thickness, μm	Z_0, single line	Z_0, line pair	C_{1G}, fF/mm	C_{12}, fF/mm	L_{11}, pH/mm	L_{12}, pH/mm	Propagation delay, ps/mm	K_C	K_L	$2a_{12}$, mV/V	b_{12}, mV/V
Microstrip:												
200	155	80.26	79.52	61.2	7.05	431.9	86.8	5.43	0.103	0.201	−53.1	76
300	155	80.26	80.04	64.0	3.65	433.3	56.6	5.41	0.054	0.131	−41.6	46
500	155	80.26	80.23	66.1	1.32	434.0	28.7	5.41	0.020	0.066	−25.3	22
Triplate:												
200	1050	84.68	79.79	60.3	27.7	560.4	176.5	7.02	0.315	0.315	0	158
300	950	81.69	79.71	67.9	17.5	542.9	111.1	6.81	0.205	0.205	0	102
500	900	80.04	79.66	76.1	7.93	533.5	50.3	6.70	0.094	0.094	0	48

Line width fixed at 100 μm for all cases, line length = 200 mm, dielectric thickness for triplate structures indicates distance between ground planes. Dielectric constant = 4. Geometries follow Figures 3-5(e) and 3-6(b), except for differing physical parameters. Note that the microstrip lines are on top of the dielectric in this example, i.e., not imbedded as in Table 3.3 and Example 3.6.

Forward coupled noise is proportional to the line length and thus is most significant for the longer lines.

SUMMARY

This chapter has presented a general discussion of the importance of the package design in preserving signal integrity in the total transmission path as well as providing the necessary power to the circuitry. It has indicated that the proper design of a printed circuit board involves tradeoff considerations relative to cost and construction of the package type itself as well as circuit performance requirements. Ultimate package design and electrical performance analysis have been greatly enhanced through the use of available computer simulation programs and basic parameter computational techniques.

APPENDIX
COUPLED VOLTAGE BETWEEN TWO
IDENTICAL TRANSMISSION LINES

In this appendix, an approximate analysis is made of the coupling of a transient between two identical transmission lines. The calculation assumes small coupling coefficients and small coupled voltages. Effects that become important with larger coupling coefficients are ignored. Such effects include the decrease of the main signal from the first line, and recoupling of noise from Line 2 back to Line 1. This approximate treatment provides some useful insight into the factors that control line-line coupling in most packaging applications. A thorough treatment based on eigenvector analysis can be found in reference [19].

In Figure 3-16 we show a single differential section that couples to the right and left to the rest of transmission line, pictured as two resistances of Z_0. We solve for the voltages across these two resistors, which we will use to form the coupled waveforms. Because of the direction of current flow as shown in the figure, the voltage across resistor R_f contributes to a forward coupled wave ($+x$ propagation direction), and that across R_b contributes to a backward coupled wave ($-x$ propagation direction), for an incident wave in Line 1 in the $+x$ direction.

The current Δi_{12} through the coupling capacitance C_{12} is the following, assuming that the voltage on Line 2 is sufficiently small so that its effect on current Δi_{12} can be neglected:

$$\Delta i_{12} = C_{12}\Delta x \frac{\partial v_1}{\partial_t} \tag{3.4}$$

With the relations $\Delta i_{12} = \Delta i_b + \Delta i_f$, $\Delta v_f = \Delta i_f Z_0$, and $\Delta v_b = \Delta i_b Z_0$, we reduce (3.4) to

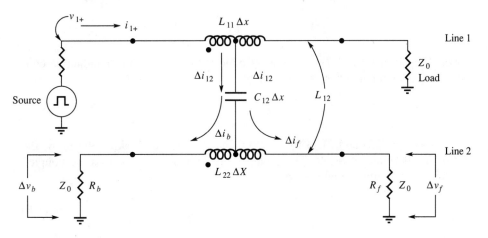

FIGURE 3-16
Differential circuit elements used in calculation of coupled noise signals.

$$\Delta v_b + \Delta v_f = K_C \frac{\Delta x}{c} \cdot \frac{\partial v_1}{\partial t} \tag{3.5}$$

Current Δi_b is defined as positive when it is directed in the $-x$ direction; thus the relation between Δv_b and Δi_b does not include a sign change. In deriving Equation (3-5), the relation for characteristic impedance includes the total capacitance of Line 1 to ground:

$$Z_0 = \left(\frac{L_{11}}{C_{1G} + C_{12}} \right)^{1/2}$$

and the capacitive coupling coefficient K_C has been defined as

$$K_C = \frac{C_{12}}{C_{1G} + C_{12}}$$

The mutual inductance between lines, L_{12}, acts as a coupling transformer as shown in Figure 3-16. The windings of the transformer are in series with the transmission lines and have matching corresponding ends as indicated by the dots. The dots enable one to determine the signal polarity. An increasing current applied to a dotted end of a line produces a positive voltage at the dotted ends of all other inductively coupled lines. Current from the main signal in Line 1, i_1, causes the following voltage difference across the differential segment Δx:

$$\Delta v_b - \Delta v_f = \Delta x L_{12} \frac{\partial i_1}{\partial t} \tag{3.6}$$

Using the relation $i_1 = v_1/Z_0$ and the propagation velocity $c = 1/[L_{11}(C_{1G} + C_{12})]^{1/2}$, we get a reduced form for Equation (3.6):

$$\Delta v_b - \Delta v_f = K_L \frac{\Delta x}{c} \cdot \frac{\partial v_1}{\partial t} \tag{3.7}$$

In deriving Equation (3.7) we have introduced the coefficient of inductive coupling, K_L, for identical lines, for which $L_{11} = L_{22}$:

$$K_L = \frac{L_{12}}{L_{11}}$$

Solving (3.5) and (3.7) for $\Delta v_f / \Delta x$ and $\Delta v_b / \Delta x$, and taking the limit, the following derivatives result. These can be integrated over x to find the coupled voltages at the output nodes of the lines.

$$\frac{\partial v_f}{\partial x} = \frac{K_C - K_L}{2c} \cdot \frac{\partial v_1}{\partial t} \tag{3.8}$$

$$\frac{\partial v_b}{\partial x} = \frac{K_C + K_L}{2c} \cdot \frac{\partial v_1}{\partial t} \tag{3.9}$$

FORWARD COUPLED WAVE. Consider first the case of the forward coupled wave. Since the lines are identical, waves in both lines travel at the same speed. The coupling region travels synchronously with the waves on both lines. Thus, the forward wave remains narrow in width, as all energy stays within the coupling region at each end of the trapezoid. The forward coupled wave builds up linearly in amplitude with distance traveled, within our approximation of small total coupled voltages. Integrating (3.8) from $x = 0$ to $x = d$, we get the following relation for the peak voltage:

$$v_f = \frac{(K_C - K_L)d}{2c} \cdot \frac{\partial v_1}{\partial t}$$

For a trapezoidal wave of equal rise and fall times on Line 1, the time derivative is nonzero only during the rise and fall times and is $\pm v_{10}/t_{\text{rise}}$, where the peak voltage of the trapezoid is v_{10}. Thus, for this case the magnitude of the peak forward coupled voltage at $x = d$ is equal to

$$v_f = \frac{(K_C - K_L)d}{2d_r} v_{10} = 2a_{12}v_{10}$$

where $a_{12} = (K_C - K_L)d/4d_r$ is the coefficient of forward coupling, and $d_r = ct_{\text{rise}}$ is the distance traveled during the rise or fall times.

BACKWARD COUPLED WAVE. With the backward coupled wave, the coupling region travels in opposite direction to the coupled wave itself. Thus, the excitation is spread out in space and time and the coupled wave duration is longer than the forward wave. The output wave at the node on the left is a superposition of wavelets coupled at earlier times that propagate and sum up at the receiving port. We form this superposition by integrating (3.9) from $x = 0$ to $x = d$, while keeping track explicitly of the travel time of a wave. The distance traveled by a wavelet includes a distance x on Line 1 to the coupling point, and a distance x traveled on Line 2 to the receiving node b. Thus, the delay time is $2x/c$. The superposition integral is then

$$v_b(t) = \frac{K_C + K_L}{2c} \int_0^d D\left(t - \frac{2x}{c}\right) dx$$

where D is defined as the time derivative of v_1 at the transmitting node at $x = 0$ on Line 1:

$$D(t) = \frac{\partial v_1}{\partial t}$$

The integration can be performed, giving the following simple expression for the backward coupled noise:

$$v_b(t) = b_{12}[v_1(t) - v_1(t - 2\tau)] \tag{3.10}$$

where $b_{12} = (K_C + K_L)/4$ is the coefficient of backward coupling, and $\tau = d/c$ is the one-pass propagation delay of the wave down the line. The computed waveshapes of Figure 3-14(b) follow directly by taking the difference of two time-shifted waveforms, as stated in (3.10).

REFERENCES

1. A. B. Glaser and G. E. Subak-Sharpe, "Integrated Circuit Engineering—Design, Fabrication, and Applications," Reading, Mass.: Addison-Wesley, 1977.
2. L. T. Olson and R. R. Sloma, "Chip Carrier Enhancements for Improving Electrical Performance," IEEE, *Proceedings 1985 35th Electronic Components Conference*, pp. 372–378 (1985).
3. A. J. Rainal, "Computing Inductive Noise of Chip Packages," *AT&T Bell Laboratories Technical Journal* 63, pp. 177–195 (1984).
4. R. E. Canright, "A Formula to Model Delta-I Noise," IEEE, *Proceedings 37th Electronic Components Conference*, pp. 354–361, 1987.
5. A. J. Blodgett and D. R. Barbour, "Thermal Conduction Module: A High-Performance Multilayer Ceramic Package," *IBM J. Res. Dev.* 26, pp. 30–36 (1982).
6. D. B. Tuckerman, "Laser-Patterned Interconnect for Thin-Film Hybrid Wafer-Scale Circuits," *IEEE Electron. Device Letters* 8, pp. 540–543 (1987).
7. R. Pound, "Packaging Links Fast GaAs Dice to High-Speed Systems," *Electronic Packaging and Production*, pp. 70–74 (August 1985).
8. W. T. Weeks, "Calculation of Coefficients of Capacitance of Multi-conductor Transmission Lines in the Presence of a Dielectric Interface," *IEEE Transactions on Microwave Theory and Techniques* MTT-18, pp. 35–43 (1970).
9. S. Ramo and J. R. Whinnery, "Fields and Wave in Modern Radio," 2d ed., New York: John Wiley and Sons, 1960.
10. R. A. Chapman, "Schaum's Outline of Theory and Problems of Transmission Lines," New York: McGraw-Hill, 1968.
11. W. C. Johnson, "Transmission Lines and Networks," New York: McGraw-Hill, 1950.
12. C. S. Chang, "Electrical Design of Signal Lines for Multilayer Printed Circuit Boards," *IBM J. Res. Dev.* 32, pp. 647–657 (1988).
13. R. F. Bonner, J. A. Asselta, and F. W. Haining, "Advanced Printed-Circuit Board Design for High-Performance Computer Applications," *IBM J. Res. Dev.* 26, pp. 297–305 (1982).

14. M. S. Pittler, D. M. Powers, and D. L. Schnabel, "System Development and Technology Aspects of the IBM 3081 Processor Complex," *IBM J. Res. Dev.* 26, pp. 2–11 (1982).
15. IBM Advanced Statistical Analysis Program, ASTAP Manual SH20-1118-0, White Plains, N.Y.: IBM Corp.
16. E. E. Davidson, "Electrical Design of a High Speed Computer Package," *IBM J. Res. Dev.* 26, pp. 349–361 (1982).
17. R. E. Matick, "Transmission Lines for Digital and Communication Networks," New York: McGraw-Hill, 1969.
18. C. W. Ho, "Theory and Computer-Aided Analysis of Lossless Transmission Lines," *IBM J. Res. Dev.* 17, pp. 249–255 (1973).
19. C. S. Chang, "Transmission Lines," in *Circuit Analysis, Simulation, and Design,* edited by A. E. Ruehli, pp. 292–332, Advances in CAD for VLSI Series, vol. 3, part II, Amsterdam: North Holland, 1987.
20. C. S. Chang, G. Crowder, and M. F. McAllister, "Crosstalk in Multilayer Ceramic Packaging," *Proceedings 1981 IEEE International Symposium on Circuits and Systems,* Chicago, 1981, April 27–29, vol. 1, pp. 6–10.

PROBLEMS

3.1. A new printed circuit board is to be fabricated out of a fluorocarbon dielectric material whose dielectric constant is 2.0. It will be used in a digital system whose electrical waveforms have rise and fall times of 200 ps (2×10^{-10} seconds).

Assume that the cross section has the dimensions shown in Figure 3-5(a). For estimation purposes, assume that the capacitances scale in direct proportion to the dielectric constant. The inductances, of course, remain the same. Use the inductance/unit length given in Table 3.1.

(a) Estimate the characteristic impedance Z_0, the propagation delay per unit length τ/d, and the propagation velocity c.

(b) Find the reflection coefficient, if lines are terminated with 50- or 90-Ω terminating resistors.

(c) For what line length does the propagation delay equal $\frac{1}{2}$ of the rise time? This is the condition for maximum backward coupled (near end) noise.

(d) Find the line length for which the propagation delay is equal to 1 ns (1×10^{-9} seconds).

3.2. A rectangular copper conductor, 0.1 mm wide and 25 mm long, is fabricated on a printed circuit board starting with a 2-ounce copper sheet (an industry standard-size sheet that weighs 2 ounces per square foot).

Find the dc resistance of the rectangular conductor.

3.3. In the copper bus bar of Example 3.2, assume that the supply voltage is 5 V and that the dc voltage at which the circuits are specified to operate is ± 10 percent of that value. Of the 10 percent, ± 5 percent is allocated to the dc voltage distribution drop.

If the bus bar distribution is allowed to be 50 percent of the drop, what is the maximum current that can be supplied to the circuits and still be within specifications?

3.4. Explain why spacing two signal lines closer together can be both advantageous and detrimental to a design.

3.5. Discuss the effects of making the spacing smaller between the following layers in a multilayer board:

(a) Signal/power.

(b) Power/power.

3.6. Calculate the reflection coefficients for the driving and receiving end for the configuration of Example 3.5. Include the cases when the receiver end is open-circuited and when it is short-circuited.

3.7. What advantage would the use of Teflon dielectric coaxial cable (dielectric constant = 2.1) have over the use of twisted pair signal wires? Assume the wires are bundled in a pack of 12 pairs per cable. Which would have the fastest propagation velocity?

3.8. For the two-line structures of Figures 3-5(e) and 3-6(b), calculate the following quantities:

(a) Characteristic impedance, and propagation time per unit length, τ/d.

(b) Peak forward and backward coupled noise voltages, assuming all terminations are at the characteristic impedance, so that reflections do not have to be considered. Use a line length of 1280 mm and compare the results with those of Example 3.6.

(c) Which line structure gives the least noise?

3.9. For the three-line structures of Figures 3-5 and 3-6, estimate the noise coupled between the two outer lines, assuming the center line is grounded. Ignore the noise signal coupled through the center line. Use a line length of 1280 mm.

CHAPTER
4

PRINTED CIRCUIT BOARD SIGNAL LINE ELECTRICAL DESIGN[1]

CHI SHIH CHANG

INTRODUCTION

The printed circuit board (PCB) is essential in bringing together all electronic components in a computer system [1–4] as discussed in the previous chapters. It is constructed with several electrically conductive layers separated by dielectric layers as discussed in Chapter 2. Some of the conductive layers are used to

---

[1] This chapter is a revised edition of a paper from the IBM Journal of Research and Development. This material is copyright 1988 by International Business Machines Corporation; reprinted with permission. The section on skin effect has been omitted to shorten the length of this chapter. A few sections have been extended, and a problem set have been added, for classroom use.

104

provide ground and power supply voltages. The remaining conductive layers are etched into well-defined lines for electrical signal interconnections among the integrated circuit chips. The layer-to-layer interconnections are accomplished through the plated through-holes. For high-density PCBs, we may also use the programmable vias for interconnections between two adjacent conductive layers.

In most PCBs the signal lines on a signal layer run parallel, in either the x or the y direction. For the purpose of explanation, we shall assume that they run in the y direction. If we follow the path of a long y direction signal line, we may find that it has other signal lines on one or both sides. There are also regions where both immediate left- and right-hand sides are vacant. In the first several sections of this chapter we shall discuss the electrical design of a simple case where the long signal line has no immediate adjacent lines throughout its entire length [5–7], and we shall also neglect the effect of the parallel lines beyond the immediate neighborhood.

For PCBs using glass-epoxy insulation material with a relative dielectric constant ϵ_r equal to 4, an electrical signal can travel with a delay of 6.67 ps/mm. For a typical interconnection line 100 mm long, it will take 0.667 ns for the signal to reach the receiving end. Depending on the input impedance of the receiver circuit, some of the electrical energy may reflect backward and return to the sending end at 1.333 ns. If the off-chip driver circuit at the sending end has a rise time shorter than 1.333 ns, we have to pay close attention to this reflected energy. This can be carefully studied by treating this signal line as a transmission line as discussed in Chapter 3. A uniform transmission line has a well-defined characteristic impedance and signal propagation delay [5–6].

When two or more signal lines are placed in close proximity, an active signal on one line will cause coupled noise on the adjacent lines [5,6,8–11]. As the length of the parallel section increases, the backward (i.e., backward to the near end) coupled noise increases in magnitude and its pulse width remains constant, equal to the active signal rise time. Beyond a certain length for the parallel section, when the time of flight across this section is equal to, or greater than, one-half of the signal rise time, the magnitude of the backward coupled noise remains constant, while the pulse width becomes longer, equal to twice the time of flight (Chapter 3). This magnitude is called the saturated backward coupled noise, and strongly depends on the edge-to-edge spacing between the adjacent lines and the distance from the closest voltage plane. If necessary, we may limit the length of the parallel section to limit the backward coupled noise to a fraction of its saturated value.

Because of the finite inductance of the on-chip voltage distribution and the chip carrier lead wires (Chapter 3), a latch circuit may experience voltage disturbance when the internal logic circuits or off-chip driver circuits are switching. This is called delta-I noise [11,12]. It is important that the combination of the delta-I noise, the coupled noise, and other dc voltage drop on the power supply and signal line at the clocking time is kept below the noise margin of the latch circuit (Chapter 3). Otherwise, false switching may happen and cause improper logic function of the chip. In this chapter, we shall focus on signal line studies.

TRANSMISSION LINE CHARACTERISTICS OF A SIGNAL LINE

As discussed earlier, a signal line over a voltage reference plane is to be treated as a transmission line for fast rise time signal propagation (Chapter 3). This pair of parallel conductors has a loop inductance of L pH/mm, and a signal-to-reference capacitance of C pF/mm. The telegrapher's equation for a transmission line has a traveling wave solution as follows [5–6,13–14]:

$$V(y,t) = f(\tau y - t) + g(\tau y + t) \tag{4.1}$$

$$Z_0 I(y,t) = f(\tau y - t) - g(\tau y + t) \tag{4.2}$$

where

$$\tau = \sqrt{LC} = 3.33\sqrt{\epsilon_r} \quad \text{(ps/mm)} \tag{4.3}$$

$$Z_0 = \sqrt{L/C} \quad (\Omega) \tag{4.4}$$

Here, the voltage waveform $V(y,t)$ is a function of position y and time t. It consists of a forward-traveling component, $f(\tau y - t)$, and a backward-traveling component, $g(\tau y + t)$. The current waveform, $I(y,t)$, is related to the voltage waveform with proportionality constants of Z_0 for the forward-traveling component, and $-Z_0$ for the backward-traveling component, as discussed in Chapter 3. Both the forward- and backward-traveling components have a signal propagation time constant τ that is proportional to the square root of the relative dielectric constant, ϵ_r. It takes the traveling wave 3.33 ps to propagate 1.00 mm in vacuum. In the glass epoxy with $\epsilon_r = 4$, the propagation time becomes 6.67 ps/mm.

In the computer system design, the CPU cycle time is one of the most important parameters (Chapter 1). Some of the signal paths within one CPU cycle may span many logic gate stages at close proximity. Some of them may have only two logic gate stages at a significant distance. For the signal paths of the latter case, the PCB propagation time constant τ may become an important design parameter. For example, if the time of flight across the PCB signal line is 60 percent of the total signal path delay, a reduction of ϵ_r from 4.0 to 2.5 will reduce the total signal path delay by 12.5 percent.

As stated earlier, Z_0 of Equation (4.4) is a proportionality constant between the forward-traveling voltage and current waveforms. It is called the characteristic impedance of the transmission line. Assume we send a 10-mA current ramp with a 1.0-ns rise time at the input end of a 300-mm-long PCB signal line in a dielectric medium with $\epsilon_r = 4$. At the end of the 1.0-ns ramp, the wave front of the forward-traveling wave is halfway along the 300-mm-long signal line. It will take another 1.0 ns for the wave front to reach the far end, and it will then be reflected. The reflected wave is a backward-traveling wave; its wave front will return to the input end at 4.0 ns. Using a high-impedance probe at the input end we will sense a voltage ramp of 1.0 ns rise time, which remains at constant magnitude until $t = 4.0$ ns [6]. If this magnitude is 800 mV, we know that this signal line has a characteristic impedance of 80 Ω. Conversely, if we want to get a 2.0-V ramp at the input end of the 80 Ω signal line, we have to design a driver circuit delivering 25 mA output current.

Note that the driver circuit design is strongly influenced by the PCB signal line characteristic impedance. For a 50-Ω PCB signal line, the driver circuit has to deliver 40 mA output current to obtain a 2.0-V ramp. That represents a 60 percent increase in driving current requirement, and will cause a 60 percent increase in transient current through the PCB voltage distribution system if the signal rise time is kept constant. Therefore, the 80-Ω line is preferred over the 50-Ω line design with respect to the transient current consideration. Usually, a PCB signal line is designed with a characteristic impedance of 40 to 90 Ω. In the remaining part of this chapter, we shall present various design considerations in the choice of signal line characteristic impedance.

CHARACTERISTIC IMPEDANCE OF A SIGNAL LINE

The typical epoxy-impregnated glass cloth used in PCBs has a thickness of 40 to 100 μm. For yield and reliability considerations, we would use two or more sheets in the laminate. The signal line width may range from 20 to 200 μm. Due to process considerations, the minimum edge-to-edge spacing between parallel signal lines is usually one to two times the line width. The line thickness is about 0.3 to 0.6 times the line width. In this section, we shall discuss the geometrical consideration in the signal line impedance design [5–7]. Combining Equations (4.3) and (4.4), we have

$$\sqrt{\epsilon_r} Z_0 = \frac{3.33}{C/\epsilon_r} \quad (\Omega) \qquad (4.5)$$

Figure 4-1 shows the cross section of a stripline in a homogeneous medium above a voltage reference plane. The medium has a relative dielectric constant ϵ_r. At a height H above the reference plane is a signal line of rectangular cross section, having width W and thickness T. The signal-line-to-reference-plane capacitance per unit length is a two-dimensional electrostatic problem. It will

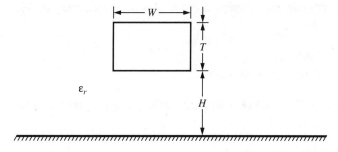

FIGURE 4-1
Stripline cross section in a homogeneous medium.

remain at the same value when H, W, and T are all scaled by the same factor. Therefore, we may use the line width as the parameter for normalization and state that the capacitance per unit length is a function of the H/W and T/W ratios. Furthermore, the capacitance is linearly proportional to the relative dielectric constant. Therefore, we have

$$C = \epsilon_r f(H/W, T/W) \tag{4.6}$$

This capacitance can be calculated numerically [15]. The signal line characteristic impedance can then be obtained from Equation (4.5). The numerical results are shown in Table 4.1 for various H/W ratios at $T/W = 0.5$. The data in Table 4.1 can be approximated as follows:

$$\sqrt{\epsilon_r} Z_0 = \frac{377 \times (H/W)}{1 + 2.62(H/W)^{3/4}} \tag{4.7}$$

The third column in Table 4.1 shows the results of Equation (4.7). The error is shown in Column 4; it is less than 1 percent when $H/W < 4.5$. The inverse of the right-hand side of Equation (4.7) contains the following factor:

$$f(W/H) = (W/H) + 2.62 \times (W/H)^{1/4} \tag{4.8}$$

The first term represents the parallel plate contribution, and the second term gives the approximate fringe field contribution to the signal line capacitance and impedance. Note that the fringe field contribution becomes the dominant term when $H/W > 0.28$.

Now we will use Table 4.1 to illustrate how the material and process considerations will determine the upper and lower bounds of the characteristic impedance. Assume that manufacturing can routinely make PCBs with 100-μm signal line widths with acceptable yield, and the epoxy-impregnated glass cloth has a thickness of 70 μm with $\epsilon_r = 4.0$. Using Equation (4.7) together with the correction percentage in the fourth column of Table 4.1, we obtain the characteristic impedance, as shown in Table 4.2, for two- to eight-ply laminate. Note that the 4-times increase in laminate thickness results in a 1.63-times increase in Z_0. A 100-Ω signal line impedance will require an eight-ply laminate, making for a very thick PCB. In addition, a thick laminate will require a large spacing between signal lines to contain the coupled noise (to be discussed later in this chapter). The alternatives are:

1. Reducing signal line width
2. Using new cloth of low dielectric constant

On the other hand, if $Z_0 < 60 \, \Omega$ is desired, different approaches will be required:

1. Reducing thickness of glass-epoxy cloth
2. Increasing signal line width at the expense of wiring density

TABLE 4.1
Signal line impedance and its approximation when T/W = 0.5

H/W	$\sqrt{\varepsilon_r}Z_0$, Ω	Equation (4.7) Ω	ΔZ_0, %	Slope, Equation (4.10)	Z_0 ratio ($H_2 = 2H_1$) Exact	Z_0 ratio ($H_2 = 2H_1$) Approximate	Z_0 ratio ($H_2 = H_1$) Exact	Z_0 ratio ($H_2 = H_1$) Approximate
0.10	25.62	25.72	0.38	0.76	0.731	0.746	0.573	0.587
0.15	34.46	34.66	0.58	0.71	0.755	0.765	0.597	0.610
0.20	42.01	42.27	0.63	0.67	0.775	0.781	0.618	0.629
0.25	48.66	48.93	0.55	0.64	0.790	0.794	0.635	0.644
0.30	54.55	54.85	0.55	0.61	0.804	0.805	0.651	0.658
0.35	59.92	60.19	0.45	0.59	0.815	0.814	0.664	0.670
0.40	64.82	65.06	0.37	0.57	0.825	0.823	0.676	0.680
0.45	69.34	69.54	0.29	0.56	0.833	0.830	0.687	0.689
0.50	73.55	73.69	0.20	0.54	0.841	0.836	0.697	0.698
0.75	91.03	90.87	−0.17	0.49	0.867	0.862	0.733	0.731
1.00	104.50	104.14	−0.34	0.46	0.883	0.879	0.759	0.754
1.50	124.70	124.25	−0.36	0.41	0.903	0.901	0.791	0.785
2.00	139.90	139.47	−0.31	0.39	0.914	0.915	0.810	0.806
2.50	152.00	151.80	−0.13	0.37	0.920	0.925	0.824	0.821
3.00	161.90	162.21	0.19	0.36	0.926	0.933	0.834	0.832
3.50	170.30	171.27	0.57	0.35	0.932	0.939	0.843	0.841
4.00	177.90	179.30	0.79	0.34	0.934	0.944	0.849	0.848
4.50	184.80	186.53	0.94	0.33	0.936	0.948	0.853	0.854
5.00	190.70	193.13	1.27	0.33	0.938	0.951	0.859	0.859
7.50	213.80	219.63	2.73	0.31	0.947	0.963	0.875	0.876
10.00	230.70	239.62	3.87	0.30	0.951	0.969	0.884	0.887
15.00	255.00	269.68	5.76	0.29	0.953	0.977	0.893	0.898

TABLE 4.2
**Laminate thickness
and signal line
impedance for stripline
structure with $\varepsilon_\gamma = 4$**

Layers	H/W	Z_0
2	1.4	60.6
3	2.1	71.3
4	2.8	79.0
5	3.5	85.2
6	4.2	90.4
7	4.9	94.8
8	5.6	98.7

The above discussions bring out the tradeoffs a designer will face in the choice of materials, process yield and reliability, wiring density, and PCB thickness.

The data in Table 4.1 are for $T/W = 0.5$ only. For other T/W ratios, we may make first-order engineering estimates of Z_0, using an approximate effective line width, as follows:

$$W(\text{effective}) = W \times (1 + T/W) \div 1.5 \qquad (4.9)$$

Numerical calculation indicates that, for the typical PCB structure with $T/W = 0.3$ to 0.6, and $H/W = 0.5$ to 5, the error in Z_0 is less than 3 percent when $W(\text{effective})$ in Equation (4.9) is substituted for W in Equation (4.7) [16].

The ratio of the percentage variation of Z_0 to the percentage variation of H/W can be derived from Equation (4.7), and is as follows:

$$\frac{\delta Z_0/Z_0}{\delta(H/W)/(H/W)} = \frac{1 + 0.655 \times (H/W)^{3/4}}{1 + 2.62 \times (H/W)^{3/4}} \qquad (4.10)$$

The result is shown in Column 5 of Table 4.1. As an illustration, we assume that the effective signal line width $W(\text{effective})$ has 8 percent variation, and the glass-epoxy cloth thickness between the signal line and the reference plane has 6 percent variation. The root-sum-square of the above two variations yields 10 percent variation for H/W. If the design is in the neighborhood of $H/W = 4$, then Z_0 will have 3.4 percent variation. If the design is in the neighborhood of $H/W = 1$, then Z_0 will have 4.6 percent variation. Note that variation in dielectric constant in Equation (4.7) will further worsen the Z_0 tolerance.

EFFECTS OF MICROSTRIP
AND TRIPLATE STRUCTURES

In some of the PCB structures, a signal line is placed on top of the glass-epoxy dielectric. The top and the two side surfaces are surrounded by air, forming an inhomogeneous dielectric medium. This is called the microstrip line [7]. Most of the electromagnetic wave is in the glass-epoxy medium. However, there is still a significant fraction of wave energy in the air dielectric. The combined traveling wave in the inhomogeneous medium is not a pure transverse electromagnetic (TEM) wave [5]. For the usual PCB structure, we take the quasi TEM waves approximation and apply the transmission line treatment as before. From laboratory measurement, we know that the propagation time constant lies between that of the air medium and the glass-epoxy medium. An effective dielectric constant will be derived for a frequently encountered microstrip structure in this section.

Since the permeability μ of the dielectric medium is the same as that of the air, the inductance per unit length of the signal line is the same as that in the homogeneous medium. The capacitance per unit length can be obtained by solving the electrostatic field problem [15]. Numerical calculations are made for a microstrip line on the glass-epoxy dielectric with $\epsilon_r = 4$, and having its top and two side surfaces surrounded by air. The results indicate that the fringe field contribution can be closely approximated by $1.37 \times (W/H)^{1/6}$. That is:

$$C = 4 \times \epsilon_0 \times [W/H + 1.37 \times (W/H)^{1/6}] \qquad (4.11)$$

The above approximation has less than 1 percent error when $0.1 < (W/H) < 10$. From Equations (4.8) and (4.11), we obtain an effective relative dielectric constant as follows:

$$\epsilon_r(\text{effective}) = \frac{4 \times [(W/H) + 1.37 \times (W/H)^{1/6}]}{(W/H) + 2.62 \times (W/H)^{1/4}} \qquad (4.12)$$

Note that Equation (4.8) has less than 1 percent error, of opposite polarity to that in Equation (4.11), when $0.1 < (W/H) < 4.5$. This yields a ± 2 percent error in Equation (4.12). Note that in the typical PCB structure, where $H/W = 0.5$ to 5, that is, $W/H = 0.2$ to 2, Equation (4.12) gives an effective relative dielectric constant in the range of 2.54 to 2.77.

In other PCB structures, the signal line may be sandwiched between two reference planes, as shown in Figure 4-2 [5–7]. This is called the shielded stripline, and will be referred to as the triplate structure in this chapter to emphasize the presence of the second reference plane. The introduction of this second reference plane will increase the electrostatic capacitance of the signal line and reduce the characteristic impedance. The parallel plate contribution in the expression similar to Equation (4.8) increases by a factor of $1 + H_1/H_2$, which is 1.5 when $H_2 = 2H_1$, and 2.0 when $H_2 = H_1$. Numerical calculation indicates that the fringe field contribution has little change when $H_2 = 2H_1$, and has a 7 percent increase when $H_2 = H_1$. Therefore, in the triplate structures, the fringe field contribution becomes the dominant term when $H_1/W > 0.48$ for $H_2 = 2H_1$, and when $H_1/W > 0.64$ for $H_2 = H_1$. For $T/W = 0.5$, the ratios of signal line impedances can be approximated by the following formulas:

$$\frac{Z_0(H_2 = 2H_1)}{Z_0(H_2 = \infty)} = \frac{1 + 2.62(H_1/W)^{3/4}}{1.5 + 2.62(H_1/W)^{3/4}} \qquad (4.13)$$

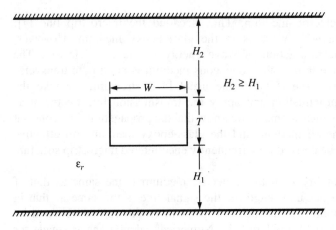

FIGURE 4-2
Shielded stripline (triplate) cross section.

$$\frac{Z_0(H_2 = H_1)}{Z_0(H_2 = \infty)} = \frac{1 + 2.62(H_1/W)^{3/4}}{2 + 2.8(H_1/W)^{3/4}} \qquad (4.14)$$

Table 4.1 shows the ratios of Equations (4.13) and (4.14) in Columns 7 and 9 respectively. They are good approximations of the numerical calculation results shown in Columns 6 and 8. The second reference plane at $H_2 = 2H_1$ or H_1 causes a line impedance reduction of 16 or 30 percent respectively when $H_1/W = 0.5$. And the reduction becomes 7 or 17 percent respectively when $H_1/W = 3.0$. For other T/W ratios, we may use W(effective) in Equation (4.9) to substitute for W in Equations (4.13) and (4.14). Numerical calculation indicates that the error in Z_0 is less than 3 percent for the typical PCB structure, where $T/W = 0.3$ to 0.6 and $H_1/W = 0.5$ to 5 [16].

MATCHED-LOAD IMPEDANCES OF PARALLEL SIGNAL LINES

Figure 4-3 shows the vertical cross section of a triplate structure with three parallel signal lines sandwiched between two voltage reference planes in a homogeneous medium. All three signal lines have the same width W and thickness T, and are at the same height H_1 above the bottom reference plane. The distance to the top reference plane is equal to, or greater than, the distance to the bottom reference plane; that is, $H_2 \geq H_1$. For a stripline structure, we have $H_2 = \infty$. The signal line in the middle has the same edge-to-edge spacing S from both neighboring lines. Table 4.3 shows the numerical calculation results for the elements in the three-by-three capacitance and inductance matrices [15] for the stripline structures with $T = 0.5W$ and $H_1 = 2.0W$. They are positive-definite [17–18] symmetrical matrices, which can be proved from stored energy considerations [19].

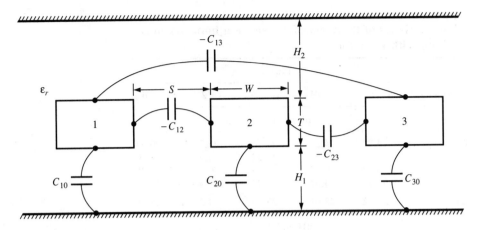

FIGURE 4-3
Cross section of three coupled lines in a homogeneous medium.

Note that the off-diagonal elements in the capacitance matrix are all negative. In the inductance calculation [15], it is assumed that all current flux is confined to the conductor surface; that is, the internal inductance is not included. When the adjacent signal lines are brought closer to the middle line, the surface current density will redistribute to reach the minimum energy state, thus the self-inductance is reduced slightly due to this proximity effect. C_{22} is the sum of three capacitances, C_{20}, $-C_{12}$, and $-C_{23}$. It is equivalent to the capacitance between the second line and the combined voltage reference, consisting of Lines 1, 3, and the original reference plane. The C_{22} value increases drastically as the spacing between adjacent lines is brought down from ∞ to W.

When Lines 1 and 3 are left floating, the capacitance between Line 2 and the reference plane is the parallel combination of three branches: (1) C_{20}, (2) series capacitance of C_{10} and $-C_{12}$, and (3) series capacitance of C_{30} and $-C_{23}$. The result is defined as C_{2F}, and is included in Table 4.3. From a physical structure viewpoint, we may treat C_{2F} as the capacitance between Line 2 and the reference plane when the space originally occupied by Lines 1 and 3 is replaced by dielectric material of extremely high dielectric constant, that is, $\epsilon_r \to \infty$. Since the high dielectric regions only occupy relatively small areas, the increase in C_{2F} is very small in comparison with the increase in C_{22}. Since the self-inductance L_{22} is calculated with the adjacent lines floating, as is C_{2F}, we have $Z_{02F} \equiv \sqrt{L_{22}/C_{2F}}$ as the open-circuited impedance of the middle signal line. The results are shown in Table 4.3. Note that the open-circuited impedance is only slightly smaller than that of a single line, that is, when adjacent lines are absent.

In the actual PCB wiring, the near ends of the two adjacent lines are not floating. They are likely to be connected to other sections of signal lines before being terminated at driver or receiver circuits. In the following we shall define the matched-load impedance of the coupled transmission lines [10]. For the lossless coupled transmission lines, which consist of n parallel signal lines

TABLE 4.3
Characteristics of three coupled lines in a homogeneous medium with $\varepsilon_r = 1.0$

Parameters	$S = \infty$	$S = 4W_0$	$S = 3W_0$	$S = 2W_0$	$S = W_0$
			Line spacing		
			pF/m		
C_{11}	23.82	24.28	24.70	25.77	29.34
$-C_{12}$	0	3.00	4.22	6.37	11.40
$-C_{13}$	0	0.57	0.73	0.96	1.35
C_{22}	23.82	24.67	25.47	27.48	34.27
C_{2F}	23.82	23.91	23.99	24.20	24.98
			nH/m		
L_{11}	466.14	465.34	464.50	462.17	453.25
L_{12}	0	58.88	81.50	118.00	181.19
L_{13}	0	18.26	27.70	46.48	91.20
L_{22}	466.14	464.71	463.20	459.13	444.84
			Ω		
Z_{02F}	139.90	139.41	138.96	137.74	133.45
$\sqrt{L_{11}/C_{11}}$	139.90	138.44	137.12	133.93	124.28
Z_{01}	139.90	138.40	137.00	133.60	123.00
$\sqrt{L_{22}/C_{22}}$	139.90	137.25	134.85	129.27	113.93
Z_{02}	139.90	137.20	134.80	129.00	112.80
			mV/V		
b_{21}	0	62.31	85.88	123.70	189.80
b_{21} (appr)	0	62.07	85.26	121.83	183.14
b_{31}	0	15.71	22.35	34.55	62.19
b_{31} (appr)	0	15.71	22.33	34.49	61.76
b_{12}	0	62.60	87.33	128.10	207.00
b_{12} (appr)	0	62.60	86.67	126.10	198.99

Notes:
1. $C_{21} = C_{23} = C_{32} = C_{12}$
2. $L_{21} = L_{23} = L_{32} = L_{12}$
3. $C_{33} = C_{11}$
4. $L_{33} = L_{11}$
5. $Z_{03} = Z_{01}$
6. $Z_{02F} \equiv \sqrt{L_{22}/C_{2F}}$

above a reference conducting plane, the characteristic impedance is an n-by-n matrix as follows [20–21]:

$$Z_0 = Y_0^{-1} = (LC)^{-1/2}L = (LC)^{1/2}C^{-1} \qquad (4.15)$$

where $(LC)^{1/2} = P\tau P^{-1}$

P = the n-by-n eigenvector matrix of the LC product
τ = the positive square root of the corresponding
eigenvalues [17–18]. It is an n-by-n diagonal matrix.

For the example shown in Table 4.3, the three eigenvalues are identical because of the homogeneous medium. Therefore, the characteristic impedance and admittance matrices, Z_0 and Y_0, are:

$$Z_0 = \frac{1}{\tau}L = \tau C^{-1} \tag{4.16}$$

$$Y_0 = Z_0^{-1} = \tau L^{-1} = \frac{1}{\tau}C \tag{4.17}$$

Note that Z_0 is the open-circuited input impedance matrix of a three-port network, which represents an infinitely long section of three coupled transmission lines, and Y_0 is the short-circuited input admittance matrix of this three-port network.

To obtain the matched-load impedance of these three coupled transmission lines, we first measure the input impedance of the first signal line using, for example, the time-domain-reflectometer (TDR) [6,7]. Then, the first line is terminated by the input impedance just obtained, and the input impedance of the second line is measured. Afterward, the second line is terminated by its input impedance just obtained, and the input impedance of the third line is measured. The third line is then terminated by its input impedance. We will repeat the above process, measuring the input impedance of Lines 1, 2, 3, 1, 2, . . . until the consecutive measurements for the same signal line are within acceptable measurement accuracy for each of the three lines. This set of three impedances—$\{Z_{01}, Z_{02}, Z_{03}\}$— is defined as the matched-load impedance. The results of numerical iterative simulation are shown in Table 4.3.

Note that the difference between Z_{02} and $\sqrt{L_{22}/C_{2F}}$ is small for $S = 4W$, but becomes unacceptably large for $S = 2W$ and $S = W$. When this difference is significant, the signal line impedance at the boundary of the sparsely and closely placed parallel wiring regions will have a large discontinuity. A quantitative assessment on matched-load impedance will be presented after the coupled noise studies.

CROSS-TALK AMONG ADJACENT PARALLEL SIGNAL LINES

A long section of three parallel signal lines over a voltage reference plane is shown in Figure 4-4. Line 1 is connected to an active signal. The near ends of Lines 2 and 3 are terminated by their corresponding matched-load impedances, as defined in the last section. Because of the long line length, the termination at the far end is immaterial in the discussion here. Since these three parallel lines can be treated as a three-port resistor network as discussed following Equation (4.17), the input ends of Lines 2 and 3 will have finite voltages, called cross-talk, due to an active signal on Line 1. With the matched-load impedance termination, the ratios of the cross-talk on Lines 2 and 3 to the active signal voltage on Line 1 are defined as the backward (i.e., backward to the near end) coupling coefficients,

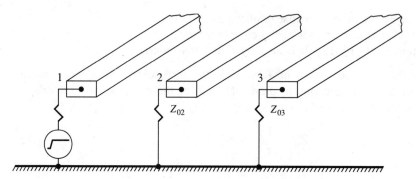

FIGURE 4-4
Near-end termination of three coupled lines.

b_{21} and b_{31}. In other words, the voltages on Lines 2 and 3 are $b_{21}V_{01}$ and $b_{31}V_{01}$, respectively, when Line 1 has a voltage of V_{01}.

In order to understand the situation of two active lines, we shall assume that Line 1 has a source impedance equal to its matched-load impedance Z_{01}, and a Thevenin equivalent source voltage [13] of $2V_{01}$, so that the voltage on the input end of Line 1 is V_{01}. Similarly, if Line 3 is the only active line, with a Thevenin equivalent voltage of $2V_{03}$ and a source impedance of Z_{03}, then the voltages on the input ends of Lines 1, 2, and 3 will be $b_{13}V_{03}, b_{23}V_{03}$, and V_{03}, respectively. From the linear superposition principle, we know that when both Lines 1 and 3 are active with Thevenin equivalent source voltages $2V_{01} = 2V_{03} = 2V_0$, the voltages on the input ends of Lines 1, 2, and 3 will be $(1 + b_{13})V_0, (b_{21} + b_{23})V_0$, and $(1 + b_{31})V_0$, respectively. It is emphasized here that when both Lines 1 and 3 are active, the voltage on Line 1 is no longer equal to V_0. The simulation results of the backward coupling coefficients are included in Table 4.3. Note that when Line 1 is the only active line, the backward coupled noise on Line 2 has the following closed-form approximation [10]:

$$V_{21}(\text{near end}) \approx b_{21}[V_0(t) - V_0(t - 2\tau D)] \tag{4.18}$$

where $V_0(t)$ = voltage on the input end of Line 1
         τD = delay time across the coupled line section of length D

$$b_{21} \text{ (approx.)} = 0.25\left(\frac{-C_{21}}{C_{22}} + \frac{L_{21}}{L_{11}}\right) \tag{4.19}$$

The second term on the right-hand side of Equation (4.19) is one-quarter of the ratio of two inductance values, and can be viewed as the inductive coupling coefficient. It is independent of the relative dielectric constant. The first term is the capacitive coupling coefficient, which is independent of the relative dielectric constant only in a homogeneous medium. Both terms are functions of the geometrical parameters. The results of these calculations are also included in Table 4.3. Note that the approximation has error within 4 percent even when the coupling coefficient exceeds 0.2 (i.e., 200 mV/V).

Equation (4.18) indicates that when the 0-to-100 percent rise time, T_r, of $V_0(t)$ is shorter than twice the delay time across the length of the coupled line section, $2\tau D$, the noise reaches its maximum magnitude, called the saturated backward coupled noise, and has a pulse width equal to $2\tau D$. On the other hand, if the length of the parallel section is shorter, such that $2\tau D < T_r$, the magnitude of the coupled noise will be smaller, with a pulse width equal to T_r. Therefore, we can reduce the magnitude of the coupled noise by reducing the length of the parallel section. This is usually accomplished by establishing a design guideline and using a computer-aided design (CAD) tool to ensure that it is strictly adhered to.

When the spacing between adjacent lines decreases, the line-to-line capacitance $-C_{12}$ increases, as does the total line capacitance C_{22}. The overall results are an increase in coupled noises and a decrease in matched-load impedances, as illustrated in Table 4.3. Even though the backward coupling coefficient b_{21} is independent of the relative dielectric constant ϵ_r in a homogeneous medium, its dependency on the geometrical parameters W, S, T, H_1, and H_2 in the triplate structure still yields four degrees of freedom. In a typical PCB fabrication, we are likely to use the manufacturable minimum signal line width, W_0, with a thickness of 0.3 to $0.6W_0$. We shall assume that $T = 0.5W_0$, and study the coupled noise of the structure with three degrees of freedom. We shall use numerical calculation results to show that these geometrical parameters have the following order of relative importance: (1) H_1/S, (2) H_2/H_1, and (3) W_0/S. Note that in the studies of the single signal line impedance in the previous several sections, the signal line width W has been used as the parameter for normalization. For coupled noise studies, we shall use the edge-to-edge spacing S as the parameter for normalization.

Assume that all three signal lines in Figure 4-3 have the same manufacturable minimum line width W_0, thickness $T = 0.5W_0$, and edge-to-edge spacing $S = W_0$ to $4W_0$ (i.e., $W_0/S = 1.0$ to 0.25). We may use a thin laminate to get a low value of line height H for low-impedance signal line design, and a high H value for high-impedance design. The three-by-three capacitance and inductance matrices are calculated [15]. The matched-load impedances, $\{Z_{01}, Z_{02}, Z_{03}\}$ as defined before, are then simulated, as are the backward coupling coefficients b_{21}, b_{31}, and b_{12}.

The two solid curves in Figure 4-5 show the backward coupling coefficient b_{21} versus the H_1/S ratio, with $W_0/S = 1.0$ and 0.25 as a parameter for the stripline structure. Note that when W_0/S is kept at 1.0 and H_1/S is reduced by a factor of four from 1.6 to 0.4, b_{21} reduces from 171 to 70 mV/V. On the other hand, when both S and H_1 are increased by a factor of four so that H_1/S is constant, and W_0/S is reduced by a factor of four from 1.0 to 0.25, b_{21} decreases from 171 to 146 mV/V. The reduction is about 20–30 mV/V throughout the range of H_1/S ratios of practical applications. When a second reference plane is introduced to form the triplate structure, similar numerical calculations are carried out for $H_2/H_1 = 2.0$ and 1.0, and these results are also shown in Figure 4-5. A few observations are summarized as follows:

1. The coupling coefficient b_{21} is heavily influenced by the H_1/S ratio, but has weak dependency on the W_0/S ratio.
2. The introduction of the second reference plane in the triplate structure will reduce the coupled noise from that of the stripline structure by about 25–35 mV/V for $H_2 = 2H_1$, and 40–60 mV/V for $H_2 = H_1$.
3. We may use linear interpolation with reasonable accuracy for W_0/S between 0.25 and 1.0, and for H_1/H_2 between 0.0 and 0.5, as well as between 0.5 and 1.0. Note that $H_1/H_2 = 0.0$ for the stripline structure.

In a typical PCB design, the total saturated coupled noise is not to exceed 200 mV per one volt of active signal swing. If a signal line has active lines on both left- and right-hand sides, then the total saturated coupled noise will be twice that of b_{21}. Therefore, we have to ensure that $b_{21} < 100$ mV/V. From Figure 4-5, we find that when $S = W_0$, the acceptable design is $H_1 < 0.65S$ for a stripline structure, and $H_1 < 0.98S$ or $1.33S$ for a triplate structure with $H_2 = 2H_1$ or H_1, respectively. From the data in Table 4.1 we find that for $H_1/W_0 = 0.65, 0.98$, and 1.33, the corresponding line impedances are $Z_0 = 42, 46$, and 46.5 Ω for $\epsilon_r = 4$, and 53, 58, and 59 Ω for $\epsilon_r = 2.5$.

Let us compare the wiring density of the stripline and triplate structures for a 60-Ω design with a restriction on coupled noise equivalent to $b_{21} \leq 100$

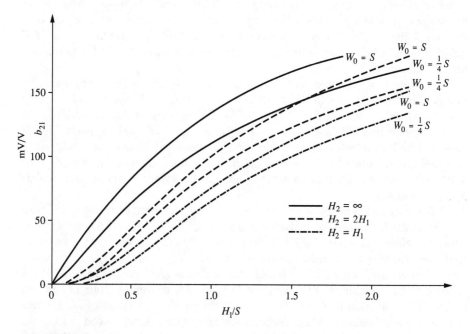

FIGURE 4-5
Backward coupling to the middle signal line in a homogeneous medium when only one adjacent line is active.

TABLE 4.4
**Wiring density comparisons between various PCBs
in the homogeneous medium for 60-Ω design with
$b_{21} \leq 100$ mV/V**

ϵ_γ	Parameter	$H_2 = \infty$	$H_2 = 2H_1$	$H_2 = H_1$
2.5	H_1/W_0	0.82	1.04	1.39
	H_1/S	0.68	0.99	1.34
	$\dfrac{S}{W_0} = \dfrac{H_1/W_0}{H_1/S}$	1.21	1.05	1.04
	$S + W_0$	$2.21W_0$	$2.05W_0$	$2.04W_0$
4.0	H_1/W_0	1.37	1.72	2.24
	H_1/S	0.74	1.03	1.37
	$\dfrac{S}{W_0} = \dfrac{H_1/W_0}{H_1/S}$	1.85	1.67	1.64
	$S + W_0$	$2.85W_0$	$2.67W_0$	$2.64W_0$

mV/V. The results are shown in Table 4.4 for $\epsilon_r = 2.5$ and 4.0. Note that, with $\epsilon_r = 4$, the acceptable center-to-center pitch between adjacent signal lines is reduced from $2.85W_0$ to $2.67W_0$ when a second reference plane is introduced to convert a stripline into a triplate structure with $H_2 = 2H_1$. On the other hand, the reduction of ϵ_r from 4.0 to 2.5 for the stripline structure will reduce the pitch from $2.85W_0$ to $2.21W_0$. In other words, under the same signal line impedance and coupled noise restriction, the low-dielectric medium has good wiring density advantages. On the other hand, the introduction of a second reference plane to form the triplate structures has only minor density advantages over the stripline structure. This is true when there is only one signal layer sandwiched between two reference planes. When there are two signal layers sandwiched between the reference planes, the effect of orthogonal lines becomes important. The triplate structure will give better electrical characteristics than the stripline. This will be discussed in the last section of this chapter.

EFFECT OF LINE SPACING
ON MATCHED-LOAD IMPEDANCES

In this chapter, we have presented the approximate formulas for the characteristic impedance of a single signal line. We may use Equation (4.9) to find the effective signal line width to be used in Equation (4.7) for the stripline structure. We may then use Equations (4.13) and (4.14) to adjust for the triplate structures. All three formulas use the imaginary parallel plate component and fringe field component. They can be easily extended to other H_2/H_1 ratios.

The presence of adjacent signal lines causes signal cross-talk, and reduces the matched-load impedances of the parallel signal lines. Figure 4-6 shows the nomographs for the percentage reduction of the matched-load impedances of the

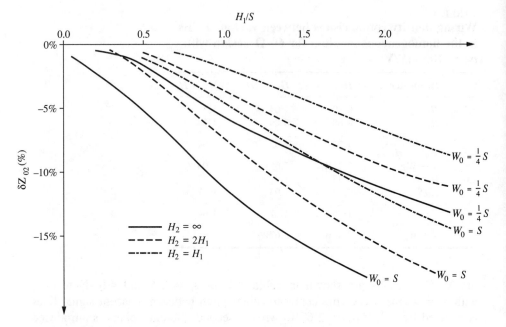

FIGURE 4-6
Reduction of the middle signal line impedance in a homogeneous medium.

middle signal line. The ordinate is defined as:

$$\delta Z_{02}(\%) \equiv 100\% \times \frac{Z_{02} - Z_0}{Z_0} \tag{4.20}$$

where Z_0 is the characteristic impedance of a single signal line with given W, T, H_1, and H_2 values, and Z_{02} is the matched-load impedance of the middle signal line. A few important points are summarized as follows:

1. The presence of adjacent lines will reduce the matched-load impedance Z_{02}. The results of Equation (4.20) are always negative. In a typical PCB design, we have to increase the laminate thickness, that is, H_1, to increase the Z_0 value by half of the percentage shown in Figure 4-6, so that the target value of line impedance will be halfway between Z_{02} and Z_0. Of course, we have to increase the value of S accordingly to contain the coupled noise.

2. The percentage reduction δZ_{02} is strongly influenced by the H_1/S ratio. Next comes the W_0/S ratio. The introduction of the second voltage reference plane in the triplate structure also has a strong effect on δZ_{02}.

3. We may use linear interpolation with reasonable accuracy for W_0/S between 0.25 and 1.0, and for H_1/H_2 between 0.0 and 0.5, as well as between 0.5 and 1.0.

In the last section, we have pointed out that, for $W_0 = S$, the backward coupling coefficient b_{21} will be less than 100 mV/V when $H_1 < 0.65S, 0.98S$, and $1.33S$ for $H_2 = \infty, 2H_1$, and H_1, respectively. From Figure 4-6 we find that the corresponding reductions in matched-load impedance are 7.5, 7.2, and 7.1 percent, respectively. That is, when the pulse, traveling on a signal line in the absence of neighboring lines on both sides, arrives at the point where both neighboring lines appear, there is a 7.1 to 7.5 percent impedance discontinuity. This will cause about a 3.7 to 3.9 percent reduction of the active voltage on the middle signal line. We may have to increase the output current of the off-chip driver circuit by about 3.8 percent to achieve the required voltage swing.

EFFECT OF ORTHOGONAL LINES
ON IMPEDANCE AND COUPLED NOISE

In the previous sections of this chapter we have discussed a special case in which there is only one signal layer. The usual PCB structure consists of two or more substructures, each of which has two signal layers placed on top of one, or sandwiched between two, voltage reference planes. A substructure with two signal layers over one voltage reference plane is illustrated in Figure 4-7. On one of the signal layers, the signal lines run in the y direction; on the other layer, they run in the x direction. We shall assume for explanation purposes that the signal lines of interest run in the y direction. With the introduction of x direction signal lines running on an adjacent signal layer, the environment for energy propagation along the y direction signal lines is changed as summarized in the following:

1. The vertical cross section of the PCB normal to the y direction is a function of y; that is, the y direction transmission line is not uniform along the propagation direction. If the dimension of, and the spacing between, the x direction lines

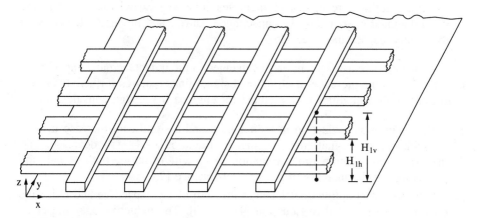

FIGURE 4-7
Two signal layers with x and y direction signal lines above a voltage reference plane.

on the adjacent signal layer are much less than a quarter wavelength (e.g., 1/20 of the wavelength), we may assume an average effect on the y direction signal line characteristics. For PCBs with $\epsilon_r = 4$, a 1.0-mm center-to-center pitch between adjacent x direction signal lines will not reach 1/20 of one wavelength until 7.5 GHz. Therefore, in the low GHz range, we may use the uniform transmission line treatment (Chapter 3).

2. The dielectric medium becomes inhomogeneous. The metal conductors of the x direction signal lines behave like a dielectric medium with infinite relative dielectric constant [10]. This will increase the signal propagation time. The impact is more extensive on the common mode (or sum mode) signal, and less on the difference mode signal. This can be visualized since energy of the difference mode propagation is concentrated in the region between the parallel y direction signal lines [5], where the influence of the medium with infinite relative dielectric constant is minimal.

Alternatively, the influence of the orthogonal lines discussed above can be expressed in terms of inductance and capacitance matrices per unit length for transmission lines under study as follows:

1. The alternating current along a y direction signal line will induce eddy current along the circumferences of the cross section of the x direction signal lines on the adjacent layer. These eddy current loops on the yz plane will reduce the self as well as mutual inductances between the y direction signal lines. Since the cross-sectional area of these x direction signal lines occupies only a small percentage of the total space surrounding the signal lines, the reduction in the inductances is usually negligible. This has been demonstrated from numerical calculation and hardware measurement [22, 23].

2. For capacitance calculation, the orthogonal lines are to be treated as part of the voltage reference. Numerical computations and hardware measurements indicate that there are significant increases in the self-capacitances, that is, the main diagonal elements in the capacitance matrix, and moderate decreases in the mutual capacitances [23–25].

A quantitative assessment on the changes to the inductance and capacitance matrices due to the presence of orthogonal signal lines requires further study. We shall only make a qualitative assessment for engineering design consideration.

1. With a negligible reduction in inductances and a significant increase in self-capacitance, the signal line impedance, Equation (4.4) will decrease, and the signal propagation time, Equation (4.3) will increase, both at half the percentage of the increase of the self-capacitance [23].

2. The backward coupling coefficient is the sum of the capacitive and inductive coupling, as indicated in Equation (4.19) [10]. The inductive coupling component has little change. The moderate reduction in the mutual capacitance and significant increase in the self-capacitance will reduce the capacitive coupling. Therefore, the backward coupling coefficient is reduced.

3. The propagation time of the common mode signal has significantly increased in comparison with that of the difference mode signal. This means that the difference mode signal arrives at the receiving end before the common mode signal. Therefore, the quiet line on the receiving end will experience an out-of-phase coupled noise [10]. The total forward coupling energy is proportional to the propagation time difference between these two modes of propagation [21]. That is, the stronger the orthogonal line effect, the worse the forward coupled noise.

4. In the triplate structure, where the signal plane pair is sandwiched between two voltage reference planes, the orthogonal lines disturb the electromagnetic field in the region midway to the farther away voltage plane. It will add a small perturbation term to the W/H_2 parallel plate contribution. The orthogonal line effect is usually small.

5. The stripline structure has only one voltage reference plane, as shown in Figure 4-7. Here, the x direction signal lines are on the layer closer to the voltage plane, and experience only a small effect due to the orthogonal y direction lines, which are further away from the voltage plane. However, if the signal lines of interest are the y direction lines on the outside layer, they will be further away from the reference plane—that is, H_1 will be larger—and also have a significant portion of their field to the reference plane disturbed. The overall effects are the following:

(a) A wider line width is needed for the y direction signal lines to compensate for the larger H_1 value to obtain the same impedance as the x direction signal lines on the inner signal layer. A large edge-to-edge spacing is also needed to limit the coupled noise. Therefore, the center-to-center pitch between the adjacent y direction signal lines has to increase, proportional to the increase in the H_1 value. This means the wiring density on the outside signal layer is reduced.

(b) There is a large reduction in line impedance due to orthogonal lines. The y direction signal lines on the outside signal layer will have a large impedance discontinuity at the boundary between sparsely and closely placed x direction wiring regions on the inner signal layer.

(c) There is a large increase in propagation delay time, especially for the common mode signal. This is particularly noticeable for the microstrip structure, since the difference mode among the y direction signal lines on the outside layer has some electromagnetic energy in the air dielectric medium. The difference in the propagation delay between the common and difference modes will increase further and worsen the forward coupled noise.

CONCLUSION

In this chapter, we have discussed the transmission line behavior of the signal lines in the PCB. The characteristic impedance of a single signal line in a stripline or triplate structure in a homogeneous medium can be approximated by using the imaginary parallel plate and fringe field contributions. The designer may use these results to choose the most desirable characteristic impedance with proper tradeoff among materials, process, wiring density, PCB thickness, driver circuit output current, and voltage distribution.

The presence of active parallel signal lines will induce cross-talk on the signal line in the middle, and reduce its matched-load impedance. Results of numerical calculation for PCBs in the homogeneous medium are presented in two nomographs. The designer may use the cross-talk data for tradeoff between the maximum allowable coupled noise and the edge-to-edge spacing between adjacent signal lines. The impedance reduction calls for an increase in the output current of the driver circuit to achieve the required voltage swing at the receiver circuit. It is illustrated that the low-dielectric-constant materials have both performance and density advantages.

In high-density and high-performance packaging, the dc resistance and skin effect resistance [5,6,14] of the signal line become important design parameters. Their effects have not been discussed in this chapter. It is noted that the total dc resistance of the signal path, including the signal line, is not to exceed 20 percent of the signal line characteristic impedance for high-performance applications, where a long signal line is always terminated by a resistor with value roughly equal to its characteristic impedance. The skin effect resistance will degrade the signal rise time and cause additional delay, which is proportional to the square of the line length, inversely proportional to the square of the characteristic impedance, and roughly inversely proportional to the square of the perimeter of the signal line cross section. For a 100-by-50-μm signal line in the 80-Ω PCB design, the critical length is 1.24 m to ensure that the additional delay is less than 0.1 ns. For a 25-by-12.5-μm signal line in the 50-Ω design, the critical length is reduced to 0.194 m.

ACKNOWLEDGMENTS

The author is deeply indebted to Dr. D. P. Seraphim. His penetrating inquiries and constant inspirations are the key motivations in pursuing this study on PCB signal line electrical design. The author also greatly appreciates the tremendous help from D. A. Gernhart in making several hundred cases of the two-dimensional capacitance calculations. Many colleagues have given positive criticism and helpful feedback in the presentation of this design approach.

REFERENCES

1. D. P. Seraphim and I. Feinberg, "Electronic Packaging Evolution in IBM," *IBM J. Res. Develop.* 25, pp. 617–629 (1981).
2. D. P. Seraphim, "A New Set of Printed-Circuit Technologies for the IBM 3081 Processor Unit," *IBM J. Res. Develop.* 26, pp. 37–44 (1982).
3. D. P. Seraphim, D. E. Barr, W. T. Chen, and G. P. Schmitt, "Electrical and Electronic Applications," in *Advanced Thermoset Composites* (edited by J. M. Margolis), New York: Van Nostrand Reinhold, 1986.

4. R. F. Bonner, J. A. Asselta, and F. W. Haining, "Advanced Printed-Circuit Board Design for High-Performance Computer Applications," *IBM J. Res. Develop.* 26, pp. 297–305 (1982).
5. R. E. Matick, *Transmission Lines for Digital and Communication Networks*, New York: McGraw-Hill, 1969.
6. C. W. Davidson, *Transmission Lines for Communications*, New York: John Wiley & Sons, 1978.
7. T. C. Edwards, *Foundations for Microstrip Circuit Design*, New York: John Wiley & Sons, 1981.
8. N. C. Arvanitakis, J. T. Kolias, and W. Radzelovage, "Coupled Noise Prediction in Printed Circuit Board for High-Speed Computer System," 7th International Electronic Circuit Packaging Symposium, 1–11 (1966).
9. A. J. Rainal, "Transmission Properties of Various Styles of Printed Wiring Boards," *Bell System Technical Journal* 51, pp. 995–1025 (1979).
10. C. S. Chang, G. Crowder, and M. F. McAllister, "Crosstalk in Multilayer Ceramic Packaging," 1981 IEEE International Symposium on Circuits and Systems Proceedings, 6-11 (1981).
11. N. C. Arvanitakis and J. J. Zara, "Design Considerations of Printed Circuit Transmission Lines for High Performance Circuits," 1981 IEEE WESCON session, 18-4 (1981).
12. E. E. Davidson, "Electrical Design of a High Speed Computer Package," *IBM J. Res. Develop.* 26, pp. 349–361 (1982).
13. D. W. Dearholt and W. R. McSpadden, *Electromagnetic Wave Propagation*, New York: McGraw-Hill, 1973.
14. S. Ramo, J. R. Whinnery, and T. van Duzer, *Fields and Waves in Communication Electronics*, New York: John Wiley & Sons, 1984.
15. W. T. Weeks, "Calculation of Coefficients of Capacitance of Multiconductor Transmission Lines in the Presence of a Dielectric Interface," *IEEE Trans. Microwave Theory Tech.* MTT-18, pp. 35–43 (1970).
16. K. T. Ho, private communications.
17. F. Ayres, Jr., *Theory and Problems of Matrices*, Schaum's Outline Series, New York: McGraw-Hill, 1962.
18. S. Lipschutz, *Theory and Problems of Linear Algebra*, Schaum's Outline Series, New York: McGraw-Hill, 1968.
19. R. Plonsey and R. E. Collins, *Principles and Applications of Electromagnetic Fields*, New York: McGraw-Hill, 1961.
20. C. W. Ho, "Theory and Computer-Aided Analysis of Lossless Transmission Lines," *IBM J. Res. Develop.* 17, pp. 249–255 (1973).
21. C. S. Chang, "Transmission Lines," in *Circuit Analysis, Simulation, and Design*, edited by A.E. Ruehli, pp. 292–332, Advances in CAD for VLSI, vol. 3, part 2, Amsterdam: North Holland, 1987.
22. P. A. Brennan, N. Raver, and A. E. Ruehli, "Three-Dimensional Inductance Computations with Partial Element Equivalent Circuits," *IBM J. Res. Develop.* 23, pp. 661–668 (1979).
23. B. J. Rubin, "The Propagation Characteristics of Signal Lines in a Mesh-Plane Environment," *IEEE Trans. Microwave Theory Tech.* MTT-32, pp. 522–531 (1984).
24. A. E. Ruehli and P. A. Brennan, "Efficient Capacitance Calculations for Three-Dimensional Multiconductor Systems," *IEEE Trans. Microwave Theory Tech.* MTT-21, pp. 76–82 (1973).
25. Q. Gu and J. A. Kong, "Transient Analysis of Single and Coupled Lines with Capacitively-Loaded Junctions," *IEEE Trans. Microwave Theory Tech.* MTT-34, pp. 952–964 (1986).

PROBLEMS

4.1. A given PCB stripline design has nominal dimensions and 3σ tolerances as follows:

$$W = 100 \pm 30 \ \mu m$$

$$T = 40 \pm 10 \ \mu m$$

$$H = 300 \pm 30 \,\mu\text{m}$$

$$\epsilon_r = 4 \pm 0.3$$

Find the nominal values of Z_0, and its tolerance due to the variation of each, as well as all of the process and material parameters.

4.2. For a PCB triplate design, the 3σ tolerances of process and materials parameters are as follows:

$$W = 100 \pm 30 \,\mu\text{m}$$

$$T = 40 \pm 10 \,\mu\text{m}$$

$$H_1 + T = 300 \pm 30 \,\mu\text{m}$$

$$\epsilon_r = 4 \pm 0.3$$

$$H_2 = 700 \pm 70 \,\mu\text{m}$$

What are the nominal value and the tolerances of Z_0?

4.3. In a given PCB stripline design, three of the signal lines run in parallel for 300 mm. The medium has a relative dielectric constant ϵ_r equal to 4. The width and thickness of these lines are 100 and 50 μm, respectively. The edge-to-edge spacings between the adjacent lines are 200 μm. The bottom side of the signal lines is 200 μm above the reference plane. The near end of the middle signal line is driven by an off-chip driver circuit with an output impedance of 64.5 Ω. It can deliver 20 mA, with 0-to-100-percent rise time of 1.0 ns, to a 64.5 Ω load (i.e., a Thevenin equivalent voltage of 2.58 V). The near ends of the remaining two signal lines are connected to long twisted-pair cables of 66.8 Ω impedance.

(*a*) If the far ends of these three lines are terminated by resistors of 66.8, 64.5, and 66.8 Ω, respectively, what are the approximate voltage waveforms from $t = 0$ to 7 ns at both ends of all three lines?

(*b*) If the far ends of all three lines are open-circuited, what are the approximate voltage waveforms from $t = 0$ to 7 ns at both ends of these lines?

4.4. A PCB triplate structure is to be designed with a pair of x- and y-direction signal layers sandwiched between two reference planes. The dielectric material sheets are 60 μm thick, with $\epsilon_r = 3.0$. The manufacturing facility can produce signal lines with nominal thickness T equal to 50 μm, line width W equal to 100 μm or wider, and edge-to-edge spacing S equal to 120 μm or larger. From process sequences, it is known that the distance between the facing surfaces of the signal layers is a multiple of 60 μm, and $H_1 + T$ is also a multiple of 60 μm. For reliability consideration, it is imperative that three or more sheets be used between electrically conductive layers for lamination (i.e., $H_1 + T \geq 180 \,\mu$m, etc.). The saturated backward coupled noise is not to exceed 160 mV/V when both left- and right-hand-side signal lines are active.

(*a*) Draw the vertical cross section of your design.

(*b*) What is the impedance of a signal line with and without adjacent signal lines on both left- and right-hand sides?

(*c*) What is the saturated backward coupled noise of your design?

(*d*) What is the maximum length of a parallel section if the backward coupled noise is not to exceed 100 mV/V?

(Note: There is no unique answer to this design problem. You have to justify your design decisions.)

CHAPTER
5

THERMAL MANAGEMENT IN ELECTRONIC PACKAGING

FRANK E. ANDROS
BAHGAT G. SAMMAKIA

INTRODUCTION

To achieve improved performance in data processing equipment, trends in electronic package designs (as discussed in Chapter 1) have moved toward larger circuit chips, higher I/O, increased circuit density, and improved reliability. Increases in chip size, I/O, and circuit density bring about improved performance by reducing electrical line lengths and therefore, point-to-point signal flight times. Such considerations are covered in Chapters 3 and 4. Greater circuit density means increased power density (W/in^2). As shown in Figure 5-1, power density has increased exponentially with time over the past 15 years and it appears that it will continue to do so in the near future.

Reliability at all levels of packaging is directly related to operating temperature. The impact of temperature, along with other environmental factors, is covered in Chapter 9. Higher operating temperatures accelerate various failure mechanisms such as creep, corrosion, and electromigration. In addition, temperature differences that occur as a system is cycled between power-off and power-

127

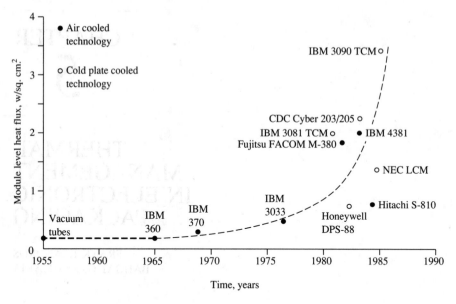

FIGURE 5-1
Module heat flux increase with time.

on conditions have a significant effect on electronic component reliability. This effect is related to fatigue in composite structures, a topic treated in Chapter 6. A large variety of materials is used in order to meet performance, density, and cost objectives. Many of the materials discussed in Chapters 10 and 11, have different coefficients of thermal expansion; therefore, on/off cycles give rise to alternating stresses at material interfaces. Thus, from a reliability viewpoint, absolute temperature, time, temperature difference, number of cycles, and cycle rate are important parameters.

The trend to increase chip power dissipation has become a major concern for the circuit designer and places significant emphasis on thermal design considerations at all levels of electronic packaging, from the chip to the system. It has prompted increased research in thermal analysis/design, as demonstrated by an increase in publications in recent years on this subject (see literature survey by Antonetti and Simons [1]).

BASIC MODES AND CONCEPTS OF HEAT TRANSFER

There are three basic mechanisms of heat transfer: conduction, convection, and radiation. These mechanisms can be extremely complex, and each is dealt with in detail in several textbooks [2,3,4,5,6]. We shall introduce them in this chapter by giving basic definitions and examples, and encourage readers to review the referenced texts.

Conduction is a mechanism associated with energy exchange at the molecular level; i.e. from molecule to molecule, and is usually associated with solids. It can also occur in fluids (gases and liquids) over very short distances. Consider the solid slab shown in Figure 5-2. The governing equation for steady, one-dimensional conduction heat transfer is Fourier's law:

$$Q = kA\frac{\Delta T}{\Delta X} \tag{5.1}$$

where Q = heat flow
A = area through which the heat is flowing
ΔT = temperature difference
ΔX = length of the heat flow path
k = thermal conductivity—a material property

In the electronics industry it is customary to think of heat flow in terms of resistances, based on an electrical/thermal analog where heat flow is considered current, and temperature difference is voltage. Then, following Ohm's law:

$$R = \frac{E}{I} \tag{5.2}$$

or from Equation (5.1)

$$R = \frac{\Delta T}{Q} = \frac{\Delta X}{kA} \tag{5.3}$$

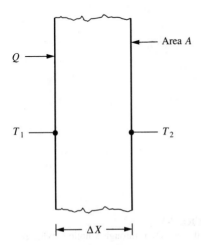

FIGURE 5-2
Conduction through a slab.

where R is the conductive thermal resistance (°C/W). Using Equation (5.1), a similar equation can be obtained (see Figure 5-3) for one-dimensional radial heat transfer:

$$R = \frac{\Delta T}{Q} = \frac{\ln(r_1/r_2)}{2\pi k L} = \frac{T_1 - T_2}{Q} \qquad (5.4)$$

where T_1 and T_2 are the temperatures at radii r_1 and r_2.

Figure 5-4 shows an example of the electrical network analog method. Although this model uses a "flip chip" module with solder joints between chips and substrate, a similar model could be shown for wire-bonded chips. The application shown is a single module attached to a card by copper pins. A cross section of the module shows that it holds a single chip attached to the substrate by means of an array of solder bumps. The chip is also heat sunk to the cap through a layer of thermal grease. Thus, heat dissipated by the chip can flow through the thermal grease to the cap and eventually to the surrounding air. Some of the heat may also be conducted through the solder bumps, the ceramic substrate, and pins until it reaches the card. Depending upon the card configuration and other components, the heat may then flow into the card or to the air, or both. Also shown in Figure 5-4 are typical values for the thermal resistances in this configuration.

The simple equations shown above are for one-dimensional steady heat transfer. For more complex geometries (three-dimensional) with internal heat sources, nonisotropic thermal conductivity, and transient heat flow, the governing equation is the energy equation:

$$\rho C_p \frac{\partial T}{\partial t} = \frac{\partial}{\partial x} k_x \frac{\partial T}{\partial x} + \frac{\partial}{\partial y} k_y \frac{\partial T}{\partial y} + \frac{\partial}{\partial z} k_z \frac{\partial T}{\partial z} + q_s \qquad (5.5)$$

where T = temperature
  ρ = material density
  C_p = specific heat

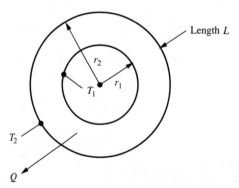

FIGURE 5-3
Radial conduction through a hollow cylinder.

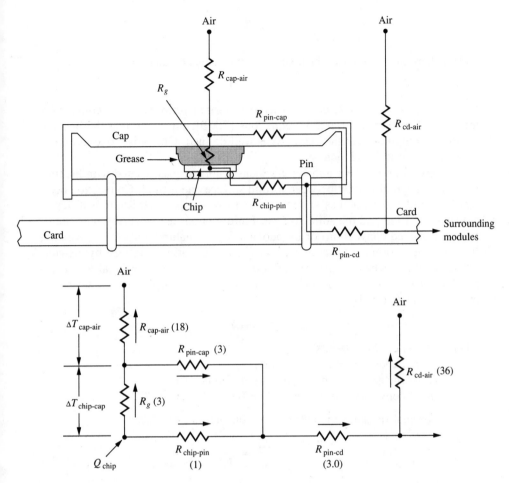

FIGURE 5-4
Typical electronic component resistance network.

q_s = volumetric heat source
k_i = thermal conductivity in the ith direction
t = time

This equation is a second-order nonlinear (since ρ, C_p, and k_i are functions of temperature) partial differential equation. The solution of this equation cannot be obtained in closed form due to its nonlinear nature. Therefore, solutions are restricted to numerical methods, the most often used being finite difference and finite element techniques [7]. The boundary conditions may be nonlinear as well. If it is assumed that the properties are constant and a steady state exists, Equation (5.5) reduces to the Poisson equation:

$$\nabla^2 T + \frac{q_s}{k} = 0 \tag{5.6}$$

and, if there are no internal heat sources, the Laplace equation:

$$\nabla^2 T = 0 \tag{5.7}$$

Equations (5.6) and (5.7) can be solved in closed form by using conventional techniques for partial differential equations [7,8,9], providing that the boundary conditions are linear and the geometry is well defined. Otherwise, one must again resort to numerical techniques.

Convection is a heat transfer mechanism normally associated with the transfer or exchange of energy between a solid surface and a fluid. Convection occurs when a fluid moves across surfaces, and there exists a temperature difference between the fluid and the solid surface. Fluid motion may occur by pressure differences (forced convection) or by buoyant forces (natural and free convection). In its simplest form convection heat transfer is qualitatively defined by Newton's law of cooling, whereby the rate of cooling is directly proportional to the temperature difference between the object being cooled and the air stream passing over the object, that is,

$$Q = hA_s\Delta T \tag{5.8}$$

where Q = heat flow (watts)
A_s = exposed surface area (cm^2)
ΔT = surface-to-fluid temperature difference (°C)
h = convective heat transfer coefficient (W/cm^2 · °C)

A thermal resistance is defined for convection as follows:

$$R = \frac{\Delta T}{Q} = \frac{1}{hA_s} \tag{5.9}$$

where R represents the convective thermal resistance, that is, the resistance to heat flow from a solid surface to a fluid immediately adjacent to the solid surface. The convective heat transfer problem is much more complex than conduction, since Newton's law of cooling is based on the assumption that the geometry and flow conditions remain constant. This implies that the heat transfer coefficient h is a function of both velocity and geometry. Therefore, if the thermal portion of the convective problem is to be solved, the velocity field must first be known.

The governing equations for the convective heat transfer problem are the conservation equations (mass, momentum, and energy), a set of nonlinear partial differential equations:

Continuity (mass balance):

$$\frac{\partial \rho}{\partial t} + \frac{\partial}{\partial x}(\rho u) + \frac{\partial}{\partial y}(\rho v) + \frac{\partial}{\partial z}(\rho w) = 0 \tag{5.10}$$

Momentum (x direction):

$$\rho\frac{Du}{Dt} + \frac{\partial P}{\partial x} = \frac{\partial}{\partial x}\left\{\mu\left[2\frac{\partial u}{\partial x} - \frac{2}{3}\left(\frac{\partial u}{\partial x} + \frac{\partial v}{\partial y} + \frac{\partial w}{\partial z}\right)\right]\right\}$$

$$+ \frac{\partial}{y}\left[\mu\left(\frac{\partial u}{\partial y} + \frac{\partial v}{\partial x}\right)\right] + \frac{\partial}{\partial z}\left[\mu\left(\frac{\partial w}{\partial x} + \frac{\partial u}{\partial z}\right)\right] + \underline{X} \quad (5.11)$$

where D/Dt is the substantial derivative

$$\frac{D}{Dt} = \frac{\partial}{\partial t} + u\frac{\partial}{\partial x} + v\frac{\partial}{\partial y} + w\frac{\partial}{\partial z}$$

Similar equations exist for the y and z directions.
   Energy:

$$\rho C_p\frac{DT}{Dt} = \frac{Dp}{Dt} + \frac{\partial}{\partial x}\left(k\frac{\partial T}{\partial x}\right) + \frac{\partial}{\partial y}\left(k\frac{\partial T}{\partial y}\right) + \frac{\partial}{\partial z}\left(k\frac{\partial T}{\partial z}\right) + \mu\Phi \quad (5.12)$$

ϕ is referred to as the dissipation function.

$$\phi = 2\left[\left(\frac{\partial u}{\partial x}\right)^2\left(\frac{\partial v}{\partial y}\right)^2\left(\frac{\partial w}{\partial z}\right)^2\right] + \left[\frac{\partial v}{\partial x} + \frac{\partial u}{\partial y}\right]^2$$

$$+ \left[\frac{\partial w}{\partial y} + \frac{\partial v}{\partial z}\right]^2 + \left[\frac{\partial u}{\partial z} + \frac{\partial w}{\partial x}\right]^2 - \frac{2}{3}\left[\frac{\partial u}{\partial x} + \frac{\partial v}{\partial y} + \frac{\partial w}{\partial z}\right]^2$$

In the foregoing

   x, y, z = spatial coordinates
   u, v, w = velocities in the x, y, and z directions, respectively
   ρ = density
   t = time
   μ = dynamic viscosity
   P = pressure
   $\underline{X}$ = body force
   C_p = specific heat
   T = temperature

   For the derivation of the conservation equations, see references [6] and [10]. These five equations with the appropriate boundary and initial conditions can, in theory, be solved for the five dependent variables: u, v, w, P, T. In practice, however, the complexity of this general set of equations renders them unsolvable in closed form. Solutions can be obtained for special cases and with simplifying assumptions.
   For the case of constant properties and incompressible flow, the equations reduce to

Continuity:
$$\frac{\partial u}{\partial x} + \frac{\partial v}{\partial y} + \frac{\partial w}{\partial z} = 0 \qquad (5.13)$$

Momentum (x direction):
$$\frac{Du}{Dt} + \frac{P}{x} = \nabla^2 u + \underline{X} \qquad (5.14)$$

Energy:
$$c\frac{DT}{Dt} = k\nabla^2 T + \mu\phi \qquad (5.15)$$

This system of equations is still nonlinear and requires the use of numerical techniques to obtain solutions. Further simplification of the above equations can be made for the cases of steady, fully developed flow in channels (see the following section). For boundary layer flow, several texts are available that address these special cases in detail [6,10,11].

CHANNEL FLOWS

Channel flows are particularly important in the thermal management of microelectronic components. This is because, typically, chips are mounted either directly on cards, or on individual modules mounted on the cards. The cards are then arranged parallel to each other and attached on the sides to form a typical package, as shown in Figure 5-5. Such an arrangement forms a number of parallel channels, each containing discretely heated components. The components are cooled either by a buoyancy-induced convective flow or by a forced convective flow produced by pumping air through the channels.

Heat transfer in a channel flow containing heated components is a function of several variables. The flow, as mentioned, could be buoyancy induced or forced. It could also, in turn, be laminar or turbulent, or it could undergo transition from laminar to turbulent at some point in the channel (see Figure 5-6). Depending on the approach to the channel, the flow could be fully developed (position-independent velocity and temperature profiles) or partly developed, and the same is true of the exit section. Other variables that determine the flow regime are the channel dimensions, module dimensions, and module heating rate. The following discussion of channel flows will be restricted to the basic physics of the flow, the solutions, and empirical correlations. No attempt will be made to present detailed derivations; instead, the reader will be referred to the appropriate references for more details.

Natural Convection

For low power dissipation rates, it may be possible to depend on buoyancy-induced flows to maintain components at acceptable temperatures. This situation is always desirable but usually impractical. The desirability stems from cost

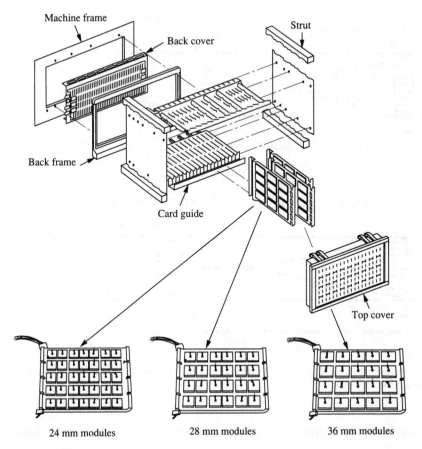

Machine frame

Strut

Back cover

Back frame

Card guide

Top cover

24 mm modules 28 mm modules 36 mm modules

FIGURE 5-5
Schematic diagram showing typical air-cooled package with some representative module-on-card arrangements.

effectiveness as well as simplicity of design. However, the ever-increasing power dissipation needs, as well as the desire for compactness, usually eliminate natural convection as a viable cooling technique, with the exception of packages operating at very low power, such as some microcomputers.

Most of the studies of buoyancy-induced flows in channels have been concerned with two-dimensional channels with simplified boundary conditions. Channels formed by symmetrically heated isothermal surfaces, symmetrically heated uniform flux surfaces, asymmetric isothermal surfaces, and asymmetric uniform flux surfaces are some examples. Three-dimensional effects due to module sizes and spacing, and to lateral effects have not been studied in detail. Typically electronic packages are made of conductive materials, and the prob-

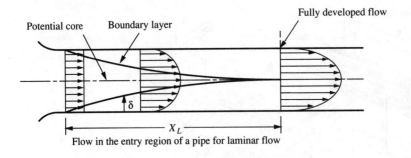

Flow in the entry region of a pipe for laminar flow

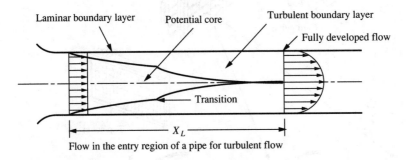

Flow in the entry region of a pipe for turbulent flow

FIGURE 5-6
Schematic diagram showing flow development and velocity profiles in channel flows, for both laminar and turbulent flows.

lem should really be approached from both a conductive and convective standpoint.

Most designers of microelectronic packages have relied on empirical data specific to their particular technology needs. Generally, heat transfer correlations for natural convection have been found to obey

$$\text{Nu} = A(\text{GrPr})^a$$

where $\text{Nu} = hx/k$ is a nondimensional heat transfer coefficient, and A and a are constants that depend on the particular configuration and boundary conditions. $\text{Gr} = g\beta\, l^3 \Delta T / \nu^2$ is the Grashof number, where β is the volumetric coefficient of thermal expansion, and $\text{Pr} = C_p \mu / k$ is the Prandtl number. These types of correlations yield the value of h in heat transfer calculations.

A detailed discussion of buoyancy-induced flows in channels will not be given here, but the reader is referred to Kraus and Bar-Cohen [12] for a complete review of the available heat transfer correlations. Other relevant studies include those by Ortega and Moffat [13], Sparrow, et al. [14], and Bar-Cohen and Rohsenow [15].

Forced Convection

Forced convection is more commonly used than natural convection in packaging applications since the power levels that can be maintained are considerably higher.

FULLY DEVELOPED LAMINAR FLOW IN RECTANGULAR DUCTS. For steady, fully developed, laminar flow in rectangular ducts, the Navier-Stokes equations (5.14) may be reduced to the following simplified form:

$$\mu \nabla^2 u = \frac{dP}{dx} \tag{5.16}$$

This simplified equation has been solved both analytically and numerically for several duct cross sections. The major concern for this type of flow is the pressure drop across the duct due to friction between the fluid and the duct. This information is required to estimate the power needs for forced air. This can be computed from the wall shear stress, as follows:

$$\tau_0 = C_f \frac{\rho V^2}{2} \tag{5.17}$$

where τ_0 is the wall shear stress, V is the mean fluid velocity, and ρ is the fluid density. The value of C_f for rectangular ducts has been determined for ducts of various cross sections as shown in Figure 5-7, from Kays and Crawford [10]. In this figure $C_f \text{Re}$ is plotted versus $1/a^*$, where $a^* = b/a$ is the aspect ratio of the channel. Thus, C_f can be computed from the figure. Here

$$\text{Re} = \text{Reynolds number, } \frac{4r_h G}{\mu}$$

$$r_h = \text{hydraulic radius} = \frac{\text{cross sectional area}}{\text{perimeter}} = \frac{A_c L}{A}$$

$$A_c = \text{cross-sectional area}$$

$$L = \text{tube length}$$

$$A = \text{total tube surface area for the length } L$$

$$G = \text{mean mass velocity, } \frac{\dot{m}}{A_c}$$

It is worth noting that the friction coefficient becomes significantly higher as the channel gets narrower. In microelectronic packaging, channels formed by parallel cards are typically very narrow, and usually blocked by modules, wires, and connectors to maximize packaging density. This greatly increases the pressure drop across the channels and the power requirements for forced air.

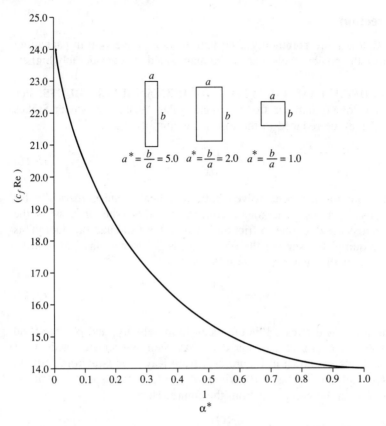

FIGURE 5-7
Friction coefficients for fully developed laminar flow in rectangular tubes. (From Kays and Crawford
[10].)

For fully developed temperature and velocity profiles, the applicable energy
conservation equation is

$$\nabla^2 T = \frac{\partial^2 T}{\partial y^2} + \frac{\partial^2 T}{\partial z^2} = \frac{u}{\alpha}\frac{\partial T}{\partial x} \tag{5.18}$$

The velocity must be determined by solving the appropriate momentum balance,
Equation (5.16). The major concern here is the heat transfer rate from the walls
of the duct to the fluid. This has been determined for a number of different ducts
for two different wall-fluid boundary conditions. Table 5.1, adapted from Kays
and Crawford [10], shows a number of solutions, with

$$Nu = \frac{hD_h}{k} \tag{5.19}$$

TABLE 5.1
Nusselt numbers for fully developed velocity and temperature profiles in tubes of various cross sections

Cross-sectional shape	b/a	Nu_H*	Nu_T
(circle)		4.364	3.66
a□ b	1.0	3.61	2.98
a▭ b	1.43	3.73	3.08
a▭ b	2.0	4.12	3.39
a▭ b	3.0	4.79	3.96
a▭ b	4.0	5.33	4.44
a▭ b	8.0	6.49	5.60
(parallel plates)	∞	8.235	7.54
(hatched channel)		5.385	4.86
(triangle)		3.00	2.35

From W. M. Kays and M. E. Crawford, [10].

* The constant-heat-rate solutions are based on constant *axial* heat rate, but with constant *temperature* around the tube periphery. Nusselt numbers are averages with respect to tube periphery.

where h = heat transfer coefficient

$$D_h = 4 \times \frac{\text{flow area}}{\text{perimeter}} = \text{hydraulic diameter}$$

k = fluid thermal conductivity

In Table 5.1 the subscript H denotes a constant axial heat rate, while T denotes a uniform wall temperature. For further details on the above solutions, the reader is referred to Kays and Crawford [10] and to Shah and London [16,17].

FULLY DEVELOPED TURBULENT FLOW IN DUCTS. For turbulent flows in ducts the viscous boundary layer is very narrow, and the velocity change from the center of the channel to any wall occurs almost independently of the other walls. Thus, the shape of the ducts is not critical to computing the pressure drop in the channel. It has been shown that by estimating an equivalent hydraulic radius for noncircular channels, one could use the known solutions for circular channels with good accuracy.

The following correlations have been determined for fully developed turbulent flows in circular channels:

$$\frac{C_f}{2} = 0.039 \mathrm{Re}^{-0.25} \qquad 10^4 < \mathrm{Re} < 5 \times 10^4 \qquad (5.20)$$

$$\frac{1}{\sqrt{C_f/2}} = 2.46 \ln(\mathrm{Re}\sqrt{C_f/2}) + 0.3 \qquad \mathrm{Re} > 5 \times 10^4 \qquad (5.21)$$

where $\mathrm{Re} = uD/\nu$ is the Reynolds number. Equation (5.21) is known as the Karman-Nikuradse equation. This is sometimes awkward to use; simplified forms of the equation are

$$\frac{C_f}{2} = 0.023 \, \mathrm{Re}^{-0.2} \qquad 3 \times 10^4 < \mathrm{Re} < 10^6 \qquad (5.22)$$

$$\frac{C_f}{2} = (2.236 \ln \mathrm{Re} - 4.639)^{-2} \qquad 10^4 < \mathrm{Re} < 5 \times 10^6 \qquad (5.23)$$

SURFACE ROUGHNESS EFFECTS. The effect of duct wall surface roughness on the coefficient of friction was investigated by Moody [18]. The entire range of Re versus friction factor f is shown in Figure 5-8 for several different wall roughness values. It should be noted that the Fanning friction factor f is defined as $f = 4C_f$. The surface roughness question is very significant for packaging of microelectronic components, since such components in a channel can be represented as a surface roughness.

HEAT TRANSFER IN FULLY DEVELOPED TURBULENT CHANNEL FLOWS. The physical principles discussed previously apply to heat transfer in turbulent channel flows, as they did to momentum transfer. The thermal and viscous boundary layers are also very narrow, and the shape of the duct can be accounted for by using an equivalent hydraulic diameter in circular channel correlations. The Nusselt number for fully developed turbulent pipe flow (circular) can be computed from

$$\mathrm{Nu} = 0.023 \, \mathrm{Re}^{0.8} \mathrm{Pr}^{1/3} \qquad (5.24)$$

where $\mathrm{Pr} = $ Prandtl number $= C_p \mu / k$. Therefore, for a two-dimensional channel, Equation (5.24) can be transformed to:

$$\mathrm{Nu} = \frac{hb}{k} = 0.0194 \left(\frac{\rho V b}{\mu}\right) \mathrm{Pr}^{1/3} \qquad (5.25)$$

where b is the distance between the two surfaces of the channel. For further details

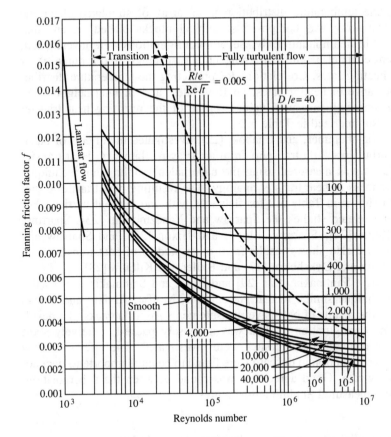

FIGURE 5-8
Fanning friction factor plot for smooth and rough tubes. (From Rohsenow and Hartnett [4].)

on heat transfer in turbulent channel flow, see the *Handbook of Heat Transfer* [19], Kays and Crawford [10], and Kraus and Bar-Cohen [12].

HEAT TRANSFER IN CHANNELS WITH DISCRETELY HEATED COMPO-NENTS. The discussion up to this point has been restricted to channels with smooth walls and uniform boundary conditions. In most electronic packages such conditions do not apply. Components mounted on cards form obstructions to the flow and change transport radically. The components are usually dissipating heat nonuniformly in space and heat rates are also unsteady in time. The card carrying the modules is nonadiabatic, so some of the heat is conducted through the card to other parts of the package. Due to these complicating effects, there have been many recent investigations related to such flows.

Moffat, Arvizu, and Ortega [20] and Arvizu and Moffat [21] report the results of an experimental and analytical study of heat transfer from an array of simulated electronic components. A superposition technique was developed for computing the temperature distribution in an array of simulated electronic components. The array was composed of cubical blocks placed on an adiabatic surface and heated nonuniformly. Several configurations were investigated, namely different channel heights, different module spacings, and a range of air velocities and heating rates. Correlations for predicting the temperature distribution for a wide range of heating rates, air flow rates, and channel heights were in good agreement with measured values. Some of the resulting heat transfer correlations are summarized in Table 5.2.

Sparrow, Niethhammer, and Chaboki [22] investigated the effects of barriers, used as "turbulators," on transport from an array of flatpacks. Heat transfer from the flatpacks was measured using the naphthalene sublimation technique, by invoking the mass to heat transfer analogy. The resulting heat transfer correlations are also shown in Table 5.2 for flatpacks with and without barriers. For more details and related studies see Sparrow and Yanezmoreno [23], Sparrow, Vemuri, and Kadle [24], Sparrow, Lloyd, and Hixon [25], and Sparrow and Tao [26]. Table 5.2 also shows results from other studies of heat transfer from flatpacks, by Wirtz and Dykshoorn [27] and Buller and Kilburn [28].

Most recently, Sammakia et al [29] and Biber and Sammakia [30] used a superposition technique similar to that presented by Moffat, Arvizu, and Ortega [20] to correlate the results of extensive measurements from flatpacks in a developing flow. The configuration studied was a densely packed, relatively short channel. Several different module sites were investigated. The effects of flow development and conduction in the card were found to be quite important, depending upon air flow rate. Good agreement was found between the superposition-based model and measured module temperatures.

TABLE 5.2
Heat transfer coefficients for seven situations of turbulent flow over modules between parallel planes having the same approach Reynolds number

Source	Case	Correlating equation*	h	C_p/row†
Kays	Smooth planes	$h = 0.023\ Re^{0.78}$	14	—
Sparrow	Dense flatpacks	$h = 0.078\ Re^{0.72}$	29	0.034
Sparrow	With barrier	$h = 0.112\ Re^{0.70}$	43	
Wirtz	Sparse flatpacks	$h = 0.324\ Re^{0.60}$	48	
Arvizu	Sparse cubes (3/1)	$h = 0.60\ \ Re^{0.56}$	60	0.120
Arvizu	Dense cubes (2/1)	$h = 0.650\ Re^{0.56}$	65	0.160
Buller	Single flatpacks	$h = 0.722\ Re^{0.54}$	61	

From A. Ortega and R. J. Moffat, [13].

* The form of each correlating equation was extracted from its parent data set as a whole, while h was found specifically for a value of Re = 3700 either using that equation or from the data. Reynolds number is based on channel height and mean velocity in the empty channel upstream of the elements.

† C_p/row is the pressure drop per row, made dimensionless by the velocity over the elements.

AIR COOLING

Air cooling is the most widely used approach in the thermal management of electronic equipment. Air cooling is cost effective, causes very little environmental impact, and imposes very little or no special requirements upon the user of the equipment (as compared to liquid cooling). Air cooling design criteria are further developed compared to liquid cooling. There is, however, a price to be paid for these advantages in the form of relatively low power dissipation capabilities and large volumetric air requirements. This is due to the low thermal capacity and density of air compared to liquids. These restrictions are not prohibitive for small- and intermediate-sized computers, but are more prohibitive for most of the high-end systems, many of which are liquid cooled.

A detailed discussion of air cooling considerations is given here to acquaint the reader with the typical constraints and requirements encountered by the thermal designer. Design requirements usually call for a controlled chip temperature rise above ambient, compact or dense packaging that results in heavily populated cards at a small card pitch (low channel heights), and limits on acoustical disturbances introduced by air blowers.

A typical system design team includes electrical, mechanical (structural), and thermal engineers. First, the application requirements are determined, such as power dissipation rates, module population and distribution on the cards, required chip junction temperatures, ambient air temperature, permissible air flow rates (determined from acoustic considerations), and channel height as determined by allowable card pitch. At this stage the thermal engineer can do a preliminary evaluation of the resulting junction temperatures to determine if the requirements can be met. This determination is based on analysis, numerical modeling, and empirical correlations. Heat sink sizes, if necessary, are determined and optimized, and the results of the analysis are then presented to the design team.

If the thermal requirements are not attainable, the design team then reexamines the application requirements to determine if they can be changed or optimized. A new set of requirements may then be obtained that helps meet the thermal objectives. Thus, typically, a series of optimization loops are carried out with all the areas of expertise involved, until a final acceptable design is reached. At that stage, it may be necessary to design, build, and test experimental hardware to determine if the predicted performance was accurate.

Computation of Chip Temperatures

Analytical and numerical techniques for chip temperature computation have been discussed earlier. Here, a discussion of empirical techniques is presented. First, a generic package design is selected for the experimental work. Typical module and card sizes are determined along with channel height and air flow rates. Then, card layouts, usually heavily populated, are chosen. The package is then powered (specially designed thermal chips are used), and chip temperatures, module cap temperatures, and air temperatures are measured. Air flow rates are also measured. Based upon that data, internal and external resistances can be determined and are plotted as a function of air flow rates. Figure 5-9 shows the resulting R_{ext} values for one column in a typical package. In that test the cards were 9 by 14 cm, the modules 2.4 by 2.4 cm, and the card pitch 1.6 cm. By determining the values of R_{ext} and R_{int} for a specific module location and air flow rate, the junction temperature can be computed. Typically, R_{int} values are determined separately and assumed to remain constant with varied air flow rate.

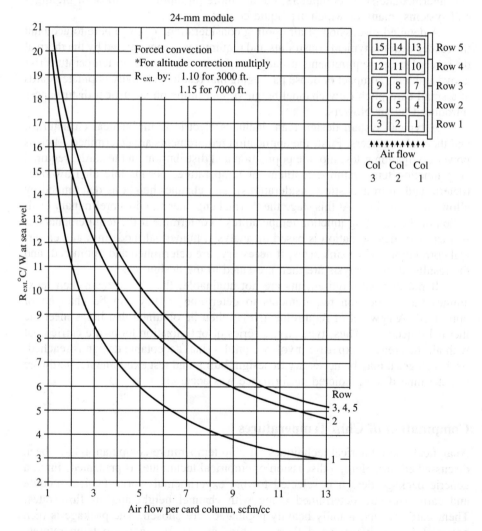

FIGURE 5-9
External resistance R_{ext} as a function of air flow rate: column 1.

The procedure used to compute chip temperatures is as follows:

$$T_{\text{chip}} = T_{\text{rm}} + \Delta T_{\text{in-rm}} + \Delta T_{\text{air}} + (R_{\text{ext}} \times P_m) + \Delta T_{\text{chip-cap}} \qquad (5.26)$$

T_{rm} is the ambient air temperature. $\Delta T_{\text{in-rm}}$ is the air temperature rise from the room ambient to entrance of the channel. This is primarily due to blower motor heat dissipation. $\Delta T_{\text{air}} = R_{\text{air}} \times P_{\text{col}}$, where P_{col} is the power dissipated by the modules below the module of concern plus $1/2\ P_m$, where P_m is the nominal power of the module of concern. R_{air} is an additional resistance resulting from the air temperature rise as it travels through the channel over heated modules. This can be obtained from Figure 5-10 as a function of air flow rate. $\Delta T_{\text{chip-cap}} = R_{\text{int}} \times P_m$ is the temperature rise in the module.

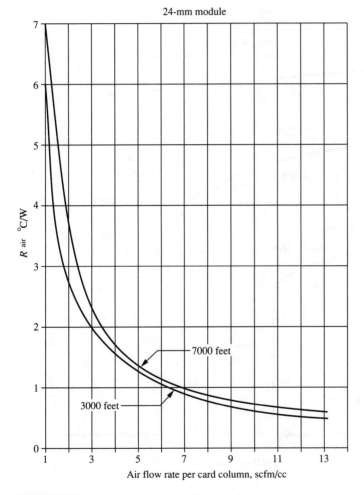

FIGURE 5-10
Air resistance R_{air} as a function of air flow rate.

Once the above information is determined, an estimate can be made of the temperature distribution for different proposed card layouts and application conditions.

LIQUID COOLING

As pointed out in earlier sections, considerable progress has been and continues to be made in integrated circuit technology. Sizes continue to decrease and power dissipation continues to increase. Although cooling techniques that use air are always desirable, particularly from a cost effectiveness standpoint, there are obvious limitations. Figure 5-11 clearly demonstrates the limits on maximum chip power and on heat flux for air-cooled systems.

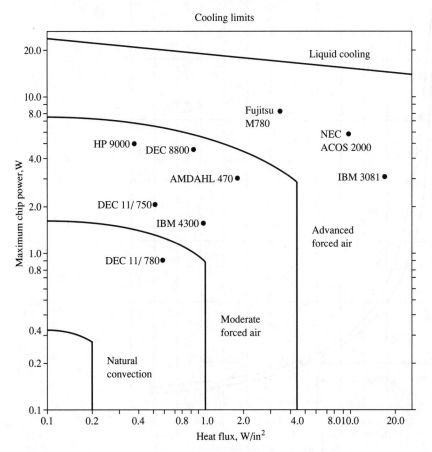

FIGURE 5-11
Cooling limits for different systems.

Liquid cooling can be classified into two general categories — single-phase cooling and two-phase cooling. Single-phase cooling may be buoyancy induced or forced. Two-phase cooling may depend upon boiling or evaporation in a variety of modes.

Single-Phase Liquid Cooling

Single-phase liquid cooling can be either direct, wherein the liquid is in direct contact with the microelectronic circuitry (as in the Cray 2 supercomputer), or indirect, coupled to a conduction mode (as in the IBM 3080 and 3090 series). Direct cooling has the advantage of minimizing the internal thermal resistance of the packages. This facilitates the dissipation of very high heat flux levels. There are, however, associated difficulties, such as reliability concerns and cost effectiveness. Due to such concerns, direct cooling has not been used widely in computer systems with the exception of the Cray supercomputer series.

Indirect single-phase liquid cooling has several advantages over direct techniques. The liquid does not come into contact with the circuitry, as shown in Figure 5-12 (*a*), (*b*), and (*c*). This reduces the reliability concerns and removes many constraints on the liquids available for use. The degree of purity of the liquid, and the range of flow rate that can be used, are not as critical. The disadvantage, as mentioned earlier, is the introduction of an internal thermal resistance

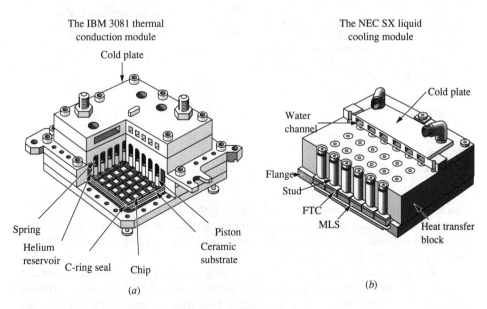

(*a*)

(*b*)

FIGURE 5-12
Schematic diagrams showing different water- and air-cooled multichip modules. (From Bar-Cohen [31].)

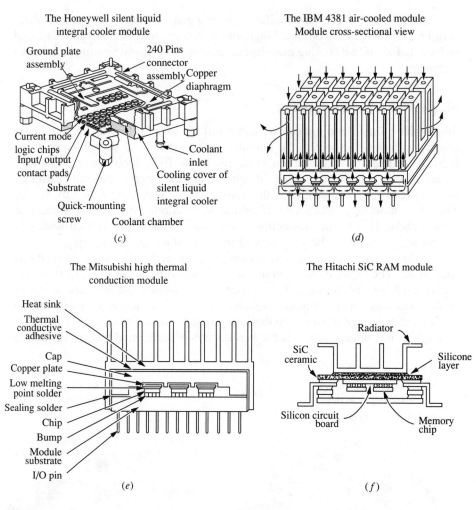

The Honeywell silent liquid integral cooler module

The IBM 4381 air-cooled module
Module cross-sectional view

Ground plate assembly

240 Pins
connector assembly
Copper diaphragm

Current mode logic chips
Input/ output contact pads
Substrate
Quick-mounting screw

Coolant inlet
Cooling cover of silent liquid integral cooler
Coolant chamber

(c)

(d)

The Mitsubishi high thermal conduction module

The Hitachi SiC RAM module

Heat sink
Thermal conductive adhesive
Cap
Copper plate
Low melting point solder
Sealing solder
Chip
Bump
Module substrate
I/O pin

Radiator
SiC ceramic
Silicone layer
Silicon circuit board
Memory chip

(e)

(f)

FIGURE 5-12
(*continued*)

based upon interface resistances and conduction effects. Figure 5-13 from [31] shows a comparison of internal and external resistances for several liquid- and air-cooled systems, as described in Figure 5-12. All six packages shown in Figure 5-12 are for multichip components.

The greatest advantage for liquid cooling (single phase, indirect) over air cooling is the lowering of the external thermal resistance. This advantage is shown in Figure 5-13. This results in much higher heat flux dissipation capabilities, as shown in Figure 5-14. The exception in Figure 5-14 is the IBM 4381 impingement air-cooled system. In that system the external thermal resistance was greatly decreased through the use of the advanced impingement type heat sink shown

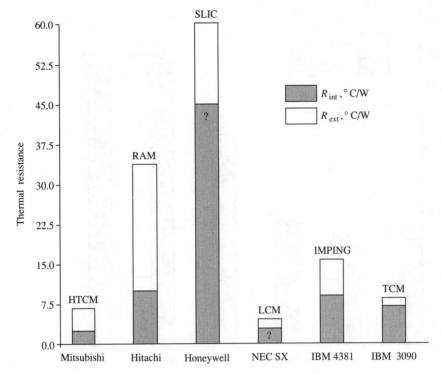

FIGURE 5-13
Total thermal resistance for multichip modules. (From Bar-Cohen [31].)

in Figure 5-12. Internally, the thermal resistance was lowered by using a thermal paste to conduct heat from the chips to the heat sink. For additional reading on single-phase liquid cooling, refer to [31–39].

Two-Phase Liquid Cooling

Two-phase liquid cooling can be applied to microelectronic components in one of many modes, some of which are shown in Table 5.3, from [40]. The resulting heat transfer capability depends upon the mode, the liquid used, and the application. As shown in Table 5.3, most past usage of two-phase cooling modes has been in specialized applications. Heat pipes, for example, discussed in some detail in a later section, are used extensively in airborne packaging situations. The same is true for subcooled pool boiling. On the other hand, forced convection subcooled boiling in channels has been used in radar and microwave applications.

Many of these two-phase modes of heat transfer are currently under investigation as potential techniques for packaging of very-high-power chips. Forced convection boiling, thin film evaporation, and jet impingement are but a few examples. As research efforts continue, current constraints such as liquid com-

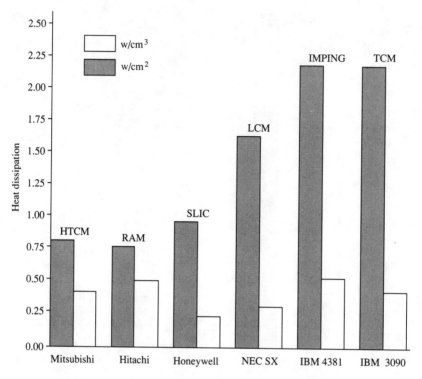

FIGURE 5-14
Heat dissipation capability for multichip modules. (From Bar-Cohen [31].)

patability, reliability, and heat transfer limitations will undoubtedly be overcome, and some of these techniques should prove successful for cooling computer systems. Additional references related to two-phase cooling of microelectronics are [41–44].

ADVANCED COOLING TECHNIQUES

In addition to cooling techniques such as natural and forced convection with gases and liquids, several other options are available to the thermal designer. They include the use of heat pipes, thermoelectric cooling, and microchannel cooling, among others. Most of these techniques, however, are not commonly used, but have applications in specific technology needs. They are discussed here briefly, and the reader is referred to additional literature, where available.

Heat Pipes

Heat pipes are devices that transfer heat by evaporation of an appropriate liquid from hot regions. The vapor is then condensed at some other, cooler location and returned to the hot region for recirculation. Figure 5-15 shows a simple heat

TABLE 5.3
Modes and applications of two-phase cooling

Mode	Liquid	Application
Forced convection subcooled boiling in channels	Water	Anode cooling for radar or microwave tubes
	Inert, dielectric	Proposed for multichip modules
Forced convection saturated boiling in channels	Various	Proposed cold plates for space systems
Subcooled pool boiling	Inert, dielectric	Submerged condenser for airborne electronics; supercomputer; testing of multichip modules for mainframe computers
Saturated pool boiling	Inert, dielectric	Proposed expandable cooling packages or packages with vapor space condensers
Thin film evaporation	Various	Proposed cold plates for space systems Proposed falling film cooling of chips
Jet impingement boiling free jet submerged jet gas jet	Various	Laboratory investigations for possible application to very high power chips
Spray cooling	Inert, dielectric	Proposed for cooling of hot components
Heat pipe	Various	Applied to specialized airborne packaging situations

A. E. Bergles et al. [40].

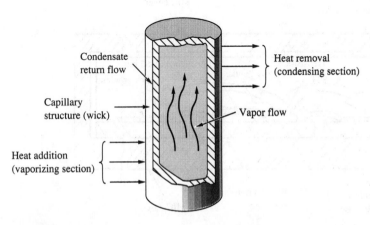

FIGURE 5-15
Straight cylindrical form of heat pipe. (From Sekhon [45].)

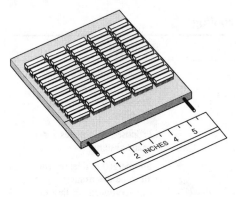

FIGURE 5-16
DIP circuit card with heat pipe thermal mounting plate. (From Sekhon [45].)

pipe, consisting of the heat addition or evaporation region, the heat removal or condensation region, and a mechanism for recirculating the condensate. This mechanism is typically a capillary structure or a wick. The role of the wick is critical since it enables the heat pipe to operate independent of gravity or orientation.

Liquid vaporization at the heated segment of the pipe raises the pressure of the liquid-vapor interface. This causes a capillary pressure to develop at the evaporation end, thereby causing the liquid to be pumped through the wick. Heat transfer rates attainable using this technique could be considerably higher than those using conventional convective techniques. Moreover, heat pipes can take any shape and are not restricted to cylindrical geometries. Figure 5-16, for example, shows a dual in-line package (DIP) circuit card with heat pipe thermal mounting plate. Another innovative use of the heat pipe principle is shown in Figure 5-17, where chips are cooled directly by evaporating the fluid over them, and the module itself acts as a heat pipe.

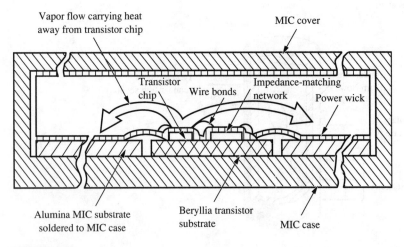

FIGURE 5-17
Section of heat pipe-cooled MIC RF transistor.

There are, however, limitations to the use of heat pipes in microelectronic packaging. Space and cost effectiveness are usually prohibitive, with the exception of very specialized cases. There are also limitations on the attainable heat transfer rates. Those are usually determined by nucleate boiling in the wick, choking of the vapor flow at sonic flows, and the capillary pumping limitation, among others. For an extensive discussion of appropriate fluids and wicking materials and shapes, refer to Kraus and Bar-Cohen [12]. A thorough discussion of specific applications to microelectronic packages can be found in Sekhon [45].

Thermoelectric Cooling

Thermoelectric cooling is based upon a principle discovered in 1834 by Peltier. Essentially, Peltier found that if two different metals are connected at two junctions (see Figure 5-18), and if the two junctions are maintained at different temperatures, t_1 and t_2, then an electric current will flow through the loop. Inversely, if the two junctions were initially at different temperatures, and a current were passed through the loop, a heat flux is generated through the loop that depends in magnitude upon the materials and the temperature difference between the two junctions.

This principle has specific advantages for the thermal packaging engineer in applications where precise temperature control is required and when temperatures below ambient room temperature are required. The principle is therefore used in the cooling of infrared detectors, optical communication systems (see Chapter 8), some avionics systems, and some shipboard applications. However, it is not commonly used for computer systems cooling.

The net heat flux from a cold junction may be computed from the following expression; for details of the derivation see Seely and Chu [43] and Kraus and Bar-Cohen [12]:

$$q = \alpha I t_c - \frac{1}{2} I^2 R - K \Delta t \tag{5.27}$$

where
α = Seebeck coefficient, V/°C
I = current, A
R = electrical resistance
K = thermal conductance, W/°C
Δt = temperature difference
t_c = temperature of the cold junction

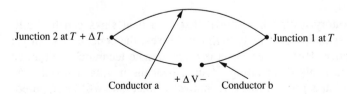

Junction 2 at $T + \Delta T$ Junction 1 at T

Conductor a $+ \Delta V -$ Conductor b

FIGURE 5-18
Schematic illustration of the Seebeck effect.

Equation (5.27) points out several relevant factors regarding the use of thermoelectric cooling in packaging. The current I must be high if a large heat transfer flux is to be achieved. Also, the electrical resistivity must be kept low to reduce the heat dissipation rate resulting from the I^2R term in Equation (5.27). Material selection is also critical, since α must be high for a high heat flux to be achieved. Finally, a low thermal conductivity is desirable to maintain a large temperature difference between the junctions. These three properties can be combined in a figure of merit, Z, to evaluate any material as a potential candidate for use in thermoelectric cooling:

$$Z = \frac{\alpha^2}{k\rho}$$

where k is the thermal conductivity and ρ the electrical resistivity. High values of Z are desirable.

The packaging engineer is also faced with further constraints on material selection from the compatibility viewpoint. Thermoelectric cooling is suitable for some very specific applications, as, for example, in situations where a large thermal gradient is to be maintained through a relatively small heat flux. Most previous usage has been in the areas of air conditioning and refrigeration (see references [45,46,47,48,49]). Due to the above mentioned considerations, as well as cost effectiveness, thermoelectric cooling has not been widely used in the computer packaging area.

Microchannel Cooling

A relatively new cooling technology has been developed that makes use of new advances in silicon microfabrication techniques. An integral compact heat sink is incorporated into an integrated circuit chip, as shown in Figure 5-19. Such a heat sink offers extremely low thermal resistances, at relatively low liquid flow rates through the heat sink. For example, a resistance of 0.1°C/W was measured for a 1-cm^2 chip, at a water flow rate of 10 cm^3/s. This can be translated to a heat dissipation rate of 600 W/cm^2 at a temperature rise of about 60°C.

The channels in the heat sink can be fabricated by using orientation dependent etching, as discussed in references [50,51,52]. Precision mechanical sawing can also be used, and was found suitable, particularly for high-aspect-ratio channels. After the channels are formed, the heat sink is completed by anodically bonding a glass plate to the top of the channels, resulting in the configuration shown in Figure 5-19.

Although laboratory testing has shown that such heat sinks can be built, and are capable of dissipating extermely high heat fluxes, the concept has not yet been reduced to practice. Part of the reason is that alternate techniques, although not as efficient, are capable of handling existing heat dissipation demands. As the power dissipation rates grow, however, innovative methods will be required, and microchannel cooling may prove successful.

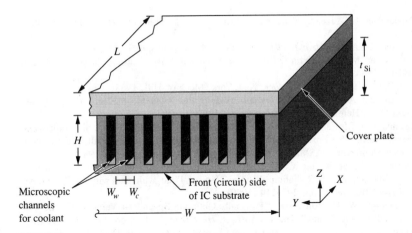

FIGURE 5-19
Schematic of the compact heat sink incorporated into an IC chip.

CONCLUDING REMARKS

Thermal management of electronic equipment will play a critical role in shaping the industry. A cursory look at the water-cooled mainframes of today (Figure 5-12) reveals the complexity, volume requirements, and additional cost imposed by the thermal management scheme.

When one considers the exponential rise predicted in power dissipation requirements (Figure 5-1), it is apparent that future designers will have to be innovative to meet this challenge. Furthermore, this increase in power dissipation is likely to occur over the whole spectrum of computer products, ranging from the high-end supercomputers to the desk-top microcomputers. New materials, manufacturing processes, and cooling techniques will certainly be required to meet the challenges of the future.

REFERENCES

1. V. W. Antonetti and R. E. Simons, "Bibliography of Heat Transfer in Electronic Equipment," *IEEE Transactions on Components, Hybrids, and Manufacturing Technology* CHMT-8, no. 2 (1985). pp. 289–295.
2. M. N. Ozisik, *Boundary Value Problems of Heat Conduction,* Scranton, Pa.: International Textbook, 1968.
3. B. Gebhart, *Heat Transfer,* 2d ed., New York: McGraw-Hill, 1971.
4. W. M. Rohsenow and J. P. Hartnett, *Handbook of Heat Transfer,* New York: McGraw-Hill, 1973.
5. W. M. Rohsenow and H. Y. Choi, *Heat, Mass, and Momentum Transfer,* Englewood Cliffs, N.J. : Prentice-Hall, 1961.
6. E. R. G. Eckert and R. M. Drake, *Analysis of Heat and Mass Transfer,* New York: McGraw-Hill, 1972.
7. G. E. Myers, *Analytical Methods in Conduction Heat Transfer,* New York: McGraw-Hill, 1971.

8. H. S. Carslaw and J. C. Jaeger, *Conduction of Heat in Solids,* London: Oxford University Press, 1947.

9. V. S. Arpaci, *Conduction Heat Transfer,* Reading, Mass.: Addison-Wesley, 1966.

10. W. M. Kays and M. E. Crawford, *Convective Heat and Mass Transfer,* 2d ed., New York: McGraw-Hill, 1980.

11. H. Schlichting, *Boundary Layer Theory,* New York, McGraw-Hill, 1968.

12. A. D. Kraus and A. Bar-Cohen, *Thermal Analysis and Control of Electronic Equipment,* Washington D.C.: Hemisphere Publishing Corp., 1983.

13. A. Ortega and R. J. Moffat, "Heat Transfer from an Array of Simulated Electronic Components: Experimental Results for Free Convection with and without a Shrouding Wall," 23d National Heat Transfer Conference, ASME, HTD-Vol. 48, pp. 5–16, Denver, Colo., 1985.

14. E. M. Sparrow, S. Shah, and C. Prakash, "Natural Convection in a Vertical Channel: I. Interacting Convection and Radiation. II. The Vertical Plate with and without Shrouding," *Numerical Heat Transfer* 3, pp. 297–314 (1980).

15. A. Bar-Cohen and W. H. Rohsenow, "Thermally Optimum Spacing of Vertical, Natural Convection Cooled, Parallel Plates," *Trans. ASME, J. Heat Transfer* 106, pp. 116–123 (1984).

16. R. K. Shah and A. L. London, "Laminar Flow Forced Convection in Ducts," in *Advances in Heat Transfer,* New York: Academic Press, 1978.

17. R. K. Shah and A. L. London, *Trans. ASME, J. Heat Transfer* 96, pp. 159–165 (1974).

18. L. F. Moody, Friction factors for pipe flow, *Trans. ASME* 66, p. 671 (1944).

19. W. M. Rohsenow and J. P. Hartnett, editors, *Handbook of Heat Transfer,* New York: McGraw-Hill, 1973.

20. R. J. Moffat, D. E. Arvizu, and A. Ortega, "Cooling Electronic Components: Forced Convection Experiments with an Air-cooled Array," 23d National Heat Transfer Conference, ASME HTD-Vol. 48, pp. 17–28, Denver, Colo. 1985.

21. D. E. Arvizu and R. J. Moffat, "The Use of Superposition in Calculating Cooling Requirments for Circuit Board Mounted Components," *Proc. of the 32d Electronic Components Conference, IEEE, EIA, and CHMT* (1982).

22. E. M. Sparrow, J. E. Niethhammer, and A. Chaboki, "Heat Transfer and Pressure Drop Characteristics of Arrays of Rectangular Modules Encountered on Electronic Equipment," *Int. J. Heat Mass Transfer* 25, pp. 961–973 (1982).

23. E. M. Sparrow, A. A. Yanezmoueno, and D. R. Otis, Jr., "Convective Heat Transfer Response to Height Differences in an Array of Block-like Electronic Components," *Int. J. Heat Mass Transfer* 27, pp. 469–473 (1984).

24. E. M. Sparrow, S. B. Vemuri, and D. S. Kadle, "Enhanced and Local Heat Transfer, Pressure Drop, and Flow Visualization for Arrays of Block-like Electronic Components," *Int. J. Heat Mass Transfer* 26, pp. 689–699 (1983).

25. E. M. Sparrow, J. R. Lloyd, and C. W. Hixon, "Experiments on Turbulent Heat Transfer in an Asymmetrically Heated Rectangular Duct," *J. Heat Transfer* 88, pp. 170–174 (1966).

26. E. M. Sparrow and W. Q. Tao, "Enhanced Heat Transfer in a Flat Rectangular Duct with Streamwise-Periodic Disturbance at One Principal Wall," *J. Heat Transfer* 105, pp. 851–861 (1983).

27. R. A. Wirtz and P. Dykshoorn, "Heat Transfer from Arrays of Flat Packs in a Channel Flow," *Proceedings 4th Annual International Electronics Packaging Conference,* pp. 318–326 (1984).

28. M. L. Buller and R. F. Kilburn, "Evaluation of Surface Heat Transfer Coefficient for Electronic Module Packages," *Heat Transfer in Electronic Equipment, HTD* 20 (1981).

29. B. G. Sammakia, C. Tai, B. Colavito, and D. Dudascik, "A Temperature Superposition Technique for Predicting the Temperature Distribution on Non-uniformly Heated Cards," IBM technical report, under preparation.

30. C. R. Biber and B. G. Sammakia, "Transport from Discrete Heated Components in a Turbulent Channel Flow," presented at the ASME winter annual meeting, Anaheim, Calif. 1986.

31. A. Bar-Cohen, "Thermal Management of Air and Liquid-Cooled Multi-Chip Modules,"

ASME/AICHE National Heat Transfer Conference, Denver, Colo., ASME/HTD-vol. 48, pp. 3–20, 1985.

32. S. Acharya and S. V. Patankar, "Laminar Mixed Convection in a Shrouded Fin Array," *Microelectronics and Reliability* 11, pp. 213–222 (1972).

33. E. Baker, "Liquid Cooling of Microelectronic Devices by Free and Forced Convection," *Microelectronics and Reliability* vol. 12, pp. 163–173, (1973).

34. A. E. Bergles, "Survey Techniques to Augment Convective Heat and Mass Transfer," *Progress in Heat and Mass Transfer* 1, pp. 331–424 (1969).

35. A. E. Bergles, R. L. Webb, G. H. Junkhan, and M. K. Jensen, "Bibliography on Augmentation of Convective Heat and Mass Transfer," HTL-19, Dept. of Mechanical Engineering, Iowa State University, 1979.

36. A. E. Bergles, et al., "Bibliography on Augmentation of Heat and Mass Transfer—II" HTL-31, Dept. of Mechanical Engineering, Iowa State University, 1983.

37. R. C. Chu, A. E. Bergles, and J. H. Seely, "Survey of Heat Transfer Techniques Applied to Electronic Packages," IBM Technical Report 100.2869, May 1977.

38. S. B. Preston and R. N. Shillabeer, "Direct Liquid Cooling of Microelectronics," *Proceedings INTER/NEPCON, P* 9, pp. 10–31 (1970).

39. E. Baker, "Liquid Immersion Cooling of Small Electronic Devices," *Microelectronics and Reliability* 12, pp. 163–173 (1973).

40. A. E. Bergles, A. Bar-Cohen, P. J. Marto, and K. P. Moran, "Liquid Cooling, Two-Phase," summary of a workshop session on research needs in electronic cooling, Boston, Mass. 1986.

41. U. P. Hwang and K. P. Moran, "Boiling Heat Transfer of Silicon Integrated Circuit Chips Mounted on a Substrate," *Heat Transfer in Electronic Equipment*, HTD-Vol. 20, ASME Winter Annual Meeting, (1981).

42. C. F. Ma and A. E. Bergles, "Boiling Jet Impingement Cooling of Simulated Microelectronic Chips," *Heat Transfer in Electronic Equipment* ASME HTD-vol. 28, pp. 5–12, Nov. 1983.

43. L. S. Tong, *Boiling Heat Transfer and Two-Phase*, New York: John Wiley & Sons, 1967.

44. S. W. Chi, *Heat Pipe Theory and Practice*, Washington, D. C.: Hemisphere Publishing Corp., 1967.

45. K. S. Sekhon, "Heat Pipe Cooling Techniques for High Density Printed Wiring Boards," Fourth Ann. Int. Electronics Packaging Soc. (IEPS) Conf. Proceedings, Oct. 1984, pp. 575–585.

46. J. H. Seely and R. C. Chu, *Heat Transfer in Microelectronic Equipment*, New York: Marcel Dekkar, 1982.

47. C. D. Hudelson, G. K. Gable, and A. A. Beck, "Development of a Thermoelectric Air-Conditioner for Submarine Applications," *ASHRAE Trans.* 70, p. 156 (1964).

48. R. E. Simons, "Thermal Conditioning System for an Optical Data Link," IBM Technical Report 00.1628-1.

49. R. L. Eichhorn, "Thermoelectric Refrigeration," 54th Annual Meeting of ASRE, Minneapolis, Minn., 1954.

50. D. B. Tuckerman, "Heat Transfer Microstructures for Integrated Circuits," Ph.D. dissertation, Stanford University, 1984.

51. D. B. Tuckerman, and R. F. W. Pease, "High Performance Heat Sinking for VLSI," *IEEE Electron Device Letters*, EDL-5, (1981).

52. K. E. Petersen, "Silicon as a Mechanical Material," *Proceedings of the IEEE*, 70, no. 5, (1982).

PROBLEMS

5.1. Assuming a circuit board card has 15 modules on it arranged as shown in Figure 5-9, perform the following calculations:

    (*a*) If $R_{int} = 5°C/W$, power dissipation is uniform at 2 W per module, air flow rate is 6 scfm/cc, ambient air temperature is 20°C, and the channel dimensions are the same as those in Figure 5-9, calculate the chip temperatures in row 4.

(b) If the air flow rate was increased to 12 scfm/cc for the same conditions as in (a), calculate the maximum chip temperature on the card.

(c) What is the minimum necessary air flow rate to maintain a maximum chip temperature of 70°C, for the conditions of part (a)?

5.2. Design an experiment based on which an algorithm can be constructed for predicting chip temperature for nonuniform power dissipation by modules in a channel flow. Assume the same card layout as described in Problem 1.

CHAPTER 6

MECHANICAL DESIGN

ERIC A. JOHNSON
WILLIAM T. CHEN
CHUN K. LIM

INTRODUCTION

Many of the limitations in electronic packaging are mechanical in nature. To increase the circuit density of a package, larger chips, with more I/Os, can be used. Unfortunately, this imposes additional strain on the chip-to-substrate interconnection. This strain arises from the difference in the coefficient of thermal expansion (CTE) between the chip and substrate, and from changes in temperature. Matching the coefficients is a possible solution to this problem but, unfortunately, the substrate-to-card mismatch is thereby increased.

Substrates are generally attached to boards with pins that are soldered into holes. By decreasing the pins' diameter and soldering them onto a pad on the board's surface, the wiring density can be increased. However, these surface mount solder joints could break if the card is flexed during assembly or field repair. Vibration and shock could also cause connection failure.

Copper circuit lines embedded in the boards can also fail through thermal expansion and flexing. Although circuit boards are designed to match copper's thermal expansion in two directions, which is done by adjusting the ratio of glass fibers to epoxy, normal board expansion can cause vias and through-holes to crack during soldering or service.

Since reliability and design limitations of electronic packages are so dependent on mechanical behavior, this chapter is devoted to applications of solid mechanics and what is generally termed "strength of materials." There is a rich body of literature covering this field and, while the fundamentals are briefly presented in this chapter, the reader should make reference to those works [1–3] to gain a more complete understanding.

Practical packaging problems are generally too complex to be dealt with using classical analysis techniques; therefore, the finite element method will be introduced. A historical treatment, which takes the reader through the several approximate methods that have evolved into the finite element method, will help provide an understanding of its correct application.

BASIC PRINCIPLES OF STRESS AND STRAIN

Stress may be defined as the force per unit area that acts upon the surface of an infinitesimal volume. On the surface of an arbitrary volume, such as that shown in Figure 6-1, the force may vary continuously, but in the limit, as dA approaches 0, the forces may be resolved into three components:

$$\sigma_{xx} = \lim_{dA \to 0} \frac{df_x}{dA} \qquad \sigma_{xy} = \lim_{dA \to 0} \frac{df_y}{dA} \qquad \sigma_{xz} = \lim_{dA \to 0} \frac{df_z}{dA} \qquad (6.1)$$

The component perpendicular to the area is the normal stress, while the two components in the plane of the elementary area are the two shear stresses.

Just as it takes three planes to define a point in cartesian space, it is necessary to define the stresses acting on three orthogonal planes to completely determine the stress field. Only six of these quantities are unique, since static equilibrium requires that $\sigma_{xy} = \sigma_{yx}$, $\sigma_{xz} = \sigma_{zx}$, and $\sigma_{yz} = \sigma_{zy}$. It is convenient to arrange these stresses into the following matrix, commonly referred to as the stress tensor:

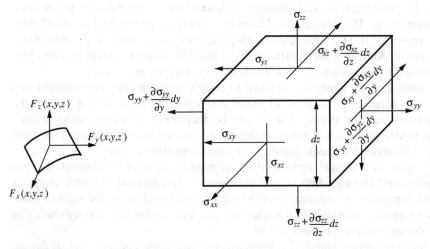

FIGURE 6-1
An element on the surface of an arbitrary volume demonstrates stress in a solid.

$$\begin{bmatrix} \sigma_{xx} & \sigma_{xy} & \sigma_{xz} \\ \sigma_{yx} & \sigma_{yy} & \sigma_{yz} \\ \sigma_{zx} & \sigma_{zy} & \sigma_{zz} \end{bmatrix} \tag{6.2}$$

The stresses in any plane can be determined by multiplying the stress tensor by the normal to the plane. In mathematics this is known as a coordinate transformation.

In statics, one must ensure that all of the forces acting on a body are in balance. The forces acting on the faces of an infinitesimal volume, such as that shown in Figure 6-1, must also be in balance; this requirement is expressed by the equilibrium equations. After multiplying the stresses in a particular direction by the areas on which they act, the following equations are obtained:

$$\frac{\partial \sigma_{xx}}{\partial x} + \frac{\partial \sigma_{xy}}{\partial y} + \frac{\partial \sigma_{xz}}{\partial z} = 0$$

$$\frac{\partial \sigma_{yy}}{\partial y} + \frac{\partial \sigma_{xy}}{\partial x} + \frac{\partial \sigma_{yz}}{\partial z} = 0 \tag{6.3}$$

$$\frac{\partial \sigma_{zz}}{\partial z} + \frac{\partial \sigma_{xz}}{\partial x} + \frac{\partial \sigma_{yz}}{\partial y} = 0$$

In the presence of stress any real material will undergo some deformation. Any relative motion between points is defined as strain. Two-dimensional or plane strain will be discussed here and the results extended to three dimensions.

Figure 6-2 shows an imaginary, thin rubber block deformed in several ways. With an eraser it is quite easy to repeat this demonstration. A square, with diagonal lines inscribed, has been drawn on the surface and is shown, undeformed, in the first illustration. When a uniform deformation is applied, as in the second illustration, the lines may stay undeformed (as segments *ab* and *cd* have done), change length only (as is the case for segments *ad* and *bc*), or even undergo both a change in length and a rotation (which is the case for segments *ac* and *bd*). Despite these deformations all of the segments remained straight, which is not the case when a nonuniform deformation is applied. The third illustration

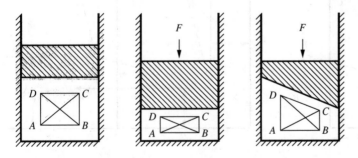

FIGURE 6-2
Deformation of a block in plain strain. A thin block of rubber, upon which a figure has been inscribed (*a*), is distorted to show the effects of uniform strain (*b*) and nonuniform strain (*c*).

of Figure 6-2 shows that under nonuniform deformation, changes in length and rotation still occur, but some segments, such as the diagonals, are no longer straight.

One of the early lessons of calculus is that a curved line segment can be represented by a number of straight segments. If one examines a sufficiently small portion of a material under nonuniform deformation the curved sides of the element can likewise be represented as straight.

Such a differential element is shown in Figure 6-3. The deformation at point A can be described by determining the change in length of the segments AB and AD, along with their rotation relative to each other. The element must be infinitesimal to define the deformation at a point, and the change in length of the segments would then also approach zero, so the normal strains are defined as:

$$\epsilon_x = \lim_{\Delta x \to 0} \left[\frac{\overline{A'B'} - \overline{AB}}{\overline{AB}} \right] \text{ and } \epsilon_y = \lim_{\Delta y \to 0} \left[\frac{\overline{A'D'} - \overline{AD}}{\overline{AD}} \right] \tag{6.4}$$

The normal strain components are dimensionless and positive when a segment elongates.

The shear strain can best be described with reference to Figure 6-3, where the distorted element has been rotated and superimposed on the original element. The sum of the angles describes the element's distortion. This distortion could also be expressed as $\text{Tan}^{-1}(\overline{D'D''}/\overline{A'D''}) + \text{Tan}^{-1}(\overline{B'B''}/\overline{A'B''})$ or, for small angles θ and ϕ, as the dimensionless number $\overline{D'D''}/\overline{A'D''} + \overline{B'B''}/\overline{A'B''}$. The shear strain for small values of θ and ϕ is therefore

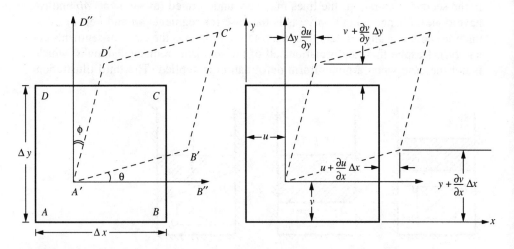

FIGURE 6-3
Plain strain. A small element of a continuous body is deformed to demonstrate the components of strain. The deformation may also be expressed in terms of the displacements u and v and their partial deratives.

$$\epsilon_{xy} = \lim_{\Delta x, \Delta y \to 0} \left(\frac{\overline{D'D''}}{\overline{A'D''}} + \frac{\overline{B'B''}}{\overline{A'B''}} \right) \quad (6.5)$$

To find the strain in terms of the displacements, consider the components u and v in the x and y directions respectively. Figure 6-3 shows the differential element with the deformations expressed using partial derivatives. By using the definitions of strain that have just been established, along with Figure 6-3, the strain may be expressed as:

$$\epsilon_x = \lim_{\Delta x \to 0} \frac{[\Delta x + (\partial u/\partial x)\Delta x] - \Delta x}{\Delta x} = \frac{\partial u}{\partial x}$$

$$\epsilon_y = \lim_{\Delta y \to 0} \frac{[\Delta y + (\partial v/\partial y)\Delta y] - \Delta y}{\Delta y} = \frac{\partial v}{\partial y} \quad (6.6)$$

$$\epsilon_{xy} = \lim_{\substack{\Delta x \to 0 \\ \Delta y \to 0}} \left[\frac{(\partial u/\partial y)\Delta y}{\Delta y} + \frac{(\partial v/\partial x)\Delta x}{\Delta x} \right] = \frac{\partial v}{\partial x} + \frac{\partial u}{\partial y}$$

where small strains have been assumed.

It was shown previously that a symmetric matrix could be assembled to completely describe the state of stress. The same is true for strain:

$$|\epsilon| = \begin{bmatrix} \dfrac{\partial u}{\partial x} & \dfrac{1}{2}\left(\dfrac{\partial v}{\partial x} + \dfrac{\partial u}{\partial y}\right) & \dfrac{1}{2}\left(\dfrac{\partial u}{\partial z} + \dfrac{\partial w}{\partial x}\right) \\ \dfrac{1}{2}\left(\dfrac{\partial v}{\partial x} + \dfrac{\partial u}{\partial y}\right) & \dfrac{\partial v}{\partial y} & \dfrac{1}{2}\left(\dfrac{\partial w}{\partial y} + \dfrac{\partial v}{\partial z}\right) \\ \dfrac{1}{2}\left(\dfrac{\partial u}{\partial z} + \dfrac{\partial w}{\partial x}\right) & \dfrac{1}{2}\left(\dfrac{\partial w}{\partial y} + \dfrac{\partial v}{\partial z}\right) & \dfrac{\partial w}{\partial z} \end{bmatrix} \quad (6.7)$$

BEHAVIOR OF REAL MATERIALS

So far, stress and strain in a solid have been described independently. In the simplest case they are related directly by the material property known as *Young's modulus*. The modulus, which is usually denoted as E, is measured using a uniaxial tension test in which a long, slender sample of some material is pulled, and the force applied to the sample and the elongation are measured. The force and deformation can be used to calculate the stress and strain and to define the ratio $E = \sigma/\epsilon$. Another consequence of deformation, which can be demonstrated by stretching a rubber band, is that the cross-sectional area is reduced as the sample elongates. This phenomenon, which occurs under both tension and compression, is known as the *Poisson effect*. The cross-sectional area reduction has been experimentally shown to be proportional to the strain in the longitudinal direction and to occur only along the principal axes. In the case of the uniaxial tension or compression test the strains along these axes are:

$$\epsilon_x = \frac{\sigma_x}{E} \qquad \epsilon_y = \frac{-\nu\sigma_x}{E} \qquad \epsilon_z = \frac{-\nu\sigma_x}{E} \qquad (6.8)$$

when the force is applied along the x axis.

A shear modulus G may be similarly defined to relate shear stress and strain.

$$\epsilon_{xy} = \epsilon_{yx} = \frac{\sigma_{xy}}{G} \qquad \epsilon_{xz} = \epsilon_{zx} = \frac{\sigma_{xz}}{G} \qquad \epsilon_{yz} = \epsilon_{zy} = \frac{\sigma_{yz}}{G} \qquad (6.9)$$

Since materials may have different properties in different directions, due to such things as grain structure, crystal lattice, or the orientation of fibers in composites, up to six equations with different moduli are needed to fully describe the relationship between stress and strain.

At low levels of strain most materials deform linearly with stress and return to their original geometry when the stress is removed. There is, however, a stress level at which a material no longer returns to its original geometry; this is known as the *yield stress*. At stresses above this level some amount of permanent deformation will remain when all external loads are removed. This is known as the plastic strain or deformation. Stress and strain are not linearly related when this occurs.

Plasticity can be most easily understood if one considers a cubic crystalline structure, which is common among metals. Figure 6-4(a) shows an *edge disloca-*

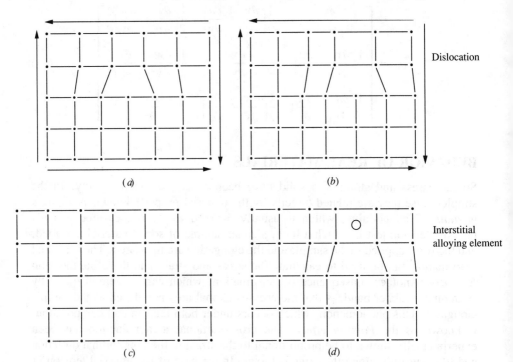

FIGURE 6-4
An edge dislocation in a cubic crystal. The edge dislocation shown in (a) can move under shear stress, as shown in (b). Eventually it reaches a boundary (c) or is pinned by an alloying agent (d).

tion in the crystal. Deformation of the crystal can cause the extra plane of atoms to displace an adjacent, bonded plane; the dislocation has, in effect, moved over. Eventually the dislocation reaches a grain boundary or the outside surface. This type of deformation will not disappear when the load on the sample is removed.

Very low stresses are necessary to cause dislocations to migrate in pure metals. However, alloying agents or impurities greatly increase the yield stress by pinning the dislocations. Furthermore, a pinned dislocation will block other dislocations that are moving in the same direction, causing them to pile up behind it. This distorts the crystal significantly so that it stores more elastic energy. If a structure is loaded past its yield point, unloaded, and then loaded again it will behave elastically up to a much higher stress than it did when first loaded. This phenomenon is known as *strain hardening*.

For dislocations to migrate it is necessary to apply a shear stress. If the crystal shown in Figure 6-4(*a*) were under a hydrostatic force (equal force along three orthogonal axes), the lattice would stretch or compress, but the distortion would not promote the movement of the dislocations. This important point demonstrates that it is the shear stress that defines the onset of yielding. In a uniaxial tensile test the sample will break on a 45-degree angle because the maximum shear stress occurs at that orientation.

For most materials the yield stress is not well defined. Some amount of plastic deformation may be present at very small stresses and the yield stress for these materials may be interpreted as defining the point below which elastic behavior dominates. In these cases the *offset method* is used to define the yield stress. Figure 6-5 shows stress-strain curves for some typical engineering materials. For engineering analyses these curves are often approximated by simple relations.

At room temperatures some materials, such as solder, undergo an initial strain followed by gradual, additional deformation when subjected to a constant load. This is known as *creep*. Almost all metals behavior in this way at absolute temperatures higher than half their melting point. Eutectic Sn-Pb solder melts at 540°K.

Creep is caused by dislocations; as the lattice gains kinetic energy, through either heating or stress, dislocations are generated and begin to move. As in the case of strain hardening, some dislocations are pinned and begin to pile up, thereby reducing the rate of deformation. When the dislocation rate and pinning rate match, a steady-state condition arises and deformation continues at a constant rate. These first two stages of creep are illustrated in Figure 6-6.

If the load is sufficiently high the creep will, at some point, increase again until fracture occurs. For engineering purposes stresses are limited so that this tertiary stage of creep is not reached within the product lifetime.

Creep fatigue is a particularly important concern since solder is a common material for both electrical and mechanical connections, as discussed in Chapter 19. The stresses in these solder joints are generally caused by *differential thermal expansion*.

Heating a material increases the kinetic energy of the atoms in the crystal lattice. This additional energy moves the atoms further apart, causing the material

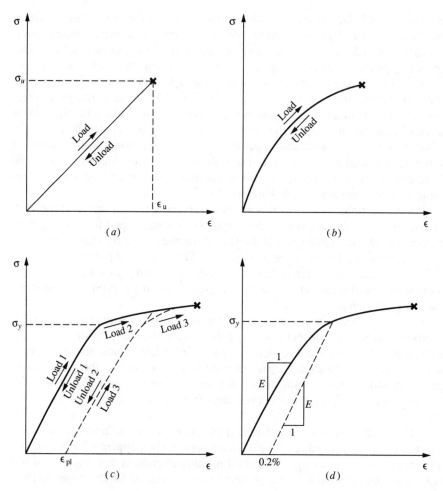

FIGURE 6-5
Typical stress-strain curves for engineering materials: (*a*) An elastic, linear material. Brittle materials, such as glass, behave in this manner. σ_u and ϵ_u represent the ultimate stress and strain. (*b*) A nonlinear, elastic material. Rubber is a good example of such a material. All deformation is recovered when external loads are removed. (*c*) A material with a linear, elastic behavior and a well-defined yield point. Below yield all deformation may be recovered but at strains above the yield strain (ϵ_y) permanent plastic deformation occurs (ϵ_{pl}). Steels behave in this manner. (*d*) most materials, such as copper, behave as shown with an initial linear portion and a poorly defined yield point. A yield stress is defined and used to mark the change from predominantly elastic deformation to plastic. Different amounts of offset are used for different materials. For copper 0.2 percent is used, while for solder 0.05 percent is appropriate.

to expand. The thermal strains that occur can create very large stresses if dissimilar materials are used. For example, connections between a circuit board, with a coefficient of thermal expansion of about 17 ppm/°C, and a ceramic substrate, which typically expands only 6 ppm/°C, must be carefully designed to avoid failure at processing or operating temperatures.

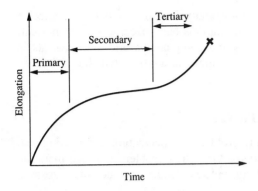

FIGURE 6-6
The three stages of creep. The elongation of a Pb-Sn solder specimen is shown as a function of time.

So far, strain has been shown to arise from stress, Poisson's effect (ν), and thermal expansion (α) associated with temperature change (T). For small strains occurring in three dimensions, the following equations relate these factors:

$$\epsilon_x = \frac{1}{E}\left[\sigma_x - \nu(\sigma_y + \sigma_z)\right] + \alpha T \qquad \epsilon_{xy} = \frac{\sigma_{xy}}{G}$$

$$\epsilon_y = \frac{1}{E}\left[\sigma_y - \nu(\sigma_z + \sigma_x)\right] + \alpha T \qquad \epsilon_{yz} = \frac{\sigma_{yz}}{G} \qquad (6.10)$$

$$\epsilon_z = \frac{1}{E}\left[\sigma_z - \nu(\sigma_x + \sigma_y)\right] + \alpha T \qquad \epsilon_{zx} = \frac{\sigma_{zx}}{G}$$

These are known as the *constitutive equations*.

Along with the equations of geometric compatibility (6.7), which relate strain to displacement, and the equilibrium equations (6.3), the constitutive equations make up the foundation of linear elasticity theory. These equations can be used to solve many idealized packaging problems where the materials behave linearly (Young's modulus is constant) and the strains are small. One reason for the small-strain requirement is that the equilibrium equation is based on the geometry of the undistorted element.

STRESS CONCENTRATIONS

While it is easy to imagine the stress field in a slender sample undergoing a tension test, intuition fails when a more difficult geometry is encountered. In all but the simplest structures it becomes important to understand the way that abrupt changes in section increase the local stress.

Up until recently, engineers used stress concentration factors to estimate the effects of geometry on stress [4]. Some stress concentration factors can be determined mathematically, but in most cases experimental studies have been used to develop tables. By using a stress concentration factor along with a generous factor of safety, structures were designed to ensure reliability; however, these parts were often more costly and massive than necessary. Under the impetus of the aerospace program more accurate stress predictions became necessary and numerical methods, such as finite element analysis, came into use. Computer packaging is an area where numerical simulations have become very important, since the size of mechanical connections and components, and the materials in use, limit the electrical performance. Finite element analysis will be introduced in a section later in this chapter.

A CHIP BONDED TO A SUBSTRATE

There are some simple, yet important, problems in packaging engineering that illustrate the use of linear stress analysis. One such problem is to estimate the thermal stresses that arise when a silicon chip is bonded to a ceramic substrate by a thin layer of adhesive.

Many low-cost packages are made by directly attaching a chip to a carrier and then using wires to make connections between the two (Figure 6-7(a)[5]). In this case it is the stresses in the chip, adhesive, and carrier that are of concern, so an appropriate model would include only these layers. A simplification that makes the problem easier to solve is to model it in only two dimensions. Although this represents a structure with infinite width, it is nonetheless useful for estimating the stresses. In the model, which is shown in Figure 6-7(b), the symbols t, E, G, α, and ν refer to thickness, Young's and shear moduli, coefficient of thermal expansion, and Poisson's ratio, respectively. The subscript 0 is used to designate the adhesive layer. F and τ represent the axial and shear forces.

Further simplification assumes the adhesive is linearly elastic and the shear stress does not vary across its thickness. It is also assumed that the top and bottom layers are subject to either extensional or compressive forces, which produce uniform stress through each layer's thickness. Finally, the temperature is assumed to be uniform in all layers.

The free-body shown in Figure 6-7c depicts an element of length x. Applying equilibrium conditions yields the equations:

$$\frac{\partial F_1}{\partial x} dx - \tau dx = 0 \qquad \frac{\partial F_2}{\partial x} dx + \tau dx = 0 \qquad (6.11)$$

where F is the force per unit width. The strain in the top and bottom layers, which arises both from thermal expansion and the internal forces, may be expressed as

$$\frac{\partial u_1}{\partial x} = \frac{F_1}{E_1 t_1} + \alpha_1 T \qquad \frac{\partial u_2}{\partial x} = \frac{F_2}{E_2 t_2} + \alpha_2 T \qquad (6.12)$$

where T is the change in temperature. In the adhesive layer the shear strain is

$$\frac{\tau}{G} = \frac{u_1 - u_2}{t_0} \qquad (6.13)$$

Differentiating Equation (6.13) and substituting from Equation (6.12) gives

$$\frac{t_0}{G} \frac{d\tau}{dx} = \left(\frac{F_1}{E_1 t_1} - \frac{F_2}{E_2 t_2} \right) + (\alpha_1 - \alpha_2)T \qquad (6.14)$$

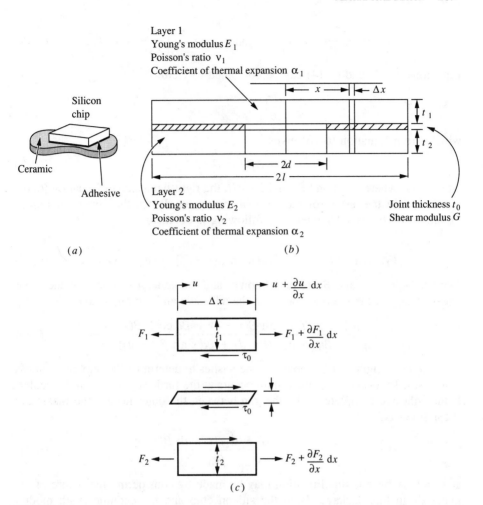

FIGURE 6-7
Thermal stresses in a first-level package. A simple package, consisting of a chip attached to a ceramic substrate by adhesive, is illustrated in (*a*). A two-dimensional model of the structure is shown (*b*) along with a free-body representing an infinitesimal portion of model (*c*).

which, after differentiating again and substituting from Equation (6.11) yields

$$\frac{d^2\tau}{dx^2} - \beta^2\tau = 0 \quad \text{where } \beta^2 = \frac{G}{t_o}\left(\frac{1}{E_1 t_1} + \frac{1}{E_2 t_2}\right) \tag{6.15}$$

This equation is one of second order and has a general solution

$$\tau = A \sinh \beta(x - d) + \beta \cosh \beta(x - d) \tag{6.16}$$

which is valid for the range $d < |x| < l$. Since the displacements at $x = d$ are known to be

$$u_1 = \left(\frac{F_1}{E_1 t_1} + \alpha_1 T\right)d \quad \text{and} \quad u_2 = \left(\frac{F_2}{E_2 t_2} + \alpha_2 T\right)d \qquad (6.17)$$

Equations (6.13) and (6.14) can be combined to yield

$$\left.\frac{d\tau}{dx}\right|_{x=d} = \left.\frac{\tau}{d}\right|_{x=d}$$

which, from Equation (6.16) gives

$$A\beta d = B \qquad (6.18)$$

At the end, where $x = 1$ and F_1 and $F_2 = 0$, the right-hand side of Equation (6.14) contains only the terms for the thermal expansion. When the general solution, Equation (6.16), is substituted, the following is obtained:

$$\frac{t_0}{G}\beta[A \cosh \beta(l - d) + A\beta d \sinh \beta(l - d)] = (\alpha_1 - \alpha_2)T \qquad (6.19)$$

The coefficients A and B are now known, and the general solution for the shear stress acting on the interface between the adhesive and the outer layers is

$$\tau = \frac{(\alpha_1 - \alpha_2)TG[\sinh \beta(x - d) + \beta d \cosh \beta(x - d)]}{\beta t_0[\cosh \beta(l - d) + \beta d \sinh \beta(l - d)]} \qquad (6.20)$$

In most engineering problems, one wishes to determine the maximum stress that arises. In this case the maximum occurs at the furthest point from the center. If the adhesive completely fills the gap between the outer layers, the maximum shear stress is

$$\tau_{\max} = \frac{(\alpha_1 - \alpha_2)TG \tanh \beta l}{\beta t_0} \qquad (6.21)$$

at $x = 1$. A further simplification may be made by considering the nature of the materials in this package. Both the silicon chip and the ceramic (with moduli of about 62 msi and 50 msi, respectively) are very rigid when compared to the adhesive (which may have a shear modulus as low as 750 psi). If $\beta l << 1$ the maximum shear stress is

$$\tau_{\max} = (\alpha_1 - \alpha_2)\frac{TGl}{t_0} \qquad (6.22)$$

If, on the other hand, the chip is thin and flexible when compared to the substrate, then $\beta = \sqrt{G_1/E_1 T_1}$ and

$$\tau_{\max} = (\alpha_1 - \alpha_2)T \sqrt{\frac{E_1 G_1 t_1}{t_0}} \tanh \beta l \qquad (6.23)$$

If the module is quite large, so that the term $\beta l >> 1$, the shear stress is

$$\tau_{\max} = \frac{(\alpha_1 - \alpha_2)TG}{\beta t_0} \qquad (6.24)$$

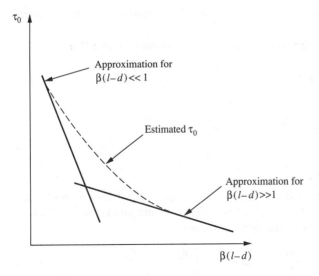

FIGURE 6-8
Maximum shear stress in a simple first-level package has been calculated for the model shown in Figure 6-7 as a function of the parameter $\beta(l - d)$. The straight lines are for the cases $\beta l \ll 1$ and $\beta l \gg 1$.

Another geometry likely to be of interest is the one with adhesive only in the outermost region. In this case the quantity $\beta(l - d)$ is very small and, once again, Equation (6.22) gives the approximate value of the shear stress. This same model could be used to estimate the shear in solder joints around the periphery of a chip attached to a ceramic substrate.

The effect of the parameter $\beta(l - d)$ on the shear stress is plotted in Figure 6-8. The actual shear stress approaches asymptotically the special cases described previously.

BENDING OF A FLEXIBLE CIRCUIT

In the previous analysis the stresses across the thickness of the chip and substrate were assumed to be constant and only the equilibrium of forces was employed. To fully satisfy equilibrium conditions, the moments that arise must also be included. However, in this case the rigidity of the chip and substrate allow the bending to be ignored. This is not the case when modeling a flexible circuit.

Flexible circuits are generally made up of flat, thin layers of plastic and copper circuitry, with an adhesive in between. Although the copper is etched into a series of lines, so that the structure is not continuous, the model developed in the previous example can be used to estimate the stresses that arise in the structure during thermal cycling.

The same assumptions used in the previous example may be applied; the only differences are that both shear and tensile forces are considered in the adhesive layer and that moments cannot be ignored.

Two additional symbols are used: M designates a moment, while V is used to indicate a shear force.

The conditions of equilibrium of forces and moments require that

$$\frac{dM_1}{dx} - V_1 + \tau_0 \frac{t_1}{2} = 0 \qquad \frac{dM_2}{dx} - V_2 + \tau_0 \frac{t_2}{2} = 0 \qquad (6.25)$$

$$\frac{dF_1}{dx} - \tau_0 = 0 \qquad \frac{dF_2}{dx} + \tau_0 = 0 \qquad (6.26)$$

$$\frac{dV_1}{dx} - \sigma_0 = 0 \qquad \frac{dV_2}{dx} + \sigma_0 = 0 \qquad (6.27)$$

The bending of beams is described in many texts, with that of Timoshenko being one of the best [1]. Provided that the deflections are small, one can relate the transverse displacement v to the applied moments:

$$\frac{d^2v_1}{dx^2} = \frac{-M_1}{D_1} \quad \text{and} \quad \frac{d^2v_2}{dx^2} = \frac{-M_2}{D_2} \qquad (6.28)$$

where

$$D_i = \frac{E_i t_i^3}{[12(1 - v_i)^2]}$$

The longitudinal displacements are

$$\frac{du_1}{dx} = \frac{(1 - v_1^2)F_1}{E_1 t_1} - \frac{6(1 - v_1^2)M_1}{E_1 t_1^2} + (1 + v_1)\alpha_1 T \qquad (6.29)$$

$$\frac{du_2}{dx} = \frac{(1 - v_2^2)F_2}{E_2 t_2} + \frac{6(1 - v_2^2)M_2}{E_2 T_2^2} + (1 + v_2)\alpha_2 T$$

As before, the shear stress in the adhesive is described by the equation

$$\frac{\tau_0}{G} = \frac{u_1 - u_2}{t}$$

while the tensile stress is

$$\frac{\sigma_0}{E_0} = \frac{v_1 - v_2}{t_0} \qquad (6.30)$$

Equations (6.25) through (6.30) involve 12 unknowns and may be reduced to a pair of sixth-order equations in σ and τ [6]. For the normal stress this may be expressed as

$$\frac{d^6\sigma_0}{dx^6} - \frac{G_0 c}{\eta}\frac{d^4\sigma_0}{dx^4} + \frac{E_0 b}{\eta}\frac{d^2\sigma_0}{dx^2} - \frac{G_0 E_0(bc - a^2)\sigma_0}{\eta^2} = 0 \qquad (6.31)$$

where the constants a, b, and c are defined as

$$a = 6\left[\frac{(1 - \nu_1^2)}{E_1 t_1^2} - \frac{(1 - \nu_2^2)}{E_2 t_2^2} \right]$$

$$b = 12\left[\frac{(1 - \nu_1^2)}{E_1 t_1^3} - \frac{(1 - \nu_2^2)}{E_2 t_2^3} \right] \quad (6.32)$$

$$c = 4\left[\frac{(1 - \nu_1^2)}{E_1 t_1} + \frac{(1 - \nu_2^2)}{E_2 t_2} \right]$$

A general solution to the set of equations is in the form

$$\sigma_0 = A_1 \cosh \beta_1 x + A_2 \sinh \beta_1 x + A_3 \cosh \beta_2 x \cos \beta_3 x$$

$$+ A_4 \sinh \beta_2 x \cos \beta_3 x + A_5 \sinh \beta_2 x \sin \beta_3 x \quad (6.33)$$

$$+ A_6 \cosh \beta_2 x \sin \beta_3 x$$

where β_1, β_2, and β_3 are the real and complex conjugate roots of the algebraic equation

$$y^3 - \frac{G_0 c}{\eta} y^2 + \frac{E_0 b}{\eta} y - \frac{G_0 E_0 (bc - a^2)}{\eta^2} = 0 \quad (6.34)$$

The equation for the shear stress is similar.

If there are no external forces or moments applied to the structure the solution can be simplified. Due to the symmetry of the structure, the shear forces on the adhesive are zero at the center, where $x = 0$. If the upper layer expands more than the lower there will be a shear force V applied to the adhesive. For all $x > 0$ this shear will be positive (it will tend to rotate the area clockwise), while for all $x < 0$ it will be negative. The shear force is an antisymmetric function and may be expressed as $f(x) = -f(-x)$; the solution must reflect that relationship. Since the cos and cosh functions are symmetric, while the sin and sinh functions are not,

$$\tau_0 = C_1 \sinh \beta_1 x + C_2 \sinh \beta_2 x \cos \beta_3 x + C_3 \cosh \beta_2 x \sin \beta_3 x \quad (6.35)$$

It can similarly be developed that the axial stresses in the top and bottom layers are symmetric with respect to x. This gives

$$\sigma_0 = A_1 \cosh \beta_1 x + A_3 \cosh \beta_2 x \cos \beta_3 x + A_5 \sinh \beta_2 x \sin \beta_3 x \quad (6.36)$$

Figure 6-9 shows the axial and shear stress distribution for a flexible circuit with two different thicknesses of adhesive. Note that the stresses remain low over most of the joint length except in the vicinity of the edge, where the stresses rise dramatically. The change in the sign of the axial stress near the edge is quite surprising. The most important conclusion drawn from this analysis is that failures depend on the thickness of the adhesive layer (t_0) relative to the size of the part (l). One would expect failures to originate near the edge.

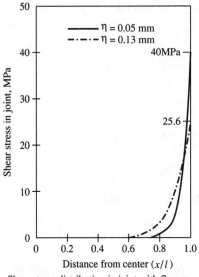

Shear stress distribution in joint with flexure.

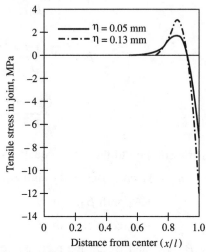

Tensile stress distribution in joint with flexure.

FIGURE 6-9
Modeling a flexible circuit. The shear and tensile stress distributions in the adhesive layer.

This model could be used to help design a flexible cable if the adhesion strength between the layers was well known, although one would still have to apply a generous safety factor to allow for the many assumptions made. A great deal of testing is called for, and an engineer would ensure that such tests were carried out. The need for experimental verification of results will be seen again in the next example.

A SURFACE-MOUNTED PACKAGE

One method for making both mechanical and electrical connections between a card and a substrate is to use solder joints between pads on the two surfaces. The model can be applied to such a package by using a technique called *smearing*, as demonstrated in the following example.

Figure 6-10 shows a leadless chip carrier and the assumptions needed to make the modeling approach tractable. The first of the illustrations shows the substrate and the solder joints that are formed around the edges. If one assumes that the solder joints are not discrete, but rather a continuous connection, the problem becomes much simpler.

As long as only shear stresses occur in the solder joints, their stiffness may be expressed by Hooke's law, $k = Gtw/h$, where tw represents the surface area of the joint on the card and h is the height of the joint. Since we are increasing the width of each joint, but wish to keep the stiffness constant, G must be adjusted.

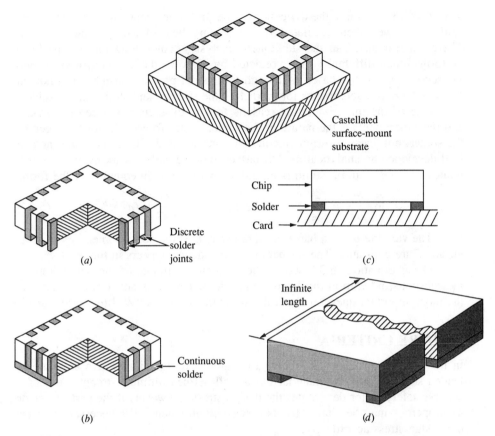

FIGURE 6-10
Modeling a leadless chip carrier. (*a*) A cut away view of an LCC, showing solder joints along the periphery. (*b*) "Smearing" the discrete solder joints produces a model with a continuous connection. One can use a two-dimensional model for such a structure, as illustrated in (*c*). The problem modeled in this case is an infinitely long substrate as shown in (*d*).

Thus, for joints with a width of 0.5 mm on 1-mm centers the shear modulus would be decreased by a factor of two. The properties of the discrete joints have, in this case, been smeared to give a continuous structure with intermediate mechanical properties.

To further simplify the problem one can treat it as if it were two-dimensional. This is the same as modeling a substrate that is infinitely long in one direction. At this point the problem may be solved in closed form.

This particular problem has been the subject of a great deal of experimentation. In a series of recent papers [7–9] Hall and his associates have observed the deflection that arises from thermal cycling of leadless chip carriers when bonded to epoxy boards. Both holography and strain-gauge measurements were used to determine the deflections. The forces and moments that act on the solder joints were then calculated by representing the card and substrate as plates. These experiments, and the subsequent analysis, demonstrate that the package performs much differently than predicted by Timoshenko's model, upon which the second example is based. Although this model provides insight, it is not, in this case, a reasonable means for predicting the stresses and strains in the solder.

One of the most important reasons for the difference between the model and the experiments is the nature of the solder itself. To account for the creep of the solder, nonlinear constitutive equations are required. Stone, Hannula, and Li [10] developed an analytical model, based upon the earlier work by Hart, which relates the stress, strain, strain rate, and temperature in an equation of the form:

$$\log \frac{\sigma^*}{\sigma} = \left(\frac{\dot{\epsilon}^*}{\dot{\epsilon}}\right)^\lambda \qquad \text{where} \qquad \dot{\epsilon}^* = \left(\frac{G^*}{G}\right)^m f e^{-Q/RT} \qquad (6.37)$$

The variable σ^* is a hardness parameter, Q is the activation energy, and λ, m, and f are constants. The temperature dependence appears in the term $\dot{\epsilon}^*$.

Using Equation (6.37) would yield a better estimate of the deflections of the substrate and chip in this example problem. This does not, however, answer the most important question about the structure, which is "Will the joints fail?"

FAILURE CRITERIA

In the simplest problems it is easy to establish design limits. If a structure is loaded statically and the material has a well-defined ultimate strength, use of a suitable safety factor determines the design stress. However, if the part creeps or its properties may be altered by its service environment, it is necessary to alter the design stress accordingly.

As an example, consider the connector shown in Figure 6-11, which shows a card forced between pair of springs in order to make electrical connections. These springs are designed (as discussed in detail in Chapter 7) to wipe across the soft plated surface (often gold) of the contact pads, pushing any dust or debris out of the way and deforming the malleable surface. The springs must maintain some minimum contact force throughout the product lifetime to prevent corrosion from eventually degrading the contact. Creep in the solder or the copper spring could cause the contact force to become less than this minimum requirement.

Figure 6-11(b) shows hypothetical curves for the contact force as a function of time and temperature. The region where the spring enters the solder is highly stressed and the solder will flow, allowing the spring to bend inside the housing. This example shows that the load generated by the deflection of the spring decreases as the solder deforms. Both the rate of deformation and the load decrease together until the system approaches equilibrium. At 150°C the equilibrium point is reached at a force below that required for electrical reliability; the connector may therefore fail.

Another type of design limit arises when trying to avoid yielding in a material. In the manufacture of cores (see Chapter 11), which are laminated

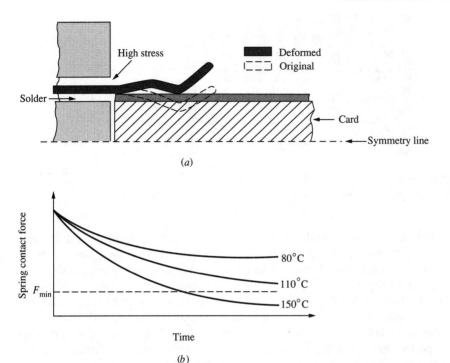

(a)

(b)

FIGURE 6-11
Mixed mode creep; A spring connector is shown soldered into a housing (*a*). Stress in the solder will cause it to creep, which decreases the contact force between the spring and pad. Increasing the temperature increases the creep rate (*b*), which could cause the contact force to become less than that required to ensure electrical integrity.

together to form a circuit board, plastic deformation of a layer containing copper lines could be disastrous. If the yield strength of the copper were exceeded during handling, the core would remain distorted and would change its dimensions when laminated. Given the extremely small size of circuit lines and holes, and their precise tolerances, this could make the core unusable. Thus, a packaging engineer could be called upon to set appropriate handling limits, which would be governed by the yield strength of the copper.

Sometimes a part must be designed so that there is a limited displacement during operation. Such is the case when thermal grease is used to conduct heat away from a chip and into a finned radiator, as shown in Figure 6-12. In this case, when a particular displacement amplitude was repeatedly exceeded, grease was forced out from between the chip and the cap. Once this limit had been established experimentally it could be applied to any number of similar packages.

The force transmitted by a member can also be a design limit and, once again, the spring connector makes a good example. Figure 6-13 illustrates a card held in place by connector springs at both ends. Since the gold surface of the

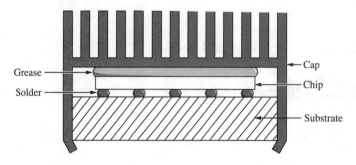

FIGURE 6-12
Thermal grease in a module. Grease is used to conduct heat away from a chip.

springs is severely deformed when a card is inserted, the connection can reliably be made and broken only a limited number of times before the plating is worn away. If the frame expands more than the cards at the operating temperature, the springs could slip across the pads, causing additional wear. To prevent this wear, it must be assured that the friction force between the spring and pads is sufficient to prevent sliding, given the relative motion between the frame and the cards.

In these last two cases failure did not occur statically, but developed during repeated power cycles. The mechanisms that produced these failures are easily identified, but failure in some cases can be unexpected. When stresses are repeatedly applied to a structure it can fail well below its ultimate strength. This phenomenon is generally referred to as *fatigue*.

FATIGUE

Fatigue failures were first observed in the mid 1800s at a time when the construction of railroads and bridges was under rapid development. One problem was

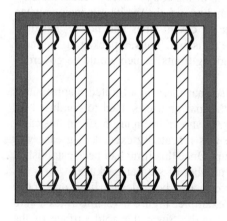

FIGURE 6-13
Springs connect cards to a frame.

the unexpected failure of railway axles after several years of service. Rankine, in 1843, attributed this to a gradual deterioration of the metal, and also demonstrated that the radius of curvature affected the resistance to repeated impacts.

Bridges with longer spans were being constructed and were expected to carry ever increasing loads. A Royal Commission inquired into the suitability of iron for bridges, concluding in 1849 that repeated application of a load of only half the static breaking load would cause a beam to fail in less than 1000 traversals.

Wohler carried out a lengthy investigation of the effect of cyclic stress on axles and developed a rotating bending machine, which is still in use today, to simulate the loading conditions. He showed (in 1871) that fatigue failures could occur below the yield limit in wrought iron and steel. In addition he showed that these materials can resist fatigue indefinitely if the stress is sufficiently low. This property is, however, seen almost exclusively in iron and low-alloy steels at room temperature. The effects of geometry and mean stress on fatigue were also studied.

The effects of fatigue were established at these early dates, but the explanation of this phenomenon is still not complete. Early work by Ewing and Humfrey, on bending in steel, showed that "slip bands" appeared on the surface if the yield limit was exceeded. The slip bands arise, for example, when dislocations in a crystal reach the outer surface of the structure, as demonstrated in Figure 6-4(c). The discontinuity that a slip-band causes at the surface becomes a stress concentration from which a crack can propagate.

Plastic deformation is therefore seen as the initiator of fatigue failure, but the mechanism has been variously attributed to wear along the slip plane, phase transformation of the material in the slip plane, and dissolved gases that enter along the slip plane.

It is now generally accepted that the rate of crack propagation governs the fatigue life of a structure since cracks form within a small fraction of the cycles to failure [11]. Crack growth depends primarily on material properties, magnitude of the stress, rate of application, temperature, and history of the part.

The cyclic stress-strain curves of Figure 6-14 illustrate how complex the behavior of some materials can be. In this case eutectic Sn-Pb solder is tested at different frequencies and temperatures with specimens of different grain sizes. Grain size may be varied by aging the samples at elevated temperature, for example, at 120°C for two hours. In each test the samples were cycled within a constant strain range and the stress was monitored, as shown in Figure 6-14. Note that the stress range shows an initial increase, due to hardening. Note also that the life increases with increasing frequency, although it appears to approach a limit at very high frequencies. Higher temperatures decrease the life and a smaller grain-size allows the sample to undergo more cycles. Tests such as these are reported extensively in the literature; however, most have been performed on bulk samples that may behave differently from the small solder connections used in electronic packages. This is an area in which a great deal of research is currently under way.

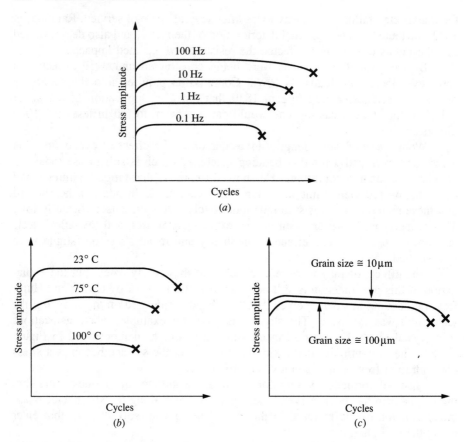

FIGURE 6-14
Behavior of eutectic Sn-Pb solder. The stress amplitude for constant strain amplitude, fully traversed fatigue tests. In (a) the frequency is varied for samples at room temperature. Part (b) illustrates the effect of temperature. Grain size is important, as (c) clearly shows.

The behavior described above is typical of most metals with low melting points. In electronic packages another complicating factor arises; loads are often caused by thermal expansion during operating cycles, and the results from isothermal tests on solder can no longer be directly applied.

There appear to be two distinct types of fatigue processes, one producing failure in 1000 cycles or less, while the other operates at lives of more than 10,000 cycles. In general the low-cycle fatigue behavior applies whenever the yield point is exceeded, while high-cycle fatigue occurs when stresses remain in the elastic region.

HIGH-CYCLE FATIGUE

It was high-cycle fatigue that brought about the failures Wohler observed in railway axles. High-cycle fatigue has been studied extensively for many years and

"S-N" curves have been measured for many materials. Since high-cycle fatigue is generally confined to stresses and strains below the yield limit it is only slightly affected by strain-rate or frequency. This makes it possible to accelerate a fatigue test by using a frequency many times higher than the structure would see in service.

As an example, consider the stresses that occur when a computer is subjected to vibrations, such as those found in a helicopter. Figure 6-15 shows a card mounted to a board inside a flight computer, and the acceleration loading that drives the card. Every flexure of the card will put the copper circuit lines through a fully traversed stress cycle. If the computer's frame had a structural resonance at 10 Hz and a service life of 4000 hours, the circuit lines would have to survive a minimum of 100 million cycles.

In a problem such as this, one would measure the vibration environment of the helicopter, or use a vibration specification supplied by the manufacturer, to determine the acceleration amplitude at the base of the computer. The computer could then be tested on a vibration table and the strain measured on the card's surface by using strain gauges. As an alternative, finite element analysis might be used to predict the strain amplitude.

Once the strain amplitude is known, the life can be estimated from handbook data or accurately determined from fatigue tests on sample cards. In the latter case the electrical resistances of the lines could easily be monitored to determine when failure occurred. At 1 kHz the lines could be tested to a billion cycles in 12 days and it would be possible to establish the S-N curve so that data would be available for the design of future systems.

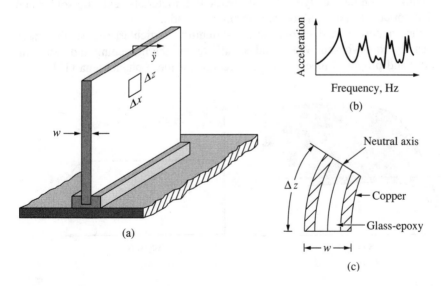

FIGURE 6-15
A computer card under acceleration loading. The card is shown in (*a*); the acceleration spectrum appears in (*b*). Flexure causes stress on the copper lines in the card (*c*).

The results of a fatigue study can be much different from calculations based on handbook values for a number of reasons. The most important reason is the effect of the grain structure of the lines, which varies according to the type of processing used, the plating rate, and temperature. The size of a sample can also affect the fatigue life and, as in the case of solder, handbook values for copper are generally determined on bulk samples. The copper purity varies with the type of processing, and any contaminants that collect in the grain boundaries can significantly reduce the fatigue life.

When more than one loading condition is applied to a structure it is still possible to make a prediction of its life. *Miner's rule* is used for this purpose:

$$\frac{n_1}{N_1} + \frac{n_2}{N_2} + \cdots + \frac{n_i}{N_i} = C \tag{6.38}$$

This assumes that the damage that occurs after n cycles, at a stress level that produces a fatigue life of N cycles, is proportional to n/N. The constant C is usually found to be between 0.7 and 2.2, which illustrates the degree of uncertainty of this method. Experimental verification is necessary whenever this is used as a design guide.

LOW-CYCLE FATIGUE

The S-N curves used in most fatigue problems are of little use when the yield point has been exceeded. Beyond that point not only is the stress-strain curve nonlinear, but it may change during testing. Figure 6-14(a) and 6-16(a) show cyclic stress-strain data of typical strain-hardening materials, eutectic solder and annealed copper, tested at a constant strain amplitude.

Other materials will exhibit strain softening, particularly those that have been heavily cold-worked. Wood and Segall studied the hardening and softening of copper and proposed mechanisms to account for these phenomena [12].

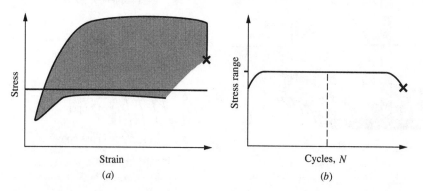

FIGURE 6-16
Fatigue of copper. The stress range was measured in this low-cycle fatigue test at constant strain amplitude. In (a) the recorder paper was indexed after every cycle so that the loops may be clearly seen. The stress range from (a) is plotted in (b) as a function of the number of cycle.

Fortunately most materials harden or soften in relatively few cycles and the stress-strain behavior is stable over almost all of the life. That makes it possible to relate the stress and strain at the middle portion of the life curve and use this as a basis for predicting fatigue.

Many engineers use the Coffin-Manson equation to relate the strain amplitude to the fatigue life. In his first paper on this subject, Manson related the amplitude of the plastic strain to the cyclic life [13]:

$$\epsilon_p = MN_f^z \tag{6.39}$$

where M and z are material constants. Coffin later hypothesized that a value of $z = -0.5$ could be adopted for most materials [14].

The elastic strain range can also be related to the cyclic life:

$$\epsilon_{el} = \frac{K}{E}N_f^\gamma \tag{6.40}$$

where ϵ_{el} is the stress range at $\frac{1}{2}N_f$ divided by E, the elastic modulus, and K and γ are other material constants [15].

These relations were determined experimentally, yet despite their simplicity, they fit data for many materials extremely well. Figure 6-17 shows the behavior of a typical material, in this case annealed 4130 steel [15].

In many cases one would rather deal with the total strain range than either the plastic or elastic ranges. In this case the two Equations (6.39) and (6.40) can be combined to give

$$\epsilon = MN_f^z + \frac{K}{E}N_f^\gamma \tag{6.41}$$

This equation is also plotted in Figure 6-17.

The four constants M, z, K, and γ can be obtained from tests made at as few as two strain amplitudes. By dividing the stress range at $\frac{1}{2}N_f$ by the elastic modulus, K and γ in Equation (6.40) can be determined simultaneously. Subtracting elastic strain range from the total strain makes it possible to use Equation (6.39) to obtain M and z.

Low-cycle fatigue is a frequent concern in electronic packaging. While processing, assembly, and shipping can all cause failures, thermal cycling is the most common cause of fatigue. Stresses arise from the power-on, power-off cycles that occur during a product's lifetime.

Figure 6-18 shows a surface-mounted ceramic module. Long, thin pins are connected to a pad on the surface of a card by solder, which wets the pins to form a fillet, as shown. With coefficients of thermal expansion of 17 and 6 ppm for the card and ceramic respectively, the pins and solder joints will be under considerable strain when the module reaches its operating temperature. Low-cycle fatigue of the solder joint is a definite possibility.

Increasing the length of the pins or decreasing their thickness will make

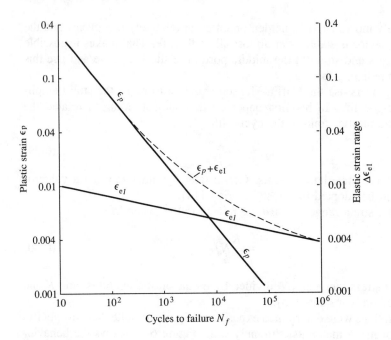

FIGURE 6-17
Low-cycle fatigue in annealed 4130 steel: relationship between ϵ_p, ϵ_{el}, and cycles to failure N_f.

them more flexible and reduce the strain in the solder joint. Unfortunately, such pins are more difficult to handle. There are also electrical penalties associated with long pins that run close together, such as induced signals, capacitance, and signal delay. To optimize the package an engineer must meet the life and reliability requirements without sacrificing electrical performance. However, predicting the strain in the solder, and from that the fatigue life, is complex. A powerful way to solve this problem is to use finite element analysis.

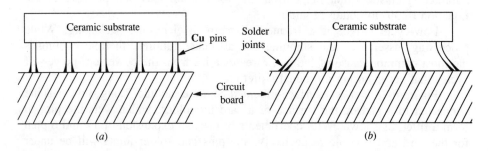

FIGURE 6-18
A surface-mounted package at room temperature (*a*) and at 85°C (*b*).

FINITE ELEMENT ANALYSIS

Finite element analysis is a procedure for breaking a complex geometry into many simple structures that may be more easily analyzed. Connecting points are chosen and the displacements at these points are equated by solving for all of the elements simultaneously. The roots of the technique lie in the analysis of trusses for bridges, which was the most demanding engineering challenge in the early 1800s. Navier may have been the first to use such an approach in his 1826 analysis of a truss problem. Since that time others have developed a systematic approach by using the *stiffness method*, notably Clebsch in 1862 and Ostenfeld in 1926 [16]. The stiffness method may be employed to obtain exact solutions for models that use only spars, and yields approximate solutions for two- and three-dimensional elements. The Rayleigh-Ritz method is the basis of modern analysis and uses the principle of minimum energy to fit shape functions to the model [17]. The term *finite element* was introduced by Clough in 1960 [18].

It is not the intent of this section to instruct readers in the finite element method, but rather to demonstrate the technique. An introduction to this subject should make one aware of its limitations and the difficulties of selecting an appropriate model. For a thorough treatment refer to references [19 and 20].

As a first example, the truss shown in Figure 6-19 will be analyzed. It is important to note the assumptions made in any finite element analysis; in this case it is assumed that the bars are uniform and linearly elastic, the ends are pin connected, and rotations are small enough to be ignored.

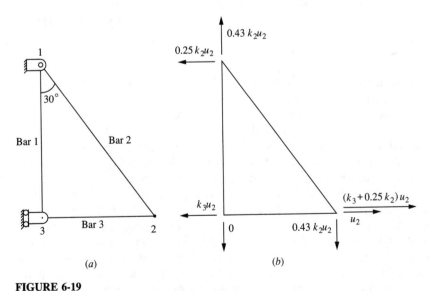

(a) (b)

FIGURE 6-19
A three-element plane truss problem. The truss is shown in (a), with joints and beams numbered. The free-body diagram in (b) shows the forces that arise from a small displacement u_2 acting on joint 2 in the x direction. Hooke's law is used to determine the reaction forces.

If the boundary conditions at nodes 1 and 3 are ignored there are six unique displacements that completely describe the truss. These are the x and y displacements at the joints or nodes. Each of these possible displacements is termed a *degree of freedom*. The free-body diagram of Figure 6-19(b) shows the forces acting upon the joints when a small displacement u_2 is applied in the x direction at node 2. Based on the geometry of the problem, bar 3 extends by an amount u_2, while bar 2 extends by $u_2/2$. Bar 1 remains undisturbed. By applying Hooke's law one can determine all the forces shown in the figure, where, for bar i, $k_i = a_i E_i / l_i$, a_i = cross-sectional area, E_i = modulus, and l_i = length. In matrix form, the forces that produce a small deflection u_2 can be written as:

$$\begin{Bmatrix} f_{x1} \\ f_{y1} \\ f_{x2} \\ f_{y2} \\ f_{x3} \\ f_{y3} \end{Bmatrix} = \begin{Bmatrix} -0.25k_2 \\ 0.43k_2 \\ k_3 + 0.25k_2 \\ -0.43k_2 \\ -k_3 \\ 0 \end{Bmatrix} u_2 \tag{6.42}$$

Similar relations can be determined for deflections of the other five degrees of freedom. Combining all of these produces the matrix equation:

$$\begin{Bmatrix} f_{x1} \\ f_{y1} \\ f_{x2} \\ f_{y2} \\ f_{x3} \\ f_{y3} \end{Bmatrix} = \begin{bmatrix} 0.25k_2 & -0.43k_2 & -0.25k_2 & 0.43k_2 & 0 & 0 \\ -0.43k_2 & 0.75k_2 + k_1 & 0.43k_2 & -0.75k_2 & 0 & -k_1 \\ -0.25k_2 & 0.43k_2 & k_3 + 0.25k_2 & -0.43k_2 & -k_3 & 0 \\ 0.43k_2 & -0.75k_2 & -0.43k_2 & 0.75k_2 & 0 & 0 \\ 0 & 0 & -k_3 & 0 & k_3 & 0 \\ 0 & -k_1 & 0 & 0 & 0 & k_1 \end{bmatrix} \begin{Bmatrix} u_1 \\ v_1 \\ u_2 \\ v_1 \\ u_3 \\ v_3 \end{Bmatrix}$$

$$\tag{6.43}$$

This demonstrates the form of the finite element problem, which may be expressed by the general equation $F = KU$.

There are some interesting observations concerning the matrix K. It is known as the *stiffness matrix* since the equation is analogous to Hooke's law. The diagonal coefficients of K are all positive, and the matrix is symmetric (which is true for linear structures). One may prove that the diagonal coefficients are positive by noting that, for each of the free-bodies analyzed above, the displacement of a node, and the nodal force that produced such a displacement, must necessarily be in the same direction. The symmetry of the stiffness matrix, for linear structures, may be demonstrated by using the principle of reciprocity, whereby the order in which forces are applied to a body are unimportant and the total work done remains the same. Consider the application of force f_1 to the body shown in Figure 6-20. The work done by the force acting upon that point is $f_1 \delta_{ij}$, where δ_{ij} is the displacement at point i due to a force acting at point j. Now, consider the application of a second force at point 2. It not only does work in moving that point ($f_2 \delta_{22}$), but also in moving point 1 by an amount $\delta_{12}, (f_1 \delta_{12})$.

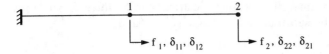

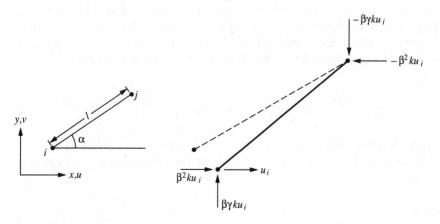

FIGURE 6-20
Two-element structure for demonstrating the principle of reciprocity: δ_{ij} is the deflection at node i due to a force at node j.

The total work done is $f_1\delta_{11} + f_2\delta_{22} + f_1\delta_{12}$. When the order in which the forces are applied is reversed, the work must remain the same. Thus $f_1\delta_{11} + f_2\delta_{22} + f_1\delta_{12} = f_2\delta_{22} + f_1\delta_{11} + f_2\delta_{21}$. The total displacement at a node i is given by $u_i = \Sigma_j \delta_{ij}$.

If one rearranges the general equation to give $\delta F = u$, where δ is the inverse of K (so that $u_i = \Sigma_j \delta_{ij} f_j$) and one makes use of the relationship between δ and u, the work done in the two cases may be expressed as $f_1 f_2 \delta_{12} = f_1 f_2 \delta_{21}$, and therefore $\delta_{12} = \delta_{21}$. Since one can show that $\delta_{ij} = \delta_{ji}$, the symmetry of δ, and also of K, is demonstrated.

While the approach used to formulate Equation (6.43) provides valuable insights into the finite element method, it becomes unwieldy when trying to assemble large matrices. Another approach to the problem, which may be easily implemented on a computer, is to determine a stiffness matrix for each element and finally combine them. Figure 6-21 shows a single bar element. As before, one can assemble the equation

$$\begin{Bmatrix} f_{ix} \\ f_{iy} \\ f_{jx} \\ f_{jy} \end{Bmatrix} = k_n \begin{bmatrix} \beta^2 & \beta\gamma & -\beta^2 & -\gamma\beta \\ \beta\gamma & \gamma^2 & -\beta\gamma & -\gamma^2 \\ -\beta^2 & -\beta\gamma & \beta^2 & \beta\gamma \\ -\beta\gamma & -\gamma^2 & \beta\gamma & \gamma^2 \end{bmatrix} \begin{Bmatrix} u_i \\ v_i \\ u_j \\ v_j \end{Bmatrix} \tag{6.44}$$

FIGURE 6-21
A bar element. The stiffness matrix for a single element of a truss is calculated by applying small displacements at the nodes. ($\beta = \cos\alpha$, $\gamma = \sin\alpha$.)

where $\beta = \cos \alpha$ and $\gamma = \sin \alpha$. The stiffness matrices for the three bars of the truss may then be found by substituting for β and γ. For bar 1,

$$\begin{Bmatrix} f_{1x} \\ f_{1y} \\ f_{2x} \\ f_{2y} \end{Bmatrix} = k_1 \begin{bmatrix} 0 & 0 & 0 & 0 \\ 0 & 1 & 0 & -1 \\ 0 & 0 & 0 & 0 \\ 0 & -1 & 0 & 1 \end{bmatrix} \begin{Bmatrix} u_1 \\ v_1 \\ u_2 \\ v_2 \end{Bmatrix} \qquad (6.45)$$

Bars 2 and 3 may be similarly treated. This technique gives the same results as in the previous example.

Equation (6.45) can be used to demonstrate the way boundary conditions are applied to a structure. The matrix can be reduced to

$$\begin{Bmatrix} f_{1y} \\ f_{3y} \end{Bmatrix} = \begin{bmatrix} k_1 & -k_1 \\ -k_1 & k_1 \end{bmatrix} \begin{Bmatrix} v_1 \\ v_3 \end{Bmatrix} \qquad (6.46)$$

but if one tries to solve this by premultiplying each side by the inverse stiffness matrix, one finds that K is singular! The bar is unrestrained and any force can displace it without distortion. This is known as *rigid body* motion.

Referring back to the truss, one must apply boundary conditions to prevent any rigid-body motion. By noting that $u_1 = v_1 = u_3 = 0$ and rearranging the equilibrium equation,

$$\begin{bmatrix} K_{11} & 0 \\ \hline 0 & 0 \end{bmatrix} \begin{Bmatrix} U_1 \\ \hline 0 \end{Bmatrix} = \begin{Bmatrix} F_1 \\ \hline 0 \end{Bmatrix} \qquad \text{where} \qquad U_1 = \begin{Bmatrix} u_2 \\ v_2 \\ v_3 \end{Bmatrix} \qquad \text{and} \qquad F_1 = \begin{Bmatrix} f_{x2} \\ f_{y2} \\ f_{y3} \end{Bmatrix} \qquad (6.47)$$

The submatrix K_{11} may now be inverted to obtain the displacements.

While pinned trusses may not seem to be related to electronic packaging problems, the formulation for the bar element, Equation (6.47), can be used to solve the example of a chip bonded to a substrate.

As a first approximation, only two elements will be used (see Figure 6-22). Due to symmetry conditions only half of the structure needs to be modeled; fixing one end of each element satisfies this requirement. The equilibrium equation for the two elements, ignoring for the moment the adhesive that connects them, is

$$\begin{Bmatrix} f_1 \\ f_2 \end{Bmatrix} = \begin{bmatrix} k_1 & 0 \\ 0 & k_2 \end{bmatrix} \begin{Bmatrix} u_1 \\ u_2 \end{Bmatrix} - \Delta Tl \begin{Bmatrix} k_1 & \alpha_1 \\ k_2 & \alpha_2 \end{Bmatrix} \qquad (6.48)$$

To include the effect of the adhesive, first note that $\tau = (u_i - u_j)G/t_0$, $f_1 = (u_2 - u_1)lG/2t_0$, and $f_2 = (u_1 - u_2)lG/2t_0$. These are the only external forces in this

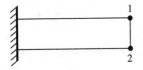

FIGURE 6-22
Two elements used to represent a chip bonded to a substrate.

problem, and they may be substituted into Equation (6.48), giving

$$\begin{bmatrix} k_1 + \beta & -\beta \\ -\beta & k_2 + \beta \end{bmatrix} \begin{Bmatrix} u_1 \\ u_2 \end{Bmatrix} = \Delta Tl \begin{Bmatrix} k_1 \alpha_1 \\ k_2 \alpha_2 \end{Bmatrix} \qquad \text{where } \beta = \frac{lG}{2t_0} \qquad (6.49)$$

Inverting K yields

$$K^{-1} = \frac{\begin{bmatrix} k_2 + \beta & \beta \\ \beta & k_1 + \beta \end{bmatrix}}{k_1 k_2 + \beta(k_1 + k_2)} \qquad (6.50)$$

and, multiplying (6.49) and (6.50), one obtains

$$\begin{Bmatrix} u_1 \\ u_2 \end{Bmatrix} = \frac{\Delta Tl}{k_1 k_2 + \beta(k_1 + k_2)} \begin{Bmatrix} k_1 k_2 \alpha_1 + k_1 \alpha_1 \beta + k_2 \alpha_2 \beta \\ k_1 k_2 \alpha_2 + k_1 \alpha_1 \beta + k_2 \alpha_2 \beta \end{Bmatrix} \qquad (6.51)$$

Since it is the shear stress τ that is of interest, the displacements of (6.51) can be substituted into the expression for τ. This gives

$$\tau = \frac{2\Delta T \beta(\alpha_1 - \alpha_2)}{1 + \beta\left[\left(\dfrac{1}{k_1}\right) + \left(\dfrac{1}{k_2}\right)\right]} \qquad (6.52)$$

where $\beta = lG/2t_0$ and $k_i = t_i E_i/l$, which can be compared directly with the previous results. The accuracy of the model can be considerably improved by using more elements. Table 6-1 compares the value of $\tau_{\max}$, determined using the finite element method, to the exact solution.

The foregoing example illustrates the first steps in any analysis using the finite element method: idealize the structure, making appropriate simplifications; identify the nodal displacements that completely describe its distortion (the degrees of freedom); and establish the equilibrium equations and solve them. To complete the analysis one must calculate the internal stresses using the appropriate stress-strain (constitutive) equations and interpret the results.

It is in the first and last steps of any analysis that the most difficulty lies. There are many computer programs available that make it simple to define the geometry of a problem. These then construct the matrices, solve the equations, and return stress or strain results. Little knowledge of mechanics is needed

TABLE 6.1
The effect of discretization on the accuracy of finite-elements results

Number of elements	$\tau_{\max}$	Percent error
1	2.3 MPa	+37
4	1.8	9
8	1.7	3

to run programs such as ADINA, ANSYS, CAEDS, or NASTRAN, but the results can be meaningless if good engineering judgment has not been applied to the problem formulation. Selecting the proper elements, choosing a reasonable element density, and defining realistic boundary conditions are critical.

SOME PACKAGING PROBLEMS

Formulating a model and interpreting the results are often the most difficult parts of a problem. Some examples from electronic packaging will show how symmetry, simplifying assumptions, and element density may be used to produce a tractable problem.

Consider a memory module constructed as follows: a ceramic substrate (discussed in Chapter 10) with a single chip attached to it by C4 solder joints, a boxlike aluminum cap that protects the chip, and an array of pins on the outer side of the ceramic that make the electrical and mechanical connections to a card. The cap is attached to the ceramic by an epoxy, which functions as a seal. A thermally conductive paste is sometimes included to transmit heat between the chip and the cap.

During operation the thermal mismatch between the ceramic and the cap causes each to deform. The coefficient of thermal expansion of ceramic is 6.5 ppm/°C while that of aluminum is 24 ppm/°C. This puts considerable stress on the epoxy holding the cap in place and causes the cap to bow upwards in the center. Motion also loads the chip-to-substrate solder connections through the paste. To limit this bowing and prevent an adhesion failure of the epoxy, the side walls of the cap had to be thinned, making it more compliant. With larger substrates it is also necessary to treat the surface of the cap to increase the epoxy adhesion.

Figure 6-23 shows the discretization of the model and the deformed geometry. Plate elements were used for the cap, ceramic substrate, and chip while beam elements were used to simulate the solder connections. Both of these element types can transmit axial and bending forces, although the plates have no rotational degrees of freedom. The thermal paste, which can transmit only compressive and tension loads, was modeled with bar elements.

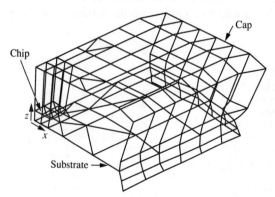

FIGURE 6-23
A finite element representation of one quarter of a 25 × 25 mm module.

Due to symmetry considerations only a quarter of the module was modeled. The substrate was fixed in the center, which corresponds to the point x, y, $z = 0$ in the figure, to prevent rigid-body motion. Since the plate elements have only three degrees of freedom it was only necessary to require that u, v, and $w = 0$ at that point. Symmetry demands that the slope of the deflection is zero along the x and y axes. Since neither plates nor bars can rotate (they have no rotational degrees of freedom), they meet this requirement. The nodes that lie on the beams along the x axis must be restrained so that only the rotations around the y axis are permitted. Similarly, along the y axis only x axis rotations are allowed. In addition, the beam and bar elements with $x = 0$ or $y = 0$ are split by the symmetry line and their thicknesses must be halved. For the elements at $x = 0$ and $y = 0$ the thicknesses must be quartered.

The same model was also used to study the flow of heat from the top surface of the chip, where the circuitry is located, to the cap through the paste. The secondary path through the chip and solder connections and into the ceramic was also included.

Another model was used to simulate handling load during manufacture of assembled printed circuit cards. Three square modules are attached to a circuit card by a 19 by 19 array of nail-headed pins brazed onto each module. The modules were attached to the card by inserting the pins into drilled holes and then filling the space that remains in each hole with solder. For the purpose of this model the pins were treated as if they were fixed at each end (all six degrees of freedom are constrained). Only a single module was modeled, and once again symmetry was employed to reduce the model's size [21].

Figure 6-24(a) shows the location of the forces applied by an equipment operator in three situations. The axial force acting on the pins was found to be greatest in load case III, at the corners of the modules. Two of these were particularly high, as Figure 6-24(b) shows. The figure also shows line data and the resultant forces from the finite element model. Experimental results using strain gages mounted on the pins confirmed the model. The three analyses agree very well.

The peak moment, shear, and axial loads predicted by the model were then used as input for a micromodel. Enlarging the head of the pin significantly reduced the stress in the brazing material. Solid elements were used here, as a three-dimensional model was necessary. This type of element, like the plate element, has only translational degrees of freedom.

A model for a more common type of package the plastic-leaded chip carrier (PLCC), is shown in Figure 6-25. Since the analyst was concerned with the behavior of the leads and not of the substrate, the elements are much smaller in the region of interest. By using a coarse discretization in some areas one can reduce the time needed to formulate a problem and save a great deal of computer time when solving the equilibrium equations.

Special provisions are included in some finite element codes to handle parametric variations. Figure 6-26 shows a drill bit used to drill the through-holes in circuit boards [22]. To obtain clean holes with a minimum of drill breakage (as discussed in Chapter 17) an engineer is faced with conflicting requirements: a

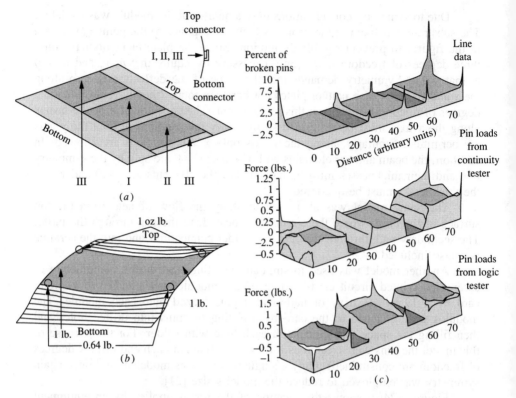

FIGURE 6-24
The macromodel for simulating handling of a card. The location of the applied forces for three handling cases is shown in (a). Handling case III produces the highest pin loads at the locations shown in (b). Line data and analytical pin loads are shown in (c).

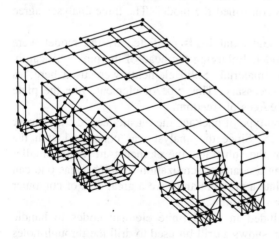

FIGURE 6-25
A plastic-leaded chip carrier.

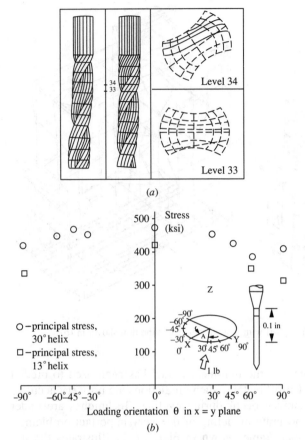

(a)

(b)

Loading orientation θ in x = y plane

O —principal stress, 30° helix
□ —principal stress, 13° helix

FIGURE 6-26
Bending stress in a twist drill. Side and cross-sectional views are shown in (a). The maximum principal stress for drills with 13- and 30-degree helix angles is plotted as a function of loading orientation (b).

deep flute is needed to transport debris out of the hole, but this flute weakens the structure. In this study the helix angle of the flute and its depth could be changed with only two commands, making it possible to rapidly determine the optimum drill geometry.

An example of a problem in which nonlinear material behavior is important relates to stresses in a surface-mounted solder joint. The stresses that arise in the solder during an isothermal mechanical test of a surface-mounted module can be calculated using linear-elastic and nonlinear constitutive equations. Since it is the solder that fails in these joints, a finite element model can be used to determine how the solder fillet's shape, the pin length, and pin diameter could be changed to decrease the stress.

For ductile materials the octahedral shear stress is often used for predicting failure for this problem. The equivalent and octahedral stresses are:

$$\sigma_{eq} = \sqrt{S_x^2 + S_y^3 - 2\gamma S_x S_y} \qquad \sigma_{oc} = \sqrt{S_x^2 + S_y^2 - S_x S_y} \qquad (6.53)$$

Manson uses octahedral shear stress and strain for predicting low-cycle fatigue

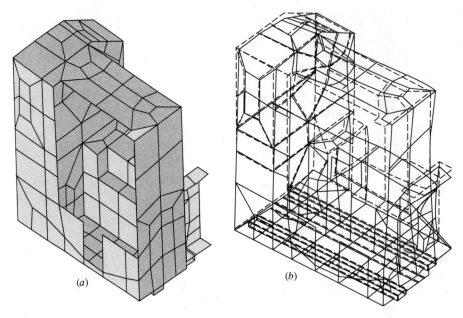

(a)   (b)

FIGURE 6-27
Dynamic analysis of a computer frame (a) with a single mode of vibration shown in (b).

life [15]. The stress calculated by the nonlinear model has been used to predict the fatigue life of this joint. Unfortunately, such predictions are not very accurate as both fatigue and creep damage occur during cycling. Techniques are under development to allow more accurate modeling of this very important problem.

The monocoque computer frame shown in Figure 6-27 illustrates the use of analysis to simulate dynamic stresses. During shipping, computers may be bumped or dropped, and hinged doors put a significant dynamic load on their supports when this occurs. By applying a transient acceleration to the base of the structure, weaknesses were identified and subsequently corrected.

CONCLUSIONS

This chapter describes a few applications of mechanics and mechanical engineering within the field of electronic packaging. Much remains to be done: new, less expensive designs are continually being sought; new materials are being introduced to improve reliability and electrical performance; and modeling and analytical techniques are being improved and used in the design stage to shorten development cycles.

REFERENCES

1. S. Timoshenko and J. N. Goodiev, *Theory of Elasticity*, New York: McGraw-Hill, 1951.
2. A. G. Green and W. Zerna, *Theoretical Elasticity*, Oxford: Clarendon Press, 1960.

3. Y. C. Fung, *Foundations of Solid Mechanics*, Englewood Cliffs, NJ: Prentice-Hall, 1965.
4. R. Roark, *Formulas for Stress and Strain*, New York: McGraw-Hill, 1965.
5. W. Chen and C. Nelson, "Thermal Stress in Bonded Joints," *IBM Journal of Research and Development* 23, no. 2, pp. 179–180 (1979).
6. C. R. Wylie, *Advanced Engineering Mathematics*, New York: McGraw-Hill, 1975.
7. P. M. Hall, *Proc. 34th Elect. Comp. Conf. IEEE*, pp. 107–116 (1984).
8. P. M. Hall, T. D. Dudderar, and J. F. Argyle, *IEEE Tr. Comp. Hybrids Mfg. Tech.* (1983).
9. P. M. Hall, *Solid State Tech.* 26, no. 3, pp. 103–107 (1987).
10. D. Stone, S. Hannula, and C. Li, *Proc. 35th Elect. Comp. Conf. IEEE*, pp. 46–51 (1985).
11. P. J. E. Forsyth, "A Two-Stage Process of Fatigue Crack Growth," Symposium on Crack Propagation, Cranfield, 1961.
12. W. Wood and R. Segall, "Softening of Cold-worked Metal by Alternating Strain," *J. Inst. Metals*, January 1958, pp. 225–228.
13. S. Manson, "Behavior of Materials under Conditions of Thermal Stress; Heat Transfer," *Symp. U. Mich. Eng. Res. Inst.*, pp. 9–75 (1953).
14. L. Coffin, "A Study of Cyclic-Thermal Stresses in a Ductile Metal," *Trans. ASME* 76, pp. 931–950 (1954).
15. S. Manson, *Thermal Stress and Low-Cycle Fatigue*, Melbourne, FL: Krieger, 1981.
16. A. Ostenfeld, *Die Deformationsmethode*, New York: Springer-Verlag, 1926.
17. W. Ritz, "Über eine neue Methode zur Lösung gewisser Variationsprobleme der mathematischen Physik," *Zeitschrift für Angewandte Mathematik und Mechanik* 135, pp. 1–61 (1908).
18. R. Clough, "The Finite Element in Plane Stress Analysis," Proc. 2nd ASCE Conf. Elec. Computation, Pittsburgh, September 1960.
19. K. J. Bathe, *Finite Element Procedures in Engineering Analysis*, Englewood Cliffs, NJ: Prentice-Hall, 1982.
20. R. Gallagher, *Finite Element Analysis*, Prentice-Hall, 1975.
21. P. Engel and C. K. Lim, "Finding the Stresses with Finite Elements," *Mechanical Engineering*, Vol. 108, No. 10, pp. 46–50, October 1986.
22. T. Niu and P. Chen, "Finite Element Analysis versus Experiments for Drills under Bending," Proc. SEM Spring Conference, June 1985, Las Vegas.

PROBLEMS

6.1. A two layer TAB film is made from polyimide attached to a copper film. Assuming that both materials are elastic, derive an expression for the direct tensile force in the copper (and direct compressive force in the polyimide), as a result of a temperature rise T degrees. Thickness, moduli of elasticity and coefficient of expansion should be denoted t, E, and α, respectively, with subscripts "p" for polyimide and "c" for copper.

6.2. In problem 1, and having found the moments and forces at the ends of the interfaces between polyimide and copper, calculate expressions for (a) the maximum resultant tensile stress in the copper, and (b) the maximum resultant compressive stress in the polyimide.

6.3. A lead leg in a surface mount chip carrier soldered to a printed circuit board takes up the thermal expansion between board and the chip carrier. If the leg can be simulated as a cantilever beam clamped at each end, calculate in terms of the bending stiffness (EI) of the leg and the relative movement (Δ) of each end, the shearing force on the solder at each end.

CHAPTER
7

ELECTRICAL
CONTACTS
AND
DESIGN

JOHN L. PIECHOTA

INTRODUCTION

In a large computing system there may be millions of interconnections, which may be a source of reliability problems in electronic packages. The interconnections may be fixed or separable as shown in Figure 7-1. Fixed interconnections such as solder joints, thermal compression bonds, or crimps are used to permanently join two parts, as discussed in Chapter 19. Separable interconnections, such as circuit card connectors and cables, are used when a circuit component may need to be removed or replaced. This chapter is concerned with separable interconnections used in electronic packaging. The subjects covered include electrical contact theory, resistance, force, geometry, materials, contamination, and wipe; friction, wear, and lubrication; packaging and process considerations; and future trends and challenges. Chapter 16 contains complementary discussions of noble metal plating, which is essential to high reliability of separable connectors.

A familiar separable interconnection is that of an electrical appliance involving a plug and a wall socket. As the plug is inserted into the socket, the metal on the plug rubs against a similar metal in the socket. A high frictional wiping action cleans the surface of oxides and thus provides for adequate electrical conduction. In this application, 120 V is available to break down any residual contamination

Fixed interconnections

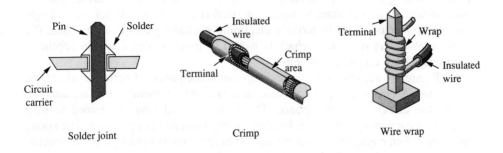

Solder joint   Crimp   Wire wrap

Separable interconnections

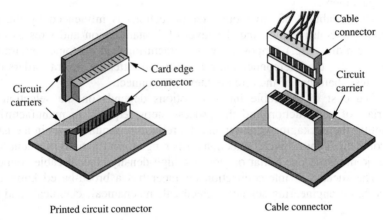

Printed circuit connector   Cable connector

FIGURE 7-1
Examples of fixed and separable connections.

left on the surfaces. In addition, the high friction loads of this connector allow the mating parts to be held in contact even when the cord is moved. This plug and its mating socket consume approximately 64 cubic centimeters of space in performing their function. A force of approximately 20 N is required to insert or withdraw this connector.

Connections within a microelectronic system must be able to do the same job. The problem is that instead of a single connection, as in the case of a plug and wall socket, a computer component can involve hundreds of interconnections, all of which must be equally reliable. The number of contacts to a card, for example, is determined by the circuits the card contains (Chapter 2). There is typically less than 5 V available to break down the residual contamination left on the surfaces of electronic connectors. Therefore, one of the keys to good connector design is to use noble metal-to-noble metal contacts that resist corrosion and stay clean. The noble metal at the contact interface must undergo local yielding along with sufficient motion to move nonconducting debris out of the contact area. The force required to insert a component having hundreds of these contacts should ideally be no greater than 100 Newtons. Therefore, the control of surface features and lubrication is important to minimize friction as well as contact wear. High-density connectors used in today's electronic packaging consume less than 0.5 cubic centimeters of space, which is over 125 times as dense as the appliance plug-and-socket connector system.

Figure 7-2 depicts some of the components used in constructing an electronic connector. A typical connector includes a spring and pin member that are attached to printed circuit or chip carriers. These springs and pins are joined to their respective carriers by soldering, brazing, or high mechanical pressure. The spring member, when engaged, applies a force to the pin member through which electric current passes. One or both members (spring and pin) are usually enclosed in a plastic housing. This housing assists in retaining and guiding the spring and pin for engagement.

The reliability of separable interconnections is influenced by the design, materials of construction, and the level of contamination and stresses (temperature, humidity, etc.) imposed by manufacturing and customer environments. Consequently, the interconnection reliability can often be several orders of magnitude less than that of the circuits they interconnect.

The cost of separable interconnections depends on the selection of the materials of construction and the process steps required in manufacturing and assembly; the packaging density; and the reliability requirements in a customer's environment. Typical interconnection costs range from fractions of a cent for fixed solder joints up to one dollar or more for high-density coaxial cable connections.

The successful interconnection engineer has a broad-based knowledge of all the basic engineering sciences (electrical, mechanical, chemical, and materi-

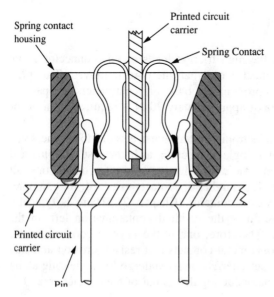

Spring contact housing

Printed circuit carrier

Spring Contact

Printed circuit carrier

Pin

FIGURE 7-2
Components of a separable connector.

al) along with experience in manufacturing processes and quality control. This requirement is necessitated by the multitude of tradeoffs (function, cost, reliability) made in a product development cycle.

Separable interconnects are designed to join two packaging members when:

1. One or both of those members may need to be removed and replaced,
2. Features need to be added to the electronic package, or
3. The use of fixed interconnections is not technically feasible.

In 1985 there were over 50 U.S. suppliers of separable connectors for the commercial and military electronic packaging industry. These suppliers offered over 600 connector product lines having a total market value of over $4.5 billion.

The design of reliable separable interconnections is the focus of this chapter—in particular, those interconnects having noble metal contact surfaces that operate typically below 5 V. A review is made of the fundamental principles involved. In-depth treatments of contact theory and related contact phenomena can be found in the references.

ELECTRICAL CONTACT

Electrical contact is the junction between two separable conductive members through which electric current passes.

When the end of a clean copper rod is pressed against a clean flat copper surface with sufficient pressure, one might expect the area of mechanical contact between these two surfaces to be that of the diameter of the rod. Such is not the case. This area is referred to as the *apparent contact area*, which is significantly larger than the area of mechanical (load-bearing) contact. The mechanical contact area is small because metal surfaces are not perfectly smooth, and therefore, when clamped together, they make asperitic contact as shown in Figure 7-3.

Electrical contact is made within these small load-bearing asperitic areas. Good electrical contact is hampered by the difficulties in maintaining clean contact surfaces. Even a contact made of high-purity gold, which is oxide resistant, has thin contaminant films deposited on its surface from ambient air in a relatively short period of time.

When two conductive surfaces mate under a sufficient load (P), local regions of metal-to-metal contact are made. These regions are commonly termed *A-spots*. The number, density, and size of these A-spots vary depending on the load applied, surface hardness, surface geometry, and the physical characteristics of oxide and contaminant films present on the surfaces. The sum of A-spot areas is the effective electrical contact area, hereafter referred to as the *effective contact area*. In general, as the applied load is increased, the extent of elastic and plastic deformation in asperities increases. Surface films on the tips of each asperity are fractured and new A-spots are generated, leading to an increase in the effective contact area (Figure 7-3). The effective contact area is an important factor that determines contact resistance.

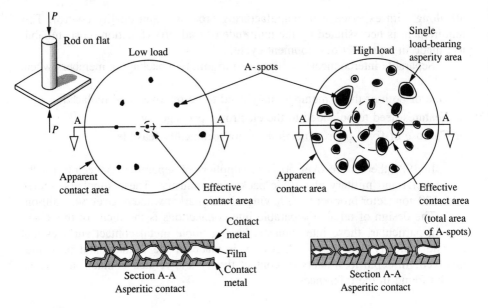

FIGURE 7-3
A-spot growth as a function of load P.

CONTACT RESISTANCE

Contact resistance (CR) is defined, from a measurement viewpoint, as the sum of the bulk resistances added in series and the resistances of parallel paths of constriction and film resistances as shown in Figure 7-4.

Bulk resistance is affected by the material's resistivity and its geometric shape. *Constriction resistance* results from the current flow being constricted through small conducting areas within A-spots. *Film resistance* is created by very thin (20 Å or less) oxide, sulfide, chloride, or other tarnishing layers present at the contact interface. These thin films conduct electrons by means of the tunneling effect. The effective mechanical contact area from a load-bearing standpoint is very small indeed, but the electrical conducting area within it is even smaller due to constriction and film resistances.

The objective of a stable contact design is to maximize the effective contact area and thereby promote stable resistance at the contact interface. To accomplish this goal, one must understand and select the optimum physical design parameters, one of which is contact force.

CONTACT FORCE

Contact force is that load P normal to the plane of contact (also referred to as normal force) that produces film fracture and penetration to allow asperitic metal-

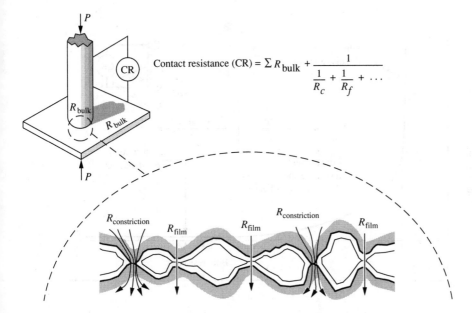

Contact resistance (CR) $= \Sigma R_{bulk} + \dfrac{1}{\dfrac{1}{R_c} + \dfrac{1}{R_f} + \cdots}$

FIGURE 7-4
Contact resistance as a function of bulk, constriction, and film resistances.

to-metal contact. Electrical contact designs are usually based on a characteristic curve for contact resistance as a function of contact force, as shown in Figure 7-5. Such a curve depends also on the resistivity of the contact materials; the thickness, hardness, and surface roughness of the contact metals; the geometry of the two mating surfaces; and, most importantly, the nature and thickness of the oxide and contaminant films present on the surfaces prior to engagement. While a minimum of 50 g contact force is recommended for a clean gold-on-gold contact system, the presence of a tough and tenacious film on either gold surface could require contact forces beyond 500 g to attain stability. In cases where a contaminant film approaches the hardness and ductility of the contact metallization, it could be extremely difficult to create enough contact force to fracture and penetrate the film.

A contact design that operates in the unstable portion of the curve in Figure 7-5 could be susceptible to contact separation due to packaging vibration or shock as well as corrosion separation. Corrosion separation occurs when corrosion products in the vicinity of the contact area migrate over time into the interface—resulting in lifting and an electrically open contact.

There is an upper bound for contact force that is dependent upon the load-bearing properties of the metals during sliding. Too much contact force can result in a loss of lubricity at asperity interfaces such that welding and tearing of contact surfaces will occur. This subject is covered in more detail in the friction, wear, and lubrication section of this chapter.

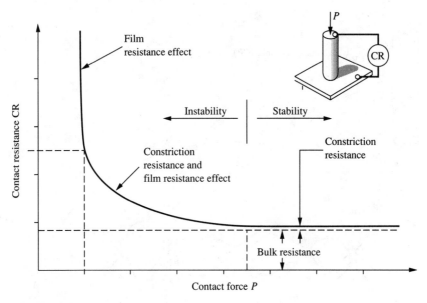

FIGURE 7-5
Contact resistance as a function of contact force.

CONTACT GEOMETRY

It has been established that electrical contact is made through A-spots and that contact force can influence the growth of A-spots in size and in number density. The maximization of the effective contact area also depends on the number of contact members (redundancy) and the geometry of the contacting surfaces.

Redundant contact points are used to increase the effective contact area and reduce the probability of a discontinuity occurring simultaneously at each point of contact. The pin and dual-tine spring receptacle in Figure 7-6 is an example of a redundant contact design. Redundancy can have its drawbacks when used in sliding contact designs. The spreading and sliding friction forces that occur when engaging a redundant sliding contact design are related to the total insertion force (Figure 7-6). Often, the contact force is minimized to reduce the frictional load of a redundant contact design. In doing so, the ability to penetrate or displace contaminate films or particulates may be reduced to the level where a single point contact design having a higher contact force may provide better electrical stability.

The geometric shape of each contact member can have a key influence on the effective contact area by increasing the compressive stress level and, thereby, the reliability of an electrical connector. The surfaces of most electrical contacts are normally plated or clad with noble metals such as gold, palladium, or their alloys to inhibit tarnishing. These materials have relatively low yield strength (for example, the yield strength of tempered steel is at least five times greater than that

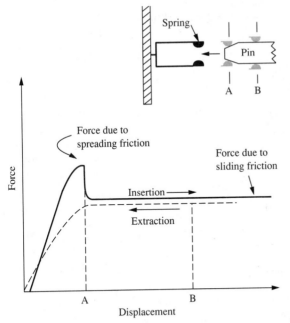

FIGURE 7-6
Typical profile of sliding connector insertion and extraction forces.

of pure gold). The asperitic contact sites of pure gold undergo plastic deformation at compressive stress levels greater than 50 MPa. The effective contact area can be very small with two flat contact surfaces under moderate load. However, if one member is changed to a hard needle-shaped profile and is mated to a very soft flat surface under that same moderate load, the contact area will become quite large.

High contact stress geometries, such as the one described above, help in maximizing the effective contact area for a given load condition. Figure 7-7 shows a comparison of the maximum compressive stress levels generated for four different contact geometries under fixed load and material conditions. These stress values are calculated by assuming elastic behavior of the metals along with an assumption of perfectly smooth surfaces (no asperities). Figure 7-7 shows that crossed-rod and sphere-on-flat geometries can create a very high localized stress at a contact interface.

In summary, redundancy of contacts and the geometric profile of the contact interface can be used to maximize the effective contact area of a connector. The proper selection of materials also aids in the above process.

MATERIALS

There are two basic parts in traditional connector systems—a plug member and a receptacle member. One of these members is normally designed to provide contact load through stored elastic energy. Each connector member is typically

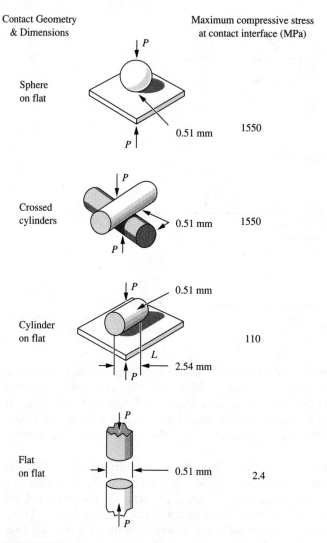

Contact Geometry & Dimensions | Maximum compressive stress at contact interface (MPa)

Sphere on flat — 0.51 mm — 1550

Crossed cylinders — 0.51 mm — 1550

Cylinder on flat — 0.51 mm — L — 2.54 mm — 110

Flat on flat — 0.51 mm — 2.4

FIGURE 7-7
Effect of contact geometry on resultant compressive stress for gold-on-gold contact metallurgy where load $P = 50$ g, the modulus of elasticity $E = 82.74$ MPa, and Poisson's ratio $v = 0.42$.

made up of a base material, a barrier material, and a contact material. Table 7.1 lists the physical properties of some typical materials used in the construction of connectors.

Base materials are selected with emphasis on the following considerations:

Conductivity—to minimize bulk resistance effects
Ductility—to aid in forming the part
Yield strength—to maximize beam deflections in the elastic range

Creep strength—to resist load relaxation with time

Hardness—to reduce the wear of the contact metallization

Base materials are typically manufactured by stamping, forming, and/or extrusion to produce the desired shape and function. The plug member (also referred to as a pin) is typically made of brass, phosphor bronze, or nickel, and its shape is produced by either stamping or forming from sheet stock or by a wire extrusion and forming process.

The receptacle member (also referred to as a spring) is typically made of alloys of copper such as beryllium copper and phosphor bronze. These metals

TABLE 7.1
Typical connector materials and their physical properties

Application	Material	Electrical conductivity (% IACS*)	Density, gm/cm³	Average hardness, kg/mm²	Percent elongation (in 50 mm)	Yield stress, MPa
Base materials	Phosphor bronze C510 alloy, spring temper	15.8	8.86	230	2–7	509
	Cu-9Ni-2Sn C725 alloy, spring temper	11.0	8.89	180	1	386
	Beryllium copper C170 alloy 165 1/2 HM temper	20.0	8.26	280	12–18	577
	Beryllium copper C172 alloy 25 1/2 HT (2hr @ 600°F)	22.0	8.25	380	2–5	931
Barrier materials	Electroplated sulfamate Ni (99.99+)	20.1	8.95	425	8–24	<500
	"A" nickel 205 wrought annealed (99.5)	18.2	8.89	100	35–55	<138
Contact materials	Electroplated "pure" gold (99.999+)	78.5	19.3	70		
	Electroplated gold w/ 0.8 Ni (99.1+)	41.2	17.4	180		
	Electroplated palladium (99.99+)	17.5	12.0	300		

* International Annealed Copper Standard

Note: The properties listed represent a compilation of values from various independent sources. The exact values and their ranges for a given material may vary depending upon the source of supply and processing conditions.

represent a compromise between good electrical conductivity and good mechanical spring and forming properties.

Barrier materials are used primarily to retard diffusion between the base and contact materials. They can also play a role in reducing the wear of the contact material, depending upon their hardness and thickness. Barrier materials are typically applied by plating or cladding to the base metal. Nickel is a typical barrier metal used in connectors.

Contact materials are selected with emphasis on the following:

• Nobility—resistance to environmental attack and adsorption of organic molecules
• Hardness—to resist wear
• Ductility—to maximize A-spot contact area and also resist cracking or spalling

While high-purity soft gold is ideal from a nobility and ductility standpoint, it lacks adequate wear properties for contact systems that undergo multiple matings under high contact stress conditions. Alloys of noble metals are commonly used because of their hardness and wear resistance. Contact materials are typically deposited by plating or cladding to the submetal structure.

By selecting a combination of contact metals with appropriate hardness, one can minimize wear on one member while maximizing the effective contact area of the total contact system. The deposition of a thin layer of soft gold over alloys of gold or palladium on one of the contact members is an example of this approach. Refer to Chapter 16 for further details on the plating of electronic connectors.

The selection of the material systems for a separable connector is critical to its overall reliability and cost. The material selection process requires synergism and tradeoffs between:

1. The desired performance properties such as bulk resistance, contact resistance, and wear life
2. The ability to be manufactured at a low cost
3. The ability to withstand the mechanical and environmental stresses of the worst-case customer environment.

Removing contamination from a contact interface is critical to the performance of a connector in a customer's environment.

CONTAMINATION AND WIPE

The primary objective of a reliable connector design should be to make it clean and keep it clean. Hermetically sealing connectors from the effects of gases and particulates in the air stream would be expensive and impractical in most cases. Many sources of film and particulate contamination of a connector's contact surfaces exist during the manufacture, storage, and operation of an electronic package. Examples of such sources are:

- Ambient air—dust, ash, sand, metals, carbon, tar, oxides, nitrates, sulfides, chlorides, and silicates
- Manufacturing process—metal chips, cotton fibers, oil mist, cement dust, paper dust, paint spray, glass fibers, solders, fluxes, and epoxy films
- People—hair, perspiration, skin, food, drinks, and clothing fibers
- Machines—paper dust, inks, ribbon lint, plastic chips, and metal filings

While cleaning steps are integrated in the connector manufacturing processes to minimize the effects of contamination, the escape of less than clean contact surfaces to a customer's site must be considered in the design of a contact system.

Because of the potential presence of films and particulates on the contact surface, simple make-or-break connector designs would be prone to intermittent failures—especially low-energy designs. Therefore *wipe* is normally built into the design of separable connectors. Wipe is defined as the relative translation under load of two contact surfaces. By sliding one surface (termed the rider) relative to another surface, a plowing action results that helps to fracture and displace surface contaminants. Figure 7-8 shows two typical methods of inducing wipe into a connector design—edge translation and impingement.

Edge translation wipe produces very high contact stresses (a function of mating angles at edge, contact force, contact materials, and contact geometries) as the spring member climbs over the edge and onto the surface of the opposing member. The wear of the contact metallization can be severe with this design if the mating angles at the edge are **not properly** controlled. The force required to insert this connector is higher than **with** impingement wipe.

Impingement wipe produces lower contact stresses (a function of contact force, materials, and geometries) during impact and translation across the surface of the opposing member. With an impingement wipe design, more care is required

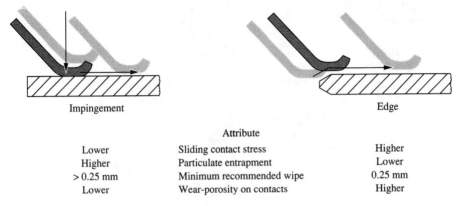

Impingement		Edge
	Attribute	
Lower	Sliding contact stress	Higher
Higher	Particulate entrapment	Lower
> 0.25 mm	Minimum recommended wipe	0.25 mm
Lower	Wear-porosity on contacts	Higher

FIGURE 7-8
Attributes of impingement versus edge wipe.

in keeping the contact surfaces free of large particulates, especially at the point of initial contact.

The extent of wipe required is largely dependent upon the design of a connector system (contact force, physical geometry, contact materials, etc.) as well as the types of contamination (thickness, morphology, etc.). In designing wipe into a sliding contact system, wear of the contact metals results from the relative motion of contact surfaces while under load. The subject of friction, lubrication, and wear becomes a necessary consideration.

FRICTION, LUBRICATION, AND WEAR

Imagine trying to engage 40 appliance plugs into their wall sockets all at the same time. It could take a force at least equivalent to the weight of a grown man to accomplish this task. This same force is required to engage 1800 electrical contacts at once in the highest density connector currently being manufactured. This force would be likely to bend and possibly break other components in the electronic packaging structure if special precautions were not taken. Therefore, the control of friction is very important in connector designs having large quantities of contacts.

In sliding two contact surfaces against each other, complex mechanical and metallurgical changes result in a change in state of those surfaces. Figure 7-9 depicts a simplified model of some of those changes in state.[1] A hemispherical

---

[1] The stresses produced at sliding contact interfaces are much more complex than depicted in this simplified model. In actuality, asperities are making contact and these asperities undergo elastic as well as plastic deformation—resulting in triaxial (three-dimensional) stress conditions being produced on and beneath the surfaces of the contact materials.

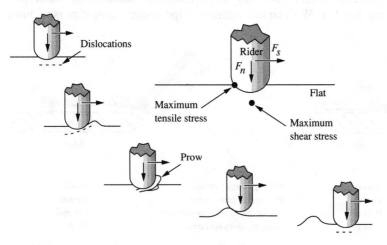

FIGURE 7-9
Sliding contact stresses and wear.

rod (rider), made of hard material, is depicted making contact with a flat surface made of a softer material. The rider, under load F_n, then moves across the surface with a resulting frictional load F_s parallel to the plane of the flat surface. As the rider makes contact with the flat surface, points of maximum tensile stress (at the trailing edge of the rider) and maximum shear stress (several atomic layers beneath the tip of the rider) are developed. The compression of the flat surface created by the rider's contact force results in dislocations being generated in the subsurface atomic structure of the flat contact material. These dislocations, by virtue of the translating shear loads of the rider, can work their way to the surface to produce a *prow* of work-hardened material at the leading edge of the rider. This prow can adhere to the rider, due to adhesive bonding of the two contact metals, and then be reattached to the flat surface and overridden. The degree of make-and-break prow formation will depend on such factors as contact material selection, surface roughnesses, lubricity at the surfaces of interacting asperities, and the applied contact force.

The translation of a rider across a flat surface can result in material transfer, the work hardening of the contact surfaces, and also the creation of wear debris. These effects give rise to a change in physical properties as well as to the topography of the contact surfaces. A photo of the wear track produced by such a rider on a flat surface is shown in Figure 7-10. Note the generation of a prow within the wear track (*a*) and the plastic deformation and compression of the gold-plated asperities inside the wear track (*b*) as compared to the undeformed asperities outside the wear track (*c*). The friction, wear, and lubricity of a sliding contact system must be predictable and controlled to prevent catastrophic galling and the resultant exposure of submetal layers. *Galling* is defined as the advanced stages of adhesive wear that tear up a surface.

Since the submetal layers are normally made of non-noble metals, the galling action allows corrosive attack around the points of contact and generates hard particles along with corrosion products. Both conditions contribute to unstable contact resistance.

An objective in a good connector design is to achieve a burnishing wear state, as opposed to galling. *Burnishing wear* is defined as the compacting and smoothing of one surface by rubbing another surface against it. Figure 7-11 shows some of the conditions under which burnishing and galling mechanisms are favored. Figure 7-12 shows a plot of wear depth as a function of the number of insertions made for the two different mechanisms. The advantage of burnishing wear is apparent. To achieve conditions that favor burnishing wear, a contact lubricant can be applied to one or both mating members. Contact surfaces have very thin films on their surfaces that act as boundary lubricants during the initial onset of sliding. These films are a natural result of air pollutants or, in some cases, result from organic impurities codeposited in the surface metallization process. But in the process of repeatedly sliding two surfaces against each other, these films are quickly removed, exposing bare metal-to-metal and leading to high adhesive wear. Therefore, a supplemental boundary lubricant is often required.

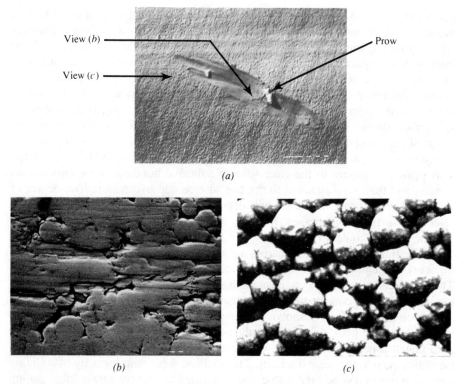

FIGURE 7-10
Wear track produced by a hemispherical rider on a gold-plated flat surface.

A supplemental lubricant helps reduce sliding friction forces, reduces adhesive transfer of materials, and, depending upon its chemistry, can also inhibit oxidation, tarnishing, and corrosion at the interface. Contact lubricants are normally highly refined natural or synthetic oils with good wetting properties coupled with low vapor pressure and thermal stability under high service temperatures. OS-124[2], a mixed isomeric five-ring polyphenylether, is one example of a synthetic oil used as a contact lubricant. The resistivity of a contact lubricant is normally very high, but the pressure of asperitic contact results in its displacement from the A-spot areas. Therefore, most contact lubricants produce a negligible film resistance effect.

One of the keys to good performance of a boundary lubricant is its cohesion to asperitic surface irregularities. Oils can "puddle" on a surface and therefore tend to reside in the valleys of an asperitic surface (refer to Figure 7-13(a)). This puddling effect can result in very high adhesive forces during the initial mating

---

[2] Trade name of contact lubricant produced by the Monsanto Corporation.

Wear mechanism	Conditions under which they are favored	Wear track showing burnishing wear
Burnishing: low adhesive transfer	• Optimum contact force • Good lubricity • Dissimilar contact metallurgies • Clean and smooth surfaces	
Galling: high adhesive transfer	• Abnormally high contact force • Poor lubricity • Similar contact metallurgies • Rough and contaminated surfaces	Wear track showing galling wear 
Fretting	• Induced by micromotion • Noble metal: formation of frictional polymer and insulators • Non-noble metal: formation of tenacious oxides and insulators	Wear spot showing fretting wear 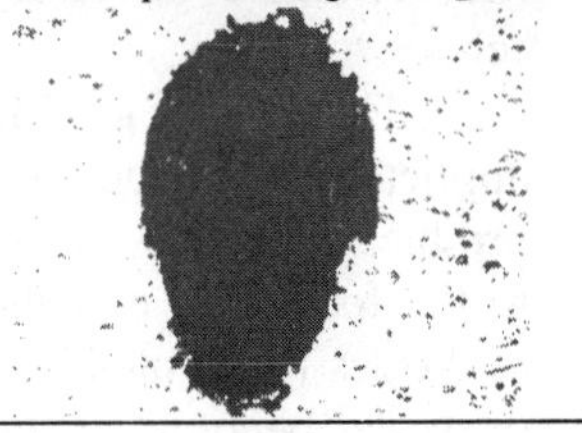

FIGURE 7-11
Typical connector wear mechanisms and conditions under which they are favored.

of a fresh (unburnished) contact surface. The key to solving this condition is to assure the meniscus height of the lubricant exceeds the height of the asperities by controlling the contact surface finish (Figure 7-13(*b*)).

A contact lubricant can also increase friction and wear by acting as a particulate collector while in service. Therefore, its advantages as a mechanical

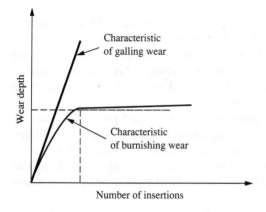

FIGURE 7-12
Wear depth as a function of the number of connector insertions for burnishing versus galling wear conditions.

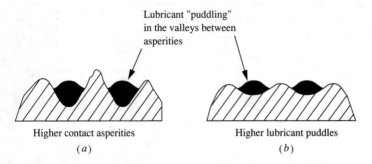

FIGURE 7-13
Contact lubricant puddling and contact surface finish.

boundary friction reducing agent must be considered in terms of the rated insertion life, tolerable insertion forces, and the expected particulate exposure during service.

The cost of gold has led to the use of less noble metals in some connector applications. In such applications, cyclic micromotion of a contact interface can produce another wear mechanism commonly termed fretting. Fretting wear leads to the formation of adhered frictional polymers (with metals such as palladium) or oxides (with metals such as tin-lead) at the contact interface. These polymers or oxides become insulators at the contact interface and result in unstable contact resistance. Fretting wear is not to be confused with fritting, which is a term applied to the electrical breakdown of tarnish films on contact surfaces that results in the formation of molten metal bridges through the film.

PACKAGING AND PROCESS CONSIDERATIONS

The performance of a connector (i.e., contact resistance stability) depends also on other design considerations. For example, the burnished wear state requires that the dimensional tolerances of the two contact members be controlled so that subsequent engagements mate in approximately the same wear track to achieve the desired results.

A contact metallization should be in a pore-free state. Initial porosity, depending on size and number density, can influence the rate of corrosion attack of the sublayer metallization. Corrosion products thus produced may over time migrate into the contact zone. It is important to minimize this type of corrosion reaction by reducing the allowed number of pore risk sites and their proximity to the contact area through good quality control.

The ability of two connecting elements to hold a fixed position to prevent disturbance of the contact interface is also important. Reliance upon friction forces alone is not adequate because the expected influences of shock and vibration during service can cause a contact system to "rock up" on contaminant films and debris in the contact area. A good design practice is to include auxiliary positive retention features into the connector system that will allow both strain relief and the prevention of macromotion of the contact interface. Positive retention prevents accidental disengagement of the connector during service and also prevents movement onto contamination or corrosion products in the vicinity of the contact area.

Poor dimensional control of the parts of a connector system is often responsible for engagement and mating in an undesired manner. Under the worst tolerance conditions, the two halves of a contact system may "stub" rather than engage, resulting in overstressing of the base material and breakage of the weakest member (see Figure 7-14). A connector might also mate on rough edges rather than the desired contact area. Therefore, the dimensional tolerances of the components and their assembly are critical to a connector's performance.

As an edge connector system is initially engaged, as shown in Figure 7-6, the friction is higher during the spreading phase of contact insertion. High friction can result in the potential for wear through the contact metallization as well as wear debris in this zone of the wear track. A good connector design will provide adequate isolation of the final contact point from these wear-through risk sites.

Most connectors use molded thermoplastic or thermosetting polymers to retain and position the contact members. Frequently, these polymers are blended with glass or fibers to enhance their mechanical properties. The composition and homogeneity of these blended polymers influence their chemical, mechanical, and electrical properties. The properties of such polymers are defined in a material supplier's literature by the use of molded test samples. But the physical shape and the melt flow patterns of a plastic connector component often vary greatly from molded test samples. Therefore, to accurately predict physical properties of a

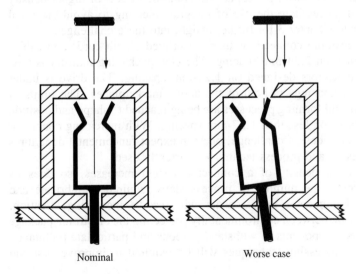

Nominal   Worse case

FIGURE 7-14
Nominal and worst-case contact alignments.

molded connector component, one should empirically characterize physical parts (impact strengths, flexural strength, solvent resistance, etc.) to assess product design and application safety factors.

Plastic polymers are often thermally and chemically stressed by the circuit packaging manufacturing process. For example, in soldering a connector to a printed circuit vehicle, the plastic body can undergo temperature changes of over 215°C. The connector body may dimensionally shrink or warp as a result of this thermal cycling. Thermal cycling can sometimes loosen the bond between the glass fibers blended into a polymer resin, resulting in ionic material from flux or solvent cleaning operations, for example, penetrating into the lattice of the polymer and thereby reducing the insulation resistance properties of the connector. It is a good practice to specify these process parameters during the connector design to assure its selected material system and physical properties remain stable through the manufacturing environment.

FUTURE TRENDS AND CHALLENGES

The basic design principles used in electrical connectors have not changed appreciably over the past 50 years. Improvements in such areas as materials and contact geometry have come about due to requirements for higher-density, higher-reliability, and lower-cost packages. The traditional stamped, formed, and plated electrical contacts used today appear crude in light of the advances made at the chip level over the past 20 years. The electrical connector is now a bottleneck to signal propagation and fidelity in an electronic package. The mating connector's electrical conductive path must be forced to shorter lengths. As signal processing speeds continue to increase, the capacitive coupling effects of higher-density contact designs will necessitate the use of integral shielding to avoid electrical interference. Refer to Chapter 3 for further insight into these challenges.

Today, the separable connections used on printed circuit cards and cables are typically designed on 2.54-mm spacing. The chip packaging industry is now using pluggable connectors designed on 1.27-mm spacing. The drive is under way for even tighter spacings in both of the above marketplaces. The limits of traditional stamping and forming processes are being reached to shape and position mechanical connectors on less than 1.27-mm spacing. With increasing connector densities being a future design challenge, one can expect fundamental departures from traditional design and process techniques currently used.

Typically, as the density of a connector system increases, so does its cost. To reduce costs, the number of process steps required to fabricate and attach an electrical connector to its carrier will receive increasing attention. Tradeoffs will be made between environmentally protecting the electronic package versus designing its components to withstand gaseous and particulate pollutants. New materials and processing techniques will be required to improve cost and performance.

The ideal separable connector of the future would utilize a common material system to join chips, chip carriers, and printed circuit carriers to one another. A challenge for the future will be the invention of this ideal material, along with its supporting design, which would have the following properties:

1. Require little if any force to engage or disengage
2. Have equivalent apparent and effective contact areas
3. Have material properties and packaging that resist environmental attack
4. Exhibit electrical properties that are practically transparent to the electrical network
5. Be capable of being packaged on 0.25-mm spacings
6. Cost less than 2 cents per connection

REFERENCES

1. F. Reid and W. Goldie, *Gold Plating Technology*, Ayr, Scotland: Electrochemical Publications, 1974.
2. R. Roark and W. Warren, *Formulas for Stress and Strain*, New York: McGraw-Hill, 1975.
3. R. Holm, *Electric Contacts*, New York/Berlin/Heidelberg: Springer-Verlag, 1981.
4. Proceedings of the Holm Conference on Electrical Contacts: Illinois Institute of Technology Dept. of Electrical Engineering, Chicago, Ill., 1967 to date.
5. W. Safranek, *The Properties of Electrodeposited Metals and Alloys*, American Electroplaters & Surface Finishers Society, New York, Elsevier, 1974.
6. D. Tabor, *The Hardness of Metals*, Oxford, England: The Clarendon Press, 1951 (Ann Arbor, Mich.: University Microfilms International, 1985).
7. F. Bowden and D. Tabor, *The Friction and Lubrication of Solids*, vols. 1 and 2, London, England: Oxford University Press, 1964.
8. M. Peterson and W. Winer, *Wear Control Handbook*, New York: ASME, 1980.
9. K. Mittal, *Surface Contamination*, Vols. 1 and 2, New York: Plenum Publishing Corporation, 1979.

PROBLEMS

7.1. What type of metallurgical wear · ιst influences galling of two contacting surfaces?

7.2. Name at least four factors whiιn promote galling wear at the interface of separable electrical contacts.

7.3. What is the maximum compressive stress, assuming elastic behavior of materials, generated by the following connector design?

> Cylinder on cylinder with axes at right angles to one another.
> Cylinder 1 plated with gold and having a diameter of 1.02 mm.
> Cylinder 2 plated with palladium and having a diameter of 0.51 mm.
> Applied contact force (normal load) of 50 g.

Assume: Poisson's ratio of 0.42 for gold and 0.33 for palladium; elastic modulus of 82.74 MPa for gold and 124.11 MPa for palladium.

7.4. What is the maximum shear stress created at the gold contact interface in Problem 7.3?

7.5. If the gold plating starts to yield at a shear stress of 55.16 MPa, what factor of safety is built into the contact design with respect to your answer to Problem 7.4?

CHAPTER
8

FIBER
OPTIC
LINK
PACKAGING

KISHEN N. KAPUR

INTRODUCTION

A glass fiber may be considered analogous to a copper wire; both carry energy, but one is optical, while the other is electrical. Glass fiber is not usually used to carry high energy (except in some applications, such as surgery), due to low efficiencies of coupling energy into and out of the fiber. Therefore, this chapter will not discuss high-energy transmission through glass fiber. In analogy with electrical conductors, we proceed from a plain wire to a transmission line (or a coaxial cable) and then to a waveguide as the frequency (or data rate of information) increases. A thin optical fiber (50 μm in diameter) actually has the bandwidth of a multigigahertz waveguide, and 100 to 1000 times less attenuation than a coaxial cable.

In the last decade fiber optic communication has grown at a rapid rate and taken over a significant portion of the telecommunication market [1]. The reason for this explosive growth is economics. It is cheaper to send information over long distances on optical fiber than on coaxial cable, due to the lower cost of fiber per meter, at high frequencies or data rates. This growth has been fueled mainly by advances in two technologies. Glass fibers have steadily improved in attenuation characteristics (see Figure 8-1) so that repeaterless links of greater than 100 km are

216

A – Low-loss high-frequency electrical cable
B – Cable television rigid (7/8" diameter) coaxial cable
C – Multi-mode optical fiber for 0.8 μm wavelength
D – Multi-mode optical fiber for 1.3 μm wavelength
E – Single-mode optical fiber for 1.3 μm wavelength
F – Single-mode optical fiber for 1.55 μm wavelength

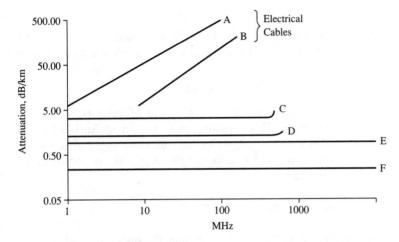

FIGURE 8-1
Optical and electrical cable characteristics.

possible today. Also, great strides have been made in semiconductor technology; in particular, advances in III-V semiconductor technology have improved the reliability of semiconductor laser diodes. However, at high frequencies the cost of launching modulated light into the fiber and detecting modulated light from the fiber is still higher than sending/receiving electrical signals over a coaxial cable. Thus, for shorter distances coaxial cable is still a viable alternative. As the cost of fiber optic transducers (transmitters and receivers) decreases with increased usage and development, the share of fiber optic communication networks will significantly increase. Fiber optic cables are also lighter in weight and smaller in diameter than comparable coaxial cables, thus giving significant handling advantage.

There is no radiation leakage from fiber optic cables, as compared to open wire transmission lines, thus reducing the EMI (electromagnetic interference) problem. Since no electrical connection is required between the transmitter and the receiver, no ground loops are created or unwanted noise coupled thru common ground as in electronic systems.

This chapter covers optical fundamentals, basic definitions, the propagation of light in the two basic fiber types (single mode and multimode), and basic connector types, including connector losses. It also covers the common electro-optical devices, frequently used packages, and electro-optical device-to-fiber coupling losses. Examples are shown, including the important topic of laser reliability and thermoelectric cooling, not treated elsewhere in the chapter. A look at the potential of fiber optic links completes this chapter.

OPTICS

Reflection and Refraction

When a ray of light hits the boundary between two optical materials with different optical indices, the ray is bent according to the following relationship (see Figure 8-2).

$$n_a \sin \Theta_a = n_b \sin \Theta_b$$

where n_x = refractive index of material $x = c/v$ (ratio of velocity of light in vacuum to that in the material). This relationship is known as Snell's law. If $n_a > n_b$, then there is an angle of incidence Θ_c at which the angle of refraction is 90 degrees; no light enters material b at this angle. Angle Θ_c is known as the *critical angle*. From Snell's law,

$$\text{Critical angle} = \Theta_c = \sin^{-1} \frac{n_b}{n_a}$$

Rays of light with angles of incidence greater than Θ_c undergo *total internal reflection* and stay within the material. This total internal reflection of light is key to transmission of light thru the fibers.

Absorption

When optical energy passes thru an optical medium, some of the energy is lost. This loss (attenuation) may be represented by the following relationship (see Figure 8-3).

$$\text{dB (decibel) loss} = 10 \log \frac{P_{\text{out}}}{P_{\text{in}}}$$

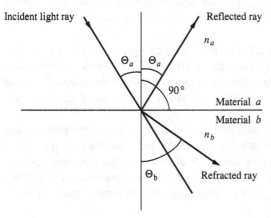

FIGURE 8-2
A ray of light incident on a material boundary.

Optical material

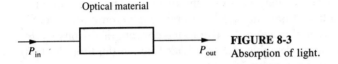

P_{in} P_{out}

FIGURE 8-3
Absorption of light.

Also, optical power is generally referenced to a 1-mW level and is given by

$$dBm = 10 \log P \qquad (P \text{ is in milliwatts.})$$

These units are used to characterize cable losses (attenuation), power outputs of light sources, and detector sensitivities.

Propagation

Let us assume that we have injected two light rays A and B into a medium a [see Figure 8-4(a)] and both Θ_A and Θ_B are greater than the critical angle. Both these rays will now propagate in the material (assumed to be a planar slab with rays in the plane of the paper and the plane of the slab orthogonal to it). These rays, if started simultaneously at the input, will arrive at the output at different times, due

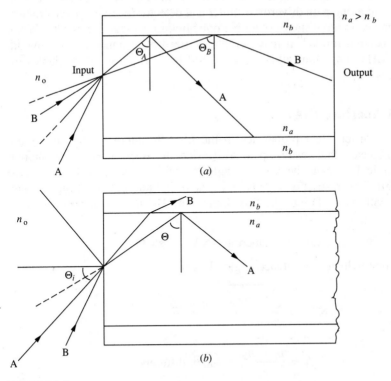

FIGURE 8-4
(a) Propagation of two light rays at different angles; (b) Angle of light acceptance into fiber.

to different path lengths taken as they pass thru the medium. The propagation of rays at various angles (allowed only at certain discrete angles on a microscopic scale by the laws of quantum physics) gives rise to the concept of various *modes* of light propagation. These modes give rise to a widening (in the time domain) of a pulsed light beam and the concept of *modal dispersion* [2].

Methods of compensating for this modal dispersion in fibers include doping the fiber to change the refractive index, making it lower near the reflective boundary. Lower refractive index near the boundary tends to speed up the light rays that are repeatedly reflected at the boundary and have longer optical paths than rays traveling along the center of the fiber. This principle is used in *graded index multimode fibers*.

If the light source is not monochromatic, energy at different wavelengths (with different refractive indices in the material) will have different delays in passing thru the material. This will cause a widening of an input light pulse as it travels thru an optical fiber. This phenomenon is known as material (chromatic) dispersion (see Reference [3] for calculation).

An optical link may be attenuation limited (not enough energy in the received signal to give a good enough signal/noise ratio for the required bit error rate), or it may be dispersion limited (when it becomes difficult to distinguish between two closely spaced light pulses in time). In real life, both these effects are present, and along with driver/transmitter and detector/receiver characteristics, they limit the maximum distance of a fiber optic link for a given bandwidth. When a link is dispersion limited, it may be characterized by distance and bandwidth product (in MHz · km), thus allowing a tradeoff between these two variables for small changes.

Numerical Aperture (NA)

As discussed earlier, light propagates in the fiber by internal reflections. Any incident light beam that results in a ray (inside the fiber) making an internal incidence angle less than the critical angle Θ_c will escape to material b [see Figure 8-4(b)]. This restriction results in a cone of light acceptance from medium (usually air with $N_0 = 1$) into the fiber input end. The term *numerical aperture* is defined:

$$\text{Numerical aperture (NA)} = \sin \Theta$$

where Θ is one-half the acceptance angle. In general

$$\text{NA} = \frac{\sqrt{n_a^2 - n_b^2}}{n_0} \approx \frac{n_a \sqrt{2\Delta}}{n_0}$$

where

$$\Delta = \frac{n_a - n_b}{n_a} = \text{Index difference}$$

In the case of graded index fiber, where $n_a(r)$ is the refractive index at point r,

NA is given by

$$\mathrm{NA}(r) = \frac{\sqrt{n_a^2(r) - n_b^2}}{n_0}$$

CABLES

The absorption curve of currently available glass fiber is shown in Figure 8-5. Region A is useful, as the low-cost silicon photodetectors are available for this region, together with the ordinary low-cost gallium arsenide light-emitting diodes. The use of low-cost compact disc lasers has also been proposed in this region [4]. The region has high losses (4 dB/km); however, the low-cost components make this an attractive region for short distances. Region B is characterized by low losses (0.7 dB/km). Moderately priced components, mainly using gallium arsenide indium phosphide, are available. Region C has the lowest fiber losses (0.3 dB/km); however, the components are costly and not readily available. Excellent graded index fiber is available for the B region (1300 nm) with almost no dispersion.

In addition to its diameters (core and cladding), attenuation, and material dispersion, the fiber is also classified by the mode of light propagation (single mode or multimode). Single-mode fiber is a step index fiber, wherein the core and cladding boundary is characterized by a step in the index of refraction. There is only one mode of propagation [2], thus one path length and no modal dispersion. Multimode fiber has a large-diameter core and allows many modes

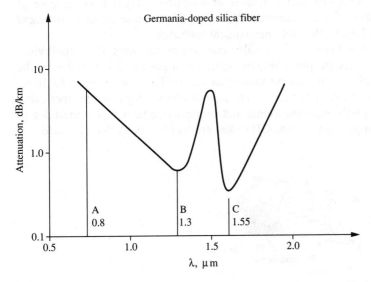

FIGURE 8-5
Fiber absorption curve (three useful regions A, B, and C).

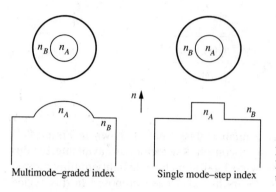

FIGURE 8-6
Refractive index profiles for fibers (not to scale).

Multimode–graded index   Single mode–step index

of light to propagate. This large number of modes gives a large modal dispersion, and step index multimode fiber is of very limited use (it has a low MHz · km product). However, when the refractive index of the core is varied radially so as to compensate for modal dispersion, excellent results are obtained (e.g., 50-μm fiber with a 1500-MHz · km product at 1300 nm; see Figure 8-6). The objective in a graded index multimode fiber is to satisfy

$$\int n \, ds = \text{constant}$$

where $n = n(r) =$ core refractive index (some function of radial distance)
$ds =$ light path distance

Due to its larger diameter (an order of magnitude larger than the core of the single-mode fiber), the multimode fiber simplifies connector and alignment problems (due to larger allowable mechanical tolerances).

The cable (see Figure 8-7) usually consists of an outer PVC (polyvinyl chloride) jacket, fibers for providing strength, and a plastic jacket enclosing the fiber. There is also a soft coating known as the buffer coating over the fiber. Multiple fiber and ribbon fiber cables are also available. Most of the fiber cable usage to date is in telecommunications, and single-mode fiber is preferred due to its high GHz · km product. A 200 GHz · km product has been demonstrated.

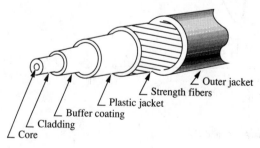

Outer jacket
Strength fibers
Plastic jacket
Buffer coating
Cladding
Core

FIGURE 8-7
Fiber optic cable.

CONNECTORS AND SPLICING

Connectors

Due to the very small diameter of glass fibers as compared to copper conductors used in coaxial cables, a means of alignment is incorporated in optical fiber cable connectors. Precision machining or molding is required, which results in higher costs for the optical connector than for electrical connectors, in general. The required alignment tolerances are ~ 1 μm for single-mode connectors and 5 to 10 μm for multimode connectors.

Tolerance requirements have dictated connector designs: (*a*) the mating cone type, (*b*) the precision sleeve type, and (*c*) expanded beam (optical) type (see Figure 8-8). The fiber-to-fiber losses are given in Figure 8-9 [5]. The losses in the connector vary from less than 1 to 4 dB. Connectors with satisfactory lives of thousands of mating cycles are available. Multiple, ribbon, and mixed (optical and electrical) connectors are also available.

Splicing

The fibers can be spliced by bringing two fiber ends (polished or as cleaved/un-polished) together in a fixture with alignment microgrooves. The fiber ends also may be fused together to provide a splice. A typical mechanical splice has a loss of <1 dB, and a typical fused splice has a loss of < 0.3 dB.

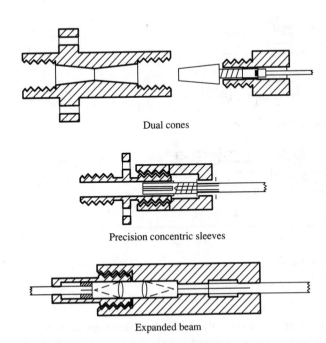

Dual cones

Precision concentric sleeves

Expanded beam

FIGURE 8-8
Sample fiber optic connectors.

1. NA mismatch losses.

    NA (transmitter) > NA (receiver)

    $$\text{Loss (dB)} = 10 \log \left(\frac{\text{NA (receiver)}}{\text{NA (transmitter)}} \right)^2$$

    No loss when NA (receiver) ≥ NA (transmitter)

2. Diameter mismatch losses.

    Diameter (transmitter) > diameter (receiver)

    $$\text{Loss (dB)} = 10 \log \left(\frac{\text{diameter of receiver}}{\text{diameter of transmitter}} \right)^2$$

    No loss when receiver diameter is larger of the two.

3. Axial misalignment (lateral) losses.

    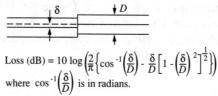

    $$\text{Loss (dB)} = 10 \log \left(\frac{2}{\pi} \left\{ \cos^{-1}\left(\frac{\delta}{D}\right) - \frac{\delta}{D}\left[1 - \left(\frac{\delta}{D}\right)^2\right]^{\frac{1}{2}} \right\} \right)$$

    where $\cos^{-1}\left(\dfrac{\delta}{D}\right)$ is in radians.

4. Reflection (Fresnel) losses.

    $$\text{Loss (dB)} = 10 \log \left[1 - \left(\frac{n_a - n_o}{n_a + n_o}\right)^2 \right]$$

    n_a = reflective index of core
    n_o = reflective index of gap

5. Gap losses

    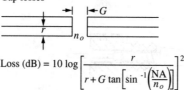

    $$\text{Loss (dB)} = 10 \log \left[\frac{r}{r + G \tan\left[\sin^{-1}\left(\frac{NA}{n_o}\right)\right]} \right]^2$$

6. Axial misalignment (angular) losses.

    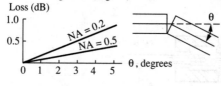

FIGURE 8-9
Fiber(step index)-to-fiber connector losses.

EMITTERS

Two types of light sources are in use today, the LED (light-emitting diode) and the LD (laser diode). Both these devices are semiconductor diodes used in the forward biased mode. Due to glass fiber characteristics (low attenuation losses at 800, 1300, and 1550 nm), only emitters in the 800, 1300, and 1550 nm regions

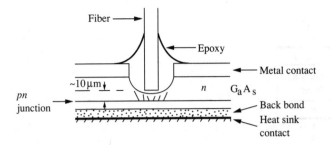

FIGURE 8-10
Surface emitting LED (Burrus diode—see reference [6]).

are useful.* Lasers and light-emitting diodes made from GaAlAs (gallium aluminum arsenide) for 800 nm and GaAsInP (gallium arsenide indium phosphide) for 1300 and 1550 nm are used (see Figures 8-10 and 8-11). The LED has low output (~1 mW) and a broad output light beam that acts as a lambertian source, with output proportional to the cosine of angle from the normal. The lasers provide high output (1 to 5 mW) in a conical beam of elliptical cross section.

The coupling efficiency of the LED is low (1 to 5 percent), but the coupling efficiency obtainable with laser diodes is high (50 to 80 percent). The spectral width of radiation from an LED is 50 to 100 nm, and that from a laser 10 nm (single-mode lasers may have a spectral width of less than 1 nm). Thermoelectric coolers are generally used with lasers to improve reliability. Due to the large temperature coefficient a monitor photodiode is used with lasers to keep the output constant.

---

*It should be noted that the 800 nm region is the result of the low cost availability of gallium aluminium arsenide LEDs and lasers as emitters and silicon pin diodes as detectors, and matching their spectral response curves to the fiber attenuation characteristics.

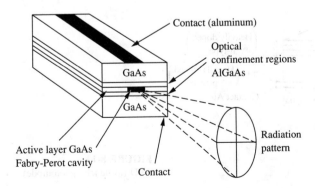

FIGURE 8-11
An AlGaAs laser diode.

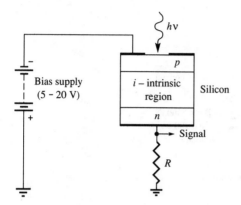

FIGURE 8-12
A PIN photodiode as a detector.

DETECTORS

There are two types of detectors in wide use, PIN photodiode and APD (avalanche photodiode) (see Figures 8-12 and 8-13). The PIN photodiode is characterized by a wide intrinsic region between the p and the n region. Most of the photons are absorbed in this region, creating electron-hole pairs, which are swept to the n and p regions by the electric field created by the reverse bias. The reverse bias also helps in reducing the device capacitance, thus enhancing its frequency response. A typical silicon PIN photodiode may have a responsivity of 0.5 A/W at 800 nm, a reverse leakage current of a few nA, and a junction capacitance of 1 pF. InGaAs PIN photodiodes are generally used in the 1300-nm region.

In APD the p-n junction is also reverse biased. Photon-induced carriers created in the intrinsic region are accelerated by the high electric field in the p-n region to cause impact ionization. These additional electron-hole pairs increase the device responsivity. The reverse bias required for impact ionization is relatively high (50 to 400 V) and has to be fairly well regulated to keep the gain constant. The advantage of APD is its high responsivity to light (10 to 500 times that of a PIN photodiode). APDs are mostly used in high-bandwidth, long-distance

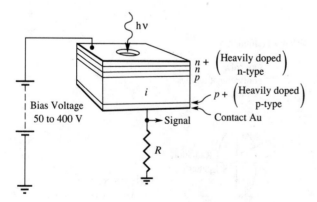

FIGURE 8-13
APD (avalanche photodiode).

links. Silicon and germanium APDs are available, but the InGaAs units are not as readily available, since they are more difficult to make.

The responsivity R of a photodetector is defined as follows:

$$R = \frac{I_p}{P_0}$$

where I_p is the average photocurrent, and P_0 is the average incident optical power on the photodetector.

PACKAGING AND DEVICE COUPLING

Packaging

The emitter and detector chips are mounted like conventional semiconductor chips. The backbond may be gold-tin, gold-germanium, tin-lead solder, or epoxy bond. The top of the chip is wire bonded by ultrasonic or thermal compression (TC) bond. The chip may then be epoxy molded, as in the low-cost LED, or may be packaged in a hermetic package, such as TO-18 and G-9. Some chips are also sold in packages with fiber pigtails, which allow the user to take advantage of a manufacturer's knowledge of efficient coupling of the fiber to the emitter. Laser packages also include the monitor photodiode and can be obtained with built-in thermoelectric coolers (TECs). Some packages are depicted in Figure 8-14.

Low-cost and low-bandwidth chips are sometimes packaged with driver, amplifier, and pulse-shaping integrated circuits in a molded package. Some of these molded packages are designed to be mounted in bulkhead connector adapters.

In addition to the molded packages, there are two other packages in wide use today. The dual-in-line package (DIP) with a connector (see Figure 8-14) is widely used for LED-based transmitters and most of the receivers. The butterfly package (see Figure 8-14), due to its low thermal resistance, is almost exclusively used with thermoelectrically cooled lasers. The thermoelectrically cooled package is generally made hermetic to avoid moisture condensation and resulting corrosion. A thermistor and a monitor photodiode are also mounted in the package. The thermistor monitors the temperature and provides control signals to TEC drive circuits to maintain the laser temperature at the desired level. The monitor photodiode monitors the laser output and provides drive signals to the laser current driver to keep the optical output constant.

Great care is required in shielding the detector circuits from undesired electromagnetic and electrostatic fields from nearby circuits or external sources. At high frequencies (greater than 100 MHz) the length of device leads may significantly affect the performance; long, unshielded leads will have higher unwanted noise coupled into detector circuits. Other inherent noise sources (e.g., thermal noise in a resistor, shot noise from a current source, dark current in a detector, reflections of signals in unmatched transmission lines, phase noise in a laser), together with unwanted noise pickup, determine minimum detectable power for a particular error rate at a given bandwidth.

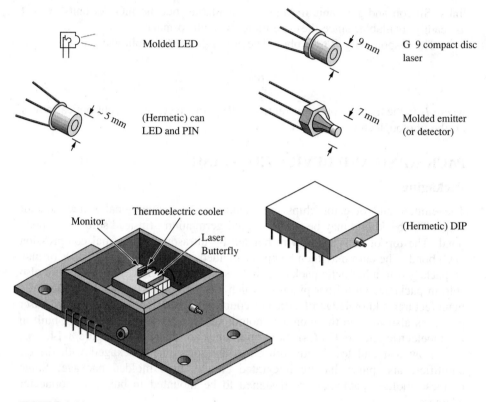

FIGURE 8-14
Fiber optic emitter and detector packages.

Device Coupling

One of the most important subjects in the optical area is emitter-to-fiber coupling. The coupling may be (*a*) direct emitter-to-fiber coupling (with or without index-matching adhesive [7]) or (*b*) emitter-to-lens (single or double convex, ball, rod or graded index, GRIN)-to-fiber coupling [8, 9]. Some useful formulas for (*a*) are given in the following:

LED (surface) to step index fiber.

$$P_{\text{fiber}} = \begin{cases} P_{\text{LED}} \, \text{NA}_{\text{fiber}}^2 & \text{for} \quad A_E < A_C \\ P_{\text{LED}} \, \text{NA}_{\text{fiber}}^2 \frac{A_C}{A_E} & \text{for} \quad A_E > A_C \end{cases}$$

where A_E is the (circular) LED emitting area and A_C is the fiber core area. P_{fiber} is further reduced by multiplication by $1 - R$, where R is the reflection coefficient, given by

$$R = \left(\frac{n_a - n_0}{n_a + n_0} \right)^2$$

LED (surface) to graded index fiber. A graded index fiber may be defined as a fiber with a radial gradient in the core refractive index, given by

$$n(r) = n(0) \left[1 - 2\Delta \left(\frac{r}{r_c} \right)^{\alpha} \right]^{1/2}$$

where r = radius and r_c = core radius

$$\Delta = \frac{n(0)^2 - n_b^2}{2n(0)^2} \approx \frac{n(0) - n_b}{n(0)}$$

and α is a constant (≈ 2).

In this case,

$$P_{\text{fiber}} = 2\Delta P_{\text{LED}} n(0)^2 \left[1 - \frac{2}{\alpha + 2} \left(\frac{A_E}{A_c} \right)^{\alpha/2} \right]$$

There is also a reflection loss in this case.

EXAMPLES

Example 1. Consider a 20-m-long plastic fiber optic link; the transmitter is a 1-mW LED (surface) and the receiver, a PIN diode. The LED is directly coupled to the fiber (but without any index-matching adhesive) and the detector is coupled via an index-matching adhesive to the fiber. Also given are

$$\lambda = 800 \text{ nm}$$

$$n_a \text{ (core)} = 1.495$$

$$n_b \text{ (cladding)} = 1.402$$

$$\text{Fiber attenuation} = 400 \text{ dB/km}$$

$$\text{LED diameter} = 0.5 \text{ mm}$$

$$\text{Detector diameter} = 1.0 \text{ mm}$$

$$\text{Fiber diameter} = 1.0 \text{ mm}$$

$$R \text{ (detector responsivity)} = 0.5 \text{ A/W}$$

Calculate
(a) Fiber NA
(b) Fiber acceptance angle
(c) Fresnel loss at emitter coupling
(d) Power coupled into fiber in dBm

(*e*) Cable loss in dB

(*f*) PIN photodiode current

Solutions

(*a*)
$$NA = \frac{(n_a^2 - n_b^2)^{1/2}}{n_0} = \left[\frac{1.495^2 - 1.402^2}{1}\right]^{1/2} = 0.519 \approx 0.5$$

(*b*)
$$\text{Acceptance angle} = 2\sin^{-1} NA = 62.6°$$

(*c*)
$$\text{Fresnel loss factor} = 1 - R = 1 - 0.039 = 0.961$$

Since
$$R = \left(\frac{n_a - n_0}{n_a + n_0}\right)^2 = \left(\frac{1.495 - 1.0}{1.495 + 1.0}\right)^2 = 0.039 \approx 4\%$$

(*d*)
$$\text{Power in fiber} = P_{\text{LED}}(1 - R)\, NA^2 = (1 \times 10^{-3})(0.96)(0.5)^2$$
$$= 0.24 \text{ mW} = -6.2 \text{ dBm}$$

(*e*)
$$\text{Cable loss} = 0.020 \text{ km} \times 400 \text{ dB/km} = 8 \text{ dB}$$

(*f*) Power at receiver (no coupling loss assumed, because index-matching adhesive used):

$$P_{\text{rcvr}} = -14.2 \text{ dBm} = 38 \times 10^{-6} \text{ W}$$
$$I_{\text{PIN}} = 38 \times 10^{-6} \text{ W} \times 0.5 \text{ A/W} = 19 \times 10^{-6} \text{ A}$$

Example 2. A laser (Figure 8-15) is mounted on a thermoelectric cooler (TEC), which is itself mounted on a natural convection air-cooled heat sink. The following data is given:

$$\text{Thermal resistance chip to heat sink} = K_{CH} = 1.0°\text{C/W}$$
$$\text{Thermal resistance chip to TEC} = K_{CT} = 1.0°\text{C/W}$$
$$\text{Thermal resistance TEC to heat sink} = K_{TH} = 0.5°\text{C/W}$$
$$\text{Thermal resistance heatsink to air} = K_{HA} = 5.0°\text{C/W}$$
$$\text{Laser drive (2 V at 0.15 A)} = W_{LD} = 0.3 \text{ W}$$
$$\text{Laser output} = W_{LO} = 0.1 \text{ W}$$

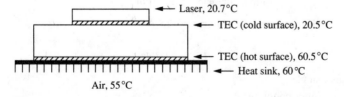

FIGURE 8-15
Laser, TEC, and heat sink assembly.

$$\text{TEC drive (1 V at 0.8 A)} = W_{TD} = 0.8 \text{ W}$$

$$\Delta T \text{ across TEC when pumping 0.2 W} = \Delta T_P = 40.0°\text{C}$$

$$\text{Air temperature} = T_A = 55.0°\text{C}$$

(a) Calculate the laser chip temperature.

(b) Calculate the chip temperature with the chip directly mounted on the heat sink, without the thermoelectric cooler.

Solutions

(a) Total power to be dissipated by the heat sink $= W$

$$W = W_{\text{LD}} - W_{\text{LO}} + W_{\text{TD}} = 1 \text{ W}$$

$$\therefore \Delta T(\text{sink-air}) = 1 \text{ W} \times K_{\text{HA}} = 1 \text{ W} \times 5°\text{C/W}$$

$$= 5°\text{C}$$

$$\therefore \text{Heat sink temperature} = T_A + \Delta T(\text{sink-air})$$

$$= 55°\text{C} + 5°\text{C} = 60°\text{C}$$

$$\Delta T(\text{TEC-sink}) = 1 \text{ W} \times K_{TH} = 1 \text{ W} \times 0.5°\text{C/W} = 0.5°\text{C}$$

$$\therefore \text{TEC hot surface temperature} = T_{\text{sink}} + \Delta T \text{ (TEC-sink)} = T_H$$

$$= 60°\text{C} + 0.5°\text{C} = 60.5°\text{C}$$

$$\Delta T_P \text{ (TEC)} = 40°\text{C (given)}$$

$$\therefore \text{TEC cold surface temperature} = T_H - \Delta T_P \text{ (TEC)} = T_C$$

$$= 60.5°\text{C} - 40°\text{C} = 20.5°\text{C}$$

$$\text{Watts from laser to TEC} = W_L = W_{LD} - W_{LO}$$

$$= 0.3 \text{ W} - 0.1 \text{ W} = 0.2 \text{ W}$$

$$\Delta T \text{ (chip-TEC)} = W_L \times K_{CT} = 0.2 \text{ W} \times 1°\text{C/W} = 0.2°\text{C}$$

$$T \text{ (chip)} = T_C + \Delta T \text{ (chip-TEC)} = 20.5°\text{C} + 0.2°\text{C}$$

$$= 20.7°\text{C}$$

(b) Heat to be dissipated to air $= W$

$$W = W_{LD} - W_{LO} = 0.3 \text{ W} - 0.1 \text{ W} = 0.2 \text{ W}$$

$$\text{Thermal resistance, chip-air} = K_{CH} + K_{HA} = K_{CA}$$

$$= (1 + 5)°\text{C/W} = 6°\text{C/W}$$

$$\therefore \Delta T \text{ (chip-air)} = W \times K_{CA} = 0.2 \text{ W} \times 6°\text{C/W} = 1.2°\text{C}$$

$$\therefore \text{Chip temperature} = T_A + \Delta T \text{ (chip-air)}$$

$$= 55°\text{C} + 1.2°\text{C} = 56.2°\text{C}$$

Example 3. Assume that the laser diode in Example 2 degrades with time, needing additional drive current to maintain a fixed output as it ages. A life test done at 60°C gives an average lifetime of 10,000 hours. Failure is defined as a 50 percent increase in drive current. The acceleration factor based on the Arrhenius relationship (degradation rate $= R_T = C e^{-E_a/kT}$) is given below.

$$\text{Acceleration factor} = \frac{R_{T1}}{R_{T2}} = e^{-E_a(1/T_1 - 1/T_2)/k}$$

where E_a = activation energy in electron volts = 1.0
  k = Boltzmann's constant = 8.6×10^{-5} eV/K
  T = temperature in degrees Kelvin

(a) Calculate the lifetime of the laser, operating under conditions given in Example 2, part (a).

(b) Calculate the lifetime of the laser, operating under conditions given under Example 2, part (b).

Solution

(a) $$\frac{R_{T1}}{R_{T2}} = e^{-(1/333 - 1/293.7)/k} = 1.07 \times 10^2$$

Life (20.7°C) = $10^4 \times 1.07 \times 10^2 = 1.07 \times 10^6$ hours

(b) $$\frac{R_{T1}}{R_{T2}} = e^{-(1/333 - 1/329.2)/k} = 1.5$$

Life (56.2°C) = $10^4 \times 1.5 = 15,000$ hours

APPLICATIONS AND FUTURE POTENTIALS

Fiber optic links have approximately 10,000 times the information capacity of ordinary telephone lines. As we learn to use the technology of fiber optics, and as new products resulting from the current activity in research and development become available (such as extremely high bandwidth optical links using phase modulation techniques), the information revolution will take a quantum jump. Picturephones, personalized newspapers, the Library of Congress available at your terminal, worldwide TV education, business and entertainment networks, and so forth, may become a reality.

Developments in material science may lead to fluoride fibers, with a thousand times less attenuation than the current germania-doped silica fiber. Progress in making multi-quantum-well structures holds the promise of making improved detectors and emitters (and in particular the lasers). Advances in optical integrated circuits may lead to arrays of low-cost optical links. Wavelength multiplexing has the potential of increasing the information-carrying capacity of existing fiber optic links. Already, as the components become available, the trend towards 1550 nm (for high-performance links) is taking shape, as the losses at 1550 nm are about half of those at 1300 nm, which in turn are about one-fifth of those at

800 nm. Lithium niobate optical integrated switching networks are under development.

The main use of fiber optic links presently is in telecommunications. Local area networks (links between computer terminals and computers) are increasingly using fiber optics. As the costs of transmitters and receivers, and to some extent connectors, are reduced (fiber optic cable already costs less today than a high-bandwidth electrical cable), fiber optic links between and within mainframe computers will be used because of size, weight, freedom from electromagnetic interference, and elimination of noisy electrical ground loops. As the technology of optoelectronic integrated circuits (OEIC) develops, optical links between printed circuit boards, cards, modules and ultimately chips will become viable. The development of optical waveguides on the substrates will greatly assist in the growth of short length optical links and in the development of optical computing. Fiber optic link technology has been used in instrumentation, in sensor applications, and in the making of a gyroscope. It has the potential of making possible a one million connections per square inch connector, three-dimensional electronic packaging (all interconnections in one direction being optical), passive multiport couplers, and a whole array of new products [10–15].

REFERENCES

1. S. E. Miller and A. G. Chynoweth, *Optical Fiber Telecommunications,* New York: Academic Press, 1979, ch. 1.
2. M. K. Barnoski, *Fundamentals of Optical Fiber Communications,* 2d ed. New York: Academic Press, 1981.
3. E. C. Jordan, *Reference Data for Radio, Electronics, Computer, and Communications,* 7th ed., Indianapolis: H. W. Sams & Co. Inc., 1985, ch. 22.
4. R. L. Soderstrom, T. R. Block, D. L. Karst, and T. Lu, "The Compact Disc (CD) Laser as a Low Cost, High Performance Source for Fiber Optic Communication," in *FOC/LAN 86,* Boston, Mass.: Information Gatekeepers, Inc., 1986, pp. 263–264.
5. *Designer's Guide to Fiber Optics,* Harrisburg, Pa.: AMP Inc., 1982.
6. C. A. Burrus and R. W. Dawson, "Small Area High Current Density GaAs Electroluminescent Diodes and a Method of Operation for Improved Degradation Characteristics," *Applied Phys. Lett.* 1, pp. 97–99 (August 1978).
7. G. Keiser, *Optical Fiber Communications,* New York: McGraw-Hill, 1983, ch. 5.
8. A. Nicia, "Lens Coupling in Fiber Optic Devices: Efficiency Limits," *Applied Optics* 20, no. 18, pp. 3136–3145 (September 1981).
9. K. Kawano and Osamu Mitomi, "Coupling Characteristics of Laser Diode to Multimode Fiber Using Separate Lens Methods," *Applied Optics* 25, no. 1, pp. 136–141 (January 1986).
10. CSELT staff, *Optical Fiber Communication,* New York: McGraw-Hill, 1981, part 5.
11. S. R. Forrest, "Optical Detectors: Three Contenders," *IEEE Spectrum,* 23, no. 5, pp. 76–84 (May 1986).
12. W. T. Tsang, *Lightwave Communication Technology, Part E: Integrated Optoelectronics,* vol. 22 of *Semiconductors and Semimetals,* New York: Academic Press, 1985.
13. A. M. Glass, "Optical Materials," *Science* 235, pp. 1003–1009 (February 1987).
14. V. Narayanamurti, "Artificially Structured Thin-Film Materials and Interfaces," *Science* 235, pp. 1023–1028 (February 1987).
15. T. Suzuki, T. Ebata, K. Fukuda, N. Hirakata, K. Yoshida, S. Hayashi, H. Takada, and T. Sugawa, "High-Speed 1.3 Micron LED Transmitter Using GaAs Driver IC," *IEEE Journal of Lightwave Technology* LT-4, no. 7, pp. 790–794 (July 1986).

16. C. Hentschel, *Fiber Optics Handbook,* Palo Alto, Calif.: Hewlett Packard, 1983.
17. R. E. Nahory, M. A. Pollack, W. D. Johnston, Jr., and R. L. Barns, *Applied Phys. Lett.* 33, p. 659 (1978).
18. L. D. Hutcheson, P. Haugen, and Anis Husain, "Optical Interconnects Replace Hardwire," *IEEE Spectrum,* 24 no. 3 pp. 30–35 (March 1987).

PROBLEMS

8.1. Calculate (*a*) $n(0)$ (or the core's axial refractive index) and (*b*) the cladding refractive index, given that the graded index core diameter is 62.5 μm, the NA is 0.29, and the index difference (Δ) equals 0.02.

8.2. Calculate the power coupled into the fiber in Problem 1 by a 1-mW (1300 nm) LED with a radiating area 30 μm in diameter. Assume no index-matching adhesive and $\alpha = 2$.

8.3. Calculate the mechanical splice loss for a step index fiber with a 0.25-mm gap, given $n(\text{core}) = 1.45$, core diameter = 50 μm, and NA = 0.21. Assume no axial angular or lateral misalignment.

8.4. Calculate the NA loss when a 62.5-μm fiber (NA = 0.29) from the transmitter is connected to a 50-μm (NA = 0.21) fiber.

8.5. Calculate the current due to Johnson (thermal) noise in the input resistor of a photodetector amplifier, given that the temperature = 25°C, resistance = 50 Ω, and bandwidth = 200 MHz. The mean square current due to Johnson noise is given by the following:

$$\overline{i^2} = \frac{4kTB}{R} \quad i \text{ in amps}$$

where k = Boltzmann's constant = 1.4×10^{-23} J/K
  T = Absolute temperature in degrees Kelvin
  B = Bandwidth in cycles/s
  R = Resistance in ohms

8.6. Assuming a detector responsivity of 0.5 A/W, calculate the equivalent power input indicated by thermal noise in Problem 5.

8.7. Assuming a signal/noise (S/N) ratio of 6 is required to meet a BER of 10^{-9} (one incorrect bit received per 10^9 bits received), calculate the minimum power required by the detector in Problem 6. (BER is a probability similar to an error function; see reference [7], chapter 7.)

8.8. Calculate the maximum capacitance (due to packaging) allowed in the input stage of a photodetector amplifier. Assume the amplifier bandwidth (frequency bandwidth over which the response is within 3 dB of maximum) is input circuit–limited. Also assume the equivalent parallel resistance and capacitance due to the detector and input amplifier devices are 100 Ω and 1.5 pF respectively. The bandwidth f is equal to $\frac{1}{2}\pi RC$ and given as 600 MHz.

8.9. A detector chip is backbonded to a copper block. The top end of the chip is wire bonded to a 1-mm-thick ceramic (with a relative dielectric constant $K = 9$ and a square bonding pad of 4 mm on a side). The capacitance (in pF.) of a parallel plate capacitor may be given as $C = 0.0884KA/T$, where A is in square centimeters; K is relative dielectric constant; and T is dielectric thickness in centimeters. Calculate the packaging capacitance associated with the bonding pad.

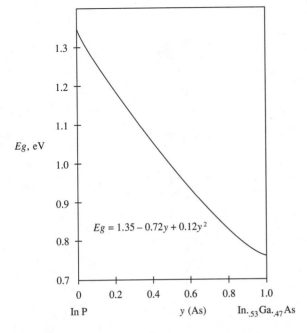

Eg, eV

$Eg = 1.35 - 0.72y + 0.12y^2$

In P y (As)   In$_{.53}$Ga$_{.47}$As

FIGURE 8-16
In$_{(1-x)}$ Ga$_{(x)}$ As$_{(y)}$ P$_{(1-y)}$ alloy diagram (lattice matched to InP, or $x = 0.47y$).

8.10. A single-mode fiber has a material dispersion of 17 ps/nm/km at 1500 nm. Calculate (*a*) the increase in the pulse width and (*b*) the bandwidth for a 2-km link, using a 1550-nm laser with a spectral bandwidth of 10 nm and Gaussian pulse inputs. Hint: Pulse width (ns) $= 0.44$/bandwidth (3 dB optical or 6 dB electrical in GHz); see reference [16] pp. 37, 63, 74, and 87.

8.11. Calculate (*a*) the band gap required for a 1.3-μm emitter and (*b*) the alloy composition of the quaternary alloy (see Figure 8-16 and reference [17]) of InGaAsP for the emitter in part (*a*).

CHAPTER
9

RELIABILITY AND TESTING

LYLE L. MARSH
ROSS D. HAVENS
SHOW-CHI WANG
JOHN A. MALACK
HOLMES B. ULSH

INTRODUCTION

Rapid advances in technology have made reliable products commonplace in the electronics industry, and because of customer dependency on information processing, quality and reliability are fundamental requirements in the choice of vendor. Sustained success is dependent on one's ability to provide products and services superior to others. Therefore, an important goal is to have each new generation of product offer service and reliability superior to the product it replaces. This ambitious goal is only achieved through careful planning and the involvement of individuals across the entire organization. This chapter acquaints readers with a fundamental approach to the reliability of electronic packages. The term *reliability* will be defined from different people's perspectives, ranging from a customer to a person operating a machine on the manufacturing line. In between is an engineer, who must address reliability from within the framework of his own particular discipline.

We will begin by separating the total electronic package into three basic components: the *chip*, the *chip carrier*, and the *printed circuit card* or *board*. All of these components can be manufactured and tested under different conditions and in separate facilities. The reliability of each component, therefore, is independent of the others and must be studied separately. After accounting for interconnections, the reliability of the total package can be considered to be the combined result of the reliabilities of each of the components.

Packages discussed are those typically found in commercial data processing systems (central processing unit, display, printer, keyboard, etc.). These products are characterized by high-volume, low-cost marketing objectives. This excludes products designed for military or special-purpose (space, underwater exploration, etc.) application. However, the concepts apply to all such products where dependable operation in the customer's environment is an essential objective.

Regardless of the package to be manufactured, successful results require the interaction of three major functions: (1) product design (provides basic function), (2) manufacturing process (produces product repeatably), and (3) product test (both in-process testing to screen defects to isolate their cause and correct the process, and stress testing to evaluate long-term reliability). When properly integrated, these organizations and functions produce a reliable product. In this chapter, we will explore this relationship in detail, with a special emphasis on testing and the manufacturing process.

Effective interaction among the major functions requires coordinated efforts from different teams. Each team has specific tasks in relation to the goal of producing a competitive product. The development team has responsibility for providing manufacturing with the product performance design, and with the process to build the package consistent with targets. Manufacturing has responsibility for producing product according to this prescribed design and process, in volumes that meet a planned schedule. A team of quality specialists monitors the manufacturing team for compliance to process procedures, and to assure that the product is defect free or otherwise meets the performance and reliability objectives. They have an added task of assuring that changes to the process continue to meet requirements imposed by the development team. Reliability engineers provide appropriate stress (life) tests and reliability models by which product reliability may be estimated. These models are constructed from stress test data using carefully chosen test conditions that accelerate each potential failure mode.

The manufacturing and quality teams share the responsibility of yield management. While there is some overlap of responsibility, manufacturing groups are generally responsible for achieving acceptable yields (thus cost control), while the quality organization is responsible for managing yields from the standpoint of defect elimination. An integral part of defect elimination is failure analysis, which provides information to both organizations concerning the source of the reduced yield. This analysis is typically performed by quality with assistance, if necessary, by development engineers.

With the exception of design, which is treated separately in Chapters 1–8, this chapter will touch on each of these areas of responsibility. The subject matter is very broad and encompasses many disciplines. This chapter provides, in capsule form, a treatise that combines these disciplines and introduces the reader to the concepts employed. Further study of the reference material is encouraged.

ENVIRONMENTAL CONSIDERATIONS AND PRODUCT TESTING

Definition of the test methodology to discover defects starts early in the product design phase, is carried through development along with the product, and ultimately is carried into manufacturing. The objective is to eliminate both time zero and latent defects. Latent defects typically develop from the time zero condition with variation in operating voltage bias and product operating environments. Each defect type requires separate attention.

At this point, it is appropriate to introduce, to a limited degree, the features of the product operating environment. The application environment, or climate, is more than the thermal-mechanical stresses generated during operation of the product. This climate plus all the other generic stresses can combine to alter or degrade product performance. Critical among these environmental factors are those of atmospheric origin. Examples of these are atmospheric pollutants and humidity, which can lead to chemical stresses and corrosion.

Typically, atmospheric pollutants are encountered in concentrations distributed over time as well as geographically. The application environment may vary between mobile and fixed locations, or between an uncontrolled outdoor environment and a controlled indoor environment. Thus, the product may see a wide range of environments, the total range of which must be understood in order that the stresses that will be imposed by the environment may be taken into account. Since the intrinsic reliability will vary with market severity, the product reliability can be expected to be distributed accordingly. Thus, important material and design issues must be resolved that address the anticipated volume distribution in each market category as defined by environmental severity. In turn, these decisions affect the selection and duration of tests to evaluate product performance.

Corrosion and Environmental Issues: Overview of Hermeticity and Package Reliability

The package environment has two immediate sources. The first is the ambient environment, external to the package. This is the operating environment in which the package is placed. The second is the internal environment that results from ad/absorbed moisture and residual ionics from incomplete or inadequate rinsing. The chemistry of any surface films (including internal surfaces) determines the reactivity of such films and the failure time distribution.

Package design, including materials selection, is important from the viewpoint of hermeticity; that is, the avoidance of easy transport paths for moisture, pollutants, or other chemically active species to the exposed package circuitry. Corrosion and failure of the package may be the result of such transport. A stylized version of a nonhermetic package is shown in Figure 9-1. Here, the time for

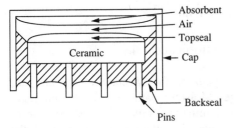

Absorbent
Air
Topseal
Cap
Backseal
Pins

FIGURE 9-1
A nonhermetic package and Memis's model for moisture permeation.

moisture to permeate the sealants and establish an internal equilibrium with the external relative humidity is modeled by Memis [1].

$$RH = RH_0 + (RH_{ext} - RH_0)(1 - e^{-t/\tau})$$

$$\tau = \tau_0 \times e^{H/RT}$$

where RH = relative humidity
   RH_0 = initial RH inside the nonhermetic package
   RH_{ext} = RH external to the package
   t = time
   τ = time constant for the diffusion constant
   H = activation energy
   R = gas constant
   T = temperature (degrees K)
   τ_0 = a constant

Pollutants in the form of moisture, etc., are adsorbed into the top seal. This seal becomes a source of the pollutant in any corrosion process. If the top seal adhesion to the substrate (ceramic or metallized layers) is poor, sometimes described as "nonwetted," a region of interfacial separation is created. Pollutants segregate to such regions and establish a local environment for corrosion. A metal sealant provides hermeticity, while polymers provide transport paths and contribute to the nonhermeticity. Defects within the polymer or at metal/polymer interfaces such as pinholes or cracks increase this loss of hermeticity.

Circuit design is more complex. Given that the designers have dealt with electrical performance features adequately, the circuit design has further importance because of design interactions with process limitations. The specified line width of the circuit may be beyond the resolution capability of the photolithographic technique available. Similarly, it may be beyond the ability of the process technology for metal deposition techniques. These process/design incompatibilities create the potential for line defects. Such defects can result in line "opens" or, under the proper environmental conditions, loss of insulation resistance.

Once the design/process parameters have been successfully managed, only environmental issues remain. The essential features of corrosion-type failure mechanisms are property degradation due to

1. Loss of insulation resistance by metal migration
2. Localized line corrosion
3. Bridging between adjacent conductive corrosion deposits.

Such failure mechanisms need a moisture path, surface contamination, and, for metal migration to occur, a bias voltage. Moisture paths are most commonly caused by assembly operations, cracks, delamination, pinholes in passivation or other coatings, and poor step coverage. Typical contaminants can include the normal atmospheric pollutants such as SO_2, H_2S, S_8, NO_2, and HCl as well as process residuals, such as soluble chlorides, phosphoric acid, alkalis, etc. The rates of such corrosion processes are accelerated by such factors as moisture, temperature, voltage, combined external interfacial contamination levels, and the design features themselves, which include conductor spacing and material properties.

To this point, the association of moisture and ionic contamination with various corrosion processes has been established. Let's examine this failure proposal in greater detail. The case for metal migration (dendritic growth) will illustrate the essential features applicable to most corrosive mechanisms. Figure 9-2 depicts the primary growth path.

From an electrochemical viewpoint, the most important feature of the model is a continuous moisture film between the adjacent circuit elements. Film continuity is a function of the relative humidity and is reflected in the leakage current i between the anode and the cathode. In turn, this leakage current is affected by ionic contamination and the film pH. The potential for these effects is illustrated in Figure 9-3, which shows an adsorption isotherm on alumina [2] at ambient temperature. The process reflected by this isotherm is monolayer coverage, multilayer colony formation, and coalescence of colonies into a continuous film. Leakage current parallels this behavior. Electrochemical corrosion processes become possible with continuous film formation. With contamination present, continuous film formation occurs at lower levels of humidity.

Marderosian [3] early demonstrated the strong interaction between moisture and contamination in promoting dendritic migration (cathode to anode) between adjacent circuit segments under a dc voltage. In this case, Figure 9-4 illustrates the effect of relative humidity on leakage current and time to failure using potassium

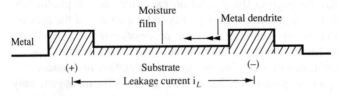

FIGURE 9-2
Primary growth path for electrochemical metal migration.

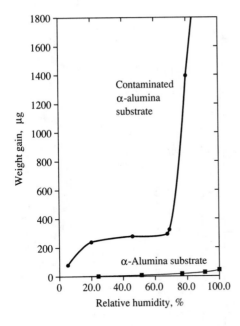

FIGURE 9-3
Absorption isotherms of contaminated versus uncontaminated α-alumina substrates.

chloride as the contaminant. The major point to be drawn from the data is that corrosion or metal migration will not occur below a threshold RH. The value of this threshold is roughly equivalent to the RH required to form a continuous cathode/anode moisture path. This corresponds to a leakage current approximating 0.005 μA. Figure 9-5 is a summary plot of the failure time versus RH for various levels of surface contamination. As the level of contamination is reduced, the time to failure for any given RH dramatically increases. This data suggests that, for a clean substrate, even RH = 100 percent (not immersion) will not support a metal migration process. The importance of surface cleanliness prior to each encapsulating step in the process, and then again of the external surfaces upon completion, in preventing corrosion and insuring reliability is thus demonstrated.

Test Considerations

Any testing is viewed as an overhead item. That is, all costs incurred during testing must be borne by the product in addition to the cost of manufacture. The need is to test stringently enough to isolate the root cause of defects, and then to eliminate them in future product by implementing appropriate process controls. The goal is to insure zero defects at the lowest cost.

To accomplish this, those who set testing strategy must decide where in the process sequence to conduct tests. They must also decide upon the types of defects to be found. Points to consider include testing costs, product cost incurred to the point of test, and the scrap and possible rework costs. The later in the process

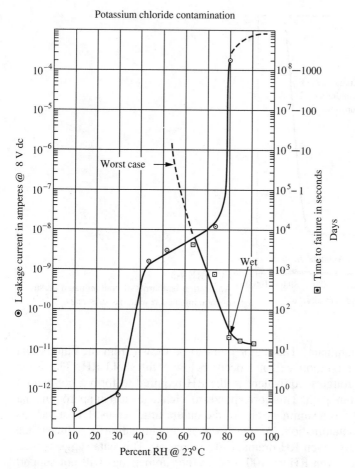

Potassium chloride contamination

FIGURE 9-4
Effect of relative humidity on leakage current and time to failure for KCl-contaminated circuit elements.

one performs the test, the more complicated and expensive the test becomes. It is certain that a final test must be made at the end of the line. The test strategist must decide what and how many earlier tests must also be made, and when they should be performed. Choices must meet the goal of eliminating defects while still containing the overhead represented by the test.

The concept described above is illustrated in Figure 9-6. A circuit card is depicted at three stages of completion. Step 1 represents the signal and voltage "core" levels of manufacture, while step 2 has combined the cores into a "finished" raw card composite. In step 3 the individual components are mounted on the card and soldered in place. The curve shows how the relative cost of testing the product increases at the various stages of product manufacture.

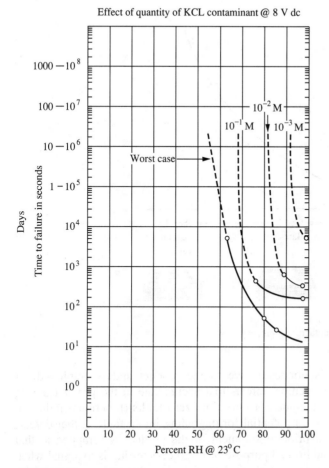

FIGURE 9-5
Effect of varying contaminant concentration on time to failure.

Test Applications

There are various reasons for subjecting a product to testing. Aside from the need to find and eliminate manufactured defects, these reasons include the need to prove the adequacy of a new design, to verify that processing is under control, to predict the performance of the product for a lifetime of conditions imposed by field service, and to provide a safe and fast alternative to field data analysis for product/process improvements. Various types of testing have evolved to cover this range of concerns. They can be separated into two general categories:

1. In-process testing
2. Stress testing

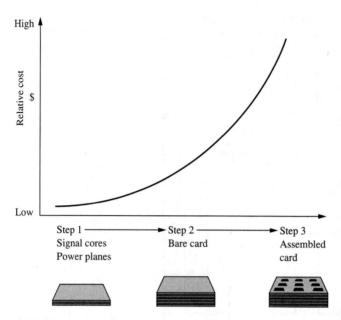

FIGURE 9-6
Relative test cost through manufacturing process.

In-process testing is nondestructive to the product and is employed for the primary purpose of screening defects that accumulate in the product during manufacture. This would include not only time zero faults (functional failures), but any latent defect having a potential for degrading into a fault. Either defects thus detected are repaired prior to shipment or the product is scrapped at that point in the process. An additional purpose of in-process testing is to signal when the process is drifting out of control. Defects found in these tests are carefully tracked to trigger timely engineering action when it is necessary.

Stress testing, on the other hand, is often destructive to the product and, while it may be performed at various process stages, it is most frequently employed on the finished product. Stress testing serves the multiple purposes of assessing the adequacy of product design and/or process controls and predicting the product's performance in the field under the anticipated stresses of the application environment. Both the in-process test and the stress test data are used in feedback fashion to process development and product design organizations to bring about future product improvements.

For both types of tests, the technique employed is driven by the type of defect to be detected and where in the process detection is desired. In some instances, a combination of in-process and stress testing techniques may be required. This is a process known as *stress screening*. Certain defect types may be undetectable under normal end-of-line product conditions and may require the product to be stressed in such a manner that defect detection is enhanced without damaging good product. An example of such a test is high-voltage testing of dielectric, in which voltages chosen are less than the dielectric strength of the material, but high enough to create a capacitive discharge (spark) if the dielectric is defective. Alternatively, the stress can involve preconditioning of the product to enhance defect detection in a subsequent test sequence. Preconditioning could involve thermal or mechanical stressing, humidity exposure, or cleaning or drying with or without a voltage bias across nodes under test.

Despite the type of testing employed, it is imperative that the test parameters chosen (electrical, environmental, and mechanical) match the defect mechanism of interest. This can be a tedious process, since defects may be subtle and very fragile in nature, thus avoiding detection with the wrong stimuli. Each test is designed with a specific purpose in mind, that can affect product performance in some way. Thus, the testing approach can have a profound effect on reliability.

Stress Testing

The industry has developed a network of tests that fall under this category. Most companies engaged in electronic packaging, however, have adopted their own battery of such tests rather than employ a standardized program such as one might hope to find in an ASTM menu. This is due in part to the need to tailor the test to the particular product and its intended application. Protection of corporate assets is another contributing factor. Despite the diversity of individual applications, each test selects one or more of the stress conditions the product will actually encounter in the field, then exaggerates the magnitude such that the product is exposed to more stress during the test than it will ever see in normal use. The stress may be applied continuously, or it may be cycled, pulsed, or otherwise repeated to simulate a specific field exposure.

Examples of specific stress tests are given in Table 9.1. The examples may apply to both first- and second-level products. Each of the tests are in common use in a variety of forms. While a brief discussion on two such tests is given below, a thorough review of each test is beyond the scope of this chapter. The tests listed represent only a sampling of stress tests in use. A selection of references on the subject ([4–12]) is provided for further study.

Remember that a stress test is performed for a specific purpose. In each case, a concern for the integrity and functionality of the product exists. That concern may be for the effect of the stress on structural integrity, electrical continuity, or insulation resistance within the product over a prolonged period of time. For example, temperature variations the product will see during power on/off cycles will result in internal strain being applied to the conductors. These strains are caused by the differences in coefficients of thermal expansion between the conductor and the insulating material to which it is bonded. An effective stress test must consider the temperature extremes the product will actually see, then choose a set of temperatures for thermal cycling that exaggerate both the low-temperature and the high-temperature ends of the range. Figure 9-7 shows an example of one such temperature profile for a typical thermal cycle test. The product depicted is

TABLE 9.1
Examples of stress tests that might be employed for specific design/defect mechanism concerns

Concern	Mechanism	Stress test
Contact failure	Corrosion thru pinhole or wear	1. Contact resistance test in environmental chamber using artificial dust and perspiration 2. Contact resistance test during vibration in presence of artificial dust
Conductor failure	Metallization layer failure thru thermal-mechanical shock or fatigue	1. Thermal cycle while monitoring change in resistance (fatigue) 2. Thermal shock while monitoring change in resistance (shock) 3. Vibration or torque testing while monitoring change in resistance (fatigue)
	Chip metallization layer failure through solid state electromigration	1. Applied current in combination with elevated temperature (monitor resistance)
Insulation failure	Electrochemical migration between conductors	1. Insulation resistance test in environmental chamber at elevated temperature/humidity and voltage (may include atmospheric pollutants)
Solder joint failure	Fracture of solder joint thru creep, fatigue, or shock	1. Thermal cycle/thermal shock as for conductor failure 2. Mechanical load application while monitoring change in resistance

designed for office use and is expected to see temperature extremes between 15°C at power off and 50°C with power on. The test temperatures chosen are between −30 and +75°C. The test temperatures correspond to shipping temperatures, as might be attained in the hold of an aircraft at high altitude and then inside a boxcar in the hot sun. In this way, we get the double advantage of evaluating shipping concerns while performing power on/off simulations. Wider δT's are possible, but care must be taken that undue or unpredictable stresses are not induced. For example, FR-4 epoxy resin (the base dielectric in some PC boards) goes through a glass transition state at approximately 120°C. Above this temperature, the coefficient of thermal expansion changes along with material properties, and the stress produced on the conductors would not be consistent for a given increment of temperature. Conductor resistance is monitored at intervals throughout the test, and maximum allowable resistance changes, referenced to time zero, are set as pass/fail criteria. The resistance may be read at ambient between cycles, or the readings may actually be taken "in situ" at the high and low temperature

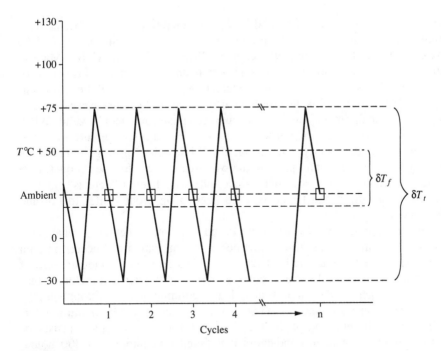

FIGURE 9-7
Typical thermal cycle test profile for a $-30/+75°C$ test. Readings may be taken at ambient or in situ (at temperature extremes).

extremes. Readings taken in situ must be normalized back to ambient conditions to rid the data of error due to coefficients of thermal resistance. In situ readings are more effective than ambient readings because damaged conductors often exhibit subtle resistance changes that become "visible" only when the conductor is under varied amounts of tension or compression.

The number of cycles required at the test temperatures to equate to the number of anticipated cycles at field temperatures can be calculated from the Coffin-Manson [13] relationship for metal fatigue, which can be written as

$$\frac{N_f}{N_t} = \left(\frac{\epsilon_t}{\epsilon_f}\right)^2$$

where the ratio of the number of cycles in the field to the number of cycles in the test is equal to the square of the ratio of the strain of the test to the strain of the field. Since strain is proportional to temperature gradient δT

$$\epsilon = \alpha \delta T$$

Then we have

$$\frac{N_f}{N_t} = \left(\frac{\delta T_t}{\delta T_f}\right)^2$$

This relationship also represents the degree of acceleration provided by the temperature extremes imposed on the product in test. Acceleration factors will be discussed in more detail later in the chapter. Where use of the Coffin-Manson equation is concerned, however, it is important to note that no correlation of field failures derived from fatigue of nondefective product has ever been proven. Therefore, while it remains in common use in the defect discovery process, attempts to equate thermal cycles to wear-out mechanisms may not be valid. When selecting test temperatures, remember that the wider the temperature extremes of the test, the larger the acceleration factor. While selecting large temperature deltas may be expeditious and provide the worst-case stress for the product, care must be taken that the test does not induce otherwise benign "defects"; that is, anomalies within the product that have no bearing on product life, to fail and result in a defect overcall.

Similar approaches can be used to address other defect mechanisms. Corrosion short circuits, caused by electrochemical migration and occurring both internally and externally to the package, are addressed through another form of environmental testing. In this case, temperature/humidity tests that include a bias voltage applied across the circuits at risk are commonly used throughout the industry. Insulation resistance is read at time zero (ambient) then in situ at intervals throughout the duration of the test. Figure 9-8 depicts a typical resistance trace for a reading between conductors that failed precipitously at 300 hours.

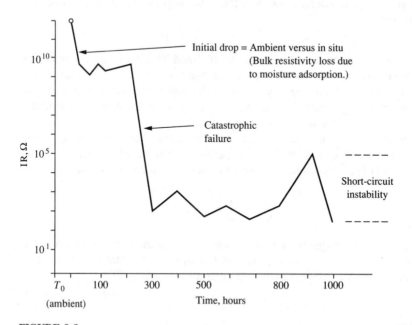

FIGURE 9-8

Typical trace of a single insulation resistance reading on a printed circuit card tested at 50/80/15 V dc.

This is characteristic of such failures, as is a tendency toward extreme fragility. Extreme care must therefore be taken through return to ambient and in failure analysis to prevent loss of fault identification. Readings were taken every 24 hours until 96 hours, then every 100 hours thereafter.

Insulation resistance tests vary in conditions, again dependent on the field stresses expected, and may even include the intrusion of atmospheric pollutants such as sulfur in some of the more sophisticated applications. For second-level products, 50°C/80 percent RH and 15 V (dc) are in common use. Conditions in use for first-level products are typically 85°C/80 percent RH/30 V dc. One test gaining acceptance in the industry is the highly accelerated stress test (HAST), which adds humidity under pressure as a stress factor. While temperature and voltage are necessary stimuli, the use of sufficient humidity to provide enough moisture at the defect site to support an electrochemical reaction is the dominating feature of these tests. Unless the defect site can provide its own moisture (possibly through solution entrapment during processing), we must rely on the diffusion of externally applied moisture into the defect site in order to activate the cell. Humidities ranging from 70 to 90 percent are most practical. Both lower and higher humidities have been successfully used, but the lower ones don't provide sufficient acceleration, while the higher ones are fraught with control problems to avoid condensation on the product. Temperatures from 25 to 85°C are common, and bias voltages may range from 5 to upwards of 100 V. Each stimulus must be evaluated for its effect on failure acceleration, as well as whether or not the exaggerated stimulus alters the mechanism of failure that would occur in the field. Temperature/humidity/voltage tests are even more likely to produce "failures" that don't translate to genuine field exposures than thermal cycle tests if stimuli chosen are too strong. Due to its importance, we will devote more discussion to this subject later in the chapter.

Once the test approach is implemented, analysis of the test data thus obtained requires some familiarity with reliability concepts, since models that forecast performance are based on this data.

QUANTITATIVE METHODS AND RELIABILITY

The ability to organize, manipulate, and derive conclusions from large information bases is one of the true benefits of the data processing industry. This ability has enabled those analyzing information about service performance to construct complex, multiple-parameter models ranging from process optimization as viewed by yield management, to models dealing with the interaction of manufacturing operations with service performance objectives. How well this activity converts raw information into a statistical measure of product quality under application stresses is the concern of reliability management. This approach is an indirect way to define reliability. Reliability is a measure of how well a product meets its performance objectives for the intended product life, while considering design, materials, process operations, and cost.

Some quantitative reliability concepts and definitions will be reviewed below. Their application in the analysis of problems will be illustrated. Before proceeding, however, it is important to point out that there are a number of axioms governing a reliability program. The following basic functions must be an integral part of day-to-day operations:

1. Design for reliability—understand how design factors influence susceptibility of risk sites (regions with a nonzero probability of being defective) to stress amplification.
2. Control reliability through proper manufacturing control.
3. Reliability assurance augmented by in-process test programs and end-of-line functional/stress testing.
4. Reliability assurance augmented by information feedback from test programs and field monitoring programs. Failure analysis feedback to design, development, manufacturing, and quality organizations is critical to attaining the reliability assurance objective.

Overall good reliability can be accomplished by:

A. Proper design analysis
B. Selection of the best materials
C. Elimination of manufacturing errors through process control
D. Proper in-line test definition-quality control
E. Understanding and analysis of product application
F. Clean environment

Distribution Functions

It has been previously stated that reliability is the successful operation of a product in a customer's application environment. "Successful" here means that the device satisfies the consumer requirements and also those of the producer, namely, cost, performance, service life, and so on. Quantifying success is a problem and forces reliability into the statistics of estimation. Information on past behavior must be related to future, similar events by probability statements. For instance, Figure 9-9 illustrates an observation that the strength of a part is distributed about a mean value according to some distribution function. It can also be demonstrated that the application stress is distributed over a range of values according to some distribution function that may or may not be of the same type. The overlap is the area of concern from a reliability viewpoint. Components possessing these strength values are going to experience stressing equal to or greater than their intrinsic strength, and failure can be expected. If repetitive application of stress accumulates damage, then component resistance is reduced—effectively shifting distribution to lower strength values and increasing the area of overlap. The

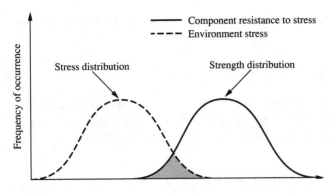

FIGURE 9-9
Intersection of stress and strength distributions.

kinetics of damage accumulation at all stress levels determine the failure time distribution.

In formulating a mathematical reliability model, the first step is to establish the statistical distribution that best describes the failure process. In our example, the statistical statement describing the overlap of the two distribution functions needs to be proposed. The second step is to estimate the parameters defining the distribution. The last step is to use the quantified distribution function to obtain a reliability estimate.

Reliability Statistics

The operational life, or more specifically, the number of hours, years, or cycles of satisfactory operation, is a random variable. The exact value depends on many factors such as variations due to manufacturing tolerances and materials and changes in the environmental conditions. Appropriate statistical models for time to failure can be constructed, and they are used to predict life, develop optimum initial burn-in procedures, and establish part replacement schedules and inventory rules. Three typical statistical models commonly used in life testing and reliability are called *exponential, Weibull,* and *lognormal.* They will be introduced in this chapter, along with methods of estimating unit failure rate and system reliability.

Reliability can be defined as the probability of a successful operation of a device in the manner and under the conditions of intended customer use during the life span of the device. Based on this definition, the following two concepts emerge:

1. Reliability is a probability that lies between 0 and 1.
2. Reliability is a function of time.

The time to failure can be considered a *continuous random variable T.* To characterize the probabilistic behavior of this random variable T, a nonnegative *probability density function f(t)* is generally used. Since T is continuous, $f(t)$ is integrable, such that

$$\int_0^\infty f(t)dt = 1$$

The quantity $f(t)dt$ is called the probability density element (or frequency element) at time t; it measures the probability that the time to failure occurs in the time increment $(t, t + dt)$. The probability contained in the time interval (t_1, t_2) is defined as

$$\Pr(t_1 \le t \le t_2) = \int_{t_1}^{t_2} f(t)dt \tag{9.1}$$

which is represented by the relative area under $f(t)$ between t_1 and t_2 (Figure 9-10).

In (9.1), if we let $t_1 = 0$ and $t_2 = t$, we have

$$\Pr(0 < T \le t) = \int_0^t f(u)du \tag{9.2}$$

This integral is a function of t, and this function, commonly denoted as $F(t)$, is called the *cumulative distribution function*. Since the random variable is time to failure, $F(t)$ represents the probability that failure occurs prior to or at time t. Note that the probability of failure at the exact time t is always 0. The probability contained in the interval (t_1, t_2) can be evaluated by using $F(t)$ as follows:

$$\Pr(t_1 \le T \le t_2) = F(t_2) - F(t_1) \tag{9.3}$$

Graphically, $F(t)$ is a nondecreasing function of t, and it is bounded by 0 and 1 (Figure 9-11).

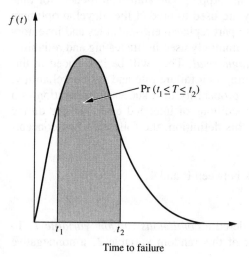

Time to failure

FIGURE 9-10
Probability density function $f(t)$.

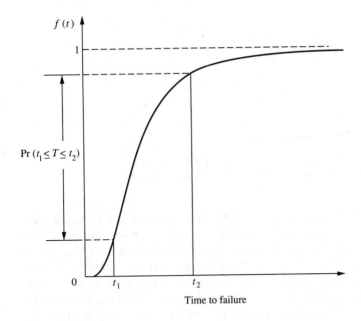

Time to failure

At any given time t, a device has either failed prior to that time or it has survived up to that time. These two events are complementary and their probabilities must add up to 1. We have already used $F(t)$ to represent the probability of failure prior to time t. Therefore, the probability of survival up to time t must be $1 - F(t)$. This probability of survival up to time t precisely matches our definition of reliability. Therefore, the *reliability function* $R(t)$ is defined by

$$R(t) = \Pr(T > t) = \int_t^\infty f(u)du = 1 - F(t) \qquad (9.4)$$

As mentioned earlier, both $F(t)$ and $R(t)$ are probabilities bounded by 0 and 1. For a highly reliable device, one may need several 9s after the decimal point to adequately quantify the reliability. Alternatively, the occurrence rate of failures can also be used to quantify reliability. It is convenient to define a function known as the *hazard function*, $h(t)$, in terms of $f(t)$, $F(t)$, and $R(t)$ as follows:

$$h(t) = \frac{f(t)}{1 - F(t)} = \frac{f(t)}{R(t)} \qquad (9.5)$$

The hazard function represents the probability of failure during a time increment, assuming that no failure had occurred prior to time t. Therefore, the hazard function, $h(t)$, can also be referred to as the *instantaneous failure rate*. Figure 9-12 represents a typical decreasing hazard function.

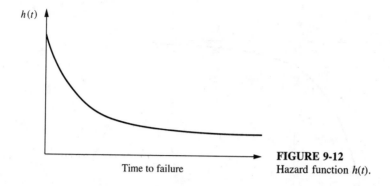

FIGURE 9-12
Hazard function $h(t)$.

The life span of a device resembles that of a human being. During the early life, a child can die of congenital or hereditary defects. Once past this period the child is almost assuredly free of such defects. Deaths from age 10 years to 30 years are mainly caused by accidents; after this period wear-out processes begin to take over and an increasing proportion of deaths can be attributed to old age. The life span of an electronic chip or component carrier can be characterized by similar periods; *early life, chance* or *intrinsic,* and *wear-out.* They are typically represented by the *bathtub curve* shown in Figure 9-13, which has been alluded to earlier. This curve, for example, might help determine the justification for a burn-in to weed out infant mortality (early life failures) or establish a replacement schedule before reaching the wear-out period.

To demonstrate the burn-in concept more fully, assume that early failures have an exponential dependence on time, so that

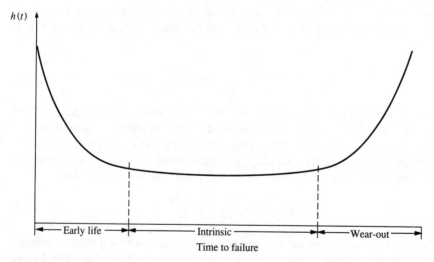

FIGURE 9-13
The "bathtub curve."

$$N_f = N_0\left(1 - e^{at^n}\right)$$

where N_f = number of failures at time t
N_0 = initial sample size
a, n = constants

It will be further assumed that the ratio $N_f/N_0 = 0.001$ at the end of 100 hours of service. If $n = 0.5$, then,

for $\qquad\qquad \dfrac{N_f}{N_0} = 0.0009 \qquad 0.0007 \qquad 0.0005$

$\qquad\qquad t\text{(hours)} = 81 \qquad\qquad 49 \qquad\qquad 25$

Thus, an operational stress for 49 hours will capture 70 percent of the failures expected in the first 100 hours of service; 81 hours will capture 90 percent. The incentive is obvious; failed units can be repaired if not too costly. Also, failure at the producer's facility is far less damaging to the producer's reputation than failure in a customer's location. Repair cost should be balanced against the cost of testing. If the test time can be reduced by increasing the functional stress, then the test time to discover early fails can be reduced or, for the same time, a greater fraction of the population can be screened.

Similar analyses can be performed to understand the intrinsic (constant) and the wear-out (increasing) failure periods. The purpose of constructing such an analysis for the intrinsic period may include other business decisions such as inventory required for spare parts. Through study of the wear-out mode, and the development of adequate models predicting its onset, product life can be anticipated.

As mentioned earlier, the hazard function $h(t)$ only evaluates the instantaneous failure rate in a time increment $(t, t + dt)$. To evaluate the cumulative failure rate, the *cumulative hazard function* is defined by the following integral:

$$H(t) = \int_0^t h(u)du \qquad\qquad (9.6)$$

By using (9.5) and (9.6), it can be easily shown that

$$H(t) = -\ln(1 - F(t)) \qquad\qquad (9.7)$$

Therefore, to quantify the reliability at time t, either $F(t)$ or $H(t)$ can be used to evaluate the probability of failure prior to time t or the cumulative number of fails prior to time t.

Various methods have been developed to estimate reliability from either test data or field data. The simplest method is *probability plotting*. Under this technique, the observed cumulative data points, either $F(t)$ or $H(t)$, are plotted on special graph paper, known as probability paper, and fitted to a straight line "by eye." This straight line is used to estimate $F(t)$ or $H(t)$.

Other estimating methods, such as the *method of least squares* and the *method of matching moments*, can also be used to estimate reliability. The most popular method in estimating reliability is known as the *method of maximum likelihood*. Maximum likelihood estimates have good consistency and are especially sound for censored or incomplete data. However, the derivation of maximum likelihood estimates is not simple and normally requires computer iterations. Various computer software packages to accomplish this task are available, and some sample outputs are given in the data analysis section of this chapter. Readers interested in estimating methods should refer to references [14] and [15].

Analysis Method:
Choice of Distribution Function

Three statistical models commonly used in modeling time-to-failure behavior of a chip or component carrier will be introduced in this section. Applications for these models are presented in the data analysis section.

The exponential distribution is the most commonly used distribution in reliability engineering. It is used to model intrinsic or chance failures when the hazard function does not vary with time. The exponential distribution has only one parameter (Θ), which represents its mean time to failure. Exponential $f(t)$, $F(t)$, $R(t)$, $h(t)$, and $H(t)$ are tabulated in Table 9.2. Note that $h(t)$ is a constant at any time t.

The Weibull distribution is probably the most versatile model in reliability applications. It has two parameters; one will determine the shape of the distribution (α) and the other will determine the scale of the distribution (Θ). It is most commonly used to model early life failures. Weibull $f(t)$, $F(t)$, $R(t)$, $h(t)$, and $H(t)$ are tabulated in Table 9.2. Note that $h(t)$ is a decreasing function of t when $\alpha < 1$, an increasing function of t when $\alpha > 1$, and a constant when $\alpha = 1$. Therefore, Weibull distributions can be used to model early life, intrinsic, and wear-out failures.

The lognormal distribution is generally used to model metal fatigue failures. For example, solder fatigue due to thermal cycling can be closely modeled by a

TABLE 9.2
Statistical models

Model	$f(t)$	$F(t)$	$R(t)$	$h(t)$	$H(t)$
Exponential	$\dfrac{1}{\Theta}e^{-t/\Theta}$	$1 - e^{-t/\Theta}$	$e^{-t/\Theta}$	$\dfrac{1}{\Theta}$	$\dfrac{t}{\Theta}$
Weibull	$\dfrac{\alpha}{\Theta}\left(\dfrac{t}{\Theta}\right)^{\alpha-1}e^{-(t/\Theta)^\alpha}$	$1 - e^{-(t/\Theta)^\alpha}$	$e^{-(t/\Theta)^\alpha}$	$\dfrac{\alpha}{\Theta}\left(\dfrac{t}{\Theta}\right)^{\alpha-1}$	$\left(\dfrac{t}{\Theta}\right)^\alpha$
Lognormal	$\dfrac{1}{\sqrt{2\pi}\sigma t}e^{-(\ln t - \mu)^2/2\sigma^2}$	$\Phi(z)$ $z = \dfrac{\ln t - \mu}{\sigma}$	$1 - \Phi(z)$	$\dfrac{e^{-z^2/2}}{\sqrt{2\pi}\sigma t[1 - \Phi(z)]}$	$-\ln[1 - \Phi(z)]$

lognormal distribution. The lognormal distribution also has two parameters; one determines the location (μ); the other determines the scale. The location parameter μ is the log of the time t_{50} for 50 percent of the population to fail, just as Θ (Weibull) was the time for $1/e$ (63.2 percent) to fail. Lognormal $f(t)$, $F(t)$, $R(t)$, $h(t)$, and $H(t)$ are tabulated in Table 9.2. Note that $F(t)$, $R(t)$, $h(t)$, and $H(t)$ for the lognormal distribution are all expressed in terms of $\Phi(z)$, a standard function often given in table form.

The data from Table 9.3 represents results from a stress test performed on a sample of size N, with n failures distributed over time. The cumulative fraction failed, $F(t)$, is $n(t)/N$. The reliability function can be obtained by $R = 1 - n(t)/N$. Similarly, the cumulative hazard function can be determined from the relation $H(t) = \ln(1/1 - F(t))$.

Figure 9-14 shows a plot of log $F(t)$ versus log (failure time). Note that a clearly nonlinear relation appears to exist between the two, with systematic deviations at longer failure times. Notice also that no failures occurred between 2100 hours and the end of the test: 3500 hours. Any model of the failure process should permit an estimate of the time to the next fail within some confidence bounds. Using this example, if that bounded estimate does not exceed 1400 hours, the assumption that all samples are mortal at the test stress, that is, are capable of failure, may not be adequate. Alternatively, the distribution function chosen may not adequately represent the data. The distribution function $f(t)$ must permit extrapolation outside the range of the experiment. A model could be selected on the basis of a fit to the data points, neglecting the early fails. This, too, is a problem, since reliability estimates tend to be more concerned with early fail projections. Figure 9-15 also shows the progressive difference between $H(x)$ and $F(x)$ as given in Table 9.3. It suggests that for small values of F the statistical model for H is a good initial estimate of $F(x)$.

TABLE 9.3
Hypothetical distribution of observed failure times

Sample size = 50
Test duration = 3500 hours

Number of fails	Number of survivors	Time	$F(t)$	$R(t)$	$H(t)$
1	49	25	.02	.98	.0202
2	48	92	.04	.96	.0408
3	47	180	.06	.94	.0618
4	46	290	.08	.92	.0833
5	45	290	.10	.9	.1054
6	44	540	.12	.88	.1278
7	43	670	.14	.86	.1508
8	42	810	.16	.84	.1744
9	41	810	.18	.82	.1985
10	40	1200	.20	.8	.223
11	39	1500	.22	.78	.2485
12	38	2100	.24	.76	.274

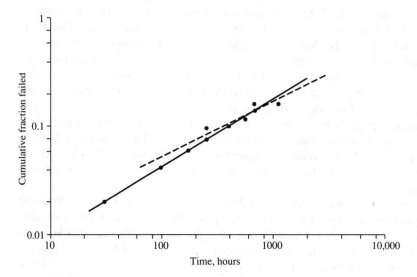

FIGURE 9-14
Illustrative example: failure-time distribution.

Acceleration of Service: The Acceleration Factor

Whenever the applied stress exceeds that anticipated for the product in field use, the test data is thought to be accelerated in time. The purpose of applying more stress than will occur during product use is to compress the field experience into a relatively small increment of time. By so doing, product reliability can be studied

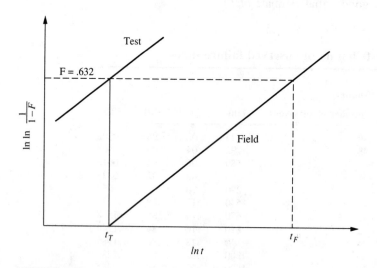

FIGURE 9-15
Schematic illustration of failure time distributions for field stresses and test stresses. The ratio t_F/t_T is the test acceleration factor.

in the laboratory in advance of product announcement. Stress testing parameters, therefore, are carefully selected such that a relationship can be drawn between the test time to failure and that anticipated for the field. Some types of stress testing and the proper selection of acceleration parameters are discussed later in this chapter. For now, we will focus on establishing the mathematical relationship between field and test conditions.

For this discussion, assume the failure analysis data are well described by a Weibull formulation. Thus,

$$F = 1 - e^{-(t/\Theta)^\alpha} \tag{9.8}$$

Here, only Θ and α are traceable to the component and its response to stress. The parameter Θ is referred to as the characteristic life or the time for F to increase to $1 - 1/e$, that is, 0.632. Figure 9-15 illustrates the variation of $\ln \ln[1/(1 - F)]$ versus $\ln t$ for stress test conditions and conditions representing field service. The ratio t_F/t_T is defined as the *test acceleration factor*, which is equal to Θ_F/Θ_T if and only if α has the same value for both the test and the field. This statement is arrived at in the following way. Equation (9.8) can be rearranged to yield

$$\ln \frac{1}{1 - F} = \left(\frac{t}{\Theta}\right)^\alpha$$

or

$$\left(\ln \frac{1}{1 - F}\right)^{1/\alpha} = \frac{t}{\Theta}$$

Thus, for a constant value of F for the field and for the stress test, we obtain

$$\frac{t_f}{\Theta_f} = \frac{t_T}{\Theta_T}$$

which equates to

$$\frac{t_f}{t_T} = \frac{\Theta_f}{\Theta_T} = \text{acceleration factor}$$

This statement claims that the field stress conditions can be accelerated in time by increasing the field stress, for example, temperature, voltage, or mechanical stress. If the acceleration factor equals 100, then every test hour is equivalent to 100 field hours. However, the act of increasing the applied stress must not alter the failure mechanism(s). Improperly chosen test parameters may even cause failures to occur in test that do not occur in field conditions. When this happens, the test may overestimate the number of field fails during a given time period.

Other conclusions can be reached. The failure mechanisms must be identical; this situation implies something about the defect distribution and further implies that the failure data should be grouped according to individual failure processes. Whenever possible a failure model should be developed around a single process; mixed failure modes needlessly complicate any statistical model.

Risk sites in the product population are presumed to possess a distribution of strengths with respect to the operation environment stress. This distribution is shown in Figure 9-16, where frequency of occurrence of risk site strength is plotted against some measure of this strength. This might, for instance, be the time to failure, with the weakest risk site having the shortest time to fail.

Sample size may have a pronounced affect on results. A random sample of 25 selected from a population might be expected to behave differently than a sample of 250. The smaller sample size group may have early, intermediate, or long-time failures censored from the test samples. In fact, if the sample size is small enough, the risk site strengths may exceed the test stress, so that no failures would be observed.

The impact of inadequate sampling of data can be disastrous. Regression analysis to model the failure statistics could be in gross error. Reliability, quality, or manufacturing process decisions based on this analysis could be equally faulty. Sampling errors typically affect the sample variance with the effect that the variance is underestimated. Statistical methods for estimating the sample size required for a given measure of confidence clearly are important for valid decision making.

One parameter in the Weibull expression is the shape parameter α. Some comments have already been made about the expected behavior when α is greater,

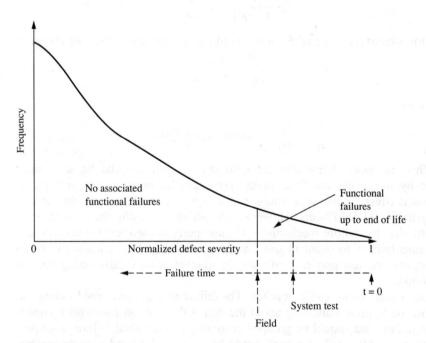

FIGURE 9-16
Defect frequency distribution measured by severity: failure time under stress.

equal, or less than unity. However, from an $h(t)$ analysis viewpoint, it is important to consider the behavior of α in a stress test versus that in the application environment. Whenever a comparison between an accelerated stress test and the field is attempted, it is implicitly assumed that the defect populations of both sample sets are identical. It is further assumed that the degradation process of both exposures is identical, differing only in kinetics. Thus, if the defect distribution is the same and the kinetics are relatable via the test acceleration factor, then the value of α has to be identical in both environments. Regression analysis of both data sets together with the added constraint of equality of slopes in the Weibull plot assures the "correct" value of α, rather than the "best fit" value.

The Weibull and other statistical formulations for the cumulative distribution function forecast values of F increasing to unity with time. Consider what happens to our analysis if that assumption is not true—that is, if not all of the product is mortal. In this case, part of the population is either defect free or contains defects that for one reason or another do not trigger the failure mechanism. In such cases, the forecast of total failures and failure rates at some fraction of product life will be overestimated. If this happens in a stress test for a newly developed product the risk for product rejection, redesign, or manufacturing process review is high. Data has to be continually examined for systematic deviations from expected behavior based on regression analysis, as measured by departure from the expected shape parameter.

Analytical tools prove useful in this situation. For instance, it can be assumed that the limited number of defects are Poisson distributed amongst the samples and discovery is distributed according to Weibull statistics. The Poisson distribution function is

$$\Phi(x, \lambda) = \sum_{k=0}^{x} e^{-\lambda} \frac{\lambda^k}{k!}$$

where for our case x is the number of defects capable of causing failure by a specific mechanism in a system sample, and λ is a parameter governing the expected distribution of defects in the sample. For instance, if $\lambda = 0.3$, then we get the following:

$X =$	0	1	2	3	4
$\Phi =$	0.741	0.963	0.996	1.00	1.00

This means that 74.1 percent of the population is defect free, 22.2 percent (96.3 − 74.1) contain one defect, 3.3 percent contain two, etc. Combined with the Weibull distribution, the equation becomes

$$N_F(\lambda, t) = \Phi(x, \lambda)N_0\left(1 - e^{-(t/\Theta)^\alpha}\right)$$

This analytical method is discussed in more detail later in this chapter. This technique recognizes that each member of the sample set can contain more than one defect (even though only one defect will be responsible for failure, excluding repair and repopulation), or even contain none at all for this failure mechanism.

Component versus System Reliability

In stress testing of new product technologies, unit and risk site failure rates are frequently employed interchangeably. Such practice arises from the necessity to equate risk site sample size to unit sample size. The issue is further complicated when field projections are made to estimate the field behavior expected over the product life. In this section these definitions are reviewed and a procedure described whereby stress test failure statistics are used to project field behavior.

System reliability can be modeled as either a series or a parallel construction. Less commonly, a combination of both is employed. In a series model, failure occurs whenever one of the system components fails. Conversely, a parallel system fails only when all of its components fail. In our terminology a unit is a card, board, module, etc., and the components are the risk sites (failure sites) whose reliability affects the reliability of the unit. An assembly of units becomes the system, and so forth. In packaging, the serial failure problem is the most common and will be discussed in some detail. Standard reliability texts can be consulted for analysis of the parallel problem.

According to probability concepts, if each successful test or failure is mutually independent for each and every risk site then the system reliability R is equal to the product of the reliabilities of each type of risk site; that is,

$$R = R(1) \times R(2) \times R(3) \times \cdots \times R(n)$$

$$R = \prod_i^n R_i$$

where R_i is the reliability at time t of the ith component of n independent components. This is known as the product rule for serial systems. To evaluate system reliability for a dependent serial system, conditional probabilities, which are quite complex, have to be used. However, there is a special case of a dependent serial system called the *weakest link* model where the maximum possible serial system reliability is attained from the minimum of the component reliabilities. For the weakest-link model, the maximum system reliability is given by:

$$R = \min R_i, \qquad i = 1, 2, \ldots, n$$

Since as before $R = 1 - F$, where F is the cumulative probability of system failure, and g_1 is the same for each type of risk site, the above expression becomes

$$1 - F = (1 - g_1) \times (1 - g_2) \times \cdots \times (1 - g_n)$$

Further, if all risk sites $(1, \ldots, n)$ are identical, then

$$1 - F = (1 - g)^n$$

Recall from the previous section that $F(t)$ is the probability that a fraction of the total sample size will fail after a time t. From this,

$$F(t) = \frac{N_f(t)}{N_0}$$

with N_0 being the sample size and $N_f(t)$ the number of failures after time t. The fraction surviving,

$$\frac{N_0 - N_f(t)}{N_0} = \frac{N_s(t)}{N_0} = R_i(t)$$

has been defined as the reliability. Now, consider the meaning of the term *failure rate*, which is the number of failures occurring during a time period.

$$N_f = N_f(t_2) - N_f(t_1)$$

$$t = t_2 - t_1$$

Thus, a definition of failure rate might be the ratio

$$\frac{F}{t} = \frac{N_f/N_0}{t}$$

or the average failure rate (AFR) for the product life (PL):

$$AFR = \frac{F \times 100 \times 1000}{PL}$$

$$= \frac{F \times 10^5}{PL} \tag{9.9}$$

where the factor 100 puts the AFR on a percent basis, and the factor 1000 is introduced to scale time to kilohours.

Actually, after a time t_1 only the survivors at t_1 can fail during the next t. Thus, N_f with respect to N_s may be of more interest than with respect to N_0. For this case

$$\frac{N_f}{N_s} = \frac{N_f/N_0}{N_s/N_0}$$

$$= \frac{F(t)}{R(t)}$$

An AFR can be defined alternately using this ratio:

$$AFR = \frac{F}{R \times PL} = \frac{F}{(1 - F) \times PL}$$

$$AFR(\%/1000 \text{ hours}) = \frac{F \times 10^5}{(1 - F) \times PL}$$

For small values of F, $1 - F \approx 1$ and we have the same expression as before.

A reasonable goal is to be able to specify a failure time distribution in a stress test that would forecast a minimum field reliability. Whenever a product reliability target can be specified, such as an AFR target at end of product life (PL), then the following results. On solving the AFR expression (Equation 9.9) for F and inserting into the serial reliability equation,

$$(1 - g)^n = 1 - F = 1 - \frac{\text{AFR} \times \text{PL}}{10^5}$$

or

$$g = 1 - \left(1 - \frac{\text{AFR} \times \text{PL}}{10^5}\right)^{1/n}$$

This says that at end of life the cumulative number of risk sites permitted to fail is known. Further, that is precisely the number permitted to fail at the end of life stress test. This information has been illustrated in Figure 9-17, which plots $\ln \ln(1/(1 - g)$ versus time, either field or test time. The point $(1/(1 - g), t)$ is known, as is the slope of the cumulative failure time plot for the failure mechanism associated with the risk site. The product passes the stress test whenever the fail time distribution falls below the predicted curve, and fails if the distribution is above the specification limit.

There is another benefit from using these equations. The reliability for a chip or component carrier can be considered as the system reliability for a system that consists of numerous types of risk sites. If the unit or intrinsic failure rate for each

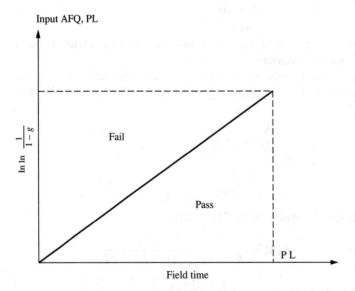

Stress test time = field time / acceleration factor

FIGURE 9-17
Arriving at pass/fail criteria for failure distribution based on reliability goal at end of product life.

risk site type is known, and if the risk sites are independent with no redundancies (parallel configurations), then the intrinsic failure rate for the carrier is given by an expression like the following (written for three components):

$$F = 1 - (1 - g_1)(1 - g_2)(1 - g_3)$$

$$= 1 - 1 + g_1 + g_2 + g_3 + \text{ second-order terms, etc.}$$

or, since F is proportional to AFR

$$\text{AFR}_{\text{carrier}} = \text{AFR}_1 + \text{AFR}_2 + \text{AFR}_3$$

In general,

$$\text{AFR} = \sum_{i=1}^{r} \text{AFR}_i$$

This is a convenient way to evaluate system failure rate based on risk site failure rates—assuming they cover the same time period, i.e., are average failure rates.

As mentioned earlier a parallel system is defined as a system consisting of n components such that system failure occurs when and only when all components fail. Equivalently, the system is successful when at least one of its components is successful. The system reliability for a parallel system with n independent components is given by

$$R = 1 - \prod_{i=1}^{n} (1 - R_i)$$

Parallel systems are much more reliable than serial systems because of the n-fold redundancy. For some products, package design may require sufficient redundancy to maximize system reliability. An example of this is the use of a bifurcated contact system. Another example is the use of two wire bonds to the same I/O site on a chip. See Chapter 7 for discussions of connector systems, and Chapter 19 for joining and interconnection.

A mixed system contains any combination of serial and parallel subsystems. By combining the expressions for the reliability of serial and parallel systems an expression for the reliability of a mixed system with independent components can be formulated. For example, consider the following mixed system with five independent components:

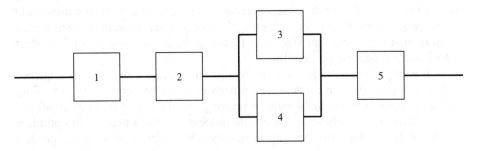

$$R = R_1R_2[1 - (1 - R_3)(1 - R_4)]R_5$$

Insertion of typical R values demonstrates that the value of R is significantly increased by the parallel redundancy versus the series configuration.

PROCESS CONTROL AND PRODUCT RELIABILITY

Previous sections of this chapter discussed the concept of reliability engineering, approaches to product testing, and related statistical methods. This section focuses on reliability in the manufacturing environment.

Process Yields, Defects, and Their Effect on Reliability

There are a few terms that have already been used in previous sections, but which should now be defined to avoid confusion in the following discussion. These terms are *process, defects,* and *yield*.

A *process* is the technique employed to produce a finished product in the manufacturing environment. In electronic packaging, this technique typically involves a sequence of mechanical or chemical operations that collectively make up the total process. The term *defect*, as used here, is any undesired and unplanned outcome of the process which, if left unrepaired, has a potential for causing a functional failure of the product. A defect may be an outcome of a single operation in the process sequence, or result from a combination of several operations. Purely aesthetic blemishes, although also not a desirable outcome, are not counted as defects here. The intent is to focus on defects as they relate to reliability. *Yield* may be defined as that percentage of the product considered usable in the next operation in the process sequence. The product of the individual yields of all process operations is referred to as the cumulative yield of the process.

It is important to remember that the next operation following the last step in the process sequence is the use of the product by the customer. Thus, the cumulative yield must take into account reliability concerns, and reliability criteria must be applied as if each step of the process sequence was the last.

Another point to consider is that it is often impractical to perform defect screening after each individual operation. Therefore, the yield of a given operation may not be counted until the product reaches a measurement point downstream in the process sequence. This increases the likelihood of encountering defects rooted in more than one operation, since one flaw may build on another as the product progresses through the process line.

It is important to understand the relationship between the process sequence and the defects that it can create if not properly executed and controlled. An ability to control the process is paramount to meeting reliability and cost objectives.

It is natural to believe that adequate process control is that which optimizes yields. If the product measures up to quality criteria imposed upon the product at the end of the manufacturing line, reliability is assumed. However, depending on product requirements, this assumption may be faulty.

There are basically two reasons why acceptable reliability should not necessarily be expected based upon quality criteria alone:

1. Our ability to define accurate criteria may be less than perfect.
2. Testers may not be 100 percent efficient at detecting every variety of defect, even when 100 percent of risk sites are tested.

Recognizing and dealing with these fundamental weaknesses is crucial to eliminating defects and controlling product reliability.

Since the potential exists for a mismatch between shipping criteria and reliability, additional strategies often must be employed to adequately protect the product. One strategy is to perform additional screening on the product, for example, at both the board level and after the board has been used in the system. Such screens may be effective, particularly while attempting to control an unacceptable defect level. However, the best approach is to avoid making the defect. A concentrated effort is required to assure that the process does not create defects that must be screened.

Figure 9-18 shows a population of defects subjected to any given screening test (a normal distribution is assumed). The shaded area to the left of the *test efficiency* line represents the portion of the distribution that escapes detection. The number of escapes, as they are called, will vary according to the test efficiency and the criteria chosen to define the defect. It is the escaping population that must be understood and eliminated to avoid a reliability impact. These defects may be detected by sufficient accelerated stress conditions (voltage, temperature, relative humidity, etc.). Subsequent failure analysis to determine a process cause is then

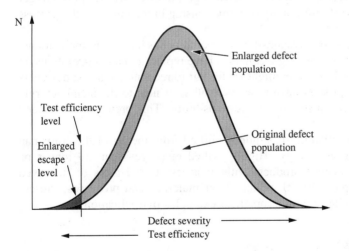

FIGURE 9-18
Effect of enlarged defect population on escape level. As defect population grows, number of escapes grows at a constant test efficiency level.

employed to define corrective actions and eliminate them from the population. Thereafter, properly chosen test parameters and defect definition will help eliminate future escapes. Chapters 21–24 provide examples of modern techniques in failure analysis as well as approaches to characterize the product as a result of changes in the process.

Looking again at Figure 9-18, it also illustrates the effect of increasing the defect population in the product before screening on the number of escapes. For any given test efficiency, the percentage of defects escaping remains relatively constant. Thus, if a process is allowed to run out of control, the enlarged defect population it produces will result in more escapes from the test.

In such cases the test becomes a process monitor as well as a defect screen. The resulting data base becomes a measure of process quality. Departure from established statistics for defect rates signals potential deterioration of process control and/or material changes. Either could trigger decreased yield and reliability.

It is necessary to continue to simplify and model the process, leading to improved monitoring and screening techniques. At the same time, remember that each defect created is a potential escape, and each escape is a potential functional failure.

Eliminating Defect Occurrence

The need to eliminate defects in manufacturing lines has always existed. However, the methods to reduce defects, and the degree to which we must go to ensure defect elimination, have changed with the increasing complexity of our products.

In a less technologically challenging environment, only a few rather straightforward actions would normally minimize defects. However, as processes become more sophisticated, there is a growing dependence on process monitoring and control to ensure a defect-free product. Increasingly, this dependence puts a larger share of the responsibility for quality of workmanship in the hands of the process engineer.

The concept of a shared responsibility for quality has important implications. Even though automatic processing techniques have replaced manual operations on today's manufacturing floor, one might easily fail to recognize that the designers and custodians of the process have now assumed seats next to the manufacturing machine operator as *co-owners* of this responsibility. This premise is expanded in Chapter 31.

Dependence for reliability has quietly shifted from the art of the craftsman to the science of the technology. Highly skilled employees can no longer be expected to produce quality products without access to the tools of improved applied science. Chapters 10–20, which cover materials and processes, present this scientific base while providing practical examples of application.

The Process Window

As noted, process engineers have a responsibility to provide the manufacturing operator with the tools needed to avoid making defects. Along with monitoring product quality, a system must be provided that monitors the process and allows process control within carefully specified bounds. By monitoring and controlling all the process parameters within certain limits, the occurrence of defects which might otherwise accumulate is eliminated. These limits are known as the process window. This process definition work is critical and must be done with intense examination of the product for defects, the determination of a defect's cause, and finally elimination of the defect through process control.

Prior to defining the process window, process development engineers must do three things:

1. Identify the candidate variables to be controlled within the process. (Define *independent* process variables.)
2. Identify all of the actions his process actually performs on the product. (Define *dependent* product variables.)
3. Establish a highly developed feedback system to supply defect information during the experimental phase of process development.

The process operation and its desired results are broken down into sets of process and product variables, such that each variable may be evaluated separately. Later, when each variable is understood, the sets can be reassembled systematically into a composite process.

Before launching a series of experiments, however, a feedback loop should be developed to indicate when the process is producing defects. In defining the process window, an engineer attempts to define the upper and lower limits of each process variable beyond which defects will occur.

Remember that the defects to be avoided may not be visible or obvious. Therefore, sensitive techniques must be established downstream from the process operation to detect all the defects to which this operation may contribute. These techniques can be temporary; i.e., for use during development, and perhaps during early hardware manufacturing stages. The fact that the techniques are not permanently required provides rationale for using otherwise prohibitive means for the sake of attaining a high defect detection rate. Such means may include sensitive analytical tools and life testing of product in simulated customer environments. Whatever method is selected, the engineer must be certain that a feedback system exists to detect as accurately as possible any process-caused anomaly that could cause a functional failure. Once established, the process window replaces this technique and itself becomes the basis of a practical feedback system for the process operator.

The foundation for experimental design is provided by first identifying the independent process variables and their dependent product variables. In a series of controlled experiments, the effect of varying each process variable individually on each product parameter will be evaluated. From this data, a set of process control limits is defined that corresponds to those points on the variable scale where defects begin to occur at an unacceptable rate. Consistent with the definition for defects as stated earlier, the term *unacceptable* is defined as that level at which they possess a real potential for causing a functional failure. This usually gives two set points: (1) a minimum set point, below which one population of defects is unacceptably large, and (2) a maximum set point, above which another population of defects becomes unacceptable.

Figure 9-19 depicts observed defect levels as a function of time for a photoresist expose system. As one increases expose time from zero to infinity, the first region of defects resulting from underexposure decreases and reaches an acceptable level at time t. Continuing to expose past time t, the defect level remains at or below this minimum until another population of defects begins to appear at time t'. This is a different population from the first in that it is derived from overexposure, whereas the first population was derived from underexposure. Chapter 22 provides an example of the use of scanning electron microscopy (SEM) for this particular application.

The region of expose times lying between t and t' in Figure 9-19 constitutes the process window for this parameter. However, a safety factor is applied before establishing the practical limits of operation, which are set safely inside of t and t'. The amount of safety factor employed depends on how fast the defect rate rises as one progresses outside the process window and the precision of the data. Once established for one parameter, this procedure is repeated for each process parameter until the window for each is defined. Remember that the ability to establish adequate quality criteria and apply process set points that are well away from danger zones depends heavily on the sophisticated failure analysis feedback system.

Thus far we have ignored the possibility of two or more process parameters *interacting* with each other to produce a different effect than they would if they

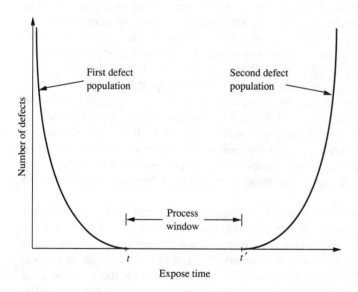

FIGURE 9-19
Definition of expose time process window.

were operating independently. Interaction is present when results of one cell of an experiment can only be reproduced when all other variables are held constant. When this occurs, it is not sufficient to independently study each variable. Any interactive process variables must be evaluated in a matrix of experiments. While evaluation of these higher-order effects can complicate the job very quickly, most processes require that this work be done if adequate control is to be achieved. Such experimental design is quite detailed, however, and is beyond the scope of this book. Factorial experimental designs, such as *Hadamard* matrices, are required, which are well documented in the literature (see references [16], [17], and [18]).

The Critical Process Variable: Measuring the Unmeasurable

By correlating process variables to product parameters that could only be seen through the additional means of our temporary feedback system (such as scanning electron microscopy), the existence of defects can now be predicted even while they are forming. Such predictions are possible even though the defect may not be detectable after the operation is complete. Now, defects that previously accumulated on each piece produced (due to lack of detectability) can be completely avoided, thus providing a measure of control never before possible.

The ability to predict the outcome becomes even more important in cases where the results of the operation cannot be seen or evaluated until later in the process sequence. Such processes warrant special attention since they represent the highest risk of operating at high defect levels for long periods of time. Because of their importance in this respect, such operations should be labeled as critical processes, and particular effort must be made to ensure that these processes remain under control.

The use of an automatic aperture-priority camera helps to illustrate this point. The photographer has no way of knowing whether the film has been correctly exposed until the film has been developed. In the meantime, the entire roll of film has been exposed without having received feedback on the film exposure process operation. If there are 24 pictures on the roll, this means 24 units of product have been committed without knowing the outcome. Despite this fact, the photographer can proceed with confidence because the designer of the camera's exposure process has recognized the need for operating within a process window, and provided a light meter monitor and controls within the camera to adjust shutter speed for any given aperture setting, light conditions, and film speed. The camera's engineers have correlated all of these process variables with the resulting product variable.

The benefits of this capability can be very significant, particularly when a product might not be tested until several operations down the line from where the defect was actually created. If there were no process monitors to sense that something was wrong, several days or weeks of production may be committed at the unacceptable defect level before feedback finally indicated a problem.

In summary, this section emphasized a strategy for manufacturing zero defects. It is based on the premise that to produce the most reliable product it is necessary to simplify and analytically model the process. This minimizes defects and at the same time leads to improved monitoring techniques for the process. Through the use of thoroughly developed process monitors and controls, an unprecedented era of quality performance can be established. The only limitation is an ability to understand the processes, and how they are fitted with appropriate measures of operator control.

TEST METHODS, PLANNING, AND DATA ANALYSIS

Reliability and quality have been discussed thus far from the viewpoint of the separate organizations, including development and manufacturing. A common thread running through the discussion is the need for feedback extracted from test data analysis. To emphasize this theme, the final section considers test methods and data analysis techniques as an essential ingredient for achieving reliability objectives.

Reliability Testing

Stress tests have been developed to reproduce the failure modes, occurring through degradation, that were observed on field returns of chip carrier packages, printed circuit packages, and the cabling interconnecting circuit packages or output devices. Aside from mechanical damage resulting from handling, the major degradation mechanisms are corrosion, thermal-mechanical fatigue, and connector contact degradation induced by wear or the accumulation of airborne particulates in the contact area. See Chapter 7 for electrical contact and design considerations.

Corrosion can take many forms ranging from a conductive bridge between adjacent circuit elements to localized corrosion attacking a circuit segment or the creep (spreading) of corrosion into the contact area of a connector. The existence of corrosion processes implies the presence of humidity and either gaseous pollutants from the environment or surface contamination from residual process chemicals. Under certain conditions corrosion can even occur internal to the package, essentially independent from the external environment.

Thermal-mechanical fatigue is another primary failure mode and results from the cracking or fracture of a circuit element such as a solder connection or plated through-hole. As was shown in Table 9.1, such failures are induced by thermal cycling, torquing, and vibration. Vibration can also accelerate wear at contact points within a connector, causing a failure through loss of precious metal (used to reduce electrical resistance and retard corrosion in the contact). Other connector failure mechanisms can be activated by vibration, such as polymerization (by friction) of lubricants in the contact area. Each of these mechanisms is further aggravated by the presence of dust particles in the contact area during vibration.

In performing the test and projecting product reliability based on these stress tests, it is necessary to know the number of elements or risk sites that exist in the product and the number of risk sites tested. Risk sites such as the number of lines can be easily counted, but the insulation resistance risk sites are not easily counted in a complex, multilevel printed circuit board. Any two lines side by side constitute a risk site. Three lines side by side would have two risk sites, between lines 1 and 2 and between 2 and 3. However, on a different wiring plane lines 1 and 3 may be side by side and so constitute a risk site. Computer programs are usually needed to determine the number of insulation risk sites as well as to determine how testing is to be performed to maximize the number of sites tested.

Risk site reliability would be determined from the test results. Then the reliability of the product would be established based on the reliability of each risk site or element and the number of risk sites in the package. Remember, it was established that

$$R(\text{system}) = 1 - F(\text{system/package}) = (1 - f(\text{risk site}))^n$$

where n is the number of risk sites and f is the probability of failure of each risk site type. For example, if $f = 1/10^6$ and $n = 10^4$, then

$$F = 1 - (1 - 10^{-6})^{10^4}$$

$$= 0.0198 (\approx 2 \text{ percent})$$

If n is increased to 50,000, F is increased to 0.0487. Thus, even if the individual risk site reliability is excellent, package reliability can be marginal.

Packages of different complexity using the same technology would have different reliabilities. Technologies with different parameters such as line widths, line-to-line spacing, as well as different materials and processes, may have different reliabilities. Therefore, reliability tests are designed to relate risk site, material, and process parameters so that product and process engineers can effectively make reliability improvements on present and future products.

Risk sites that fail the stress test (prior to end-of-life equivalence) in the absence of any defect are due to faulty design, in which case the data is used to either alter or eliminate the risk site in future design iterations. Given a mature design, risk sites failing in the stress test (as well as the field) are usually less severe examples of the time-zero defects being captured by in-process tests. If failures are encountered, then this data is used in like manner to the failure analysis data obtained during process window definition to eliminate future occurrences. In addition to process correction, there may be adjustments necessary in test strategy before successfully eliminating the problem.

It is important to maximize the use of field data on current product to signal areas requiring attention on product still being developed. Risk site field performance demonstrates how these sites behave under operating conditions in the real packaging environment. All electrical and thermal-mechanical stresses are present for synergistic interactions to occur if such interactions could contribute enhanced degradation. Thus analysis of such field data permits calculation of the intrinsic failure rate of each risk site. This means that the package failure rate is the sum of the intrinsic failure rates of each risk site type in the package, weighted by the number of risk sites contained in the package. Such methodology is acceptable as long as the new product is simply an extension of the existing technology to a greater performance, rather than a different technology in the sense that the application stress on the risk site is now totally untested or geometrically altered.

The risk site in the new design is really an unevaluated site, and can present a dilemma of choices in strategy. Here is an example. Assume a package with 10,000 type A risk sites has a reliability such that 1/100 packages are expected to fail at end of life; one package per hundred contains a defect capable of causing a package and a system failure. What is the strategy to assure that the new product meets or exceeds that reliability; especially if each package, subjected to a destructive stress test, costs $1000? Try to take at least a qualitative position on this question.

Data Review: Failure Analysis

As stress test and field return data are organized for analysis, one major difference is immediately apparent. Product enters field service distributed over time; i.e., the number of stress hours is a variable, not like a typical stress test of fixed initial sample size and fixed test duration. The question becomes how to assess the field data. Somehow, attention has to be focused on product survival, as well as the incidence of failure. An analogy is that of a test with a sample size of N_0 from which samples are removed periodically (randomly), even though they have not failed. This process is called censoring. The result is a test population remarkably similar in test hour distribution to the field population with its distribution of service hours.

Table 9.4 contains an illustrative example of field behavior of an imaginary product. The data represent performance over three vintage periods, that is, three six-month periods of continuous shipping. These data have been screened to reflect a single failure mechanism. What is noteworthy is the response to the problem by corrective actions resulting from failure analysis and feedback to the appropriate process/manufacturing operations. A plot of the failure time distributions for each of these vintages is presented in Figures 9-20, 9-21, and 9-22.

Based on the progressive increase in the characteristic life value with vintage, and the fact that the shape parameter is less than unity, the following conclusions can be reached.

1. The results from failure analysis fed back into the process/manufacturing cycle resulted in steady improvement in the defect level, but not all defects were eliminated.
2. The behavior would be classified as "infant mortality": failures caused by manufactured defects that escaped from the test procedures.

TABLE 9.4
Examples of field failure data for vintage A, B, and C products

	Vintage A			Vintage B			Vintage C		
Unit number	Service time to fail	Total service hours		Unit number	Service time to fail	Total service hours	Unit number	Service time to fail	Total service hours
1		424		1	8	8	1	8	8
2	64, 852, 1103	1193		2*	36, 108	456	2	48	473
3*		3435		3		112	3		982
4	64	3435		4*	212, 657	1210	4	148	398
5*	8, 168	300		5	800	1305	5	473	1142
6	2497	2497		6	168	300	6	523	643
7	256	652		7		2410	7		143
8		5305		8		1650	8		687
9		863		9		650	9		924
10	3882	4530		10	48	48	10		1418
11		1113		11		4900	11		1932
12	1	1		12		3600	12		1247
13*	224, 657	1180		13		2870	13		600
14		3839		14		4300	14		748
15		873		15		1910	15		824
16	60	899		16		2240	16		268
17		3719		17	1510	3940	17		312
18		96		18		2100			
19		311		19		275			
20		3454							
21	12	3800							
22	48	2600							
23		5800							
24		3610							
25		4387							

* Denotes multiple fail/repair actions.

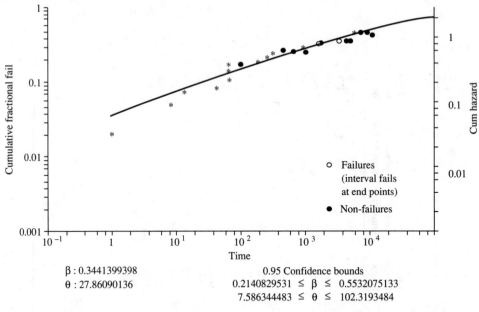

β : 0.3441399398   0.95 Confidence bounds
θ : 27.86090136   0.2140829531 ≤ β ≤ 0.5532075133
                                    7.586344483 ≤ θ ≤ 102.3193484

FIGURE 9-20
Two-parameter Weibull maximum likelihood method analysis of product field data: Vintage A.

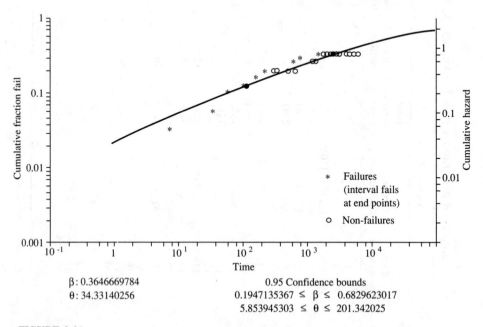

β: 0.3646669784   0.95 Confidence bounds
θ: 34.33140256   0.1947135367 ≤ β ≤ 0.6829623017
                                    5.853945303 ≤ θ ≤ 201.342025

FIGURE 9-21
Two-parameter Weibull maximum likelihood method analysis of product field data: Vintage B.

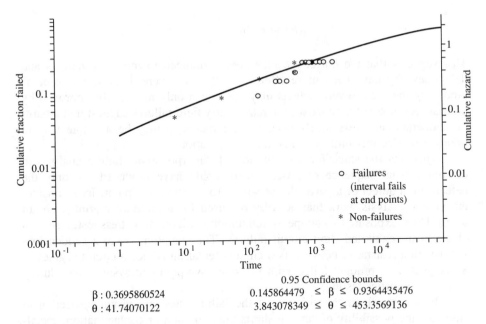

β : 0.3695860524   0.145864479 ≤ β ≤ 0.9364435476
θ : 41.74070122   3.843078349 ≤ θ ≤ 453.3569136

FIGURE 9-22
Two-parameter Weibull maximum likelihood method analysis of product failures: Vintage C.

Further reduction in defects requires additional study, particularly of the critical process steps.

Additional information can be extracted from the data with an alternative analysis; the *Poisson-Weibull-MLE* method. This method assumes defects are Poisson distributed within the population and their discovery is distributed according to Weibull statistics. The analysis provides the following results:

Poisson-Weibull-MLE			
Vintage performance	A	B	C
Shape parameter α	0.47	0.73	0.74
Characteristic life Θ	1683	472	197
Expected fraction mortal	0.63	0.47	0.27

Now, a different concept of product mortality emerges. This analysis claims that not all the product is mortal (as does the Weibull method) and forecasts that a predicted fraction of multiple failures can occur, i.e., a product can be repaired and replaced in service.

However, the data for vintages B and C reveal a concern. While the defect number has been dramatically reduced, the remaining defect severity, assessed by the time to failure, has not been proportionately reduced. This conclusion is based on a *decrease* in the characteristic life, which translates to early fails, i.e.,

$$t(F) = \Theta\left(\ln \frac{1}{1-F}\right)^{1/\alpha}$$

This suggests that the corrective measures introduced to minimize defects and defect severity have been more effective with less severe defects. The process producing the most severe defects may have been only marginally improved by process parameter review. Also, the rather early field failures suggest that a burn-in at operational stress levels prior to customer ship may be a viable way to improve service reliability for this defect population.

This sets the stage for a discussion of "interpretative" failure analysis to determine what sequence of process steps could have produced not only the defect, but a defect of a particular severity. To illustrate the point, let's consider the failure analysis work that actually occurred for a multilayer printed circuit board. Upon exposure to a temperature/humidity/voltage bias stress test, a section of the board exhibited short-circuit failures. The symptoms were a sudden loss of insulation resistance between two circuits featuring adjacent circuit segments. Subsequently, a product failed with the same symptoms at systems test during routine diagnostics testing.

Following electrical isolation of the failure sites and visual inspection to eliminate the possibility of surface shorts via corrosion or contamination, metal-lographic examination revealed the physical cause of each of the failures. The location of the failures was a risk site between a plated through-hole (PTH) and an adjacent internal signal line. Aside from involving the same risk site, a common feature of all the features was a dendritic-appearing copper deposit bridging the PTH/signal line region. The copper growth direction was cathode-to-anode with the PTH biased anodically. Other details of the analysis were not necessarily shared by all failure sites:

1. Copper deposit along apparently "dewet" glass fibers. This dewet path bridged the spacing between circuit line and PTH, making direct contact with each one.

2. Extensive deposits covering a surface of a voided region located between the shorted conductors. This voided region was located at the circuitized subassembly/sticker sheet (prepreg) interface, rather than along glass fibers.

3. Similar to (1) but along the interface defined in (2). This was distinguished from (2) in that it resembled interfacial separation or delamination more closely than it resembled a void.

4. Spiral or helical ribbon of copper deposited on cotton fiber that became entrapped during lamination. Other organic fibers had caused similar shorts.

5. Those failing stress test resistance criteria, either as a transient observation, or present as a "failure" only under environmental stress. It is common for many "failures" to disappear upon returning to normal (ambient) conditions.

Of those failures only (2) and (3) were observed in system test.

Additional temperature/humidity/voltage bias stress test studies were performed to clarify the corrosion mechanism involved. This was particularly necessary because of the experimental results reported by Lahti, Delaney, and Hines [20] which forecast that the mechanism described in (1) projected to service conditions. The results are briefly discussed as follows:

In order to support an electrochemical corrosion process a continuous moisture path must exist between electrodes. This adsorbed moisture film must be sufficiently thick as to exhibit the behavior of bulk water. This only happens above a threshold relative humidity RH_c for external or internal surfaces. Ionic contamination lowers the value of RH_c.

Given the presence of the adsorbed film along an internal interface, the electrochemistry of the corrosion process is determined by the pH of the resultant film. Acidic films provide an electrolyte that permits copper ions to be transported to the cathode and deposited. This deposition process leads to dendritic growth patterns. For films with pH values greater than approximately 6-7, the electrode reactions at and around the anode result in copper oxide formation.

The failure process that emerged at these internal risk sites was that a defect, either a voided region or a region of local interfacial separation, with sufficient adsorbed/absorbed moisture can provide the basics to form a corrosion cell. Consistent with the strategy set forth earlier in this chapter, the best solution to the problem is to eliminate the defects. Therefore, a review of the process causes surrounding the formation of these cells, and elimination at their source, was required. However, a parallel method of insuring that the defects do not become active is to reduce the moisture content of the circuit board below the threshold value.

It is also significant that the failures observed in the stress tests that involved foreign fibers required the high humidity conditions of the test to manifest themselves; such fibers were not a cause of failure in system test with the moisture content of manufactured circuit boards.

The feedback message was clear. As an accompaniment to the strategy of eliminating defects at their source, manufacturing processes must also be tailored to control the moisture content by judicious placement of baking operations and surface cleaning following wet chemical processes (circuitization processes, PTH cleaning process, etc.). At the same time, circuit board handling procedures must be sufficient to prevent contaminants from coming in contact with the product.

REFERENCES

1. I. Memis, "Water Permeation of Modules," *IBM Internal Technical Report* TR-01.1826, June 1974.
2. P. Wynblatt, et al., "Investigation on the Failure of Electronic Devices by Dendritic Growth," Quarterly Reports for 1985–87 by Carnegie-Mellon University for IBM.
3. A. Marderosian, "Humidity Threshold Variations for Dendrite Growth on Hybrid Substrates," Institute of Printed Circuits, April 1977 (IPC-TP-156).
4. G. K. Hobbs, "Development of Stress Screens," *Proceedings, Annual Reliability and Maintainability Symposium,* pp. 115–119 (1987).

5. MIL-STD-2164 (EC), "Environmental Stress Screening Process for Electronic Equipment," April 5, 1985, by the Space and Naval Warfare Systems Command, Washington D.C.
6. C. E. Mandel, Jr., "Fundamentals of Environmental Stress Screening," in *Environmental Stress Screening—A Tutorial*, Institute of Environmental Sciences, 1985.
7. B. R. Livesay, "Defect Related Failure Mechanisms in Electronic Materials Accelerated during Environmental Stress Screening," in *Environmental Stress Screening—A Tutorial*, Institute of Environmental Sciences, 1985.
8. IPC Test Methods Manual, Institute for Interconnecting and Packaging Electronic Circuits, IPC-TM-650, Evanston, Ill., 1988 (Updated yearly).
9. James J. Tomaine and Robert M. Murcko, "Predicting Reliability of LSI Printed Circuit Carriers," *Proceedings of the Fifth Annual International Packaging Conference*, pp. 83–86 (1985).
10. Frank W. Haining, James E. Hespenheide, and Roy F. Shaul, "Method of Generating Zero Risk Analysis Data," *IBM Technical Disclosure Bulletin*, 26, no. 7A (December, 1983).
11. Robert M. Murcko, "Method of Voltage Stressing Complex Carriers to Assess Reliability," *IBM Technical Disclosure Bulletin* 24, no. 4 (September, 1981).
12. Robert M. Murcko and James J. Tomaine, "A New Approach to Printed Circuit Board Temperature Cycle Testing," *Proceedings PC '82*, pp. V 31–41 (May 1982).
13. G. E. Diter, *Mechanical Metallurgy*, McGraw-Hill, 1976.
14. N. R. Mann, R. E. Schafer, and N. D. Singpurwalla, *Methods for Statistical Analysis of Reliability and Life Data*, John Wiley & Sons, 1974.
15. W. Nelson, *Applied Life Data Analysis*, John Wiley & Sons, 1982.
16. W. J. Diamond, *Practical Experiment Designs*, Lifetime Learning Publications, 1981.
17. C. R. Hicks, *Fundamental Concepts in the Design of Experiments*, CBS College Publishing, 1982.
18. D. C. Montgomery, "Design and Analysis of Experiments," John Wiley & Sons, 1984.
19. D. J. Lando, J. P. Mitchell, and T. L. Welsher, "Conductive Anodic Filaments in Reinforced Polymeric Dielectrics: Formulation and Production," *Proceedings of 1979 International Reliability Physics Symposium*.
20. J. N. Lahti, R. H. Delaney, and J. N. Hines, "Characteristic Wearout Process in Epoxy-Glass Printed Circuits for High Density Electronic Packaging," *Proceedings of 1979 International Reliability Physics Symposium*.

PROBLEMS

9.1. In a multilevel design, a total of 18 vias are used to provide interlevel connections on the following seven buses:

> Four corner buses with two vias on each bus
> Two side buses with three vias on each bus
> One center bus with four vias on it

In order for the bus to be functional, at least one via on that bus must be functional. However, in order for the substrate to be functional, all seven buses must be functional. A reliability objective of 1 ppm for fails due to via defects must be met. The vendor can only deliver substrates with via defect level less than or equal to 8000 ppm. A functional test has been developed to test all eight corner vias.

(*a*) Can we use this test as a sort tool to achieve the 1-ppm reliability objective, assuming the test is 100 percent efficient for corner vias, the probability of being defective is the same for all vias, and all vias on the same substrate are independent of each other?

(*b*) If the test is only 90 percent efficient, what will be the reliability impact for fails due to via defects?

9.2. Listed below are results from an insulation resistance test at 70°C and 80 percent relative humidity. The test was terminated after 1008 hours, and 746 samples had survived.

From	To	Fails	From	To	Fails
0	23	0	449	462	15
24	92	0	463	486	12
93	116	3	487	510	8
117	141	2	511	535	11
142	166	7	536	584	36
167	190	9	585	608	15
191	214	10	609	614	16
215	265	18	615	648	9
266	282	4	649	672	17
283	307	7	673	744	39
308	329	6	745	816	74
330	355	8	817	840	26
356	384	17	841	912	43
385	406	12	913	960	40
407	427	12	961	1008	32
428	448	6			

(a) Plot the failure time data on Weibull probability paper, and from the probability plot, estimate the times at which 50 percent and 90 percent of the population will fail.

(b) Plot the failure time data on lognormal probability paper and estimate the 50 and 90 percentile points from the plot.

CHAPTER
10

CERAMICS
IN ELECTRONIC
PACKAGING

RICHARD B. HAMMER
DOUGLAS O. POWELL
SHYAMA P. MUKHERJEE
RAO R. TUMMALA
RISHI RAJ

INTRODUCTION

Ceramics, defined as inorganic, nonmetallic materials, have played an important role in circuit packaging since the 1950s. Their extensive use can be attributed to their unique combination of mechanical, dielectric, physical, and chemical properties. Ceramics find use in chip carriers or modules for discrete devices, single integrated circuits, and hybrids (combinations of discretes and ICs), as well as ceramic cards and boards for higher-level packaging. Ceramics are used for substrates (the base on which the circuitry is placed and devices mounted), module covers and caps, sealing materials for modules, components of thick film conductors, resistors and dielectrics, and components of thin film resistors and dielectrics.

Modules constructed entirely of ceramics or ceramics in combination with metals can be made hermetic. A hermetic module is one where the interior of the module is totally isolated from the ambient atmosphere, thus making the contained circuitry immune to the effects of harmful atmospheres. Modules constructed from plastics or sealed with organic polymers, in general, cannot be made hermetic.

This chapter has been divided into two parts. The first part describes material usage and material requirements in electronics and different techniques for substrate processing. The second part presents a brief outline of ceramic science with particular emphasis on the chemical processing and sintering of ceramics. Further information can be obtained from various published works [1–4].

PART 1
MATERIALS AND PROCESSING

MATERIAL USAGE AND REQUIREMENTS

Substrates

Materials to be used as substrates for integrated circuit mounting and hybrid circuit deposition should have a number of properties:

Good insulating characteristics for electrical isolation of circuit lines and components.

Low dielectric constant for low capacitive loading of circuit lines and low capacitive coupling. The importance and interrelation of these characteristics are discussed in Chapters 3 and 4 which cover electrical design concepts in electronic packaging and signal line electrical design.

High thermal conductivity to help remove heat from the circuitry. Thermal design factors are described in Chapter 5.

High strength and toughness for ruggedness and processibility. The mechanical aspects significant to packaging are covered in Chapter 6.

High dimensional stability (all ceramics excel here).

Stability at the processing temperatures used for module fabrication.

Thermal expansion near that of materials to be attached to the substrate (e.g., silicon).

Acceptable cost.

Many ceramic materials meet the resistivity and temperature stability requirements and also excel in one or more of the other properties, but there is not a perfect material. The choice of material for a particular application requires tradeoffs among the properties to achieve optimization with respect to a specific set of application requirements. Table 10.1 lists the properties of the most commonly used ceramic packaging materials.

Aluminum oxide (alumina, Al_2O_3) is by far the most common ceramic used in electronic packaging [2]. The compositions used range from about 90 percent to over 99 percent purity. Thick-film processes typically employ a 96 percent alumina. The balance of the material is silicon dioxide (SiO_2, 3–3.5 percent),

TABLE 10.1
Properties of common ceramic packaging materials [5–7]

Property	Alumina	Beryllia	A1N	Glass-ceramic
Electrical resistivity, ohm · cm	$> 10^{14}$	$> 10^{14}$	$> 10^{13}$	$> 10^{14}$
Dielectric constant	9.5	6.8	8.5	4.5–6.5
Bending strength, MPa	300	200	300	150–240
Thermal expansion, ppm/°C	6.5	7.5	4.0	2.5–6.5
Thermal conductivity, W/mK	22	260	100–260	0.8–2

magnesium oxide (MgO), and calcium oxide (CaO), which in combination with some of the alumina form a glass phase in the boundaries between the alumina grains. The composition of this glass phase is critical to the processing of thick-film materials (particularly resistors), since the glasses in the substrate and ink interdiffuse during the thick-film sintering to achieve good bonding.

Alumina represents an optimal combination of properties for many applications. Its electrical resistivity, dielectric loss, and strength are excellent, while its thermal conductivity is much higher than that of plastics and most common ceramics, and is as good as that of some metals. Its relatively high thermal conductivity is needed in many packages to help remove the heat dissipated by active components. However, in high-power applications its thermal conductivity may not be adequate. Alumina's coefficient of thermal expansion is low compared to plastics, which allows small to medium size silicon chips to be mounted directly to the substrate, either with back bonding or "flip chip" mounting. But, when the chips become large (greater than 6 to 8 mm) the thermal expansion mismatch results in excessive stresses, and alternate mounting methods or materials must be used. Alumina has a moderate dielectric constant, and is adequate for all but the highest-performance (speed) circuitry (such as high-speed digital circuits or microwave circuits). Alumina has the lowest cost among the high-performance ceramics.

Beryllium oxide (beryllia, BeO) finds use in substrates where thermal conductivity is of utmost importance, usually in power hybrids, or power semiconductor packages [2]. Beryllia powder can be extremely toxic, and requires special safeguards during the manufacture of beryllia substrates. Because of this special handling required, the high purity necessary to achieve high thermal conductivity, and higher processing temperatures, beryllia substrates cost about ten times as much as alumina.

Aluminum nitride (AlN) is a newer material of interest for electronic packaging [5,6]. Its thermal conductivity is extremely high for a ceramic (it can be as high as beryllia) and its thermal expansion is more closely matched to that of silicon than is alumina. AlN is being developed as a replacement for beryllia in high-power packages, and for packaging large chips where the thermal expansion of alumina is a problem. The cost of AlN is between those of alumina and beryllia.

Glass-ceramics are another class of materials used in packaging for substrates, package sealing, and thick-film ink components [2,7]. Glass-ceramics are glasses that can be transformed into crystalline materials by suitable heat treatments. The melting temperatures of the resulting crystalline materials are usually higher than those of the starting glasses, which gives the final material good thermal stability during subsequent melting or sintering of the same glass ceramic (e.g., for multilayer thick-film dielectrics). Glass-ceramics generally have lower permittivity than the other ceramics used in packaging, and both the permittivity and thermal expansion can be adjusted over a moderate range to tailor the properties for specific applications. The major drawback to glass-ceramics in many applications is their very low thermal conductivity.

Caps and Covers

The main requirement for materials to be used as module covers and caps is that they match the thermal expansion of the material to which they are bonded, or else some compliance must be built into the cover/cap design to accommodate the thermal stresses that result. The easiest way to meet this requirement is to make the cap/cover of the same material as the substrate or module.

Sealing Materials

Ceramic materials are used to seal caps and covers to modules and to seal pin feed-throughs in modules.

Ceramic module sealing is typically used when both the substrate and cover are ceramic. The sealing materials are based on glasses with low melting temperatures since the sealing temperature must not degrade the devices within the module. Typical sealing temperatures are in the range of 400 to 500°C. The glass is normally ground into a powder, then mixed into a paste with organic binders and solvents. The paste is applied to the seal areas by screening, stenciling, or extrusion, and the mated substrate and cover (with mounted IC enclosed) are heated to flow the glass and form a hermetic seal.

Glass is also used to fasten and seal I/O pins into metal case modules. Short tubes of the glass sealing material are placed over the pins. Then the case, with the pins and preforms, is placed in a fixture for joining. The joining is accomplished by melting the glass which then bonds to the pins and case. Since this operation is performed before any devices are mounted in the module, the sealing temperature is only limited by the properties of the case and pins, and can be much higher than for module sealing.

Property requirements of ceramic sealing materials include:

Ability to form a hermetic seal

Electrical insulating property to prevent shorting

Sealing temperatures compatible with other module materials

Ability to wet the mating materials at the sealing temperature

Thermal expansion coefficient matched to that of the materials to which it must seal

CERAMIC PROCESSING

Dry Pressing

Dry pressing is suitable for forming almost any monolithic (no internal structure) shape, from flat plates to complex, multiple cross section parts [3,8,9]. It cannot make parts with undercuts, reentrant openings, or thin sections. It is typically used to produce ceramic substrates, covers, and caps. Parts made by dry pressing typically have tighter dimensional tolerances than parts made by other techniques. The best obtainable tolerances are ± 0.5 to 0.3 percent or ± 12 μm, whichever is greater. The surface roughness attainable is in the 0.6 to 1.0 μm range, which aids the adhesion of fired thick-film materials, but is too rough for most high-density thin-film applications. Dry pressing is the lowest-cost precision ceramic forming technique for packaging applications.

The dry-pressing process consists of compacting a ceramic powder under pressure in a die cavity that has all the features of the finished part. The pressed parts are then sintered in a furnace employing a time-temperature profile that will result in the desired final density. Alumina ceramics are typically sintered at 1500 to 1650°C for times of a few hours at the maximum temperature. Figure 10-1 shows the four stages of the pressing cycle, and Figure 10-2 shows an overall process flow for dry pressing.

During sintering the ceramic composite undergoes 15–20 percent linear shrinkage (40–50 percent volume shrinkage). Repeatably attaining the required dimensional tolerances on the sintered parts requires that the green density of the pressed piece be uniform within each part and consistent from part to part, and that the shrinkage during sintering be repeatable and predictable. The green density requirements are met by filling the die cavity with the same amount of material every time, distributing the powder uniformly throughout the cavity, and pressing to the same final thickness. Since the die filling is done on a volumetric basis, the starting powder must be free flowing, or the volume and distribution will vary from press stroke to press stroke. The typical ceramic powders used have particle size distributions covering 0.1 to 10 μm, and tend to be very fluffy and have extremely poor flowability. Because of this, the powders are agglomerated before pressing into roughly spherical particles with sizes in the 50 to 200 μm range. These particles have the consistency of fine sand, and flow easily and repeatably. The packing of these particles under compression leaves a void structure in the

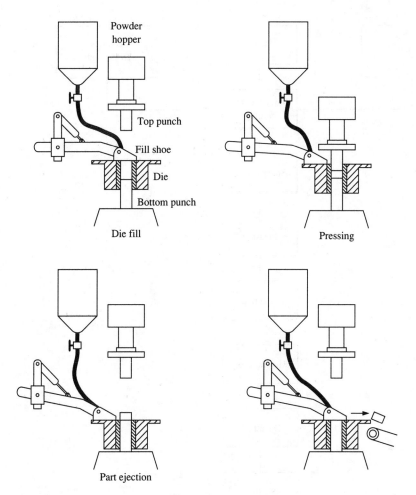

FIGURE 10-1
Dry pressing of ceramic substrates from powders.

fired ceramic which in the practical applications is never sintered to 100 percent density.

The process used for agglomeration is spray drying. The ceramic powders are mixed into a slurry (usually with water) containing small amounts of binders (to hold the agglomerates together) and other additives (to obtain optimized dispersions). The slurry is forced through an atomizer in a cylindrical drying chamber to form spherical drops. Hot air is blown through the chamber to dry the droplets. The process must be adjusted so that the droplets are dry before they contact any of the surfaces of the drying chamber. Cyclones are typically used to separate the undersized particles from the desired dried product. To obtain good pressing characteristics, it is necessary to adjust the slurry solids loading and feed

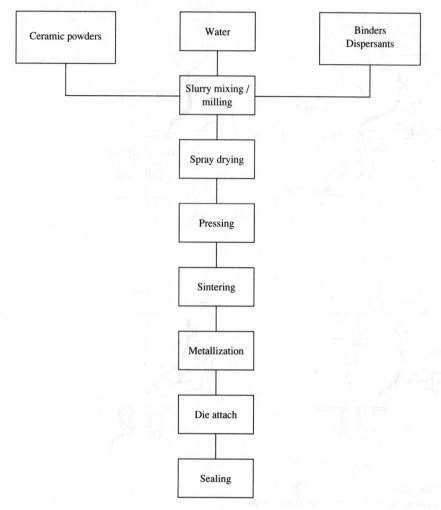

FIGURE 10-2
Dry pressing process flow.

rate, and the air temperature and flow rate so that the resulting particles are solid rather than hollow.

Figure 10-3 shows a typical density versus time curve for a sintering process. The important feature is the asymptotic approach to a limiting density (the limiting density is likely to be different for different sintering temperatures and different green body powder packing structures). The most straightforward way to achieve consistent sintering shrinkage is to insure that the sintering time-temperature profile reaches the flat portion of the densification curve. It is also important not to extend the sintering time too long to prevent the possibility of excessive grain growth. Excessive grain growth causes significant gradients in internal stresses and decreases toughness of the material, especially to impact stresses.

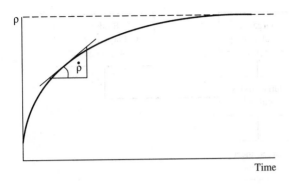

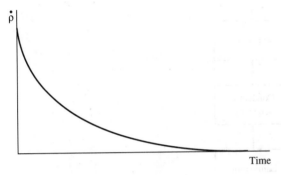

FIGURE 10-3
Density versus time for sintering
of ceramics. Note the asymptotic
approach to a limiting density and
the reduction in densification rate with
time.

Tape Casting and Laminating

Tape casting (also known as doctor blading) is used to make ceramic substrates with very smooth surfaces (0.08–0.25 μm average roughness) for thin-film hybrids, and the individual dielectric layers for multilayer ceramic packages [3,9,10]. It is also used to make multilayer ceramic capacitors.

Figure 10-4 shows a schematic of the tape casting process, and Figure 10-5 shows a process flow chart for making a multilayer ceramic substrate or chip

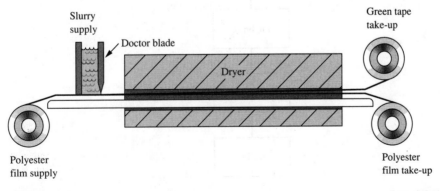

FIGURE 10-4
Schematic of tape casting process for making flexible ceramic green sheets.

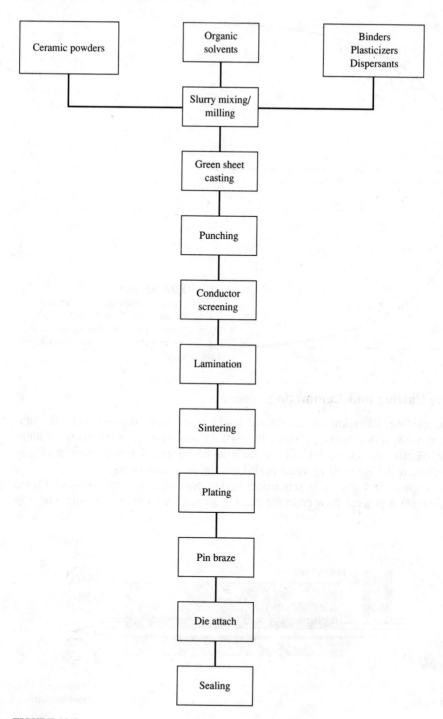

FIGURE 10-5
Tape casting process flow.

carrier. Ceramic powders of the desired composition are mixed and ground with organic binders, solvents, and plasticizers to form a readily flowable slurry. The slurry is then cast onto a smooth surface (polyester film) with the thickness of the slurry layer being controlled by the "doctor blade." The slurry layer is dried to remove the organic solvents, leaving a flexible "green sheet" consisting of the ceramic powder, binders, and plasticizers. The green sheet is then cut into pieces of a convenient size for processing. Vias, cutouts, and tooling holes are punched into the green sheet and conductor ink screened onto the sheets to fill the interlayer vias and form the circuit patterns. The sheets for all the layers of the finished substrate are then stacked, placed in a heated press, and laminated together under pressure. The laminates are cut to the proper size, and are ready for firing. The substrates are sintered in a reducing atmosphere to prevent the oxidation of the metal circuit lines. During sintering the ceramic powders and the conductor material densify and form a monolithic piece. After sintering the exposed metal is plated both for protection and to allow attachment of components and pins by brazing, soldering, or wire bonding. Pins are then brazed onto their I/O pads on the substrate, and the substrate is ready for device attachment and sealing.

Control of the starting material characteristics and all steps of the tape process is essential if dimensional tolerances of the sintered substrates are to be consistently met. The key to maintaining tolerances is to obtain uniform and repeatable sintering shrinkage throughout each part. The first step is to obtain uniform green density in each part, and the second step is to achieve a consistent amount of densification during sintering. Sintering rate and ultimate density are strongly affected by the starting particle size distribution and the particle packing in the green state. Particle packing is also a function of the particle size distribution, the characteristics of the slurry (stability and loading), and the drying operation. Slurry stability is affected by the surface area and surface chemistry of the powders used to make the slurry.

Selection of suitable binders and plasticizers for the tape-casting process requires the consideration of many requirements:

The green sheets must be tough and flexible but easy to punch.

The glass transition temperature must be manipulated to give tack-free sheets at room temperature, yet allow softening and some flow at the lamination temperature and pressure.

The green sheets must have good dimensional stability.

The binders must burn out of the body completely during the firing operation.

The combination of solvents, binders, plasticizers, and dispersants must be capable of forming stable dispersions of the ceramic powders.

The slurry-mixing operation must be developed and controlled to provide a consistent slurry with the desired casting and drying properties. For example, premilling the ceramics, binders, and dispersants before mixing in the solvents and plasticizers will usually result in very different slurry properties than mixing the ceramics with the solvents and then adding the binders, plasticizers, and dispersants. The rheology of the slurry should be adjusted so that it flows easily under the shear stresses applied by the doctor blade and under the pressure head provided by the casting reservoir, but the wet cast film should not tend to spread out under its own weight. Drying conditions (time-temperature profile, local solvent vapor pressure over the film, etc.) must be controlled to prevent distortion of the film during drying and minimize density gradients through the thickness of the film. Also, residual stresses in the dried films should be avoided as these can tend to cause curling in thin laminates or warping of thicker ones during the sintering process.

The conductor ink sintering characteristics and ultimate thermal expansion coefficient must be matched to those of the ceramic body to prevent such problems as delamination, via wall separation, via cracking, or warping during sintering and subsequent cooling. The amount of shrinkage of the conductor inks during sintering should be about the same as that of the ceramic, and sintering should proceed at comparable rates at the sintering temperatures employed. It is important that the coefficient of thermal expansion of the sintered conductors match that of the sintered ceramic closely to prevent stresses from developing during cooling after sintering. The high temperatures (1550 to 1650°C) required for sintering alumina necessitate the use of refractory metals for conductors. Tungsten and molybdenum are the two most commonly used.

Roll Compaction

Roll compaction is an alternate method for making thin sheets of ceramic materials when the sheets do not need to be laminated together [3,9]. Figure 10-6 is a schematic of the roll compaction process. Spray-dried powders, similar to those employed in dry pressing, are fed between two compacting rolls and emerge

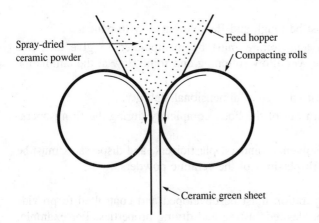

Spray-dried ceramic powder

Feed hopper

Compacting rolls

Ceramic green sheet

FIGURE 10-6
Schematic of roll compaction process for making ceramic sheets.

after compaction as a continuous sheet of green ceramic. The powders contain more binders than used for dry pressing but less than used for tape casting. This results in a sheet that can be cut and punched to form the required shapes, but not a flexible and laminatable sheet. Sintering is accomplished as in the other processes.

SUBSTRATE PROCESSING

There are three approaches for fabricating circuit patterns on ceramic substrates. These are thick-film screening, thin-film photolithography, and plating. Thick-film processing can produce conductors, resistors, and capacitors. Thin films can be used for conductors and resistors, while plating is limited to conductors only. Typically, the thickness of elements in thick-film and plating technologies is 10 to 25 μm or greater. For thin films, thickness is generally around 1 to 8 μm. Thin film and plating can achieve line widths down to about 25 μm, whereas thick film is limited to about 125 μm. All three processes can be employed to fabricate multilayer circuit structures. Thick- and thin-film processing are described here; Chapter 13 covers dry processes, while Chapters 16 and 17 cover seeding and plating fundamentals.

Thick-Film Processing

In thick-film processing, an emulsion-coated screen or metal stencil is prepared with openings in the emulsion or stencil to define the circuit pattern. The screen or stencil is placed over the substrate and a squeegee is drawn across it to force a thick-film ink through the openings onto the substrate, forming circuit patterns [11]. Printing can be done with the screen or stencil in contact with the substrate (contact mode, more common with stencils) or with the screen or stencil held a small distance above the substrate (off-contact mode). In the off-contact mode the squeegee forces the screen down onto the substrate during the printing pass, and the snap-back of the screen after printing helps achieve clean separation of the ink from the screen. Figure 10-7 is a schematic of the screen printing process. The screen is removed, the wet ink is allowed to level (5–10 minutes), and then the ink is dried at 100 to 150°C to remove the vehicle solvent. The thick film material is then fired in a furnace at temperatures between 600 and 1000°C. During firing the organic binder phase is removed by burning or pyrolysis, and the film sinters (densifies) and bonds to the substrate [12,13]. The bonding process itself is not well understood. Most likely a contributing factor is interdiffusion of the ink glassy material with the glassy material of the substrate. The elements of interdiffusion and adhesion are covered in Chapters 25, 26, and 27.

Thick-Film Inks

A thick-film ink consists of three main components [14,15]. The functional phase contains the base conductive or dielectric powders. The binder phase consists of

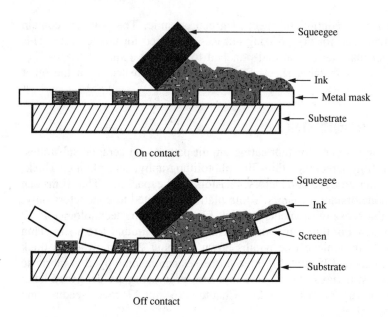

On contact

Off contact

FIGURE 10-7
Screen printing process for defining thick-line circuitry.

glass or metal oxide powders, and the carrier vehicle consists of resins, solvents, and modifiers. Ink rheology is critical to controlling the screening process and obtaining the required definition of circuit elements.

The solvents are low-volatility organic compounds with low ambient vapor pressures but high volatility at 100 to 150°C. They must wet the powder components, but not react with them or chemically attack the screen. They typically consist of terpenes or high molecular weight esters. The resin components impart flow characteristics to the ink, and serve as a binder before firing. They are typically terpene or cellulosic resins that burn off well in an oxidizing atmosphere. In the case of nitrogen-firable inks, acrylic or nitrocellulose resins that decompose on heating are employed.

The binders are either glasses or metal oxides, such as copper oxide, or mixed glasses and oxides. The most widely used glasses are borosilicates, lead borosilicates, and bismuth-lead borosilicates. Glasses, as noted above, are responsible for both adhesion to the substrate and cohesion of the fired film. Glass migrates to the film-ceramic interface during the firing operation, and in many cases the glass reacts with the substrate to form a chemical bond. The glass and substrate surface compositions are both critical to achieving good bonding. As a minimum the molten glass must readily wet the substrate surface. Reaction and interdiffusion are desirable for improved bonding, but reactions should not form crystalline phases, which can degrade the bond strength. The softening and melting points of the glass determine the firing temperature. The coefficient of thermal expansion should be adjusted to match that of the substrate. A problem that can occur with glass-bonded inks is glass migration to the top surface of the film, which degrades wire bonding and soldering. Particle size is important since it affects the fired film density and the amount of glass flow during firing.

Adhesion to the substrate in oxide bonded systems is the result of chemical bonding. The oxides form aluminates after reacting with the alumina ceramic, for example

$$Cu_2O + Al_2O_3 \rightarrow 2CuAlO_2$$

at high temperatures. Film cohesion is the result of sintered bonds between the metal particles, rather than embedding in a glass matrix.

Rheological properties are very important because the ink must transfer rapidly and cleanly from the screen to the substrate, must level well, must not spread excessively, and must have a uniform thickness from print to print. Ink rheology is affected by the particle sizes of the functional and binder phases, the vehicle chemistry and viscosity, and the interactions of the vehicle materials with the powder surfaces. Particle size distribution, morphology, surface area, and density are very important. Large particles disrupt the fired film and affect the electrical and mechanical properties, while too small particles can result in film blistering. Some aspects of ink rheology and thixotropy are discussed in Chapter 11; in this case the inks are for a different application but the fundamentals are consistent.

Conductor, resistor, and dielectric inks that are to be used together must be compatible. That is, they must have similar firing temperatures, and must be suitable for firing in the same furnace atmosphere.

Conductors. Metal powders make up the electrically active component of conductor inks. Typical conductor metals are gold, silver, copper, molybdenum, and tungsten, as well as platinum/gold, palladium/silver, and platinum/palladium/silver alloys. Gold, silver, copper, and the alloys are used in traditional hybrid circuitry. All of them but copper, are fired in air atmospheres. Copper, because of its susceptibility to oxidation, must be fired in a nitrogen atmosphere [16]. Adhesion, solderability, wire bondability, resistivity, migration resistance, solder leaching, and cost are factors to consider in making a selection. The refractory metals (tungsten and molybdenum) are used when cofiring with alumina, beryllia, or AlN is required, as when making monolithic, multilayer packages.

Resistors. Thick-film resistor inks are commercially available that cover a sheet resistivity range of from 1.0 Ω to 10 MΩ.[1] The functional phases of resistors are composed of electrically conductive ceramic powders. The most commonly employed are ruthenium dioxide, bismuth-ruthenium oxide and tantalum nitride [17,18]. The oxides are used with air-fired conductor inks, and the nitride with

---

[1] Sheet resistivity is bulk resistivity ($\Omega \cdot$ cm) divided by the nominal thickness of the film (typically 0.0025 cm for thick film resistors).

nitrogen-fired inks. Modifiers are added to control the temperature dependence of resistance and are proprietary in nature [19].

In thick-film resistor processing, it is important to consistently produce stable components with the desired resistance and a low temperature coefficient of resistance (TCR).

The resistance and TCR are sensitive to the ink firing profiles, and inter-diffusion and reaction of the conductor and resistor materials often affect the resistance in the region of the conductor-resistor interface. The interaction of conductors and resistors can lead to the formation of third phases with incompatible thermal expansions, which can result in tensile stresses at the resistor terminations. If the stresses are large enough to cause cracking, severe instability of resistance values usually results.

TCR is defined as the slope of the fractional resistance deviation ($\Delta R/R$) versus temperature curve (see Figure 10-8). The TCR limits specified for an ink are the slopes (one positive, one negative) of the two lines passing through the reference point on the curve that bound the resistance deviation. In practice, the maximum deviation usually occurs at the extremes of the specified temperature range, so that the average TCR can be taken as the maximum and minimum slopes of lines through the reference point and points at the temperature extremes. Typically, the reference temperature is 25°C and the temperature range is −55°C to +125°C.

Resistors are designed so that the average as-fired resistance is lower than the required resistance value to optimize the yields. The final resistors can than be

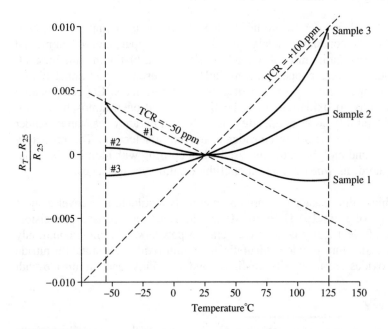

FIGURE 10-8
Typical resistance versus temperature data for thick-film resistors.

trimmed with a laser or abrasive jet to remove material and increase the resistance value to very high precision, about 1 percent or better. The final resistance is typically about 1.5 to 2.0 times the original value.

Thick-film dielectrics Thick-film dielectric inks have three different uses. They are used to isolate circuit lines in multilayer structures, to construct thick-film capacitors, and as protective coatings for thick-film components.

Dielectrics for multilayer applications employ low-permittivity ceramic powders as the functional phase. The glass and powder must be chosen so that the thermal expansion of the resulting film is very closely matched to that of the substrate to prevent stresses that can cause the substrate to bow. It is also desirable for the film to be stable when subjected to multiple firings, which are common in multilayer applications [20].

Thick-film inks for capacitor dielectrics usually include a high-permittivity ferroelectric compound, often based on titanates (such as barium titanate). They also usually contain a Curie point modifier. The glass binder phase acts as a series capacitor, and consequently, the composite permittivity is greatly reduced from that of the pure functional phase.

For protective applications the functional and binder components are the same material, which is usually a low-melting glass. They are typically employed to protect resistors and conductors from silver migration or other harmful environmental effects. Reliability considerations and test techniques are discussed in Chapter 9, "Reliability and Testing."

Thin-Film Processing

Thin films are formed by either evaporating or sputtering the desired materials onto the substrate in a vacuum as presented in Chapter 13. The films are then patterned using photolithography and etching. This technique is applicable to conductors, resistors, and dielectrics [21–26]. The thin-film process is adaptable to both single-layer and multilayer applications. A typical thin-film structure consists of an interface/adhesion layer (optional), a resistor material layer (optional), an interface/etch mask layer (optional), a conductive layer, and a solder mask layer (optional). Adhesion/interface layers are usually chromium, titanium, or nickel. Tantalum, tantalum nitride, nichrome, or a chromium-silicon dioxide composite are often-used resistor materials. Copper, aluminum, and gold are common conductor materials. Etch masks and solder masks are chosen from materials that can be etched selectively with respect to the layers either above or below them. Conductive layers are typically 1 to 10 μm thick, and the other layers range from 5 to 100 nm thick [25]. Figure 10-9 shows a typical processing flow diagram for thin-film circuitry.

Multilayer circuitry can be produced by alternating circuitry layers with dielectric layers. Considerations for electrical design of interconnections are provided in Chapters 3 and 4. Where connection is required between two or more layers, metallized through-holes are formed in the dielectric layer. These holes, com-

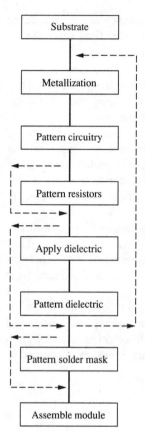

FIGURE 10-9
Thin-film process flow.

monly called vias, can be formed by photolithography and etching, or screened using thick-film processing.

Polyimides are commonly used as a dielectric material because of their low permittivity, high thermal stability, planarizing characteristics, and ease of processing [27–29].

Packaging Examples

Figure 10-10 shows a typical hybrid substrate employing thick-film circuitry (conductors and resistors), chip capacitors, discrete semiconductors, and integrated circuits. Such substrates are typically dry-pressed or roll-compacted 96 percent alumina. High-power applications may employ beryllia or aluminum nitride instead of alumina to improve heat dissipation. Similar circuitry can be produced by using thin films rather than thick films. Substrates for thin-film hybrids are usually produced by tape casting because the finer circuit geometries require the use of smoother surfaces. The hybrid substrate must undergo further packaging

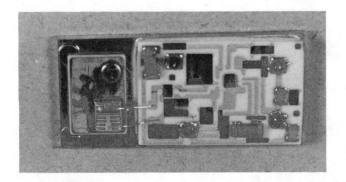

FIGURE 10-10
Typical hybrid circuit on ceramic substrate.

to provide means for connection to a printed circuit card and environmental protection.

Figure 10-11 shows a type of metal package in which hybrid substrates are often mounted. The pins are sealed in place and electrically isolated from the case by glass rings. The hybrid substrate is bonded to the bottom of the case, wire bonds connect the hybrid circuitry to the pins, and then a metal cover is welded or soldered on top to complete the hermetic package.

Figure 10-12 shows a ceramic dual in-line package (DIP) [30,31]. It is constructed from three layers of tape laminated together and contains internal wiring. The bottom layer contains a metallized pad in the center to which the integrated circuit chip is back bonded. The middle layer contains a cutout in the center, which forms the cavity in which the IC is placed, and has metal circuit lines running from the edges of the IC cavity to the edges of the substrate. Wire

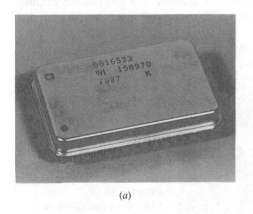

(a)   (b)

FIGURE 10-11
Metal package for ceramic substrate: (*a*) top view and (*b*) bottom view. Note the glass seal between pins and metal case.

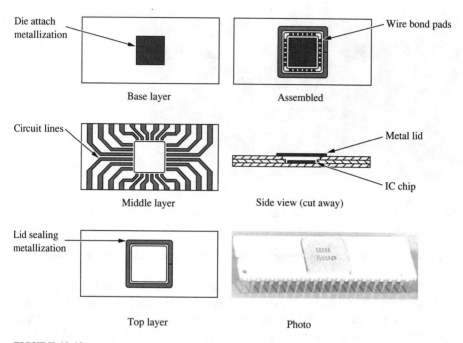

Base layer

Assembled

Middle layer

Side view (cut away)

Top layer

Photo

FIGURE 10-12
Ceramic dual in-line package (DIP). Note pins brazed to the side of the module.

bonds connect the IC to the circuit lines at the cavity, and I/O pins are brazed to the ends of the lines along the sides of the package. The fundamentals of joining and interconnections are covered in Chapter 19. A ring of metallization runs around the cavity in the top layer. A metal cap is soldered to this ring to complete the package after die and wire bonding.

Figure 10-13 shows a CERDIP, which is a lower-cost form of the ceramic DIP [30,32]. Internal circuitry is replaced with a stamped metal lead frame, and

FIGURE 10-13
CERDIP module. Note the ceramic sealing layer between the top and bottom ceramic slabs.

two ceramic plates form the top and bottom of the package. A paste made from a low-melting glass powder is applied to the periphery of the mating surfaces of both pieces of ceramic, and the lead frame embedded in the paste on the bottom piece. The IC is bonded to the bottom ceramic plate, wire bonds connect the IC to the lead frame, and the ceramic cover is placed over the assembly. Sealing is accomplished by heating the module to a temperature at which the sealing glass melts, and the glass on the top and bottom plates flows together.

Figure 10-14 shows a multilayer ceramic pin grid array (PGA) package [30,31]. PGAs are constructed with three to five layers of laminated tape, and contain internal wiring and vias to connect wiring on different levels. Construction is similar to that used for ceramic DIPs, except the pins are brazed in a square grid pattern on either the surface opposite the die cavity (cavity up) or on the same surface (cavity down), instead of just along two edges. The use of a PGA allows more pins to be placed in a smaller area than with a DIP, thus allowing smaller packages with shorter circuit lines. The cavity-up configuration allows the entire bottom surface to be used for pins, maximizing the number of module I/Os. The cavity-down configuration allows a heat sink to be attached to the module to improve thermal dissipation.

Figure 10-15 shows another type of PGA. This module employs a dry-pressed ceramic and a single layer of circuitry [25]. The circuitry is formed using either thin- or thick-film methods. The pins are swaged into holes in the ceramic and soldered to the circuit lines, rather than being brazed to the surface. The module shown also employs a chip attachment method referred to as "flip chip," which uses solder bumps on the chip to make solder connections to the substrate circuit lines, rather than using wiring bonding [33]. This is also discussed in more detail in Chapter 19. The advantage of this type of module over the standard, multilayer PGA is its lower cost.

Figure 10-16 shows ceramic leadless chip carriers (CLCC) [30,31]. This type of module has no I/O pins and is designed for surface soldering or socket mounting. The CLCC is constructed with from one to three layers of laminated

FIGURE 10-14
Ceramic pin grid array (PGA) package. (*a*) Cavity-up configuration; (*b*) cavity-down configuration.

FIGURE 10-15
Metallized ceramic substrate
on dry-pressed substrate with
swaged pins and flip-chip IC
bonding. (*a*) module with chip
before capping and (*b*) after
capping.

tape. The half-round *castellations* are metallized with a solderable alloy, and
designed to form easily inspectable solder fillets. The I/O pads are placed along
all four edges of the module rather than only two as with DIPs, and the I/O pad
spacing is usually either 0.040 or 0.050 inches instead of the 0.100 inches used
for DIPs. These combine to increase the I/O density and decrease the signal line
lengths when compared to DIPs. Because it is directly soldered to the next level of
packaging without any compliant members in between, the thermal expansions of
the module and next level must be closely matched to prevent joint failure during
thermal cycling. Because of this, CLCCs are often mounted on ceramic cards or
carriers, or on printed circuit cards with expansion-controlling metal cores.

FIGURE 10-16
Ceramic leadless chip carriers.

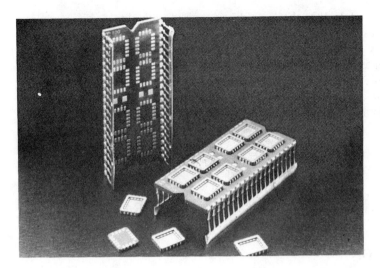

FIGURE 10-17
CLCCs on ceramic motherboard.

Figure 10-17 is an example of several CLCCs containing memory chips mounted on a ceramic DIP carrier to form a high-density memory module.

Figure 10-18 shows a multichip, multilayer ceramic module [34]. This represents an extension of the single-chip multilayer PGA. Here the module contains circuitry to interconnect the many chips on the module and reduce the

FIGURE 10-18
Multichip, multilayer ceramic (MLC) substrate.

total number of module I/Os that would be required with single-chip modules, as well as decreasing signal wiring lengths. This leads to increases in system packaging density and improves system performance as discussed in Chapter 1, "Packaging Architecture."

Such modules may contain in excess of 30 circuitry layers, and over 100 integrated circuit chips. The pinning processes are covered in complementary chapters, Chapter 19, "Joining and Interconnections," and Chapter 20, "Mechanical Processes."

PART 2
CERAMIC SCIENCE

INTRODUCTION

The discussions of the modern scientific and technological issues in the development of new or improved ceramic substrates for electronic packaging fall mainly into two categories. One is the development and application of new materials, including both ceramics and metals for high-performance substrates. The second category of issues is related to improvements and innovation in processing technology.

Alumina has been so widely used because both the industry and the ceramists are familiar with the processing of alumina. The similar thermal expansion coefficients of alumina and molybdenum, combined with the high-melting point of molybdenum, have been a key factor in the development of multilayer technology. However, there is now a greater understanding of the processing of alternate materials that have even lower dielectric constants (see Table 10.1 near the beginning of this chapter) and can be sintered at lower temperatures. Low sintering temperatures mean that the ceramic can be cofired with high-conductivity metals, for example copper. Cordierite glass-ceramic is an example of one such ceramic material. Another example is a glass-based dielectric that is filled with particles of alumina. One major advantage of glass-based systems is that they can be etched and patterned, and gradually built up to several layers by the same technique used in the IC technology. From a materials standpoint, this approach offers an opportunity to tailor the dielectric layers in order to optimize their thermal and electrical properties. The materials can be tailored by controlling the glass composition, and also by controlling the volume fraction, particle size, and particle morphology of the ceramic filler. This research is currently in its infancy.

The other set of issues is related to the understanding and innovation in processing of ceramic substrates. Since sintering is the main technique for consolidating ceramics, many of the substrate processing issues relate to the science of sintering. The second section in this part of the chapter discusses the fundamental aspects of sintering. The important point is that the sintering behavior is critically dependent on the size and the distribution of the ceramic particles in the starting powder, and on the packing of these particles in the green microstructure. If sintering additives are used to promote sintering, then they are often left behind at grain interfaces. If the dielectric properties of the interfaces are very different from that of the ceramic crystallites, then they can have a significant effect on the macroscopic properties of the dielectric. A good example of the importance of sintering aids is the use of oxides to promote sintering in aluminum nitride. The dichotomy arises because the sintering aids degrade the thermal conductivity of aluminum nitride.

The green microstructure highlights the importance of particulate science in the processing of ceramic substrates. We can envisage an ideal microstructure consisting of particles that are all of the same size and are packed in a close-packed array, like atoms in a crystal. If such structures could be produced, their sintering behavior would be highly controlled and predictable. In practice, even if monodispersed particles could be produced in large quantities, the procedure for packing them in ordered arrays presents a problem. If the powder compact contains ordered domains that are misoriented with respect to each other, then the interdomain interfaces become packing defects which can evolve into cracklike flaws during the sintering process [35]. Even if the ideal goal of obtaining single crystallike arrangements of monosized particles is unattainable, we must develop methods which afford control in the particle consolidation process. The ability to disperse the particles so that they do not form agglomerates and then to pack them without large-scale packing defects is a very worthwhile objective [35,36]. Schemes can be devised wherein particles are randomly packed in such a manner that packing defects that extend across several particle diameters are avoided. This scheme would also require good control of the particle size distribution so that crystalline packing of the powder could be suppressed. Particle size control must be combined with the use of colloidal techniques for the consolidation of powders into green compacts. Colloidal methods allow control over dispersion of the powder so that agglomeration-type defects can be avoided. The principle behind the dispersion is either to control the charge on particle surfaces by changing the pH, or to attach polymeric macromolecules to the surface of the particles so that they are sterically repelled from each other. Again, the scientific application of colloidal science to ceramic processing is in its infancy and the reader is referred to the literature [37,38] for a discussion on colloidal chemistry.

An understanding of the interaction between ceramic particles and organics is important because organics are often used in green state processing. In colloidal processing organic polymers are used for steric dispersion. But even more important is the use of polymer binders in the tape-casting process. Without the polymers the flexibility of design and fabrication of the multilayer ceramic substrate would not be possible. The binder solutions used in tape casting consist of four main constituents: a solvent, a plastic or a binder to impart strength, a plasticizer for introducing flexibility, and a dispersant to prevent flocculation during processing. The mechanical properties and the dimensional stability of the binder are of key importance, since the green tape is handled extensively before the final sintering step. The binder is removed from the green body just before the final sintering step; this is a very complex process that depends on the molecular structure of the binder, which dictates how it will pyrolyze during the burnout, and on the permeability of the porous microstructure of the green body. In most instances these procedures are empirically optimized using binder compositions that are commercially available. Fundamental research in this area is quite scarce. The reader is referred to [39,40] for further details in this area.

An alternate method for the fabrication of ceramic substrates is the synthesis by chemical routes at low temperatures. The starting materials in this case are metal alkoxides or inorganic salts. The techniques are particularly suitable for the preparation of films since the solutions can be cast into thin films and then directly transformed into ceramics. Since this type of processing is carried out at lower temperatures, there is less probability of contamination. Recently, there has been considerable interest in this technique for the synthesis of ceramic materials. The next section in this chapter is an in-depth discussion of this topic.

The issue of metal-ceramic interfaces is of great general interest in electronic packaging. In spite of its importance, there is little fundamental understanding of these unusual interfaces. Very recent observations of an interface between single-crystal sapphire and single-crystal niobium by high-resolution electron microscopy show that the interface between the metal and the ceramic is atomically sharp [41]. That is, the crystal lattices from either side of the interface are continuous right up to the plane where the two crystals meet. The structure is shown in Figure 10-19. An interesting consequence of this observation is that bonding between metals and ceramics, at least in some cases, can be atomic, and therefore very strong. Measurements of mechanical strength of metal-ceramic interfaces, which up to this point are very scarce in the literature, support the idea of strong bonding

FIGURE 10-19
Crystalline structure of the metal-ceramic interface.

between metals and ceramics when the ceramic is an oxide and the metal is a transition metal such as niobium or molybdenum [42,43]. Bonding between silicate glasses and metals has been widely used in industry for many years, which suggests that they bond well to each other, but direct measurements of the adhesion strength are not yet available [44].

While the work on interfaces between metals and oxide ceramics is limited, there is practically no work on bonding between metals and nonoxide ceramics such as aluminum nitride and silicon carbide [45]. Since these ceramics oxidize readily and form a protective oxide layer, the bonding may take place first with the oxide layer. Interestingly, the problem of direct bonding between metals and carbides is relevant to the mechanical properties of stainless steels and nickel-base superalloys. In these metals, fractures nucleate first at the interfaces between precipitates of carbides and the metal. Experience suggests that the carbide-metal interfaces are more resistant to fracture than the interfaces between oxide inclusions and the metal, which leads us to believe that the interfaces between *pure* carbides and metals may be even stronger than oxide-metal interfaces.

Ceramics are brittle materials and are therefore susceptible to failure from the largest flaw. Thus ceramics follow extreme value statistics, for example Weibull statistics [46]. (See also Chapter 9 where Weibull statistics are applied to make failure predictions.) This flaw contributes to a loss in reliability in ceramic materials. Time-dependent growth of flaws by stress corrosion is also of concern, in that it can lead to long-term failure at stresses much smaller than the stress required for catastrophic failure. For a given material, and for a given mechanism of failure, the fracture stress varies inversely as the square root of the flaw size. The issue of strength is, therefore, intimately related to the origin of the flaws in the ceramic substrate. The schematic in Figure 10-20 (courtesy F.

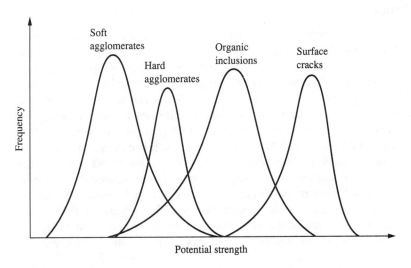

FIGURE 10-20
Factors producing flaws in ceramics and their relative severity.

F. Lange) shows the factors which can produce flaws in the ceramic, and their relative severity. The flaws can be produced during the sintering process due to the presence of agglomerates or from large pores as a result of organic inclusions. These flaws are often more degrading to strength than surface flaws that may be caused by grinding and handling. Thus the issue of strength and reliability is related intimately to the green state processing of ceramics.

CHEMICAL APPROACHES IN CERAMIC PROCESSING

Background

The key to the improvement of the properties, performance, and long-term reliability of a ceramic material is to control its structure and microstructure at a very fine scale (50–1000 Å) during the early stages of fabrication. One of the limitations of the conventional approaches of ceramic processing is the use of raw materials, i.e., powders, of several micron sizes that cannot be modified during processing to impart controlled bulk and interface microstructures and homogeneity at a molecular level. However, the future electronic and optoelectronic materials demand a precise control of structure, microstructure, and interface structure at ultrastructure level for the precision and reliability of their performances. In recent years, chemical routes that offer significant advantages in achieving these goals have received considerable attention in producing advanced ceramics and glasses [37,47–50].

Two chemical approaches that have become most popular in advanced glass-ceramic and composite fabrication are the sol-gel process and the colloid process.

Sol-Gel Process

The sol-gel process referred to here is based on the chemical polymerization of organometallic compounds. Alkoxysilanes, metal alkoxides, and reactive metal salts in nonaqueous solvents at low temperatures (around room temperatures) react to produce metal oxide gels, which are subsequently converted to glass/ceramics [51]. The basic concept of this approach is to take advantage of the chemical polymerization of organometallic compounds in solutions to produce inorganic polymeric units or the skeleton of high-temperature inorganic materials such as oxides, nitrides, or carbides, and subsequently to consolidate the polymeric building blocks to dense glass/ceramics after the removal of residual organics. A variation of this approach consists of reacting appropriate pure organic monomers or partially polymerized monomers with the organometallic polymeric units to produce a new class of materials called organically modified silicates (ORMOSILs) having a combined structure of the inorganic and organic components [52].

Some of the unique features and advantages of the sol-gel process are:

High purity in the final product can be achieved because of the purity of starting compounds and low processing temperature [53].

The sol-gel process provides high homogeneity, good stoichiometry, and compositional control [49,50,54].

The process produces very fine particles with a narrow size distribution leading to low-temperature sintering and microstructural control [55].

The possibility also exists for producing new glasses/ceramics and mixed organic-inorganic materials [37,52].

Sol-gel microporous materials can be used for catalyst support, insulators, and membranes [49,56].

The process allows the fabrication of biphasic or triphasic composites [47,57].

Thin films and thin self-supporting glass-ceramic sheets and fibers are produced directly from the sol-gel [58–60].

Sol-gel can be used for optical glass fibers and gradient index lenses [53, 60–62].

The low-temperature processing is of significant importance in fabricating composites of glasses/glass-ceramics, embedded metals, and plastics in sheets and thin-film forms. Sol-gel processes potentially are capable of fabricating multilayered structures of diverse materials, such as dielectrics, capacitors, resistors, and conductors. Recently, the processing of electronic substrate materials, such as cordierite ($2MgO \cdot 2Al_2O_3 \cdot 5SiO_2$) glass-ceramic, and capacitor materials, such as $BaTiO_3$ and $Pb(Z,Ti)O_3$, in bulk and thin-film form indicate the potential applications of the sol-gel processes in microelectronic packaging and thin-film devices [50].

The science of the sol-gel process for making glasses/glass-ceramics has been investigated somewhat intensively. The following subsections discuss the various aspects of the sol-gel processing of a model glass-forming system such as silica. The underlying principles apply to other glass-forming systems to a certain extent. However, the electronegativity of the central atom in metal alkoxides and the crystallization tendency of the gel should be considered in each particular metal oxide system. The sol-gel process for the fabrication of glasses/glass-ceramics consists of the following steps:

1. Synthesis of homogeneous sol by the polycondensation of metal alkoxides, and subsequent transformation to wet gel
2. Drying of wet gel
3. Removal of organics and hydroxyl groups
4. Densification of the porous gel to glass/glass-ceramics

A typical schematic diagram of the process is shown in Figure 10-21.

SYNTHESIS OF SOL AND SOL-TO-GEL TRANSFORMATION. The gel formation by the hydrolytic polycondensation of metal alkoxide can be represented by the following two steps:

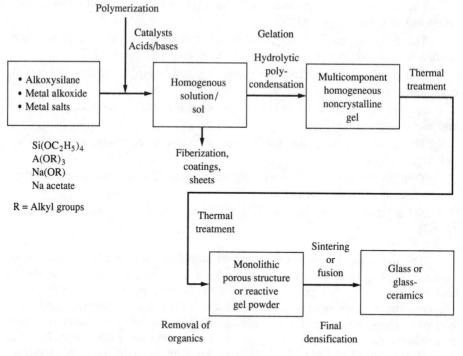

FIGURE 10-21
Sol-Gel process for glass/glass-ceramic fabrication using metal alkoxides.

1. Controlled hydrolysis of the metal alkoxide molecule in alcohol solvents:

$$M(OR)_n + xH_2O \rightarrow M(OH)_x(OR)_{n-x} + xROH \qquad (10.1)$$

2. Polycondensation of hydrolyzed species leading to gel formation:

$$2M(OH)_x(OR)_{n-x} \rightarrow (M\text{-}O\text{-}M) + H_2O + ROH \qquad (10.2)$$

where R = alkyl group and M = metal atom (e.g., Si, Al, Ti, B).

In the case of multicomponent oxide systems, different metal alkoxides such as $Si(OR)_4$ and $Ge(OR)_4$ can be copolymerized by hydrolytic polycondensation to form a mixed network having a structure like (Si-O-Ge-O-Si) where the distribution of the second network former, as example Ge^{+4}, can be controlled by the process parameters [63]. The polymeric units produced at the sol stage link together and grow with time and temperature, and subsequently a continuous network forms at the gel stage. The structure and microstructure of the gel obtained depend on several physicochemical factors that control the hydrolytic polycondensation reactions of metal alkoxides. These are:

Catalyst used: acid/base
Molar ratio of H_2O to $M(OR)_n$

Nature of solvents

Chemical nature of metal alkoxides

Temperature

The nature of catalyst and the concentration of water play important roles in the hydrolytic polycondensation process, which in turn controls the physicochemical structure of the gel produced. Several studies have been done on the kinetics of the polycondensation process of alkoxysilane, particularly of tetraethoxysilane (TEOS) [64–66]. Small angle x-ray scattering results [64,65] indicate that depending on the water content, the hydrolytic polycondensation of TEOS can lead to the formation of polymeric species ranging from polysiloxane chains to colloidal particles. The hydrolysis of TEOS with a four-molar ratio of water yields primarily linear or randomly branched polymers at pH 1 and discrete, more highly branched, polymeric clusters at pH 7. (See Figure 10-22 from Brinker and Scherer [64].) These differences in structure occur because at low pH (< 2.5) hydrolysis involves a mechanism of electrophilic attack on alkoxide oxygen, and at intermediate to high pH, hydrolysis and polycondensation involve nucleophilic attack of OH on silicon [64–66]. The structure of the gel produced at different pH conditions can strongly influence the densification and the crystallization process and thus the structure of the resultant glass/glass-ceramic [67]. The molar ratio of H_2O to $Si(OR)_4$ also plays an important role in determining the molecular configuration of the polymeric species, which dictates the rheology of the sol. An understanding and control of the rheology of the sol are required for the fabrication of thin glass sheets, coatings, or fibers from the sol. Experimental works [59] show that when the molar ratio of H_2O to $Si(OC_2H_5)_4$ is less than four in acid-catalyzed solution, the solution becomes spinnable and the fabrication of fibers or thin sheets can be achieved. Fabrication of fiber is extremely important since these are currently pulled from expensive processes employing high-temperature melts.

The spinnability of acid-catalyzed low-water-content solution is attributed to the formation of linear polymer. Sakka and Kamiya [68] determined the in-

Microstructural variations
due to processing chemistry

Acid catalyzed - entangled primarily linear
molecules crosslinked at junctions

Base catalyzed - agglomerated
branched clusters

FIGURE 10-22

Schematic representation of gelation in acid- and base-catalyzed silica gels prepared from tetraethoxysilane hydrolyzed with 4 moles H_2O per mole $Si(OC_2H_5)_4$ (after Brinker and Scherer [64]).

TABLE 10.2
Dependance of gelling time on the chain length of R (= SiOR) (After Schmidt [66])

Compound	Gelling time from the start of hydrolysis, hours
$Si(OC_2H_5)_4$	2
$Si(OC_4H_9)_4$	32
$Si(O(CH_2)_6CH_3)_4$	25
$Si(n-C_4H_9O)$ $(t-C_4H_9O)_3$	75
$Si(sec-C_4H_9O)_4$	500

trinsic viscosity and number average molecular weight of acid-catalyzed solutions having different molar ratios of H_2O to $Si(OC_2H_5)_4$. They used the Mark-Houwink equation, $[\eta] = KM_\eta^a$ where $[\eta]$ is intrinsic viscosity, M_n is number average molecular weight, and the magnitude of a is a measure of the hydrodynamic volume of the polymeric species, i.e., the main chain rigidity and branches. The a values of the solutions having molar ratios of H_2O to $Si(OC_2H_5)_4$ of 20, 5, 2, and 1 were found to be 0.34, 0.2, 0.64, and 0.75 respectively. According to Sakka and Kamiya the highest values of a (0.5 to 1.0) in low-water-content solution indicate the formation of linear polymers, whereas values smaller than 0.5 and close to zero indicate disks or sphere particles.

The nature of metal alkoxides influences the hydrolytic polycondensation rates. In the case of alkoxysilane, the polycondensation rate decreases with increase of alkyl chain length (see Table 10.2). This change in the hydrolytic polycondensation rates is due to (1) change in electron density at the silicon atom because of different electron-pumping effects of different alkoxides and (2) the difference in steric hindrance of different sizes of alkyl groups [66]. The chemical nature of the metal alkoxide also influences the homogeneity of multicomponent gels, because the molecular complexity (i.e., the degree of polymerization of the metal alkoxide) and the polycondensation rate depend on the nature of the central atom as well as the alkyl group [69] (see Table 10.3). It is evident from Table 10.3 that the degree of polymerization of metal alkoxides increases with the increase of ionic radii and with the decrease of the electronegativities of the central metal atom [69].

DRYING OF WET GELS. The viscous sol produced by the partial polymerization of metal alkoxides is molded to produce a desired shape. The initial sol, a Newtonian liquid, transforms to gel on aging. During gelation the viscoelastic sol (liquid) changes to viscoelastic solid (wet gel) which transforms into a microporous elastic solid gel during evaporation of the solvent from the wet gel. When a liquid evaporates from microporous or spongy solids such as gels, a capillary force due to the interfacial tension of the liquid puts a high stress on the solid during drying. This drying stress can cause collapsing of the pores, leading to high

TABLE 10.3
Molecular complexities of Group IV element ethoxides (after Reference [69]

(i) Compound	C(OEt)$_4$	Si(OEt)$_4$	Ge(OEt)$_4$	Sn(OEt)$_4$	Pb(OEt)$_4$	Ti(OEt)$_4$	Zr(OEt)$_4$	Hf(OEt)$_4$	Th(OEt)$_4$
(ii) Electronegativity	2.50	1.74	2.02	1.72	1.55	1.32	1.22	1.23	1.11
(iii) Covalent radii (A)	0.77	1.11	1.22	1.41	1.47	1.32	1.45	1.44	1.55
(iv) Degree of polymerization	1.0	1.0	1.0	4.0	—	2.4	3.6	3.6	6.0

shrinkage and cracking. The nonuniform capillary pressure due to evaporation of solvents from capillary pores of nonuniform sizes causes stresses that could lead to the fragmentation of monolithic gels. Even if the monolithicity is maintained, the entrapment of organics and water inside the collapsed pores during air drying can lead to bloating of the glass at the onset of viscous flow during sintering process.

The fundamental aspects of the drying mechanics are discussed by Zarzycki [70,71] and Scherer [72]. A simple analysis [70] shows that the capillary force or pressure developed during evaporation of solvents can be expressed as follows: For a cylindrical capillary of radius r,

$$\Delta P = \frac{2\gamma\cos\theta}{r} \tag{10.3}$$

where γ = interfacial tension of evaporating liquid, and θ = wetting angle.

The magnitude of capillary force depends on the size of the capillaries and the surface tension of the evaporating liquid. Thus the capillary pressure, or drying stress, increases without limit as r approaches zero. Figure 10-23 (after Zarzycki [71]) shows the variation of ΔP as a function of the pore radius for water (γ = 0.073 N/m) when θ = 0. It is evident from Figure 10-23 that the magnitude of the capillary stress across the gel structure can vary a considerable extent, depending on the range of the pore diameter. When the differential stresses thus developed during the drying are greater than the mechanical strength of the network, the gel will tend to crack. Hence, all measures which will (1) reduce the capillary stresses and (2) increase the tensile resistance of the gel-monolith

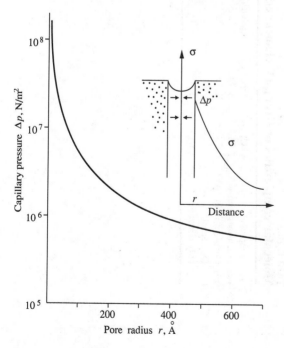

FIGURE 10-23
Capillary pressure (ΔP) as a function of the pore radius for water-filled capillaries. Inset shows the stress distribution in the vicinity of the pore [71].

will help in maintaining the monolithicity of large gel pieces. Several possible ways of achieving these goals are [70,73]:

Strengthening the gel structure

Making pore diameter uniform and larger

Reducing the surface tension of the liquid

Removal of the liquid under supercritical conditions where the liquid vapor interface does not exist

Removal of the solvent by freeze-drying

The following two drying approaches have been shown to be successful in producing large gel monoliths: (1) supercritical drying, and (2) use of some drying control chemical additive (DCCA) to make the pores of uniform sizes and strengthen the gel network [70,71–76].

Supercritical drying technique [70,71,74–76]. When a liquid changes imperceptibly to a gas without passing through a phase boundary, it is called supercritical or hypercritical drying. This takes place when a liquid is heated above its critical temperature (T_c) and above its critical pressure (P_c) as would happen in an autoclave. The change would be illustrated as: liquid $\rightarrow$ fluid above T_c $\rightarrow$ vapor. At a critical point the macroscopic densities of both the liquid and vapor phases become identical and liquid transforms to a fluid, eliminating the liquid/vapor interfacial effect and circumventing the drying stress.

The supercritical drying process consists of placing gel monoliths submerged in the solvent in an autoclave containing an excess solvent to develop an overpressure. The system is heated to increase the pressure to above the critical pressure of the solvent (e.g., 63.8 bar for ethanol). The internal pressure rises quickly due to the positive thermal expansivity of ethanol. When a pressure above P_c is reached, the temperature is increased at constant pressure to a value several degrees above T_c, the critical temperature (243°C for ethanol). At this stage the liquid is in the fluid regime. The fluid is now slowly purged by depressurizing and flushing with dry inert gas (Ar or He). The system is cooled slowly to room temperature after the flushing out of the solvent. The gel thus produced is called aerogel. It is highly porous and remains monolithic even with pore diameters of different sizes. Since no capillary pressure due to surface tension exists during supercritical drying, no shrinkage or collapsing of the pores occurs. It should be noted that instead of developing overpressure by heating the solvent an overpressure of inert gas such as N_2, Ar or He can be applied in the beginning to increase the pressure above P_c. This approach has also been used to prepare large silica gel monoliths [75].

Drying under atmospheric conditions using drying control chemical additives (DCCA) during gel formation [77,78]. This approach is based on the use of certain chemical compounds such as formamide or oxalic acid, which modify the hydrolytic polycondensation in such a way that a narrow pore size distribution and/or uniform larger pores develop after gelation; both effects reduce the drying stresses by eliminating or lowering the stresses generated due to porosity. The formation of pores of uniform size lowers the differential stresses because the local variation of capillary pressure P due to the variation of pore sizes across the structure is minimized. The larger pore size also considerably reduces the capillary pressure and, thus, the stress due to porosity (see Figure 10-23).

DENSIFICATION PROCESS. The densification process consists of two steps: (1) removal of residual organics and hydroxyl groups, (2) sintering.

Removal of residual organics and hydroxyl groups. Residual organics (such as alkoxy groups or alcohols physically and chemically bonded with the gel) and free molecular water can be removed relatively easily by thermal treatment in air or in oxidizing atmosphere in the temperature range of 100–350°C. The removal of hydroxyl groups up to around 1000 ppm can be achieved by passing dry oxygen/helium in the temperature range 300–1000°C, depending on T_g of the oxide system and pore structures. However, the removal of hydroxyl groups to the few-ppm level requires special chemical treatments with reactive gases such as CCl_4, Cl_2, or F_2 [53,79].

The effectiveness of the removal of residual organics and hydroxyl groups depends strongly on the pore size and pore morphology of the dried gels. Similarly, the kinetics of the sintering process depends on the molecular and pore structure of the gels. The physicochemical structures of dried gels are determined by the drying conditions including mechanical stress, temperature, and atmosphere during drying. The removal of residual organics and hydroxyl groups from supercritically dried gels is achieved before the onset of densification by viscous flow so that no bloating is observed during densification at higher temperature, whereas the gel monoliths dried slowly in air usually show a bloating phenomenon during densification when viscous flow occurs [71,74]. This is primarily due to entrapment of gases and water in collapsed pores during air drying. The Brunauer-Emmett-Teller (BET) isotherms of the aerogel and the air-dried gel were found to be distinctly different [74]. The isotherm observed with the air-dried gel falls into a category which is generally found with microporous solids showing a large number of extremely small (~ 50 Å) pores, whereas the isotherm observed with aerogels falls into a different category that is generally common among porous solids having tubular or ink-bottle type pores. Thus the drying stress can significantly change the pore volume and pore morphology [74].

Besides the influence of the drying process, the pore structure and the structure of dried gels depend on the processing parameters such as the pH and the concentration of water and solvents, which control the nucleation and growth of polymeric species [80]. Systematic studies [74] on the pore structure and microstructure of SiO_2 aerogel prepared at pH 8 indicate that the gel is composed of primary spherical particles in the range 50–100 Å, which cluster to form large (1000 Å) spherical aggregates loosely bound together. A transmission electron

AS–Autoclaved

0.1 µm

FIGURE 10-24
Transmission electron micrograph of supercritically dried silica gel prepared from Si(OCH$_3$)$_4$ with 5 moles H$_2$O per mole Si(OCH$_3$)$_4$ at pH 8 showing elementary particles and ultraporosity [74].

micrograph of such an aerogel prepared by the hydrolytic polycondensation of Si(OCH$_3$)$_4$ at a high pH of 8 with 5 moles of water per mole Si(OCH$_3$)$_4$ is shown in Figure 10-24. The scanning electron micrograph of the same gel is shown in Figure 10-25. Two types of porosity as revealed by TEM and BET surface area analysis exist in as-dried gels: macroporosity (1000 Å scale) and ultraporosity (100–200 Å). Hence the microstructures of the as-dried base catalyzed aerogel appear to be composed of primary particles (50–100 Å), which cluster to form large (~ 1000 Å) spherical aggregates (see Figures 10-24 and 10-25). A complete densification of the SiO$_2$ aerogel monolith to transparent low-OH-content glass is achieved by following a certain heat treatment schedule up to 1300°C [74].

0.1 µ m

FIGURE 10-25
Scanning electron micrograph of SiO$_2$ aerogel showing loosely bound large aggregates and macropores [74].

SINTERING. After the removal of organics and hydroxyl groups, the gel is still substantially porous; only a minor shrinkage occurs due to condensation reaction involving the formation of the Si-O-Si linkage by the removal of H_2O and ROH. The main driving force for sintering is the high surface energy of the porous gel; and in the case of glasses, the densification occurs by Newtonian viscous flow at around T_g, the glass transition temperature. However, the densification becomes complicated with nonglass forming systems when other dynamic factors operate during sintering: (1) The gel tends to crystallize during sintering. A competition between the sintering kinetics and the crystallization kinetics occurs. The crystallization changes the parameters such as viscosity which play a strong role in the sintering process. (2) The concentration of hydroxyl groups, which plays a strong role in determining the viscosity and the crystallization of glasses, also changes during the thermal densification process unless it is removed before the densification process. The fundamentals of densification kinetics are discussed by Zarzycki [70,71] and Brinker and Scherer [64]. Two models to represent the texture of the porous gels were used: (a) the closed-pores model proposed by MacKenzie and Shuttleworth [81], (b) the open-pores model developed by Scherer [82]. The details of these models and their applicability to the sintering kinetics of gels of different types are discussed by Zarzycki [70,71], Scherer [82], and Brinker and Scherer [64].

During densification, certain multicomponent silicate gel systems and non-silicate metal oxide gel systems tend to crystallize. The successful conversion of gel to glass depends on a competition between phenomena that lead to densification, that is, sintering, and those that enhance crystallization. The time-temperature-transformation (TTT) diagrams that have been determined for silica gels [70] represent a convenient method of studying crystallization versus glass formation. A knowledge of (1) viscosity change as a function of time and temperature and (2) the nucleation mechanism in multicomponent gels is needed to predict the progress of densification and crystallization processes occurring simultaneously during gel-to-glass transformation [83].

The crystallization kinetics and the nature of crystallization, however, are strongly influenced by the nature and distribution of the metal oxide building blocks and thermal history. Since the gel is produced at around room temperatures by the polymerization process and subsequently transformed to glass/glass-ceramics by densification at around T_g, the time and temperature involved in sintering are not always sufficient for attaining the equilibrium structure. The structure and microstructure of gel-derived glass/glass-ceramics, particularly in the multicomponent system, could be significantly different from the structure of melt-cooled glasses/glass-ceramic, in which case the melt structure and cooling conditions dictate the glass structures. The structural differences can originate from the following phenomena: (1) the orderliness and the distribution of network formers or modifiers in gels, (2) a preferential reaction between two monomers/oligomers, in which case a polymerized molecular complex can produce an ordered/disordered structure of a compound that can persist after densification [83].

Colloid Process

The colloid process [36,55,84–88] is based on the preparation of stable dispersion colloidal particles in an aqueous or nonaqueous liquid medium. The colloidal sol is either transformed to a gel by destabilization or compacted to make a green body with minimum pore volume and uniform pore diameter. High packing density and consequently low shrinkage are achieved by using colloidal particles in the 0.05–1.0 μm range. The colloidal particles are prepared by any one of the following techniques:

1. From inorganic metal salts by chemical reactions in solution [88]
2. Hydrolysis of metal alkoxides [55,87]
3. Vapor phase oxidation of metal halides/organometallic compounds [85,86]

In the case of the preparation of glasses, noncrystalline nonaggregated sols of colloidal particles are transformed into monolithic gel bodies having a large number of interparticle contacts. Subsequently, the drying and densification are achieved by following similar steps as mentioned before in the sol-gel process.

In the case of colloidal processing of crystalline ceramics, the prime requirements are (1) packing of unagglomerated colloidal particles of narrow size distribution (0.1 μm or less) to a compact having a uniform distribution of pores and (2) as many particle-particle contacts as possible (approaching close packing), and (3) the selection of the thermochemical variables such that densification rather than particle coarsening occurs and that pores are pinned to grain boundaries. This approach offers a method of making uniform, fine-grained (like the size of colloids) microstructures at substantially lower temperatures. However, one of the practical limitations of this approach is the formation of irregular aggregation or flocculation of the particles, which precludes dense packing and uniform pore size. Moreover, particle size has a strong influence on the aggregation; when particles are too small ($\sim$ 50 nm) they adhere to each other upon contact and the forces between them are very large [84]. Even with larger colloidal particles ($\sim$ 100 nm) the problem in interparticle bonding leading to irregular agglomerates in concentrated sol exists. Hence, the surface modification of particles with organics and the selection of appropriate particle sizes are the means to avoid agglomeration/interparticle bonding before compaction.

Different methods are used to achieve compacted molded shape: [100,103]

1. Colloidal filtration
2. Mechanical compaction and extrusion
3. Centrifugal packing

Sintering of colloidal compacts of monosized particles has been investigated by several researchers [35,36,55,87]. Sacks, et al. [87] studied the effect of ordered vs. random packing of silica spheres of uniform size in the diameter range 0.1–1 μm. The degree of uniformity of the pore diameter is very important in controlling the sintering temperature required to achieve 100 percent theoretical density and the stresses that are developed during sintering. Well-dispersed suspensions of silica particles prepared at high pH produce highly ordered compacts having a narrow pore size distribution but still contain defects analogous to vacancies, dislocations, grain boundaries, etc., in crystals. Flaws and pores originated from these defects persist to higher densities after the densification of the ordered portion and require higher temperatures and/or longer sintering time for their elimination [35,87].

However, once a dense packing and uniform pore size distribution are achieved in a certain scale, the sintering of fine colloidal powders can be achieved at considerably lower temperatures when compared with the conventional powders. Dense uniform compacts of monosized TiO_2 are sintered to greater than 99 percent of theoretical density at 1050–1100°C [55], a temperature much lower than the 1300–1400°C reported to sinter conventional TiO_2 powders to 97 percent of theoretical density. Hence, the key to the success of colloidal processing for producing fine-grained ceramics lies in producing uniform distribution of fine pore sizes in a green body. It is less important to achieve an orderly arrangement of particles than to produce a uniform distribution of pore sizes [35,36].

SINTERING OF CERAMICS

Background

Sintering may be defined as a process whereby a cluster of particles can be consolidated into a dense solid by diffusional mechanisms. The diffusional transport process may be visualized in different ways. It can be considered as a change in the shape of the particles, by transporting matter from one point on the surface of a particle to another, in such a way that the particles fit more snugly with their neighbors. As a result the external volume of the particle cluster decreases. When the particles fit perfectly the cluster is said to have reached theoretical density. The remarkable feature of this mechanism is that diffusional transport can occur very quickly at temperatures considerably lower than the melting point of the particles [89,90]. The key point is that the size of the particles should be small so that the diffusion distances are short. As will be shown later, the rate of sintering, in fact, scales with the fourth power of the particle size; the smaller the particle size, the faster the rate of sintering.

The geometrical configuration of the ensemble of particles in the powder compact also has important consequences in the sintering process. The idealized situation, where the structure of the powder compact is assumed to be that of a close-packed single crystal, is unrealistic. Even if all particles could be made to have exactly the same size, like atoms in a crystal, it would be difficult if not impossible to pack them into a perfect crystal of macroscopic dimensions. In the real situation the structure is likely to have defects such as grain, or domain, boundaries. If the particles are not all the same size, then the packing structure is likely to be random and, therefore, more homogeneous. Green state processing methods such as dry pressing, tape casting, and injection molding can have a profound influence on the packing configuration.

Inhomogeneities and defects in the packing density raise the question of incompatibility in the sintering rate in different regions of the powder compact. In extreme situations the incompatibility can lead to the formation of cracklike defects in the ceramic, which degrade the reliability of the sintering process. This issue will be addressed from two standpoints. The first is with regard to the magnitude of the incompatibility stress that can be generated in inhomogeneous sintering. The knowledge of the stress will provide insights in the use of hot isostatic pressing and hot pressing to suppress the formation of processing flaws. The second point relates to the scale of the homogeneity in the packing microstructure. The question must be asked whether or not there is a threshold size of inhomogeneity below which sintering defects will not form.

In contrast to ceramics, packing defects in the sintering of glass are less important because the incompatibility stresses arising from such defects can be easily relaxed by viscous flow. The sintering mechanism in glass is also different; pore closure in glasses is achieved by viscous flow rather than solid state diffusion, and the sintering rate is less sensitive to the particle size. The sintering behavior of glasses and gels is reasonably well understood and is predictable on the basis of models which were originally proposed by Frenkel [91] and by Mackenzie and Shuttleworth [81]. These models have been refined and applied to gels as well as to glasses with bimodal particle size distributions in several papers by Scherer [82,92].

The last topic discussed in this chapter is the sintering of composites, such as fiber-reinforced glasses and ceramics. The dispersed phase in the composite creates a situation of incompatible sintering. Here we will present a fairly general conclusion that sintering of ceramic-ceramic composites is likely to be difficult, but composites where the matrix is a glass should sinter well. At the end of this chapter references are given for additional reading [93,94].

Two Particle Models for Sintering

The role of diffusion kinetics in the sintering process can be considered by assuming the powder pack to have an idealized structure. For simplicity we assume that the structure is simple cubic, although other periodic configurations, such as face-centered cubic or hexagonal close-packed can also be considered. The influence of particle size and diffusion coefficients on sintering, which is of main interest to us, is adequately predicted by the simple analysis presented here.

The sintering of a three-dimensional cubic structure, shown schematically in Figure 10-26, can be modeled by considering the unit cell drawn in the heavy line. The change in the geometry of the pore as it shrinks is shown at the bottom in Figure 10-26. Atoms are transported into the pore from the adjacent grain boundaries by grain boundary diffusion. The atoms arrive at the triple grain junctions and are then distributed around the pore by surface diffusion. When the pore is large, surface diffusion is slow and cannot keep up with the arrival of

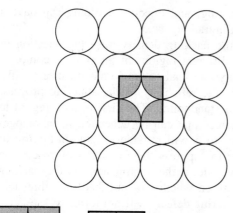

(a)   (b)   (c)   (d)

FIGURE 10-26
An idealized model for the sintering of spheres arranged in a simple cubic arrangement. The interparticle pore will change its shape as it sinters, as shown in (a) through (d).

atoms into the pore, but eventually the transport process becomes grain-boundary-diffusion limited and the pore begins to shrink in a quasi-equilibrium shape. This transition from a nonequilibrium to the equilibrium shape of the pore is shown from (b) to (c) in Figure 10-26. The interesting point about this transition is that it is accompanied by a change in the topology of the pore structure. As long as the pores have convex surfaces as in (a) and (b), they remain interconnected throughout the material, so that the ceramic appears permeable to the external environment. When the pores assume a concave surface curvature then the pores become totally enclosed within the material. The shape of the pores in the quasi-equilibrium case is controlled by the dihedral angle α. It is a state variable and is related to the ratio of the surface energy γ_s and the grain boundary energy γ_b, as follows:

$$\cos\alpha = \frac{\gamma_b}{2\gamma_s} \tag{10.4}$$

The angle α is susceptible to change when impurities segregate to the grain boundary and/or to the surface, since that can influence the interface energies. Note that the sign of the radius of curvature, r, of the pore depends on the dihedral angle α. For example, the pore shown in Figure 10-26, which has a fourfold coordination, assumes flat surfaces when $\alpha = \pi/4$. If α is greater than $\pi/4$, then the pore surface becomes concave (r positive). If it is less than $\pi/4$, then the shape becomes convex (r negative). The condition on α for a transition from concave to convex pores changes with the coordination number of the pore.

The magnitude and the sign of r are extremely important in the sintering, process since the driving force for sintering, p_o, is given by

$$p_0 = \frac{2\gamma_s}{r} \tag{10.5}$$

where r is positive if the pore surface is concave as shown in Figure 10-26(c) and (d). Equation (10.5) includes the effect of the dihedral angle since the magnitude of r depends on α. Generally, the smaller the dihedral angle the smaller the magnitude of the sintering pressure p_0. If r becomes infinity (flat surface) then the sintering pressure goes to zero and the pore becomes stable. The stability of pores depends on the dihedral angle and on the coordination of the pore. For example, the four-sided pore shown in Figure 10-26 will be stable if α is less than $\pi/4$; a three-sided pore will be stable if α is less than $\pi/3$, etc. Geometrical scaling requires that r and the particle size a should be proportional to each other, so that

$$r = fa \tag{10.6}$$

where f is a shape function that depends on α.

Figure 10-26 shows that pores are filled by the transport of atoms from the grain boundary into the pore. While this process can take place by lattice diffusion and by boundary diffusion, in most practical situations boundary diffusion is much faster, and we may ignore the contribution from lattice diffusion. In that case the equation for the densification rate is of the following form:

$$\frac{d\rho}{dt} = -(1 - \rho)B\frac{\delta D_b}{a^3} \cdot \frac{p_0 \Omega}{kT} \tag{10.7}$$

Here δD_b is the boundary width times the grain boundary self-diffusion coefficient, and ρ is the relative density. B is a weak function of ρ and may be assumed to be a constant; it is dimensionless. Ω is the atomic volume, k is Boltzmann's constant, and T is the temperature in degrees K. The quantity p_0 is the sintering pressure described by Equations (10.5) and (10.6). Particle size is denoted by a. Note that combining Equations (10.5) to (10.7) gives that $d\rho/dt$ varies inversely with the fourth power of the particle size. Such a strong dependence provides a strong incentive for reducing the particle size in powder compacts.

In Equation (10.7) we note that the rate of densification is proportional to the grain boundary diffusion coefficient δD_b. In ionic materials, such as zirconia, and partially ionic materials, for example alumina, grain boundary diffusion occurs fast enough at temperatures approaching approximately 70 percent of the melting point to provide rapid sintering. In covalent materials, however, solid state diffusion is too slow to be useful. In such instances the sintering rate can be enhanced by using additives that form a fluid phase in the interparticle regions [95]. Densification can now take place by the dissolution of the crystal into the fluid at the grain boundary and reprecipitation out of the fluid in the pore region [96]. The mechanism is illustrated schematically in Figure 10-27. In contrast to the solid state case, here it is desirable that the dihedral angle shown in Figure 10-27, α_l, should be close to zero, since the fluid must wet the particle-particle interface in order to enhance matter transport. The driving force for liquid phase

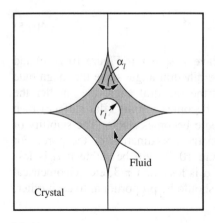

FIGURE 10-27
In liquid phase sintering the driving force for densification arises from the negative pressure due to the pore r_l.

sintering, p_l, is given by the radius of the spherical pore, r_l, and the surface tension of the fluid γ_l:

$$p_l = \frac{2\gamma_l}{r_l} \tag{10.8}$$

The rate of densification in the presence of the fluid is now given by

$$\frac{d\rho}{dt} = -\frac{(1 - \rho)B_l\bar{c}}{\eta a^3} \cdot p_l \tag{10.9}$$

where η is the viscosity of the fluid phase, and $\bar{c}$ is the solubility of the ceramic in the liquid at the sintering temperature.

In liquid phase sintering, the atomic structure of the solid-liquid interface is particularly important, since a process similar to crystal growth takes place. Crystal growth from a supersaturated fluid is often controlled by the density of steps on the surface since they provide sites where atoms can become attached to the crystal. It is likely that in liquid phase sintering the rate of solution precipitation is also determined by the density of the steps. In that case the sintering rate becomes interface controlled, as opposed to diffusion controlled, and the equation for densification rate becomes:

$$\left(\frac{d\rho}{dt}\right)_{int} = -(1 - \rho)B_l'\frac{k_l'\bar{c}}{a} \cdot p_l \tag{10.10}$$

The main difference between Equations (10.9) and (10.10) is the particle size dependence; it is much weaker for interface control. The parameter k_l' in Equation (10.10) is the interface velocity for a driving force of one kT. Since diffusion and interface reaction are both necessary to complete the solution precipitation cycle, the slower of the two equations will be rate controlling in the sintering process.

Phenomenology of Sintering Behavior

The measurement of shrinkage with time generally yields a curve of the type shown in Figure 10-3. The sintering rate decreases with time. This nonlinearity can be partly explained in terms of the $1 - \rho$ term on the right-hand side of Equations (10.7), (10.9), and (10.10) since the densification rate must necessarily go to zero as ρ approaches one.

However, the data are often even more nonlinear than the simple equation would suggest. The reason is that the effective particle size increases during densification as a result of grain growth. Since the densification rate is very sensitive to particle size, grain growth has a very strong decelerating effect on the densification rate [95]. The extent of the change in particle size is shown in pictures of alumina taken at three stages of sintering (Figure 10-28). The grain size increases from 0.7 μm to 2.2 μm when the density increases from 0.6 to 0.93; this would lead to a 100-fold decrease in the sintering rate. Thus modeling and understanding of the sintering process must include the study of grain growth.

Although the importance of grain growth has been generally recognized in the literature, the manipulation of microstructure to suppress grain growth remains a relatively unexplored area of research in ceramic materials. Note that grain growth is likely to be less important in liquid phase sintering if the process is interface controlled, since then the rate of sintering is much less sensitive to grain size.

The final exercise in this section will be to make a rough estimate of the densification viscosity of a powder compact. This is defined as the ratio of the average sintering pressure to the average densification rate. Later in this chapter it

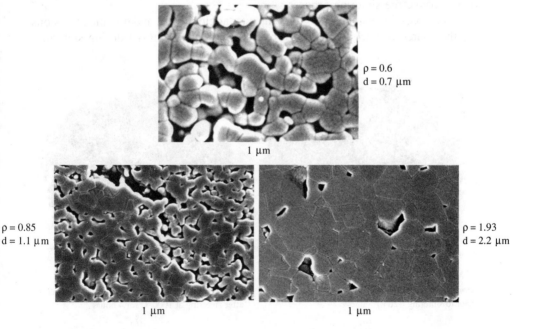

FIGURE 10-28
The change in the distribution and the morphology of pores in the sintering of alumina at 1773 K. The average particle size of the starting powder was 0.3 μm. The grain size d increases with increasing density ρ. (Ref. [100] courtesy American Ceramic Society.)

will be shown that the typical value of the sintering pressure in ceramic sintering when the particle size is approximately 1 μm is about 1 MPa. A realistic value of the sintering time would be about 2 hours or 7×10^3 s. The total densification strain is given by $\Delta\rho = \ln(1/\rho_g)$, where ρ_g is the green density; assuming $\rho_g = 0.5$ gives $\Delta\rho = 0.7$, which leads to an average rate of sintering of about 10^{-4} s^{-1}. The average sintering viscosity is, therefore, approximately 10^{10} Pa · s or 10^{11} poise. An estimate of this viscosity can be obtained by relating it to glass. The annealing point of glass is defined at 10^{13} poise and the softening point at $10^{7.6}$ poise. Therefore, it can be concluded that the sintering of ceramics is a highly viscous process.

The Importance of Packing Geometry

Real sintering does not have the idealized arrangement of powder particles, such as that shown in Figure 10-26, even if the particles are all of the same size. A more typical situation is shown in Figure 10-29. It shows the sintering of a two-dimensional raft of glass spheres. It can be noted that the hexagonally close-packed domains sinter very quickly. Once that has happened, the interdomain regions control the sintering rate. The influence of domain size is not shown here, but it has been observed that if the domains are larger than about thirty particle diameters, cracks develop in the interdomain boundaries. These cracks often survive the sintering process.

In three-dimensional powder compacts, agglomerates and organic inclusions are the source of packing defects [96–102]. Organics burn out during sintering,

Monolayer of monosized 140 = μm glass spheres prior to sintering (20×)

Monolayer of monosized 140 = μm glass spheres after sintering at 1000 °C for 4 hours

(a)

(b)

FIGURE 10-29
Early stages of sintering of a two-dimensional raft of glass spheres (Ref. [35]).

leaving behind large pores that are very difficult to sinter. Packing defects can also be introduced during the fabrication of the powder compact; for example, dry pressing of powders can create shear type packing flaws.

While there is not yet a quantitative understanding of the connection between irregular packing geometries and the macroscopic sintering rate, it is now recognized that large packing defects such as interdomain boundaries and agglomerates can give rise to sintering defects. These defects probably do not have a significant effect on the sintering rate, but they can seriously degrade the strength of the ceramic material. Recent theoretical and experimental work has shown that only very large packing defects develop into flaws. Generally, in ceramics, the critical size of the packing defect is about 100 μm; packing defects smaller than this will most likely not lead to sintering cracks.

Some possible techniques exist for limiting the scale of the inhomogeneities in packing densities to less than 10 μm. Colloidal methods, tape casting for example, have the potential for eliminating large agglomerates and inclusions. The use of controlled bimodal or multimodal particle size distributions can also reduce the size of the single-crystal domains that are likely to form if mono-sized particles are used. Model experiments with two-dimensional structures have shown that the domain size decreases as particles that are 50 percent smaller are added to monodispersed particles. These results would suggest that packing inhomogeneities can be controlled by careful control of particle size distributions and colloidal mixing.

Hot Isostatic Pressing

The previous section discussed the generation of incompatibility stresses as a result of inhomogeneous sintering in the powder compact. The magnitude of these stresses has been calculated in a model where two regions in the compact, one surrounding the other, are assumed to sinter at different rates. The stress can be expressed as a multiple of the sintering pressure, and lies in the range of one to ten times the sintering pressure. Recently, the sintering pressure has been directly measured in experiments where sintering is carried out under a uniaxial stress [100]. The total pressure for sintering can be varied by changing the uniaxial stress, and the extrapolation of the densification data to zero applied load gives a measure of the sintering pressure. In alumina, with a starting particle size of 0.7 μm, the sintering pressure is found to lie in the range of 1 to 2 MPa. In titania the pressure is a factor of three lower, probably because the surface energy of titania is approximately one-third that of alumina.

It is interesting to compare the measurements of the sintering pressure with the values estimated from Equations (10-5) and (10-6). Assuming that the shape factor $f = 1$, using γ_s for alumina to be 2 Jm^{-2} and $a = 0.7$ μm, and substituting into Equation (10-5), we find $P_o = 5.7$ MPa. This agrees with the measured value to within a factor of about four. The comparison would be better if the details of the shape of the pore were taken into account.

The measured value of the sintering pressure and the theoretical stress analysis suggest that the maximum value of the internal stress developing in the compact during differential sintering will be about 20 MPa. In principle, a tensile stress of this magnitude can be generated in the solid due to inhomogeneous sintering. Therefore, if sintering were carried out under a superimposed hydrostatic pressure of 20 MPa then all stresses in the compact would remain compressive and the formation of processing flaws would be suppressed. Conventional hot isostatic pressing (HIP) is carried out at pressures of up to 200 MPa. The discussion here would suggest that HIP at much lower pressures should be equally effective in enhancing densification and increasing reliability.

Sintering of Composites

The sintering of composites is necessarily inhomogeneous because the dispersed phase is often fully dense. An example would be a composite of alumina and whiskers of silicon carbide. The whiskers would constrain the sintering of surrounding alumina. Theoretical calculations show that the internal stress caused by the constraint will stop the sintering process if the volume fraction of the dispersed phase exceeds about 10 percent, and if the matrix is a polycrystalline ceramic. In these instances one must resort to pressure-assisted sintering by hot isostatic pressing, hot pressing, or sinter forging to achieve full densification.

An interesting case arises when the matrix phase is a glass [92,101] instead of a ceramic. In that case the analysis predicts that the incompatibility stress generated in the glass will be quickly relaxed by shear flow and sintering will be unaffected by the presence of a nonsintering dispersed phase [101]. Recent experiments [102] have borne out these predictions. The next question is how much of the dispersed phase can be added to the composite before it becomes difficult to sinter. A systematic study of this problem has yet to be carried out. The answer is likely to depend on the wetting angle between the glass and the ceramic phase, and the morphology of the ceramic particles. If the wetting is poor, then glass will become isolated in the interparticle regions and sintering of pores will be prevented. Sintering of fibrous composites is likely to be more difficult than the sintering of particulate composites because the fibers will develop an interconnected network at lower volume fractions. The sintering pressure must force this network to disintegrate before the composite can consolidate to full density.

REFERENCES

1. J. B. Blum and W. R. Cannon, *Advances in Ceramics*, vol. 19, *Multilayer Ceramic Devices*, Columbus, OH, American Ceramic Society, 1986.
2. R. C. Buchanan, *Ceramic Materials for Electronics*, New York: Marcel Dekker, Inc., 1986.
3. J. A. Mangels and G. L. Messing, *Advances in Ceramics*, vol. 9, *Forming of Ceramics*, Columbus, OH, American Ceramic Society, 1984.
4. E. A. Giess, K. N. Tu, and D. R. Uhlmann, *Materials Research Society Symposia Proceedings*, vol. 40, *Electronic Packaging Materials Science*, Materials Research Society, 1985.

5. W. Werdecker and F. Aldinger, "Aluminum Nitride—An Alternative Ceramic Substrate for High Power Applications in Microcircuits," *IEEE Transactions on Components, Hybrids, and Manufacturing Technology* CHMT-7, no. 4, pp. 399–404 (December 1984).
6. Y. Kurokawa, H. Hamaguchi, Y. Shimada, K. Utsumi, H. Takamizawa, T. Kamata, and S. Noguchi, "Development of Highly Thermal Conductive AlN Substrate by Green Sheet Technology," *Proceedings of the 36th Electronic Components Conference*, IEEE, pp. 412–418 (1986).
7. B. Schwartz, "Conference Studies Next Generation of Electronic Packaging," *American Ceramic Society Bulletin* 65, no. 7, p. 1032 (July 1986).
8. W. D. Kingery, *Introduction to Ceramics*, first edition, New York: John Wiley & Sons, 1960, pp. 39–45.
9. E. P. Hyatt, "Making Thin, Flat Ceramics—A Review," *American Ceramic Society Bulletin* 65, no. 4, pp. 637–638 (1986).
10. W. G. Burger and C. W. Weigel, "Multi-Layer Ceramics Manufacturing," *IBM Journal of Research and Development* 27, no. 1, pp. 11–19 (1983).
11. A. Cundy and D. Horne, "Automation and Thick Film Printing of Hybrid Circuitry," *Hybrid Circuit Technology*, pp. 21–27 (September 1986).
12. L. Miller, *Thick Films Technology and Chip Joining*, New York: Science Publishers, Inc., 1972.
13. J. Sergent, "Hybrid Circuit Industry and Its Technology," *Semiconductor International*, pp. 104–108 (September 1986).
14. R. Vest, "Materials Science of Thick Film Technology," *American Ceramic Society Bulletin* 65, no. 4, pp. 631–636 (1986).
15. A. Rose, "Composition and Manufacturing of Thick Film Inks," *Hybrid Circuit Technology*, pp. 29–34 (September 1986).
16. R. DeLaney, "The Risks, the Reliability of Copper Thick Film Technology," *Hybrid Circuit Technology* 2, no. 4, pp. 19–23 (April 1985).
17. J. Pierce, D. Kuty, and J. Larry, "The Chemistry and Stability of Ruthenium Based Resistors," *Solid State Technology* 25, no. 10, pp. 85–93 (October 1982).
18. J. Pierce, R. Johnson, H. Schmidt, and J. Larry, "New Thick Film Resistors for Potentiometer Applications," DuPont Electronics Materials Division Publication E-37335, October 1980.
19. R. Headley and R. Cote, "Tracking of Resistance and TCR in Thick Film Resistors–Distinctions, Definitions, and Derivations," DuPont Electronics Materials Division Publication E-29761, October 1979.
20. B. Taylor, R. Getty, J. Henderson, and C. Needes, "Air and Nitrogen Fireable Multilayer Systems: Materials and Performance Characteristics," *Solid State Technology* 27, no. 4, pp. 291–295 (April 1984).
21. W. DeForrest, *Photoresist Materials and Processes*, New York: McGraw-Hill, 1975.
22. L. Thompson, C. Willson, and M. Bowden, "Introduction to Microlithography," *ACS Symposium Series* 219, Washington, D.C.: American Chemical Society, 1983.
23. J. Roberts, "Resists Used in Lithography," *Chemistry and Industry*, pp. 251–257 (April 15, 1985).
24. P. Marcoux, "Dry Etching: An Overview," *Hewlett-Packard Journal* 33, pp. 19–23 (August 1982).
25. D. Bendz, R. Gendey, and J. Rasile, "Cost/Performance Single Chip Module," *IBM Journal of Research and Development* 26, no. 3, pp. 278–285 (May 1982).
26. M. Hill, "Thin Film Patterning Techniques for Hybrid Circuits," *Semiconductor International*, pp. 112–113 (September 1986).
27. J. Reche, "Polyimides in Hybrid Circuit Processing," *Semiconductor International*, pp. 116–117 (September 1986).
28. J. Shurboff, "Polyimide Dielectric on Hybrid Multilayer Circuits," *Proceedings of the 33rd Electronic Components Conference*, IEEE, pp. 610–615 (1983).
29. C. Ho, D. Chance, C. Bajorek, and R. Acosta, "The Thin Film Module as a High Performance Semiconductor Package," *IBM Journal of Research and Development* 26, no. 3, pp. 286–296 (May 1982).

30. J. J. Farrel, "Large-Scale Cost-Effective Packaging," *IEEE Micro* 5, no. 3, pp. 5–10 (June 1985).

31. R. Bowlby, "DIP May Take Its Final Bows," *IEEE Spectrum* 22, no. 6, pp. 37–42 (June 1985).

32. F. K. Moghadam, "Development of Adhesive Die Attach Technology in Cerdip Packages," *Solid State Technology* 27, no. 1, pp. 149–157 (January 1984).

33. L. Miller, "Controlled Collapse Reflow Chip Joining," *IBM Journal of Research and Development* 13, no. 2, pp. 239–250 (March 1969).

34. A. J. Blodgett and D. R. Barbour, "Thermal Conduction Module: A High Performance Multilayer Ceramic Package," *IBM Journal of Research and Development* 26, no. 1, pp. 30–36 (January 1982).

35. E. Liniger and R. Raj, "Packing and Sintering of Two-Dimensional Structures Made from Bimodal Particle Size Distributions," *J. American Ceramic Society* 70, no. 11, pp. 843–49 (1987).

36. I. Aksay, "Microstructural Control through Colloidal Consolidation," in *Advances in Ceramics*, vol. 9, edited by J. A. Mangels and G. Messing, Columbus, OH: American Ceramic Society, 1984.

37. L. L. Hench and D. R. Ulrich, editors, *Ultrastructure Processing of Ceramics, Glasses, and Composites*, New York: John Wiley & Sons, 1984.

38. R. F. Davis, H. Palmour, and R. L. Porter, editors, *Emergent Process Methods for High Technology Ceramics*, New York: Plenum Press, 1984.

39. G. Y. Onoda and L. L. Hench, editors, *Ceramic Processing before Firing*, New York: John Wiley & Sons, 1978.

40. R. A. Gardener and R. W. Nufer, "Properties of Multilayer Ceramic Green Sheets," *Solid State Technology*, pp. 38–43 (May 1974).

41. M. Florjancic, W. Mader, M. Ruhle, and M. Turwitt, "HREM and Diffraction Studies of an Al_2O_3/Nb Interface," *J. de Physique*, p. C4–129. Vol. C4 (1985). Proceedings of International Conference on the Structure and Properties of External Interfaces, Irsee, W. Germany, Aug 19–23, 1984.

42. G. Elsner, T. Suga, and M. Turwitt, "Fracture of Ceramic-to-Metal Interfaces," *J. de Physique* 46, Colloque C4, supplement to Vol. 46, no. 4, p. C4–597 (April 1985).

43. A. G. Evans, M. Ruhle, and M. Turwitt, "On the Mechanics of Failure in Metal-Ceramic Bonded Systems," ibid., p. C4–613.

44. M. P. Borom and J. A. Pask, "Role of Adherence Oxides in the Development of Chemical Bonding at Glass-Metal Interfaces," *J. American Ceramic Society* 49, no. 1, pp. 1–6 (1966).

45. K. Suganuma, T. Okamoto, M. Shimada, and M. Koizumi, "A New Method for Solid-State Bonding between Ceramics and Metals," *J. American Ceramic Society* 66, p. 117 (July 1983).

46. C. Lipson and N. J. Sheth, *Statistical Design and Analysis of Engineering Experiments*, New York: McGraw-Hill, 1973, p. 36.

47. C. J. Brinker, D. E. Clark, and D. R. Ulrich, editors, *Materials Research Society Symposia Proceedings*, vol. 32, *Better Ceramics through Chemistry I*, New York: North-Holland, 1984.

48. J. Zarzycki, editor, *Proc. 3rd Int. Workshop on Glasses and Glass-Ceramics from Gels*, New York: North-Holland, 1986.

49. L. L. Hench and D. R. Ulrich, editors, *Science of Ceramic Chemical Processing*, New York: John Wiley and Sons, 1986.

50. C. J. Brinker, D. E. Clark, and D. R. Ulrich, editors, *Materials Research Society Symposia Proceedings*, vol. 73, *Better Ceramics through Chemistry II*, Pittsburgh: Materials Research Society, 1986.

51. S. P. Mukherjee, "Sol-Gel Processes in Glass Science and Technology," *J. Non-Crystalline Solids* 42, pp. 477–488 (1980).

52. H. Schmidt, "Organically Modified Silicates by the Sol-Gel Process" in Ref. 47, pp. 327–335.

53. S. P. Mukherjee, "Ultrapure Glasses from Sol-Gel Processes," in *Sol-Gel Technology*, edited by L. Klein, Park Ridge, New Jersey: Noyes Publications, 1988.

54. S. P. Mukherjee, "Homogeneity of Gels and Gel-Derived Glasses," *J. Non-Crystalline Solids* 63, p. 35 (1984).

55. E. Barringer, N. Jubb, B. Fegley, R. L. Pober, and H. K. Bowen, "Processing of Monosized Powders," in Ref. 37, p. 315.
56. G. Carturan, G. Facchin, G. Navazio, V. Gottardi, and G. Cocco, "A Novel Approach to Preparation of Support for Glass-Titania-Pd Catalyst in Hydrogenation of Olefins," in Ref. 37, pp. 197–206.
57. R. Roy, S. Komarneni, and D. M. Roy, "Multi-phasic Ceramic Composites Made by Sol-Gel Techniques," in Ref. 47, pp. 347–359.
58. H. Dislich and E. Hussmann, "Amorphous or Crystalline Dip Coatings, Obtained from Organometallic Solutions: Procedures, Chemical Processing, and Products," *Thin Solid Films* 77, pp. 129–139 (1981).
59. S. Sakka, K. Kamiya, K. Makita, and Y. Yamamoto, "Formation of Sheets and Coating Films from Alkoxide Solutions," *J. Non-Crystalline Solids* 63, pp. 223–236 (1984).
60. S. P. Mukherjee and J. Phalippou, "Glassy Thin Films and Fiberization by the Gel Route," in *Glass—Current Issues*, edited by A. F. Wright and J. Dupuy, Boston: Martinus Nijhoff Publishers, 1985.
61. S. P. Mukherjee, "Gradient Index Lens Fabrication Processes," paper presented at Topical Meeting on Gradient Index Optical Imaging Systems TQAI (1–5) held at Honolulu, Hawaii, May 4–5, 1981, Optical Society of America.
62. M. Yamane, J. B. Caldwell, and D. T. Moore, "Preparation of Gradient-Index Glass Rods by the Sol-Gel Process," in Ref. 50.
63. S. P. Mukherjee, "Gels and Gel-Derived Glasses in the Silica-Germania System," in Ref. 47, pp. 111–118.
64. C. J. Brinker and G. W. Scherer, "Relationship between the Sol-to-Gel and Gel-to-Glass Conversions," in Ref. 37, pp. 43–59.
65. K. D. Keefer, "The Effect of Hydrolysis Conditions on the Structure and Growth of Silicate Polymers," in Ref. 47, pp. 15–24.
66. H. Schmidt and H. Scholze, "Mechanisms and Kinetics of the Hydrolysis and Condensation of Alkoxides," in *Glass—Current Issues*, edited by A. F. Wright and J. Dupuy, Boston: Martinus Nijhoff Publishers, 1985.
67. S. P. Mukherjee and S. S. K. Sharma, "Structural Studies of Gels and Gel-Glasses in the Silica-Germania System Using Vibrational Spectroscopy," *J. American Ceramic Society* 69, pp. 806–810 (1986).
68. S. Sakka and K. Kamiya, "The Sol-Gel Transition in the Hydrolysis of Metal Alkoxides in Relation to the Formation of Glass Fibers and Films," *J. Non-Crystalline Solids* 43, pp. 31–46 (1982).
69. D. C. Bradley, R. C. Mehrotra, and D. P. Gaur, *Metal Alkoxides*, London: Academic Press, 1978.
70. J. Zarzycki, "Sol-Gel Preparative Methods," in *Glass—Current Issues*, edited by A. F. Wright and J. Dupuy, Boston: Martinus Nijhoff Publishers, 1985.
71. J. Zarzycki, "Monolithic Xero- and Aerogels for Gel-Glass Processes," in Ref. 37, pp. 27–42.
72. G. W. Scherer, "Drying Mechanics of Gels," in Ref. 50, pp. 225–230.
73. R. K. Iler, *The Chemistry of Silica*, New York: Wiley, 1979, p. 536.
74. S. P. Mukherjee, J. F. Cordaro, and J. C. Debsikdar, "Sintering Behavior of Silica Gel Monoliths Having Different Pore Structures," *J. American Ceramic Society* to be published in 1988.
75. J. G. Van Lierop, A. Huizing, W. C. P. M. Meerman, and C. A. M. Mulder, "Preparation of Dried Monolithic Silica Gel Bodies by an Autoclave Process," in Ref. 48, pp. 265–270.
76. S. P. Mukherjee, "Supercritical Drying in Structural and Microstructural Evolution of Gels: A Critical Review," to be published in *Proc. of 3rd Int. Conf. on Ultrastructure Processing of Ceramics, Glasses and Composites*, edited by J. D. MacKenzie and D. R. Ulrich.
77. R. D. Shoup, "Controlled Pore Silica Bodies Gelled from Silica Sol-Alkali Silicate Mixtures," pp. 63–69 in *Colloid and Interface Science*, vol. 3, edited by M. Kerker, New York: Academic Press, 1976.
78. L. L. Hench, "Use of Drying Control Chemical Additives (DCCA) in Controlling Sol-Gel Processing," in Ref. 49, pp. 52–64.

79. E. M. Rabinovich, D. L. Wood, D. W. Johnson, Jr., D. A. Fleming, S. M. Vincent, and J. B. MacChesney, "Elimination of Cl_2 and Water in Gel-Glasses," *J. Non-Crystalline Solids* 82, pp. 42–49 (1986).

80. D. W. Schaefer and K. D. Keefer, "Fractal Aspects of Ceramic Synthesis," in Ref. 50, pp. 277–288.

81. J. K. MacKenzie and R. Shuttleworth, "Phenomenological Theory of Sintering," *Proc. Phys. Soc.* (London) 62, pp. 833–852 (1949).

82. G. Scherer, "Sintering of Low Density Glasses (I and II)," *J. American Ceramic Society* 60, pp. 236–239, 243–246 (1977).

83. S. P. Mukherjee, "A Comparison of Structures and Crystallization Behavior of Gels, Gel-Derived Glasses and Conventional Glass," in Ref. 50, pp. 443–460.

84. R. K. Iler, "Inorganic Colloids for Forming Ultrastructures," in Ref. 49, pp. 3–20.

85. G. W. Scherer and J. C. Long, "Glasses from Colloids," *J. Non-Crystalline Solids* 63, pp. 163–172 (1984).

86. E. M. Rabinovich, D. W. Johnson, J. B. MacChesney and E. M. Vogel, "Preparation of High Silica Glasses from Colloidal Gels Part 1," *J. American Ceramic Society* 66, pp. 683–688 (1983).

87. M. D. Sacks and T. Y. Tseng, "Preparation of Silica Glass from Model Powder Compacts (I and II)," *J. American Ceramic Society* 67, pp. 526–537 (1984).

88. E. Matijevic, "Monodispersed Colloidal Metal Oxides, Sulfides, and Phosphates," in Ref. 37, pp. 334–352.

89. R. L. Coble, "Diffusion Sintering in the Solid State," in *Kinetics of High Temperature Process*, edited by W. D. Kingery, Cambridge, Mass.: Tech. Press of MIT, 1959.

90. R. L. Coble, "Diffusional Models for Hot-Pressing with Surface Energy and Pressure Effects as Driving Forces," *J. Appl. Phys.* 41, pp. 4798–4807 (1970).

91. J. Frenkel, "Viscous Flow of Crystalline Bodies under the Action of Surface," *J. Phys. (Moscow)* 9, pp. 385–391 (1945).

92. G. W. Scherer, "Viscous Sintering of a Bimodal Pore Size Distribution," *J. American Ceramic Society* 67, pp. 709–715 (1986).

93. G. C. Kuczynski, editor, *Sintering Processes*, New York: Plenum Press, 1980.

94. D. Kolar, S. Pejovnik, and M. M. Ristic, editors, *Sintering—Theory and Practice*, Amsterdam: Elsevier Scientific, 1982.

95. R. M. German, *Liquid Phase Sintering*, New York: Plenum Press, 1985.

96. R. Raj, "Creep in Polycrystalline Aggregates by Matter Transport through a Liquid Phase," *J. Geophys. Res.* 87, pp. 4731–4739 (1982).

97. R. J. Brook, "Pore Boundary Interactions and Grain Growth," *J. American Ceramic Society* 52, pp. 56–57 (1969).

98. R. K. Bordia, "Sintering of Inhomogeneous or Constrained Powder Compacts: Modelling and Experiments," Ph. D. thesis, Cornell University, Ithaca, N.Y., June 1986.

99. R. Raj and R. K. Bordia, "Sintering Behaviors of Bi-modal Powder Compacts," *Acta Metall.* 32, pp. 1003–1019 (1984).

100. K. R. Venkatachari and R. Raj, "Shear Deformation and Densification of Powder Compacts," *J. American Ceramic Society* 69, pp. 499–506 (1986).

101. R. K. Bordia and R. Raj, "Analysis of Sintering of a Composite with a Glass or Ceramic Matrix," *J. American Ceramic Society* 69, C55–C57 (1986).

102. V. Ducamp, "Thermodynamic Aspects of Sintering of Glass and Glass-Ceramic Composites," M.S. thesis, Cornell University, Ithaca, N.Y., 1987.

PROBLEMS

10.1. With linear shrinkage (fractional) defined as

$$S_\ell = \frac{\ell_i - \ell_f}{\ell_i}$$

where ℓ_i = linear dimension before firing (sintering)
ℓ_i = linear dimension after firing

and volume shrinkage S_v defined similarly, derive an expression relating volume shrinkage to linear shrinkage, assuming that shrinkage is isotropic (i.e., the same in all directions.)

10.2. Assume a ceramic sintering process that achieves a fired density of 3.90 ± 0.02 g/cm^3 for a ceramic powder with a particle density of 3.96 g/cm^3, and that nominal linear shrinkage of dry pressed parts produced using this process is 17 percent. What must be the density and density tolerance achieved in the pressing operation to achieve linear tolerances of ±0.5 percent in the final product? Ignore any contributions of fugitive binders.

Hints: Solving Problem 1 is a prerequisite for solving this problem. Using a Taylor series approximation for the value of a function of two variables may be helpful:

$$f(x + \Delta x, y + \Delta y) = f(x, y) + \left(\Delta x \frac{\partial}{\partial x} + \Delta y \frac{\partial}{\partial y}\right) f(x, y) + \cdots$$

or

$$\Delta f \simeq \Delta x \frac{\partial f}{\partial x} + \Delta y \frac{\partial f}{\partial y}$$

10.3. Calculate the capacitance of a 0.050-by-0.050-inch thick-film capacitor when the dielectric has a thickness of 0.0008 inch and a relative dielectric constant (ϵ_r) of 9.6.

10.4. A thin-film resistor measures 150 Ω at 25°C and 151.5 Ω at 100°C. Calculate its TCR in ppm/°C.

10.5. Calculate the length of a 500-Ω thin-film resistor fabricated from a material with a sheet resistivity of 100 Ω and having a width of 200 μm.

10.6. A particular application requires a thick-film resistor with a value of 500 Ω ± 1 percent over the temperature range of −55 to 125°C. If the resistor can be trimmed to ±0.1 percent, and drift of the resistor value over life is limited to ±0.5 percent, what is the required TCR of the resistor?

CHAPTER 11

POLYMERS AND POLYMER-BASED COMPOSITES FOR ELECTRONIC APPLICATIONS

GEORGE P. SCHMITT
BERND K. APPELT
JEFFREY T. GOTRO

INTRODUCTION

Polymer science is one of the most active areas of applied science. The field of polymer science was essentially established during the early part of this century. However, many principles of rubber elasticity were already practiced in the 19th century. Fundamental understanding of polymers grew rapidly around the time of World War II, with extremely rapid growth following the war. Excellent recent texts treat the subject generally [1–5], and from the viewpoints of the chemist [5–7], and of the physicist [8–13]. The application of polymers in electronics has been covered in chapters of texts and review articles [14–18]. For those desiring a critical understanding of the subject matter of this chapter, the references cited above and throughout the text provide entry points to suit a variety of technical backgrounds.

Polymers are materials whose usefulness comes from their high molecular weight, achieved by linking together many small molecules. Indeed, the word *polymer* means "many units." Most polymers in use today are synthetic, but polymers such as unvulcanized natural rubber, cellulose, and silk occur naturally.

Other high-molecular-weight materials are also found in nature, for instance lipoproteins and complex saccharides, but these substances are not formed by repetitive linkage of identical small molecules, and thus do not exhibit the characteristics typical of long chain synthetic polymers.

There are two classes of synthetic polymers, thermoplastics and thermosets. Thermoplastics are high-molecular-weight linear or branched polymers, which may either be crystalline or amorphous, depending on their chemical compositions. Polyethylene, polystyrene, polypropylene, polycarbonate (Lexan®), and polyethylene terephthalate (Mylar®) are examples of thermoplastic polymers. Thermoplastics are processible by the application of heat and pressure, and in the absence of degradation, no chemical reaction takes place during processing. If a thermoplastic polymer is crystalline, then the polymer must first be melted in order for flow to occur. Amorphous thermoplastic polymers must be heated above their glass transition temperature, T_g, in order for softening and flow to occur. Upon cooling, thermoplastics either crystallize or transform from the rubbery to the glassy state as the temperature falls below T_g.

Thermosetting polymers are typically low-molecular-weight polymers (oligomers) that undergo large chemical and physical changes during processing. Epoxies, Bakelite® (phenolics), and Formica® (melamine formaldehyde) are examples of thermosetting polymers. Upon application of heat, these multifunctional resins chemically react, causing an increase in the molecular weight, and form a highly branched, interconnected (cross-linked) network. Under certain well-defined conditions, these polymers undergo gelation and macroscopic flow ceases. Further chemical reaction leads to full conversion of all reactive groups and produces a polymer with high hardness, high heat distortion temperature, and good chemical resistance. Further application of heat and pressure will not cause a thermosetting polymer to flow.

Thermoplastic polymers are supplied to the molder at their final molecular weight and are melted in processing. The high molecular weight causes the viscosity to be very high, and large amounts of energy must be expended to process the highly viscous polymer melt. In contrast, typical thermosetting polymers are supplied to the molder at low molecular weights and thus allow part of the processing to take place at much lower viscosities. As the resin reacts during curing, the viscosity rapidly increases as the network builds and approaches the final cure state.

Our purpose is to present a brief overview of polymer applications in electronic packaging. A detailed discussion will focus on the state-of-the-art technology used to manufacture complex multilayer printed circuit boards. The main emphasis will be on the materials and processes employed to fabricate these boards. A wide range of scientific disciplines will be described, including thermochemistry, kinetics, rheology, chemical analysis, and mechanical analysis.

The chapter begins with a discussion of the application of polymeric materials in chip carriers (first-level packaging). Then the application of a variety of thermosetting polymers in printed circuit boards (second-level packaging) is extensively covered. The chapter concludes with a description of a variety of techniques used for materials and process characterizations. The methods described will be applicable to the investigation of circuit board materials, as well as high-performance resins used for advanced composites (for example, in aerospace applications).

POLYMERS IN FIRST-LEVEL PACKAGING

The focus of this section will be the use of polymers in first-level packaging. This will not include any of the polymeric applications in device (chip) fabrication. The major application of polymers in chip fabrication is in the area of micron and submicron photoresist technology, which is treated in the chapter on lithography. There are also excellent texts [14,15] that cover this subject in detail.

Chip Carrier

First-level packaging encompasses the technology required to get electrical signals into and out of the chip. The chip is most often mounted on a substrate as part of a component, called a module or chip carrier. A number of types of chip carriers are commonly in use: flatpacks, pin grid arrays (PGA), and dual in-line packages (DIP). More recent types are plastic-leaded chip carriers (PLCC), lead-less chip carriers and small-outline integrated circuits (SOIC). These components are illustrated in Figure 11-1. The latter three types of carriers are utilized in surface mount technology (SMT). The SMT chip carriers are soldered onto pads on the surface of the card or board. Surface mount carriers are usually vapor phase soldered and thus require a thermally resistant circuit board material. The PGA and DIP carrier leads, on the other hand, are inserted into the plated through-holes and are typically attached by wave soldering methods, at temperatures lower than vapor phase soldering.

ENCAPSULATION OF CHIP CARRIERS. For the DIP carrier, the chip is typically connected to the lead frame via wire bonding. Subsequently, the chip and lead frame are encapsulated in a polymeric insulator. The encapsulating material serves as a dielectric insulator and shields against environmental degradation. These encapsulating materials are typically thermosetting polymers, with epoxy-based resins most widely used. Other thermosetting polymers, such as silicones and unsaturated polyesters, are employed to a lesser degree.

One reason for the widespread utilization of epoxies is their reactivity with a variety of curing agents [19]. This allows resin systems to be formulated with a wide range of properties. The inherent strength of epoxies is high, with neat (unfilled) resins reaching 12,000–14,000 psi in tension. Epoxy-glass cloth laminates can reach tensile strengths of 40,000 psi or more.

Cure shrinkage of epoxy systems is very low; with typical fillers it can be on the order of 1 to 2 percent. This low cure shrinkage is important in maintaining close tolerances during molding, and the low shrinkage produces less residual stress in the cured part, thus improving long-term dimensional stability. The

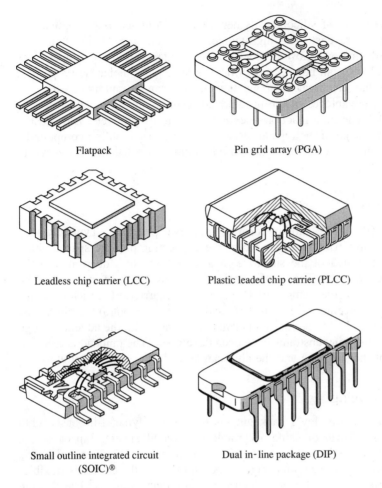

Flatpack

Pin grid array (PGA)

Leadless chip carrier (LCC)

Plastic leaded chip carrier (PLCC)

Small outline integrated circuit
(SOIC)®

Dual in-line package (DIP)

FIGURE 11-1
Various types of chip carriers commonly in use.

end use temperatures for epoxy-molded carriers can be as high as 125–150°C, depending on the selection of reactants, fillers, and modifiers. Silicone polymers are usually required if higher operating temperatures are encountered.

Epoxies wet and adhere well to many types of solid surfaces. They are useful encapsulant materials due to a combination of factors including low cure shrinkage, a polar nature, and the ability to be blended with a wide variety of modifiers to impart the desired mechanical properties.

Epoxies are also attractive from a process point of view. Fast cure kinetics can easily be attained, and only the most rapid of other chemical systems can exceed the cure speed of epoxies. Epoxy monomers and prepolymers are commercially available in a wide variety of viscosities, and thus are easily tailored to meet a particular molding requirement.

The primary use of silicone polymers is in high-temperature applications that warrant the extra cost. The high-temperature endurance of this family of polymers exceeds that of almost all electrical encapsulants. End use temperatures in the range of 200°C can be attained. Silicones have good dielectric properties, wet and encapsulate electrical components well, and can be rapidly cured.

One of the advantages of the unsaturated resins is the extremely fast cure kinetics possible. The cure reaction does not generate by-products. Low cost is the principle advantage of the typical unsaturated polyester molding compounds. Generally, they have inferior strength and environmental resistance compared to epoxies and silicones.

Pin Grid Arrays

Pin grid arrays (PGA) are another form of first-level packages. Typically they are defined as a circuitized substrate where the connecting pins are spaced either uniformly over the area of the substrate or at least two rows deep around the perimeter of the substrate. This configuration takes advantage of the area of the substrate to maximize the number of I/Os (pins) per carrier. Historically, chips were mounted in the center of the substrate and lines fanned out to the pin heads. Consequently, chips had to be located between the pins. With the advent of large memory chips, flush pin substrates have been developed. The pinhead is recessed in the surface of the substrate and the chip can be as large as the substrate.

Thin-Film Packaging

Another approach to first-level packaging is to use a polymer film-based chip carrier (PFBCC). Instead of using a ceramic or molded plastic chip carrier, a polymeric film carrier is employed as shown in Figure 11-2. The polymeric carrier has several significant advantages. The thin film in the carrier is flexible, and more importantly has a lower dielectric constant than ceramics. The flexible nature of these packages has already been exploited extensively in the form of cables and tape-automated bonding (TAB). TAB chip carriers are composed of a relatively fragile lead frame, which is supported by the polymeric carrier. The chip is bonded to the lead frame and the chip/lead frame/polymer film carrier is bonded to the second-level package. The TAB technology was stimulated by the continuous processing technology of the photographic film industry. Thus, the polymeric carrier is metallized (sputter or vapor deposition) and circuitized via

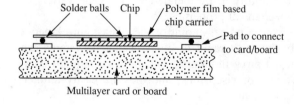

Solder balls Chip Polymer film based chip carrier

Pad to connect to card/board

Multilayer card or board

FIGURE 11-2
Schematic of the polymer film based chip carrier (PFBCC).

photolithography. The carrier is etched to produce free-standing leads, and the chips are bonded to the frame. All these processes can be handled in roll form, either by processing a roll through one process at a time (e.g., metallization) or through several process modules in sequence (e.g., photoresist apply, bake, expose, develop).

PFBCC may be viewed as an extension of TAB. The circuitry is entirely supported by the polymeric carrier. The chips are typically mounted in the center, with lines fanning out to the connection pads on the perimeter. Pads can be arranged so that the I/O approaches that of the PGA. By introducing vias and double-sided and multilayer structures, the packaging density of the PFBCC can be increased significantly.

The polymeric film carriers are frequently based on polyimides, with Kapton® being the most prominent polyimide. Polyester films are commonly used (mainly polyethylene terephthalate, Mylar®) and as of late a new generic group of materials, liquid crystalline polymers, is finding applications as well. The choice of carrier is largely governed by the process parameters like metallization and chip bonding. Polyimides and their properties are discussed in later sections.

ENCAPSULATION OF PIN GRID ARRAY CHIP CARRIERS. All first-level packages are encapsulated to protect chip circuitry and first-level circuitry from corrosion and handling damage. The packages may be protected by a cap (metal, ceramic, or plastic), which in turn needs to be attached to the chip carrier. For ceramic chip carriers, modified epoxies provide a semihermetic seal against the environment. The epoxies are applied either in liquid form or as preforms. They are commonly based on bisphenol-A epoxy and an anhydride. Flexible anhydrides (usually aliphatic) and reactive diluents (aliphatic epoxies and dialcohols) may be employed as well to yield an encapsulant that transmits little stress between ceramic and cap. The cure kinetics and rheology need to be controlled closely to guarantee a good seal around the pins, and seal between the cap and the ceramic without covering the back side of the ceramic (where the chip is attached). The same principle applies to plastic caps. Metal caps have the advantage of better chip-cooling capacity especially when thermal grease or metallic springs are used to transmit the heat from chip to cap.

Thermal greases are highly thixotropic suspensions of a heat conductor (e.g., metal oxide, boron nitride) in a liquid polymer. Silicone oils (dimethylene siloxane and diphenylene siloxane, for example) or hydrocarbon oils at 20 to 30 percent concentration are used together with small amounts of thixotropes (fumed silica, clays, soaps, etc.) to make pastes that act like thermal bellows. This is accomplished by filling the gap between chip and cap, as shown schematically in Figure 11-3. The grease will ensure that the thermal path is maintained during expansion and contraction due to thermal cycling of the cap, chip, and substrate. Loss of thermal grease in the gap leads to an interruption of the thermal path and subsequent overheating.

The specific rheological behavior and the grease-mixing process are still largely characterized using empirical methods. While the molecular weight and

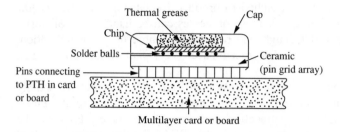

FIGURE 11-3
Schematic showing the application of thermal grease between the chip and the cap to ensure proper heat dissipation.

the molecular weight distribution of the oils are important, the dominant factor may be the control of the mixing process. Likewise, the wetting of the particle surfaces by the oils and the nature of the agglomerates being formed during a given shear history is still poorly understood. A prevailing model involves entangled polymer chains into which the particles are embedded, and hydrogen bonding between the silica reinforcing the structure [20]. During shearing these reinforcing bonds supposedly break, and chains become unentangled, with the structure rebuilding when shearing stops.

Ceramic or metal caps may also be soldered directly to the carrier to provide hermetic seals, if necessary. These chip carriers can also be encapsulated with plastics, similar to DIP. Most common are epoxy encapsulants in the form of transfer molding compounds or, lately, injection-molding resins.

The above encapsulants may be contrasted with topseals. Topseals encapsulate the chip (or circuitry) directly while leaving the rest of the carrier accessible. The topseals may be polyimides, polyamidimides, or silicone type materials. The topseal may also be applied between chip and ceramic to exclude corrosive materials and act as alpha radiation barriers. Ceramics and solders may contain enough natural radioactive isotopes to erase a memory cell of a chip. Alpha particles are low-energy particles that can be stopped by thin films; thus, polymers containing heavy elements (like silicon) are not necessarily required to impart protection.

CIRCUIT BOARD FABRICATION

Process Flow

The basic process flow for printed circuit board manufacturing will be outlined in this section. Specific details may be found in later sections of this chapter, or in other chapters of this text. Only glass-cloth-reinforced circuit boards will be discussed here. Low-cost boards based on paper reinforcement, specialty boards using organic fibers (Kevlar or woven PTFE), and low-expansion materials (quartz cloth, copper-invar-copper) will not be addressed.

The manufacturing process for all printed circuit boards begins by impregnating glass cloth with a laminating resin using a treater tower. This process

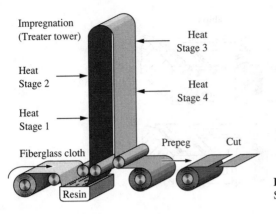

FIGURE 11-4
Schematic of the impregnation process.

is shown schematically in Figure 11-4. The impregnation process involves coating a woven glass fabric with an epoxy resin solution, carefully evaporating the solvents, and partially curing the resin to a prescribed state called the B-stage. This yields an easily handled, stable material termed prepreg. Prepreg is cut into sheets of the desired size and stored. To produce a *core*, several plies of prepreg are sandwiched between sheets of treated copper foil and laminated at elevated temperature and pressure (Figure 11-5), fusing and curing the resin. The resulting core is a copper-clad, fully cured epoxy-glass composite. The cores are then circuitized using photolithographic processes.

The circuitization process involves application of a photosensitive polymer (termed *photoresist*) to the core, exposing the photoresist through a mask with the desired circuit pattern, and developing the resist using a suitable solvent. The copper circuitry can be fabricated using either a subtractive or an additive process. These two processes are compared in Figure 11-6. For low to medium circuit densities, the subtractive process is the most common. This process typically uses "negative" photoresists, which crosslink upon irradiation with UV light. The photoresist is exposed to UV light through a mask having the desired circuit pattern. The UV light crosslinks the photoresist and the resist in the unex-

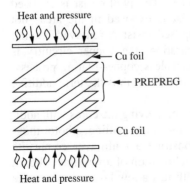

Heat and pressure

— Cu foil

— PREPREG

— Cu foil

Heat and pressure

FIGURE 11-5
Schematic of the lamination process.

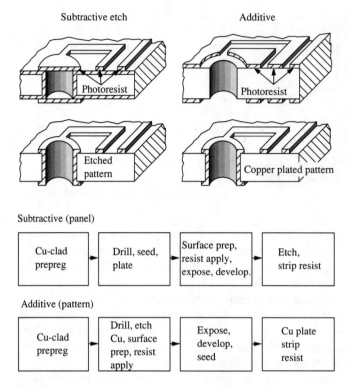

FIGURE 11-6
Comparison of subtractive circuitization and pattern plate (additive) circuitization processes.

posed areas is removed with a solvent. This process is termed *developing*. The crosslinked resist protects the underlying copper. The exposed copper is etched and the photoresist is removed. The copper foil under the resist is not attacked during etching, thus producing the desired copper circuitry.

For fine lines and high circuit densities, an additive process is used to plate the circuitry onto a bare epoxy glass laminate. The photoresist is exposed and developed in a similar manner, but the resist is exposed in the areas not desired as circuitry. After developing, the unexposed resist is removed leaving channels in the resist layer. After surface preparation, the copper circuitry is plated into the channels in the resist using an electroless copper-plating process. In this process, the copper is added to the circuit channels, thus the term additive plating.

The composite circuit board is fabricated by interleaving the cores with additional sheets of prepreg and copper foil. Lamination, hole drilling, photolithography, and plating processes are repeated to construct a multilayer composite (Figure 11-7). Although the process flow for the fabrication of a multilayer board is fairly simple, even multilayer circuit boards with just a few layers may require

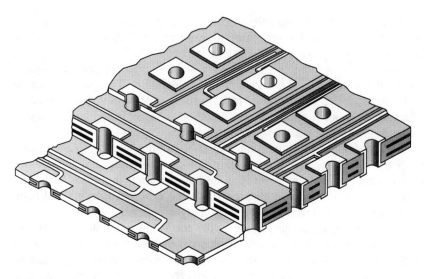

FIGURE 11-7
Schematic of a multilayer printed circuit board.

a hundred process steps. These are necessary to ensure interfacial cleanliness, to develop adequate adhesion of copper circuitry to the base resin, to achieve resist adhesion and proper resist performance, and to maintain the integrity of the interplane joint in the plated through-holes.

As a consequence of the complex sequence of highly diverse processing steps, the composite materials are exposed to a number of very aggressive environments. These include aqueous etching/plating solutions, aqueous solutions or organic solvents for developing and stripping photoresists, various cleaning and degreasing agents, and high-temperature soldering operations. Each process has a high degree of complexity with the possibility of many types of interactions. To control the quality and manufacturing process, the complexity of the technology does not allow the use of empirical testing. It is necessary to take a fundamental approach to understand the basic physical and chemical aspects of the materials and processes.

Impregnation

Circuit board fabrication begins with impregnation of glass cloth with a thermosetting resin solution. To enhance adhesion between the resin and the glass, the woven glass fabric is coated with a coupling agent, typically an aminosilane [18]. After adequate mixing and aging, the catalyzed resin solution (termed *varnish*) is pumped into a dip tank. The glass cloth is immersed in the resin solution using a variety of methods to ensure good wetting of the fiber bundles. The coated glass cloth is passed between metering rolls to control the resin pickup. The web then travels through a multizoned tower where the solvent is volatilized and the resin is partially reacted to a predetermined extent.

The web emerges from the tower as a stable, tack-free prepreg with a glass transition temperature, T_g, above room temperature. The prepreg can be cut into the desired size or stored in roll form. Prepreg is normally stored at room temperature under humidity control, although some formulations require refrigeration to prevent advancement during storage.

Penetration of the resin solution into the glass bundles with complete wetting of each fiber is necessary to ensure good electrical insulation after lamination. Controlled solvent removal and resin reaction have to occur in the treater tower. For reactive systems (in this case epoxies) the balance between drying and resin advancement dictates the tower temperatures and web speed.

A common problem during drying is the formation of bubbles of solvent vapor. These bubbles occur in the resin rich areas or between the strands in the fiber bundles. Under a light microscope, the bubbles trapped in the fiber bundles have long cylindrical shapes and are thus termed *cigar voids*. When the resin is in the fluid state, the bubbles are easily expelled. Cigar voids are much more difficult to remove due to spatial constraints. The rapidly increasing viscosity encountered in the later stages of the treater tower magnifies this difficulty. Bubble formation may also occur during lamination when the resin temperature exceeds the boiling point of the entrapped solvent. This leads to laminate voids, which may also pose a serious reliability concern. Prepreg must exhibit reproducible conversion (or degree of B-stage) below the gel point of the resin, to allow for additional flow in the lamination press. Thus, the treating operation has to balance solvent removal with chemical advancement of the resin.

Additional complications arise when cosolvents with widely different boiling points are used or when removal of one solvent precipitates the curing agent from the resin. With a dicyandiamide (DICY) cured epoxy resin, a potential problem is the formation of crystalline DICY if the temperature profile is not optimized. DICY crystals rotate plane-polarized light and are easily observed in prepreg samples using cross-polarized light microscopy. DICY crystals are clearly visible in the micrograph shown in Figure 11-8.

Control of the impregnation process is difficult and in the past has been largely empirical. Recently, work by Colucci [21] has resulted in a more fundamental understanding of the impregnation process. Work along these lines will enable impregnators to have more control over the impregnation process.

Lamination

Copper-clad laminates are fabricated by stacking sheets of prepreg between pretreated copper foils of the desired thickness. The copper-prepreg-copper layups are placed between polished steel planishing plates and placed in the laminating press. A schematic of the layup is given in Figure 11-5. If copper-clad laminates

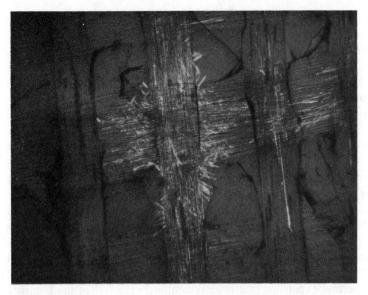

FIGURE 11-8
Optical micrograph of prepreg showing DICY crystals on the fiber bundles. Transmitted light using crossed polarizers.

are not desired, separator sheets or lubricants ensure removal of the plates from the cured laminate.

When the temperature rises in the press, the prepreg resin liquefies, wets the copper, and intermixes with resin from other prepreg sheets. As the cure reaction progresses, the resin gels, retaining the macroscopic shape. Further time in the press allows complete reaction of all functionalities.

Lamination conditions vary with resin type, thicknesses of the laminate, and thermal capacity of the press. Typical lamination conditions for epoxy prepregs utilize a press temperature of 170–180°C at a pressure of 250–1000 psi for 30–60 minutes at temperature. The cool-down cycle is optimized to minimize residual stresses and warpage. Polyimides and other high temperature polymers may require lamination at 200°C or more for complete cure. Hot water and steam heated presses do not perform well much above 175°C, so frequently only a partial cure is obtained after lamination at this temperature. Full cure is reached by postcuring the laminates in an oven at higher temperatures.

Photosensitive Coatings: Photoresists and Solder Masks

The detailed discussion of the technology of photosensitive coatings may be found elsewhere in this book. *Photosensitive coating* is a general term that does not allude to any specific application. Photoresists are the most common photopolymers employed in second-level packaging. These are used as temporary masks to produce circuit patterns on laminated cores.

Second-level packaging mainly employs negative-acting resists. Typically, these are liquid resists and are applied by spray, curtain, or dip coating, and silk screen printing. Dry film negative photoresists are frequently replacing liquid resists because of some very distinct process advantages. These include solvent-free application (hot roll laminated onto the circuit board), no bake required prior to expose, and controlled thicknesses (up to 75 microns). Recently, dry film positive-acting resists have become available that offer potentially higher resolution.

Photosensitive dielectrics or permanent photoresists are processed essentially the same as other photoresists, except that they are not stripped after circuitization, thus becoming a permanent part of the circuit package. The permanent nature of these coatings demands that they be compatible with the manufacturing process (for example, copper plating and drilling) and meet the performance requirements of the final circuit package. Due to the thermal excursions experienced by typical multilayer boards during processing and service life, the permanent resist will have to possess thermal expansion coefficients similar to the rest of the laminate to ensure that no adhesive failure occurs. Photopolymers are also used as solder masks. After application of the resist, the circuit features to be soldered are exposed using UV light. The solder mask serves as a protective coating to minimize handling damage as well as cover and protect the circuitry not to be soldered.

Obviously, each photosensitive coating must be tailored for the specific application. Here, some polymer aspects are highlighted. Liquid resists have some common parameters that are important for process control. In general, the application technique is sensitive to the viscosity of the resist solution. Therefore, resists are supplied in concentrated form, allowing dilution to the appropriate viscosity. Adjustments are limited because the concentration and the wet thickness control the dry thickness of the resist. Occasionally two coats of resist are applied to ensure pinhole-free coatings. The viscosity of the resist is a function of the molecular weight of the resin and its concentration. The molecular weight also affects the dissolution rate during developing.

Molecular weights and molecular weight distributions can be monitored conveniently either by intrinsic viscosity or size exclusion chromatography (SEC) measurements. The resin and photosensitizer can be characterized by spectroscopic techniques (FT-IR, NMR, UV). Photoreactions can be investigated and monitored via the same spectroscopic techniques. Additional techniques have been developed to characterize the photoreactions in situ by integrating an exposure source into an analytical tool (photo-FTIR [22,23], photo-DSC [24,25], and photodielectrometry [26]).

Dry film resists derived their name from the fact that they are virtually solvent free. Typically, a reactive monomer acts as the solvent for sensitizers/initiators, additives, and film-forming polymers. The photoresist is dissolved in a casting solvent, coated onto a carrier film and dried. The resist is supplied in roll form with a separator sheet to prevent adhesion to the backside of the carrier. The resist is hot roll laminated to the substrate (circuit board). Good resist adhesion to the substrate is attained using this method. The molecular weight distribution of the polymer controls the flexibility and tackiness of the resist film and the adhesion of the laminated film, and impacts the photospeed of the resist. The ultimate glass transition temperature and adhesion after photolytic curing are also affected by the molecular weight.

Drilling

After lamination of a multilayer circuit board, there must be means to interconnect various levels of circuitry and components mounted on the board. Interconnections are commonly made by mechanically drilling through the composite, followed by copper plating either electrolytically or using autocatalytic (electroless) deposition. Drilling and copper plating will be covered in detail in other chapters, with the polymer aspects of drilling briefly discussed here. Through-hole drilling requires the removal of both cured epoxy resin and glass cloth as the drill cuts through the board. One of the critical factors governing hole quality is the heating of the hole wall and resin chips [27]. During drilling of an epoxy-glass laminate, the epoxy immediately in contact with the drill may be heated above the T_g of the resin. The rotation of the drill causes the resin to be subjected to large shear stresses. The now rubbery ring (due to heating above T_g) acquires a large molecular orientation tangential to the hole. The chip debris, also heated above the glass transition, may be smeared (or redeposited) on the interior of the hole wall. If the epoxy smear covers a copper plane, then electrical continuity will be lost if the smear is not adequately removed. The polymer in the rubbery ring will cool rapidly as the drill is removed, with the subsequent quenching of molecular relaxations resulting in residual stresses in the hole barrel. After drilling, annealing the laminate at a temperature above T_g will allow the molecular orientation to relax.

The aggressive cutting action of the drill also causes resin-glass fiber fracture along the barrel of the hole. These fractures can be attributed to the pulling and pushing action of the drill on the glass fibers in the heated resin during drilling. Degradation of the glass-resin interface is detrimental, providing a location for process solutions to become trapped, thus creating a reliability concern.

Drilling is a complex combination of polymer physics, mechanics, and engineering. The ultimate goal is to design a drilling process that minimizes smear and damage to the interior of the hole, while maintaining an adequate process flow from a manufacturing point of view.

Dimensional and Humidity Effects

The integrity of the epoxy–coupler–glass cloth interface is extremely important when producing high-reliability laminates to be used in uncontrolled environments. Under conditions of high humidity and applied voltage stress, many couplers appear to lose their properties [28–31]. The result is the creation of a copper shorting filament plated along the glass surface. If this bridges two conductors (for example, a plated through-hole and power plane), a short is obtained from one circuit element to another. Accelerated testing under conditions of high humidity and applied voltage stress is a functional test of coupler quality.

The cured resin strongly bonded to the glass cloth acts as a composite with anisotropic mechanical properties. Residual stresses have been shown to affect the dimensional stability of the composite laminate [32–35]. One source of these stresses is the glass fabric itself. Tension on the web in the treater tower (in the warp or machine direction) causes the fill yarns to be flattened, while the crimp angle in the fill yearns is increased. Tension in warp yarns causes them to acquire a more circular cross section. During lamination, the tension in the yarn induced during impregnation can be relaxed as the resin is heated and softens into a viscous liquid. The flowing resin also exerts a force on the glass cloth, along with the pressure applied by the press. Temperature variations across the surface of the laminate, resin flow in areas of dense circuitry, and hole drilling all create stress-induced dimensional change [36,37].

Cross-ply lamination can compensate for glass cloth tension [38], but in most cases tracking such factors is not straightforward. Consequently, detailed correlation is lacking between dimensional change and the factors most predictive models require to describe the deformation (orthotropic contraction, warp, twist, and other high-order strain functions). The general effects are discernible, with the complex multilayer fabrication process monitored and controlled using empirically determined computer models. Highly precise measurement techniques ensure that circuitry in each layer will exactly register to others in the composite. Moisture and temperature significantly affect dimensional stability; therefore, prepregs, cores, and subcomposites are often temperature and humidity stabilized prior to critical process steps.

POLYMERS IN SECOND-LEVEL PACKAGING

Epoxy resin/woven glass cloth laminates have been the most widely used substrate in mulitlayer printed circuit boards for the last 25 years. Epoxy resins achieved commercial significance in the 1950s, and have enjoyed a considerable growth in applications and the variety of resins available [19]. Glass cloth technology also developed very rapidly during World War II. Both the glass cloth and epoxy resins used in electronic components have undergone substantial development to accommodate the demanding applications in the electronics industry.

Epoxy Resins

The type of laminate that has found the most acceptance is designated FR-4 (fire-retardant epoxy-glass cloth). The typical epoxy resin used is the diglycidyl ether of 4, 4'-bis(hydroxphenyl)methane, commonly referred to as diglycidyl ether of bisphenol-A (DGEBA). Fire retardancy is imparted by including approximately

15–20 percent bromine through the addition of a brominated DGEBA in the resin formulation. Some resin formulations include higher-functionality epoxies, either epoxidized cresol novolacs (ECN) or other multifunctional resins. The chemical structures of typical epoxy-based resins are summarized in Figure 11-9. The purpose of multifunctional epoxies is to increase the crosslink density, thereby raising the glass transition temperature T_g and improving chemical resistance. The

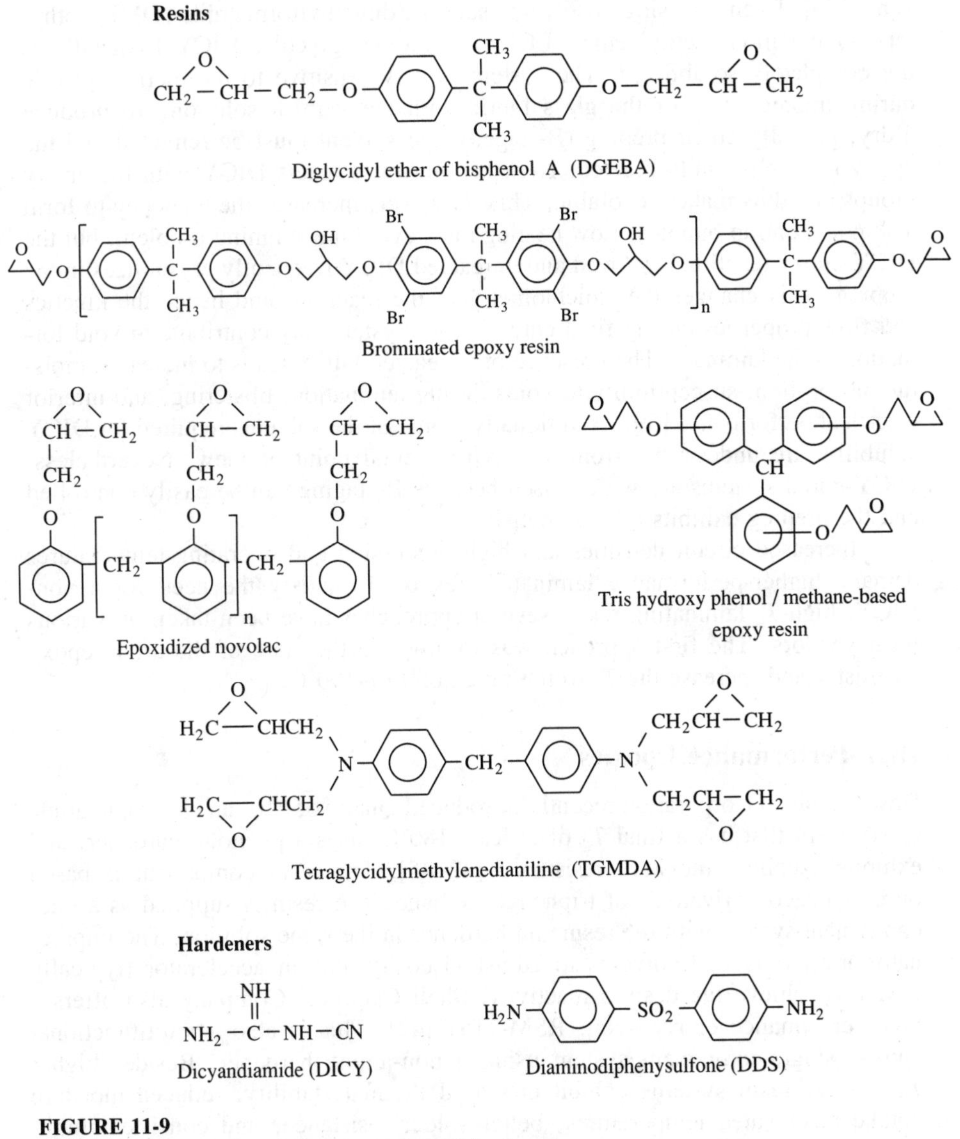

Resins

Diglycidyl ether of bisphenol A (DGEBA)

Brominated epoxy resin

Epoxidized novolac

Tris hydroxy phenol / methane-based epoxy resin

Tetraglycidylmethylenedianiline (TGMDA)

Hardeners

Dicyandiamide (DICY)

Diaminodiphenylsulfone (DDS)

FIGURE 11-9
Chemical structures of typical epoxy resins and hardeners.

glass transition temperature T_g of the final laminate is in the range of 100–130°C, depending on the formulation and process conditions.

A catalyst is used to increase the polymerization rate, usually tertiary amines (0.01–5 percent). Typical catalysts are benzyldimethylamine (BDMA), tetramethylbutane diamine (TMBDA), 2-methyl imidazole (2MI), and other tertiary amines [19,39–41]. Diaminodiphenyl sulfone (DDS) and dicyandiamide (DICY) are the most commonly used curing agents. One major drawback of the FR-4 system is the poor solubility of DICY in common impregnation solvents [39]. Even by using cosolvents such as dimethylformamide (DMF), ethylene glycol monomethyl ether (EGME), or other glycols, DICY frequently is not completely solubilized. The system is very sensitive to the thermal profile during impregnation of the glass fabric with the varnish solution. To produce a dry, partially cured prepreg (B-staged), the solvent must be removed and the epoxy groups partially reacted. High heating rates reacts DICY with the epoxy groups and thus makes it soluble. This, however, increases the tendency to form a skin and entrap bubbles. Low heating rates avoid the skinning problem, but the DICY-epoxy reaction is slowed and unreacted DICY frequently crystallizes in the prepreg. This changes the stoichiometry of the reaction, and hence the kinetics and flow properties during final cure. These crystals may contribute to void formation in the laminate. The presence of unreacted DICY leads to increased moisture absorption, susceptibility to voids during lamination, blistering, and inferior electrical performance [39]. Additionally, some of the solvents required for DICY solubility are undesirable from an environmental point of view. Nevertheless, DICY-cured systems are widely used because B-staging can be easily controlled and the prepreg exhibits a long shelf life.

Increased circuit densities and high processing and operating temperatures demand higher-performance lamination resins. To satisfy the need for a non-DICY, high-T_g laminating resin, several approaches have been taken by various resin vendors. The first approach was to advance the state of the art in epoxy chemistry and increase the T_g to the range of 170–190°C.

High-Performance Epoxies

Dow Chemical Corporation recently introduced Quatrex 5010, an electronic grade epoxy resin that has a final T_g of at least 180°C, uses a phenolic hardener, and exhibits excellent moisture resistance [42–45]. The resin component is based on epoxidized derivatives of triphenyl methane. The resin is supplied as a one-component system with the resin and hardener in the same solution. The impregnator must only add solvents to adjust viscosity and an accelerator (typically 2-MI) to achieve the desired reactivity. Shell Chemical Company also offers a high-performance epoxy resin, RSM-1151 [42]. This is also a multifunctional epoxy, single-component system using a non-amine hardener. Besides higher T_g's, these resin systems exhibit increased thermal stability, reduced moisture uptake at elevated temperatures, better solder resistance, and consistent high-temperature peel strengths.

Another epoxy-based high-T_g resin system is based on tetraglycidyl methylenedianiline (TGMDA) and diaminodiphenyl sulfone (DDS) (Figure 11-9). This resin has gained wide acceptance in the aerospace industry [46] but is not widely used for electronic applications.

Cyanate Resins

Cyanate resins are typically derived from bisphenol-A and were first introduced to the circuit board industry by Weirauch [47]. Mitsubishi Gas Chemical developed a commercial resin by mixing dicyanates with methylene dianiline bis-maleimide. The bis-maleimide triazine (BT) type resins may be blended with epoxies to achieve the desired laminate properties. Typical formulations contain 50 percent BT resin (approximately 10 percent bis-maleimide and 90 percent bis-cyanate) and 50 percent of various types of epoxy resins [48]. Figure 11-10 shows (a) the chemical components and (b) the major reaction pathways for the BT/epoxy resin system. These resins are usually catalyzed with zinc octanoate or other transition metal carboxylates. Advantages include high heat resistance due to the ring structure developed during curing, low dielectric constant, and excellent electrical insulation resistance after moisture absorption [28].

Similar to the BT resins are a class of laminating resins consisting of mixtures of bisphenol-A dicyanate and epoxies [49]. These polymerize during lamination, forming triazine ring structures and oxazoline rings. Suitable formulation yields a resin system capable of fully curing at standard circuit board lamination conditions. These laminates have properties similar to the BT/epoxy blends previously discussed.

Interez has developed a high-performance laminating resin consisting of pure dicyanates [50,51]. The network is formed through the cyclotrimerization of the cyanate functionality [52,53]. In general, cyanate resins and their blends behave quite differently from conventional epoxy resins during impregnation and lamination. Impregnation typically starts with triazines (oligomerized dicyanates), which have very high functionality. Consequently, crosslinking progresses very rapidly, and quickly leads to gelation.

Polyimides

Polyimides are polymers with high-temperature stability and good mechanical properties, and they are suitable for a variety of packaging applications. Despite drawbacks such as high water absorption, only moderate dielectric properties, and high cost, polyimides have established a firm position in the electronics industry.

Polyimides can be classified into three types, according to the polymerization mechanism required to achieve the fully cured polymer. These types are *condensation, addition,* and *polymerization of monomeric reactants* (PMR).

Historically, the first polyimides developed were the condensation type, and these are utilized extensively in electronic packaging. Kapton® is a condensation polyimide formed from the two-step condensation of pyromellitic dianhydride

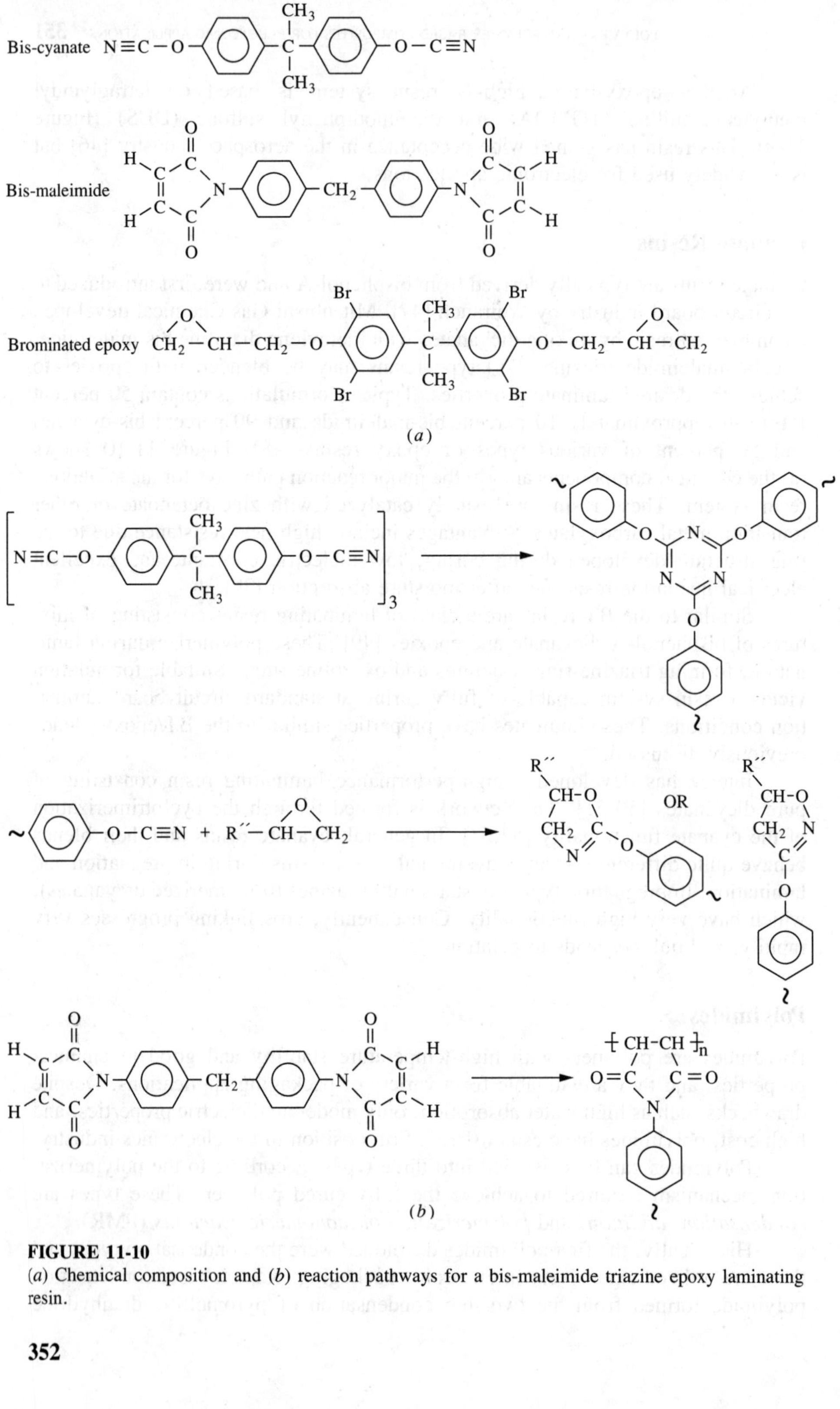

(a)

(b)

FIGURE 11-10
(a) Chemical composition and (b) reaction pathways for a bis-maleimide triazine epoxy laminating resin.

and 4,4'-bis(aminophenyl)oxide (or oxydianiline). This combination is commonly referred to as PMDA-ODA polyimide and is shown in Figure 11-11. Films of PMDA-ODA and other condensation polyimides can be strong and extensible. As coatings, they produce superior leveling layers that can withstand high-temperature metal deposition. Other formulations are available, providing a variety of properties for specific end users [54]. However, since the curing process produces water, condensation polyimides are not suitable for applications requiring thick cross sections, such as multilayer laminates.

For laminating, the addition type of polyimide is used. Addition polyimides (based on bis-maleimides, BMI) are commercially available from various vendors [55]. These utilize aromatic amines (usually methylene dianiline, MDA) to enhance the cure reaction [56,57]. While these resins are well suited for lamination, and have higher end-use temperatures than epoxies, they are not as thermally stable as the condensation polyimides previously discussed. The chemical structures of typical addition polyimides are summarized in Figure 11-12.

The third type of polyimide, PMR, utilizes the methyl ester-acid instead of the anhydride to condense with the aromatic diamine. This allows the imide to form at lower temperatures, compared with anhydride imidization. The polyimide formed at this stage has a lower molecular weight than a condensation polyimide, but the molecular weight is controlled by endcapping the chains with the ester-acid of a norbornene derivative. At the temperature of final cure, the norbornene endcaps combine in an addition reaction. The temperature resistances of the PMR-type polymides are similar to those for condensation polyimides, and therefore they are used in the same types of applications.

Among the new promising systems for rigid laminates are the hybrid polyimide-siloxanes, which combine many of the advantages of both individual components [58]. Polyimides absorb water readily, are attacked by alkaline reagents, and are not easily bonded when fully imidized. The polyimide-siloxane copolymers show promise in overcoming these drawbacks.

The use of flexible circuits is a growing portion of the industry. Multilayering is becoming more common with hybrids between rigid and flexible constructions utilized. Polyesters and polyimides are the most frequently used materials for flexible circuit insulation layers. Polyethylene terephthalate and similar polyesters are the most widely used general substrates.

Polyimides are more expensive than other films, but their film strength, high temperature stability, good dielectric properties, and resistance to a wide variety of organic solvents allow widespread application in electronic packaging.

Fluoropolymers

Fluoropolymers appear most promising for applications where propagation speeds and low signal noise are the driving factors. The low dielectric constant of fluoropolymers derives from their symmetrical structure, which in turn leads to minimized dipoles. This structure, however, is also responsible for the difficulties

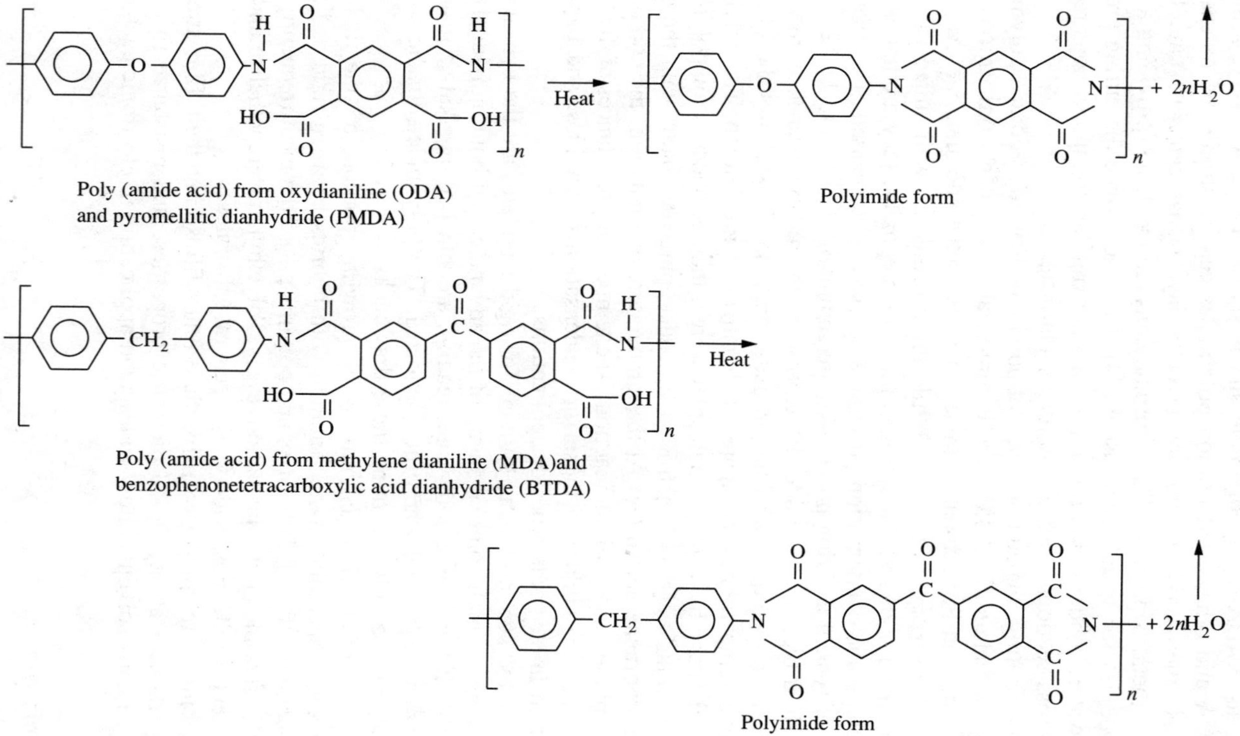
Heat
Poly (amide acid) from oxydianiline (ODA)
and pyromellitic dianhydride (PMDA)
+ 2nH2O
Polyimide form
Heat
Poly (amide acid) from methylene dianiline (MDA)and
benzophenonetetracarboxylic acid dianhydride (BTDA)
+ 2nH2O
Polyimide form

FIGURE 11-11
Chemical structures of PMDA-ADA and BTDA-ODA condensation-type polyimides.

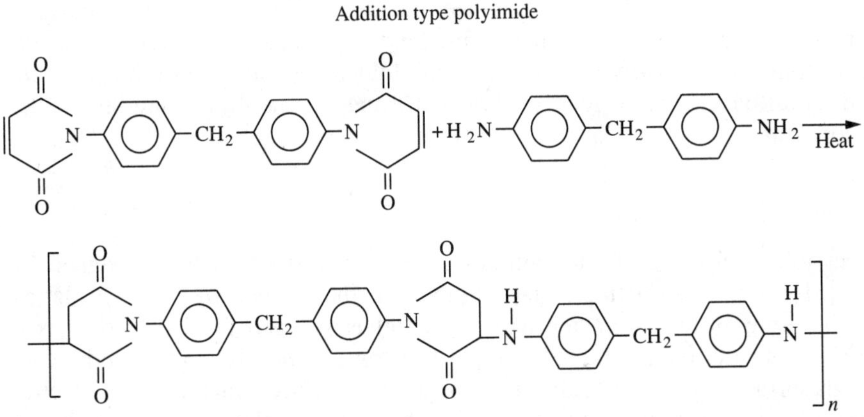

Addition type polyimide

FIGURE 11-12
Chemical structures of a typical addition-type polyimide.

in processing these polymers with conventional solvents, chemicals, and temperatures.

Typical polymers of this category are polytetrafluoroethylene (PTFE), which has extremely low dielectric loss, and remains unchanged over wide ranges of temperature, humidity, and frequency [59,60]; polychlorotrifluoroethylene (CTFE); and polyperfluoropropylene. These polymers are crystalline and demand high-temperature processing due to their high melting points. Adhesion presents a twofold challenge when fabricating multilayer structures. First, since these low dielectric constant materials are thermoplastics, composites have to be fabricated in a way that avoids melting the subcomposite (i.e., using polymers or adhesives with lower melting points). In most applications, these polymers need to be reinforced to control the thermal expansion. Secondly, adhesion to copper is critical, and the fluoropolymer surface typically requires special treatment with plasma or very strong bases to achieve acceptable peel strengths.

A number of materials are currently offered, ranging from PTFE filled with ceramics or glass (Rogers Corporation), glass fabric reinforced PTFE (Oak Corporation), and Kevlar®-reinforced fluoropolymers (Chemfab Corporation). A different approach was taken by the Gore Company [61,62]. Gore markets porous woven PTFE fabrics impregnated with several different resins. Thermoset/PTFE materials offer the advantage of conventional thermoset processing while dielectric constants below 2.8 can be obtained.

Engineering Thermoplastics

Currently efforts are under way to utilize engineering thermoplastics for printed circuit boards. Resins under evaluation include polysulfones, polyethersulfones, and polyetherimides [63]. These resins are commercially available in both filled and unfilled grades. The use of engineering thermoplastics for circuit packaging is a new and rapidly growing technology. Present thermoplastics applications encompass relatively simple injection-molded, double-sided printed circuit boards. Multilayer circuit boards fabricated using thermoplastics are still under development.

Copper Foil

Copper foil is produced by a continuous electrodeposition method invented by Yates [51]. A slowly rotating metal drum acts as the cathode with copper plating onto the drum surface from the bath. The foil is stripped from the drum as it emerges from the bath. The side of the foil toward the drum is smooth, with a rough surface toward the bath. To enhance the surface roughness, an electrochemically produced layer of metals and metal oxides is plated on the rough side of the foil.

The composition and degree of roughness varies depending on the resin to be bonded and the processing conditions. Rough or absorbent adherends are readily bonded together. Smooth, impervious materials are much more difficult to bond, and these types of surfaces are typically encountered in printed circuit substrates.

After circuitization the copper circuitry is smooth, but many kinds of foreign substances must be removed if adequate surface preparation is to be obtained. Water rinsing, sometimes with scrubbing, and often in conjunction with detergents or other chemical agents, is used to remove particulates or water-compatible stains. Oils and greases are displaced by solvents in liquid or vapor phase. Environmental restrictions in recent years have curtailed the use of many common organic solvents. Clean-room processing and factory automation will undoubtedly reduce much of the need for cleaning in the future.

Smooth surfaces, even if clean, usually cannot be bonded unless roughening or chemical treatment of the adherend surfaces precedes adhesive application. Roughening of the underlying substrate is achieved by laminating the resin against the oxide surface of the copper foil and subsequently etching the copper, leaving the resin with the topography of the oxide surface [65]. An epoxy surface may be roughened using a molten alkali etch, resulting in a surface topography yielding good adhesion [66]. Polyimides can be caused to revert to the more compliant permeable amide-acid state, which enhances bondability [67].

Metal surfaces too smooth to be bonded can be roughened by abrasion, but more frequently the metal surfaces on power or ground planes are prepared for bonding by chemical modification. Alkaline oxidation of copper provides an oxide surface which is more polar and has increased surface irregularity, compared to the originally smooth copper. Chelates such as benzotriazoles bond well to copper oxides and are stable at elevated temperatures (approximately 200°C). These chelates allow good adhesion to organics, have high cohesive strength, and serve as corrosion inhibitors when used as coatings [68].

MATERIALS
AND PROCESS CHARACTERIZATION

In this section a brief description of methods used to characterize thermosetting polymers will be presented. These methods are equally applicable to all types of thermosets, but the discussion will highlight how these methods are utilized to characterize the materials and processes used in printed circuit board fabrication. When appropriate, material parameters will be discussed, when these place limits on the process or desired end use properties.

Thermal Analysis

Differential scanning calorimetry (DSC) is an analytical technique that yields both qualitative and quantitative information and allows determination of the cure kinetics on a more basic level [69]. DSC monitors the energetic state of materials by recording the energy required to maintain a precise temperature equilibrium between the sample and a reference as a function of temperature or time. As a consequence, transition temperatures (melting points, glass transition temperatures), heat capacities, and the progress of chemical reactions can be observed. Figure 11-13 shows a typical thermogram (heat flow as a function of temperature) for an epoxy laminating resin. At lower temperatures, the endothermic deflection in the baseline defines T_g, the point where the amorphous organic material softens. As the temperature rises, the resin undergoes an exothermic crosslinking reaction. The peak area is proportional to the heat of reaction, which reflects the initial composition of the material. The peak shape, temperature of peak onset, and the temperature of peak maximum are specific to a given resin formulation (resin, hardener, and catalyst).

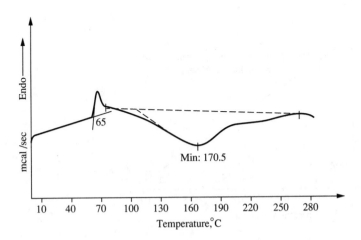

FIGURE 11-13
Typical DSC thermogram for a FR-4 epoxy laminating resin. Heating rate was 20°C/min.

Heats of reaction obtained from DSC can be employed to monitor varnish reactivity, composition, and extent of B-stage advancement. Cure conditions can be optimized using heat of reaction and T_g data. In addition, kinetic mechanisms and kinetic parameters can be determined using DSC when other analytical methods become intractable due to the insolubility of thermosets at high degrees of conversion. The cure must be characterized when the material is a very low viscosity liquid (impregnation process) up to a high-modulus material with essentially infinite viscosity. Using DSC, the T_g was found to vary linearly with conversion up to about 95 percent for a typical FR-4 epoxy resin [70]. Conversion was computed from the heat of reaction of samples that had been reacted directly in the calorimeter, using vacuum-dried varnish as the normalizing standard. At very high conversions, tightening of the network by only a few residual reactive groups causes the T_g to rise more rapidly than suggested by the heat of reaction. Therefore, a high degree of cure is required to develop the ultimate properties of the epoxy (modulus, coefficient of thermal expansion, and chemical resistance). Thus, T_g measurements are useful in evaluating the cure state in laminates after processing.

Quality control monitoring of the laminate thoughout the process is also possible using T_g measurements. In many of the process steps required to fabricate circuit boards, the laminate is exposed to aqueous or organic solvent solutions. Moisture and some solvents easily diffuse into epoxy and plasticize the resin causing T_g to decrease as shown in Figure 11-14. Moisture uptake also causes swelling of the laminate, and the resulting expansion induces stress in the laminate. These stresses can only be totally removed by drying at temperatures above or close to

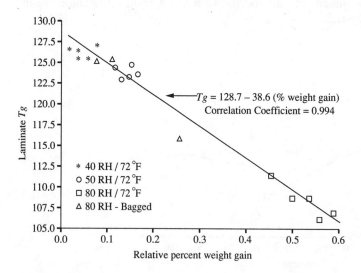

FIGURE 11-14
Glass transition temperature as a function of relative percent weight gain for FR-4 resin exposed to various temperature/humidity conditions.

T_g. Thus, DSC can be useful in designing annealing and stabilization procedures after wet processing.

Infrared Spectroscopy

Calorimetric methods are somewhat limited in that they provide information only on the overall kinetics. A spectroscopic method is required to investigate specific cure reactions. Fourier transform infrared spectroscopy (FTIR) using a programmable heated cell provides a rapid method to monitor individual chemical reactions during curing. The curing of a bis-maleimide triazine/epoxy thermosetting resin is a good system to demonstrate the utility of combining DSC and FTIR measurements to elucidate the cure pathways.

The thermogram of BT/epoxy prepreg shown in Figure 11-15 exhibits two distinct yet overlapping peaks. The combined heat of reaction was determined to be 60 cal/g. The shape of the exotherm suggests several reactions, with rapid and complete reaction expected only when the temperature is close to the second T_{max}.

Heated sample cell FTIR was used to investigate the chemical origin of the observed peaks in the DSC. The bis-maleimide triazine epoxy system has very complex curing mechanisms [71], and only the cyclotrimerization reaction will be presented here to illustrate the utility of the FTIR method. The cyclotrimerization of bis-cyanate to form the triazine rings is a major reaction in the BT system. The aromatic cyanate vibration at 2270 cm^{-1} was used to monitor the consumption of cyanate, and the —C≡N— vibration at 1580 cm^{-1} was utilized to monitor the formation of the triazine ring structure.

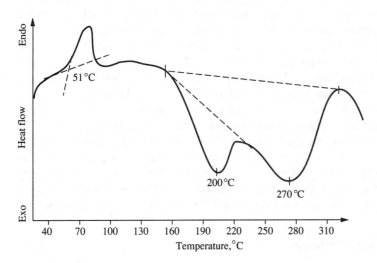

FIGURE 11-15
DSC thermogram for a BT/epoxy thermosetting resin. Heating rate was 20°C/min.

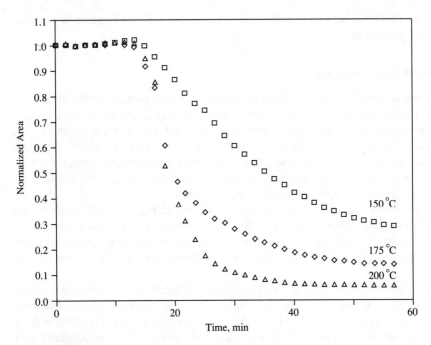

FIGURE 11-16
Normalized aromatic cyanate vibration versus time for a 10°C/min ramp and isothermal hold at the indicated temperatures.

Figures 11-16 and 11-17 show the normalized cyanate and triazine vibrations respectively. In this experiment, the sample was heated at 10°C/min to the indicated temperature and held isothermally. The rapid decrease of the cyanate vibration begins at about 150°C, which corresponds to the initial exothermic deflection in the DSC thermogram in Figure 11-15. The reaction rate and final conversion is a strong function of the reaction temperature. Only at 200°C does the consumption of cyanate appear to level off. The rate of triazine ring formation at 150°C is nearly identical to the rate of cyanate consumption, indicating that cyclotrimerization is the predominant reaction mechanism at this temperature. At higher reaction temperatures (177 and 200°C), the rate of cyanate consumption is greater than the rate of triazine formation, indicating that the cyanate functionality is reacting with epoxy groups (see Figure 11-16). Due to the complex curing mechanism of the BT/epoxy resin, a combination of DSC and a spectroscopic method was required to gain a sufficient understanding of the cure pathways.

Rheological Measurements

The lamination process is an area of intense interest. Laminate quality largely depends on the strength of the adhesive bond between the individual layers. Bonding is accomplished only if the prepreg can flow sufficiently to wet copper

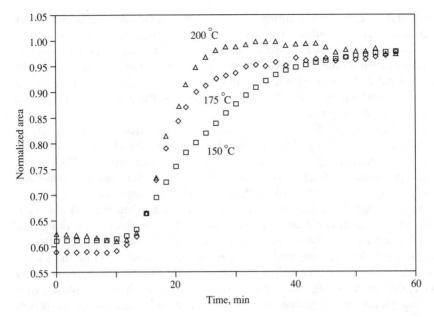

FIGURE 11-17
Normalized triazine ring vibration versus time for a 10°C/min ramp and isothermal hold at the indicated temperatures.

surfaces and mix with other prepreg. Excessive flow, on the other hand, is unacceptable since it will reduce the dielectric thickness between circuit layers. Thus, the optimum lamination process is a balance of the temperature dependence of the reaction kinetics. Therefore, the processing and resultant properties of the composite during lamination are controlled by the initial prepreg conversion and the effective temperature profile.

Traditionally, varnish reactivity is monitored using the *stroke cure* test and the extent of reaction measured via *flow testing*. The stroke cure is an empirical test that relates to the time to reach the gel point, a measure of the reaction rate of the varnish. Flow testing monitors prepreg conversion via the amount of macroscopic flow that occurs between the B-stage and the gel point, where macroscopic flow ceases. The flow test is an empirical test and is limited to comparing prepreg with similar resin loading, initial conversion, and reactivity. A more detailed discussion of the use of flow testing to characterize the lamination process may be found elsewhere [72–74].

The lamination process causes physical and chemical changes to the B-staged prepreg as heat and pressure are applied. Heating causes softening and flow in the resin, along with initiating the chemical crosslinking reaction. Crosslinking continuously changes the physical structure of the resin leading to changes in the viscosity and moduli. In order to control the lamination process, an understanding of how the chemical crosslinking reaction affects the rheological properties is necessary. There are a variety of rheological methods available to measure the viscosity and moduli during curing [75–78]. One useful method is oscillatory parallel plates. A sinusoidally varying strain with a fixed frequency is applied to a resin disk. By measuring the magnitude and phase angle of the resulting stress (in this case torque), the dynamic moduli and viscosity are calculated. The sample and parallel plates are enclosed in a temperature-controlled and nitrogen-purged environmental chamber. The air temperature in the chamber is programed to yield various heating rates.

Several heating profiles are used to investigate the effects of heating rate on a particular resin's viscosity (for a given lot of prepreg); Figure 11-18 shows the viscosity/time relationship for a wide range of temperature profiles. The heating rate has a pronounced effect on the minimum viscosity. Between the fastest and slowest heating rates there is more than an order of magnitude difference in the minimum viscosities. The *process window*, or time when the resin is fluid, is narrower for the faster heating rates. This leads to a shorter gel time. The minimum viscosity steadily increases as the heating rate decreases. The flow is governed largely by the magnitude of the minimum viscosity. While the slower heating rates allow more time before gelation, the level of the minimum viscosity is higher, thus yielding less flow. The ultimate goal is to obtain a viscosity/time profile that yields an acceptable time to gel, along with maintaining the proper

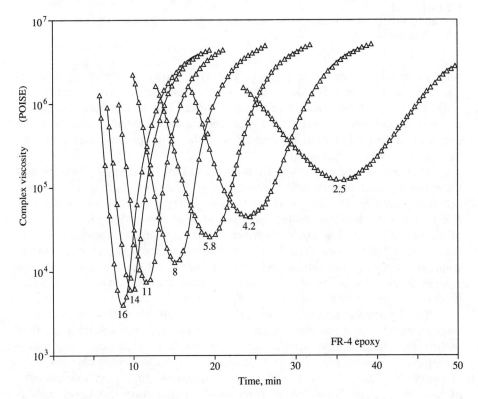

FIGURE 11-18
Complex viscosity as a function of time for various heating rates. (Heating rates are noted in °C/min.)

amount of flow to achieve good wetting of the adjacent epoxy laminate and copper circuitry.

COMPOSITE LAMINATE CHARACTERIZATION

Residual Stresses and Warpage

The mechanical and physical properties of the laminated composite play an important role in the subsequent processing and performance of the circuit board. As outlined previously, during lamination, the layup is heated to an elevated temperature, maintained at the curing temperature for a specified length of time and then cooled to room temperature. During cooling residual stresses develop in the initially stress-free laminate. Similar stresses are introduced by changes in temperature and moisture content. Residual stresses are a function of many parameters, such as ply orientation and stacking sequence, curing process, fiber tension and content, and other variables. They can reach values sufficient to cause matrix cracking, delamination, and premature failure under external loading.

Residual laminate stresses have been analyzed theoretically using lamination theory and laminate material properties [79,80]. Experimentally, they have been measured and determined using embedded strain gage techniques [81–83]. It has been shown by viscoelastic analysis that residual stresses are also a function of the cool-down cycle [84]. For a given temperature drop over a specified length of time, there is an optimum cool-down cycle that minimizes residual stresses. Recent work on the characterization of a circuit board prepreg showed that residual stresses and warpage could be calculated and compared favorably with experimental results [85,86]. It was found that warpage occured as expected in asymmetric laminates. Symmetric laminates exhibited warpage when theoretical considerations predicted that no warpage should occur. Theoretical calculations indicated that the observed warpage could be accounted for by considering a slight misalignment of one of the plies in the layup. Shadow Moiré techniques [85] have proved to be very useful in characterizing the out-of-plane deflections for these types of laminates. Figure 11-19 shows the experimental Moiré pattern for an asymmetric layup. The expected warpage is observed and compares well with the predicted fringe pattern given in Figure 11-20. Material factors such as fiber tension, defects in the weave, fiber misalignment, and matrix variability during cure all must be carefully controlled to assure a uniform, flat laminate.

Micromechanics

The techniques of micromechanics have been widely utilized to study the mechanical interactions between the constituents of the composite. They also are useful to investigate the structural and mechanical reliability of circuit boards. Micromechanics play an important role in establishing how the macroscopic properties can be controlled by the geometric and material properties of the constituents. To meet exact mechanical and thermal performance specifications, printed circuit boards are designed with a glass content sufficient to provide a

FIGURE 11-19
Shadow Moiré pattern for $[0_6/90_6]$ laminate (grating of 16.7 lines/cm, 42.5 lines/in.).

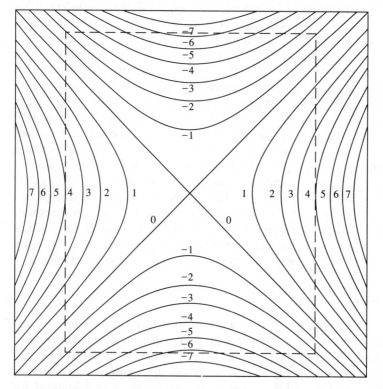

FIGURE 11-20
Predicted Moiré fringe pattern for $[0_6/90_6]$ laminate. (Fringe orders marked correspond to a ruling of 16.7 lines/cm; relative deflection of points of neighboring fringes is 0.6 mm.)

thermal match with the in-plane copper circuits of the board. With increasing circuit densities, the size of the lines and interconnections are so small that microscopes are required to inspect these features. Micromechanical testing is often employed to evaluate these small design elements. In cases where this is not feasible, analytical techniques based on the finite element method have proven invaluable. Such models require elastic property estimates, which account for the considerable attention given to determining the mechanical properties of composites.

Multilayer circuit boards experience many thermal cycles during the various stages of fabrication. Therefore, the thermal properties of the composite are important. One technique used to investigate thermal expansion is thermomechanical analysis (TMA). The composite sample is placed in contact with a very sensitive displacement transducer (usually a linear variable differential transformer, LVDT) and subjected to a linear heating cycle. The displacement is plotted as a function of temperature. Figure 11-21 shows two TMA traces for a laminated composite. Figure 11-21(a) is the expansion in the x-direction (in-plane). Below T_g the expansion is in the range of 2.0×10^{-5} inch/inch/°C, or 20 ppm/°C. Above T_g the expansion is not much larger due to the constraint imposed in the x-y plane by the woven glass cloth. In contrast, the z-expansion shows a very marked temperature dependence. Figure 11-21(b) shows the z-axis expansion for the same composite. The sub-T_g expansion is in the range of 65 ppm. Without the constraint of the glass cloth in the z-direction, the expansion is thus about a factor of three higher. The effect of the glass fabric is even more evident in the z-expansion above T_g. The z-expansion increases substantially (in the range of 250 ppm) due to the absence of fiber constraints in the z-direction.

The large z-expansion above T_g plays a role in the stresses developed in a composite circuit board during various thermal excursions above T_g. Finite element modeling (FEM) techniques provide a powerful tool in analyzing the mechanical behavior of a complex structure. In using this technique, the entire structure is divided into a finite number of elements. A set of local stiffness equations is then obtained for each typical element, and through the uniqueness of nodal points, global stiffness equations are assembled. With the advent of powerful data-processing equipment, the equations can be solved using numerical methods. This technique has been used extensively to characterize composite materials.

The FEM technique is useful to investigate the effect of temperature on small components of the circuit board, such as a plated through-hole (PTH). In modeling the PTH, due to symmetry with respect to the center of the board, only one half of the PTH needs to be modeled. As previously discussed, a symmetric layup is required in order to eliminate warpage after lamination. Because of the thermal expansion mismatch between the PTH barrel and the z-direction expansion of the laminate, stresses (or strains) are generated in the copper barrel when the board is subjected to a temperature change. Thermal strains have been calculated for this case using a finite element model. To test the accuracy of the model, a simple technique was developed to measure the barrel strain and laminate expansion. A thin cross-sectioned PTH sample was placed on a hot plate, which sat on an x-y

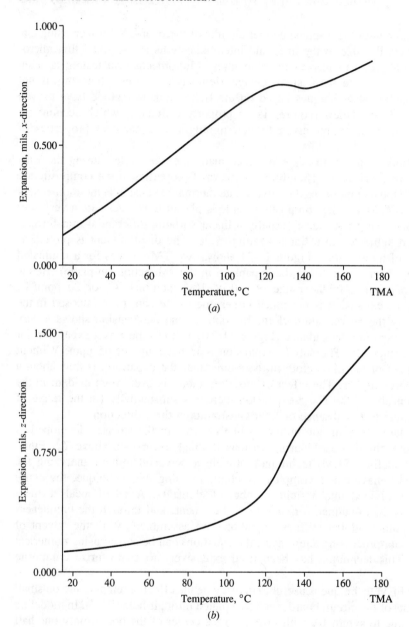

FIGURE 11-21
(a) Thermal expansion for FR-4 laminates in the x-y plane. Data is for expansion in the warp direction. (b) Thermal expansion in the z direction for an FR-4 laminate.

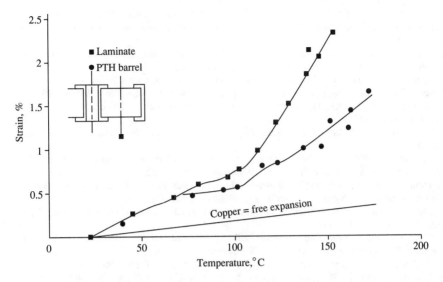

FIGURE 11-22
Measurement of thermal strain (in the z direction) in printed circuit board sections when equilibrated at various temperatures above and below T_g.

micrometer stage. At various temperatures, measurements were made across the barrel and on the thickness of the laminate between the PTHs. The displacement measurements were used to calculate corresponding strains at each temperature. The thermal strains are plotted as a function of temperature in Figure 11-22. For comparison, the unconstrained z-expansion for copper is plotted. The upper curve is the expansion of the laminate between two PTHs. The middle curve is the expansion of the PTH barrel. As was previously demonstrated (Figure 11-21(b)), the large z-expansion of the laminate above T_g plays a significant role in the strain induced in the PTH. For practical purposes, the PTH acts as a rivet in the laminate and large strains are induced in the PTH during thermal excursions above T_g. The thermal strain data was in substantial agreement with values calculated using the finite element model. Both the model and the experimental results indicate the need for sufficient copper ductility in the PTH to avoid significant cracking problems.

REFERENCES

1. F. Billmeyer, *Textbook of Polymer Science*, 3rd ed., New York: John Wiley & Sons, 1984.
2. F. Rodriguez, *Principles of Polymer Systems*, New York: McGraw-Hill, 1970.
3. F. Bovy and F. Winslow, *Macromolecules, an Introduction to Polymer Science*, New York: Academic Press, 1979.
4. P. Heimenz, *Polymer Chemistry, the Basic Concepts*, New York: Marcel Dekker, 1984.
5. G. Odian, *Principles of Polymerization*, 2d ed., New York: John Wiley & Sons, 1981.
6. S. Rosen, *Fundamental Principles of Polymeric Materials*, New York: John Wiley & Sons, 1982.

7. P. Flory, *Principles of Polymer Chemistry*, Ithaca, N.Y.: Cornell University Press, 1953.
8. R. Seymour and C. Carraher, Jr., *Polymer Chemistry: An Introduction*, New York: Marcel Dekker, 1981.
9. J. Mark, A. Eisenberg, W. Graessley, L. Mandelkern, and J. Koenig, *Physical Properties of Polymers*, Washington, D.C.: American Chemical Society, 1984.
10. L. Sperling, *Introduction to Physical Polymer Science*, New York: John Wiley & Sons, 1986.
11. J. Ferry, *Viscoelastic Properties of Polymers*, 3d ed., New York: John Wiley & Sons, 1980.
12. J. Aklonis and W. MacKnight, *Introduction to Polymer Viscoelacticity*, 2d ed., New York: John Wiley & Sons, 1983.
13. I. Ward, *Mechanical Properties of Solid Polymers*, 2d ed., New York: John Wiley & Sons, 1983.
14. E. Feit and C. Wilkins, editors, *Polymer Materials for Electronic Applications*, American Chemical Society Symposium Series 184, Washington D.C., 1982.
15. L. Thompson, C. Willson, and M. Bowden, editors, *Introduction to Microlithography*, American Chemical Society Symposium Series 219, Washington D. C., 1983.
16. T. Davidson, editor, *Polymers in Electronics*, American Chemical Society Symposium Series 242, Washington, D.C., 1984.
17. L. Thompson, G. Willson, and J. Frechet, editors, *Materials for Microlithography*, American Chemical Society Symposium Series 266, Washington D.C., 1984.
18. E. P. Plueddemann, *Silane Coupling Agents*, New York: Plenum Press, 1982.
19. H. Lee and K. Neville, *Handbook of Epoxy Resins*, New York: McGraw-Hill, 1967.
20. Cabot Corporation, "Cab-O-Sil®2: Properties and Function," product bulletin, Tuscola, Il. 1985.
21. M. J. Colucci, "Void Formation in Glass Reinforced Epoxy Resin Prepregs," M.S. thesis, University of Texas, 1984.
22. S. J. Fuerniss, A. J. Sommer, and B. K. Appelt, "Characterization of Negative Resists via Photo FT-IR," *Proc. North Amer. Thermal Anal. Soc.* 15, p. 435 (1986).
23. R. Snyder and J. Gotro, "The Use of FT-IR to Characterize Photosensitive Thermosets," *J. Appl. Spectroscopy* 41, p. 476 (1987).
24. B. K. Appelt, "Characterizing Photoresists by Thermal Analysis", *Polm. Eng. Sci.* 23, p. 125 (1983).
25. J. E. Moore, *UV Curing: Science and Technology*, edited by P. Pappas, Norwalk, Conn.: Technology Marketing Corp., 1980.
26. J. D. Reid and W. H. Lawrence, "Dielectric Properties of a Commercial Photoresist: Dependence on Ultraviolet Light Exposure Time," *J. of Radiation Curing*, (April p. 4, 1986).
27. R. E. Weiss, "The Effect of Drilling Temperature on Multilayer Board Hole Quality," *Circuit World* 7, no. 3, p. 8 (1977).
28. D. Lamdo, J. Mitchell, and T. Welsher, "Conductive Anodic Filaments in Reinforced Polymeric Dielectrics: Formation and Prevention," paper presented at IEEE symposium, 1979.
29. J. N. Lahti, R. H. Delaney, and J. N. Hines, "Characteristic Wear-Out Process in Epoxy/Glass PC for High Density Electronic Packaging," in *Proceedings of the Technical Program* I. Rel. Physics Symposium, 1979.
30. C. J. Tautscher, "Measuring Cleanliness of Printed Circuit Wiring—What Really Counts," in *Proceedings of the Technical Program*, National Electronic Packaging and Production Conference, Anaheim, Calif., March 1–3, 1983.
31. L. L. Marsh, R. J. Lasky, D. P. Seraphim, and G. S. Springer, "Moisture Solubility and Diffusion in Epoxy and Epoxy/Glass Composites," *IBM J. Res. Develop.* 28, pp. 655–661 (1984).
32. H. T. Hahn and M. J. Pagano, "Curing Stresses in Composite Laminates," *J. Composite Materials* 9, p. 91 (1975).
33. H. T. Hahn and M. J. Pagano, "Curing Stresses in Composite Laminates," *J. Composite Materials* 10, p. 91 (1976).
34. R. H. Marloff and I.M. Daniel, "Three-Dimensional Photoelastic Analysis of a Reinforced Composite Model," *Exp. Mechanics* 9, pp. 156–162 (1969).

35. C. C. Chamis, "Lamination Residual Stresses in Cross-plied Fiber Composites," in *Proceedings of the Technical Program*, 26th Annual Conference of the Society of the Plastics Industry, Reinforced Plastics/Composites Division, 1971.
36. I. M. Daniel, T. Liber, and C. C. Chamis, "Measurement of Residual Stresses in Boron/Epoxy Laminates," in *Proceedings of the Technical Program, Composite Reliability*, Special Technical Publication 580, American Society for Testing and Materials, 1975, p. 340.
37. I. M. Daniel and T. Liber, "Lamination Residual Strains and Stresses in Hybrid Laminates," in *Proceedings of the Technical Program, Composite Materials: Testing and Design (Fourth Conference)*, ASTM Special Technical Publication 617, American Society for Testing and Materials, 1977, p. 331.
38. I. M. Daniel and Y. Westman, "Stresses Due to Environmental Conditioning of Cross Ply Graphite/Epoxy Laminates," in *Proceedings of the Technical Program*, Fourth Internal Conference of Composition Materials, August 1980, vol. 1, pp. 529–542.
39. M. Shah, F. Jones, and M. Bader, "Residual Dicyandiamide (DICY) in Glass Fibre Composites," *J. of Materials Sci. Letters* 4, p. 1181 (1985).
40. R. Bauer, *Epoxy Resin Chemistry*, American Chemical Society Symposium Series 114, Washington D.C., 1979.
41. R. Bauer, *Epoxy Resin Chemistry II*, American Chemical Society Symposium Series 221, Washington D.C., 1983.
42. M. Brody, *Printed Circuits Fabrication* (December 22, 1985).
43. R. Urscheler, "New Epoxy Laminating System with High Heat Distortion Temperature Performance," *Proceedings of the Printed Circuit World Convention III*, Washington D.C., 1984.
44. A. M. Pignerl, E. C. Galgoci, R. J. Jackson, and G. C. Young, *Proc. of NEPCON* West, p. 22, Anaheim, CA, 1987.
45. C. A. Haper and S. M. Lee, *Int. SAMPE Electronics Conf.* vol. 1, p. 657 (1987).
46. J. Moacanin, M. Cizmecioglu, S. Hong, and A. Gupta, "Mechanism and Kinetics of the Curing Process in a Resin System", p. 83 in *Chemorheology of Thermosetting Polymers*, edited by C. May, American Chemical Society Symposium Series 227, Washington D.C., 1983.
47. K. K. Weirauch, "A New Thermosetting Polymer for Use in Printed Circuit Boards," IPC Technical Paper TP-066, Chicago, Il, 1975.
48. M. Gaku, "BT Resin Widens the Choice of Laminate Materials," *Electronic Packaging and Production*, 30 (December 1985).
49. D. Shimp, *Polymeric Materials Science and Engineering*, vol. 54, American Chemical Society, New York 1986.
50. F. A. Hudock and S. J. Ising, "Cyanate Ester Resins for Printed Wiring Board Caminates," IPC Technical Paper TP-618, San Diego, (1986).
51. D. A. Shimp, "Thermal Performance of Cyanate Functional Thermosetting Resins," *Int. SAMPE Symp.* 32, p. 1063 (1987).
52. K. A. Jensen and A. Holm, in *Chemistry of Cyanates and Their Derivatives*, Vol. 1, edited by S.Patai, New York: John Wiley & Sons, 1977.
53. M. Bauer, J. Bauer, and G. Kuhn, "Kinetics and Modeling of Thermal Polycyclotrimerization of Aromatic Dicyanates," *Acta Polym.* 37, p. 715 (1987).
54. K. Mittal, editor, *Polymides: Synthesis, Characterization, and Applications*, vol. 1 and 2, New York: Plenum Press, 1984.
55. V. St. Cyr, "Polyimide Notebook," *PC Fabrication*, p. 24 (February 1986).
56. M. Bargain, A. Combet, and P. Grosjean, U.S. Patent 3,562,223 (1971).
57. H. Stenzenberger, M. Herzog, W. Romer, R. Schelblich, and N. Reeves, *British Polymer Journal* 15, p. 2 (1983).
58. J. Summers, C. Arnold, R. Bott, L. Taylor, T. Ward, and J. McGrath, "Synthesis and Characteristics of Novel Polyimide Siloxane Segmented Copolymers," *Polymer Preprints*, vol. 27-2, p. 403, American Chemical Society, 1986.
59. T. A. Smith, "Drilling and Plating of Teflon/Glass Printed Circuit Boards," in *Proceedings of the Technical Program*, National Electronic Packaging and Production Conference, Anaheim, Calif., March 1–3, 1983, p. 293.

60. *3M Sourcebook, Electronic Products Div.*, publication No. 225-4S, St. Paul, Minn.: 3M Center.
61. D. D. Johnson, U.S. Patent 4,680,220 (1987).
62. D. D. Johnson, "Laminate Spurs High-Speed Digital Processing," *Electronic Packaging and Production* 27, p. 80 (1987).
63. D. C. Frisch, "New PCB Substrates Expand Packaging Engineer's Choices," *Electronic Packaging and Production*, p. 194 February, (1985).
64. Charles Yates, Canadian Patent 359,618 (1936).
65. H. L. Rhodenizer, J. J. Grunwald, and W. P. Innes, U.S. Patent 3,666,549 (to McDermid Corporation).
66. G. V. Elmore and K. C. David, "Mechanism of Bonding Electroless Metal to Organic Substrates," *J. Electrochemical Soc.* 116, p. 1455 (1969).
67. H. G. Linde, "Adhesive Interface Interactions Between Primary Aliphatic Amine Surface Conditioners and Polyamic Acid/Polyimide Resins," *J. Polymer Sci. Polm. Chem.* 20, p. 1031 (1982).
68. G. W. Poling, "Inhibition of the Corrosion of Copper and Its Alloys," Final Report INCRA Proj. No. 185, February 1979.
69. R. B. Prime, "Thermosets," in *Thermal Characterization of Polymeric Materials*, edited by E. A. Turi, New York: Academic Press, 1981.
70. B. K. Appelt and P. A. Cook "Defining B-Stage and Total Cure for a DICY Based Epoxy Resin by DSC," in *Analytical Calorimetry*, vol. 5, p. 57 edited by J. F. Johnson and P. S. Gill, Plenum Press , 1984.
71. J. T. Gotro, B. K. Appelt, and K. I. Papathomas, "Thermal Characterization of a Bis-Maleimide/Bis-Cyanate/Epoxy Thermosetting Resin for Composites," *Polymer Composites* 8, p. 39 (1987).
72. B. K. Appelt, J. T. Gotro, T. L. Ellis, G. P. Schmitt, and J. P. Wiley, "Composite Lamination-Analysis and Modeling," *Proc. Soc. Plastics Eng., ANTEC 85*, p. 289 (1985).
73. J. T. Gotro, B. K. Appelt, T. L. Ellis, G. P. Schmitt, and J. P. Wiley, "Composite Lamination-Analysis and Modeling," *Polymer Composites* 7, p. 91 (1986).
74. J. Gotro, B. Appelt, M. Yandrasits, and T. Ellis, "Thermoanalytical Investigation of Composite Lamination," *Polymer Composites* 8, p. 222 (1987).
75. J. M. Dealy, *Rheometers for Molten Plastics* New York: Van Nostrand Reinhold, 1982.
76. C. A. May, editor, *Chemorheology of Thermosetting Polymers*, American Chemical Society Symposium Series 227, Washington, D.C., 1983.
77. A. V. Tungare, G. C. Martin, and J. T. Gotro, "A Rheological Analysis of the Cure Behavior of Epoxy Resins," *Proc. Soc. Plastics Eng., ANTEC 87*, p. 330 (1987).
78. A. V. Tungare, G. C. Martin, and J. T. Gotro, "Chemorheological Characterization of Thermoset Cure," *Polm. Eng. Sci.*, 28, p. 1071 (1988).
79. C. C. Chamis, "Lamination Residual Stresses in Cross-plied Fiber Composites," in *Proc. of 26th Annual Conference of SPI, Reinforced Plastics/Composites Division* 17-D (1971).
80. I. M. Daniel and T. Liber, "Lamination Residual Stresses in Fiber Composites," IITRI Report D6073-I, for NASA Lewis Research Center, NASA CR-134826, 1975.
81. I. M. Daniel, T. Liber, and C. C. Chamis, "Measurement of Residual Strains in Boron/Epoxy and Glass/Epoxy Laminates," in *Composite Reliability*, Special Technical Publication 580, American Society for Testing and Materials, 1975, p. 340.
82. I. M. Daniel and T. Liber, "Effect of Laminate Construction on Residual Stresses in Graphite/Polymide Composites," *Exper. Mechanics* 17, p. 21 (1977).
83. I. M. Daniel and T. Liber, "Lamination Residual Strains and Stresses in Hybrid Laminates," in *Composite Materials: Testing and Design (Fourth Conference)*, Special Technical Publication 617, American Society for Testing and Materials, 1977, p. 331.
84. Y. Weitsman, "Residual Thermal Stresses Due to Cool-down of Epoxy-Resin Composites," *J. Applied Mechanics* 46, pp. 563–567 (September 1979).
85. I. G. Zewi, I. M. Daniel, and J. T. Gotro, "Residual Stresses and Warpage in Circuit Board Composite Laminates," *Proc. Soc. for Experimental Mechanics*, fall meeting, p. 14 (1985).
86. I. G. Zewi, I. M. Daniel, and J. T. Gotro, "Residual Stresses and Warpage in Woven-Glass/Epoxy Laminates," *J. Experimental Mechanics* 27, p. 44 (1987).

PROBLEMS

11.1. What are the principal features in the DSC-thermogram of a thermoset?

11.2. Describe effect of moisture on printed circuit boards.

11.3. Rank the coefficents of thermal expansion (CTE) of circuit board materials (epoxy, copper, glass cloth). How do the individual CTEs affect the board CTEs?

11.4. You are designing a new circuit card using an experimental thermoset resin and a new copper. Metallurgical analysis of the copper reveals that the critical strain (in tension) for fracture is 2.0 percent. The T_g of the resin is 150°C. Careful TMA experiments on the resin/glass composite yield the following data:

z expansion:

$$\alpha_g = 60 \times 10^{-6} \text{ in./in./°C} \qquad \text{(below } T_g\text{)}$$

$$\alpha_e = 250 \times 10^{-6} \text{ in./in./°C} \qquad \text{(above } T_g\text{)}$$

x expansion:

$$\alpha_g = 20 \times 10^{-6} \text{ in./in./°C} \qquad \text{(below } T_g\text{)}$$

$$\alpha_e = 40 \times 10^{-6} \text{ in./in./°C} \qquad \text{(above } T_g\text{)}$$

The critical geometrical constraints are as follows:

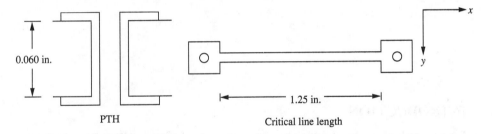

PTH   Critical line length

To solder components, the manufacturing engineer needs to heat the card from room temperature (25°C) to 230°C (vapor phase solder). The dwell time is long enough for the card to reach equilibrium. Can you use this resin and copper in this card? Show details.

12

LITHOGRAPHY IN ELECTRONIC PACKAGING

GERALD W. JONES
JANE M. SHAW
DONALD E. BARR

INTRODUCTION

Resists are high-molecular-weight compounds that define a pattern for subsequent processing. The metal patterning processes for printed circuit boards (PCBs) or ceramic substrates usually involve selective addition of metals or other elements to the surface by plating, or selective subtraction of the base material by etching. For integrated circuits (ICs) the addition takes the form of metal evaporation and silicon oxidation. Resists derive their name from their ability to protect the area under the pattern from subsequent processing. Although there are many commercially available resists, they fall generally into two basic classes: those that require irradiation to define a pattern and those that do not.

The type of resist that does not require irradiation for image formation uses stencil screens that contain patterns. The ink (resist) is squeezed directly through a very fine mesh screen pattern made of either silk, polyester, or stainless steel onto the substrate to define a circuit. The ink can then be cured by using either heat or light to harden it. From this point on, the ink resist can be used in the same manner as photoactive resists (photoresists) discussed below in the photolithography section. Screen resists are commonly used today for applications where line widths are in the range of 10 mils and greater. These resists are cheaper but have limited resolution capability. The materials and rheology factors for this type of resist are discussed in Chapter 11.

Photoresists are used to image lines less than 10 mils wide. PCB fine-line imaging involves widths 5 mils and below, whereas IC fine lines are 2 μm wide or less. The resist thickness is adjusted accordingly: thinner resists produce finer lines.

The circuit density and electronic speed of PCBs, ceramic substrates (Chapter 10), and ICs has increased dramatically in the last decade, and this, to a large extent, was made possible by the advances in photoresist technology to define smaller and smaller features. Today's smallest memory chip feature is about 100 times smaller than the diameter of a human hair. These advances are even more remarkable when one realizes that to make an integrated circuit, the photolithographic process must be repeated perhaps 20 times. The photoresist pattern for each sequential step must be exactly aligned with the previous pattern, which puts stringent requirements on the imaging tools.

PHOTOLITHOGRAPHY

Lithography, or "writing on stone," was discovered in the late 18th century. The process involved treating a stone or other suitable imaging material with an ink-attracting hydrophobic substance. The ink was repelled in nondesign areas by moistening the stone with water and the ink image was transferred by pressing paper against the patterned stone. The term *photolithography* generally refers to a process whereby images are created by exposure to light. Present day photolithography uses a polymeric coating that is imaged with light and developed to form a desired pattern, creating open and protected areas. Photolithography is also used to make plates for printing newspapers and magazines. Its applications in electronic packaging are described in this chapter.

Radiation sensitive polymers, which are used for photoresists, are about 1 million times less sensitive than conventional photographic film. Silver halides, which are used in photographic emulsions, are sensitive to the near ultraviolet and blue end of the visible spectrum, i.e., about 0.300 to 0.450 μm (Figure 12-1). Photographic emulsions can be sensitized to the entire visible region by incorporating sensitizing dyes that absorb light and transfer energy to the silver halide grains. The same approach can be applied in photoresists; that is, dyes can be added to the formulations to shift the spectral sensitivity to maximize photo yield. Since photoresists cannot approach photographic film speeds, intense ultraviolet lamps are used to achieve shorter imaging times. The resolution of silver halide photographic films is determined by the emulsion grain size. A large grain size produces faster speed, but also a rougher line edge. Photoresist resolution is affected by the radiation wavelength, by the polymer molecular size, and the development process.

The portion of the electromagnetic spectrum (Figure 12-1) between 0.200 and 0.450 μm is used to expose photoresists. The amount of energy contained in one mole of photons (an einstein) of a given wavelength is expressed in Equation

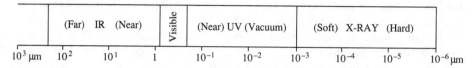

FIGURE 12-1
Electromagnetic spectrum. The spectral region most important for photolithography ranges from x-rays to the near uv. For very fine lines in the range of 1 μm, x-rays are used, and for 10 μm and above, uv is used. The spectral energy is inversely proportional to the wavelength. Energy must be considered with respect to bond making or breaking.

(12.1). An einstein of 200-nm light contains 143 kcal of energy, whereas light of 365 nm would have only 78.3 kcal per einstein.

$$\text{Energy (kcal)} = \frac{2.86 \times 10^4}{\text{wavelength (nm)}} \tag{12.1}$$

The high energy of 200-nm light can cause the rupture of carbon-to-carbon bonds (82.6 kcal/mole) and thus would degrade the resist. The use of higher-energy radiation, such as x-rays and electron beams, will be discussed later in relation to integrated circuit fabrication.

PHOTORESISTS

Photoactive resists or photoresists can be classified as either negative or positive acting. Figure 12-2 shows the imaging steps involved with both types of photoresists. Upon exposure to light, negative resists become more insoluble in the developer and positive resists become more soluble. The positive resists are rarely used to produce patterns for printed circuit boards. Generally, positive resists are costly and are not suited for the board process because of their application mode (spin coating) and their thickness (1–5 μm). However, positive-acting photoresists provide excellent resolution and are used to fabricate components that require resolution on the order of 1–15 μm.

Positive Resists

Wherever light strikes a positive resist, the sensitizer (usually a diazoketone) undergoes reactions to form an alkali-soluble product (see Figure 12-3). The unexposed resist is not soluble in the mildly alkaline solutions used as developers; therefore, only the irradiated area is washed away when developed.

Both positive and negative resists are activated by light, which causes them to change chemically. In the case of positive resists, the reactions that form the alkali-soluble species are intramolecular and therefore quite fast. For this reason, positive systems are not influenced by radical scavengers, such as oxygen, and no special precautions to exclude oxygen are necessary. With negative resists special precautions are necessary to exclude oxygen during exposure. Positive

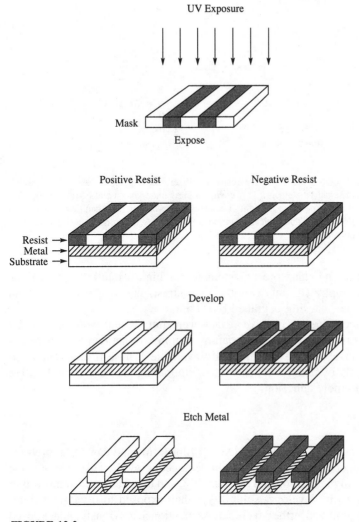

FIGURE 12-2
Schematic depicting a positive- and a negative-acting resist. A positive resist becomes soluble in the developer upon exposure, whereas a negative resist becomes insoluble. The areas left when the developed resist is removed can either be plated with a metal (additive process) or etched (subtractive process) to form circuit lines and features.

resists, unlike negative resists, can be reexposed to open new windows in the coating. This allows multiple exposures without reapplying resist.

Another important difference between positive and negative resists is their response to the developer. The aqueous alkaline developer used to dissolve the light-exposed regions in positive systems converts the newly formed acid into a water-soluble salt. The aqueous developer does not penetrate the areas that are not exposed, because these regions are hydrophobic. Positive resists distort very little

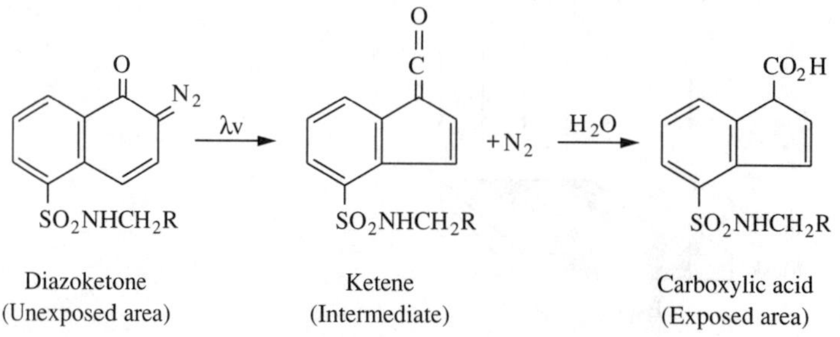

FIGURE 12-3
Chemical transformation of a typical positive resist sensitizer. Most positive resists have alpha-diazoketones as the light sensitive element. This chemical class undergoes a classic ring contraction (Wolff rearrangement) when exposed to uv light. The resulting carboxylic acid can then be removed by a caustic aqueous developer. The typical quantum efficiency for this process is 1 because the rearrangement is intramolecular.

during development and therefore are suitable for high-resolution applications such as microlithography (IC fabrication). In contrast, negative resists swell in the organic developer, causing degraded resolution.

Until quite recently, all positive resists were liquids and were applied by techniques such as roller coating, spin coating, dipping, or spraying. Positive dry film photoresists have become commercially available. These films are applied in sheet form directly by using a hot roll, thus eliminating the bake cycle to drive off solvents and promote adhesion.

Negative Resists

Negative resists are available in liquid and dry film forms and can be tailored for development by either organic solvents or dilute aqueous alkali. This type of resist undergoes radical polymerization and crosslinking, and thus increases in molecular weight and decreases in developer solubility, when exposed to ultraviolet light. Unexposed resist is removed by the developer. As mentioned previously, nonexposed positive resists are unaffected by the developer, but negative resists swell, and the resulting stresses can produce adhesion problems and/or distorted images. With aqueous developing resists a proper balance of acidity and molecular weight in the polymeric resin system is required for facile development. Distortion due to developer penetration in negative resists is possible even with aqueous development.

Negative resists polymerize by intermolecular reactions, and the lifetime of the reactive radical is sufficiently long to allow side reactions involving other radicals or oxygen. These undesirable reactions reduce the photo efficiency and increase the exposure time. Negative systems, therefore, need to be protected from oxygen during exposure for optimal performance. This is accomplished by exposure in an inert atmosphere or, in the case of dry films, by a protective polyester cover sheet that excludes oxygen.

FABRICATION OF INTEGRATED CIRCUITS

Most resists are engineered for near UV response, because of mask, optical lens, and transmission characteristics of the light source. A typical negative resist system for less demanding applications is Kodak's KTFR (Figure 12-4), which incorporates a bis-azide sensitizer in an unsaturated matrix. Upon exposure, the bis-azide reacts to form a crosslinked matrix that is insoluble. The unexposed resist is developed away leaving the exposed pattern. For high-resolution applications, when patterns of two μm or less are desired, positive photoresists are used. These positive resists contain novolak resins and diazo-napthaquinone photosensitizers.

Process Sequence for Positive-Acting Photoresist

The process steps in using a positive photoresist include coating the resist on the substrate, baking, exposing to radiation, developing, and postbaking. The resist pattern acts as a mask for further processing such as oxidation and metal etching. As described below, each processing step is important in determining the integrity and ultimate resolution of the patterned resist circuit.

Application. The positive-acting resists used for high-resolution applications are usually applied by spin coating, but multipass spray coating is also useful and can provide adequate thickness control on ceramic substrates. In spin coating, an adhesion promoter is applied to the substrate first to prevent film delamination during subsequent processing. The substrate is then covered with a photoresist

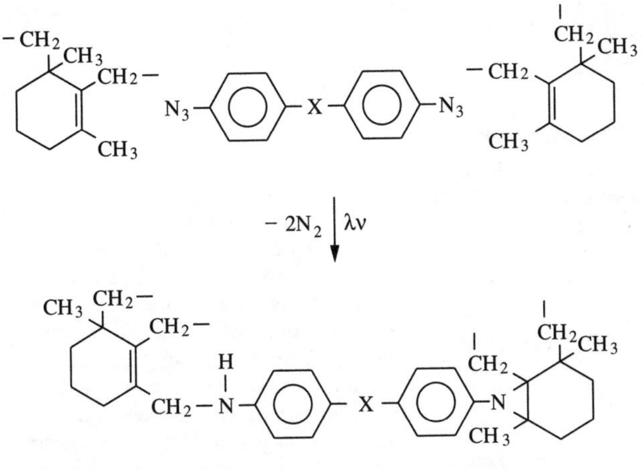

FIGURE 12-4

Crosslinking reaction of negative-acting KTFR resist. Light activates the diazide sensitizer, which likely forms a diradical nitrene, which can either insert into an allylic carbon-hydrogen bond as shown above or react with unsaturation to form a bis-aziridine. In either case the result is a high-molecular-weight polymer insoluble in solvent developers such as 1,1,1-trichloroethane.

solution and spun at speeds up to 4000 rpm to achieve a very thin and highly uniform film.

Prebaking. After coating, the photoresist is baked to remove solvent and promote adhesion to the substrate. The baking temperature is kept between 80 and 90°C because the photoactive diazo compound will thermally degrade above 100°C. The bake step is important because the development rate of the positive photoresist depends on the residual solvent remaining in the film.

Exposure. After baking, the photoresist is exposed through a mask or artwork. This transfers the mask circuit pattern to the photoresist. The various exposure techniques used in this step include contact, proximity, and projection printing (Figure 12-5). Contact printing is used extensively in electronic packaging where dimensions are large, masks are easy to generate, and mask damage such as scratches and nicks can be repaired. For high-resolution applications, where the mask is expensive and difficult to make, optical projection printers are usually used. The exposure energy is carefully monitored and measured in millijoules (mJ). The dose (mJ/cm^2) is the product of the incident light intensity and the exposure time in seconds.

(a) Contact

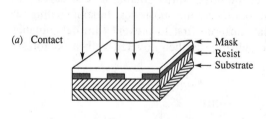

Mask
Resist
Substrate

(b) Proximity

Mask
Resist
Substrate
Space →

(c) Projection

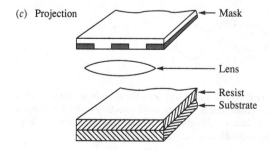

Mask

Lens

Resist
Substrate

FIGURE 12-5
Three common exposure techniques.
(a) Contact exposure is the most common photoresist printing method because of excellent artwork reproduction and low exposure tool cost. (b) Proximity printing involves keeping a small gap, e.g. 10 mils, between the artwork and the imaging materials. The artwork does not get damaged as easily as in contact printing, but collimated light is necessary. (c) Projection printing is used in high-volume areas where high resolution and close tolerances are necessary. Projection printers are quite costly and are focused within a narrow range.

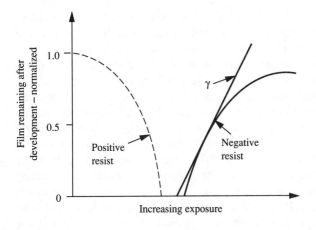

FIGURE 12-6
Contrast (gamma) in positive
and negative photoresists.
Graphically, contrast is depicted
as the slope of the above
curves. A photoresist with high
contrast is highly differentiated
in exposed versus unexposed
areas; that is, a slight increase in
exposure produces a large resist
response.

Development. The *contrast* of resist systems is determined by measuring the film thickness after development as a function of exposure. Figure 12-6 shows the resist thickness decreasing with exposure dose for a positive resist until all of the material is removed. Conversely, the thickness of a negative film will increase until the entire exposed area is totally crosslinked. The slope of this curve measures the resist response to development and defines the contrast.

Postbake. After the pattern is developed, the resist is baked to remove residual solvent and increase substrate adhesion. Dimensional thermal stability is important because any image flow will reduce pattern resolution and integrity.

Resist removal. After pattern transfer to the substrate the resist is removed by solvents or plasma sputtering, and the substrate is cleaned prior to subsequent chemical treatment and the next photoresist application.

HIGH-RESOLUTION ALTERNATIVES TO OPTICAL LITHOGRAPHY

Optical lithography is the workhorse of the semiconductor industry. In 1987 one megabyte (1 Mb) memory chips were produced commercially with 1-μm dimensions. The ultimate theoretical resolution achievable by optical patterning is 0.35 μm. Because of alignment and processing tolerances effects in a manufacturing environment the practical limitation is about 0.5 μm. Resolution less than 0.5 μm requires other forms of radiation that are not limited by diffraction effects, such as electron beams and x-rays.

Electron Beam Lithography

Polymeric resists can be patterned by a very small beam of electrons (100 Å); this is analogous to raster scanning a TV screen. No mask is required when using e-beam systems because the information for the raster pattern is stored in digital form. E-beam systems can resolve features smaller than 0.25 μm because an electron wavelength is typically 0.01 Å and diffraction effects are negligible. Polymers can be chosen in which either scission or crosslinking reactions predominate to produce negative or positive images. Poly-(methylmethacrylate) is a high-resolution positive-acting material, and is widely used throughout the industry.

Although electron beam lithography is not limited by diffraction effects, the high-energy electrons can be backscattered as much as 5 μm by a silicon substrate. The scattered electrons can reduce the contrast. Electron beam writing time is longer than optical imaging time because of computer time needed to correct the dose due to the backscattered electrons and the sequential electron beam writing process. E-beam lithography is currently used to produce low-volume specialty logic chips, where mask generation cost for optical lithography is high. The high resolution and alignment capability of e-beam tools have found their greatest use in fabricating high-quality masks for optical lithography.

X-Ray Lithography

X-rays, with a wavelength of 8 Å, do not suffer from diffraction or backscattering effects as do optical or electron beam exposures. However, x-rays cannot be focused like electrons; therefore, the pattern must be transferred to the polymeric resist by proximity printing. The polymer is exposed to x-rays transmitted through a mask, which has the same dimension as the desired chip pattern. This process places large demands on mask fabrication and alignment capability. X-rays will not penetrate the common chrome/glass mask used in optical lithography. Therefore, a gold absorber pattern supported on a thin (2 μm) transparent silicon nitride membrane is used for the x-ray mask. Polymeric materials, in general, lack x-ray sensitivity (only 10 percent of the incident x-rays are absorbed by a polymer film). Sensitive polymer materials must be designed or high-energy x-ray sources such as synchrotrons must be used to achieve short exposure times. The resolution capabilities of optical, e-beam, and x-ray lithography are compared in Figure 12-7. Continued improvement in lithographic materials and tools is required to meet the demand for high-speed, low-cost integrated circuits.

FABRICATION OF PRINTED CIRCUIT BOARDS

In this section negative dry film photoresists are discussed and the photolithographic process steps to build a printed circuit card or board are demonstrated.

Artwork

At the simplest level, a draftsman draws the design layout in a format much larger than the final product. Visualization is easier if the traces are several times their actual size. The layout is reduced to proper size by photocopying with a high-

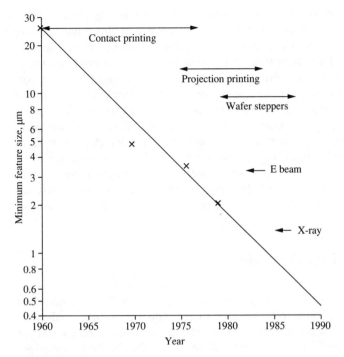

FIGURE 12-7
Resolution capability of various imaging systems. Submicron lithography is practiced today and features continue to decrease as the requirements increase. Tomorrow's technology will test the limits of resolution versus wavelength.

resolution camera. The artwork is then recorded on photographic film or glass plates which serve as masters to generate working copies. The working copies are usually made on either diazo or silver halide polyester films that are susceptible to dimensional changes due to variations in temperature and humidity. Diffusion in polymers is discussed in Chapters 25 and 26. If circuit board dimensions are to be controlled to plus or minus 1 mil (0.001 in.) in 10 inches, the artwork environment must be kept within plus or minus 4°C and plus or minus 9 percent relative humidity. If dimensional control is needed, glass plates must be used, but they are very expensive compared to films.

At a more sophisticated level, the design is created on a computer-aided design (CAD) system. The automated design system drives a photoplotter which writes on glass plates or polyester films. These systems are quite expensive and usually take several hours to draw the layout. Laser photoplotters take computer data and write directly on an emulsion, producing a finished product in minutes.

Surface Preparation

Copper panels, on which patterns are made, must be prepared for photolithographic processing by first removing contaminants and excessive surface oxides.

Contaminants may be removed by cleaners that contain wetting agents and detergents to emulsify soils such as fingerprints. Surface oxides are removed with hydrochloric or sulfuric acid. The panel surface is then roughened either mechanically or chemically. Mechanical treatments often use brushes impregnated with abrasive particles that scour the surface and remove a few microinches of copper, or a pumice slurry is used either as a spray or in conjunction with brushing to roughen the surface. A pumiced surface is preferred for fine lines because it does not produce surface grooves as does mechanical brushing. Grooves create pathways for chemicals to seep under the resist and cause copper etching in areas that should be protected. Figure 12-8 illustrates results of the two surface treatments. A very light chemical etch is sometimes used to roughen the surface. When chemical treatments are used, the ionic residues must be completely removed; otherwise, they will interfere with the adhesion of the resist to the copper panel.

Application of Dry Film Photoresist

Dry film photoresist is commonly used for circuit boards. A dry film photoresist is a three-layer composite: (1) a polyester support sheet for the photosensitive material, (2) a layer of photoactive monomer mixed with a polymeric binder and other materials, and (3) a polyolefin cover sheet which prevents the photoresist from sticking or "blocking" when it is wound into a roll for handling or storage. Dry films are more expensive than liquid systems, but they offer advantages such as uniform thickness, fewer processing steps, and elimination of organic solvents.

Dry films are applied to copper surfaces with a hot roll laminator. The polyolefin sheet is removed during lamination and the hot rollers contact the polyester support sheet to press the resist onto the copper surface under pressure. The heated rolls cause the resist to flow and fill the roughness created by the surface preparation. If the heating is not adequate, the resist will not adhere well to the copper surface, but an overheated resist may react thermally and begin polymerization before light exposure. Boards are sometimes preheated to increase the lamination speed, especially if their thermal mass is large. Some dry film manufacturers incorporate adhesion promoters to increase the adhesion between the resist and copper surface. These agents are believed to migrate to the resist-copper interface during lamination and provide a chemical bond between the two surfaces.

Exposure

Panels must be registered exactly the same way each time with respect to the photo tool for proper alignment in a multilayered structure. This is usually accomplished by placing a panel containing holes over registration pins and then placing the artwork (polyester film or glass plate) over the same pins. Registration, as discussed in Chapters 17 and 20, is important for double-sided boards that contain internal planes because subsequently the boards will be drilled. The circuit traces

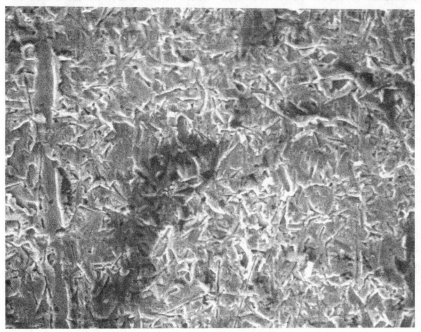

FIGURE 12-8
Comparison of the pumiced (left) and mechanically abraded copper surfaces (1200×). Mechanically scrubbed copper surfaces have a directional appearance due to the panel and brush orientation. Pumice treatment provides better surface topography for resist adhesion.

on both sides and on internal planes must all be aligned to very close tolerances to insure connections in the z direction between the various levels by plated through-holes (PTHs). Once the registration step is completed, a vacuum is applied. This process forces the mask into very close contact with the polyester cover sheet. The resist is then exposed for a preselected time.

The resist contains chemicals called initiators, which absorb the ultraviolet radiation of the mercury lamp and decompose into radicals that initiate polymerization. During the polymerization reaction, monomers react to form polymers of varied molecular weights. The resulting polymers in the exposed areas are insoluble in developer and remain intact during development. The developer removes unexposed low-molecular-weight material, thus producing resist patterns. These patterns define areas where copper can be plated (additive processes, Chapter 16) or etched (subtractive processes, Chapter 18).

The exposure process appears quite simple, but it is not. For instance, establishing the correct radiation dose is not an easy task. This can be done in several ways. The first method involves inspecting the developed pattern to see if the surface is glossy and the lines are well defined. An underexposed resist will be attacked by the developer and the surface will appear dull. An overexposed resist will "bloom" or grow outside the defined area, causing poor resolution.

Inspection, as described above, is usually adequate for boards having coarse circuit geometries, but for finer lines and spaces a more refined technique is preferred. Resist line width measurement is a more precise procedure. In this technique, the resist is given a series of different exposures and the developed lines are measured. The proper exposure is the dose that reproduces the mask dimensions. If a mask of varying lines and spaces is used, the resist resolution can also be determined. Resolution refers to the smallest lines and spaces that a resist can reproduce.

A third method, which is perhaps the most accurate, but also the most difficult, involves exposing a large number of boards to small changes in dose and monitoring the finished product yield. The proper dose produces the best yield and reflects the effects of other operations such as plating or etching.

Once the proper exposure is determined by one of the above methods, the challenge is to keep the dose constant. The resist on each piece should receive the same exposure, otherwise the process consistency and yield will be affected. The two common techniques for measuring exposure dose use either a step wedge or a radiometer. There are two step wedges commonly used in the industry, the 21- and 41-step tablets. A photograph of both is shown in Figure 12-9. Each step on a 21-step wedge differs from the adjacent step by 0.15 optical density unit, and on a 41-step wedge each differs by 0.05 unit. Optical density is defined according to

$$OD = \log \frac{I}{T} \tag{12.2}$$

where I is the incident light and T is the transmitted light. Thus, the amount

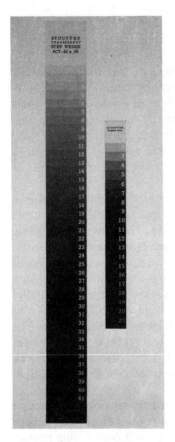

FIGURE 12-9
Two commonly used step tablets. The Stouffer 21 and 41
step tablets are used throughout the industry for exposure
consistency. The general feature of all tablets is that they
attenuate the light by increasing the opacity of successive
steps. Once the optical density is known, the dose needed to
bring the resist to the point of just surviving the developer can
be calculated.

of light passing through a particular step will be half that passing through a step
0.3 unit less:

$$0.30 = \log \frac{100}{T}$$

$$1.70 = \log T$$

$$50 = T$$

A 21 wedge increases 0.3 unit for every two steps and a 41 wedge increases 0.3
unit for every six steps.

Once the proper exposure has been determined, a step wedge is placed
between a clear glass plate (to simulate the glass artwork) and the photoresist.
The resist is then exposed and developed to reveal the step wedge image. All
resist above a certain step will be developed off and all steps below will be
intact. The last remaining step provides a measure of exposure dose. A concern
of the step wedge method is the developer dependence. If the developer is too

aggressive, the wedge reading will be lower than normal because more resist will be removed. Thus, the wedge method monitors both exposure and development.

A radiometer is an instrument that measures light by converting light energy into electrical energy. In many radiometers this is accomplished with a silicone-based photodiode. Some instruments are thin enough to fit inside exposure tools having only a 1-in. clearance. These thin units generally only integrate the total dose. However, the lamp intensity can be approximated by dividing the dose by the exposure time. Energy is typically expressed in terms of millijoules (a millijoule is a milliwatt-second), and dose, which is energy per unit area, is reported as mJ/cm^2.

Collimated light exposure sources are required for higher resolution, because they produce parallel light rays. Figure 12-10 shows that nonparallel rays strike the mask at an angle and cause unwanted polymerization. The consequence is line growth, or blooming, as it is sometimes called, in the case of negative resists. Any medium or space between the artwork and the resist, such as the polyester support sheet or panel depressions, can lead to blooming. Collimation is measured in units of degrees half angle as illustrated in Figure 12-11. The more nonparallel the light, the larger the half angle. It is possible to have collimated light (zero degree half angle) and still not have rays perpendicular to the resist plane. The deviation is measured by the declination angle (Figure 12-11). Collimated light can reproduce the mask features more exactly, but it can also image artwork defects and dirt that would not print with noncollimated sources. Collimated light is necessary when printing by the proximity method.

Proximity printing avoids the expense of glass master repair and inspection by suspending the artwork about 5–10 mils above the resist surface. The expensive glass master does not contact the resist cover sheet as it does during contact printing, and thus scratches and contamination are avoided. There is some resolution loss with proximity printing, even with units where collimation half angle

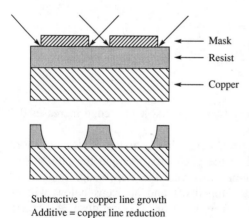

Subtractive = copper line growth
Additive = copper line reduction

FIGURE 12-10
Effect of uncollimated light on resolution. Light which strikes a resist at an oblique angle can undercut the artwork mask and initiate unwanted reactions. In the case of negative resists, line broadening occurs leading to an oversized etched or undersized plated lines. This effect becomes greater as the space between the mask and photoresist increases.

Mask
Resist
Copper

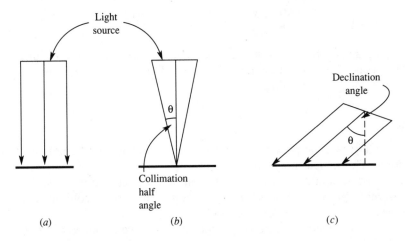

FIGURE 12-11
Depiction of (*a*) collimated light, (*b*) uncollimated light, and (*c*) declination angle. Collimated light has all rays parallel and is measured by half angle as shown in (*b*). Truly collimated light has a 0-degree half angle. Declination angle is the deviation of incident light from the perpendicular. Thus, it is possible to have collimated light with a declination angle as in (*c*).

is 1 degree or less. The glass savings must be weighed against the increased cost of collimated proximity printers.

Projection printing is used extensively for chip imaging but is not a useful alternative for board lithography. This technique requires the focal plane to be within a few mils of the surface across the image area. Chips are about the size of a pencil eraser, and boards can be as large as IBM's 3091 mainframe boards (24 by 28 in.). These boards may contain more than 20 layers and cannot be held to the tight z axis tolerance required for projection printing. The optics required for projection printing are very expensive when compared to either contact or proximity printing.

The same laser technology used to produce artwork is now being applied to photoresist imaging. An argon ion laser, emitting in the 480-nm (green light) visible spectrum region, is used to write directly on photoresist without artwork. The laser raster scans an 18- by 24-in. panel in 30 seconds with a 1-mil spot. This technology makes new demands on photoresists because the resists must now be sensitive to green light instead of uv radiation. Uv-sensitive resists can be safely handled in yellow light but visible laser resists are so sensitive that they must be kept in red light. Visible lasers will continue to dominate until more economical and stable uv lasers are commercially available.

Commercial argon ion lasers produce about 5 to 6 watts of power, which is optically reduced to about 2 W at the resist surface. A quick calculation shows that at a dose of about 20 mJ/cm^2, such a laser can image an 18- by 24-in. (2787 cm^2) board in 30 seconds.

$$\text{Dose} = \frac{2000 \text{ mW} \times 30\text{s}}{2787\text{cm}^2} = 21.5 \text{ mJ/cm}^2 \tag{12.3}$$

This dose requires a resist that is about five times more sensitive than conventional dry films. Direct laser imaging allows quick turnaround times for prototype boards because artwork fabrication is eliminated.

Development

The exposed negative-acting photoresist is given time to complete the polymerization reaction. The radical polymerization process continues for a short time after initiation, and 30 minutes is usually sufficient to complete the process. The resists are designed to produce a visual image in the film by turning color in areas where light strikes, thus avoiding double or no exposure.

There are three general classes of photoresist developers: aqueous, semiaqueous, and organic solvent. A resist is formulated to develop in only one of these systems. Aqueous developers are alkaline solutions usually containing sodium carbonate. Semiaqueous developers use water and organic chemicals such as cellosolve or butyl carbitol. Organic solvent developers usually use 1,1,l-trichloroethane. Resists developed in organic solvents show the greatest processing latitude, followed by semiaqueous and finally aqueous type. This is because aqueous resists have acid functionality imbedded within the polymeric backbone and, therefore, tend to swell more than organic solvent resists.

Prior to development, the protective polyester support sheet is removed. This sheet served to exclude oxygen during the polymerization step and also protected the dry film resist from dirt and physical damage. Development is accomplished by placing the board on a conveyor that passes through a spray chamber. The developer composition and temperature is selected for each resist according to the manufacturer's specifications. The time required to remove the resist is referred to as the *time to clear*, and the point in the developer where the resist is removed is called the *break point*. Usually a board is exposed to the developer for twice the time-to-clear to ensure that resist residues are totally removed.

Circuitization

The resist was placed on the board to form a pattern that defines a printed circuit upon circuitization. There are three general methods (Figure 12-12) to form circuits on a board: subtractive, semiadditive, and fully additive.

The subtractive method utilizes copper already on the board. The standard copper thickness is 1.4 mils and is referred to as one-ounce copper. (The term *one ounce* refers to the weight of one square foot of copper that is 1.4 mils thick.) The resist covers the copper in areas that are to become circuit traces. The board is passed through an etching chamber containing copper-oxidizing agents such as acidic cupric chloride or hydrogen peroxide/sulfuric acid as discussed in Chapter 18. The chemistry converts metallic copper to water-soluble complexes, which

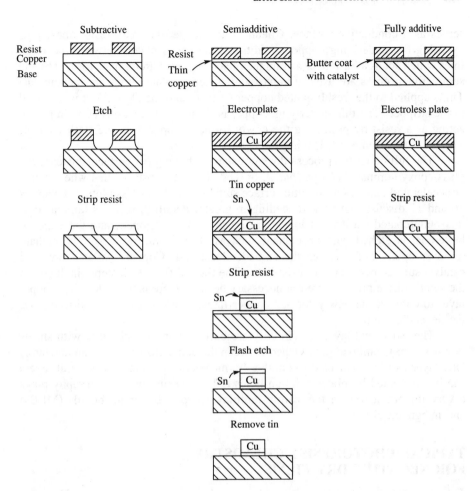

FIGURE 12-12
Three common PCB circuitization methods. The subtractive method is most widely used by the industry to make features 5 mils and greater. The additive methods can routinely produce lines in the 3–5 mil range.

are then removed by spray action. The copper under the resist is protected from the oxidant and is not altered. The final process involves removing the resist in a stripper. The stripper chemistry swells the resist until stress causes debonding. This debonding is the result of Case II diffusion, discussed in Chapter 26. The loose resist is removed from the board surface by mechanical means, leaving the desired circuit traces. This process is termed *subtractive* because copper is removed or subtracted from the part.

The semiadditive process involves both copper addition and subtraction. The resist is usually placed on a very thin copper layer (5 μm or less), which serves as a conductive surface. Channels are imaged between resist lines, thus exposing the underlying copper. Additional copper is plated between the resist lines which act as a form to constrain the plated copper. Circuit lines formed by the subtractive process are usually trapazoidal, whereas plated lines are rectangular. Tin is applied to the freshly plated copper surface, and the photoresist is removed in the stripper. The thin underlying copper is flashed—that is, etched with the tin acting as a resist by protecting the newly formed copper from the flash etchant. The tin is then removed, leaving the desired circuitry.

The fully additive process does not start with a copper surface. Instead, the glass-epoxy laminate is *buttercoated*, or seeded with a substance that when etched or accelerated exposes a catalytic surface suitable for electroless plating. Chapters 16 and 17 discuss plating and seeding in greater detail. A resist is laminated to this buttercoated surface, channels are imaged and developed, and the exposed buttercoat is etched with an etchant such as chrome/sulfuric acid. The etchant removes epoxy and exposes the embedded catalyst. Copper is plated onto the catalyst surface between the resist lines. The tin and flash etch steps, included in the semiadditive process, are not necessary because there is no underlying copper layer to connect the newly formed copper lines. The resist is stripped to expose the desired circuitry.

The individual layers of circuitry are built on top of each other with sheets of glass-epoxy laminate sandwiched between the conductors to act as an insulator. One layer can be connected to another by means of via holes and several layers can be connected by plated through-holes (PTHs). Thus, photolithography plays a very important role in fabricating multilayered printed circuit boards (MLBs) and integrated circuits.

TYPICAL PHOTORESIST CHEMISTRY FOR NEGATIVE DRY FILMS

The chemistry of all photoresists is beyond the scope of this chapter. However there are general principles that are followed for the major commercial systems. The most widely used photoresists fall into three major classes: (1) negative-acting dry films, (2) negative-acting liquids, and (3) positive-acting liquids. A chemical scheme will be presented for only the negative-acting dry films; however, the other categories are briefly addressed in the "Fabrication of Integrated Circuits" section.

Dry films do not flow into small depressions as well as liquids and, therefore, do not conform as well to the surface topography, allowing solution to undercut the resist. The dry film formulator must, therefore, devise a resist system that will conform as well as possible to the substrate but still not exude from a tightly wound roll containing perhaps 1000 feet of resist. The delicate balance between desirable properties is achieved by selecting the correct polymer backbone or binder. As with most resists, molecular weight is very important for achieving good physical properties such as modulus, resistance to developer, and adhesion. Pendant carboxyl groups are added to the backbone polymer to aid aqueous development in nonexposed areas and to improve resist to copper adhesion. Most binders in dry film resists are not photochemically active but become entangled in the three-dimensional network as the reactive monomer polymerizes. They are also good film-forming agents and act to contain the less viscous components.

The reactive monomers are usually acrylates that polymerize by a free radical mechanism and are capable of forming three-dimensional networks. Usually the monomers absorb ultraviolet energy below the polyester protective sheet and photomask cutoff limit (about 320 nm). Therefore, they cannot be exposed directly and must be initiated by another molecule.

Absorbed radiant energy is transferred to the initiator through bimolecular collisions. Triplet excited state radicals are thought to be involved in the reaction. The reaction rate slows significantly when the polyester cover sheet is removed and oxygen diffusion is allowed. Oxygen is a ground state triplet molecule and can react with excited state triplet species to terminate free radical polymerization. With free radical initiation, the excited state molecule can either fragment or hydrogen abstract to produce radicals that add to the monomer and start polymerization.

Certain dyes are added to dry films for contrast and identification purposes. The contrast dyes do not change chemically and usually absorb sufficiently far away from the blue region (yellow absorbs blue) so as not to affect photospeed. Dyes added for imaging are called leuco dyes and react by photooxidation to form conjugate colored dyes, giving a printout or latent reverse image of the mask. The visible printout image informs the operator that the part was exposed and thus prevents reexposure.

Additionally, dry films can contain adhesion promoters, thermal stabilizers, plasticizers, antiblocking agents to prevent resist sticking, and chain transfer agents to facilitate polymerization. It is, therefore, not an easy task to formulate a successful dry film photoresist. Figure 12-13 outlines some typical ingredients found in an aqueous developable dry film photoresist. The resist is composed of a binder, styrene/maleic anhydride/maleic acid polymer (V, 60 percent); a reactive monomer, trimethylolpropanetriacrylate (IV, 28 percent); a photooxidant, 2,2′-bis-(o-chlorophenyl)-4,4′,5,5′-tetraphenylbiimidazole (I, 2 percent); a leuco dye (III, 0.4 percent); and an electron-donating initiator (II, 0.4 percent). Photooxidant (I) absorbs uv energy and disassociates into radical (Ia). Imidazole radicals (Ia) can abstract a hydrogen atom from tertiary amine (II) thus forming radical (IIa) which is capable of initiating polymerization of monomer (IV). Imidazole radicals (Ia) can also photooxidize leuco dye (III) to the conjugate color form (IIIb). The binder becomes entrapped in the polymer network formed when the triacrylate monomer (IV) polymerizes. The entrapped binder is thus rendered insoluble in 1% sodium carbonate developer and a negative resist image results. The conjugate color form of leuco dye (III) differentiates exposed and unexposed areas.

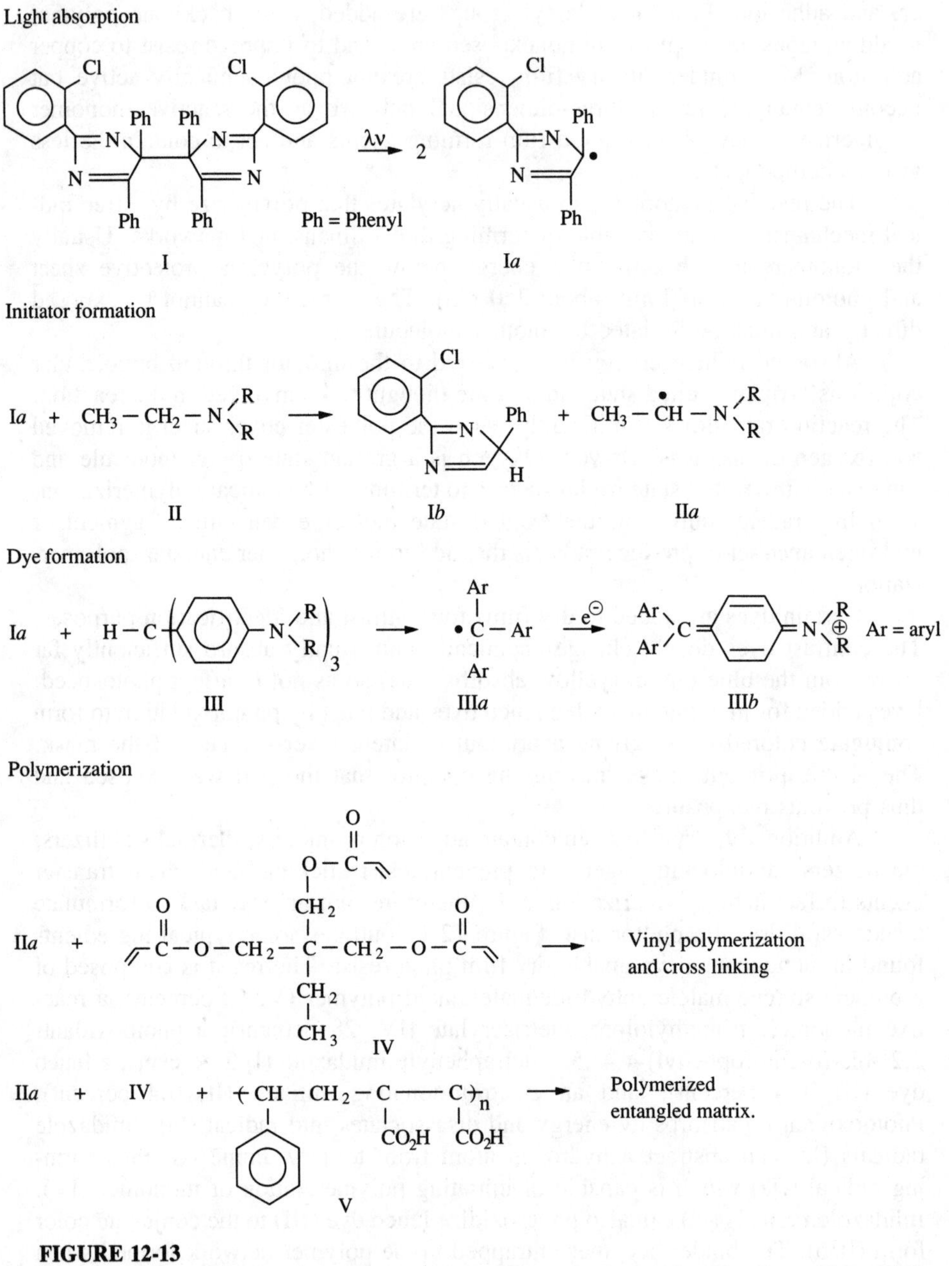

Light absorption

Initiator formation

Dye formation

Polymerization

FIGURE 12-13
Typical reactions occurring during polymerization of negative-acting dry film photoresists. U. S. Patent 4,298,678 (1981).

REFERENCES

R. A. Colclaser, *Microelectronics Processing and Device Design*, New York: John Wiley & Sons, 1980.

W. DeForest, *Photoresist: Materials and Processes*, New York: McGraw-Hill, 1975.

E. D. Feit and C. W. Wilkins, *Polymer Materials for Electronic Applications*, American Chemical Society Symposium Series 184, Washington, D.C., 1982.

H. Freitag, editor, "Packaging Technology," *IBM J. Res. Develop.* 26, no. 3, pp. 275–396 (May 1982).

K. I. Jacobson and R. E. Jacobson, *Imaging Systems*, New York: John Wiley & Sons, 1976.

S. S. Labana, *Ultraviolet Light Induced Reactions in Polymers*, American Chemical Society Symposium Series 25, Washington, D.C., 1976.

S. P. Pappas, editor, *UV Curing: Science and Technology*, vol. I, Norwalk, Conn.: Technology Marketing Corporation, 1980.

S. P. Pappas, editor, *UV Curing: Science and Technology*, vol. II, Norwalk, Conn.: Technology Marketing Corporation, 1985.

PROBLEMS

12.1. Why are resist systems necessary in the fabrication of chips and associated packaging in the industry?

12.2. Explain why a resist is called positive or negative and give examples of the chemistry associated with positive and negative resist systems.

12.3. The light source you are using has an intensity of 10 mW/cm^2. Your resist has a sensitivity of 300 mJ/cm^2. How long will you have to expose your resist to light?

12.4. A resist was found to have optical densities of 0.65 and 0.325 at 365 and 405 nm respectively. Calculate the percentage of light that would be absorbed at each wavelength in this film.

12.5. An inexperienced exposure operator did not apply vacuum during contact printing of a negative dry film photoresist. The positive artwork had 6-mil opaque and 6-mil clear areas. The exposure unit had a collimation half-angle of 16.5 degrees. How would the resulting copper lines compare to the glass master dimensions if one board were processed additively and the other board subtractively?

12.6. Consider two artworks; one is positive and has 6-mil opaque lines and 3-mil clear spaces, and the other is negative and has 3-mil opaque lines and 6-mil clear spaces. The positive artwork is printed with a negative resist and the negative artwork is printed with a positive resist. Both parts are subtractively processed in a normal manner. What are the copper line dimensions for each board?

CHAPTER
13

VACUUM METALLIZING IN PACKAGING

RUSSELL T. WHITE JR.
RONALD S. HORWATH
JEROME J. CUOMO

INTRODUCTION

Background

Vacuum metallization is a technology essential to the fabrication of integrated circuits. Without the ability to uniformly deposit a wide variety of conductive and insulating materials, in extremely narrow and thin layers, with good topological coverage, VLSI would not be the reality it is today. The utilization of vacuum metallization in electronic packaging has not, however, been as extensive as that in chip fabrication. Examples of vacuum metallization processes that have been developed for packaging will be given in this chapter and also are discussed in Chapter 10. More packaging-related applications will no doubt be developed in the future if the trend toward dry processing is to be realized.

Two metallization processes are described in some detail here to provide the reader with the necessary fundamental and practical information. These processes are *vacuum evaporation* and *sputtering*. Further information can be gained from references at the end of the chapter.

An Example

The circuitry on a ceramic chip carrier provides an intermediate level of interconnect between an epoxy/glass circuit board and a chip (see Chapter 10, Figure

394

10-18). Some means of forming conductive circuit lines on the surface of the ceramic chip carrier is needed. Vacuum evaporation or sputtering has been used successfully to deposit various thicknesses of adherent, electrically conductive metal for this purpose. When the inherent adhesion of the desired metal to the ceramic is poor, an advantage of vacuum deposition is the ability to deposit a thin layer of adhesion-promoting material between the ceramic and the metal. Standard photolithographic techniques can be used to form the fine lines, 25–75 μm (0.001–0.003 inches) wide, to which the chip is soldered to allow communication from the input/output devices to the integrated circuits.

Advantages

More traditional methods of making electrically conductive circuit elements for packaging components include the use of laminated metal foils and/or wet chemical seeding and plating (as discussed in Chapters 16 and 17) for organic composite circuit boards. Some of the advantages of vacuum metal deposition processes are (1) cost saving through less expensive, easily automated, highly reproducible processes; (2) less environmental effects because of the nature of the vacuum processes; and (3) the use of materials not usable in more traditional techniques.

Comparison of Vacuum Metallization Processes

Two important metallization processes are evaporation and sputtering. Both are discussed in detail in this chapter.

Deposition of a metal by evaporation in its simplest form involves heating a metal, in a vacuum chamber, to a temperature at which it changes from either a solid or liquid phase to a gaseous phase. This gaseous phase travels from the source to some cooler surface where it condenses back again to form the deposited film.

Deposition of metal by sputtering is a somewhat more complex and interesting process. It involves the establishment of a glow discharge (plasma) in a low-pressure inert gas in a vacuum chamber, the acceleration of ions from the glow discharge across a voltage gradient, and transfer of energy from these accelerated ions to atoms of a solid metal surface of a *target*. The target is the source of the metal to be deposited. Energized atoms of target metal are ejected and travel through the space in front of the target to the cooler surface of the substrate where they condense to form the deposited film.

Table 13.1 is a summary of some of the important characteristics of the two deposition processes.

Limits

The limits of vacuum metallization in packaging applications can be illustrated by comparison with its applications in chip fabrication.

TABLE 13.1

	Evaporation	Sputtering
Mechanism of production of depositing species	Thermal energy	Momentum transfer
Deposition rates	Can be very high (up to 750,000 Å/min)	Low (up to 24,000 Å/min)
Depositing species	Atoms	Atoms
Uniformity for large flat substrates	Poor, unless complex multiple sources	Good
Ability to coat:		
a. Complex shaped object	Poor, line-of-sight	Fair, but nonuniform
b. Into small blind holes	Poor	Poor
Kinetic energy of depositing species*	Low ($\approx$0.1 to 0.5 eV)	Can be high ($\approx$1 to 100 eV)
Bombardment of substrate by inert gas ions*	Not normally	Yes or no depending on configuration
External substrate heating during deposition*	Yes, normally	No, not generally

* Described later in this chapter.

The use of vacuum metallization in electronic packaging differs from that for chips in several ways. The most obvious difference is size. Silicon wafers now are being processed with diameters up to several inches, so that several thousand chips can be metallized in one production run of a large vacuum system.

While certain packaging components have sizes similar to chips, others, such as circuit boards for the largest computers, are several feet square (i.e. 24 by 28 inches). Due to this larger size, a single circuit board may constitute a run of the vacuum system, thereby limiting the throughput capabilities. This is not necessarily a problem if the production requirements of the chips and the packaging components are similarly proportioned. Otherwise large numbers of vacuum systems may be required, with associated costs and space requirements.

The difference in the thickness of the metal layer is an obvious one. Whereas with chips, a thickness of 1000 to 10,000 angstroms of metal is usually sufficient, thicknesses of 80,000 angstroms (ceramic chip carrier) to 500,000 angstroms (circuit board) are usually required for increased current-carrying capacity. Thus the time required to produce a circuit is proportionally longer.

Another big difference is vacuum compatibility and heat resistance of the materials used. Since most vacuum metallization processes involve heat generation in one form or another, the thermal stability of the material being coated is important. The silicon wafer, and the metals and insulators that make up the chip, are generally compatible with the vacuum environments and are able to withstand exposures to high processing temperatures. This is not necessarily the case with packaging components. Ceramics are similar to chips in their vacuum compatibility and heat resistance, but, on the other hand, a typical circuit board is made of epoxy polymer resins with fiberglass reinforcement. In order to maintain dimensional stability, this material generally should not be taken above 150°C for extended periods. Absorbed gases, residual solvents, and moisture present in certain materials can contaminate the vacuum environment, especially when heated during deposition. Polyimides used in packaging applications today are capable of withstanding temperatures up to 350°C for extended process times and are usually relatively free from volatile contaminants because of the high-temperature curing operation they undergo.

Another Example

A potential approach for overcoming the thickness limitation discussed above involves a combination of vacuum deposition of a thin "seed" layer followed by electrolytic or electroless plating to final thickness. This is a good example of a process that takes advantage of the strengths of vacuum and wet processes to achieve a result that neither can do as well alone. The combination of wet seeding and plating of metals onto nonconductive polymer surfaces has been done for years (as described in Chapter 17). The utilization of vacuum metallization for seed layer deposition in order to achieve adherent plating on materials that do not work with wet seeding can open important new areas of application.

Cost Considerations

Once a vacuum metallization process is proved to be technically achievable, a careful cost analysis of this process (see Chapter 29) compared to more traditional methods (if applicable) is necessary. An important consideration is the effective cost, in capital dollars, per each unit of throughput. High throughput, production vacuum metallizers generally cost between $500,000 and $4,000,000 or more. The system must be carefully designed for maximum throughput, with minimum downtime, to achieve the lowest effective cost per part. This is especially important with large packaging components, where small numbers of parts can be processed per run and the capital cost of the system, as well as the cost of running the process, must be recovered on relatively few parts instead of being spread over large numbers of parts as in chip manufacturing.

Reference Materials

One of the most complete sources of information on vacuum processes is *The Handbook of Thin Film Technology* [1]. Several other very good reference books are included in the reference list [2,3,4]. In addition, a wealth of information about almost any vacuum technology can be obtained from, or through, the American Vacuum Society's publications, courses, and bibliography lists [5,6].

VACUUM EVAPORATION

Background

Thin metal films were being vacuum deposited as early as 1887 [7]. Much of this early work was done for research interests. It was not until the advent of large-scale, reliable vacuum equipment that vacuum metal deposition became a routine manufacturing process.

Today, the applications of vacuum evaporation range from the deposition of aluminum on paper for shiny gift wrap, to the deposition of exotic, multicomponent materials for integrated circuit and electronics manufacturing. The functions of these films range from purely aesthetic (such as Christmas tree tinsel) to utilitarian (such as corrosion protection, light filtering and reflection, or the fabrication of electrical circuit lines and electronic devices). There are a multitude of applications of this technology, and certainly many more are yet to be imagined.

Vacuum Levels

In this chapter, pressure or vacuum levels are stated in units of *torr*, 1 torr being equal to 1 mm Hg. Vacuum evaporation is generally carried out in the 10^{-5} to 10^{-9} torr range. Sputtering is carried out in the 10^{-1} to 10^{-4} torr range. The reasons that these vacuum levels are used is discussed below. It should be noted that the term *high vacuum* often is used as a generic term to describe any vacuum level below about 10^{-3} torr.

Molecular Movement in the Vacuum Chamber

Gas atoms and molecules at normal atmospheric pressure are packed so closely together that no one of them can travel very far without colliding with a neighbor. The term *viscous flow* is used to describe this pressure region, which is usually identified as greater than 0.5 torr. As pressure is reduced, the distance that gas molecules can travel between collisions is increased. The term *molecular flow* is used to describe molecular motion where, because of low particle densities, molecules travel somewhat independently from one another. This occurs at pressures less than 0.005 torr. At very high vacuum levels, gas molecules have the possibility of traveling the entire length of the vacuum chamber, from wall to wall, without colliding with another gas molecule. This possibility is very important to deposition by vacuum evaporation. The pressure range between viscous flow and molecular flow is described as *transitional flow*, between 0.005 and 0.5 torr. Throughout this pressure region, as the pressure is decreased the crowding becomes less, and gaseous particles have the ability to move longer distances without colliding with neighbors. This actual length of free movement is very important to deposition processes and can be estimated as a value called the mean free path, as described in the following section.

Mean Free Path

A very important parameter in vacuum metallization is the mean free path, the average distance that a molecule travels in the vacuum chamber without colliding with another molecule in the gas phase. The mean free path (mfp) λ is calculated by

$$\lambda = \frac{kT}{p\pi\sigma^2\sqrt{2}} \tag{13.1}$$

where T is the temperature in kelvins, σ is the molecular size, p is the pressure, and k is Boltzmann's constant. The mfp is a mean value, which means that some molecules will travel longer distances and some shorter before collision. As will be discussed later, the mfp can be longer than the distance between the metal vapor source and the substrate. In such a case, the metal can traverse that distance without once striking another gas molecule. This is important, since the number of collisions experienced by a traveling metal atom, for example, can significantly affect the quality and characteristics of the final deposited film.

Wall Impingement Rate

The rate at which the gas in the chamber impinges on a given surface area inside the chamber, including the substrate being metallized, is an important parameter. This is especially important if the gas is reactive to the depositing metal. Each time a gas molecule strikes the exposed fresh metal surface, it has a chance of reacting, and an undesired addition to the depositing metal film may result. The following equation allows one to calculate an approximate impingement rate N, defined as the number of gas molecules of mass m striking a square centimeter of exposed surface in the vacuum chamber as a function of gas temperature and pressure.

$$N = \frac{P}{(2\pi mkT)^{1/2}}\mathrm{cm}^{-2}\mathrm{s}^{-1} \tag{13.2}$$

Knowing the metal deposition rate, and the fraction of molecules that react upon impingement, the extent of incorporation of contaminant gas in the deposited film can be estimated.

An Example of a Basic Process

As the name implies, vacuum evaporation involves the phase change of a metal, in a vacuum, from solid or liquid to vapor by heat input, transfer of the vapor from the source to the substrates, and condensation of that vapor to a solid. In a typical process, substrates are first cleaned to remove any trace of oils, salts, or other contaminants. The substrates then are mounted in the upper portion of a vacuum chamber as shown in Figure 13-1. A quantity of the metal to be deposited, copper

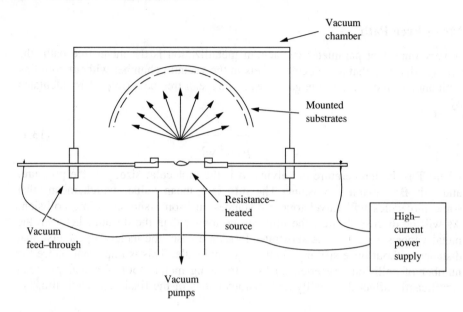

FIGURE 13-1
Schematic of vacuum system configuration used for evaporation deposition.

for example, is placed in a depression in a strip of tungsten foil that is mounted between two electrodes in the lower portion of the chamber. The chamber is closed and pumped down to a high vacuum (10^{-6} torr or better). Sufficient time is allowed for the removal of contaminating gases. Electric current is then passed through the tungsten foil via the electrodes, causing the foil to become very hot. The copper on the tungsten foil melts. The temperature is adjusted to achieve a desired evaporation rate. Atoms evaporate from the surface of the molten metal and, because of the vacuum, travel in straight lines to condense on a cooler surface, either the substrates or the shielding located above. When the desired deposition thickness is reached, the current is turned off and the evaporation source is allowed to cool. The chamber is then vented to atmospheric pressure and the substrates removed.

The Need for a Vacuum

It is important that vacuum evaporation is accomplished at very low pressures. At sufficiently low pressures, atoms of evaporated metal can travel the complete distance from the evaporation source to the substrate without sustaining a single collision with another atom. Poor vacuum would result in three possible problems. The first is the problem of loss of energy from too many collisions with cold gas atoms. The second problem is change in the directions of the evaporated atoms by the collisions. The third problem is probably the most important one. If the gas in the chamber happens to be a reactive gas, such as oxygen or water vapor, the collisions between this gas and the evaporating metal may result in a chemical reaction, and a different material may be deposited than was evaporated (this is sometimes intentionally done in a process called reactive evaporation).

In general, the lowest achievable pressure is maintained during the evaporation process to avoid these types of problems. Gloves are always worn when handling items that will be exposed to the vacuum since a small amount of fingerprint oil or other such contamination can make the ultimate pressure orders of magnitude higher, especially since the heat that is inherent in the evaporation process causes the oils to vaporize. Many contaminants, oils in particular, can destroy adhesion between the deposited metal and the substrate. Special precautions are taken to avoid moisture adsorption on items to be exposed to the vacuum. Water molecules tend to adsorb or absorb very tenaciously to certain materials, and then desorb or outgas when heated by the evaporation process. This is especially troublesome since water vapor is highly reactive to most evaporating metals.

Even with a long mean free path, and few gas phase collisions, the presence of a small amount of a reactive contaminant can still cause significant problems. This is especially true with low deposition rates, since there is more time for reactions to occur. At pressures as low as 10^{-7} torr, there is a sufficient amount of background gas to produce impingement rates of several monolayers per second onto the surface of a growing film. Water vapor desorbing from the surfaces inside the chamber during deposition may react with, and become incorporated in, the depositing film by this impingement, and the film's electrical conductivity may suffer. For this reason, the inside surfaces of the chamber are often baked by separate radiant heaters prior to the start of deposition. Another technique that is sometimes used to reduce the amount of condensible materials in the chamber during deposition is the use of cryogenic pump or trap. The extent to which steps are taken to rid the chamber of contaminants is heavily dependent on the intended function of the coating and the nature of the substrates being coated. Generally, less precautions need to be taken when metallizing gift wrapping paper, for example, than when making submicron integrated circuit devices.

The Phenomenology of Deposition/Evaporation

The calculation of deposition thickness and variations can be approached as one would approach calculation of light intensity and uniformity, per Lambert's laws of illumination and emission. The three important variables are distances from the source, angle of the receiving surface, and angular emission characteristics of the metal atoms as they leave the evaporating surface.

GEOMETRIC CONSIDERATIONS. A useful concept for the relationship between source-to-substrate distance and deposited thickness is that of a *point source*. An ideal point source of evaporating metal can be treated mathematically the same way as a point light source would. The upper or lower portions of some evaporation sources act much like point sources (Figure 13-2) especially when the source-to-substrate distance is large compared to the source size. The metal

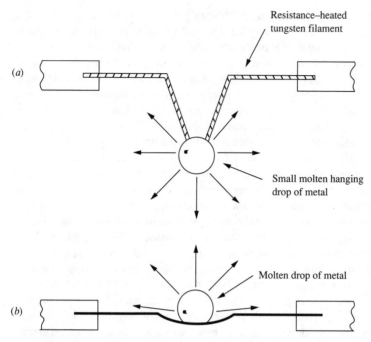

FIGURE 13-2
Examples of evaporation situations that closely approximate point sources.

evaporated from an ideal (tiny) point source in the center of a hollow sphere would be deposited uniformly over the inside surface of the sphere. The surface area of a sphere is proportional to the square of the radius. Therefore, deposited thickness is proportional to the inverse square of the distance from the source.

The angling of the receiving surface relative to the arriving flux affects the deposition thickness by effectively changing the area receiving that flux. This is described by a simple cosine function.

EMISSION OF ATOMS FROM A SURFACE. Atoms emitted from a flat evaporating surface (or a sputtering surface, discussed later) do not come out uniformly at all angles. Instead, angles perpendicular to the surface are preferred. The flux distribution closely approximates a cosine function. Figure 13-3 illustrates

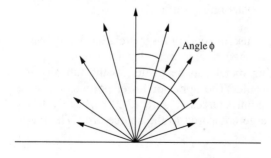

FIGURE 13-3
Cosine emission of atoms from a flat surface. Emission is proportional to the cosine of the angle from the perpendicular.

that the amount of material leaving the flat surface at angle ϕ is proportional to the cosine of that angle. This is a very important fact, especially when using a large, flat evaporation surface, such as a large pool of molten metal or an evaporating solid [1].

MODELING OF THE DEPOSITION PROCESS. The relationships described above can be combined as Equation (13-3), yielding deposition thickness as a function of distance from the source d, angle of the receiving surface θ, angle of emission from the source ϕ, and a proportionality constant a.

$$\text{Thickness} = \frac{a \cos \theta \cos \phi}{d^2} \qquad (13.3)$$

These types of calculations provide important insights into the strengths and weaknesses of the vacuum evaporation metal deposition process. From these discussions, the coverage of irregular surfaces can be understood. Figure 13-4 illustrates examples of deposition onto different surface features and some of the problems that can be encountered when trying to deposit a conductive path over previously formed features. Computer models have been used to predict and optimize the step coverage that results from different system geometries [9].

When good step coverage and deposition uniformity are needed, substrate motion relative to the evaporation source is often used. A special substrate holder, called a planetary, moves the substrates relative to the evaporation source(s). It should be noted that sufficient step coverage may not be achievable with vacuum evaporation, even with the use of planetaries, because of the line-of-sight nature of the process.

Evaporation Source Types

Some common types of evaporation sources are described in the following.

RESISTANCE HEATING. Resistance heating of a container holding the material to be evaporated (evaporant) was the earliest method used for vacuum metallization. In this method, large electrical currents are passed through a refractory metal container, usually made from tungsten. A major problem of this method is the interaction between the evaporant and the container. Aluminum, for example, alloys with tungsten, forming brittle intermetallic compounds that cause the container to fracture easily. It is this alloying or wetting action that allows simple tungsten wires to act as containers. Some metals (e.g. aluminum) melt, spread out, and cling to the wire as they evaporate. Figure 13-5 shows several of the common forms of resistance heated containers.

Another problem in resistance heating is that some of the container material may evaporate and be deposited along with the intended metal. Still another problem is the finite size of the container. It is difficult to deposit large quantities of metal without opening the system to refill the container, unless some automatic method is used. There have been several ingenious methods invented to accom-

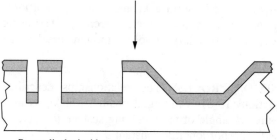

Perpendicular incidence

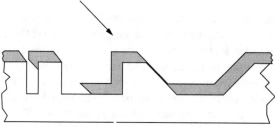

Single-angle incidence

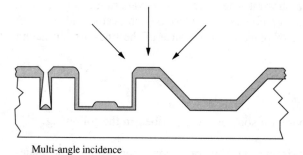

Multi-angle incidence

FIGURE 13-4
Examples of different types of coverage due to different arrival angles.

plish this [1]. One of the simplest ideas is shown in Figure 13-6. The metal to be evaporated is continuously fed, in wire form, into the resistance-heated container.

ELECTRON BEAM. The interaction between the evaporant and the container has historically limited the choice of depositing metals and the usefulness of the evaporation process. The introduction of the water-cooled copper hearth (container) and direct electron beam heating of the evaporant removed many of these limits. In this approach a high-current electron beam is generated and focused by electric and magnetic fields to impinge directly on the material to be evaporated, thereby causing local heating and evaporation. Because the heating occurs primarily at the surface of the metal being evaporated, the boundary between this metal and the vessel holding it can be kept relatively cool by water

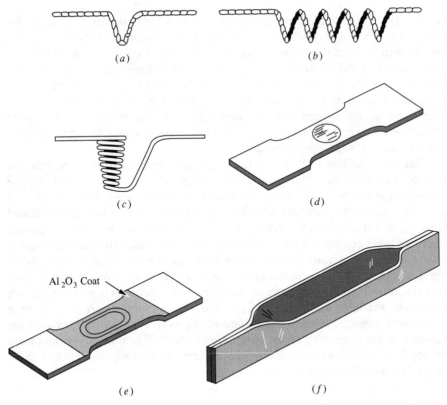

FIGURE 13-5
Wire and metal foil evaporation sources: (a) hairpin source; (b) wire helix; (c) wire basket; (d) dimple foil; (e) dimpled foil coating; (f) canoe type. (Permission: Maissel and Glang, page 1–40.)

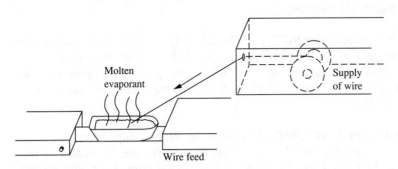

FIGURE 13-6
Schematic of automatic wire feed evaporant source.

cooling. This method prevents undesired reactions from occuring between the molten metal and the container by providing a solid interface between the two, with the metal essentially acting as its own container. Contamination of the film by container material is greatly reduced. The electron beam can be electronically scanned across the surface of a large source of metal to provide more control of deposition uniformity. In order for the electron beam to operate properly, the pressure in the system must be kept very low. Therefore, electron beam evaporation sometimes cannot be used when low pressures are difficult to achieve, as in the case of outgassing substrates.

Recently, hollow cathode electron sources have been used to provide the electrons that heat the evaporant. These are plasma arc devices that operate generally at pressures higher than conventional electron beam sources, 10^{-4} to 10^{-5} torr, and therefore do not suffer from the high-vacuum constraints mentioned above. In this method, high currents of electrons are generated from a plasma (described later) confined inside a refractory metal tube through which inert gas is flowing. The plasma heats the tube to temperatures at which thermal emission of large numbers of electrons occurs inside the tube. These electrons in turn liberate more electrons from the plasma inside the tube. Ultimately, there is an abundance of electrons available at the outlet of the tube. These electrons are directed by a confinement magnetic field to the surface of the evaporant, where they heat the evaporant in the same way as the conventional electron beam. The development and application of hollow cathode type devices is currently an active area of plasma and metallization research.

Special System Configurations

The discussions so far have been focused on single-source evaporation systems, evaporating one material at a time. The following are some examples of more elaborate systems and their uses.

MULTIPLE EVAPORATION SOURCE. Occasionally more than one evaporation source is installed in an evaporation system. This is done to provide better coverage and uniformity, to deposit two or more subsequent layers of metal during one vacuum cycle, or to codeposit two different metals simultaneously to make an alloy or mixture.

HIGH-THROUGHPUT SYSTEMS. Examples of high-throughput systems are load-locked systems (Figure 13-7) or in the case of flexible substrates, a roll-coater system. In a load-locked system, loading, initial pumpdown with preheat, deposition, cooling, venting to atmosphere, and unload are done at different parts of a large system simultaneously. One of the limits of this type of system is the finite amount of time required to pump down and degas one carrier full of substrates. If the deposition process is such that it could be done very quickly, the throughput of this system may well be gated by the pumpdown cycle. One way to avoid this problem is to use magazine type sample handling. A large number

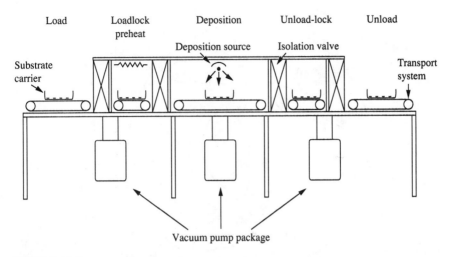

FIGURE 13-7
Schematic of in-line type deposition system. Parts are loaded at one end, proceed through a load-lock into the metallization chamber, and then through another load-lock to be unloaded.

of substrates may be pumped down at one time, then moved into a staging area in the next section of the vacuum chamber whence they are rapidly fed through the deposition zone, one after another, while another large batch of substrates is being pumped down and made ready for deposition.

ROLL-COATER METALLIZING. Extremely high throughputs can be achieved when metallizing thin flexible substrates by processing them in a roll-coater format. Figure 13-8 illustrates a roll-coater metallizer utilizing multiple evaporation sources, each with wire evaporant feed, coating a wide roll of material. This type of system is often used to coat gift wrap paper or plastic food wrap. The roll can travel as high as 500 feet per minute or more.

SPUTTERING

Background

The process presently known as sputtering was originally reported by Grove in 1852 [10], but it was not until the 1960s that increasing numbers of articles about it began appearing in the literature. In the past 20 years there has been a large number of publications, patents, and applications in this field. A good description of the history, theory, and applications of this technology can be found in the *Handbook of Thin Film Technology* by Maissel and Glang [1].

Applications of sputtering were originally thought to be limited, primarily because of low deposition rates, excess thermal and radiation damage to substrates, and poor uniformity of deposition. Recently the development of high-rate, lower-temperature sputtering systems designed for high throughput and uni-

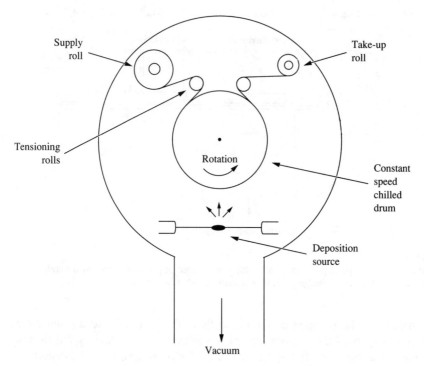

FIGURE 13-8
Schematic of a roll metallizer.

form deposition, together with many of the unique characteristics of sputtering, has regenerated acceptance and application of this technology.

Not only is sputtering a useful manufacturing process in industry today, it also is a very interesting process from a scientific viewpoint. It provides an opportunity to investigate glow discharges, plasmas, ionization, ion bombardment, atom/molecule interaction, crystal structure, crystal growth, adhesion, atomic scale chemistry, alloying, and other subjects.

Sputtering offers added flexibility in materials fabrication that is not available in the thermal processes. The sputter processes provide energetic bombardment at the deposition surface that can be utilized to modify the properties of the materials being deposited.

Basic System and Process

The essential elements of an early vacuum system designed for sputtering are shown in Figure 13-9. A typical operating cycle consists of loading clean substrates into the vacuum chamber, evacuating the chamber to a low pressure to remove contaminants (10^{-5} torr), backfilling the chamber to the operating pressure with an inert gas such as argon (usually in the 10^{-3} torr range), initia-

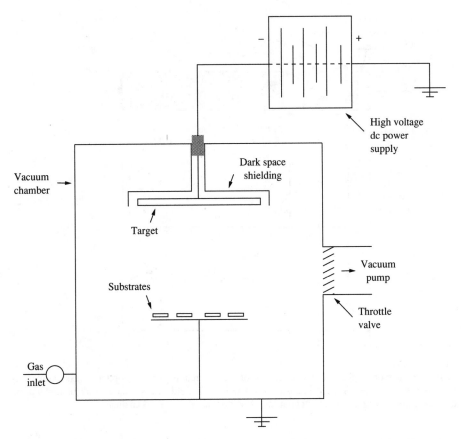

FIGURE 13-9
Schematic of simple system for sputter deposition (diode configuration).

tion and sustaining of a glow discharge or plasma by application of a high negative voltage to the cathode/target assembly, and sputter depositing of the target material onto a substrate. After the deposition process is complete, the system is vented to atmospheric pressure and the substrate removed.

A discussion of the sputtering process can begin with a simplified model of a self-sustaining glow discharge. Figure 13-10(a) shows two electrodes spaced some distance apart in a low-pressure gas. For this discussion, argon will be assumed as the gas since it is the one most commonly used. If a sufficiently large dc voltage is applied between the electrodes, the gas breaks down and current flows from one electrode to the other. As the term *glow discharge* implies, various regions of the space between the electrodes give off bright light, Figure 13-10(b). This glow discharge has several distinguishing features that are indicative of the events occurring between the electrodes.

Of significant importance is the region identified as the Crookes dark space. The largest change in the electric field strength between the electrodes occurs

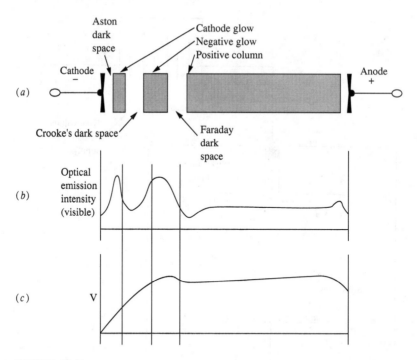

FIGURE 13-10
Schematic view of dc glow discharge. Names of the different regions (a), optical emission intensity (b), and effective electric field intensity (c). The electric field intensity is a result of nonuniform charged particle distribution throughout the glow discharge. (Public domain.)

across this space. The nonuniform electric field gradient between the electrodes is a result of a nonuniform distribution of charged particles in this space. Figure 13-10(c) is a graphical representation of the distribution of electric potential between the electrodes. The cathode is maintained at a high negative voltage relative to the anode. Electrons leave the cathode surface and accelerate across the Crookes dark space. Depending on the pressure, these electrons have a certain probability of colliding with and ionizing argon atoms, thereby liberating additional electrons and forming positively charged argon ions.

The edge of the *negative glow* nearest the target is the point at which sufficient amounts of ionization and atom excitation are occurring to give off visible light due to relaxation of excited atomic states. This negative glow portion of the discharge is composed of approximately equal numbers of free electrons and ions in a very dynamic state of continuous ionization, recombination, and motion. There is little electric field gradient across this region and it is, as a whole, approximately neutral, thereby allowing motion of the electrons and ions to occur by diffusion. A region such as this is called a *plasma*.

The key event of the sputtering process occurs when an argon ion diffuses close to the border of the negative glow that is nearest the cathode. At that time the positively charged argon ion is attracted toward the negative cathode surface. The ion accelerates and gains kinetic energy. When it hits the cathode, energy is transferred to the cathode's atomic matrix. If the energy is sufficiently high, atoms of the cathode material are ejected. This process can be likened to the action occurring when a cue ball hits an organized assembly of balls in a game of billiards. Studies have concluded that the main collision interactions occur near the surface of the cathode material, typically in the first three monolayers [11,12]. Along with the ejection of cathode atoms, secondary electrons also are ejected from the cathode by the bombarding argon ions. These electrons accelerate, strike argon atoms, and, as previously described, provide a continuation of the process; hence the term *self-sustaining glow discharge*.

Since the process just described results in removal of material from the surface of the cathode, it could not be continued very long before the cathode was destroyed. From a practical standpoint, the surface of the cathode is generally covered with a sheet or a plate of the material to be sputtered. This renewable surface, which is either bolted or bonded to a water-cooled backing plate, is called the *target* (it is the target of the argon ion bombardment). The term target will be used to refer to the sputtering surface for the rest of this chapter. Cooling of the target, usually by water channels behind it, is very important in high-rate sputtering. More than 70 percent of the power delivered to the target from the power supply ends up as heat energy in the target and must be removed.

For the most part, target atoms are ejected as individual neutral atoms or small groups of atoms. The angular ejection pattern approximates a cosine function as described in the vacuum evaporation section. These atoms travel through the space in front of the target and deposit on a substrate positioned there. Since a finite pressure of argon gas must be used to sustain the plasma, and since this results in a shorter mean free path compared to that of vacuum evaporation, the distance between the target and the substrate is often shorter. If no reactive species are present in the vacuum chamber, the sputtered atoms deposit on the substrate with the same composition as the target material, with perhaps a small amount of buried argon gas depending on the process conditions. The relatively high energy imparted to the sputtered atoms by the ejection process provides them with sufficient surface mobility to orient themselves on the substrates. The microstructure of the deposited material may not, however, be exactly as that of the target. The properties of the deposited films are highly dependent upon process parameters such as target voltage, pressure, system geometry, and power.

Magnetron Sputtering

Application of magnetic fields to sputtering has resulted in revolutionary changes in this technology. As discussed previously, it is the action of the electrons that initiates the plasma, which in turn causes the sputtering process to occur. The electrons gain kinetic energy as they accelerate away from the negatively charged target. One of the main functions of the electrons is to collide with and ionize argon atoms. Because the cross sections of these particles are small, and space between them is large due to the low pressure, the probability of collisions is low. High sputtering pressures were used in the past to increase this probability, but they generally led to poor film properties because of the shorter mean free paths. Magnetron sputtering is a means of making the electrons collide with argon atoms more efficiently while maintaining a low pressure. Magnetron sputtering also significantly improved several of the other early problems associated with sputtering: low deposition rates, excess thermal and radiation damage, and poor uniformity of deposition. The way that magnetron sputtering improved some of these problems is discussed below.

As shown in Figure 13-11, magnets are positioned behind the target in such a manner as to form a closed path of magnetic field lines in front of the target surface. As electrons try to accelerate away from the negative target surface, they move through perpendicular magnetic fields and therefore have forces applied to them. These forces change the directions of the electrons such that they are turned back toward the target, where they are repelled again.

From Figure 13-11, it can be seen that the effect of the magnetic field is to trap the high-energy electrons in a region in front of the target. This results in this region of space having a high concentration of high-energy electrons

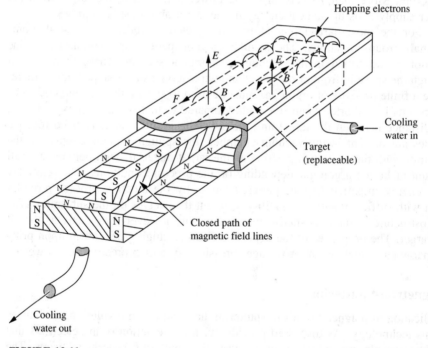

FIGURE 13-11

Cut-away view of magnetically enhanced (magnetron) sputtering target/cathode assembly. Electrons trying to leave the negatively charged target surface are trapped by the $E \top B$ fields. Improved ionization probability occurs along with localized target sputtering beneath the trapping region.

and the highest probability of collisions with argon atoms. Since the argon ion concentration is the highest in this region, the target surface immediately adjacent will experience more argon ion bombardment and therefore more sputtering.

The pattern of high target material removal will closely approximate the shape of the magnetic path, resulting in a nonuniform erosion pattern on the surface of the target. The rectangular magnetron target pictured in Figure 13-11 will end up having a racetrack-shaped groove in its surface. Sputtering from this target continues until the groove becomes too deep; then it must be replaced. This is in some respects an inefficient use of the target material, but the benefits are generally worth it. It is important to note that, compared to non–magnetically enhanced sputtering, magnetron sputtering provides much higher deposition rates, lower target voltages, less electron bombardment of the substrates and therefore less substrate heating, lower operating pressures, and better control of deposition uniformity over large areas.

Figure 13-12 shows a typical method for achieving uniform deposition over large areas in production-type equipment. Rectangular magnetron sputtering targets that are quite narrow in width (e.g., 10 cm) can be long in length (2–3 meters). Uniformity of deposition can be achieved in a system such as that shown in Figure 13-12 when the substrates are continually moved past the sputtering target in a direction perpendicular to its length axis. Uniformity of deposition

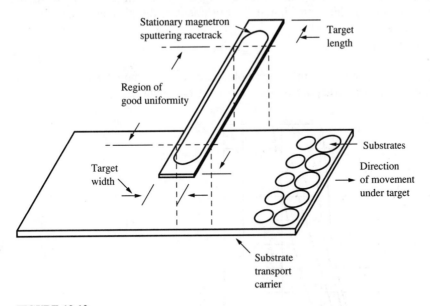

FIGURE 13-12

Schematic of one type of high-volume magnetron sputtering configuration. Parts to be coated are continually transported past the target. The direction of motion is perpendicular to the long dimension of the target, thereby producing uniformity of deposition in that direction. Uniformity in the direction of the long axis of the target is achieved by symmetry of the target shape, provided the target is sufficiently longer than the sample path is wide.

is typically good to a few percent along the length of the target except near the corners of the racetrack. Therefore, all substrates moved at a constant speed in front of the center portion of the target will have uniform deposition.

Magnetron sputtering suffers from difficulty with depositing magnetic materials at high rates, since the magnetic field lines are blocked and electron trapping becomes less efficient.

Process Parameters

Important parameters that affect the sputtering process and influence the film properties are deposition rate, deposition thickness, sputter gas pressure, and substrate bias.

DEPOSITION RATE. In a sputtering system operating at a given pressure, whether magnetron or not, the rate of ejection of target atoms, and therefore the deposition rate, is a function of the power applied to the target. Power to the target is given by the product of the voltage applied to the target and the current flowing from the power supply to the target. This current represents the electron flow through the plasma to the grounded chamber walls or a separate positively charged anode and ultimately back to the other terminal of the power supply. A typical current-voltage relationship for a magnetron system is shown in Figure 13-13. This relationship, after initial establishment of the plasma, generally follows the form:

$$I = bV^n \tag{13.4}$$

where I is the current, V is the target voltage, and b is some constant. The more efficient the electron trapping is, the higher the exponent n.

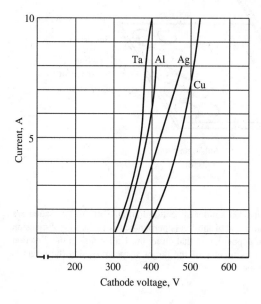

FIGURE 13-13
Typical current-voltage characteristics of high-rate magnetron sputtering targets. (Permission: Leybold-Heraeus.)

In a sputtering system operating in the normal voltage/current region, which is a large range, the deposition rate is approximately linear with respect to the power of the target.

DEPOSITION THICKNESS. The thickness of metal deposited on a substrate is determined by the deposition rate and length of time the deposition occurs. Deposition onto static substrates is simply the rate times the time. Deposition onto substrates that are continuously moved past a target, as is often the case in manufacturing systems, is more involved. To achieve a desired thickness of metal, the substrate must be transported through the deposition zone in front of the target in a given amount of time. This time together with the length of the deposition zone in the direction of movement define the transport speed of the substrates, and ultimately the throughput capability of the system. Higher deposition rates and longer deposition zones (multiple targets) can provide higher throughputs. For a given gas pressure the following empirical formula is helpful in deciding on the approximate sputtering conditions to achieve a desired thickness in a system where the substrates are moved one or more times past one or more targets.

$$\text{Thickness} = \frac{(d)\ (\text{power})\ (\text{number of targets})\ (\text{number of passes})}{\text{transport speed}}$$

where d is an empirically determined proportionality constant.

GAS PRESSURE. As the pressure in the sputtering chamber increases, the likelihood that the electrons will ionize argon atoms improves. Therefore, for a given voltage, the ion density increases and therefore the current increases. Most modern sputtering systems operate with current-regulated power supplies; therefore, as the pressure of the system increases, at a given current, the voltage decreases. Constant current implies the same number of ions striking the target (neglecting changes in secondary electron emission). But each ion has lower energy because of the lower target voltage so that the deposition rate decreases.

Higher pressure can affect the deposition in other ways. At sufficiently high pressures, above approximately 130 millitorr, more than half of the sputtered target atoms may be reflected back to the target surface by collisions with the gas atoms in the chamber, thus effectively lowering the deposition rate. For these reasons it is difficult to precisely predict the effect of pressure on deposition rates.

Aside from the effects on deposition rate, sputter gas pressure can significantly alter the properties of the deposited material. As described earlier, additional collisions between the sputtered atoms and the sputter gas can reduce surface mobilities of the former. Higher pressures may also cause the inclusion of sputter gas atoms in the depositing film [13]. Changes in the sputter gas pressure in general are more likely to affect the properties of the deposited film than changes in the deposition rate.

SUBSTRATE BIAS DURING DEPOSITION. A process variable that is often not used is substrate bias. Substrate bias is the application of a voltage, usually negative, on the substrate during the deposition process. In order to accomplish this, additional substrate fixturing is required to maintain electrical isolation from other parts of the chamber. A few hundred volts of negative voltage applied to a depositing metal surface, if sufficiently close to the plasma, can result in significant ion bombardment of the surface receiving the deposition. This ion bombardment can alter the crystal structure and other physical properties of the deposited layer. It can have a favorable or unfavorable effect on the ability of the depositing metal to cover irregular substrate surfaces. This ion bombardment also can cause significant additional substrate heating, which may or may not be desirable.

Additional Considerations

Other aspects of sputter deposition that warrant discussion include sputtering yield, multicomponent target sputtering, residual stresses, substrate heating, and substrate coverage. Discussion of these aspects of sputtering provide insight into potential uses of sputtering and into potential problem areas.

SPUTTER YIELDS. One of the advantages of sputtering over other deposition techniques is its ability to deposit a wide variety of materials including refractory materials, alloys, and insulators, as well as metals. The metals that are commonly sputtered have similar sputter yields, usually within a factor of four. The sputter yield is a term that measures the number of target atoms ejected for each bombarding sputter gas ion that strikes the target (atoms per ion). The yield depends on the energy and mass of the bombarding ion, but is useful for comparing the rate at which different materials will be deposited when exposed to the same ion bombardment conditions. Table 13.2 lists sputter yields for various metals under bombardment by argon ions of two energies.

TABLE 13.2
Sputter yield of argon ions*

Target material	Bombarding energy	
	200 V	600 V
Ag	1.6	3.4
Al	0.35	1.2
Au	1.1	2.8
Co	1.6	1.4
Cr	0.7	1.3
Cu	1.1	2.3
Mo	0.4	0.9
Pd	1.0	2.4
Si	0.2	0.5
Ti	0.2	0.6
W	0.3	0.6

*From reference 14

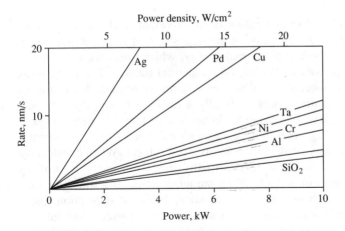

FIGURE 13-14
Typical deposition rates as a function of power applied to a target. Note that power is expressed as overall watts supplied to the target (*a*) and effective power as watts/cm² (*b*). (Permission: Leybold-Heraeus.)

Table 13.2 shows that with the same type of target configuration, operating at 600 volts, copper sputters at about 1.77 times the rate of chromium. The same type of data is shown in Figure 13-14 in graphical form, with sputter yields for different material plotted against target powers. It is interesting to note that in vacuum evaporation, tungsten must be heated to about 3230°C to have an evaporation rate similar to aluminum heated to only about 1220°C; while in sputtering, under similar sputter conditions, tungsten sputters at a rate that is only a factor of 2 lower than aluminum.

MULTICOMPONENT TARGETS AND ALLOYS. The difference in sputter yields might lead one to think that sputtering from a target made of a 50/50 alloy of copper/chrome might result in the deposition of a film that is 1.77 times richer in copper, since at 600 V, the relative sputter yields of these two metals are 2.3 and 1.3 respectively. This is indeed the case, but only for a very short time the first time the target is used. When the target is first bombarded with sputter ions, more copper is ejected than chrome, but this causes the surface of the target to be depleted of copper. Soon an equilibrium is reached, and the subsequent deposited film is essentially the same chemical composition as the bulk target. This is a phenomenon that can cause difficulties in chemical analysis techniques that rely on sputtering to clean or remove layers from the surface of a multicomponent material prior to surface analysis. The message here is that, although sputter yields of different materials are not the same, alloys can be sputtered to produce deposited materials of the same composition as the target. In contrast, evaporation from a molten source can produce a vapor, and therefore a deposited film, of different composition.

SUBSTRATE HEATING. Substrate heating is an important, and potentially limiting, factor in sputtering. Accompanying the deposition of target atoms onto a substrate is a substantial amount of thermal energy, which raises the temperature of the substrate by an amount depending on its thermal mass. When using sputtering to deposit target material on silicon or ceramic substrates, for example, heating has not been a major concern. Usually a rather high substrate temperature is desirable to achieve the required properties. External heating of these substrates, prior to or during deposition, along with its effects on the microstructure and adhesion of the deposited film, is an involved subject. When considering using sputtering in conjunction with less thermally stable substrates, controlling sputtering heating becomes important. When the temperature of a substrate must be kept below 150°C, such as with some epoxies, limits of deposition thickness as a means of avoiding substrate overheating are encountered. Without special cooling, or film thickness limitations, substrate temperatures can reach over 400°C.

The control of substrate temperature is not a simple problem. It is generally difficult to effectively cool substrates in a vacuum, especially when high-volume production is required. With the low pressures used in sputtering, convection cooling is negligible. This means that cooling must be accomplished by either conduction or radiation. Conduction of heat from a substrate is difficult unless the substrate is in good thermal contact with a heat sink; therefore, most substrate cooling in a vacuum occurs by thermal radiation, and radiation becomes effective only at relatively high temperatures.

Most of the heat input to substrates during vacuum evaporation comes in the form of latent heat of condensation of the evaporated target material. There is only a small amount of kinetic energy carried by each atom, and some heat will come in the form of radiant heat from the evaporation source. Careful design of the evaporation elements and shielding can reduce the radiant heating. Sputtering, on the other hand, has several inherent sources of substrate heating. Again, careful and creative design can minimize them. Sources of heat input to the substrate during the sputter deposition process include:

a. Latent heat of condensation of the sputtered atoms
b. Kinetic energy of the sputtered atoms from the ejection process
c. Particle and electromagnetic radiation from the plasma (ion, electron, uv, ir)
d. Infrared radiation from hot objects in the chamber (the target surface, shielding, etc.)
e. Bombardment of the substrate by reflected neutral sputter gas atoms

Depending on the system configuration, target material, and sputter gas pressure, the combined heat input to the substrate can be more than four times the heat of condensation of the depositing material. This means that unless the substrate has sufficient thermal mass, or some elaborate cooling means is provided, or the deposition thickness is sufficiently small, substrate overheating may occur.

In one type of magnetron sputtering system for depositing copper films, as each copper atom arrives at the substrate, so does about 14 eV of thermal energy [15]. Put in more meaningful terms, 14 eV/atom is equivalent to about 4.5 calories per micron of deposited copper per square centimeter of substrate surface. This type of heat input can cause a significant temperature rise in low-thermal-mass substrates.

Figure 13-15 illustrates that significant substrate heating can occur at typical magnetron deposition rates when cooling is accomplished by thermal radiation only [16]. Equation (13.5) relates sputter parameters and substrate characteristics to temperature rise of a thin flat substrate that receives deposition on one side, but is allowed to radiate from both sides [16]—thus the factor of 2 in the cooling part of the equation.

$$\frac{dT}{dt} = \underbrace{\frac{1}{\tau\rho C_p}}_{\substack{\text{thermal} \\ \text{mass}}} \quad [\underbrace{kER}_{\substack{\text{heat in by} \\ \text{deposition}}} + \underbrace{2\sigma\epsilon\,(T_1^4 - T_2^4)}_{\substack{\text{heat out by} \\ \text{radiation}}}] \tag{13.5}$$

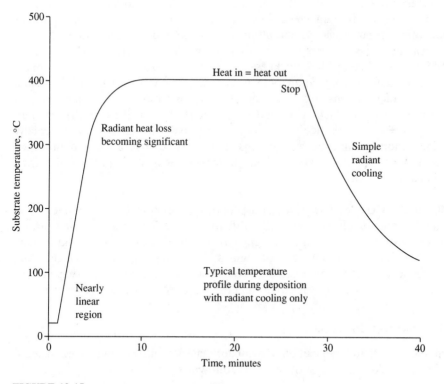

FIGURE 13-15
Typical temperature profile of heating of a substrate by sputter deposition when cooling of the substrate is accomplished by thermal radiation alone.

$$T_{2\,max} = \left(\frac{kER}{2\sigma\epsilon} + T_1^4\right)^{0.25} \tag{13.6}$$

where τ = thickness of substrate
  ρ = substrate density
  C_p = substrate specific heat
  k = proportionality constant
  E = energy per atom
  σ = Stefan-Boltzmann constant
  ϵ = substrate emissivity
  T_1 = temperature of surroundings
  T_2 = temperature of substrate
  R = sputter deposition rate

Equations (13.5) and (13.6) provide insight into some potential solutions to the overheating of low thermal mass substrates. A reduction in the maximum temperature reached by the substrate can best be accomplished by changing the parameters represented in the equation. This is more easily said than done for most of the parameters except deposition rate (R). One big disadvantage of using this parameter to reduce substrate temperature is its connection to the time required to deposit the desired film thickness. In a manufacturing process, this time represents product throughput capacity, and ultimately capital and process dollars.

Another possible perspective provided by these equations is that certain combinations of deposition thickness, substrate thermal mass, and reasonable deposition rates are not compatible with temperature constraints of particular substrates. In situations like these, one solution is to use some other forms of cooling, such as conduction. Conduction is generally much more efficient but, as explained earlier, difficult to achieve inside the vacuum system, especially in high-volume manufacturing systems. One exception is a method in which rolled flexible substrates are held under tension against a chilled roll during metallization.

SUBSTRATE COVERAGE. Another aspect of sputtering that can be looked upon as an advantage or a limitation is its ability to cover irregular surfaces and surface recesses. Evaporation is a line-of-sight process. It does not coat around corners and leaves shadows of unmetallized regions. Sputtering, on the other hand, does coat around corners to some extent. The extent is dependent on the shape of the target, the sputter gas pressure used, and the presence or absence of sputter bias.

Other Sputter Processes

Technology of sputter deposition of insulator materials and in situ reacted material has been developed, as described in the following:

RF SPUTTERING. RF sputtering has the ability to deposit nonconductive insulator materials such as Al_2O_3. This is accomplished by the application of ac instead of dc power to the target. Usually this alternating voltage is in the megahertz radio frequency range, thus the term *RF sputtering* (13.56 MHz is an industrial frequency often used). The attachment of the power supply, here called a generator, to the target is more complex than in the dc case. A tuning or matching network is used between the target and the generator to allow power transmission without excess power reflected back to the generator.

A layer of nonconductive material, such as a plate of Al_2O_3, is bonded onto a conductive backer plate. The generator produces alternating voltage on the backer plate, typically $+/-$ 500 V or more, several million times per second. Getting the plasma started is sometimes more difficult in an RF system, but once it is started, the electrons in the plasma respond to the rapidly changing voltages and oscillate back and forth through the plasma region. The energies gained by the electrons are sufficient to cause ionization of the gas atoms in their paths, producing more electrons and thereby sustaining the plasma, often without the help of secondary electron emission from the target.

At the frequencies used, the voltage reverses itself so quickly that not all of the electrons have the chance of reaching either the target surface or, in the opposite cycle, an anode or the vacuum chamber walls. The key to the sputtering action is that some electrons do strike the surface of the nonconductive target and accumulate to produce hundreds of volts of charge. Even conductive targets can be made to accumulate a voltage if they are isolated from the generator by a blocking capacitor. Other conductive surfaces on which the electrons land can be grounded.

The ionized gas atoms in the plasma (argon for example), having much greater masses than the electrons, are not significantly affected by the rapidly changing electric fields. Instead, as in the dc case, they are attracted to the resulting negatively charged target, accelerate toward it, and cause sputtering of the target material.

REACTIVE SPUTTERING. Reactive sputtering has a more direct analog in reactive evaporation, but the process control and flexibility often make reactive sputtering the preferred choice. One of the most common uses of this process is to sputter material from a pure target, allow it to react either in the gas phase or on the depositing surface with a reactive gaseous species purposely introduced in the chamber during the deposition, and in effect deposit a new material. An example of this is the sputtering of titanium in the presence of nitrogen in order to deposit titanium nitride, a very hard gold-colored coating with exceptional wear properties. Substrate bias is usually used in this process to improve the extent of reaction and to increase the hardness of the deposit.

CONCLUSION

There is little doubt that an increased use of vacuum metallization processes in packaging manufacturing will occur. The advantages include material choices, flexibility, less environmental contamination, cost savings, and increased automation possibilities. The combination of vacuum and wet processes, such as dry seeding followed by wet plating, offers added choices. On the other hand, consideration must be given to vacuum compatibility of the substrate, the ability of the substrate to withstand certain processing conditions (such as heat input), the product throughput as a function of required deposition thickness and substrate size, and the ability of the product to support the relatively high capital cost of the required equipment.

REFERENCES

1. L. I. Maissel and R. Glang, editors, *Handbook of Thin Film Technology*, New York: McGraw-Hill, 1970.
2. J. L. Vossen and W. Kern, editors, *Thin Film Processes*, New York: Academic Press, 1978.
3. B. Chapman, *Glow Discharge Processes*, New York: John Wiley & Sons, 1980.
4. J. F. O'Hanlon, *A User's Guide to Vacuum Technology*, New York: John Wiley & Sons, 1967.
5. J. L. Vossen, editor, *Bibliography on Metallization Materials and Techniques for Silicon Devices*, New York Thin Film Division, American Vacuum Society, 1974–1982.
6. American Vacuum Society, New York.
7. R. Nahrwold, *Ann. Physik* 31, p. 467 (1887).
8. M. Knudsen, *Ann. Physik*, 52, p. 105 (1917).
9. A. Blech, D. B. Fraser, and S. E. Haszko, *J. Vac. Sci. Technology* 15, no. 1, p. 13 (1978).
10. W. R. Grove, *Phil. Trans. Roy. Soc. London* 142, p. 87 (1852).
11. Wright, *Am. J. Sci.* 13, p. 49 (1877).
12. D. E. Harrison, N. S. Levy, J. P. Johnson, and H. M. Efron, *J. Appl. Phys.* 39, p. 2742 (1968).
13. H. F. Winters and E. Kay, *J. Appl. Phys.* 38, p. 3928 (1987).
14. N. Laegreid and G. K. Wehner, J. Appl. Phys. 32, p. 365 (1961).
15. J. A. Thornton and J. L. Lamb, *Thin Solid Films* 119, p. 87 (1984).
16. R. T. White and K. J. Blackwell, internally published work, IBM Corporation, Endicott, N.Y. (1986).

PROBLEMS

13.1. Calculate the mean free paths at 10^{-6}, 10^{-5}, 10^{-4}, and 10^{-3} torr for argon gas at 25°C. (Assume 3.67 Å as effective molecular diameter.)

13.2. In a vacuum evaporation system, copper is being deposited at 200 Å per second, depositing metal with nearly bulk density (usually the case), and there is a background pressure of 10^{-5} torr water vapor and a gas temperature of 25°C. (*a*) Assuming 10 percent of the water molecules that impinge on the growing film react to form the contaminent CuO, what will be the molecular percent of this contaminant in the final film? (*b*) What will be the contaminent level at a deposition rate of 1 Å/s?

13.3. Assuming an ideal point source (metal evaporating equally in all directions from a very small point) of evaporating metal, what thickness of metal will be deposited on a surface perpendicular to the arriving metal at 10, 20, and 30 cm from the source when 10 g of aluminum is completely evaporated? (The density of aluminum is 2.7 g/cm^3)

13.4. Assuming an ideal point source as in the previous question, with 10 g of aluminum being completely evaporated, what is the thickness of metal deposited on a sample

30 cm from the source when the surface of the sample is at 0°, 30°, 45°, and 80°angles with respect to the arriving metal?

13.5. Assuming a very small, flat evaporation source (i.e., subliming solid), what will be the relative deposited thicknesses on (small) substrates that have surfaces perpendicular to the arriving metal, equidistant from the source, at 0°, 30°, 50°, and 80°angles from the perpendicular to the surface at the source location?

13.6. In the normal operating range of a sputtering system, doubling the power to the target causes what effect on sputter deposition rate, approximately?

13.7. Which of the processes described in the chapter might be the best choice to deposit a 50/50 metal alloy on a large flat substrate? Why?

CHAPTER
14

PLASMA PROCESSING

WALTER E. MLYNKO
STEPHEN R. CAIN
FRANK D. EGITTO
FRANCIS EMMI

INTRODUCTION

A plasma is a collection of charged species which is, on the whole, electrically neutral. It is generally assumed to be an ionized gas; however, plasmas can exist in liquids and solids as well. Examples of plasmas are fluorescent light bulbs and other glow discharges, the flame of a match, and even our sun. A plasma environment can interact with surfaces to etch, deposit, and change the nature of the surface. It can be used to deposit materials and selectively etch fine patterns in many materials.

There are many advantages to using plasma instead of wet chemical methods. For example, plasma is clean. This advantage is one reason that plasma is used in the manufacture of integrated circuits. A plasma does not leave residual ions which can contribute to mobile charge in a semiconductor or dendritic growth in a circuit board. Plasma processes are inherently self-contained with little chance of human exposure. The plasma only appears when power is supplied. Contrast this with a chemical etchant such as sulfuric acid, which maintains activity before and after use (until neutralized).

The reactive species in a plasma environment can be tailored to preferentially etch certain materials and not others. The ability of a plasma to etch anisotropically makes plasma processing valuable in etching fine features in a solid or producing vertical texture in a surface.

Reactions can occur in a plasma that would otherwise be impossible at room temperatures. Thermodynamically improbable reactions can be made to occur. Plasma creates a kinetically limited mixture of very reactive radicals and ions.

Plasma requires little electricity, material, or maintenance compared with liquid methods, and thus, the operating costs are small. However, the capital costs can be very large. A vacuum-tight chamber with electrical feed-throughs and a pumping package is generally more expensive than equipment for liquid processing. Also, because they are not understood very well, plasma processes tend to involve a substantial amount of development time and engineering. Although research has provided basic physical/chemical information, the manufacturing scale-up is still much an art.

PLASMA FUNDAMENTALS

Overview of Plasmas

Plasma is sometimes called the fourth state of matter. Philosophers in ancient Greece thought that the entire universe was made of four substances: earth, water, air, and fire. Today we might think of them as solid, liquid, gas, and plasma, though strictly speaking, plasma is not a distinct state of matter. As energy is added to a substance, the temperature of that substance increases until a change of phase occurs. As the substance reaches the transition temperature for melting and vaporization, the temperature stays constant while heat is supplied to melt or vaporize. However, when ionization begins, the transition is not abrupt as it is for phase changes, but occurs gradually. As more energy is added, the temperature increases and ionization proceeds. The added heat is partitioned between the ionized species and the neutrals without an abrupt change in state.

Two parameters commonly used to describe a plasma are the electron temperature T_e and the electron density n_e. Although less common, the ion temperature T_i and the ion density n_i can also be used. It turns out that the ion density is approximately the same as the electron density because of the tendency towards electrical neutrality. The fraction of the gas that is ionized is typically about 10^{-6} to 10^{-3} for processing plasmas.

Though T_e can be hundreds of thousands of degrees, the electrons do not generally transfer that energy to nearby surfaces. Unless a surface is biased positively to attract electrons, the electrons are repelled from it. However, electrons can do useful work within the plasma by colliding with atoms and molecules to produce useful chemical reactions. The electron energy is commonly referred to in units of electron volts (eV). One eV is equivalent to 11,600 degrees Kelvin. (This is strictly true only when the electron energy distribution has a Maxwell-Boltzmann form.) The electron energy of plasmas used for processing in the electronics industry is usually about 2 to 10 eV. Some typical electron energies are displayed on the plasma map in Figure 14-1. This map includes such naturally occurring plasmas as the stars, earth's ionosphere, and fire.

How do we know if a plasma exists in some container? To answer this question, let us define the Debye length, a kind of yardstick that is specific to

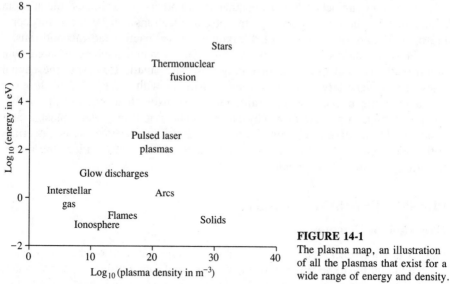

FIGURE 14-1
The plasma map, an illustration of all the plasmas that exist for a wide range of energy and density.

plasma. The Debye length is derived from treatments of charged particle interactions in a plasma or ionic solution. We refer the reader to physical chemistry or statistical mechanics textbooks for discussions of the Debye-Huckel theory of plasmas (or ionic solutions). The Debye length is

$$\lambda_D = \left(\frac{kT_e \epsilon}{2n_e e^2} \right)^{1/2} \tag{14.1}$$

where k is Boltzmann's constant, ϵ is the permittivity of the plasma (very nearly equal to that of free space), and e is the electronic charge. The Debye length is the distance at which the effective potential due to a single charged particle drops to $1/e$ of its original value. A plasma exists if the average spacing between charged particles is much smaller than λ_D, and if λ_D is much smaller than the container that houses the plasma.

In the laboratory, plasmas are generated typically with a simple electrical discharge at low pressure, also called a glow discharge. A glow discharge may be created by connecting a dc power supply to an evacuated chamber, shown in Figure 14-2. The glowing positive column of the discharge is a plasma, the dark regions near the electrodes are sheaths.

A magnetic field can be added to increase the plasma density. One common example of the use of a magnetic field is in magnetron sputtering [3,4,9]. (Also see Chapter 13.) The magnetic field makes electrons follow a helical path, which decreases the tendency of the electrons to escape from the discharge to the walls of the chamber. This effect increases both the electron and ion density and makes the discharge more efficient.

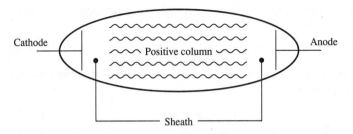

FIGURE 14-2
The most common laboratory plasma, a dc glow discharge.

Basic Plasma Physics

The physics of plasmas generally deals with describing the behavior of large numbers of charged particles in the presence of electric and magnetic fields.

IONIZATION EQUILIBRIUM. If we have a plasma in equilibrium—for example, a gas at high temperature in an insulated oven—the concentrations of electrons, ions, and neutrals are related by the Saha equation:

$$\frac{n_{A+} n_e}{n_A} = K(T, \text{ system}) \tag{14.2}$$

where n_{A+}, n_e, and n_A are the concentration of positive ions, electrons, and neutrals, respectively. K is an equilibrium constant that depends on the temperature and the quantum mechanical details of the atoms. This equation permits us to solve for the degree of ionization at a given temperature. The Saha equation can be approximated [1] by:

$$\frac{n_i}{n_n} = \left(\frac{(5.4 \times 10^{12})^{3/2}}{[n_n \exp(U_i/kT)]} \right)^{1/2} \tag{14.3}$$

where T, n_i, n_n, and U_i are the temperature, ion density, neutral density, and ionization energy in cgs units, respectively. At room temperature and 1 torr of air $n_n = 3 \times 10^{19}$, $T = 300$ K, and $U_i = 14.5$ eV. The resulting ionization is 10^{-122}, a very small amount indeed. The obvious result of this calculation is that we don't have equilibrium plasmas for processing materials near room temperature.

IONIZATION. Fortunately, most plasmas used for processing applications are not in equilibrium, and the amount of ionization depends on the kinetics of creation and loss of charged species. The primary mechanism of ionization in processing plasmas is by electron impact collision:

$$e^- + A \rightarrow A^+ + 2e^- \tag{14.4}$$

where A is a neutral species and A^+ is the ionized version of A. The rate of this reaction depends on the particular system but can often be described by a simple rate expression:

$$R = kn_e n_A \tag{14.5}$$

where k is unique for every system and depends on process parameters. An alternate form of this equation is:

$$R = \nu(\text{e-n, ionization})n_e \tag{14.6}$$

where $\nu(\text{e-n, ionization})$ is the collision frequency for ionization by electron-neutral (e-n) impact. Its units are the same as for frequency, s^{-1}.

Other mechanisms of ionization such as photoionization:

$$A + h\nu \rightarrow A^+ + e^- \tag{14.7}$$

and charge exchange:

$$A^+ + B \rightarrow A + B^+ \tag{14.8}$$

occur with far less frequency.

CHARGED PARTICLE LOSS. To balance the creation rate of charged particles, there are various loss processes. The primary one, however, is diffusion to the chamber walls. Charged particle diffusion in process plasmas occurs by ambipolar diffusion [1]. This means that both electrons and ions diffuse out of the plasma together. The diffusion rate of electrons is much faster than that of positive ions. However, if many electrons left the plasma, an enormous positive charge would build up. Thus, the electrons are retarded from diffusing at their "free" rate. In a sense, the electrons have to wait for the positive ions to catch up.

Electrons and ions must diffuse together. The resultant diffusion rate is midway between that of the electrons and ions. The electrons are slowed and the ions are sped up. The rate of diffusion is given by Fick's first law [2]:

$$J = -D_a \frac{dn}{dx} \tag{14.9}$$

where D_a is the ambipolar diffusion coefficient and dn/dx is the local gradient of the charged species concentration, either electron or ion.

CONDITION FOR PLASMA. To have an active plasma, the ionization must equal or exceed the loss by diffusion. This relationship can be expressed as:

$$\nu_{(\text{e-n, ionization})} \geq \frac{D_a}{\lambda^2} \tag{14.10}$$

where λ is some characteristic size of the chamber.

SHEATHS. The region between a plasma and an adjacent solid surface is known as a sheath. In a sheath region, there are generally many more of one type of charge than another, distributed in space. For this reason they are also sometimes called space charge sheaths. The physical dimensions of sheaths are generally about one Debye length but can be much larger when large electric fields occur, such as when high frequencies are used to sustain the plasma [10]. Sheaths are generally studied as a part of plasma behavior. Knowledge of the sheath behavior can also be used to diagnose the plasma by using Langmuir probes [8,11]. A sheath plays an important role because there is a voltage drop across it that accelerates positive ions toward the solid surface and electrons into the plasma. This voltage drop can have a major effect on chemical reactions occurring on substrate surfaces. Ions accelerated across the sheath also can impart enough energy to a solid surface to eject particles (sputtering), as discussed in Chapter 13.

ELECTRICAL CONDUCTION IN PLASMAS. Because a plasma has mobile charged particles, it seems natural that plasmas conduct electricity. The conduction current through a plasma is given by Ohm's law:

$$j = \sigma E \qquad (14.11)$$

where j is the current density (A/m^2), σ is the conductivity (mho/m) of the plasma, and E is the electric field (V/m) imposed on the plasma. The conductivity can be divided into contributions from electrons and ions:

$$\sigma = e(n_e \mu_e + n_i \mu_i) \qquad (14.12)$$

where e is the electronic charge (coulombs), and μ_e and μ_i are the electron and ion mobilities, respectively. This expression can be simplified, however, because mobility is inversely proportional to the mass of the particle. Thus, the electron mobility is much larger than the ion mobility, and the conductivity of a plasma is approximately

$$\sigma = e n_e \mu_e \qquad (14.13)$$

Most of the current in a plasma is carried by electrons, analogous to the way that most of the current in a semiconductor is carried by electrons rather than holes. (Note, however, that in the sheath regions, electrons may not be the primary current carriers. Since surfaces tend to be negative with respect to all plasmas, ions can carry a substantial amount of current through the sheath regions.)

Plasma Chemistry

BASIC PROCESSES. The plasma can generate reactive species from relatively inert parent molecules via inelastic collisions with electrons. Such reactions are illustrated below for a fictitious molecule A.

$$A + e^- \rightarrow A^* + e^- \qquad \text{excitation}$$

$$A + e^- \rightarrow A^+ + 2e^- \qquad \text{ionization}$$

$$A + e^- \rightarrow B + C + e^- \qquad \text{dissociation} \qquad (14.14)$$

$$A + e^- \rightarrow B^+ + C + 2e^- \qquad \text{dissociative ionization}$$

$$A + e^- \rightarrow B^* + C + e^- \qquad \text{dissociative excitation}$$

Here, A^* represents electronically excited A. The terms B and C refer to fragments of the parent molecule, A.

Electron density and energy, as well as the residence time and fragmentation chemistry of the parent molecule(s), determine the efficiency of excitation, ionization, and dissociation into reactive fragments. The electron energy distribution can be approximated by a Maxwell-Boltzmann distribution, as illustrated for a typical gas in Figure 14-3(a). The shape of the distribution is a function of the plasma process parameters (e.g., gas composition, power, pressure) as well as machine configuration.

Dissociation energies are 4–10 eV, while ionization energies are typically about 10 eV. Thus, it is usually those electrons at the high-energy region of the distribution that produce dissociation and ionization; the process is inherently somewhat inefficient. Now, one may be tempted to think that higher electron velocities will in general lead to more production of ions, excited species, and dissociation products. To a point, this hypothesis is true. However, at very high electron energies, the reaction probability falls off, as illustrated by the cross section in Figure 14-3(b). The regions of the cross-sectional curve may be understood as follows. At very low electron energies, the collision is not sufficiently energetic to give rise to reaction, be it ionization, excitation, or dissociation. With increasing average energy, a larger number of electrons will have sufficient kinetic energy to cause the reaction; thus the cross section increases. At very high impact energy the electron passes by (through) the molecule too rapidly to perturb the electronic orbits; little reaction takes place, and the cross section decreases.

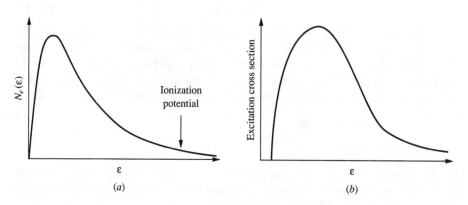

FIGURE 14-3
(a) Electron energy distribution, $N_e(\epsilon)$ is the number of electrons within the energy range ϵ to $\epsilon + d\epsilon$. (b) Excitation cross section as a function of electron impact energy ϵ.

GAS MIXTURES. Mixtures of gases are commonly used to enhance the production of a particular reactive fragment. A frequently used mixture is oxygen (O_2) and tetrafluoromethane (CF_4); therefore, this mixture will be discussed in more detail. Oxygen added to CF_4 enhances the production of atomic fluorine in the plasma. The observed increase in fluorine atom concentration in this mixture has been proposed to be the result of either the reaction between atomic or molecular oxygen and the dissociative products of CF_4 [12,13] or the reaction between electronically excited metastable oxygen molecules and CF_4 [14]. A dependence of species concentration in CF_4-O_2 mixtures on the electron energy distribution in the plasma is also possible [13]. Addition of CF_4 to O_2 leads to increased production of reactive oxygen atoms. Hence, these mixtures result in increased etching rates over those achieved by using the pure gas plasmas.

OPTICAL EMISSION FROM THE PLASMA. A frequently used technique for measuring relative (although not absolute) number densities of atoms and molecules in the plasma is monitoring the intensity of emitted light (optical emission spectroscopy). Light is emitted when an electron decays from an excited electronic orbital to a lower orbital. The frequency of this light serves as a fingerprint of the atom or molecule from which it originated. The intensity is a function of the number of excited atoms or molecules, which in turn is a function of the ground state number density, electron density, and the electron energy distribution. Figure 14-4 shows an optical emission spectrum for a CF_4-O_2 plasma.

Of great interest is the variation in reactive species concentrations with changes in plasma process parameters. Since the electron energy distribution can vary with parameter changes, one cannot be sure that observed changes in spectral

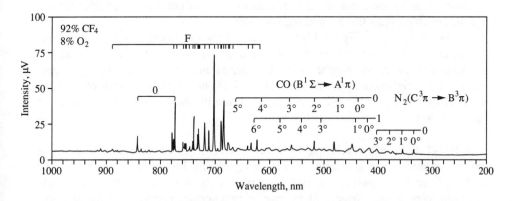

FIGURE 14-4
Optical emission spectra for a CF_4-O_2 discharge. (Reproduced from W. R. Harshbarger, R. A. Porter, T. A. Miller, and P. Norton, *Appl. Spectroscopy* 31, (1977), p. 201, by permission of The Society for Applied Spectroscopy.)

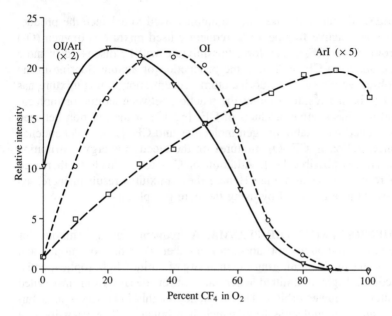

FIGURE 14-5
Oxygen, argon, and actinometrically corrected emission intensities as functions of gas feed composition.

intensities are due solely to changes in ground state number densities. A technique used to eliminate the dependence of emission intensities on the electron density and electron energy distribution is known as actinometry. Here, a small amount of chemically inert gas with an excitation cross section similar to the species of interest is added to the plasma. The dependence on electron density and energy distribution is factored out when one takes the ratio of emission intensities for the species of interest to the excited inert species.

As an example, Figure 14-5 compares emission intensities for atomic oxygen, argon (the actinometer), and the ratio of the two as functions of feed gas composition. Since the number density of argon atoms is constant in this experiment, the variation in argon atom emission intensity is an indication of changes in electron energy and/or electron density in the plasma, as discussed above for gas mixtures.

ETCHANT TRANSPORT. Once the reactive fragments are generated, they must be transported to the substrate surface, before being lost to recombination. The etch rate can be limited by the rate of arrival of etchant at the surface, or by the reaction rates at the surface, or by rate of product desorption from the surface. A good illustration of the effect of reactant flux to the surface on etching rate is shown in Figure 14-6, which plots the etch rate of polyimide as a function of linear gas flow velocity for etching downstream of an O_2-CF_4 microwave plasma. In this instance, as the flux of reactants to the sample increases, so does the etch rate.

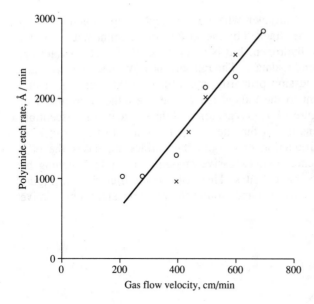

FIGURE 14-6
Effect of flow velocity on etch rate in a microwave system. (Reproduced from V. Vukanovic, G. A. Takacs, E. A. Matuszak, F. D. Egitto, F. Emmi, and R. S. Horwath; *J. Vac. Sci. Technol. A 4* (1986), p. 698, by permission of the American Vacuum Society.)

Surface Chemistry

REACTIONS ON THE SURFACE. With a knowledge of the fundamental chemistry, such concerns as selectivity (etch rate ratios for different materials) and surface diagnostics may be addressed. Since polymers are the materials most commonly etched or modified for electronic packaging, they are used extensively as examples here.

For clarity, focus on a simple polymer, polyethylene, in an O_2-CF_4 plasma. A sequence of reactions is depicted in Figure 14-7. Fluorine migrates from the

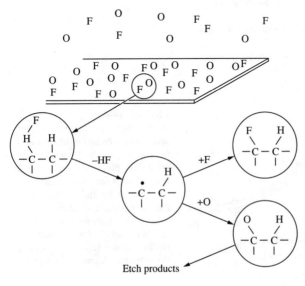

Etch products

FIGURE 14-7
Reactions at the polymer surface.

plasma to the surface of the polymer where it abstracts a hydrogen, leaving behind a radical, which may be attacked by the species in the immediate vicinity. (Oxygen also can abstract a hydrogen, but is much less effective than fluorine, as indicated from bond strength data.) Though any nearby species can attack the radical, the ensuing discussion primarily is confined to the active etchants, O and F. Etchant attachment to the radical leaves the site either fluorinated or oxygenated, as shown in Figure 14-7. Oxygenation of the site leaves it susceptible to degradation, while fluorination of the site leaves it passive to further etching processes. Thus, while introduction of F onto the surface increases the etch rate via enhanced radical generation, excessive amounts of F retard etching by competing with O for surface radical sites. These effects are illustrated in Figure 14-8, which shows the atomic oxygen and atomic fluorine concentrations relative

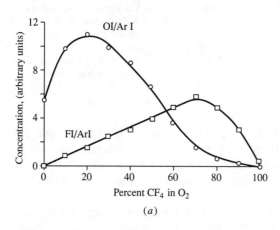

(a)

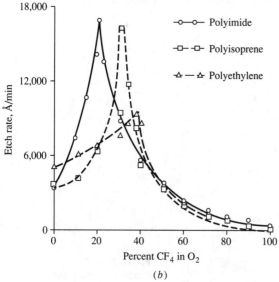

(b)

FIGURE 14-8
Actinometrically corrected oxygen and fluorine emission intensities (a) and polymer etch rates (b) for an O_2-CF_4 plasma in a planar diode reactor. (Part (a) was originally presented at the fall meeting of the Electrochemical Society, Inc., held in New Orleans, Louisiana, in 1984. Part (b) was reproduced from reference 15 by permission of the American Vacuum Society.)

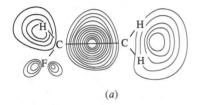

(a)

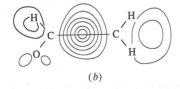

(b)

FIGURE 14-9
Electron contour maps for fluoroethane (a) and oxyethane
(b). Note the difference in the electron density in between
the carbons.

to argon, and the etch behavior of polyethylene, polyisoprene, and polyimide as
a function of the O_2-CF_4 gas feed composition.

Differences between the fluorinated and oxygenated intermediates are effectively accounted for by using molecular orbital theory [15]. A particularly useful
indicator is the "overlap population," which measures the degree of bondedness
between two atoms in a molecule (the greater the overlap population, the better
the bonding). Hydrogen abstraction followed by attachment of F to the radical
leaves the C–C bond virtually unaltered. On the other hand, attachment by O
weakens the bond by about 25 percent, facilitating degradation of the polymer.

What distinctive properties of oxygen enable it to weaken the C–C bonds?
The most striking feature is that oxygen has one less electron than does fluorine.
(F has seven valence electrons, O has six.) This deficiency is manifested in C–C
bonding type orbitals, as illustrated by the electron density contour plots in Figure
14-9. Note that the electron density between the two carbons is significantly lower
in an oxygenated species than in a fluorinated species, which accounts for the
relative weakness of the C–C bond. This argument is based solely on electron
count.

In extending these concepts to the etching of olefinic groups (C–C double
bonds), one may be tempted to assume a hydrogen abstraction type mechanism.
Indeed, the C–C overlap populations show the same trend. Though oxygen significantly weakens the C–C double bond (for the same reasons as discussed above),
the bond retains partial double bond character and is still significantly stronger
than a C–C single bond. Therefore, etching through a hydrogen abstraction type
mechanism seems highly unlikely. This situation is similar for phenyl groups.

Etching reactions of olefins are still possible, but through a different
mechanism. A well known class of olefinic reactions is radical addition [16],
shown below.

$$\begin{array}{c}>\!C\!=\!C\!<\; + \; F \; \longrightarrow \; -\!\overset{\displaystyle\cdot}{\underset{\displaystyle |}{\overset{\displaystyle |}{C}}}\!-\!\overset{\displaystyle |}{\underset{\displaystyle |}{C}}\!-\!F\end{array}$$

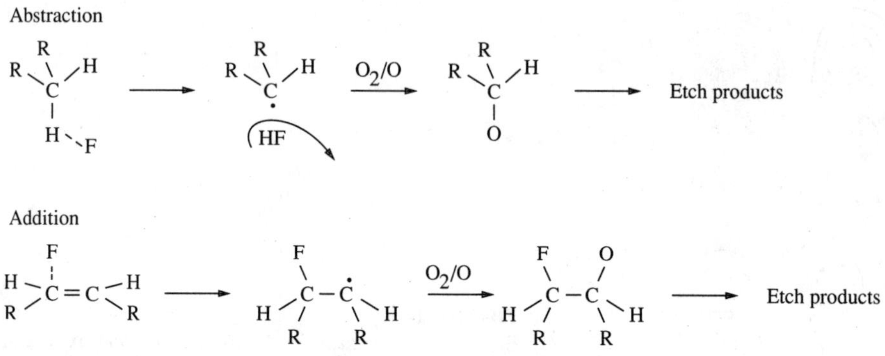

Abstraction

Addition

FIGURE 14-10
Difference in the dominant mechanisms. Note that during abstraction, fluorine leaves the surface as HF, but during addition, fluorine remains bound to the polymer.

The addition prepares a radical site for attack by oxygen, as did the hydrogen abstraction in the polyethylene case, and saturates the olefin so that the oxygenation step involves only a C-C single bond.

The two mechanisms are illustrated in Figure 14-10. As discussed above, in the case of etching a saturated polymer such as polyethylene, fluorine prepares the radical site by removing hydrogen, then leaves the reaction zone as HF. With the addition of fluorine to unsaturated groups, F remains chemically bonded to the polymer, which is revealed by subsequent surface analysis using x-ray photoelectron spectroscopy [15].

SURFACE ANALYSIS. Polymer degradation leads to volatile etch products via bond breaking and new bond formation. What remains on the surface is the unetched material, which has been modified by the plasma, yet has not reacted to a sufficient extent to be volatilized. Since modifications occur in the near surface region (the depth of modification varies), analyses that are performed to characterize changes to the material must be surface sensitive.

The most commonly used analytical techniques to determine compositional or structural changes resulting from exposure of packaging materials to plasmas include x-ray photoelectron spectroscopy (XPS), auger electron spectroscopy (AES), Rutherford backscattering spectroscopy (RBS) and Fourier transform infrared spectroscopy (FTIR). These techniques are described in more detail in Chapters 23 and 24. Each technique yields a unique type of information related to surface modification.

XPS and AES are the most surface sensitive of the techniques mentioned here; the analysis depth is generally on the order of 10–30 Å. These techniques are generally used for determining elemental compositions. Figure 14-11 shows a plot of the elemental composition of polyimide, determined by XPS, as a function of treatment in various O_2-CF_4 plasmas. XPS also gives some information on

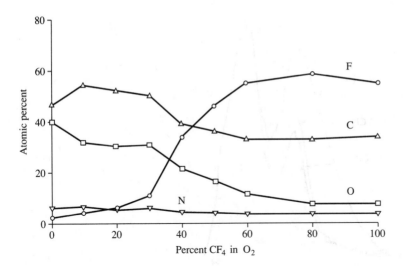

FIGURE 14-11
Surface composition of O_2-CF_4 plasma treated polyimide as a function of feed gas composition. (This figure was originally presented at the fall meeting of the Electrochemical Society, Inc., held in New Orleans, Louisiana, in 1984.)

chemical bonding. For example, the carbon spectra for polyimide exposed to a plasma with increasing amounts of CF_4 in O_2 are given in Figure 14-12. These spectra, normalized to the hydrocarbon (C–H) peak intensity at 285.0 eV, show that with higher CF_4 concentration in the gas feed, a greater contribution of higher binding energy carbon is present. Additional peaks are attributed to covalent-type CF_x bonds [17]. Similar high-resolution spectra may be obtained for other elements present in plasma-treated polyimide (O, F, N). Since there are problems performing AES on nonconducting samples, this technique has not been used as widely as XPS in the analysis of plasma-modified polymers. Although XPS and AES are quantitative techniques, they are not as accurate as RBS.

RBS, as discussed in detail in Chapter 23, uses an ion beam of known energy (typically 1–3 MeV) to probe the elemental nature of the sample. The incident ions are scattered, then energy-analyzed and counted to determine the elemental composition of the sample. The depth of analysis is on the order of micrometers; hence, RBS is not as surface sensitive as XPS or AES. With RBS, elemental information as a function of depth may be obtained in a nondestructive manner. For example, Figure 14-13 shows RBS spectra of polyimide films exposed to an O_2–CF_4 plasma for 1.5, 7, and 30 minutes. The intensity of the peak due to fluorine increases with increasing exposure time to the plasma. In addition to a greater total content of fluorine, the spectra reveal differences in the depth of fluorination. Figure 14-14 shows a plot of depth of fluorination of polyimide as a function of plasma exposure time. With both XPS and AES, elemental depth profiling is usually obtained in a destructive manner by monitoring the

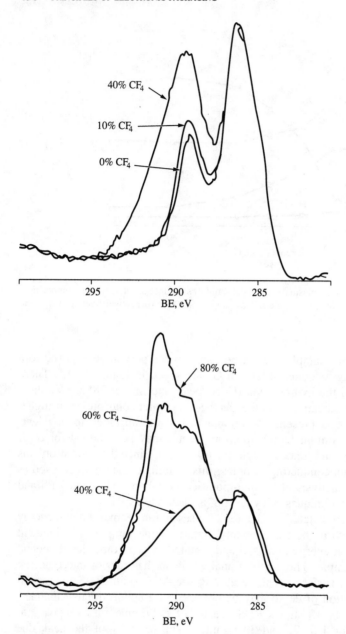

FIGURE 14-12
Carbon 1s spectra of O_2-CF_4 plasma treated polyimide. BE refers to the binding energy, and is given in eV. The peaks were normalized to the hydrocarbon peak at approximately 285 eV. (This figure was originally presented at the fall meeting of the Electrochemical Society, Inc., held in New Orleans, Louisiana, in 1984.)

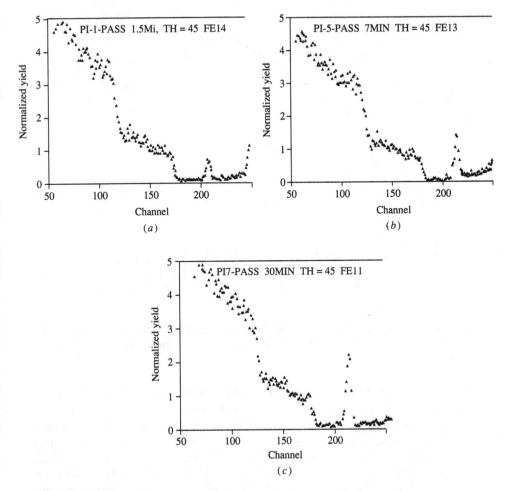

FIGURE 14-13
RBS analysis of polyimide treated for (*a*) 1.5 min, (*b*) 7 min, and (*c*) 30 min in an O_2-CF_4 plasma. Note the increase in the fluorine peak at the 210 channel.

composition as a function of sputter time by using an ion beam directed at the sample surface. Many potential problems may arise with this depth-profiling technique (see Chapter 23).

Further information regarding the nature of bonding is obtained by using reflectance IR spectroscopy. With this technique, the depth of analysis is between 200 Å and 2 μm, depending on experimental conditions. Samples must be on a reflecting substrate. Molecules absorb infrared radiation leading to vibrational and rotational transitions. Spectra yield quantitative information regarding functional groups. The structural information obtained using IR complements the compositional data received from XPS, AES, and RBS. The information obtained with the

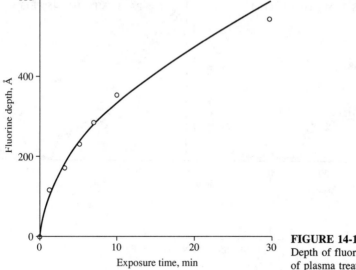

FIGURE 14-14
Depth of fluorination as a function of plasma treatment time.

techniques mentioned above, combined with an understanding of reactions in the gas phase and at the gas-solid interface, is used to delineate the various etching mechanisms.

CHEMICAL KINETICS OF SURFACE REACTIONS. Since etching occurs at a gas-solid interface, the nature of the etchant-surface adsorption should be addressed. Etchants pass from the plasma into a *reaction zone* in close proximity (a few angstroms) to the surface. Once in the reaction zone, the etchants may either establish a quasi equilibrium between adsorption and desorption, or initiate a series of surface reactions which give rise to the final products. It is convenient to view this reaction zone as a surface that chemisorbs the etchants prior to reaction. Henceforth in this section, *reaction zone* and *surface* will be used interchangeably.

Consider now polymer etching in a CF_4-O_2 plasma, where the active etchants are generally accepted to be atomic fluorine, F, and atomic oxygen, O. Suppose that a potentially reactive site, say a phenyl ring, on the surface chemisorbs i atoms of F and j atoms of O to form some locally modified site. This modified site is susceptible to subsequent reaction to form the final products, as illustrated in Equation (14.16):

$$\text{Polymer} + i\text{F} + j\text{O} \rightleftharpoons \{\text{polymer-}i\text{F-}j\text{O}\}^* \rightarrow \text{products} \qquad (14.16)$$

The intermediate, $\{\text{polymer-}i\text{F-}j\text{O}\}^*$, is an activated complex that forms the product. The particular values of i and j, along with the nature of the initial site on the polymer surface determine what the reaction products are. Only certain values of i and j in the activated configuration give rise to volatile, hence removable, products. Other combinations of i and j lead to compounds that may be passive to further etching. The etch rate is directly proportional to the number of the activated sites on the surface. An etch rate expression may be obtained by summing over all reactive configurations, weighted by the reciprocal of their life times τ. Hence,

$$\text{Rate} = \Sigma \frac{n(i, j)}{\tau(i, j)} \tag{14.17}$$

where $n(i, j)$ is the number of activated configurations, {polymer-iF-jO}*, on the surface. Of course, the sum only runs over those values of i and j that give rise to volatile products. As a hypothetical example, suppose that a phenyl ring degrades if one atom of F attaches itself to a carbon, and one atom of O attaches itself to an adjacent carbon. The sum would then include all possible configurations where an O atom is adjacent to an F atom on the phenyl ring. In practice, this sum may be approximated by a single term, so that Equation (14.17) becomes

$$\text{Rate} \approx n_{\text{poly}^*}/\tau_{\text{poly}^*} \tag{14.18}$$

In Equation (14.18), n_{poly^*} is the total number of chemically reactive (with respect to etching) configurations on the surface. In the above example, n_{poly^*} would include all configurations where a phenyl ring is bound to F and O such that at least one oxygen is attached to a carbon with fluorine attached to an adjacent carbon. An average lifetime, τ_{poly^*}, for all of the active configurations is used.

Terminology alternative to that used in Equations (14.17) and (14.18) is typical in the literature dealing with chemical kinetics. Reaction rates are expressed in the form,

$$\text{Rate} \approx k_{\text{poly}^*}[\text{poly*}] \tag{14.19}$$

where k_{poly^*}, the reaction *rate constant*, is inversely proportional to the lifetime, τ_{poly^*}. Reactant concentration [poly*] is directly proportional to the number of reactants n_{poly^*}. It should be noted at this point, that the rate constant is temperature dependent. For simple decomposition reactions of the type discussed above, the rate constant is generally greater (shorter lifetime) at higher temperatures. This constant is well approximated by an exponential in $1/T$:

$$k = k_o \exp \frac{-E_a}{k_b T} \tag{14.20}$$

Such an expression incorporates the effect of reaction barriers through the *activation energy*, E_a. Chemical reactions that occur with difficulty have large barriers to the reaction and high activation energies. In Equation (14.20), k_o is a temperature-independent preexponential factor, and k_b is the Boltzmann constant.

The concentration of activated sites may be related to the gaseous etchant concentrations via the techniques of statistical mechanics. For those interested in more detail, we recommend consulting a textbook on statistical mechanics or thermodynamics.

ION BOMBARDMENT. The foregoing discussion has concentrated on the chemical aspect of the plasma-etching reactions. However, the etching behavior of a given material depends not only on the number of arriving neutrals, but also on the flux and energies of ions, electrons, and photons.

Bombardment by energetic positive ions has been shown to increase etch rates in certain gas-solid systems and to decrease etching rates in others [18]. Hence, any one explanation of the effect of ions cannot be universal. Typically, ion bombardment has been observed to increase the etch rates of organic materials used in packaging. However, within a given gas-solid system, opinions are diverse as to which ion enhancement mechanism dominates.

A good example of ion-enhanced chemistry is the etching of Si using XeF_2 only, Ar^+ only, and both simultaneously [19]. XeF_2 is dissociatively chemisorbed onto the silicon surface producing two fluorine atoms (the chemically reactive etchant); the Xe evaporates. The etch rate obtained using XeF_2 with ion bombardment together is much greater than the sum of the rates obtained with either alone. Because of the synergistic effect, it is clear that physical sputtering alone is not the source of ion enhancement.

Since the concentration of ions in typical etching plasmas is roughly 10,000 times smaller than the concentration of neutral radicals, the degree of enhancement may seem surprising. It has been proposed that the magnitude of the effect is largely due to the relatively higher mobility of the ions and space-charge enhanced ambipolar diffusion [20]. The relatively large kinetic energy that ions reach in traversing the sheath regions may compensate for their smaller density.

Two additional effects that an ion can have upon striking a surface are the removal of surface contaminants that could otherwise retard etching and the enhanced formation of etch product by creating a higher number of reactable sites via disorder in the surface [21]. In addition, this modified surface may have a different affinity for the etchant species than the unmodified surface. Acceleration of the product formation step (i.e., formation of a product molecule weakly bound to the surface) has also been proposed as a mechanism for ion enhancement in Si-F systems [18]. The desorption of etch product molecules can be affected by ions via local thermal effects (further weakening of chemically weakened bonds) or by detrapping [21] of product molecules.

PRODUCT FLUX. Desorption and transport of the final products complete the etching process. As noted above, desorption can be enhanced with bombardment by energetic ions. The rate of desorption is determined primarily by the vapor pressure of the etch products, which may be found in an appropriate handbook. For example, chlorine-containing gases are more suited to etching aluminum than are fluorine-containing gases (aluminum chlorides possessing higher vapor pressures than aluminum fluorides). The etch products for polymer etching in O_2-CF_4 plasmas are primarily HF (for polymers possessing some saturated functionality), CO, and CO_2, all high-vapor-pressure gases.

Substrate temperature can affect the nature of the etch product. Consider, for example, aluminum etching using Cl_2 without plasma. At 33°C, Al_2Cl_6 is the primary etch product, while at higher temperatures, $AlCl_3$ is the major product [22].

REACTOR CONFIGURATIONS AND APPLICATIONS

The most commonly used electronic packaging materials are epoxy, polyimide, photoresists, cermet (a chrome/SiO_2 resistor), chromium, copper, and ceramics. Of these, only copper and ceramics cannot be plasma-etched practically. The types of processes for which plasma is useful include pattern replication, surface modification (chemical or physical), and contaminant removal. Typical applications include removal of residual material from drilled through-holes ("drill smear removal," Chapter 17), surface modification to promote metal-substrate adhesion, and stripping of photoresist (Chapter 12).

As discussed previously, the nature of plasma-etching processes can be purely chemical, purely physical, or a combination of the two. The mode is determined by the configuration of the reactor. The choice of which type of process to employ is dictated by the product to which it is applied. For example, when replication of fine-line patterns is required, purely chemical and hence isotropic etching processes are not suitable. Such requirements are encountered commonly in IC fabrication, where dimensional control is the major driving force for plasma implementation. Generally, packaging applications involve the use of larger features in thicker films with lower aspect ratios than in IC manufacture. The major process concerns are high rates over larger substrate areas to accommodate high throughput. Hence, very different machine configurations may be required.

Downstream Etching

Several variations of downstream etching apparatus are shown schematically in Figure 14-15. The barrel etch configuration, with grounded etch tunnel (Figure 14-15(a), was one of the first reactor designs used for electronic component manufacture (ca. 1960); it was used for resist-stripping applications in IC fabrication. Figure 14-15(b) shows an alternate design most commonly employed when using microwave plasmas. Here, species generated in the plasma are transported to the remotely located samples. The third example, shown in Figure 14-15(c), is merely a variation on the barrel etcher. Samples can be placed in what is known as the secondary plasma region. In each of these systems the samples are remote from the plasma. Thus, they are not surrounded by a sheath and hence are not subjected to highly directional energetic ion bombardment. However, they lend themselves to highly selective etch processes. Etching is isotropic (nondirectional); these configurations are not commonly used for fine-line pattern delineation, but are widely applicable for surface treatments or blanket film removal.

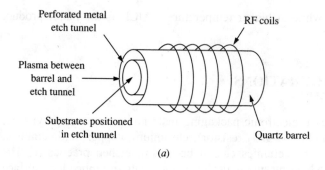

(a)

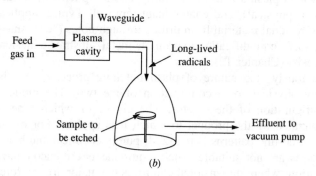

(b)

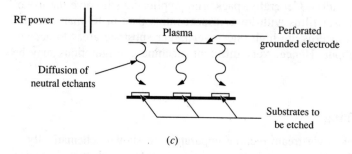

(c)

FIGURE 14-15
Downstream etching configurations.

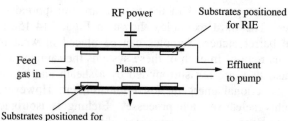

FIGURE 14-16
Planer diode etching system.

Conventional Planar Diode

This configuration is shown in Figure 14-16. Here the sample sits directly in the plasma, typically on the grounded electrode (externally biased substrates constitute the "triode" configuration). Substrates are exposed to ions that are accelerated through the sheath, but in this case the voltage drop is relatively low. Some degree of anisotropy can be achieved. Process pressures are on the order of hundreds of millitorr.

Reactive Ion Etching (RIE)

For reactive ion etching, the substrate is placed on the driven electrode. The surface area of this electrode is much less than that of other surfaces in contact with the plasma. As a result, the sheath voltage adjacent to the substrate is large. This condition reduces the possibility of sputtered contaminants from grounded surfaces depositing onto the etched material. Etching in this mode can be highly directional, but because of the more energetic ion bombardment, sputtered contamination can be a problem. The contamination can act as an etch mask, leading to residual unetched material. Typical process pressures are less than 0.1 torr in order to increase the mean free path of the ions.

Inert Gas Sputter Etching

The configuration and principles that govern sputter deposition (Chapter 13) apply to this etching technique. Argon is used almost exclusively. Of course, in this instance, the sputtering target is replaced by the substrate to be etched. Etching is physical and highly directional, but not selective.

Ion Beam Etching

Here, the substrate region is separated from the ion source, an ion gun. Either inert or chemically reactive ions can be used. The latter technique is known as reactive ion beam etching (RIBE). As for inert gas sputter-etching processes, ion beam etching is highly directional.

ENGINEERING AND SCALE-UP OF PLASMA PROCESSES

Manufacturing Considerations

CHARACTERISTICS OF A PROCESS. The qualities of the final product of plasma processing are generally dependent on several characteristics of the process itself. They are selectivity, uniformity, and reproducibility.

Selectivity. The tendency for a plasma-etching process to etch some materials faster than others is called selectivity. For example, the ability to etch the organic portion of a printed circuit board without etching any of the copper cladding is an important example of selectivity. In this case the selectivity is virtually infinite, the organic etches so much faster than the copper. In other examples, however, the selectivity is not as large and may require monitoring.

Uniformity. All plasma reactors are nonuniform! The question is how much nonuniformity we can tolerate in a process. Some reactors have only a 1 percent variation across the active surface; other have 100 percent or more. The etching uniformity can determine the ultimate size of the features being etched. For example, if the etching variation is 100 percent and five μm of material must be removed, then in order to insure that five microns are removed everywhere, some locations must be etched 10 μm. The feature size on the product must be large enough to accept the extra 5 μm.

Reproducibility. A plasma reactor must often be *conditioned* in order to operate reproducibly. Conditioning is done by operating the plasma prior to when it is actually used to process parts. Sometimes this conditioning is a preheat to bring the temperature up to a nominal value. Manufacturers of plasma equipment also recommend a burn-in period, where a new reactor is run several times to bring it to normal. After a plasma machine is used for a while, it may be necessary to clean or condition it in some other way so that it operates properly. In these treatments, the goal is to insure that subsequent runs are reproducible. It is not fully understood why some of these treatments work.

PROCESS WINDOW. When a process for manufacturing is good enough to implement, the key process parameters and measurables associated with it are formally recorded in the form of a process window. The characteristics of selectivity and uniformity discussed above, as well as the key parameters on the machine, such as pressure, temperature, and flow, are specified, and the maximum excursions in all of these variables are recorded and referred to as a manufacturing process specification.

Scaling Manufacturing Systems

Laboratory experiments with plasmas establish the major operating conditions that are needed to achieve the desired result. However, the manufacturing-scale machine is usually larger, and thus the conditions are generally different. There is not a one-to-one correspondence between the small-scale plasma and the larger one. Experiments must be done on the larger machine. This duplication of effort increases expense and development time.

Machine Similarity and Scaling

As mentioned above, a manufacturing system is quite different from a laboratory system. Manufacturing machines are generally large, automated, fixed in size, and limited in geometry by the needs of throughput and cost. The scaling of a process from a laboratory reactor to a manufacturing machine usually has some surprises, since process conditions used in the laboratory do not usually apply to the larger-scale machine.

The usual engineering approach to any kind of scaling is to do a dimensional analysis [23]. By this approach the known variables and constants associated with the plasma reactor would be manipulated to derive combinations or groups of variables appropriate for experimental design [12b]. The use of combined variables would reduce the number of experiments required to develop a plasma process. However, this approach is not commonly used. Instead, the process development is repeated in the large machine, which is time-consuming and expensive. Ideally, the operating conditions in a large manufacturing machine would be similar to those in the laboratory.

Example of Scaling Model Systems

Since plasma processes are inherently complex, let us start this discussion with work done on simpler systems, such as the basic dc discharge.

DC DISCHARGES. Some of the earliest work on dc discharges was to determine the voltage needed to produce an electric discharge through a gas, known as the breakdown voltage. The general relationship for breakdown voltage as a function of pressure and electrode spacing came to be known as Paschen's law [5]:

$$V_B = f(nd) \qquad (14.21)$$

where n is the neutral density and d is the electrode spacing. This relationship was alternately expressed as:

$$pd = \text{constant for a given gas and system} \qquad (14.22)$$

where p is the gas pressure.

A commonly used similarity parameter for a dc glow discharge is the ratio of electric field to neutral gas pressure (actually gas density), usually expressed as E/p. For example, the electron-neutral collision frequency can be expressed as a function of E/p:

$$\nu(\text{e-n}) = nf(E/p) \qquad (14.23)$$

where $\nu(\text{e-n})$ is the electron-neutral collision frequency, E is the electric field, p is the discharge pressure, and $f(E/p)$ is some function of E/p. Often, data on some property of a gas discharge plasma is plotted versus the parameter E/p.

Some more recent work [24] considers the detailed diffusion effects in an actively maintained discharge to produce a scaling law of the form:

$$Ip^3R^2 = f(E/p) \qquad (14.24)$$

where p is the gas pressure, I is the discharge current, and R is the discharge tube radius. This law is a generalization of others which considered the product pR a constant if I/R was constant.

The above parameters arose from the desire to minimize the number of experiments and the need to summarize data on dc glow discharges in a universal form. For example, it was much easier to do a single experiment by varying E/p than to do two experiments where first E was varied, then p varied.

RF DISCHARGES. Just as in the dc discharge, we may use pd and E/p as proper variables to describe the initial breakdown of a gas in an RF field. An additional variable, the RF wavelength, can be included in two sets of variables [5]:

$$Ed, pd, p, \lambda$$

or   (14.25)

$$Ed, E/p, p, \lambda$$

Either set of variables can be used to describe the breakdown of a gas in an RF field and can be conveniently mapped on a graph [12b].

These proper variables, as Brown [5] called them, have also been extended to actively maintained RF discharges [25,26]. Similar to that for a dc discharge, the condition for plasma maintenance in an RF discharge can be expressed as:

$$\nu(\text{ionization}) \geq \frac{D_a}{\lambda^2} \tag{14.26}$$

where $\nu(\text{ionization})$ is the ionization frequency, D_a is the ambipolar diffusion coefficient, and λ is the characteristic length of the discharge chamber. The electron density distribution is seen to depend on pd, the product of pressure and electrode spacing, as in Paschen's law [Equation (14.21)]. Also, a quantity analogous to E/p also exists for RF discharge; this quantity is the effective electric field, given by

$$E_{\text{eff}} = \frac{E}{(1 + \Omega^2/\nu^2)^{1/2}} \tag{14.27}$$

where Ω is the radian frequency of the applied field.

SHEATHS. The presence of sheaths and the large ion bombardment component in RF discharges requires new scaling concepts. Since ion bombardment is responsible for enhanced surface chemistry, the similarity between two plasma reactors depends on their having similar sheath properties. A major complication in this comparison is that the sheaths exist only at surfaces, whereas the quasi-neutral plasma depends on the volume between the electrodes. The sheaths scale with surface area, whereas the plasma scales with volume. Also, sheaths can wrap around the back side of electrodes, changing the characteristics of otherwise very similar reactors [27]. Because of this geometrical difference, it may not be possible to scale the sheaths and the plasma volume equally.

Another example of the complication involved in scaling ion bombardment is the range of energies that ions may attain. There may be ions of a broad range of energies striking the surfaces, depending on the frequency of the applied field, the ion mass, and sheath properties [28,29]. A general scaling parameter for this relationship has been described [30], having the form

$$(\text{RF frequency} \times \text{sheath thickness})^2 \times \text{ion mass/sheath voltage}$$

This parameter determines the shape of the distribution of energies of bombarding ions. The shape can range from monoenergetic to bimodal. Since the etching enhancement by ion bombardment depends on energy [31], this shape may be important in achieving similarity.

SCALING WITH CHEMICAL REACTIONS

General considerations. There are some general observations about plasma processes and plasma etching in particular that are useful for similarity considerations. These are:

Primary reactive species density. If one species is doing most of the chemistry, then it provides a good monitor, providing you can measure it. For example, fluorine is the major etching species for silicon. Thus, the silicon etch rate can be monitored by measuring the fluorine concentration. The relative fluorine density can be estimated by optical emission [32].

The atomic ratio concept [33,34]. The surface reaction rate may depend on the ratio of two atomic species rather than a single species. Example: If you are trying to etch silicon with F atoms, a larger F/C ratio will make the etching go faster. For instance CF_4 will etch silicon faster than C_2F_4.

Power density in RF plasmas. Power is usually expressed in watts per surface area of electrode rather than plasma volume. This convention has arisen because the electrode area has a larger effect on how much power should be used than does the overall plasma volume. This may be because large amounts of power are dissipated by ions accelerating toward the surfaces across sheath regions.

The loading effect. As the amount of material being etched in a plasma is increased, the etch rate decreases correspondingly. An example is plasma etching silicon. The more silicon being etched in the plasma chamber, the slower the etch rate.

The flow rate/residence time effects. The faster the flow rate of etchant gas, the faster the etch rate, up to a maximum. Beyond the maximum, the gas flows through the chamber too fast to have a chance to dissociate and etch [33]. Thus, there is generally an optimum residence time.

Similarity and scaling. The entire business of similarity and scaling is usually done with *similarity parameters*, which are groups of variables and constants that describe something about the system. These similarity variables are usually dimensionless. They are a major extension of the proper variable concept discussed by Brown [5]. Two different systems are said to be similar if their similarity variables are the same. For instance, a scale model of an airplane in a wind tunnel is encountering turbulence similar to a full-size airplane if the Reynolds number is the same for the scale model as it is for the full-size plane. The Reynolds number is a similarity variable for turbulence. Many scaling variables commonly used in chemical engineering are outlined in reference [23].

Reactor models. Recent modeling work has been applied to the Chemical Vapor Deposition (CVD) process [35,36]. Modeling has been extended to plasma-activated CVD using basic transport and reaction kinetics analysis. The added complexity of the plasma is generally dealt with by making a major assumption about the contribution that the plasma makes to the chemical reactions. For example the generation rate of reactive radicals can be assumed proportional to the average electron density. Then the diffusion and convection equations can be solved for the plasma deposition profiles [37]. Effects of diffusion and convection are particularly important if a neutral reaction dominates the deposition process [38].

Plasma etching technology has also benefited from reactor modeling. Extensive computer simulation of silicon etching in a CF_4 plasma [39] showed that, indeed, gas phase chemistry is dominated by neutral reactions, and verified the importance of various transport processes in the reaction mechanism. The major importance of transport processes, especially gas phase diffusion, in controlling the etch rate profile in plasma etching, has led to predictive etch rate models in which the etch rate profile is shown to be a balance between reaction and transport rates [40–42]. Additional modeling efforts have addressed in more detail both the charged particle behavior and the behavior of reactive neutrals [43].

REFERENCES

General references on plasma

1. F. F. Chen, *Introduction to Plasma Physics*, New York: Plenum Press, 1974.
2. J. L. Shohet, *The Plasma State*, New York: Academic Press, 1971.

Electrical discharges

3. B. Chapman, *Glow Discharge Processes*, New York: John Wiley & Sons, 1980.
4. J. L. Vossen and W. Kern, editors, *Thin Film Processes*, New York: Academic Press, 1978.
5. S. C. Brown, *Introduction to Electrical Discharges in Gases*, New York: John Wiley & Sons, 1966.
6. E. Nasser, *Fundamentals of Gaseous Ionization and Plasma Electronics*, New York: Wiley Interscience, 1971.

Plasma production in a laboratory

7. R. Jones, "Laboratory Plasma Devices" *Physics Reports* 61, no. 5 (1980).

Plasma diagnostics

8. R. H. Huddlestone and S. L. Leonard, editors, *Plasma Diagnostic Techniques*, New York: Academic Press, 1965.

Additional references

9. G. N. Jackson, "RF Sputtering," *Thin Solid Films* 5, pp. 209–246 (1970).
10. H. S. Butler and G. S. Kino, "Plasma Sheath Formation by Radio Frequency Fields," *Physics of Fluids* 6, no. 9, p. 1346 (1963).
11. B. E. Cherrington, "The Use of Electrostatic Probes for Plasma Diagnostics—A Review," *Plasma Chemistry and Plasma Processing* 2, no. 2, p. 113 (1982).
12. *a.* E. Kay, J. Coburn, and A. Dilks, in *Plasma Chemistry III*, edited by S. Verprek and M. Venugopalan, vol. 94 of *Topics in Current Chemistry*, Berlin: Springer, 1980, p. 1.
    b. D. L. Flamm, V. M. Donnelly, and D. E. Ibbotsen, "Basic Chemistry and Mechanisms of Plasma Etching" *J. Vac. Sci. Technol. B* 1, p. 23 (1983).
13. C. J. Mogab, A. C. Adams, and D. L. Flamm, "Plasma Etching of Si and SiO_2—the Effect of Oxygen Additions to CF_4 Plasmas," *J. Appl. Phys.* 49, p. 3803 (1978).
14. G. K. Vinogradov, P. I. Nevzorov, L. S. Polak, and D. I. Slovetsky, "Kinetics and Mechanisms of Chemical Reactions in Nonequilibrium—Plasma Etching of Silicon and Silicon Compounds," *Vacuum* 32, p. 529 (1982).
15. S. R. Cain, F. D. Egitto, and F. Emmi, "Relation of Polymer Structure to Plasma Etching Behavior: Role of Atomic Fluorine," *J. Vac. Sci. Technol. A* 5, p. 1578 (1987).
16. For reviews, see
    a. P. I. Abell, in *Free Radicals*, vol. 2, edited by J. K. Kochi, New York: John Wiley & Sons, 1973, p. 63.
    b. E. S. Huyser, *Free Radical Chain Reactions*, New York: Wiley Interscience, 1970.
    c. D. H. Hey, in *Advances in Free Radical Chemistry*, vol. 2, edited by G. H. Williams, London: Logos Press and Academic Press, 1967, p. 47.
17. *a.* H. J. Leary, Jr. and D. S. Campbell, in *Photon, Electron, and Ion Probes of Polymer Structure and Properties*, edited by D. W. Dwight, T. J. Fabish, and H. R. Thomas, ACS Symposium Series 162, Washington, D.C., 1981.
    b. A. Dilks, *Electron Spectroscopy: Theory, Techniques, and Applications*, vol. 4, edited by C. R. Brundle and A. D. Baker, London: Academic Press, 1981.
    c. D. T. Clark, W. J. Feast, D. Kilcast, and W. K. R. Musgrave, "Applications of ESCA to Polymer Chemistry. III. Structures and Bonding in Homopolymers of Ethylene and the Fluoroethylenes and Determination of the Compositions of Fluoro Copolymers," *J. Polym. Sci. Polym. Chem.* 11, p. 389 (1973).
18. H. F. Winters and J. W. Coburn, "Plasma-Assisted Etching Mechanisms: the Implications of Reaction Probability and Halogen Coverage," *J. Vac. Sci. Technol. B* 3, p. 1376 (1985).
19. J. W. Coburn and H. F. Winters, "Ion- and Electron-Assisted Gas-Surface Chemistry—An Important Effect in Plasma Etching," *J. Appl. Phys.* 50, p. 3189 (1979).
20. M. J. Kushner, "A Kinetic Study of the Plasma-Etching Process. II. Probe Measurements of Electron Properties in an RF Plasma-Etching Reactor," *J. Appl. Phys.* 53, p. 2923 (1982).
21. J. Dieleman, F. H. M. Sanders, A. W. Kolfschoten, P. C. Zalm, A. E. deVries, and A. Haring, "Studies on the Mechanism of Chemical Sputtering of Silicon by Simultaneous Exposure to Cl_2 and Low-Energy Ar^+ Ions," *J. Vac. Sci. Technol. B* 3, p. 1384 (1985).
22. H. F. Winters, "Etch Products from the Reaction on Cl_2 with Al(100) and Ca(100) and XeF_2 with W(111) and Nb," *J. Vac. Sci. Technol. B* 3, p. 9 (1985).
23. R. H. Perry and D. W. Green, editors, *Perry's Chemical Engineers' Handbook*, 6th edition, New York: McGraw-Hill, 1984.
24. Cherrington, B. E. "Modeling of Low-Pressure Gas Discharges," *IEEE Transactions on Electron Devices* ED-26, no. 2, p. 148 (1979).
25. A. T. Bell, "Spatial Distribution of Electron Density and Electric Field Strength in a High-Frequency Discharge: Criteria for Similarity," *Ind. Eng. Chem. Fundam.* 9, no. 1, p. 160 (1970).

26. J. A. Thornton, "Plasma-Assisted Deposition Processes: Theory, Mechanisms and Applications," *Thin Solid Films* 107, pp. 3–19 (1983).
27. W. E. Mlynko and D. W. Hess, "Electrical Characterization of RF Glow Discharges Using an Operating Impedance Bridge," *J. Vac. Sci. Technol. A* 3, no. 3, p. 499 (1985).
28. J. W. Coburn, and E. Kay, "Positive Ion Bombardment of Substrates in RF Diode Glow Discharge Sputtering," *J. Appl. Phys.* 43, p. 4965 (1972).
29. R. T. C. Tsui, "Calculation of Ion Bombarding Energy and Its Distribution in RF Sputtering," *Physical Review* 168, no. 1, p. 107 (1968).
30. M. J. Kushner, "Distribution of ion energies incident on electrodes in capacitively coupled RF discharges," in *J. Appl. Phys.* 58, no. 11, p. 4024 (1985).
31. J. W. Coburn, H. F. Winters, and T. J. Chuang, "Ion-Surface Interactions in Plasma Etching," *J. Appl. Phys.* 48, no. 8, p. 3532 (1977).
32. J. W. Coburn and M. Chen, "Optical Emission Spectroscopy of Reactive Plasmas: A Method for Correlating Emission Intensities to Reactive Particle Density," *J. Appl. Phys.* 51, p. 3134 (1980).
33. B. N. Chapman and V. J. Minkiewicz, "Flow Rate Effects in Plasma Etching," *J. Vac. Sci. Technol.* 15, no. 2, p. 329 (1978).
34. J. W. Coburn and E. Kay, "Some Chemical Aspects of the Fluorocarbon Plasma Etching of Silicon and Its Compounds," *IBM J. Res. Develop.* 23, no. 1, p. 33 (1979).
35. K. F. Jensen and D. B. Graves, "Modeling and Analysis of Low Pressure CVD Reactors," *J. Electrochem. Soc.* 130, no. 9, p. 1950 (1983).
36. S. Middleman and A. Yeckel, "A Model of the Effects of Diffusion and Convection on the Rate and Uniformity of Deposition in a CVD Reactor," *J. Electrochem. Soc.* 133, no. 9, p. 1951 (1986).
37. G. Turban and Y. Catherine, "A Kinetic Model for Radio Frequency Plasma-Activated Chemical Vapour Deposition," *Thin Solid Films* 48, pp. 57–65 (1978).
38. F. Weling, "A Model for the Plasma Activated Chemical Vapor Deposition Process," *J. Appl. Phys.* 57, p. 4441 (1985).
39. D. Edelson and D. L. Flamm, "Computer Simulation of CF_4 Plasma Etching of Silicon," *J. Appl. Phys.* 56, p. 1522 (1984).
40. J. F. Battey, "The Effects of Geometry on Diffusion Controlled Chemical Reaction Rates in a Plasma," *J. Electrochem. Soc.* 124, no. 3, p. 437 (1977).
41. R. C. Alkire and D. J. Economou, "Transient Behavior during Film Removal in Diffusion Controlled Plasma Etching," *J. Electrochem. Soc.* 132, no. 3, p. 648 (1985).
42. M. Dalvie, K. F. Jensen, and D. B. Graves, "Modelling of Reactors for Plasma Processing I: Silicon Etching by CF_4 in a Radial Flow Reactor," *Chemical Engineering Science* 41, no. 4, p. 653 (1986).
43. See, for example, entire issue of *IEEE Transactions on Plasma Science* PS-14, no. 2 (1986).

PROBLEMS

14.1. What is the Debye length of a single electron-ion pair in a cubical room that is 20 ft on a side? A crack in a package that is 1 μm on a side? (Assume room temperature.)

14.2. At what temperature would you have to heat up the air in an oven in order for the fraction ionized to be 10^{-5}, the same as in a typical processing plasma?

14.3. What is the minimum ionization frequency to maintain a plasma for a chamber with a radius of 5 cm if the ambipolar diffusion coefficient is 0.1 m^2/s?

14.4. What are the advantages and disadvantages to using oxygen plasma etching to remove the organic binding material from old photographic film to recover the silver?

14.5. Suppose that you have just discovered a plasma process to deposit antireflection coatings on the inside of hollow optical fibers. Where the fiber material is rare or difficult to make, use of hollow fibers is much less expensive than use of solid ones. Suppose that you deposited under conditions where the Thiele modulus, a similarity variable, was 1.5. The Thiele modulus is given by:

$$\phi = L\left(\frac{2k}{Dr}\right)^{1/2}$$

where L is the length of the fiber, k is the reaction rate constant for the surface reaction, D is the diffusion coefficient, and r is the inside radius of the fiber. In the original experiment, 10-cm-long fibers with an inside radius of 1 μm were used.

Suppose that you wanted to scale the process to big plasma machines for fibers that were 10 m long. What inside diameter fiber would you have to use in order to maintain similarity? If you can adjust the reaction rate constant k by changing the temperature of the deposition, by what factor would you have to adjust k in order to perform this process on 10-m-long fibers with 1-μm radius and still maintain similarity?

14.6. Why is ion-assisted etching (e.g., RIE) highly directional compared to purely chemical plasma etching (e.g., microwave)?

14.7. Consider the following kinetic data obtained for O_2–CF_4 etching of two polymers, A and B.

Etch rate, Å/min				
A	B	[O]	[F]	Temperature, K
2,500	3,600	2	3	300
6,200	9,000	5	3	300
8,800	12,500	7	3	300
18,500	27,000	5	9	300
10,300	15,000	5	5	300
5,500	13,500	5	5	275

where the O and F atom concentrations are in arbitrary units. Assume a rate expression of the following form:

$$R = k[F]^i[O]^j$$

From the above data, find the reaction orders, the rate constant, and the activation energy. Suppose that for a given process, polymer A must etch at least 5 times as fast as polymer B. Under what conditions might this selectivity be realized?

CHAPTER
15

LASER
APPLICATIONS
IN ELECTRONIC
PACKAGING

CHRISTOPHER R. MOYLAN

INTRODUCTION

Electronic components manufacturing uses technically sophisticated processes to form patterns of metal lines, semiconductors, and insulators. Early in the development of these technologies, screening processes borrowed from the art and printing industries were used to make these patterns. Since then, patterns have become smaller, and radiation has been utilized to pattern resists, which are used in subsequent processing to obtain the microscopic patterns. The subsequent processing is actually a lengthy and complex series of steps (Figure 15-1), which includes postbaking, development, rinsing, etching or deposition, removal of photoresist, and final cleans and residue rinses. Research and development laboratories have expended extraordinary efforts perfecting these lithographic processes because of their vital role in the entire electronics industry. The technical achievements have been spectacular, but the economic costs have been high.

As the dimensions of electronic devices and their interconnections shrink, new techniques for fabricating them are required. As the size of packages increases, their cost follows, and yields required to make their manufacture economically feasible rise dramatically. In the packaging world, therefore, as our products

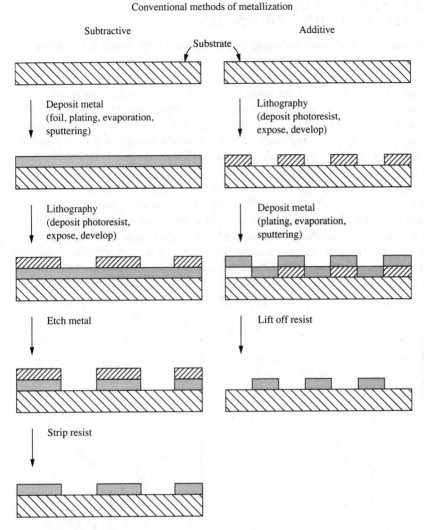

FIGURE 15-1
Conventional methods of metallization.

tend toward infinite size with infinitely small features on them, there is great
pressure to develop defect-insensitive processes that can deal with tiny features.
The characteristics of laser light make it an appealing energy source to use in
such processes. In this section, the distinctions between various laser-induced
processes are explained, then the types of lasers best suited for each process are
enumerated. The most important packaging materials are briefly listed with the
steps of their handling best suited for laser processing, and some examples are
described.

WHAT LASERS CAN DO
THAT OTHER TECHNOLOGIES CAN'T

Two characteristics of laser radiation make it particularly suitable for advanced processing applications: directionality (spatial coherence) and monochromaticity. The directionality enables us to focus the light onto a particular microscopic site rather easily, or project an image down to microscopic dimensions. The monochromaticity allows us to select which of several simultaneously irradiated species will absorb the light. There are in general two types of laser-induced processes: pyrolytic (photothermal), where chemical reactions are initiated by adding vibrational energy to molecules or materials, and photolytic (photochemical), where electrons in molecules are excited to higher-energy orbitals, leading to reactions. The former type uses the laser as a highly localized heat source. The latter uses it to perform photochemistry on a particular molecular species; the products of the photoreaction then complete the processing (Figure 15-2).

It is often not immediately obvious which type of event is occurring with a new process observed in the laboratory. There are, however, certain diagnostic experiments that clearly identify a laser process as either pyrolytic or photolytic. The most obvious one is to measure the rate of the process as a function of laser fluence or intensity (flux). Rates of pyrolytic processes scale with the amount of

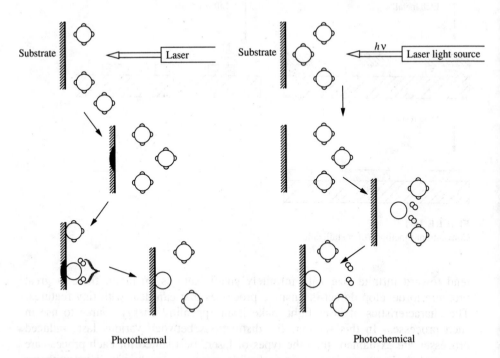

Photothermal   Photochemical

FIGURE 15-2
Photothermal versus photochemical laser processing.

heat deposited, whereas rates of photolytic ones depend on the number of photons supplied. If the rate of the process can be measured at two different wavelengths, one can vary either the total amount of laser energy absorbed or the number of photons absorbed while keeping the other quantity constant. Identification of the parameter that must be held constant to give equal rates at the two wavelengths reveals whether the process is photothermal or photochemical. For the latter class, determination of the process rate as a function of photon flux or fluence shows the number of photons required to induce the rate-limiting photochemistry. If the dependence is linear, it's a one-photon process. If it goes as the intensity squared, it's a two-photon process, etc.

There are, of course, other ways of making this distinction, dependent upon the specific materials being irradiated. For example, one should know the absorption spectrum of each liquid, gas, or translucent solid being used, and the reflectance spectrum of each opaque solid material. This knowledge is usually sufficient to determine whether the laser process is pyrolytic or photolytic. If, for example, a gas phase species being used is transparent at a given wavelength, but the substrate material absorbs strongly, a laser at that wavelength will heat the surface without inducing photoreactions in the gas phase, and any chemical process that occurs must be thermally initiated. If one has the choice, should a photothermal or photochemical process be designed? Unsurprisingly, the answer depends upon the circumstances. A pyrolytic process is often preferable because the chemistry is more predictable. Thermal reaction parameters and products are known for many more systems than are photochemical ones. Modeling is often easier, because heat flow through materials is well understood. A greater choice in laser sources is possible with photothermal processes, because materials can be heated at many wavelengths, whereas they perform photochemistry at only a select few. Visible laser sources in particular are not usually capable of initiating photochemical reactions, but are commonly used for localized heating; they are more easily aligned and their optics are cheaper and more widely available.

On the other hand, thermally sensitive materials are better suited to photochemical processing. Projection patterning can be done with greater resolution using photolytic processes, because there is no loss of resolution due to spreading of heat in the material. Processing of transparent substrates must be done photochemically, as must any processing that involves chemistry different from ground electronic state (thermal) chemistry. If the desired laser-induced reaction is to occur in a liquid or gaseous sample above a solid surface, it must be done photochemically, because it is generally impractical to heat a fluid in an effort to achieve some localized effect on an adjacent solid. And one is not necessarily restricted to products different from those observed in a pyrolytic process; photochemical decompositions often yield the same fragments. Also, if infrared lasers are used, reaction via the lowest threshold energy pathway is guaranteed, yielding thermal products. This last remark is an appropriate point at which to shift to a discussion of the types of laser sources available for designing packaging processes.

TYPES OF LASERS
AND WHEN TO USE THEM

Laser light sources can be conveniently divided two ways: by the wavelength regime at which they emit, and the manner in which the energy is delivered (pulsed or continuous wave). The particular advantages of each type are commented upon here [1].

Wavelength Regimes

INFRARED. Infrared (ir) lasers can be used to heat many materials, but metals are an important exception. Useful packaging metals like copper absorb so little infrared light that they are used as ir mirrors. For photochemical applications, ir wavelengths correspond to molecular vibrational energies, so ir photochemistry produces thermal products. Infrared photolysis is therefore a method of creating thermal products photochemically, as mentioned above. But optical materials suitable for infrared light can be difficult to use, being hygroscopic (like KCl) or opaque (like Ge). The longer wavelengths limit the maximum achievable resolution, sometimes to dimensions unacceptable for particular applications, even in packaging.

VISIBLE. Any material that is not white, silver, or transparent can be heated with a visible laser to initiate a pyrolytic process. Furthermore, those materials that do not absorb visible light can often be dyed so that they do. Visible lasers are therefore very convenient for use in photothermal processing. But little photochemistry can be accomplished at these wavelengths. Since the beam can be seen, it is easy to align, and optics are widely available and reasonably inexpensive. The large number of available lasers that are in the visible range makes it the only regime that is essentially completely accessible via one source or another.

ULTRAVIOLET. Most excited-state photochemistry must be initiated with uv light. The strictest resolution requirements necessitate this as well. Most materials become more absorptive as the photon energy increases, so photothermal reactions can usually be driven with ultraviolet light as well. The available wavelengths in the uv region are the most limited of the three regimes, however.

Energy Output Mode

CONTINUOUS WAVE. If a process requires one steady action to be induced by laser light, a continuous wave laser is the obvious choice. For example, if a line on a substrate is to be drawn by scanning the substrate with respect to the beam, a continuous wave laser should be used.

PULSED. If the process has several steps that have to occur in a repeating sequence, a pulsed laser is more suitable. Pulsed lasers are also required for processes that need a high peak power, because their output is concentrated into short bursts of much higher intensity than the output of continuous wave lasers.

Naturally, the optimal output mode may not be available. If a particular wavelength or power is required, one may have to make do with the other type of laser. It is nevertheless important to note that materials processes can be divided into those that are inherently continuous and inherently stepwise, and it is desirable to use a light source that matches those characteristics if at all possible.

PROCESSING PACKAGING MATERIALS WITH LASERS

The major categories of materials used in packaging of semiconductor devices are metals, polymers, and composite insulators (including ceramics) [2]. The amount and variety of laser activity on these materials follows that sequence in descending order. Accordingly, we start here with some discussion of laser studies on metals, and follow with the progressively shorter sections on polymers and ceramics.

Metals

DEPOSITION. Patterning of thin film metal interconnect wiring can be done with lasers both additively and subtractively, and each method can be performed both pyrolytically and photolytically to varying degrees. The advantage in metal deposition is that it can be a single-step process; the advantage in etching is that one is guaranteed to start with the lowest possible (bulk) resistivity, since the metal is deposited in conventional blanket modes of evaporation or sputtering.

Laser-induced metal deposition can be performed from precursors in all three phases: solids, liquids, and gases. The gas-phase processes are most appealing because only they are truly dry processes. Accordingly, they will be discussed first.

Dry laser processing has the advantage of being quite straightforward. Typically, the sample to be laser-patterned is mounted in a vacuum chamber facing an optical window. A controlled amount of a desired compound in the vapor phase is added, and may be flowed through the system if desired. Either the light source is focused onto the surface, or it irradiates a mask pattern projected onto the surface (Figure 15-3). Patterning can occur in several stages if desired, by pumping out one gas and allowing another one to enter the chamber. After processing, the chamber is vented and the sample removed. The relative importance of gas-phase photochemistry (as opposed to surface thermal or photolytic reactions) can be determined rather easily by mounting the sample sideways (parallel to the laser beam but not touched by it). The parallel orientation restricts

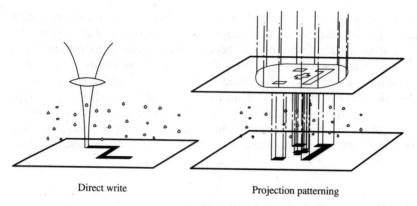

Direct write

Projection patterning

FIGURE 15-3
Laser patterning methods.

the process to its gas-phase photochemical component, which can be compared with the direct irradiation results. The perpendicular orientation is quite defect-insensitive, because any particulates that have landed on the irradiated areas tend to be superheated and volatilized off the surface by the laser.

The metals that have been photothermally deposited, that is, by laser chemical vapor deposition (LCVD), appear to have been chosen in general not because of their use as conductors but more because carbonyl or alkyl compounds of these metals are easily available. Despite the flurry of activity in laser deposition of metals in recent years [3–6], the search for new and better precursor compounds is being pursued by only a few researchers. It has been shown that two parameters of a precursor compound are critical in determining its use for pyrolytic deposition. The first is its vapor pressure. For many useful metals, few compounds with respectable vapor pressures exist. Deposition rates for copper, for example, so far are always limited by the vapor pressure of the inorganic precursor [7], and deposition chambers often must be heated to increase the vapor pressure to usable levels. What is needed are essentially chemical vapor deposition (CVD) reagents that will be used for localized rather than blanket deposition. So, any precursor mixture that has been proven successful for CVD is a good candidate for LCVD, and investigators looking for new compounds for a particular metal are well advised to check the CVD literature. In each case, deposition occurs by heating a surface until a gas phase compound decomposes to yield a material of choice. In CVD, the entire sample is heated and coated. In LCVD, only laser-heated regions undergo deposition.

The second key parameter is decomposition temperature. If a precursor compound decomposes at a low temperature, low laser intensities can be used and more substrates can be treated without damage. But if the decomposition temperature is too low, metal tends to deposit over a broad area with very little provocation and deposition becomes difficult to control. The optimum decomposition temperatures are in the neighborhood of 150–200°C, and optimum vapor pressures are about 50 torr. (Above this approximate pressure regime, a higher vapor pressure does not increase the deposition rate because the increased number of precursor molecules in the gas phase is offset by their decreased ability to diffuse to the microscopic heated zone [4,7].)

Based on these criteria, it is unsurprising that the bulk of the reported metal deposition involves either aluminum, zinc, or cadmium deposition from the corresponding methyl derivatives, or chromium, molybdenum, tungsten, iron, or nickel deposition from parent carbonyl compounds. There has also been a fair amount of work on tungsten deposition from tungsten hexafluoride with added hydrogen [8]. One might ask why this precursor is used rather than hexacarbonyl, since HF is formed as a by-product; the answer is apparently that the carbonyl decomposition is not completely clean, and some disproportionation occurs, leading to formation of tungsten carbide [9].

In packaging, metals like cadmium, tungsten, and iron are not particularly useful. What one needs are highly electrically conductive metals, metals that adhere well to nonmetallic materials, and metals that do not corrode to be used as protective layers. Therefore, the three most useful metals to deposit would be gold, copper, and chromium. Chromium deposition from the hexacarbonyl compound has been attempted more often using the photochemical method [5], so we use copper and gold deposition as our examples for photothermal metal deposition.

Pyrolytic deposition of copper from the gas phase was first performed by Jones et al. [10] using copper hexafluoroacetylacetonate—abbreviated Cu(hfac)$_2$. Among other observations, it was found that heating the reaction chamber resulted in significantly thicker spots of copper. Heating the cell caused the vapor pressure of the inorganic compound to increase (it is only 3 mtorr at room temperature) so that many more molecule-surface collisions were occurring. It has been shown that if transport of precursor molecules to the hot surface is rate limiting, rather than the decomposition kinetics in the adsorbed phase, such a pressure effect does not obtain. At these pressures, the increase in concentration of gas phase molecules is offset by the decrease in the diffusion coefficient, and no net increase in reactive flux occurs. The observation that increasing the vapor pressure increases the deposition rate is therefore a diagnostic for surface-limited kinetics.

Complete metallization of copper for packaging applications would not be practical as the process has been reported. Scan rates are only several micrometers per second, and the length of typical thin film wires would make process times extremely long. Scan rate concerns for complete patterning are typical for such "direct write" processes as LCVD; a strong motivation for photochemical deposition is the potential to project an ultraviolet image onto a surface and grow the entire wiring pattern directly (see below).

The kinds of limitations seen in the copper LCVD process exist, but to a lesser extent, in the gold deposition process developed by Baum and Jones

[11]. Here the precursor is a true organometallic, dimethyl gold acetylacetonate, $(CH_3)_2Au(acac)$, or one of its fluorinated derivatives (the trifluoro or the hexafluoro acac derivatives). The unfluorinated parent compound has a higher room temperature vapor pressure (8 mtorr) than does the copper compound, and also exhibits a lower decomposition temperature. Deposition rates are therefore higher for gold, and thicker spots may be prepared.

In summary, LCVD of conductive metals has been shown to depend primarily on the following parameters: precursor compound vapor pressure (the higher the better), precursor compound decomposition temperature (the lower the better, up to a point), laser intensity (the deposition rate appears to go through a maximum), substrate properties (optical absorptivity and thermal diffusivity), and total cell pressure (any buffer gas or leak reduces the rate drastically).

Copper and gold may also be pyrolytically deposited from condensed phases. While copper has been prepared by laser-induced decomposition of a paste containing copper formate [12], the bulk of the work on these metals involves localized heating of a substrate immersed in a solution containing the precursor of interest. This technique has become known as laser-assisted plating [13], and is actually a set of four similar techniques, described as follows:

Laser-assisted electroplating. Standard electroplating solutions are used, and a laser is focused on some area that is in good electrical contact with the voltage source. Plating occurs several orders of magnitude faster where the laser heating occurs than elsewhere.

Laser jet plating. Rather than immersing the substrate in electroplating solution, a jet of the liquid is trained on the area to be metallized. The laser is directed down the jet colinearly. Even greater rate enhancements over standard electroplating are possible with this method, at some sacrifice in resolution.

Laser-assisted electroless plating. No external voltage is needed; autocatalytic plating baths are used. As in the first method, plating can occur everywhere but happens much faster where the laser heats the substrate.

Thermal battery effect. The standard electroplating solution is used without the external voltage source. Localized laser heating is sufficient to disturb the equilibrium and cause plating to occur locally, and etching to occur everywhere else that is in electrical contact with the area being plated.

All four of these methods are photothermal methods. Any metal that can be electroplated can be deposited by at least two of these techniques. On occasion, the thermal battery effect may work in the opposite direction of that desired and become a photothermal etching method from solution.

Pyrolytic deposition of metals also can be initiated from solid phase precursors. In particular, Fisanick, Gross, and coworkers have demonstrated extremely fast deposition of gold lines from an organogold polymer [14]. The polymer is applied to the surface, and irradiated with a visible laser in air. The organic parts of the polymer burn away, leaving gold deposits. After the unirradiated polymer is removed by standard methods, gold lines remain. Although the thickness of such lines has been limited, scan rates are extremely fast, approaching those needed to make direct writing of metals useful for complete circuitization in packaging.

We now turn to photochemical deposition of metals. Photolytic deposition, also known as photochemical vapor deposition (PCVD), can only be accomplished from gas phase precursor compounds, because excited solids tend to relax and become hot, and the products of solution photochemistry would probably not be localized sufficiently to make deposition practical.

The precursor compounds that are useful for LCVD, as mentioned above, are obvious choices for PCVD as well since the key factor is getting a high concentration of molecules in the gas phase. Yet no photochemical deposition of useful packaging metals has yielded high-purity material. In most cases, photochemically deposited metal is loaded with carbon. Carbon contamination of metals deposited from carbonyl compounds is a general phenomenon, but the same problem is observed with the acac derivatives. Photolytic deposition of metals for packaging, therefore, remains a wide-open field with potentially enormous return on any successful experiment. If projection deposition of useful metals becomes a reality, a great deal of costly lithographic patterning can be eliminated.

ETCHING. The characteristics that make metals like copper, gold, and chromium useful in packaging applications make them difficult to etch. In deposition, the difficulty lies in finding volatile compounds that contain these metals. In etching, one must form reasonably volatile metal-containing compounds directly from bulk metal and ablate them. The number of compounds that metals like gold form is few anyway; the number that are volatile is even smaller and the number of volatile ones that can be prepared by irradiating the solid metal in the presence of some other species is even smaller.

It has been shown that copper may be patterned by laser-assisted chemical etching using halogen gases [15]. The process works as follows. The corrosive gas reacts with the copper everywhere it is exposed. The laser beam heats the copper halide surface locally, ablating the etch product and exposing new copper to react with the gas. While the entire sample is slowly corroding, the irradiated areas etch much faster due to the effect of the laser. If the sample is etched quickly enough and the gas is removed, patterns can be formed before the entire sample is irretrievably corroded.

The knowledge that copper reacts with halogens to form removable etch products has been used to create a photochemical etching method [16]. In this process, the sample is exposed to bromotrichloromethane (CCl_3Br), which does not spontaneously react with the copper. When irradiated by an excimer laser, the gas photolyzes to produce bromine atoms and chlorinated radicals that react with the surface. The laser thus initiates photochemistry as well as removing etch products. Since the starting gas is not corrosive, real packaging parts could be patterned in this manner without fear of contamination or destruction.

Except for the aforementioned result, there has been essentially no published work on photochemical etching of copper, gold, or chromium from the gas phase. Like photochemical deposition of these metals, this is an area ripe for exploitation in the packaging world. Excimer laser patterning is easily capable of achieving the dimensions of packaging structures; only the chemistry remains to be worked out.

There are scattered reports of laser-induced etching in solution. Copper has been etched in a nickel electroplating solution with an argon ion laser, and gold has been etched in aqueous iodine with a krypton ion laser [5]. In principle, since laser-assisted electroplating is known to work, laser-assisted electroetching should also be achievable. Furthermore, the principle behind laser enhancement of any process involving an activation energy should be applicable to etching; if a particular solution is known to etch a metal, localized heating of the substrate in the solution should accelerate the reaction. A number of materials have been laser-etched using this approach [5], but not metals useful for packaging.

Polymers

Polymers may easily be patterned by ultraviolet light in a process that has become known as *ablative photodecomposition* [3]. This method, which employs an excimer laser as the light source, was first demonstrated on poly(ethylene terephthalate) by Srinivasan and Leigh [17]. Since then, excimer laser patterning of a variety of polymers has been accomplished. The process is characterized by an apparent lack of melting and charring, as well as by relatively low translational temperatures (velocities) of the ejected molecular fragments. These observations suggest that the etching mechanism is purely photochemical, involving no internal conversion of the electronic energy prior to decomposition of the polymer chain. On the other hand, it has been observed for polyimide etching at three different excimer wavelengths that the amount of deposited heat per pulse required to induce etching (as determined by the observed threshold fluence and the known absorption coefficients) is identical within experimental error [18]. This observation implies that the decomposition is a pyrolytic one; otherwise, the existence of a wavelength-independent threshold absorbed energy would have to be dismissed as an amazing coincidence. A controversy has therefore built up about the mechanism of uv laser ablation of polymers, with some work supporting photolytic etching and other results suggesting pyrolytic etching.

In most cases, where two camps participate in a sustained debate backed up on both sides by experimental evidence, the answer to the question "Is it A or B?" is usually "Both," "It depends," "Neither," or "A and B are actually the same thing." The photochemistry adherents now tend to argue that at 193 nm photochemistry is occurring, while at longer wavelengths the process is indeed photothermal. Srinivasan and coworkers have obtained SEM photographs of material ejected during ablation of poly(methyl methacrylate) and found that at 193 nm, the material resembles shredded coconut, whereas at 248 nm (and higher fluence) beads of obviously melted polymer are found [19]. The melting implies that a substantial amount of heat was deposited in the material (thermal mechanism), whereas the shredding is consistent with clean, photochemical bond scission.

Although this particular laser process is difficult to classify as either completely thermal or completely photochemical, there is no question about whether it works. Etching of polymers with patterned excimer laser light proceeds at hundreds of angstroms per pulse, and these lasers run at hundreds of pulses per second. Fluences required are low (on the order of 100 mJ/cm^2). A low threshold power for a projection patterning process translates into significantly higher throughput, because the laser beam does not have to be focused as tightly to reach that fluence, so that a larger area of the substrate can be patterned at one time. In particular, if a regular pattern such as an array of holes needs to be formed, low-power etching provides wide-area drilling cheaply. In short, potential applications for laser patterning of polymers in packaging are numerous, regardless of the nature of the mechanism.

An alternative method for patterning polymers has been demonstrated by Bjorkholm and coworkers [20]. This technique fits into the "direct write" category. Frequency-doubled argon ion laser light is focused onto the polymer surface, and the beam is scanned with respect to the surface. The authors find that the threshold power for etching PMMA is essentially zero with this method, and conclude that the mechanism is photochemical.

Ceramics and Composites

It is perhaps not surprising that very little laser patterning of ceramics has been reported. Oxide materials are by nature very resistant to heat, so low-power pyrolytic processes are not expected to be effective in patterning them. Inorganic oxides in general have very strong bonds and do not undergo photochemical reactions. Laser applications in composites have therefore been limited to brute force thermal processing using high-power lasers, such as burning serial numbers into ceramic parts. Holes can be drilled in glass-epoxy circuit board materials using focused laser beams as well [21].

Lasers have also found applications in ceramics with regard to materials preparation before sintering. For example, a lengthy report on formation of silicon nitride and silicon carbide particles for use in ceramic fabrication has appeared [22]. The authors find that when a CO_2 laser is used to photolyze gases such as silane, ammonia, methane, and ethylene, particles are produced that are more suitable for ceramic fabrication than those prepared by any other method.

CONCLUSION

Extensive work that is applicable to electronic packaging has been done in the areas of pyrolytic metal deposition and uv laser etching of polymers. Complete patterning by either uv projection deposition or etching of metals remains to be achieved despite its obvious desirability. Thus, applications of lasers in which the laser is used primarily as a heat source have been well exploited, but methods in which the laser is used to induce specific and useful chemical reactions, particularly at gas/solid interfaces, still need to be invented and developed. When those methods are developed, we anticipate that lithographic processes, reactive ion etching, and evaporation and sputtering (all costly processes) will no longer be necessary in packaging. Packages will be formed by complete laser patterning, either by deposition or etching. Clearly, there is exciting work to be done.

REFERENCES

1. M. J. Weber, editor, *CRC Handbook of Laser Science and Technology*, Boca Raton, Fla.: CRC Press, 1982.
2. E. A. Giess, K.-N. Tu, and D. R. Uhlmann, editors, *Electronic Packaging Materials Science*, Pittsburgh: Materials Research Society, 1985.
3. D. Bauerle, editor, *Laser Processing and Diagnostics*, Berlin: Springer-Verlag, 1984.
4. D. J. Ehrlich and J. Y. Tsao, "A Review of Laser-Microchemical Processing," *J. Vac. Sci. Technol. B* 1, pp. 969–984 (1983).
5. Y. Rytz-Froidevaux, R. P. Salathe, and H. H. Gilgen, "Laser Generated Microstructures," *Appl. Phys. A* 37, pp. 121–138 (1985).
6. J. G. Eden, "Photochemical Processing of Semiconductors: New Applications for Visible and Ultraviolet Lasers," *IEEE Circ. Dev.* 1, pp. 18–24 (1986).
7. C. R. Moylan, T. H. Baum, and C. R. Jones, "LCVD of Copper: Deposition Rates and Deposit Shapes," *Appl. Phys. A* 40, pp. 1–5 (1986).
8. T. F. Deutsch and D. D. Rathman, "Comparison of Laser-Initiated and Thermal Chemical Vapor Deposition of Tungsten Films," *Appl. Phys. Lett.* 45, pp. 623–625 (1984).
9. D. K. Flynn, J. I. Steinfeld, and D. S. Sethi, "Deposition of Refractory Metal Films by Rare-Gas Halide Laser Photodissociation of Metal Carbonyls," *J. Appl. Phys.* 59, pp. 3914–3917 (1986).
10. F. A. Houle, C. R. Jones, T. Baum, C. Pico, and C. A. Kovac, "Laser Chemical Vapor Deposition of Copper," *Appl. Phys. Lett.* 46, pp. 204–206 (1985).
11. T. H. Baum and C. R. Jones, "Laser Chemical Vapor Deposition of Gold: Part II," *J. Vac. Sci. Technol. B* 4, pp. 1187–1191 (1986).
12. R. B. Gerassimov, S. M. Metev, S. K. Savtchenko, G. A. Kotov, and V. P. Veiko, "Laser-Induced Decomposition of Organometallic Compounds," *Appl. Phys. B* 28, p. 266 (1982).
13. R. J. von Gutfeld, "A Review of Laser Enhanced Plating and Etching for Electronic Materials," *Denki Kagaku* 52, pp. 452–459 (1984).
14. M. E. Gross, G. J. Fisanick, P. K. Gallagher, K. J. Schnoes, and M. D. Fennell, "Laser-Initiated Deposition Reactions: Microchemistry in Organogold Polymer Films," *Appl. Phys. Lett.* 47, pp. 923–925 (1985).
15. W. Sesselmann, E. E. Marinero, and T. J. Chuang, "Laser-Induced Desorption and Etching Processes in Chlorinated Cu and Solid CuCl Surfaces," *Appl. Phys. A* 41, p. 209 (1986).
16. C. R. Moylan and T. J. Chuang, "Photochemically-Induced Etching of Copper," in *Laser*

Processes for Microelectronic Applications, J. J. Ritsko, D. J. Ehrlich, and M. Kashiwagi, eds., Electrochemical Society, Pennington, NJ, 1988. Proceedings Volume 88-10, pp. 123–129.

17. R. Srinivasan and W. J. Leigh, "Ablative Photodecomposition: Action of Far-Ultraviolet (193 nm) Laser Radiation on Poly(ethylene terephthalate) Films," *J. Am. Chem. Soc.* 104, pp. 6784–6785 (1982).

18. J. H. Brannon, J. R. Lankard, A. I. Baise, F. Burns, and J. Kaufman, "Excimer Laser Etching of Polyimide," *J. Appl. Phys.* 58, pp. 2036–2043 (1985).

19. R. Srinivasan, B. Braren, D. E. Seeger, and R. W. Dreyfus, "Photochemical Cleavage of a Polymeric Solid: Details of the Ultraviolet Laser Ablation of Poly(methyl methacrylate) at 193 and 248 nm," *Macromolecules* 19, pp. 916–921 (1986).

20. J. E. Bjorkholm, L. Eichner, J. C. White, R. E. Howard, and H. G. Craighead, "Direct Writing in Self-Developing Resists Using Low-Power cw Ultraviolet Light," *J. Appl. Phys.* 58, pp. 2098–2101 (1985).

21. W. R. Wrenner, "Laser Drilling PC Substrates," *Circuits Manufacturing* 17, p. 28 (1977).

22. J. S. Haggerty and W. R. Cannon, in *Laser-Induced Chemical Processes*, edited by J. I. Steinfeld, New York: Plenum Press, 1981.

PROBLEMS

A. Photochemical Etching

Imagine that we wish to etch material M by using a laser to photolyze molecule AX_4 in the gas phase to produce etchant X_2, have that react with M at the surface to form product MX_2, and have the laser also remove the product from the surface to expose a new layer of material M.

15.1. What kind of laser would be best for our process?

15.2. Assume that we are projecting patterned light through a window into a vacuum chamber and onto a substrate. The AX_4 gas will be allowed to fill the chamber to some pressure, and then the laser will be turned on. If the cross section for light absorption by AX_4 is 2×10^{-19} cm^2, the quantum yield for the photochemical reaction is 0.83, and 90 percent of the light that hits the window gets through it, what fraction of the laser fluence incident on the window will be left to impinge on the substrate if we put 50 torr of our precursor in the cell and leave 5 cm between the window and the substrate?

15.3. Now imagine that we wish to etch M so as to produce features that are 10 μm in size. Etchant X_2 is very reactive, and will either hit the surface and react with M immediately, or will react with either AX_4, AX_2, or another X_2 to form various chemically benign molecules. Therefore, we must confine the etchant molecules so that they encounter a collision with either the surface or another gas phase molecule before they travel a distance larger than our feature dimension. The mean free path of an etchant (λ) is given by the following expression:

$$\lambda = \frac{kT}{\sqrt{2}\pi d^2 P}$$

If we want to keep our sample as far away from the window as possible to avoid getting etch product debris on it, and we want to allow as much laser fluence to hit the surface as possible so that all the MX_2 gets removed, what pressure of

AX$_4$ should we use in our chamber? Assume that X$_2$ has an effective molecular diameter d of 3 Å.

B. Photothermal Deposition

Imagine that we wish to deposit material M from gas phase precursor MX$_2$ by using a laser to heat a spot on the substrate and inducing the precursor to decompose at the hot spot, with fragments X flying away to leave M behind. We would like to form a line by scanning the laser across the surface in a continuous fashion. The substrate is a solid, opaque material with a reddish color.

15.4. What kind of laser should we use?

15.5. It has been shown that the rate of such a process can be expressed as $D[R]/(l + d)$, where D is the diffusion coefficient, [R] is the concentration of the reactant molecules in the gas phase, l is a sort of mean free path parameter, and d is the diameter of the laser focal spot on the surface. The parameter l is defined as $2D/\epsilon v$, where ϵ is the efficiency of the surface pyrolysis reaction (the fraction of molecules striking the surface that result in a deposited M atom) and v is the root-mean-square velocity of precursor molecules in the gas phase.

Inspection of the rate expression reveals that it has two limiting forms. In the first, $d \gg l$. This means that the pressure is high, so that the diffusion coefficient and mean free path for MX$_2$ are both low. This is the *diffusion-limited kinetics* regime, where the rate of the process is the rate at which precursor molecules strike the surface. The second limiting case is where $l \gg d$; at these low pressures, the rate is *surface kinetics–limited*, so that the efficiency of the pyrolysis reaction (ϵ) determines the deposition rate.

Now imagine that MX$_2$ exists as a solid compound, and that the vapor of that solid is the supply of gas phase precursor. Let the vapor pressure of MX$_2$ be given by the following expression:

$$\log P \text{ (torr)} = 15 - \frac{5 \times 10^3}{T}$$

where T is in degrees Kelvin. The diffusion coefficient is expressed in cm^2/s. It is given by the following expression:

$$D = \frac{3}{8\sigma^2[R]} \sqrt{\frac{kT}{\pi m}}$$

This formula includes [R], the concentration of molecules in the gas phase. The vapor pressure expression above gives us that information in pressure units (torr) rather than in concentration units (molecules/cm^3). Using the ideal gas law, derive the expression for the diffusion coefficient as a function of pressure P rather than as a function of concentration [R].

15.6. Now let the molecular mass of MX$_2$ be 300, and its effective molecular diameter (σ) be 5 Å. Assume that 1 out of 10 gas-surface collisions results in a deposited M atom, and let the laser focal spot diameter be 10 μm. Pressure expressed in cgs units (dyne/cm^2) is 1.33×10^3 times the value of the pressure expressed in torr. The root-mean-square velocity of a gas phase molecule is given by the following formula:

$$v = \sqrt{\frac{3kT}{m}}$$

By determining l as a function of temperature, determine the temperature that divides the diffusion-limited regime from the surface kinetics–limited regime. What conditions will prevail at room temperature?

CHAPTER
16

METAL
DEPOSITION

PERMINDER BINDRA
SOLOMON L. LEVINE
WALTER T. PIMBLEY
ROBERT R. SCHAFFER
JONATHAN REID
WILLIAM L. UNDERKOFLER
RAYMOND T. GALASCO

INTRODUCTION

Metal deposition can be performed on metallic or nonmetallic substrates by electrochemical or chemical means from solutions. In electronic packaging, metal deposition has two principal applications:

Metal finishing. This is the deposition of a thin film of a metal on a metallic or nonmetallic substrate to achieve desired surface properties without altering the bulk properties of the substrate.

Integrated circuits. The deposition of fine lines of a metal, usually copper, on a dielectric material to provide the interconnections between chips and other devices, to distribute electric power, and to generally provide a suitable operating environment for electric currents.

The deposition of a metal from a solution of cations, M^{Z+}, superficially appears to take place by a simple discharge reaction of the type:

$$M^{Z+} + ze^- \rightarrow M^0 \tag{16.1}$$

In actual fact, electrolytic deposition of metals is a highly complex process. Metal ions exist in solution not only as cations, but also as complex anions, and the deposition of the metal is likely to be a function of the following:

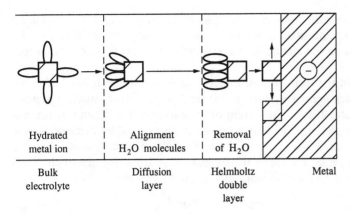

Hydrated metal ion	Alignment H_2O molecules	Removal of H_2O	
Bulk electrolyte	Diffusion layer	Helmholtz double layer	Metal

FIGURE 16-1
Schematic representation of the mechanism of metal deposition.

Processes in solution such as mass transfer.

The charge transfer reaction, e.g., the slow rate of detachment of the hydration sheath in the case of a metal cation in solution, or of the ligands when the metal exists as a complex ion in solution.

Nucleation of discrete centers of the metallic phase. The nucleation behavior is a function of the nucleate-substrate interaction.

The growth of the continuous metallic phase. Additives in solution affect the growth mechanism either directly via incorporation in the lattice, or indirectly by inhibiting the growth of certain facets with respect to others.

A schematic of the various stages of metal deposition is depicted in Figure 16-1. The overall rate of metal deposition reaction can be controlled by any one stage or the combination of two or more stages. For a systematic study of metal deposition, it is necessary to delineate the mechanism of the various stages.

In this chapter, the phenomena of nucleation and crystal growth are treated first. This is followed by a consideration of electroless metal deposition and electrolytic metal plating. Finally, some of the engineering aspects of metal deposition are described.

FUNDAMENTALS

Nucleation

Nucleation is the formation and cohesion of the very first particles of one phase from another. The phenomenon of nucleation is of particular importance in metal deposition.

In this section, the theory of homogenous nucleation and growth in vapors is adapted to the description of nucleation in electrodeposition of metals, but first, certain concepts necessary for the theory of nucleation are presented.

Most phase transformations proceed discontinuously through the formation and subsequent growth of small embryos or nuclei of new phase. When nuclei appear at active sites on the electrode, the nucleation process is termed heterogeneous.

For example, in the formation of metals, stable nuclei are formed after the application of an overpotential η, through the random growth of small aggregates or clusters of the metal atoms. The origin of an activation barrier in this process is easily seen by regarding the clustering of metal atoms as thermodynamic fluctuations. Following Gibbs [1,2], who was the first to treat quantitatively the stability of a new phase as a result of the fluctuation process, the electrochemical free energy of formation of a stationary cluster A_n of size n is

$$\overline{\Delta G} = \pm ze\eta n + \sigma H n^{2/3} \tag{16.2}$$

where z electrons are transferred per atom or molecule, e is the electronic charge, η the overpotential, σ the surface free energy, and H is a shape factor. For spherical clusters

$$H = 4\pi \left(\frac{3}{4\pi} \frac{M}{\rho N} \right)^{2/3} \tag{16.3}$$

where N is Avogadro's number, M the molecular weight, and ρ the density. The second term in Equation (16.2) represents the excess free energy necessary to form the surface of the cluster. This quantity is always positive. The first term, which represents the electrochemical free energy change for the transfer of n species from the aqueous into the metallic phase, is negative—the positive sign is taken for cathodic and the negative sign for anodic processes. As a result, $\overline{\Delta G}$ exhibits a maximum

$$\overline{\Delta G}^* = \frac{4H^3\sigma^3}{27z^2e^2\eta^2} \tag{16.4}$$

at a critical cluster size

$$n^* = \frac{8H^3\sigma^3}{27z^3e^3\eta^3} \tag{16.5}$$

Clusters smaller than this size are unstable and tend to disintegrate or dissolve; while for clusters larger than the critical nucleus, spontaneous growth takes place; see Figure 16-2.

The equilibrium number density N_n of clusters A_n is given by the Boltzmann type distribution function

$$N_n = N_0 \exp\left(-\frac{\overline{\Delta G}}{kT} \right) \tag{16.6}$$

where k is the Boltzmann constant and N_0 is defined simply as a normalization constant [3]. Clearly N_n has the minimum value for $n = n^*$, and A_n^* is the least probable species in the system.

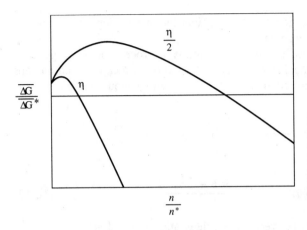

FIGURE 16-2
Plot of the electrochemical free energy of formation of clusters $\overline{\Delta G}$ against the cluster size n for overpotentials η and $\eta/2$. $\overline{\Delta G}$ and n are normalized to $\overline{\Delta G}^*$ and n^* of the critical cluster at the overpotential η.

The work of formation of the critical nucleus represents an activation barrier that must be overcome before phase change can occur. The dependence of this nucleation barrier on the overpotential is easily seen in Equation (16.4). For a spherical nucleus, the formula $r^* = (M3n^*/N\rho4\pi)^{1/3}$ and Equation (16.5) lead us to the electrochemical analogue of the Gibbs-Kelvin equation:

$$r^* = \frac{2\sigma M}{z \epsilon \eta \rho N} \tag{16.7}$$

The work of formation of a nucleus is then

$$\overline{\Delta G} = \frac{1}{3}\sigma 4\pi r^{*2} = \frac{16\pi\sigma^3}{3z^2 e^2 \eta^2}\left(\frac{M}{\rho N}\right)^2 \tag{16.8}$$

Therefore, as the overpotential is increased, the number of molecules or atoms necessary to form a cluster of critical size becomes smaller, and the nucleation barrier decreases.

All the clusters of the new phase that form must, in the process of their random growth, pass through the critical cluster A_n^*, and the rate of nucleation is equal to the net rate at which nuclei capture an atom to become aggregates of supercritical size. If the nuclei are in thermodynamic equilibrium with the single atoms and every collision of an atom with a nucleus results in a stable cluster, one obtains the equilibrium rate expression first proposed by Volmer and Weber [4]:

$$I_r = bS_{n^*}N_{n^*} \tag{16.9}$$

where I_r = rate of nucleation per unit volume, b = number of atoms impinging on a unit surface per unit time, and S_{n^*} is the surface area of a critical cluster.

Equation (16.9), in conjunction with Equations (16.6) and (16.8), gives for I:

$$I_r = bS_{n^*}N_0 \exp\left(\frac{-\overline{\Delta G}^*}{kT}\right) \tag{16.10}$$

Equation (16.10), with Equation (16.8), predicts the strong dependence of the rate of nucleation on the overpotential.

The electrodeposition of metals is essentially a process of heterogeneous nucleation. The electrochemical free energy of formation of a cap-shaped nucleus on a substrate can be described in terms of the homogeneous energy barrier, $\overline{\Delta G^*_{\text{hom}}}$, thus:

$$\overline{\Delta G^*_{\text{het}}} = \overline{\Delta G^*_{\text{hom}}} f(\theta) \tag{16.11}$$

where

$$f(\theta) = \frac{2 - 3\cos\theta + \cos^3\theta}{4} \tag{16.12}$$

and θ is the angle of contact between the nucleus and the substrate.

Crystal Growth

The evolution of any metastable state involves two consecutive steps: (1) surmounting the energy barrier that defines this condition, and (2) subsequent passage to a state of lower energy and greater stability.

Nucleation and growth are examples of these two steps, respectively, and historically they were considered to be independent of each other, nucleation being the more difficult and the slower.

Kossel [5] was the first to consider the simultaneous process of nucleus generation and subsequent growth, and proposed a natural connection between the two with his detailed model of the mechanism of crystal growth. This model, which is depicted in Figure 16-3, was developed further by Stranski [6].

The basic idea is that of two-dimensional nucleation followed by layer growth. A growth unit of atomic dimensions arrives at the surface, is adsorbed, and diffuses over the surface planes until it strikes the edge of the advancing layer, known as a *step* (see Figure 16-3). It then travels along this step until it finds the characteristic position of high coordination known as a *kink*, where it stops. This sequence of arriving from the ambient phase, diffusing across the

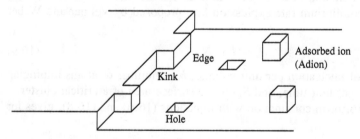

FIGURE 16-3
Kossel model of a crystal surface.

surface, colliding with a step, and finding the kink site is repeated several times by subsequent growth units.

In this way, the steps advance along the surface towards the edge of the crystal where they disappear, thus completing a layer. After building up one such layer, nucleation must again occur and a further layer is thus commenced.

This atomic model clearly suggests the unification of nucleation and growth ideas, since the rate of addition of new material to a growing surface will be a function of the frequency of surface nucleation, the step height, and its rate of advance. A closer examination of the model reveals many restrictions on this oversimplified picture.

A growth unit, for example, is not necessarily disposed to immediate incorporation in the growing base once it has arrived on location by diffusion, because it may dissolve or diffuse away or because it may not have settled at a favorable point. In the latter instance it must migrate to a more favorable point of attachment. It appears that each step line is in equilibrium with growth units adsorbed close to it, while those further away are in equilibrium with the ambient phase. The two processes are linked by surface diffusion, which in the case of growth from the vapor phase is the rate-determining step. Consideration of such restrictions are the central theme of the adsorption theories of growth of Volmer [7], Brandes [8], and Becker and Doring [9].

Electrocrystallization

It is quite possible that the study of electrocrystallization is one of the oldest parts of experimental electrochemistry. Faraday's laws were established in 1834 for application to the electrodeposition of tin, lead, and antimony onto platinum [10].

Electrocrystallization can take place in a number of situations and the most important [11] are:

1. The new phase may be formed on an inert substrate by electrodeposition from ions in solution.
2. The phase may be deposited on a substrate of the same material by electrodeposition as in (1).
3. The new phase may be formed from the parent substrate by electrodissolution followed by subsequent precipitation by reaction with the solution; alternatively, ions may pass from the substrate through the new phase to the solution interface.
4. The new phase may be formed electrochemically from a phase that is itself formed on a substrate as in (3).

Categories (1) and (2) refer principally to cathodic deposition, while categories (1), (3), and (4) refer to anodic electrocrystallization. Clearly, a process similar to (2) must always be part of the other categories and, under limiting conditions, could correspond to the growth of perfect crystal planes. The other categories will normally include the nucleation of growth centers and are more strictly analogous to crystallization from the vapor.

Although the basic model for electrolytic crystal growth is considered to be the same as that for crystal growth from the vapor, the analogy between the two processes is by no means complete. The principal difference arises from the presence of the metal-solution double layer in the electrolyte process. Other important differences have been enumerated by Bockris and Damjanovic [12] and Bockris and Razumney [13]:

The driving force in the case of crystal growth from the vapor is simply given by the supersaturation, while in the case of electrolytic growth an externally controllable force, the applied overpotential, provides the required driving force.

The rate of incorporation of particles at the crystal is slower in the electrocrystallization case than in that of growth from the gas phase because of the higher diffusion coefficient in the gas phase (10^{-1}–1 cm^2/s) compared with that in solution (10^{-5} cm^2/s).

The particles involved in electrodeposition are charged and hydrated, and a sequence of charge transfer and dehydration steps must occur. The energetics of these steps are dependent upon the applied potential difference. Passage across the double layer can be rate-determining in the electrochemical case. This could be due to the electrorestrictive distortion caused by the high field strength (10^7 V/cm) in the double layer.

Surface diffusion is likely to be slower in electrocrystallization than in the deposition from the vapor phase because of adsorption on the electrode of water molecules and ions, and because of the hydrated nature of the diffusing species.

These differences affect both metal deposition and crystal growth of non-metallic phases. In general, the tendency is for the electrocrystallization case to be slowed down and affected by more factors, at the same degree of supersaturation, and hence to be a more complicated phenomenon than that of crystal growth from the vapor.

ELECTROLESS METAL PLATING

Techniques

IMMERSION PLATING. When an iron substrate is immersed in a solution of copper sulfate or silver nitrate, the iron dissolves while the copper or the silver is plated out onto the surface of the substrate. This is called immersion plating. The process can be explained mechanistically in terms of the electrochemical series, part of which has been reproduced in Table 16.1. The electrochemical series predicts the course of a reaction when two redox systems with different values for $E°$, the standard redox potential, are brought into contact. The most electropositive redox system will be reduced (deposition) while the most electronegative will

TABLE 16.1
Standard redox potentials at 25°C

Reaction	Potential (V) vs. Standard Hydogren Electrode (SHE)
$Ag^+ + e^- \rightleftharpoons Ag^0$	+0.799
$Cu^{+1} + e^- \rightleftharpoons Cu^0$	+0.520
$Cu^{2+} + 2e^- \rightleftharpoons Cu^0$	+0.337
(pH = 0) $HCOOH + 2H^+ + 2e^- \rightleftharpoons HCHO + H_2O$	+0.056
$Ni^{2+} + 2e^- \rightleftharpoons Ni^0$	−0.250
$Fe^{2+} + 2e^- \rightleftharpoons Fe^0$	−0.440
(pH = 0) $H_2PO_3^- + 2H^+ + 2e \rightleftharpoons H_2PO_2^- + H_2O$	−0.504
(pH = 14) $HCOO^- + 2H_2O + 2e^- \rightleftharpoons HCHO + 3OH^-$	−1.070

undergo oxidation (dissolution). Table 16.1 shows that the standard redox potential for iron ($E^0 = -0.44$ V vs. SHE). Therefore, the reactions occurring on the surface of the iron substrate are as follows:

Anodic

$$Fe^0 \rightarrow Fe^{2+} + 2e^- \qquad (16.13)$$

Cathodic

$$Cu^{2+} + 2e^- \rightarrow Cu^0 \qquad (16.14)$$

or

$$2Ag^+ + 2e^- \rightarrow 2Ag^0 \qquad (16.15)$$

The surface of the metallic substrate consists of a mosaic of anodic and cathodic sites, and the process will continue until almost the entire substrate is covered with copper. At this point, anodic oxidation virtually ceases; and hence, plating is stopped. The oxidation imposes a limit on film thickness obtained by this technique, with average values of only 1 μm being observed. Other limitations on this method are that the metallic coatings produced usually lack adhesion and are porous in nature, the solution quickly becomes contaminated with ions of the base metal, and the process is only applicable to obtaining noble metal deposits.

ELECTROLESS PLATING. In electroless plating, the electrons for the cathodic reaction are provided by the oxidation of a reducing agent. This reaction must occur initially on a metallic substrate (activated substrate if the substrate is non-conducting) and subsequently on the deposit itself. The redox potential for the reducing agent has to be more negative than that for the metal being plated. The selection of the reducing agent is further limited in that the electrochemical reactions must only occur on the substrate and not cause homogeneous reduction (decomposition) of the solution. An example of solution decomposition is the addition of formaldehyde ($E^0 = 0.056$ vs. SHE at pH = 0) to silver nitrate solution, which causes spontaneous precipitation of metallic silver. A summary of

TABLE 16.2
Components of electroless copper plating baths

Reducing Agents	Complexants	Stabilizers	Exaltants
Formaldehyde	Sodium potassium tartrate (Rochelle salt)	Oxygen	Cyanide
Dimethylamine borane (DMAB)	Ethylenediaminetetraacetic acid (EDTA)	Thiourea	
Sodium hypophosphite	Glycolic acid triethanol amine		

the reducing agents that may be employed for electroless copper plating is given in Table 16.2. Theoretical conditions for the electroless plating of a metal may be determined from the Pourbaix [14] diagrams for the metal and the principal element in the reducing agent. Pourbaix diagrams (as discussed in Chapter 18) show the ranges of potential and pH over which various ions, oxides, pure metals, etc., are thermodynamically stable.

Potential-pH diagrams for the copper-water system and carbon-water system are shown in Figure 16-4(a) and (b), respectively. At potentials greater than $+0.337$ V versus SHE in acid solution, the copper ion is thermodynamically stable. For deposition of the metal, the potential of the substrate must be lowered below this value into the region where metallic copper is stable. In the carbon-water system, oxidation of the formaldehyde to formic acid or formate anion occurs at potentials below $+0.337$ V versus SHE over the entire range of pH. Thus, the electrochemical processes that could theoretically occur on the metal surface in acid solution are:

Cathodic

$$Cu^{2+} + 2e^- \rightarrow Cu^0 \qquad\qquad E^0 = +0.377 \qquad (16.16)$$

Anodic

$$HCHO + H_2O \rightarrow HCOOH + 2H^+ + 2e^- \qquad E^0 = -0.056 \qquad (16.17)$$

The two diagrams are superimposed in Figure 16-4(c). The region of potential and pH over which plating is theoretically possible is shaded.

COMPOSITION OF ELECTROLESS COPPER-PLATING SOLUTIONS. Copper is the most common metal electrolessly plated. Thermodynamic conditions for electroless copper plating are described in Figure 16-4(c), but successful electroless plating cannot be guaranteed by simply adding a solution of the reducing agent to one containing metal ions. In actual fact, local changes in pH can lead to precipitation of the metal in bulk solution instead of on the intended substrate. To overcome this difficulty, complexants depress the free metal ion

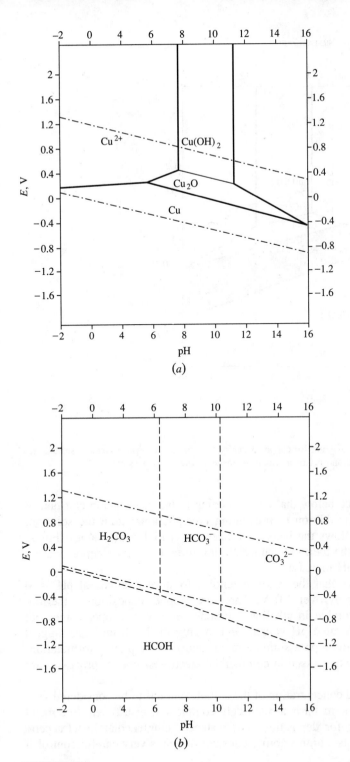

FIGURE 16-4
(*a*) Potential-pH diagram for copper-water system. (*b*) Potential-pH diagram for carbon-water system.

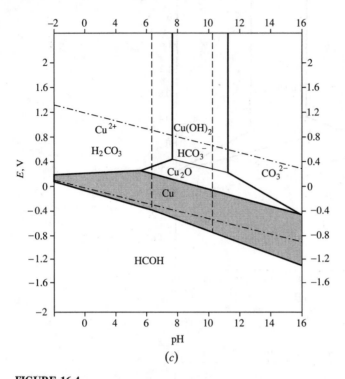

FIGURE 16-4

(*c*) Combined potential-pH diagram for copper-water system and carbon-water system. Shaded area indicates region of potential and pH over which plating is theoretically possible.

concentration to a value below that corresponding to the dissociation constant of the metal complex. In addition to preventing precipitation within the solution, the complexant also allows the bath to be operated at higher pH values. Figure 16-4(*c*) indicates that the thermodynamic driving force for copper deposition becomes greater as the pH increases.

Table 16.1 shows that the redox potentials for formaldehyde at pH = 0 and pH = 14, differs by over 1.0 V. Therefore, copper deposition is thermodynamically more favorable in alkaline solutions. However, copper comes out of solution at these elevated pH values, and to prevent this from happening, a complexant such as sodium potassium tartrate is added to the plating formulation. Some common complexants used in commercial electroless copper processes are listed in Table 16.2.

The selection and concentration of the complexant has to be considered very carefully, because if the metal is too strongly complexed, insufficient free metal ions will be available for deposition. For instance, ethylenediaminetetraacetic acid (EDTA), which has a high stability constant, requires very careful control if plating is not to cease.

Oxidation of the reducing agents employed in electroless copper plating invariably involves the formation of either hydrogen or hydroxyl ions. Conse-

TABLE 16.3
Compositions and operating parameters for commercial electroless copper baths

1. Copper sulfate	12 g 1^{-1}
Formaldehyde	8 g 1^{-1}
Sodium hydroxide	15 g 1^{-1}
Sodium potassium tartrate	11 g 1^{-1}
EDTA	20 g 1^{-1}
pH	11
Temperature	25°C
2. Copper sulfate	
DMAB	
3. Copper sulfate	
Sodium hypophosphite	

quently, the pH of the plating solution alters during plating, and thus affects the rate of deposition and composition of the deposit. Therefore, buffers are added to stabilize the pH of the solution. These include carboxylic acids in acid media (which also act as complexants) and organic amines in alkaline solutions.

Electroless plating formulations are inherently unstable and the presence of active nuclei such as dust or metallic particles can lead over a period of time to heterogeneous "decomposition" of the plating bath. The presence of the complexant in the correct concentration does not prevent this decomposition from occurring. To circumvent this problem, stabilizers such as oxygen are added to the bath in small concentrations. The stabilizers adsorb on the active nuclei and shield them from the reducing agent in the plating solution. If, however, the stabilizers are used in excess, metal deposition may be completely prevented even on the substrate itself.

The rate of plating is sometimes inordinately lowered by the addition of complexants to the bath. Additives that increase the rate to an acceptable level without causing bath instability are termed exaltants or accelerators. These are generally anions, such as CN^-, which are thought to function by making oxidation of the anodic process easier.

In summary, a typical electroless plating formulation may contain (1) a source of metal ions, (2) a reducing agent, (3) a complexant, (4) a buffer, (5) exaltants, and (6) stabilizers. Typical examples of formulations for industrial processes are given in Table 16.3.

The Mixed Potential Theory

The Wagner and Traud [15] theory of mixed potentials has been verified for several corrosion systems [16,17]. According to this theory, the rate of a faradaic process is independent of other faradaic processes occurring simultaneously at the electrode and thus depends only on the electrode potential. Hence, the polarization curves for independent anodic and cathodic processes may be added to predict the overall rates and potentials that exist when more than one reaction occurs simultaneously at an electrode. Wagner and Traud [15] demonstrated that the dissolution of zinc amalgam is dependent on the amalgam potential but independent of the simultaneous hydrogen evolution process.

The concept of mixed potentials also can be applied to an electroless plating process. For example, in the case of electroless copper plating, the anodic reaction is the oxidation of the reducing agent, methylene glycol or hydrated formaldehyde, in acid media:

$$2H_2C(OH)_2 + 2OH^- \rightarrow 2HCOOH + H_2 + 2H_2O + 2e^- \qquad (16.18)$$

$$(R^0 \rightarrow R^{z+} + ze^-)$$

And the cathodic reaction is the reduction of the metal complex:

$$[CuL_6]^{(6n-2)-} \longleftrightarrow Cu^{2+} + 6L^{n-}$$
$$\downarrow {\scriptstyle 2e^-} \qquad\qquad\qquad (16.19)$$
$$Cu^0$$

$$(M^{z+} + ze^- \rightarrow M^0)$$

The current potential curves for the anodic and the cathodic partial reactions in an electroless plating process are shown schematically in Figure 16-5. It is clear form Figure 16-5 that a necessary condition for electroless plating to occur is that the equilibrium potential for the reducing agent, E_R^0, is more cathodic than the corresponding potential, E_M^0, for the metal deposition reaction. At equilibrium,

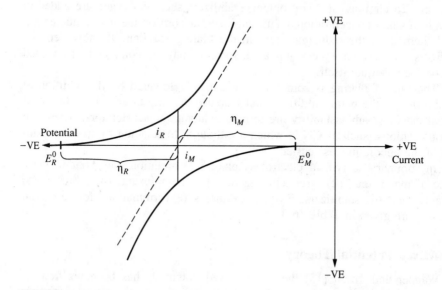

FIGURE 16-5
Current-potential curves for the partial processes involved in electroless plating. Dotted line indicates current-potential curve for the two simultaneous processes. E_R^0 and E_M^0 are equilibrium potentials.

the Wagner-Traud postulate applies and the plating rate i_{plating} is given by

$$i_{\text{plating}} = |i_R| = |i_M| \tag{16.20}$$

where i_R and i_M are the anodic and cathodic partial currents (with opposite signs). The potential associated with this dynamic equilibrium condition is referred to as the mixed potential E_{MP}. The value of the mixed potential lies between E_R^0 and E_M^0 and depends on parameters such as exchange current densities i_R^0 and i_M^0, Tafel slopes b_R and b_M, and temperature. The mixed potential corresponds to two different overpotentials,

$$\eta_R = E_{\text{MP}} - E_R^0 \tag{16.21}$$

and

$$\eta_M = E_{\text{MP}} - E_M^0 \tag{16.22}$$

If a current is passed through the cell, the measured current-potential curve is the algebraic sum of the partial current-potential curves for each electrode reaction, as shown in Figure 16-5 by the dotted curve.

APPLICATION OF THE MIXED POTENTIAL THEORY. In practice, the concept of mixed potentials is studied by constructing Evan's diagrams [18]. In Figure 16-6, a schematic Evan's diagram is constructed for the partial polarization curves shown in Figure 16-5. The coordinates of intersection of the anodic and the cathodic E versus $\log i$ relationships give the value of the mixed potential E_{MP}

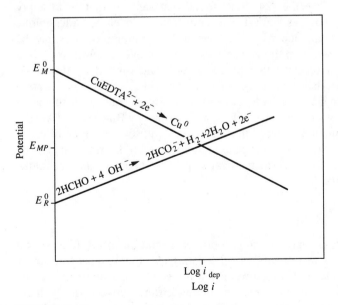

FIGURE 16-6
A schematic Evans diagram for the two partial processes of the additive copper bath.

and the plating rate in terms of a current density i_{dep}. This suggests that the dynamic relationship between the mixed potential and the copper plating rate can be obtained from the individual electrode processes if their current-potential relationships are known. Such polarization curves have been obtained by one or more of the following methods: (1) by applying the steady state galvanostatic or potentiostatic pulse method to each partial reaction separately, (2) by applying potential scanning techniques to a rotating disk electrode, (3) by measuring the plating rate from the substrate weight gain as a function of the concentration of the reductant or the oxidant [19,20]. The plating rate is then plotted against the mixed potential to obtain the Tafel parameter [21]. These methods suffer from the usual limitations associated with the theory of mixed potentials. For example, extrapolation of the polarization curve for the catalytic decomposition of the reducing agent to the plating potential is not valid if the catalytic properties of the surface change with potential over the range of interest. It also is not valid if the rate-determining step, and hence the Tafel slope, for any process changes in the potential range through which the polarization curve is extrapolated. Further, at least one of the two partial reactions involved in electroless metal plating is invariably diffusion controlled. Therefore, the weight gain method cannot be used to ascertain the plating mechanism unless the electrochemically controlled partial reaction is first identified.

A further limitation to the extrapolation of polarization curves and to the application of the mixed potential concept to electroless plating, which is often not realized [22], is that the two partial processes are not independent of each other. For corrosion processes, such a limitation was first discovered by Andersen et al. [23]. Bindra and Tweedie [24] have found that same limitation to apply to electroless plating systems as well. These authors have developed a technique based on the mixed potential theory by which electroless plating processes may be classified according to their overall mechanisms. For example, electroless plating of metals can involve a reaction that proceeds at a rate limited by diffusion. Donahue [25] has shown that the plating rate of copper, in a certain concentration range, in a copper-formaldehyde bath, is determined by the rate of diffusion of copper ions to the plating surface. The technique described by Bindra and Tweedie [24] allows a clear distinction between those reactions whose rate is controlled by the rate of diffusion of reactants to the plating surface and those reactions whose rate is limited by some slow electrochemical step. The theory of this technique is described in Appendix A.

Electroless Copper Plating

The purpose of electroless copper plating in printed circuit manufacture is to provide a means of initiating metallization on insulating substrates, most commonly in the plated through-hole, but sometimes also on the substrate surface.

The application of electroless copper plating differs between the additive and subtractive construction processes. In the subtractive process, an insulating substrate (usually epoxy-glass) with surface copper layers and drilled holes is coated with a thin layer of electroless copper. This provides a conducting layer over the surface and in the holes for electroplating. The function of the composite plated layer is to provide electrical connection to inner circuit planes exposed in the hole wall. The circuit board then is finished by a masking and etching process to create surface circuitry.

In the additive process, the entire copper layer is deposited by electroless plating. The surface circuitry is defined by a mask applied before plating, and the lines and holes are plated simultaneously. Additive electroless plating can produce uniformly thick deposits in small-bore, high-aspect-ratio holes where electroplating would give nonuniform tapered deposits. In addition, finer surface circuitry can be produced than by etching processes. A discussion of the various processes also is provided in Chapter 12.

OPERATING CHARACTERISTICS. To initiate plating, the substrate surface must be rendered catalytic by treatment with a seeder, usually adsorbed palladium as discussed in Chapter 17. Once initiated, plating continues on the copper surface, which is itself catalytic (autocatalysis).

In practice, all electroless baths manifest some instability in the form of plating on tank surfaces or on uncatalyzed areas of the substrate (extraneous plating). Ultimately, the bath must be removed and the tank etched. However, modern baths with careful control are capable of operating for many weeks before tank etching is required.

On the other hand, too much stability is undesirable. Overstabilization results in depressed plating rate, incomplete coverage (plating voids) and ultimately cessation of plating (passivity). Electroless systems are highly sensitive, and satisfactory operation can be obtained only within narrow ranges of operating parameters. Tight process control is necessary.

Electroless systems also are sensitive to contamination. Depending on the source, contaminants can destabilize or overstabilize the bath. Purity of feed chemicals and cleanliness of the operating environment is of paramount importance.

Structure and properties needed in the deposit are dependent on the application. In panel electroplating applications, the thick electroplated layer provides the bulk of the mechanical properties and electrical conductivity. The electroless layer need only be good enough to provide a sound interface in the copper to copper-plated joint. In additive plating, good copper structure and mechanical and electrical properties are important, since the entire deposit is plated electrolessly. The current state of the art in electroless copper chemistry is represented by numerous patents [26–49].

COMMERCIAL PRACTICES. Commercial plating vendors provide electroless formulations and pretreatment systems capable of meeting most applications. Selection is based on process and product requirements: thickness required, process time/plating rate, bath loading, batch or continuous operation, automatic or manual control, environmental considerations, etc.

Low-build electroless plating is used in continuous process lines where the product is transferred directly and immediately from electroless plating to the electroplating tank. The electroless deposit is thin (15–30 microinches) since it needs only to provide initial electrical conductivity for electroplating. The process time for electroless deposition is typically about 10 minutes. Such thin deposits cannot survive process interruption without surface oxidation. Low-build baths are tailored for high stability and reliability, long life, and operation with minimal analysis and control.

High-build electroless baths are used where there is a need for process interruption between electroless plating and electroplating (because of parts stocking, intervening photomasking, etc.). Typically, deposits of greater than 100 microinches are plated in 30 minutes or less. At these increased plating rates, operation becomes more critical, with increased requirements for analysis and control. Chemical consumption is faster and automatic controllers and continuous replenishment are often used. When resuming processing, parts are recleaned using acids and etchants to deoxidize and prepare the surface to accept electroplating. The thicker deposit makes cleaning with some surface removal possible without cutting down to the substrate surface.

Additive electroless plating imposes the most stringent process requirements. Because the entire deposit is plated electrolessly, priority is given to product quality requirements. Good deposit metallurgy and good mechanical and electrical properties are essential. In addition, the bath must be very stable to avoid extraneous plating on the product that could result in line-to-line shorting. As a consequence, low plating rates, and hence long plating times, are accepted. Typical rates are 50 to 100 microinches per hour. Depending on thickness required, the plating cycle can be as long as 24 hours.

Additive baths adjusted for stable plating on copper usually suffer from difficulties in initiating plating on the palladium-catalyzed substrate surface. To initiate plating, the bath can be temporarily destablized during the initiation "take" period, or alternatively, take can be established in a separate, more active strike bath, and the piece then transferred to the stable bath for thickness build.

ELECTROLYTIC METAL PLATING

Production of most circuit boards and cards, as well as the surface finishing of connectors used in computer components, involves electroplating processes. All electroplating processes utilize solutions containing dissolved salts that react, during passage of an electrical current through the bath, to form the desired metal deposit. Generally, electroplating processes are fast and inexpensive, form desirable metallurgy, and are easily controlled in comparison with competing electroless plating processes.

Fundamentals

All electroplating processes involve the concepts of Faraday's law, current efficiency, and cell overpotential. They are described below and in more detail in references [50–53].

Faraday's laws of electrolysis [10] state that the total amount of chemical change occurring at the electrodes of a cell is directly proportional to the amount of electricity (coulombs) passed:

$$W = \frac{QA_w}{Fn} = \frac{ItA_w}{Fn}$$ (16.23)

where W = weight of deposit in grams
Q = coulombs (amp-seconds)
I = current flow in amps
t = time in seconds
F = Faraday's constant; 96,500 coulombs
n = number of electrons transferred
A_w = Atomic weight

Current efficiency is the proportion of current passed in an electrode reaction to total required for all reactions. For a plating solution or process that is 95 percent efficient, 95 percent of the current is used to deposit the metal ion and 5 percent for hydrogen evolution or some other cathodic process. With respect to all the electrolytic or Faradaic processes occurring in solution, the sum of all efficiencies must be 100 percent.

Every reaction at the electrode has a standard potential; the portion of the reaction at the cathode,

$$M^{2+} + 2e^- \rightarrow M^0$$ (16.24)

is called a half-cell reaction. The standard potential E^0 is measured versus the *normal* or *standard hydrogen electrode* (NHE or SHE), for which $E^0 = 0.00$ V by definition.

The equilibrium set up between the metal and its ions in solution is a measurable parameter called electrode potential [51] defined by the Nernst equation:

$$E = E^0 + \frac{RT}{nF} \ln \frac{a(\text{products})}{a(\text{reactants})}$$ (16.25)

where E = observed electrode or Nernst potential in volts
R = gas constant
T = absolute temperature
a = activity

E^0, n, and F have been previously defined. For many practical purposes, the activity can be approximated by concentration.

When a potential is applied to an electrode, ions or dipoles (atoms or molecules having equal, but opposite, electrical charges separated by some finite distance; water is a dipolar molecule) of the opposite charge line up near the electrode surface as shown in Figure 16-7. This line-up of charges is known as the electrical double layer; it is basically a single-plate capacitor having a capacitance of up to 100 μF/cm^2.

During plating or passage of faradaic current across an electrochemical cell, the cell potential departs from its equilibrium value by an amount characterized as the overpotential or cell polarization.

The overpotential causes a reaction such as that given in Equation (16.24) to proceed at an appreciable rate. The overpotential is calculated

$$\eta = E_{cell} - E \qquad (16.26)$$

where η = overpotential in volts
E_{cell} = cell potential when current is flowing
E = Nerstian potential from Equation (16.25)

In general, plating systems are chosen to plate at an appreciable rate at the lowest possible overpotential, thus reducing power supply voltage requirements. The overpotential is a sum of several overpotentials including terms for activation

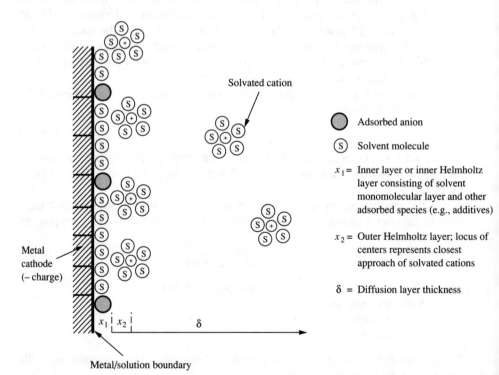

FIGURE 16-7
Model of electrical double layer.

overpotential; concentration overpotential; resistance, or ohmic, overpotential; and hydrogen overvoltage.

Activation overpotential is the portion associated with overcoming the reaction's activation free energy; i.e., the shift in energy level required to move ions across the double layer to form the deposit. It also has been called the charge transfer overpotential. The activation overpotential is related to the current by the Butler-Volmer equation:

$$i = i_O \left[\exp^{(-\alpha F n \eta_{act}/RT)} - \exp^{-((1-\alpha)nF \eta_{act}/RT)} \right] \qquad (16.27)$$

where η_{act} = the activation overpotential
$\quad i$ = net current density
$\quad i_O$ = exchange current density (the current flowing at the equilibrium or rest potential where no actual reaction is occurring)
$\quad \alpha$ = transfer coefficient

For a simple cathodic reaction, Equation (16.27) is the Tafel equation:

$$\eta_{act} = \frac{RT}{\alpha n F} \ln i_O - \frac{RT}{\alpha n F} \ln i \qquad (16.28)$$

indicating that over a large range, the activation overpotential is a logarithmic function of the applied current density. In some cases, η_{act} is a small value and can be neglected (unless the current density is very large) such as in the plating of Ni, Co, or Fe.

A more significant portion of the overpotential is the concentration overpotential. As the reduced species (i.e., the deposit) forms for some finite time, the concentration of ions near the cathode (in the diffusion layer) depletes. As this occurs and a concentration gradient builds up (i.e., concentration polarization occurs) as shown in Figure 16-8, the Nernstian potential becomes

$$E = E^0 + \frac{RT}{nF} \ln C_O + \eta_{conc} \qquad (16.29)$$

which can be reduced to

$$\eta_{conc} = \frac{RT}{nF} \ln \frac{C_e}{C_O} \qquad (16.30)$$

where C_e is the concentration of metal ion at the electrode surface and C_O is the concentration of metal ion in the bulk electrolyte.

For a cathodic process such as plating, $C_e \ll C_O$ and the concentration overpotential is negative. It is interesting to note that additives often tend to increase the concentration overpotential.

The rate of mass transfer from the bulk solution through the diffusion layer to the electrode surface determines the concentration overpotential. There are three modes of mass transfer:

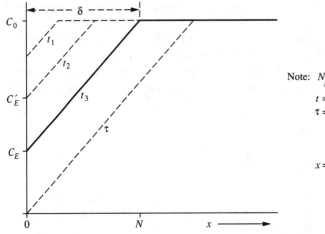

Note: $N_M = \dfrac{D(C_O - C_E)}{\delta}$

t = Time

τ = Time when species is being reduced as soon as it reaches electrode surface and current is limiting

x = Distance from electrode surface

FIGURE 16-8
Concentration gradient.

Diffusion—the natural, but slow and random, diffusion of the ion through the solution to the diffusion layer and then to the electrode under the influence of a concentration gradient. The region next to the electrode where the concentration differs form the bulk solution concentration by over 1 percent is the diffusion layer (as shown in Figure 16-8) and can be 15,000–200,000 times thicker than the double layer.

Ionic migration—i.e., the movement in the electric field, which is also a slow process.

Convection—both thermal and forced, via stirring of some sort. This also tends to reduce the concentration overpotential by reducing the diffusion layer thickness, making C_e and C_O from Equation (16.30) closer to being equal.

Another contribution to the overpotential is the resistance or ohmic overpotential—that portion required to overcome the ohmic resistance (iR drop between the anode and cathode) of the plating electrolyte.

The overpotential is comprised of several other minor constituents. The plating rate and efficiency is a tradeoff between minimizing the overpotential and maximizing the chemical reaction kinetics. Some of these are summarized in Table 16.4. As is apparent in looking at this table, the interactions of many items related to plating rates, the process parameters, and deposit properties must be investigated to provide the optimum process for a given part and metal deposit.

One other potential must be mentioned in discussing the plating of metals—the hydrogen overvoltage. In both acid and alkaline solution, the Nernstian potential for hydrogen evolution is more positive than those for many of the commonly plated metals. For example, at a pH of 10, this potential is -0.59 V.

TABLE 16.4
Minimizing overpotential/maximizing reaction kinetics

Item	Action
Reduce ohmic drop and resistance overpotential	Increase electrolyte concentration Adjust anode/cathode area ratios and spacing (better current density distribution)
Aid mass transfer and reduce mass transfer overpotential	Increase metal ion concentration Forced convection (stir/agitate) Thermal convection (heat)
Increase reaction kinetics	Heat; stir Increase current density
Reduce concentration overpotential	Increase bulk metal ion concentration Lower current density

However, the activation overpotential for this reaction is high; this inhibits the hydrogen reaction and allows the metal deposition to occur.

Plating Techniques

Electroplating requires a current at the electrode in order to provide energy for the charge transfer reaction and charge the double layer. Energy can be provided by using a constant current (galvanostatic), a constant voltage (potentiostatic), or a periodic reversal of current or voltage (pulse). The total current (i_t) passed during plating processes is the sum of Faradaic (i_f) and double-layer charging (i_{dl}) components:

$$i_t = i_f + i_{dl} \qquad (16.31)$$

Double-layer charging current normally are negligible in comparison to Faradaic currents during steady-state galvanostatic or potentiostatic plating. During pulse plating, double layer charging currents may equal or exceed Faradaic currents, depending on the pulse parameters chosen.

CONSTANT VOLTAGE (POTENTIOSTATIC). In potentiostatic plating, the concentration of reducible species at the electrode surface is a function of the applied potential chosen for the plating operation. The current that occurs at the chosen potential is determined by the concentration of reducible species in the diffusion layer region and can therefore decrease with time and vary with convection conditions. Current efficiency decreases as applied potential increases if species in solution other than the ion being deposited become reducible.

CONSTANT CURRENT (GALVANOSTATIC). In galvanostatic plating, the surface concentration of reducible species is a function of a known current or flux and the resulting concentration gradient. As the concentration of the metal ion at the cathode is depleted, the cell potential may increase and the overall current efficiency may decrease if additional species in solution become reducible at the electrode. In general, galvanostatic plating is more useful when plating must occur over a long period of time since it allows simple precise control of the amount of metal deposited.

PULSE PLATING. In galvanostatic or potentiostatic plating, the continuously applied current or voltage can lead to edge thickness buildup, porosity, roughness, and even, on occasion, nodular plating. The same average current or voltage applied in periodic pulses provides a more uniform, smooth deposit with less porosity [54–58]. In fact (due to the smaller grain sizes and, therefore, tighter packing of the crystals for a denser deposit), thinner pulse-plated deposits show the same porosity as 30 percent thicker plated dc deposits.

Figure 16-9 shows the types of pulse plating, ranging from normal to periodic reversal pulse plating. Key parameters in each are the *duty cycle R_d*, defined as

$$R_d = 100\% \frac{t_{on}}{t_{pulse}} \tag{16.32}$$

where t_{on} is defined as shown in Figure 16-9(a) and t_{pulse} is the total pulse cycle time or the sum of t_{on} and t_{off}; and the *average current density i_{av}*:

$$i_{av} = \left(\frac{R_d}{100}\right) i_p \tag{16.33}$$

where i_p is the peak current density, usually 2 to 10 times higher than the average.

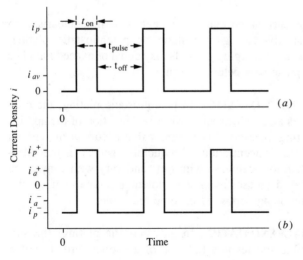

FIGURE 16-9
Typical pulse plating wave forms: (a) normal pulse; (b) periodic reversal.

Applications of Plated Metals

Perhaps the first commercial plating application was a silver plating solution (similar to those in use today) described in an 1840 patent; common platings in the early 1900s were Ni, Ag, Cu, Au, and "brassing." This section will provide a short introduction to the common electroplated metals [59–68] and their use in electronic packaging. The general references can provide additional detail as well as articles from various journals. Additional data on the properties of plated metals has been compiled by Safranek [69].

DECORATIVE/PROTECTIVE COATINGS—Cr, Ni, AND Cu. Although commercially important, it is interesting to note that good commercial Cr plating processes have only been available since about 1925. Even with improvements over the years, Cr plating costs are relatively high because of the low (10–25 percent) current efficiencies for the plating (high voltages are required to obtain current densities to plate at reasonable rates). Also, the throwing power (ability to plate non-uniform areas evenly) of most Cr baths is poor.

Nickel is one of the most commonly plated metals, due to the ease of both electro- and electroless plating and to its ability to be easily plated as an alloy. Thin deposits of Ni on metallized plastics or other base materials are very bright, durable, and ductile.

A major use of Ni plating on electronic connectors (see Chapter 7) is as an underlayment for a precious metal plating over the base copper, Be-Cu, or phosphor bronze.

The Ni layer serves two purposes:

Cu can diffuse through gold deposits; the Ni acts as a barrier to this diffusion, which in turn allows a thinner (less costly) layer of gold to be used, and maintains a corrosion-free surface.

Many precious metal plating baths are sufficiently acidic or basic so as to dissolve small amounts of Cu prior to the initiation of plating. These baths are also contaminated by Cu; since this does not occur with Ni, the lifetime of the precious metal plating solution is increased.

Electroplated Ni can vary widely in its properties—hardness, ductility, stress, and tensile strength are functions of the plating solution. A comparison of these properties for some Ni baths is shown in Table 16.5.

Copper is perhaps the most commonly plated metal because of its decorative, protective, and functional uses. Other than silver, less costly and much more abundant Cu has the highest electrical conductivity (1.59 milliohm-cm for Ag versus 1.67 for Cu) of any metal. This property leads to the major use of plated Cu on steel wire, in circuitry on glass-epoxy for printed circuit boards (PCB) and the through-hole interlayer connections on these PCBs.

In air, Cu forms a mixed oxide and sulfide tarnish that is not easily wet with solder. For most electronic applications, Cu is protected via tin or tin/lead

TABLE 16.5
Comparison of properties for Ni baths

Solution type	Hardness (Vickers)	Tensile stress, MPa	Elongation, % in 2 in.	Residual stress, MPa
Watts	140–160	380	30	125
Chloride	230–360	690	20	275–345
Sulfamate	250–350	620	20–30	3.5
Hard (sulfate)	350–500	1030	5–8	300

(solder) plating or dipping, or is protected by some other means of corrosion inhibition.

Copper electroplating solutions containing oxalate, nitrate, cyanide, chloride, acetate, and fluosilicate salts were used in the past, but the only solutions widely used today are alkaline copper pyrophosphate and acidic copper sulfate/sulfuric acid or copper fluoborate/fluoborate acid [73–75]. Typical operating conditions and compositions of these baths are shown in Table 16.6. Each bath type may be used for electrodeposition over a wide range of current densities, but deposit metallurgy may vary. The circuit processing industry needs a relatively thick, ductile, and rapidly formed deposit, and, as a result, is trending toward exclusive use of copper sulfate/sulfuric acid (acid copper) baths.

In this section, emphasis is placed on electrodeposition from acid copper baths; much of the material is, however, applicable to deposition from alternate copper electroplating baths. Although acid copper bath operating conditions,

TABLE 16.6
Bath operating conditions and compositions

	Sulfate	Fluoborate	Pyrophosphate
[Cu] g/L	15 to 60	60 to 120	20 to 40
Common anions, g/L	150 to 300 $SO_4^=$ 0.03 to 0.15 Cl^-	200 to 400 BF_4^- 15 to 30 $BO_3^=$	150 to 250 $P_2O_7^{4-}$ 15 to 30 NO_3^-
pH	−0.3 to 0.5	0.2 to 1.7	8 to 9
Specific gravity	1.1 to 1.4	1.1 to 1.4	1.1 to 1.2
Resistivity (ohm-cm)	2 to 5	5 to 10	10 to 25
Operating temperature,°F	55 to 120	65 to 120	110 to 140
Anodes	Phosphorized	Oxygen free or rolled electrolytic	Oxygen free or rolled electrolytic
Cathode current (ASF)	20 to 300	75 to 350	10 to 75
Anode current (ASF)	5 to 50	50 to 200	10 to 50

* ASF = Amps per square foot

precise composition, and trace organic additive content have continually evolved, the bath and procedure still strongly resemble the historical descriptions and patents of the 1830s [76].

The primary function of sulfuric acid in acid copper baths is to provide a high solution conductivity, typically about 0.5 mho · cm^{-1}. Specific conductivity increases almost linearly with acid concentration from 25 to 150 g/liter [77]. Concentration increases beyond 150 g/l yield progressively smaller conductivity changes owing to bisulfate species formation. Increased acid concentrations generally reduce deposit grain or crystal size [78], but cause little change in cathode polarization.

Copper sulfate is a convenient source of cupric ions. Concentration increases tend to reduce cathode polarization and may increase solution resistance if bisulfate species are formed. Too low a copper concentration leads to hydrogen evolution and reduced current efficiency at the cathode as current density is increased. Anode passivation or excessive polarization at high current densities occurs at high total sulfate concentrations. This is due to precipitate film formation at the anode when the solubility product of copper sulfate is exceeded.

Mechanism and kinetics. Copper is largely present as Cu^{2+} in all acid copper plating baths. Trace concentrations of Cu^{+} species often are present in complexes formed with organic addition agents. The overall reduction reaction may be considered as a three-step process:

$$Cu^{2+} + e^- \rightarrow Cu^+ \tag{16.34}$$

$$Cu^+ + e^- \rightarrow Cu^0 \tag{16.35}$$

$$Cu^0 \rightarrow Cu \text{ (lattice)} \tag{16.36}$$

Steps 1 and 2 are redox processes, while step 3 involves incorporation of the neutral atom into a lattice position. Under conditions typical of acid copper plating baths, the cupric ion reduction step has been shown to be relatively slow and rate-limiting in comparison to cuprous ion reduction or lattice incorporation steps [79]. Rotating disk electrode studies [80,81] have shown that step 2 is highly reversible, and the homogeneous disproportionation of Cu^{+} is insignificant. At overpotentials much smaller than those normally employed, it is generally thought that electrocrystallization steps limit the overall rate [82].

The total overpotential at the cathode can be divided into three parts corresponding to each reaction step. This overpotential results from both departures of interfacial species concentrations from equilibrium values as well as actual charge transfer overvoltages.

Assuming rapid Cu^{2+} diffusion and mass transfer, the steady-state current-potential relationship for copper deposition at typical practical conditions is approximated by

$$i = i_0 \exp -\left(\frac{\alpha n F \eta}{RT}\right) \tag{16.37}$$

Experimental data from ln i versus E plots has shown α to be approximately 0.5 for cathodic slopes. Values of i_O have ranged from 0.02 to 30 mA/cm^2 for various copper crystalline substrates [83–85].

It is not clear whether the reported values vary due to actual redox reaction rate differences or to changes in growth site density and surface character. Exchange current density in acid copper baths may be expressed in terms of bulk Cu^{2+} activity using the equation

$$i_O = nFAk_OC_O\exp^{-}\left(\frac{\alpha nF\eta}{RT}\right) \qquad (16.38)$$

where k_O = redox rate constant for $Cu \rightleftharpoons Cu^{2+} + 2e$
  A = electrode surface area
  C_O = bulk solution concentration of Cu^{2+}

Exact analysis of current-potential relations for pulse or ac current conditions are complicated by the existence of two electrode reactions [86].

To obtain desirable metallurgy and avoid hydrogen evolution, acid copper baths are usually operated at less than 50 percent of the limiting current i_M^D;

$$i_M^D = \frac{nFADC_O}{\delta} \qquad (16.39)$$

where D is the Cu^{2+} diffusion coefficient.

High limiting currents for a given bath composition are achieved by decreasing the diffusion layer thickness δ by vigorous agitation. Air sparging and panel motion are most frequently used, but solution jets directed at the cathode allow substantially higher currents. Increased cupric ion concentration clearly yields a proportional increase in limiting current.

The point of zero charge of copper metal in sulfuric acid solutions is approximately 0.08 V versus SHE [87]. Copper metal rest potentials in an acid copper bath vary in a Nernstian manner with Cu^{2+} concentration and are influenced by organic additive absorption. Cathode potentials are typically 100 to 200 mV negative of the rest potential during plating.

Acid copper bath chemistry. Acid copper baths can be divided between "high-speed" and high-throw" compositions [87,88,89]. The high-speed type is used to plate circuit board products without holes or with relatively low (1:1) aspect ratio holes. These baths operate at current densities of 100 to 200 mA/cm^2 and contain 20 to 50 g/L sulfuric acid and 50 to 150 g/L Cu. Baths are formulated to minimize hydrogen evolution even at very high current densities. In these baths, current changes vary sharply with overpotential (owing to the high Cu^{2+} concentration). The relatively low hydrogen ion concentrations yield poor solution conductivity. As a result, deposit thickness decreases sharply on surfaces with unusually large anode-to-cathode distances, such as in the center of holes. The deposit thickness formed using high-speed baths may be approximated by using a primary current distribution. These baths are usually operated at temperatures from 25 to 40°C.

High-throw baths are designed for plating circuit boards with aspect ratios of up to 20:1. These baths are formulated to minimize solution resistance and exhibit a small change in current with interfacial overpotential. Current densities of 10 to 50 mA/cm^2 and compositions containing 10 to 25 g/L Cu and 150 to 250 g/L sulfuric acid are typical of high-throw baths. Operating at or below room temperature usually improves thickness distributions as a result of the increased interfacial resistance. When modeling the deposit thickness distribution in a high-throw bath, one must consider secondary current distribution, since both kinetic and geometric factors are in operation.

Nearly all acid copper baths contain chloride ion at 30 to 100 parts per million. Chloride precipitates excessive heavy metal accumulations and influences deposition kinetics through adsorption at the cathode [90].

The organic additives used in acid copper baths contain chemicals commonly classified as "brighteners," "levelers," and wetting agents or "carriers." The additives adsorb competitively at the copper cathode and strongly influence deposit metallurgy and thickness distribution. Although the effective concentration range of these additives is often as large as a factor of ten, inappropriate concentrations or component concentration ratios will yield an unacceptable deposit. The overall effect of additives on metallurgy is to disrupt the large columnar grain produced in the absence of additives and to form a fine equiaxial grain structure.

Carriers are usually polyether or polyoxyether molecules with molecular weights from 2000 to 6000. Polyethylene glycol and polypropylene glycol have traditionally been used as carriers. These polymers form an adsorbed layer at the cathode which uniformly polarizes the interface and influences the adsorption of other additive components. The increased interfacial polarization improves throwing power and thus surface/hole thickness distributions [91].

Levelers are generally nitrogen-containing surfactant molecules including amines and amides that are protonated in acidic solution. Because of their positive charge, these molecules adsorb most strongly at edges and surface irregularities where a large interfacial potential drop occurs. This decreases the size of surface topography features that are present prior to plating [92].

A wide variety of molecules are mentioned in patents and the general literature as brighteners [93,94]. Most frequently, these contain pendant sulfur atoms (thiols, disulfides, mercaptans) and a functionality to increase water solubility. These molecules adsorb very strongly at the cathode owing to the pendant sulfur atom's high affinity for copper. Solution concentrations of only about 10 parts per million are required to achieve the desired degree of surface coverage. Brighteners frequently participate directly in the copper redox and/or electrocrystallization processes and thus strongly influence deposit grain structure and metallurgy.

Properties of acid copper deposits vary sharply depending on plating conditions and additive chemistry [95,96]. Vickers hardness ranges between 50 and 80, and elongations between 8 and 39 percent have been reported. Tensile strength varies from 12 to 63 kg/mm^2, and stress varies between 0.07 and 1.0 kg/mm^2. Ductility, conductivity, and deposit thickness are frequently tested following plating processes [97,98]. For circuit board applications, a low-stress deposit with high elongation is desirable. Hardness and tensile strength are not normally critical.

Since additive components decompose by chemical and electrochemical reactions, a control method is required. Usually additions are made based on total amp-hours passed and the duration of bath shutdowns. Additions are adjusted based on results of a Hull cell method in which copper is deposited on a brass plate over a range of current densities [99]. The Hull cell method is based only on deposit appearance and is therefore nonquantitative for any additive component. However, it has proven more useful than electrochemical [100] liquid chromatography and ion exchange chromatography methods, all of which have been difficult to apply in production situations (see Chapter 21 for discussions on Hull chemical analysis). The use of phosphorized copper anodes appear to slow the rate of oxidative additive decomposition at the anodes. A dense black film forms on and adheres to the anode during plating and presumably limits diffusion of additive components to the anode surface.

TIN AND LEAD PLATING. Tin is easily alloyed, and solders are the most common alloys of tin and lead that are plated. Another alloy becoming more widely used is Sn-Ni (usually 65 percent Sn). This material does not exist on thermal equilibrium phase diagrams and is produced only by electrodeposition; it is relatively inexpensive and very corrosion resistant in outdoor and marine atmospheres.

THE PRECIOUS METALS. Gold, silver, and the other platinum group metals are used not only for their well-known decorative properties, but also for their chemical and physical properties. Although expensive and relatively rare, they are noble metals and used for their electrical conductivity and their good wear and corrosion resistance properties.

The precious metals other than gold and palladium (osmium, platinum, iridium, ruthenium, and indium) have found little use in electronic packaging and are not commonly plated. The exception is the use of Pt-plated niobium or titanium anodes for gold or palladium plating, primarily to reduce the cost of using pure Pt.

There was a flurry of activity on Ru as a possible connector material; but its high cost, brittleness, and higher contact resistance have slowed such activity. The details of the use of gold and palladium are covered in the next section, on plating of electronic connectors.

It should be noted that gold usually is plated from acid cyanide solutions with deposits ranging from soft to hard, with bright ductile finishes offering excellent corrosion resistance and good electrical properties. Soft gold is defined as 24-carat or 99.99 percent minimum purity with a Knoop hardness in the 60–90 range. It is easily soldered, but has very poor wear properties. Major uses are for chip site solder pads, for strikes (flashes or thin deposits sometimes used to promote adhesion or protect a plating bath from contamination), for hard gold deposits, and as ir reflectors (for aerospace applications where gold, even in very thin layers, is the best ir reflection material). A layer of soft gold as thin as 0.1

μm will produce the gold color finish—such as on costume jewelry—but it may not last long.

Industrial hard gold is 99–99.8 percent gold alloyed with Ni or Co as the hardening agent (Fe and Th can also be used) and has the properties of being quite durable, with ductile deposits in the 170–250 Knoop hardness range. In electronic connector applications, good wear, friction, electrical, and corrosion properties are provided by layers ranging from 0.75–2.5 μm.

Decorative hard gold can range from 10 to 21 carats (41.6–87.5 percent) Au and usually is alloyed with Rh or Ir; colored (green) or antique gold may contain Ag, Cu, or Pb. White gold is alloyed with Ir. Government regulations on the gold content are quite strict.

With only a few exceptions (e.g., a sulfite process), gold is plated from acid cyanide solutions. Most commercial baths contain some proprietary ingredients to tailor the deposit (one vendor, for example, offers over 25 slightly different formulations), but all are based on the Au(I) cyanide complex.

Acid formulations (generally citric acid) allow the highest purity (99.999 + percent) gold to be plated without attacking delicate substrates such as plastics and ceramics. The disadvantage: Soft gold from such solutions can be plated at close to 100 percent current efficiency, but 99 percent gold is plated at only 25–40 percent efficiency. Also, gold anodes cannot be used to replenish the baths, so additions of gold is via potassium gold cyanide.

ELECTRONIC CONNECTOR PLATING. In electronic packaging, low-voltage, low-contact-force, high-reliability connectors are often used to connect the module or chip to the card, cards to boards, and cables to other subassemblies (see Chapter 7 for a comprehensive discussion of connectivity design). These take the form of pins in through-holes, pad-to-pad connectors in surface mounting, spring-to-pin or edge-to-tab connectors, or even relays such as in telephone gear. Several properties are desired in a connector:

Low contact and mated bulk resistance to meet circuit requirements

Corrosion resistance to prevent environmental failure

Low friction forces with good wear properties for repeated reliable insertion

Adequate spring characteristics

Cost

To meet all the requirements, the connector usually consists of a hierarchy of metals beginning with a base metal of copper, phosphor-bronze, beryllium-copper, iron-nickel alloys, or even pure nickel; except for the latter two, generally an underlayment of nickel is plated for the reasons previously noted in the discussion of Ni above. The most common top surface metallurgy to meet the corrosion, wear, and resistance properties is gold, generally hard gold.

Gold has two disadvantages: (1) hard gold has reduced solderability, which is easily overcome by adding a flash layer of soft gold; and (2) gold is expensive.

TABLE 16.7
Comparison of properties of Au and Pd

Property	Gold		Palladium
	Soft	Hard	
Density (g/cm^3)	19.32		12.02
Hardness (kg/mm^2)			
Range	60–100	130–200	200–400
Average	70	180	300
Resistivity (microhm-cm)	2.1	3.8	9.9

There have also been scattered data indicating poor resistance to creep corrosion in sulfide atmospheres.

The cost factor has led to extensive work on alternatives to gold for connector applications; palladium has found increasing use for many connector applications. Table 16.7 compares the properties of Au and Pd; cost is the most important difference. In addition to the fact that the raw metal cost is approximately one-third that of gold, the density difference allows for a lower weight of Pd to be used for an equivalent volume.

Prior to the late 1970s, Pd plating was limited by processes producing dull, porous deposits that often cracked due to entrapped hydrogen. Also, when mated against itself, Pd tended to form insulative frictional polymers by micromotion fretting corrosion. Newer plating baths produce ductile, nonporous, crack-free deposits. IBM was a leader in this area, utilizing Pd-plated connectors in equipment in the late 1970s.

More recently, Pd alloys—particularly Pd-Ag and Pd-Ni—have received much attention. The latter, when overplated with a thin (0.075–0.25 μm) soft gold flash has shown exceptional wear properties and is comparable to gold in terms of corrosion resistance and contact resistance.

ENGINEERING ASPECTS

Fluid Flow Considerations

MOTION OF THE LIQUID ELECTROLYTE. In order for metal deposition to take place, the ion species in the electrolyte must be transported to the solid surface. There are three mechanisms for this transport: ionic migration, diffusion, and bulk movement of the liquid electrolyte. The gradient of the electrochemical potential causes ionic migration, which is usually negligible due to excess supporting electrolyte. Diffusion provides most of the motion of the ionic species when bulk motion of the liquid is absent. However, bulk motion of the liquid can be much more effective in moving the ions, except in the immediate vicinity of a solid boundary. At solid boundaries, the liquid velocity is zero relative to that boundary. Therefore, in the vicinity of such a boundary, diffusion and bulk liquid motion are of interest. The study of the motion of the liquid electrolyte in the vicinity of the solid boundaries to be plated is important to the understanding of the plating process. Many of the methods of causing liquid motion are amenable to analysis, although numerical techniques quite often have to be used.

HYDRODYNAMIC CONCEPTS. The motion of liquids is governed by the Navier-Stokes equations, which are a statement of Newton's second law and of continuity for fluids. These Navier-Stokes equations, which are discussed in Schlichting's text [101], are valid for Newtonian fluids, in which viscosity is independent of the velocity shear rate and time. Almost all electrolytes have viscosities that vary only with temperature, and therefore, the Navier-Stokes equations apply to them.

Solutions to the Navier-Stokes equations must satisfy the appropriate boundary conditions. One of the geometric configurations for which a closed form solution exists is the disk electrode rotating about its axis in a liquid that is otherwise at rest. Since the flow regime is rigorously known for this geometry, the rotating disk electrode is often used as a research tool to study the electrochemistry of surface plating in the laboratory [102].

Another geometry of interest is that of the plated through-holes of a printed circuit board. The steady laminar flow that is achieved sufficiently far from the entrance of the tube is called Hagen-Poiseuille flow, and also is capable of being solved in a closed form from the Navier-Stokes equations [103]. The liquid's velocity for this flow is along the direction of the tube's axis and is given by

$$V_Z = 2V_0\left(1 - \frac{r^2}{a_r^2}\right) \tag{16.40}$$

where r is the radial coordinate in the cylindrical coordinate system, and a_r is the tube's radius. As can be seen in Figure 16-10, the velocity equals zero at the tube's surface, satisfying the boundary condition. When this solution is substituted into the Navier-Stokes equations:

$$\frac{\partial p}{\partial Z} = -\frac{8\mu V_0}{a_r^2} \tag{16.41}$$

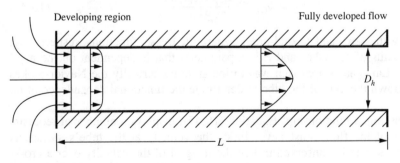

Developing region   Fully developed flow

D_h

L

FIGURE 16-10
Laminar flow in plated through-holes.

The resulting pressure gradient, $\partial p / \partial Z$, is constant, and is directed along the axis opposite to the velocity. The volume flow in the tube is given by

$$Q_V = \int_0^{a_r} 2\pi r V_Z dr = \pi a_r^2 V_O = -\frac{\pi a_r^4}{8\mu} \frac{\partial p}{\partial Z} \qquad (16.42)$$

and the velocity shear strain at the tube's boundary by

$$\frac{\partial V_Z}{\partial r}\bigg|_{r=a_r} = -\frac{4V_O}{a_r} \qquad (16.43)$$

The velocity shear at the boundary is an important parameter affecting the rate of metal deposition in mass-transport-limited plating.

In situations such as liquid flow in a tube, where body and surface tension forces do not play a role, the state of the flow can be determined from the dimensionless Reynolds number, which is the ratio of inertial to viscous forces on a volume element. For flow in a cylindrical tube, the Reynolds number is given by

$$\text{Re}_{\text{tube}} = \frac{2a_r \rho V_O}{\mu} \qquad (16.44)$$

Experimentation has shown that when the Reynolds number is significantly less than 2000, the liquid flow is laminar. However, when the number is above 2000, turbulent flow occurs, and the solution presented for the flow in Equation (16.40) is not valid. Flows in tubes of different sizes using liquids of differing properties are said to be similar if they have the same Reynolds number.

Hagen-Poiseuille flow is fully developed laminar flow in a cylindrical tube, sufficiently far from the entrance so that the flow is steady with respect to the axial distance. For plating the interior surface of the holes in circuit boards, however, the entrance region of these tubular holes is of interest. The flow in the entrance region of such tubes is not amenable to rigorous solution from the Navier-Stokes equations. Langhaar [104] used numeric methods for a solution and started with the following relationship for the axial velocity in the tube:

$$V_Z = V_O \left[\frac{I_0(\beta a_r) - I_0(\beta r)}{I_2(\beta a_r)} \right] \qquad (16.45)$$

I_0 and I_2 are the zero- and second-order hyperbolic Bessel functions, respectively; a_r is the radius of the tube, and β is a parameter that is dependent on the axial coordinate. Langhaar's approach was to integrate numerically the Navier-Stokes equations down the axis of the tube to determine the functional dependence of on Z.

The assumed relationship, Equation (16.45), displays a number of requisite properties that the flow must have. First, the velocity at the tube's surface is zero. The flow rate is conserved in that the integral of the velocity over a cross-sectional area is a constant independent of β and Z. As β goes to zero, the velocity approaches that for Hagen-Poiseuille flow, Equation (16.40). Finally, as

β goes to infinity, the velocity approaches a constant V_O over the cross section except for an infinitely thin boundary layer at $r = a_r$. This is the flow situation at the entrance plane of the tube.

The functional dependence of β on the axial coordinate, as obtained by numerically integrating the Navier-Stokes equation along the axis of the tube, is shown in Figure 16-11. Also shown is the velocity along the tube's axis as a function of the axial distance from the entrance plane. This distance from the tube's entrance is shown using a unitless distance factor:

$$Z_d = \frac{\mu Z}{a_r^2 \rho V_O} = \frac{2Z}{Re_{tube} a_r} \tag{16.46}$$

Figure 16-12 shows a number of velocity profiles for different distance factors from the tube's entrance to where fully developed Hagen-Poiseuille flow is achieved. Fully developed flow occurs at a point where the distance factor equals 0.2. The velocity shear at the tube's wall, from Equation (16.45) is

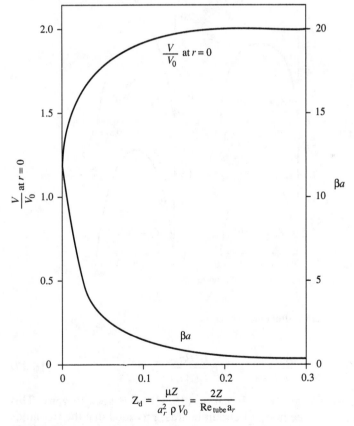

FIGURE 16-11
Effect of axial distance of β.

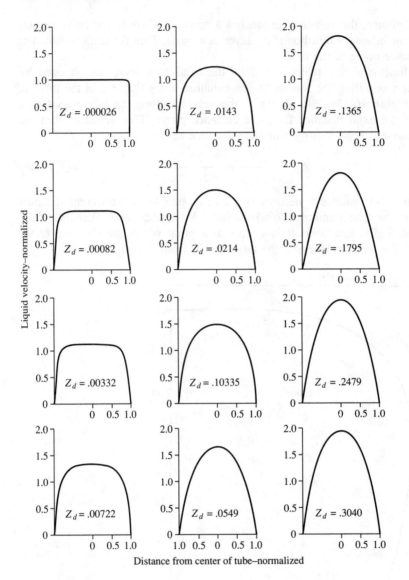

FIGURE 16-12
Velocity profiles at various distances from tube entrance.

$$\frac{\partial V_z}{\partial r}\bigg|_{r=a_r} = -\beta\frac{I_1(\beta a_r)}{I_2(\beta a_r)} \qquad (16.47)$$

This shear approaches that given by Equation (16.43) as β goes to zero. This gives the flow condition in the holes of circuit boards, provided that the Reynolds number is low enough to assure laminar flow.

This type of analysis is important in determining the solution flow process parameters and their relationship to the metal deposition process. For example, copper electroplating in blind vias (single-open-ended holes) and through-holes (PTHs) is performed to provide electrical interconnections between separated circuit planes of a printed circuit panel. Compared to surfaces, it is difficult to provide adequate solution flow in through-holes. Methods for inducing flows in holes include reciprocating panel movement, gas sparging, and forced solution flow by pumping [105,106,107]. For flow induced by reciprocating panel movement, the corresponding laminar flow in the holes can be predicted and criteria for defect-free processing established [108].

Plating Distribution

From Faraday's law, a uniform thickness distribution implies a uniform current distribution on the cathode. During electrodeposition, the cathode and anode may each be considered equipotential surfaces [109] separated by the plating solution. To complete the electrochemical circuit, charge is transferred through the solution by movement of ions. While current flows, the cell potential is distributed across the solution between anode and cathode surfaces. The potential field (Φ) at steady state in solution follows the Laplace equation with respect to cell geometry. Solution of the Laplace equation with appropriate boundary conditions for insulators and electrodes allows calculation of cathodic current density i from

$$i = -\kappa \frac{\partial \Phi}{\partial m}\Big|_{m = \text{cathode position}} \qquad (16.48)$$

where κ is the solution conductivity [110]. Different types of current distributions result depending on boundary conditions at the electrodes (see Table 16.8).

Primary current distribution is a result of ohmic potential drops as current flows through the electrically resistive solution, and is strictly a function of geometry and path length.

Secondary current distribution results from the charge transfer (or kinetic) potential drops at the electrode surfaces. Thus, the system geometry (e.g., elec-

TABLE 16.8
Different types of current distribution for electrodeposition

Current distribution	Controlling overpotential for distribution
Primary	Ohmic (cell geometry effects)
Secondary	Ohmic
	Charge transfer (reaction kinetic effects)
Tertiary	Ohmic
	Charge transfer
	Mass transfer (solution transport effects)

trode sizes and relative spatial positions) and deposition kinetics (e.g., degree of polarization) both influence the cathodic current density distribution. In the Tafel kinetic region at the cathode (where a linear ln i-versus-potential relation exists), higher average current densities will result in a less uniform current distribution [111].

Tertiary current distributions result from mass transfer effects leading to cupric ion depletion at the cathode surface. Poor quality deposits result when $i > 0.5i_M^D$, the limiting current. Thus, solution agitation must be maintained sufficiently to avoid operating under mass transport limiting conditions.

Copper electrodeposits of uniform thickness on circuit boards require that both interfacial potentials and anode-cathode solution resistance drops be equivalent at all plated surfaces. As a result, cell geometry is designed to yield a relatively uniform electric field at all surfaces. The total applied potential drop is distributed between anode and cathode interfacial regions and the bulk solution. The bulk solution potential drop is determined by solution conductivity and the electric field resulting from relative anode-cathode positions. The solution potential drop generally is large compared to interfacial resistance. The potential drop at the anode is generated by concentration overpotentials, cupric ion migration through the characteristic "black film," and the dissolution charge transfer resistance. The cathode potential drop is mainly associated with charge transfer resistance for cupric ion deposition, and ion movement through any film of adsorbed organics. Since anode and cathode polarizations usually are less than 200 mV and copper plating baths are relatively conductive, a power supply output of only 3–5 V is usually adequate.

Uniform thickness distributions across surfaces and through-holes are required for copper circuitization. These thickness distributions usually are described in terms of throwing power, which is defined as the percentage thickness ratio for panel edge to panel center, or, for holes, from hole center to panel surface. Generally, throwing powers above 70 percent for holes and 90 percent for surfaces are minimal requirements.

Calculation of current distribution on panel surfaces is not trivial for copper electrodeposition. The secondary current distribution generally has nonlinear boundary conditions requiring numerical solution. In practical systems, complicated cell geometry (for panel plate) and cathode geometries (for pattern plate) make routine numerical analysis difficult, if not impossible. Limited analysis of simple systems, coupled with experimental data, mainly is used to predict high and low current density areas [112,113]. Empirical experiments are performed with shields, thieves, auxiliary electrodes, rack design, and system geometry to improve distribution [114–117]. For panel plating, thieving metal racks often are used to divert current from high-current panel edges. For pattern plating, shielded racks (all noncontact points dip-coated with dielectric material) are recommended to eliminate excessive copper plating waste on racks. With good designs, standard deviations of less than 10 percent in surface thickness distribution are achievable for current density ranges of 30–70 mA/cm^2.

For through-holes in panels, modeling of thickness distribution has been performed for simple electron transfer kinetics [118–120]. A Wagner type number χ specific to through-holes was introduced as

$$\xi_{T_\kappa} = \left(\frac{\alpha_c FL^2}{\kappa RTD_h}\right)i \tag{16.49}$$

This dimensionless number represents the ratio of ohmic (solution) resistance to charge transfer (kinetic) resistance. For high throwing power, charge transfer resistance must dominate ($\xi_{T_\kappa} \ll 1$). Therefore, at a given hole geometry, lower plating current densities and higher solution conductivities will increase throwing power. Recently, modeling of plating thickness distribution in through-holes was extended to handle more complicated electrode kinetics, including additive effects [121]. Both models establish low throwing power as the major factor limiting plating current density for high-aspect-ratio holes. Higher current densities for plating low-aspect-ratio holes usually are limited by the mode of solution transport and/or the effectiveness of the additives [122,123].

APPENDIX A
TECHNIQUE OF MIXED POTENTIAL THEORY

In order to achieve conditions of controlled mass transfer, a rotating disk electrode (RDE) is used. The current due to the diffusion of metal ions to such a geometric surface is given by:

$$|i_M| = B'_M(C_M^\infty - C_M^a)\sqrt{\omega} \tag{16.50}$$

where C_M^∞ is the bulk concentration of the metal ions, C_M^a is the surface concentration, ω is angular velocity, and B'_M is a diffusion parameter given by

$$B'_M = 0.62n_M F D_M^{2/3} \nu^{-1/6} A \tag{16.51}$$

The symbols in this equation are defined at the end of the chapter.

For a diffusion-controlled cathodic partial reaction, $C_M^a = 0$, and the limiting current i_M^D is independent of potential and takes the form

$$|i_M^D| = B'_M C_M^\infty \sqrt{\omega} \tag{16.52}$$

Similarly, the diffusion-limited current for the anodic partial reaction is

$$i_R = B'_R C_R^\infty \sqrt{\omega} \tag{16.53}$$

The concentration of metal ions at the surface may be expressed by the Nernst equation

$$E_M = E_M^0 + \frac{RT}{n_M F} \ln C_M^a \tag{16.54}$$

which by substituting Equations (16.50) and (16.52) becomes

$$E_M = E_M^0 + \frac{RT}{n_M F} \ln \left| (i_M^D - i_M) \right| - \frac{RT}{n_M F} \ln B_M' - \frac{RT}{n_M F} \ln \sqrt{\omega} \qquad (16.55)$$

The corresponding equation for the reducing agent also can be worked out in a similar manner.

When the anodic partial reaction is under electrochemical control, the polarization curve is described by the equation

$$E = E_R^0 - b_R \ln i_R^0 + b_R \ln i_R \qquad (16.56)$$

The anodic Tafel slope b_R is given by

$$b_R = \frac{RT}{(1 - \alpha_R) n_R F} \qquad (16.57)$$

Similarly, when the metal deposition reaction is activation-controlled, the kinetics are described by the cathodic Tafel equation:

$$E = E_M^0 + b_M \ln \left| i_M^0 \right| - b_M \ln \left| i_M \right| \qquad (16.58)$$

where

$$b_M = \frac{RT}{\alpha_M n_M F} \qquad (16.59)$$

Since each partial reaction either is under electrochemical control or under mass transfer control, the overall reaction scheme consists of four possible combinations. These will be considered next, and the dependence of E_{MP} on experimental parameters such as rotation rate ω and C_R^∞ and C_M^∞ determined.

Case 1: Cathodic Partial Reaction Diffusion Controlled; Anodic Partial Reaction Electrochemically Controlled

The diffusion-limited cathodic partial current depends on C_M^∞, D_M and ω, and its magnitude is given by Equation (16.52). Then combining Equation (16.52) with Equation (16.56), using $|i_R| = |i_M|$ from Equation (16.20), gives

$$E_{MP} = E_R^0 - b_r \ln i_R^0 + \frac{b_r}{2} \ln B_M'^2 \omega + b_r \ln C_M^\infty \qquad (16.60)$$

Equation (16.60) shows that the mixed potential is a linear function of $\ln \omega$ and $\ln C_M^\infty$ and that the Tafel slope for the anodic partial reaction may be obtained by plotting E_{MP} against either of these experimental parameters. Similar functions have been obtained by Makrides [124] for corrosion processes. Case 1 is represented graphically in the symbolic diagram of Figure 16-13. Oxygen or air frequently is bubbled through electroless copper baths to oxidize any Cu(I) species formed and thus avoid bath decomposition via Cu(I) disproportionation. Under these circumstances, there is another cathodic current due to oxygen reduction,

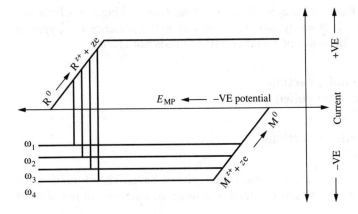

FIGURE 16-13
Graphic representation of Case 1 (see Appendix A): cathodic partial reaction diffusion controlled; anodic partial reaction electrochemically controlled.

and the total cathodic current is the sum

$$i'_M = i_M + i_{O_2} \tag{16.61}$$

where i_{O_2} is the current for oxygen reduction. At the plating potential, i_{O_2} is diffusion limited; that is, it is independent of the electrode potential and therefore equivalent to a cathodic current applied externally. Under the circumstances, Equation (16.61) becomes

$$\left| i_M^{D'} \right| = (B'_M C_M^\infty + B'_{O_2} C_{O_2}^\infty) \sqrt{\omega} \tag{16.62}$$

and Equation (16.60) becomes

$$E_{MP} = E_R^0 - b_r \ln i_R^0 + b_r \ln(B'_M C_M^\infty + B'_{O_2} C_{O_2}^\infty) + \frac{b_R}{2} \ln \omega \tag{16.63}$$

Therefore, the diagnostic criteria for Case 1 remain unchanged in the presence of oxygen in the plating bath.

Case 2: Cathodic Partial Reaction Electrochemically Controlled; Anodic Partial Reaction Diffusion Controlled

This is the converse of Case 1; the cathodic and anodic partial reactions are described by Equations (16.53) and (16.58) respectively. Combining these with the help of Equation (16.20) yields

$$E_{MP} = E_M^0 + b_M \ln \left| i_M^0 \right| - \frac{b_M}{2} \ln B_R'^2 \omega - b_M \ln C_R^\infty \tag{16.64}$$

Once again, E_{MP} is linearly dependent on $\ln \omega$ and $\ln C_R^\infty$.

The slope of E_{MP}-versus-ln ω plot is negative, thus making it easily distinguishable from Case 1. It easily can be shown that in the presence of oxygen in the plating solution, the form of Equation (16.64) does not change.

Case 3: Both Partial Reactions Electrochemically Controlled

Case 4: Both Partial Reactions Diffusion Controlled

Such cases rarely are encountered in electroless plating baths. Therefore, detailed relationships between the mixed potential and other parameters are not worked out.

APPENDIX B

SOME "TYPICAL" PLATING BATH FORMULATIONS

Type II Acid hard gold

Gold (as potassium gold cyanide)	4 to 12 g/L
Citric acid	20 to 70 g/L
Potassium citrate	50 to 90 g/L
Cobalt (as sulfate)	1 to 3 g/L
Temperature	29 to 35°C
pH (adjusted with citric or phosphoric acid)	3.2 to 4.0
Current density	0.8 to 2 A/dm^2

From E. C. Rinker and R. Duva, *USP 2,905,601* (1959).

0.5 to 1.0 g/L Ni (from potassium nickel cyanide) can be used in place of cobalt.

99% (min) gold with Knoop hardness of 91 to 200.

Max. impurity level 0.1%(not including Co or Ni).

Semibright high-speed nickel

Nickel sulfate	225 to 375 g/L
Nickel chloride	30 to 60 g/L
Boric acid	30 to 40 g/L
Sodium lauryl sulfate	0.1 to 0.5 g/L
Sodium saccharin	0.5 g/L
Temperature	45 to 65°C
pH (adjusted with hydrochloric acid)	1.5 to 4.5
Current density	2.7 to 10.7 A/dm^2

60/40 tin-lead solder plating

Stannous tin (from fluoroborate)	12 to 20 g/L
Lead (from fluoroborate)	8 to 14 g/L
Free fluoroboric acid	350 to 500 g/L
Peptone	2 to 7 g/L
Temperature	15 to 38°C
Current density	1.5 to 2.5 A/dm^2

From: *USP 3,554,878.*

REFERENCES

1. J. W. Gibbs, *Commentary*, New Haven, Conn.: Yale University Press, 1936.
2. J. W. Gibbs, *Collected Works*, London: Longmans Green & Co., 1918.
3. J. Frenkel, *Kinetic Theory of Liquids*, New York: Oxford University Press, 1946.
4. M. Volmer and A. Weber, "Nucleus Formation in Supersaturated Systems," *Z. Physik. Chem.* 119, p. 277 (1926).
5. W. Kossell, "Existenzbereiche von aufbau—und Abbauvorgängen auf der KristallKugel," *Ann. Physik.* 33, p. 651 (1938).
6. I. N. Stranski, "Zur Theorie des Kristallwachstums," *Z. Physik. Chem.* 136, p. 259 (1927).
7. M. Volmer, *Kinetic der Phasenbildung*, Leipzig-Dresden: Th. Steinkopff-Verlag, 1939.
8. H. Brandes, "Zur Theorie des Kristallwachstums," *Z. Physik. Chem.* 126, p. 196 (1927).
9. R. Becker and W. Doring, "Kinetische Behandlung der Keimbildung in übersättigten Dämpfen," *Ann. Physik* 24, p. 719 (1935).
10. A. F. Holleman and E. Wiberg, *Lehrbuch der Anorganischen Chemie*, Berlin, W. Germany, Walter de Gruyter, 1985.
11. M. Fleischmann and M. Thirsk, *Advances in Electrochemistry and Electrochemical Engineering*, vol. 3, Chapter 3, edited by P. Delahay, New York: John Wiley & Sons, 1963.
12. J. O'M Bockris and A. Damjanovic, *Modern Aspects of Electrochemistry*, vol. 3, edited by Bockris and Conway, London: Butterworth 1964.
13. J. O'M Bockris and G. A. Razumney, *Fundamental Aspects of Electrocrystallization*, New York: Plenum Press, 1967.
14. M. Pourbaix, *Atlas of Electrochemical Equilibrium in Aqueous Solutions*, English edition, Houston, TX: Pergamon Press, 1974.
15. C. Wagner, and W. Traud, "The Interpretation of Corrosion Phenomena by Superimposition of Electrochemical Partial Reactions and the Formation of Potentials of Mixed Electrodes," *Z. Electrochem.* 44, p. 391 (1938).
16. J. V. Petrocelli, V. Hospadaruk, and G. A. Dibari, "The Electrochemistry of Copper, Nickel and Chromium in the CorrodKote and CASS Test Electrolytes," *Plating*, 49, p. 50 (1962).
17. E. J. Kelley, "The Active Iron Electrode; 1. Iron Dissolution and Hydrogen Evolution Reactions in Acidic Sulfate Solutions," *J. Electrochem. Soc.* 112, p. 124 (1965).
18. M. Paunovic, "Electrochemical Aspects of Electroless Deposition of Metals," *Plating* 55, p. 1161 (1968).
19. F. M. Donahue and C. U. Yu, "The Electrochemistry of Electroless Deposition of Copper," *Electrochem. Acta* 15, p. 237 (1970).
20. A. Molenaar, M. F. E. Hodrinet, and L. K. H van Beek, "A Study of the Mechanism of the Electroless Deposition of Nickel," *Plating* 61, p. 238 (1974).
21. S. M. El-Raghy and A. A. Aho-Solama, "Kinetics of Electroless Copper Plating With EDTA as the Complexing Agent for Cuprous Ions," *J. Electrochem. Soc.* 126, p. 171 (1979).
22. F. M. Donahue, "Interactions in Mixed Potential Systems," *J. Electrochem. Soc.* 119, p. 72 (1972).

23. T. N. Andersen, M. H. Ghandehari, and M. Ejuning, "A Limitation to the Mixed Potential Concept of Metal Corrosion Copper in Oxygenated Sulfuric Acid Solutions," *J. Electrochem. Soc.* 122, p. 1580 (1975).

24. P. Bindra and J. Tweedie, "Mechanisms of Electroless Metal Plating: 1. Application of Mixed Potential Theory," *J. Electrochem. Soc.* 130, p. 1112 (1983).

25. F. M. Donahue, "Kinetics of Electroless Copper Plating: III. Mass Transfer Effects," *J. Electrochem. Soc.* 127, p. 51 (1980).

26. U.S. Patent 2,874,072.

27. U.S. Patent 2,996,408.

28. U.S. Patent 3,033,703.

29. U.S. Patent 3,075,855.

30. U.S. Patent 3,075,856.

31. U.S. Patent 3,095,309.

32. U.S. Patent 3,146,125.

33. U.S. Patent 3,222,195.

34. U.S. Patent 3,259,559.

35. U.S. Patent 3,269,861.

36. U.S. Patent 3,327,972.

37. U.S. Patent 3,326,700.

38. U.S. Patent 3,383,244.

39. U.S. Patent 2,403,035.

40. U.S. Patent 3,453,123.

41. U.S. Patent 3,457,089.

42. U.S. Patent 3,492,135.

43. U.S. Patent 3,607,317.

44. U.S. Patent 3,615,732.

45. U.S. Patent 3,615,736.

46. U.S. Patent 3,649,350.

47. U.S. Patent 3,663,242.

48. U.S. Patent 3,664,852.

49. U.S. Patent 4,124,339.

50. J. O'M Bockris and A. K. N. Reddy, *Modern Electrochemistry*, vols. 1 and 2, New York: Plenum, 1970.

51. G. Milazzo, *Electrochemistry*, New York: Elsevier, 1963.

52. C. R. Palin, *Electrochemistry for Technologists*, New York: Pergamon Press, 1969.

53. A. R. Denero, *Elementary Electrochemistry*, 2d ed., London: Butterworth, 1971.

54. H. Y. Chen, "The Limiting Rate of Deposition by P-R Plating," *J. Electrochem. Soc.* 118, p. 1132 (1971).

55. K. I. Popov, "Fundamental Aspects of Pulsating-Current Metal Electrodeposition: I. The Effect of the Pulsating Current on the Surface Roughness and the Porosity of Metal Deposits," *Surf. Technol.* 11, p. 99 (1980).

56. "Fundamental Aspects of Pulsating-Current Metal Electrodeposition: II. The Mechanism of Metal Film Formation on an Inert Electrode," *Surf. Technol.* 11, p. 111 (1980).

57. N. Ibl, "Some Theoretical Aspects of Pulse Electrolysis," *Surf. Technol.* 11, p. 81 (1980).

58. J. Cl. Puippe, N. Ibl, H. Angerer, and K. Hosokawi, "The Morphology of Pulse Plated Deposits," *Plating and Surf. Fin.* 67, no. 6, p. 68 (1980).

59. F. A. Lowenheim, editor, *Modern Electroplating*, 3d ed., New York: John Wiley & Sons, 1974.

60. F. A. Lowenheim, *Electroplating*, New York: McGraw-Hill, 1980.

61. A. K. Graham, editor, *Electroplating Engineering Handbook*, 3d ed., New York: Van Nostrand Reinhold, 1971.

62. N. Allen, editor, *Metal Finishing Guidebook Directory*, Hackensack, N.J.: Plastics Pubs., annual.

63. M. Schwartz, "Deposition from Aqueous Solution: An Overview," in *Deposition Technology for Finishes and Coatings*, Park Ridge, N.J.: Noyes Pubs., 1982, pp. 385–453.
64. H. Silman, G. Isserlis, and A. F. Averill, *Protective and Decorative Coatings for Metals*, Middlesex, England, Finishing Publications, 1978.
65. J. A. Murphy, editor, *Surface Preparation and Finishes for Metals*, New York: McGraw-Hill, 1971.
66. A. F. Bogenschutz, *Surface Technology and Electroplating in the Electronics Industry*, London: Porticullis Pubs., 1974.
67. J. I. Duffy, editor, *Electroplating Technology Recent Developments*, Park Ridge, N.J., Noyes Data Corp., 1981.
68. E. Raub and K. Muller, *Fundamentals of Metal Deposition*, New York: Elsevier, 1967.
69. W. H. Safraneil, editor, *Properties of Electrodeposited Metals and Alloys*, 2d ed., Orlando, Fla.: American Electroplaters and Surface Finishers Soc., 1986.
70. U.S. Patent 946,903.
71. W. H. Safranek, in *Modern Electroplating*, edited by F. A. Lowenheim, New York: John Wiley & Sons, 1963.
72. T. Chocianowicz-Biestek and K. Sznidt, "Methods of Obtaining Bright Copper Coatings From a Cyanide Bath," *Prace Inst. Mech. Precyz*, 8, p. 45 (1960).
73. R. W. Couch and R. G. Bikales, "Plating of Printed Circuits With Pyrophosphate Copper and Tin-Nickel," *Proc. Am. Electroplaters Soc.*, 48, p. 176 (1961).
74. C. Struyk and A. E. Carlson, "Copper Plating from Fluoborate Solutions," *Monthly Review Am. Electroplaters Soc.*, 33, p. 923 (1946).
75. J. W. Dini, "Plating Through Holes in Printed Circuit Boards: Evaluation of Some Copper Baths," *Plating*, 51, p. 119 (1964).
76. L. Mayer and S. Barbieri, "Characteristics of Acid Copper Sulfate Deposits For Printed Wiring Board Applications," *Plating and Surface Finishing*, 68, p. 48 (1981).
77. S. Skowronsk and E. Reinoso, "The Specific Resistivity of Copper Refining Electrolytes and Method of Calculation," *Trans. Am. Electrochem. Soc.* 52, p. 205 (1927).
78. A. K. Graham, "A Study of the Influence of Variables on the Structure of Electrodeposited Copper," *Trans. Am. Electrochem. Soc.* 52, p. 157 (1927).
79. E. Mattson, J. O'M Bockris, "Galvanostatic Studies of Deposition and Dissolution of Copper," *Trans. Faraday Soc.* 55, p. 1586 (1959).
80. O. R. Brown and H. R. Thirsk, "Rate Determining Step in Electrodeposition of Cu on Cu from $CuSO_4$ Solutions," *Electrochem. Acta* 10, p. 385 (1965).
81. A. I. Molodov, G. N. Markosian, and V. V. Losev, "Regularities of Low Valence Intermediate Accumulation," *Electrochem. Acta* 17, p. 701 (1972).
82. T. Hurlen, G. Ottesen and A. Staurset, "Kinetics of Copper Deposition and Dissolution in Aqueous Sulfate Solutions," *Electrochimica Acta*, 23, p. 39 (1978).
83. A. Damjanovic, T. H. V. Setty, and J. O'M Bockris, "Effect of Crystal Plane on the Mechanism and Kinetics of Copper Electrocrystallization," *J. Electrochem. Soc.* 113, p. 429 (1966).
84. H. Seiter and H. Fisher, "Relation Between the Conditions of Precipitation and the Growth Form in the Electrocrystallization of Copper," *Z. Electrochem.* 63, p. 249 (1959).
85. U. Bertocci, C. Bertocci, and B. C. Larson, "Electrocrystallization of Copper," *J. Crystal Growth* 13, p. 427 (1972).
86. K. J. Bachmann and U. Bertocci, "Metal/Metal-Ion Electrodes with Two Charge Transfer Steps: Me/Me^{Z+} And $Me^{Z+}/Me^{(Z+1)+}$. Potentiostatic Transients Under Charge-Transfer and Diffusion Control," *Electrochem. Acta* 15, p. 1877 (1970).
87. F. Mansfield, "The Copper Plating Bath," *Plating and Surface Finishing*, 65, p. 65 (1978).
88. D. A. Luke, "Acid Copper Plating on PCB's," *Electronic Production* 7, p. 73 (1978).
89. E. W. Rouse and P. K. Aubel, "Analyses of Copper Refining Cell Voltages," *Trans. Am. Electrochem. Soc.* 52, p. 189 (1927).
90. W. H. Gauvin and C. A. Winkler, "The Effect of Chloride Ions on Copper Electrodeposition," *J. Electrochem Soc.* 99, p. 71 (1955).

91. L. Mayer and S. Barbieri, "Characteristics of Acid Copper Deposits for PCB Applications," *Plating and Surface Finishing* 68, p. 46 (1981).

92. L. Mirkova, S. Rashkov, and C. Nanev, "Leveling Mechanism During Bright Acid Copper Plating," *Surface Technology* 15, p. 181 (1982).

93. R. L. Sarma and S. Nageswar, "Electrodeposition of Copper in the Presence of 2-Mercaptoethanol," *J. Appl. Electrochem.* 12, p. 329 (1984).

94. S. C. Barnes, "The Effect of Thio Compounds on the Structure of Copper Electrodeposits," *J. Electrochem. Soc.*, 111, p. 296 (1964).

95. W. H. Safranek, "Structure and Property Relationships for Bulk Electrodeposits," *J. Vac. Sci. Technol.*, 11, p. 765 (1974).

96. T. E. Such, "Physical Properties of Electrodeposited Metals," *Metallurgica* 56, p. 61 (1957).

97. E. Hofer and P. Javet, "Modern Methods of Structural Studies of Electrolytic Deposits," *Microtecnic*, 17, p. 104 (1963).

98. C. Owen, "A Modified Ductility Tester for the Tensile Testing of Thin Metal Foils," *Plating* 57, p. 1012 (1970).

99. M. Sanicky, "A Versatile Plater's Tool. All About the Hull Cell," *Plating and Surface Finishing* 72, no. 10, p. 20 (1985).

100. M. Rothstein, "Control of Plating by Cyclic Voltammetry," *Plating and Surface Finishing*, 71, p. 36 (1984).

101. H. Schlichting, *Boundary Layer Theory*, New York: McGraw-Hill, 1955.

102. R. N. Adams, *Electrochemistry at Solid Surfaces*, Marcel Dekker, New York, 1969.

103. S. Goldstein, *Modern Developments in Fluid Dynamics*, New York: Dover Publications, 1965.

104. H. L. Langhaar, "Steady Flow in the Transition Length of a Straight Tube," *J. Appl. Mech.* 9, p. A-55 (1942).

105. K. Heikkla, L. Whitney, G. Peterson, and R. Pylkki, "Factors Affecting Through-Hole Plating of Printed Wiring Boards," *Electronic Packaging and Production* 24, p. 29 (1984).

106. R. Haak, C. Ogden, and D. Tench, "Evaluation of Agitation within Circuit Board Through Holes," *J. Appl. Electrochem.* 11, p. 771 (1981).

107. J. Ju, "High Speed Impinging Jet Plating: Slot Jet and Through Hole Plating," Ph.D. dissertation, University of Illinois, 1984.

108. R. Galasco, A. David, and R. Schaffer, "Panel Movement Effects on Solution Flow in Through-Holes of Printed Circuit Panels," *AICHE Journal* 33, p. 916 (1987).

109. O. P. Watts, "Cleaning and Plating in the Same Bath," *Trans. Am. Electrochem. Soc.* 27, p. 141 (1915).

110. J. Newman, *Electrochemical Systems*, Englewood Cliffs, N.J.: Prentice-Hall, 1973.

111. T. Hoar and J. Agar, "Factors in Throwing Power Illustrated by Potential-Current Diagrams," *Disc. Faraday Soc.* 1, p. 162 (1947).

112. C. Kasper, "The Theory of the Potential and the Technical Practice of Electrodeposition," *Trans. Electrochem. Soc.* 77, p. 353 (1940).

113. S. Dalby, J. Nickelsen, and L. Alting, "Metal Distribution in Electroplating," *Electroplating and Metal Finishing* 28, p. 18 (1975).

114. W. Engelmaier, T. Kessler, and R. Alkire, "Current Distribution Levelling from Auxiliary Bipolar Electrodes," *J. Electrochem. Soc.* 125, p. 209 (1978).

115. W. Bissinger, *IBM Technical Disclosure Bulletin* 27, no. 5, p. 22 (1984).

116. U.S. Patent 4,534,832, 1985.

117. U.S. Patent 4,259,166, 1981.

118. R. Alkire and A. Mirarefi, "Current Distribution within Tubular Electrodes under Laminar Flow," *J. Electrochem. Soc.* 120, p. 1507 (1973).

119. T. Kessler and R. Alkire, "A Model for Copper Electroplating of Multilayer Printed Wiring Boards," *J. Electrochem. Soc.* 123, p. 990 (1976).

120. W. Engelmaier and T. Kessler, "Investigation of Agitation Effects on Electroplated Copper in Multilayer Board Plated Through-Holes in a Forced Flow Plating Cell," *J. Electrochem. Soc.* 125, p. 36 (1978).

121. O. Lanzi III, U. Landau, J. Reid, and R. Galasco, "Effect of Local Kinetic Variations on Through-Hole Plating," *J. Electrochem. Soc.*, accepted for publication (1988).

122. A. Sonin, "Jet Impingement Systems for Electroplating Applications: Mass Transfer Correlations," *J. Electrochem. Soc.* 130, p. 1501 (1983).
123. W. Hastie, "High Speed Copper," *Circuits Manufacturing* 23, no. 6, p. 25 (1983).
124. A. C. Makrides, and M. T. Coltharp, "Preparation of Solid Electrodes for Hydrogen Overpotential Studies," *J. Electrochem. Soc.* 107, p. 472 (1960).

NOMENCLATURE

A	electrode surface area
A_w	atomic weight
A_n	stationary cluster of size n
a	solution activity
a_r	tube radius
B'_M	metal ion diffusion parameter
B'_{O_2}	diffusion parameter for oxygen reduction
B'_R	anodic reaction diffusion parameter
b	number of atoms impinging on unit surface per unit time
b_M	Tafel slope of cathodic reaction
b_R	Tafel slope of anodic reaction
C^a_M, C_e	surface concentration of metal ion
C^∞_M, C_O	bulk concentration of metal ion
$C^\infty_{O_2}$	bulk solution concentration of oxygen
C^∞_R	bulk concentration of species R
D	diffusion coefficient
D_M	diffusion coefficient of metal ion
D_h	hole diameter
E	Nernst potential
E_M	Nernst potential for reduction reaction
E_{MP}	mixed potential
E_{cell}	cell potential during current flow
E^0	standard redox potential
E^0_M	potential for metal deposition reaction
E^0_R	equilibrium potential for reducing agent
e	electron charge
F	Faraday's constant
$\overline{\Delta G}$	electrochemical free energy of formation
$\overline{\Delta G^*_{het}}$	electrochemical free energy of formation for heterogeneous nucleation
$\overline{\Delta G^*_{hom}}$	electrochemical free energy of formation in terms of homogeneous energy barrier
H	shape factor

I	current
I_r	rate of nucleation per unit volume
I_0, I_1, I_2	zero-, first-, and second- order hyperbolic Bessel functions
i	current density
i_M	cathodic partial current
i_M^D	limiting current
i_M^o	exchange current density of cathodic reaction
$i_M^{o'}$	total cathodic limiting current
i_M'	total cathodic current
i_O	exchange current density
i_{O_2}	cathodic current due to oxygen reduction
i_R	anodic partial current
i_R^o	exchange current density of anodic reaction
i_{av}	average current density
i_{dep}	deposition current
i_{dl}	double layer charging current
i_f	Faradaic current
i_p	peak current density
$i_{plating}$	plating current
i_t	total current
k	Boltzmann's constant
k_o	redox rate constant for $Cu^+ \rightleftharpoons Cu^{+2} + e^-$
L	hole length
M	molecular weight
M^0, M^{Z+}	metal of given charge
m	general coordinate dimension
N	Avogadro's number
N_M	rate of mass flux
N_N	equilibrium number density for given cluster size
N_O	normalization constant
n	number of electrons transferred; cluster size
n_M	number of electrons transferred for cathodic reaction
n_R	number of electrons transferred for anodic reaction
P	pressure
Q	electrical charge
Q_v	volumetric solution flow in tube
R	universal gas constant
R_d	duty cycle for pulse plating

Re_{tube}	Reynolds number in a tube
r	radial distance
S_{n^*}	surface area of critical cluster
T	temperature
t	time
t_{off}	pulse-off time per cycle
t_{on}	pulse-on time per cycle
t_{pulse}	pulse cycle time
V_O	average velocity in direction of tube axis
V_Z	velocity in direction of tube axis
W	weight of deposit
X	distance from electrode surface
X_1	inner Hemholtz layer thickness
X_2	outer Hemholtz layer thickness
z	charge of metal
Z	distance in direction of tube axis
Z_d	unitless distance factor
α	transfer coefficient
α_c	cathodic transfer coefficient
α_M	cathodic transfer coefficient for metal deposition
α_R	anodic transfer coefficient
β	axial coordinate parameter
δ	diffusion layer thickness
η	overpotential
η_M	cathodic overpotential
η_R	anodic overpotential
η_{act}	activation overpotential
η_{conc}	concentration overpotential
θ	contact angle between nucleus and substrate
κ	solution electrical conductivity
μ	solution viscosity
ν	kinematic viscosity
ρ	density
σ	surface free energy
τ	time when species is reduced when reaching surface
Φ	potential in solution
ξ_{T_κ}	dimensionless ratio of ohmic to charge transfer resistances
ω	angular velocity of rotating disk electrode

PROBLEMS

16.1. List the components of one electroless copper plating solution and one copper electroplating solution. Write a one-line description of the function of each component.

16.2. Define: immersion plating, electroless plating, electroplating.

16.3. List a typical continuous process sequence (plating baths) which could be used to plate a pin type connector that may be subject to repeated insertion and removal. What are some desirable properties of the protective electrodeposit?

16.4. List several factors that must be considered when designing an electroplating system.

16.5. Two copper surfaces of area 1 cm^2 are submerged 10 cm apart in an acid copper plating solution such that they are parallel and define a volume 10 cm by 1 cm by 1 cm in size.

(a) Describe the potential profile that will develop when a potential of about 1 V is applied between the two surfaces by a power supply.

(b) If the activation overpotential (η) is -0.1 V at the cathode and the Tafel equation ($i = i_0 \exp(-\alpha n F \eta / RT)$] is valid, then what is the current at the 1-cm^2 surface? (Use $i_0 = 0.001$ A/cm^2, $T = 298$ degrees Kelvin, $\alpha = 0.5$, $R = 8.3$ J/M $\cdot$ K.)

(c) What is the resistive potential drop in solution between the two parts during the above current flow? (Use solution resistivity $= 2\Omega \cdot$ CM and assume conductivity only within the defined rectangle).

(d) How much copper is deposited in 1 hour?

(e) What is the mean thickness of the deposit?

CHAPTER 17

PLATED THROUGH-HOLE TECHNOLOGY

VOYA R. MARKOVICH
FRANCIS S. POCH

INTRODUCTION

Interconnections between the various signal and voltage layers of a printed circuit board are usually accomplished using plated through-holes. This technology involves a series of process steps; mechanical drilling, hole cleaning, catalyzing for autocatalytic deposition of copper and finally copper plating. These steps follow the composite lamination processes and precede the electroless or electrolytic deposition of copper, discussed in other chapters of this book. This chapter will review the machinery and tools used in hole making, but the focus of the chapter will be the various process considerations in the plated through-hole technology. A more detailed discussion of drill machinery can be found in Chapter 20.

It is important to realize that there is a large mismatch in thermal expansion between the deposited copper in the plated through-hole and the surrounding composite laminate. The copper walls are so highly stressed that the copper/composite interface becomes a concern. The mechanics of this situation are discussed in the mechanical design chapter, while this chapter focuses on the interfacial considerations: the cleaning of the copper inner-planes and the cleaning of the hole wall and its preparation. The application of catalyst will be discussed. Analytical considerations and control techniques will also be introduced.

MECHANICAL DRILLING OF MULTILAYERED PRINTED CIRCUIT BOARDS

Drilled holes in multilayered printed circuit boards provide the means for electrical interconnection between the layers of the board. When copper-plated, these holes provide the electrical continuity that allows the signal and power carriers within the board to communicate to one another. Signal layers and power layers provide electrical continuity in the plane of the board, while plated through-holes provide continuity perpendicular to the plane of the board.

The most common way to produce holes in printed circuit boards is to mechanically drill them using a twist drill bit. Because the laminate is composed of a variety of dissimilar materials, the drill bit must cut copper, grind and break glass and transport the debris out of the hole. Wherever copper is exposed, an electrical connection will be made when the hole is copper-plated. Any copper left untouched by the drilled hole will be insulated by the epoxy/glass laminate from the plated through-hole and will not attain electrical continuity with it.

Any description of the drill process must include discussion of these four key elements: equipment, drill bits, parameters, and hole quality.

Mechanical Drilling Equipment

The piece of equipment used to drill holes in printed circuit boards is designed specifically for that task [1]. The task of the drill machine is to deliver a spinning drill bit to the product, execute a drill stroke, and repeat the sequence many times over. The drill machine is basically an x-y-z positioning system with a drill spindle. The x-y axis positions the spindle over the product and the z axis moves the spindle up and down. The spindle spins the drill bit, and when positioned it enables the drill bit to drill a hole. The machines are computer numerical controlled (CNC) with closed-loop feedback control systems. A stable drill machine maintains a minimum of vibration while positioning, though it might change position very fast. A balance between productivity and flexibility must be attained to allow the machine to drill various products efficiently, and as inexpensively as possible. Refer to Chapter 20 for a more detailed discussion of drilling equipment.

Positioning systems similar to those found on drill machines can be found on other machines used to manufacture printed circuit boards. Routing machines are used to perform vertical milling operations. Lasers sometimes replace spindles to perform machining operations such as drilling and cutting. Many types of inspection equipment have similar x-y-z positioning systems with video cameras or x-ray equipment attached to the z axis.

For productivity, the drill machine can be designed to carry several spindles to work on several panels simultaneously. The spindle package can take the form of a one- or two-dimensional matrix. The greater the number of spindles working on a panel, the faster the panel can be drilled. Multiple machine stations can each hold a panel, or more if stacked. The greater the number of panels drilled in each load, the higher the productivity. To make the drill process as cost effective as possible, the greatest number of panels must be drilled in the least amount of time. Multi-spindle/station drill machine designs are the best way to improve productivity without sacrificing quality.

A variety of support systems must back up a good drill operation. The factory must supply temperature- and humidity-controlled environments and air supplies to minimize errors due to expansion and contraction of the machine and product. Substantial effort goes into making the machines thermally and structurally stable. This effort should not go to waste because of poor environmental systems. The electrical supply must be isolated from the main line to keep it free from voltage spikes and transients. The control hardware is sophisticated and delicate, and an electrical spike can cause the machine to malfunction. Vacuum systems are necessary to transport drill debris away and to hold the panels flat on the positioning table. Sound suppression devices and adequate lighting should be used to make the work area as pleasant as possible for the equipment operators.

Drill Bits

The drill bit has a very difficult job to perform. It must carry out several tasks in a rather hostile environment. Drilling holes in fiberglass is a difficult proposition. The drill bit must not only grind its way through a laminate composed primarily of glass, but it must also cut through copper conductors and then transport the debris produced up the flutes and out of the hole. In addition, many times the holes produced are an order of magnitude (or more) longer than they are in diameter. All this makes the printed circuit board drill bit a highly compromised tool.

The drill bit is made of a material that makes the assigned task easier. Solid cemented tungsten carbide materials are used exclusively [2], because of their superior strength, rigidity, and hardness. High strength is of primary concern to keep the drill bit from breaking during use. A good rigid material like tungsten carbide keeps bending and wander to a minimum and makes for the best possible hole location. Tungsten carbide is also one of the hardest materials available. High hardness is required to maintain reasonable tool life.

Tungsten carbide materials manufactured specifically for use in printed circuit boards are required. The drill bits used to drill printed circuit boards are very small and are susceptible to breakage. The survival of the tool is paramount, so only high-quality materials should be used. Defects in the carbide matrix (and there can be many) produce weak spots and stress concentrations that can cause tool fracture. Once a crack is initiated, it will propagate immediately and the tool will fail. High-quality materials are manufactured to minimize the defects that can lead to failure under load. However, defects that do exist can be sorted out during manufacture by bend testing each drill bit to a specified stress level. If the drill bit survives the test, it will eventually be used. If it fails the stress test by breaking, the defective drill bit will not break during use and damage the product.

Special carbide materials developed for use in very small tools are readily available for the manufacture of printed circuit board drill bits. They are much stronger than more typical materials used in less demanding applications, but softer. Specialty carbides have 10 percent cobalt binders, giving them 400,000 psi transverse rupture strengths. They are stronger, but the added strength comes to the detriment of hardness. At 92 Rockwell A, they are slightly softer than their lower-cobalt competitors, but the reduction in measured hardness can be overcome with a reduction in the grain size of the carbide particles in the matrix. Submicron grain sizes help improve wear resistance by reducing the damage done to finely machined cutting edges on the first drill strokes. Smaller particles breaking away from the cutting edges result in acceptable wear resistance, despite a reduction in measured hardness. Little penalty is incurred by sacrificing hardness for strength with high cobalt content and submicron grain size in tungsten carbide materials.

A good tool for drilling printed circuit boards begins with good materials. Beyond that lies a myriad of geometries and features [3,4] that can be integrated into the two flute twist drill bit. Many features, dimensions, and tolerances have been empirically developed through time. The history of modern machining includes the trial-and-error evolution of the twist drill bit. There is still plenty of trial and error associated in finding the proper tool for any application, especially for drilling printed circuit boards. There may be differing opinions as to the best tool design to use, but the most important consideration is that the drill bit get the basic job done. Empirical process optimization plays a large role in selecting the best set of features, dimensions, and tolerances for a specific printed circuit board drilling application.

The most important dimensions of the drill bit are its length and diameter. Both dimensions are dictated by product design. After length and diameter are specified, the drill bit manufacturer can offer many alternatives to optimize the performance of the tool. These are too numerous to discuss here, but to summarize, the printed circuit board drill bit is a two-flute twist drill bit of proper length and diameter and made of high-quality solid-cemented tungsten carbide material (Figure 17-1).

Drill Process Parameters

The drill process parameters of greatest interest are the spindle feed rate and the spindle rotational speed. Both of these are actually speed parameters, though in different directions. The spindle feed rate is the speed at which the spindle and the drill bit approach and progress through the panel. In-feed is expressed in inches per minute (ipm). Spindle speed is the rate at which the spindle and tool rotate, and is expressed in revolutions per minute (rpm). Typical feed rates vary in the vicinity of 100 ipm, while typical spindle speeds range as high as 100,000 rpm.

Generally the best productivity and hole quality are attained when the drill in-feed is as high as possible without resulting in excessive drill breakage. Because drill bits used to drill printed circuit boards are so small, they must be spun at very high speeds to cut properly. Many spindles can turn up to 100,000 rpm and are frequently used in this range. The faster the drill bit is spun, the faster it can be fed into the product, resulting in better productivity and usually better hole quality.

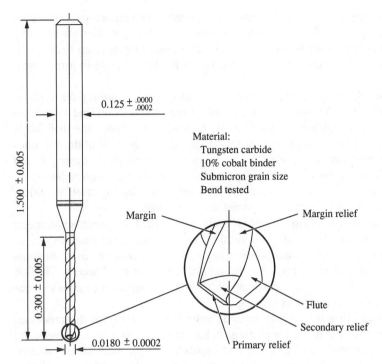

FIGURE 17-1
Typical printed circuit board drill bit.

The advance rate of the drill bit is the ratio of in-feed to spindle speed. This is expressed in inches per revolution and is a measure of the chip load on the drill bit (or, in other words, the flow of chips up the drill bit flutes). Only a limited quantity of material can flow up the drill bit flutes over a specified period of time. This maximum rate of flow varies based upon the size of the hole, thickness of the laminate, and the laminate component materials. If the chips are forced at too high a rate, the flutes will begin to load (plug) and smear epoxy on the hole wall [5]. If it is unable to clear the debris, the drill bit will eventually break.

Typical advance rates vary from 0.0005 in./rev to 0.0015 in./rev. Small drill bits below 0.010 in. in diameter are very delicate and require advance rates in the 0.0005 in./rev range to avoid breakage. These tools are used in simple, thin laminates, and they must be spun very fast, with a minimum of vibration, and fed very slowly. Larger drill bits, 0.010–0.020 in. in diameter, are used to drill composite through-holes. They are significantly stronger and can withstand advance rates in the 0.001 in./rev range, though composite cross sections are quite complex and thick.

For the best productivity and hole quality the drill bit should remain in the hole for as little time as possible [6]. Getting the drill bit in and out quickly decreases the time required to drill the panel. Also, the drill bit rubs on the hole wall slightly, and the resulting frictional effects not only contribute to drill wear but heat up both the tool and the laminate. Localized hole wall heating can soften the epoxy if the epoxy's glass transition temperature is exceeded. Reflowed epoxy will smear the copper inner-planes. To keep the amount of generated heat small, the total stroke profile should be optimized to get the hole drilled in a minimum of time [7–9].

After in-feeding, the spindle must stop briefly before pulling the drill bit back out of the hole. This short period is called the stroke dwell and lasts from 0.01 to 0.02 seconds. After dwelling at the bottom of the hole, the drill bit is withdrawn at a relatively high rate in the range of 500 ipm. Withdrawal rates five times the feed rate are common, but only when drill flute loading is not a problem. When the laminate is composed of materials that tend to stick, or aspect ratios are high enough to keep the drill bit from clearing itself of chips, withdrawal rates must be reduced to avoid drill breakage.

The number of holes (hits) a drill bit is allowed to drill is a process parameter that depends upon the hole quality required by the product. Drill changes should be as infrequent as possible to aid productivity and reduce drill bit costs. At some point drill wear may reach a level where the drill bit must be changed. The life of a drill bit depends upon the specific application and what level of epoxy smear is allowable on the copper inner-planes.

The level of allowable smear depends upon the capability of the processes used to clean the drilled holes for plating. A balance must be struck between the cost of drilling the panel and the cost and capability of the hole clean process. For printed circuit boards of any sophistication, a hole clean process is always required, even if solely to remove gross debris from the holes. More often than not, some form of chemical treatment must be performed to remove smeared epoxy resin from copper inner-planes.

In general, drill bits used to drill moderately complex cross sections can be used to remove 500 linear inches of material. For more complex laminates, this level of drill wear may result in a level of epoxy smear too heavy for the hole clean process to remove reliably.

Hole Quality

The discussion to this point has focused upon productivity concerns. The required hole quality for the particular product being drilled limits the overall process productivity. Product quality and reliability issues must be maintained in proper balance with costing issues. The drilled hole quality features of primary interest can be broken down into three main categories: drill bit breakage, drilled hole location, and plated through-hole criteria.

Remember, the primary responsibility of the drill process is to produce holes without broken drills. A functional hole is created only if the drill bit does not break. A quality product is one free of any gross defects such as those created by drill breakage and the rework that results.

The location of drilled holes is an important quality feature, because all through-holes must make contact with the appropriate copper conductors inside the laminate and avoid contact with inappropriate conductors. PTH-to-conductor contact must be made between external signal and power pads, internal signal lands, and internal power planes. Contact with external lines, internal lines, and internal power plane clearance holes, which can all be in close proximity to the PTH, must be avoided. In addition, component pins are sometimes placed in the holes, and if the hole matrix does not match the pin array very well, the component will not plug properly.

The effect of panel surface imperfections can be minimized with the use of a disposable process material entry layer, usually made of aluminum, copper, or phenolic paper [10]. The entry material may also be integrated into the panel itself, as in a protective external copper layer for example. The entry material serves as a soft layer to allow the drill bit to start a true (perpendicular) hole without being deflected by surface irregularities such as dents, scratches, or the impressions on the panel surface of the fiberglass cloth weave below.

Any early deflections become exaggerated by the time the drill bit exits the laminate. The amount of wander produced is a function of the diameter of the hole and the thickness of the panel. These two product specifications define the aspect ratio of the hole (length/diameter), and determine the most important design dimensions of the drill bit. Given length and diameter dimensions, not much can be done to eliminate the resulting drill wander. A minimum amount of random wander will always exist, because the drill bit is deflected off glass bundles in the laminate.

The most significant technique to minimize drill wander is to run a two-pass process by drilling a pilot hole first and then completing the through-hole on the second pass with one or more drill strokes. The pilot hole is drilled with a short rigid drill bit of the same diameter as the final hole. The hole produced is shallow and very nearly perpendicular to the panel surface. After the pilot holes are drilled, the program is repeated with a through drill of the proper length to complete the holes. For processes in which exit-side hole location approaches 0.005 in. deviation with only one pass, a two-pass process can cut that in half.

Depending upon the type of printed circuit board material being drilled, there are various hole quality criteria to be considered. Some holes are plated, such as signal vias and composite through-holes. Others, like drilled power plane clearance holes (usually these are etched), are not plated. Clearance holes are intended to avoid power plane continuity with the PTH. The only power plane hole quality concern is the production of copper burrs. These are minimized with the use of entry and exit materials to provide the proper backup for effective penetration and breakthrough of the copper plane.

Vias in signal cores are the simplest of plated holes and have few requirements for hole quality. The primary concern is that the via exist and that it is free of gross debris so that it can be plated properly. This is true whether the via is a blind hole on the surface of the panel or a through-hole in a signal plane.

Composite through-holes have complex cross sections and more demanding quality requirements (Figure 17-2). Epoxy smear on copper inner-planes is the most important hole quality criterion for composite through-holes. Epoxy smear

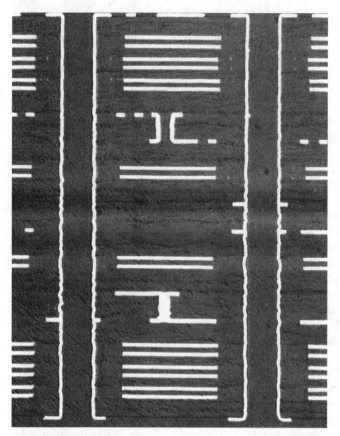

FIGURE 17-2
Typical plated through hole.

rates are related directly to drill wear. The drill and hole clean processes must be matched and balanced to produce the best quality and most economical hole possible.

The drill process is inherently destructive to the laminate; otherwise a hole could not be made. Small cracks in the hole wall will allow processing solutions to propagate into the laminate. This effect is most noticeable after copper plating; thus the term copper wicking to describe the effect. Copper wicking is promoted by excessive drill wear, but can be controlled with reduced drill bit usage and subsequent processing.

Hole wall roughness and nailheading are two quality criteria that receive more attention than they deserve. These hole wall "defects" are usually cosmetic in nature and only serve as indicators that something might be wrong somewhere else. Neither produce functional problems for the PTH, but when monitored they can alert quality control personnel to look harder at those criteria that do have an effect on the performance of the PTH.

Depending upon the complexity and design of the product, there are other hole quality criteria to consider. These are product-specific and not relevant for this discussion. For any product, however, it is important that hole quality be monitored and trends be noted. Quality products begin with quality control and constant vigilance.

The remainder of this chapter will deal with PTH quality concerns.

HOLE CLEANING

Drill Smear

One of the most important factors in achieving good mechanical and electrical connections in a multilayer circuit board is successful removal of dielectric material debris deposited on the PTH's walls during drilling. The drill smear has to be removed prior to the PTH metallization processes.

Smear is a result of heating generated by friction during drilling [8,9]. During drilling of PTHs, a drill's temperature may exceed the glass transition temperature T_g of the dielectric material (200–300° C), resulting in melting. The melts are smeared across the hole walls and copper interconnect surfaces by the penetration and extraction of the drill bit. Materials with a lower T_g will have a tendency to smear more than materials with higher T_g drilled under the same process parameters.

The smear may be several microns thick and can be seen at low magnifications on cross sections of plated through-holes (see Figure 17-3).

Smear Removal Methods

In the selection of a process for smear removal [11], the goal is to select the chemical process that will remove the smear while doing the least damage to the basic laminate material of a multilayer circuit board.

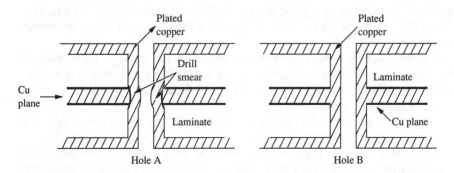

FIGURE 17-3
Plated through-holes with and without smear. Hole A has been drilled and plated directly, so the smear can easily be seen. Hole B has been cleaned with a manufacturing process and plated.

There are four basic methods for drill smear removal:

Chemical

Mechanical

Mechanical/chemical

Plasma

CHEMICAL SMEAR REMOVAL. There are five chemical methods in current use to remove drilling smears from the inner layers in multilayer board manufacturing. These are immersion in

1. Concentrated sulfuric acid (H_2SO_4)
2. Concentrated chromic acid (CrO_3)
3. A mixture of chromic and sulfuric acid
4. A mixture of potassium permanganate and sodium hydroxide for epoxy resin materials
5. Sodium naphtalene for fluoropolymers

Other materials, most of those with conventional C–H bonds, will be cleaned by (1) to (4), while for others special developments are required.

The principal advantage of sulfuric acid is that it attacks epoxy at a high rate. In contrast to chromic acid, sulfuric does not contain a heavy metal that must be kept to a very low level (1 ppm) in waste water. Most of the disadvantages of sulfuric acid are associated with process control and consistency due to its strong dependency on water concentration.

Sulfuric acid dissolves fully cured epoxy; however, water does not. If water is present, even in relatively small concentrations, in sulfuric acids, the resin becomes insoluble and either fails to dissolve or precipitates from the solution or a combination of both. Thus the acid concentration must be kept above 88 percent.

The advantages of chromic acid are bath life, reliability, and control. Chromic acid followed by a good neutralizer (like sodium bisulfate) that reduces Cr^{+6} to Cr^{+3} will produce good quality plated through-holes ready for further processing.

Chromic acid has three main disadvantages: it has difficult waste disposal specifications; a smooth surface finish is created on the epoxy resulting in poor electroless copper adhesion on hole walls; and residual Cr^{+6} is a severe poison to palladium/tin catalysts.

The use of a combination of the two acids is an effective method for epoxy etching and surface texturing.

Alkaline potassium permanganate has long been known by researchers to be an effective surface microetchant, which improves the adhesion of electroless copper. Alkaline permanganate is an effective oxidizer for a number of organic compounds (difunctional, tetrafunctional, and polyimide resin substrates). The chemistry is fairly complex, with several reactions occurring simultaneously. In alkaline solutions, permanganate ions (Mn^{+7}) slowly decompose into manganate (Mn^{+6}) ions with the release of oxygen gas:

$$4KMnO_4 + 4KOH \rightarrow 4K_2MnO_4 + 2H_2O + O_2$$

This can lead to disproportionation and a net loss of manganese from the system in the form of manganese dioxide sludge:

$$3K_2MnO_4 + 2H_2O \rightarrow 2KMnO_4 + 4KOH + MnO_2$$

The formation of manganese dioxide sludge (MnO_2) results in inconsistent performance and high operating costs. However, with the introduction of a secondary oxidizer, capable of oxidizing manganate into permanganate, this can be avoided. The rate of consumption of a regenerating oxidizer is proportional to the rate of production of manganate ions:

$$2K_2MnO_4 + [0]\cdot 0 + H_2O \rightarrow 2KMnO_4 + 2KOH + \text{reduced regenerator}$$

The regenerating oxidizer keeps the relatively expensive manganese chemistry in a balance with losses due primarily to dragout depletion and contamination.

For multilayer boards fabricated with fluoropolymers, desmearing is done with sodium naphtalene etchant. Sodium naphtalene etches the fluoropolymer [12] surface in order to render it wettable for seeding and electroless copper deposition. The reaction is self-limiting to a maximum surface penetration of about 10 μm.

Conventional multilayer boards contain woven glass laminates. After drilling and desmearing operations, exposed glass yarns in the hole barrel should be removed prior to electroless deposition. Otherwise, the entrapped yarns may act as stress-raising defects or flaws in the hole plated deposit. Also, the etched fabric in the hole wall may act as a seed catcher and possibly as a mechanical adhesion center for the copper. The glass etch is often accomplished with ammonium bifluoride followed by a hydrochloric acid or with fluoboric acid (48% wt/wt) followed by an alkaline rinse.

The mechanism of drill smear removal by using chemical methods may be summarized as follows:

1. Laminate material is chemically swelled. This is a Case II diffusion process (Chapter 26), occurring very rapidly with penetration of the solvent through the smeared organic material to the inner-plane. The expansion of the organic induces spalling from the interfaces. While this process is going on at an inner-plane interface, the ground-up combination of glass and organic is likewise detached along the hole wall.

2. A strong acid or base is used to clean up the metal interfaces. These chemicals are usually designed to balance the dissolution of both organic materials and metal. When hydrofluoric acid is used, glass buried in the inner-planes by the grinding action of the drill is removed, and some etch-back will be achieved into the fabric in the hole wall. This roughening action may be important

for forming pockets to catch the palladium seed particles. Additionally, the roughening produces a surface for mechanical linkage of the copper.

MECHANICAL SMEAR REMOVAL. The mechanical drill smear removal process can be manual or conveyorized, and can use a vapor blast or a jet of water under high pressure (300–900 psi). The preferred process is a combination of high-pressure jets of water and oscillation of a pumice slurry across the surface, deburring the inside of each hole. This is followed by rinsing and drying. The main advantages of mechanical smear removal is that there is no chemical waste.

CHEMICAL/MECHANICAL SMEAR REMOVAL. Chemical/mechanical smear removal uses a combination of the kinds of processes discussed above. First, a mechanical vapor blast process is used to remove all loose particles. In the next step, the chemical cleaning process, the epoxy smear is removed. In the final step, the vapor blast process is repeated to further clean and remove any remains of the drill smear. This is followed by a clean water wash.

PLASMA SMEAR REMOVAL. A plasma is an ionized gas, made up of free radicals, atoms, and positive or negative ions, generated under a vacuum.

The plasma process is very different from the previous processes, which rely to a large degree on swelling. The plasma actually chemically machines off molecular layers. This means that the process must have a sufficient duration [13] to remove the thickest layer of smear that has been created. Plasma attacks epoxies and is successfully used as an etchback for polyimide, multilayer, flexible circuitry. However, it does not etch glass fibers. To use plasma for PTH cleaning in the multilayer boards, it is necessary to use subsequent wet chemical processes to remove the glass fibers extending from the hole walls.

Plasma etching is controlled by gas flow, temperature, time, gas pressure, type of gas, and the loading configuration, as discussed in the chapter on plasma processes.

Equipment

Equipment requirements vary both in size and capacity depending on the volume of boards being processed and their physical dimensions. Equipment can be completely manual or totally automated and computer controlled.

Regardless of the equipment, it is important to construct the hole-cleaning equipment so that adequate flow of the solutions is achieved through the drilled holes. One of the best ways to force the solution through the PTHs is to place the board surface perpendicular to the flow direction of the cleaning fluids (Figure 17-4).

See Chapter 16 for a discussion of the agitation techniques used to assure chemical process adequacy. The processing tanks and pipes must be constructed with noncorrodible materials, and standard measures must be employed for the operator's personal protection.

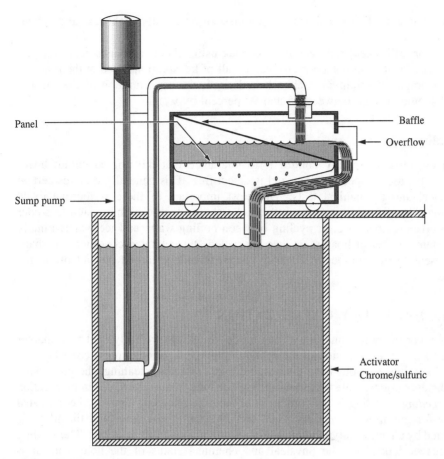

FIGURE 17-4
Schematic of chemical hole clean process equipment.

Process Control

The adequate removal of drill smear and etchback are verified on a sampling basis on special test panels, or *coupons*, which can be plated and microsectioned. Test coupons are often a part of the board design made to specified patterns, or with particular features for mechanical testing, such as peeling at the inner plane to determine bonding integrity.

Usually, microsectioning is done in x and z directions to determine PTH quality. The criteria are usually no smear, no dangling glass fibers, limited or specified etchback of the inner plane, no dielectric delamination, and no undercut or loss of adhesion between the inner plane and the organic. There is a risk of loss of adhesion when a copper oxide interface is used for adhesion and the acids selectively attack it.

The z-direction microsectioning is commonly used to evaluate smear, especially at interconnect areas. The x-direction microsectioning is used to observe protruding glass fibers, nailheads, roughness of hole walls, etchback, and delamination.

For drill smear removal and etchback using chemical/mechanical systems, daily chemical analyses are performed on all of the solutions used in the process. For example, if minimum etchback is desired, the concentration of the sulfuric acid should be kept between 85 and 90 percent by weight.

Quality

The objective of quality control is to guarantee that the printed circuit board meets its product specifications and is defect free. This generally requires certain physical and mechanical tests. The most important of these is thermal cycle testing. Usually there is a shock thermal test, for example, immersion in solder of a given temperature, or cycling between boiling water and ice water. Finally a certain number of boards are thermal cycled at accelerated conditions, perhaps representing the number of times a system would be turned on and off in the field.

THROUGH-HOLE CATALYZATION

A number of metals and alloys can be deposited electrolessly, but this chapter will focus on a discussion of the catalyzing process for electroless copper.

Previous sections have discussed drilling and hole cleaning, the processing of the epoxy-glass hole walls and the copper inner plane interfaces in preparation for *seeding*. Seeding is the formation of metallic nuclei, which are the activated sites for initiation of plating. The "seed," usually palladium and tin alloy, is created by an autocatalytic reduction of the metal ions from solutions. The seeding processes depend on the physical and chemical status of the hole wall. It is believed, but not verified, that the porosity and the size and shape of the pores in the epoxy are as crucial as the chemistry over the hole wall in establishing and maintaining good seed coverage of the surfaces. The treatments that establish the proper surface for seeding are usually proprietary, as is the actual detailed seed chemistry and physical characteristics. A general discussion of the fundamentals of the processes is given here.

There are three material components on which a seed can deposit: organic (usually epoxy), glass, and copper. Since the copper surface of the inner planes is easily prepared to accept an autocatalytic copper deposition, the seed process is balanced toward nucleating copper deposition on the glass and epoxy. Indeed, if a thick seed layer is deposited on the copper, it prevents a satisfactory copper-to-copper bond. On the other hand, if the seed coverage of the epoxy and glass is insufficient for the copper to be continuous during later deposition, the hole wall will likely fail to withstand process and environmental stresses. Thus all four processes, drill, hole clean, seed, and plate, are very strongly interactive, and a change in any one of them may influence the final properties of the plated through-hole.

The Catalyzation Processes

In electroless deposition, the driving force for reducing an ion in solution to a metal on the surface is provided by introducing a reducing agent homogeneously in the solution.

The solution is made inherently unstable; that is, the reaction between the metal ion and the reducing agent is thermodynamically favorable both homogeneously in the solution and heterogeneously at the surface. However, the rate of deposition is dependent on the density of nucleation sites. Thus the successful practice of the technique depends on the application of a suitable catalyst on the surface, making the heterogeneous reaction for nucleation and growth of the copper on the surface much more likely than the homogeneous reaction. The homogeneous reaction in solution causes deposits and precipitates of copper everywhere.

During electroless copper deposition, two electrochemical reactions are proceeding simultaneously on the surface—the reduction of copper ions to copper and the oxidation of the reducing agent, formaldehyde, to the formate ion. The potential of the plating surface is the mixed potential, the potential at which the rates of the oxidation (anodic) and reduction (cathodic) processes are equal. It is known that the rate of copper reduction is determined by the copper ion concentration at the surface and that the formaldehyde oxidation reaction is under kinetic control. Catalyzing the overall process is thus basically a matter of choosing a good catalyst for the formaldehyde reaction.

The first step of the reaction is the dissociative adsorption of formaldehyde into an adsorbed carbon-containing fragment and adsorbed hydrogen. Thus palladium, platinum, nickel, and copper, which adsorb hydrogen, can be used to catalyze electroless copper deposition, but gold, which does not adsorb hydrogen, cannot [14,15]. The formaldehyde reaction is impeded by the formation of surface oxides, and thus nonnoble metals are not good catalysts.

The catalyst most commonly used to activate surfaces for electroless copper deposition is palladium: it can both dissolve and adsorb hydrogen, and it is economically favorable due to its low mass per unit volume and surface compared to platinum. Copper is also a good catalyst providing it is kept oxide free or the oxide is stripped prior to plating. Thus the plating process will continue on copper once it is initiated.

TWO-STAGE CATALYZATION. An earlier method of seeding involved immersing the printed circuit board in a solution of $SnCl_2$ to deposit tin ions on the surface (Figure 17-5). During a subsequent immersion in a solution of $PdCl_2$, the adsorbed tin reduced the Pd^{+2} to palladium metal on the surface. There are several drawbacks in activating a surface in this way. One of the most serious ones is that the $PdCl_2$ solution electrochemically drives copper metal dissolution while forming a thick and poorly adherent palladium deposit on the copper as a replacement reaction. It is difficult to produce good inner plane joints with this process due to the loosely adherent palladium.

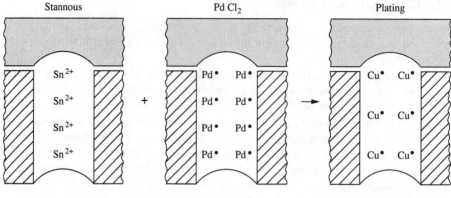

$$SnCl_2 + PdCl_2 \rightarrow SnCl_4 + Pd^0 \rightarrow \text{Copper plating}$$

FIGURE 17-5
Schematic of the two-stage catalyzation process.

SINGLE-STAGE CATALYSIS. Another method of seeding is the direct use of an optimized Pd/Sn colloid (Figure 17-6). In this case the solutions are premixed to form the desired state of the dispersed colloid [16], particle size, density, etc. In the colloid, the palladium is already in the metallic state. The structure of the colloid is known to be a metallic palladium core surrounded by a stabilizing layer of tin and associated anions (Figure 17-7).

The surface charge on the colloid particle is negative, causing the particles to stay dispersed and thus to be attracted preferentially to positive sites on glass or epoxy.

THREE-STAGE CATALYZATION. Another alternative method is three-stage seeding (Figure 17-8). In this case the application of sensitizer ($SnCl_2$) and activator ($PdCl_2$) is followed by immersion into colloidal Pd/Sn solution. Therefore,

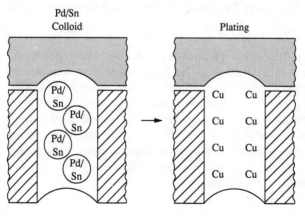

FIGURE 17-6
Schematic of the single-stage catalyzation process.

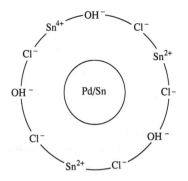

FIGURE 17-7
Diagram of single-stage colloid showing the metallic Pd/Sn core surrounded by a layer of tin and associated anions.

three-stage catalyzation is a combination of two-stage and single-stage catalyzation methods.

Once the colloid Pd/Sn is on or in the surface, the ionic tin still plays a role: it can degrade the activity of the catalyst. First, it tends to block the active palladium, and second, it might be oxidized and thus prevent further chemical

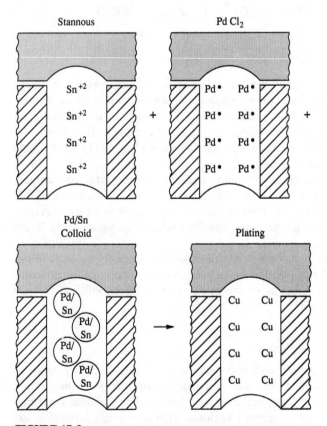

FIGURE 17-8
Schematic of the three-stage catalyzation process.

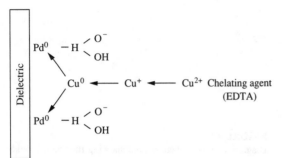

FIGURE 17-9
Reduction of Copper onto a catalyzed surface.

activation. Accordingly, thorough rinsing of the seed along with chemical removal of the tin and tin oxide increases chemical activation. During rinsing by water, tin forms the insoluble meta-stannic acid or stannic oxide.

$$H_2SnO_3 \cdot 6H_2O \xrightarrow{H_2O} SnO_2 + 7H_2O$$

$$5(H_2SnO_3 \cdot 6H_2O) \xrightarrow{H_2O} H_2Sn_5O_{11} + 10H_2O$$

$$2SnCl_2 + 2H_2O + O_2 \rightarrow 2SnO_2 + 4HCl$$

$$10SnCl_2 + 12H_2O + 5O_2 \rightarrow 2H_2Sn_5O_{11} + 24HCl$$

To enhance the chemical activity the tin and oxidized tin compounds are intentionally removed to create a surface richer in palladium. This part of the process is called acceleration [17–19]. It prepares the colloid surface for the hydrogen adsorption and formaldehyde oxidation. Common accelerators are solutions of HCl, NaOH, NH_4BF_4. HCl is commonly used since it also readily strips oxide from the copper inner plane surfaces.

The next step in the process is copper deposition on the catalyzed surface. This part of the process is still obscure, since the seed particles are so small that few practical observations can be made. However, high resolution metallography [20] appears to indicate that the particles actually coalesce during the early stages of the process.

The cupric ions are reduced by formaldehyde in a strongly basic medium. The primary product of the reaction is metallic copper. The reaction proceeds through the cuprous $(+1)$ state as shown in Figure 17-9. Both Pd^0 and Cu^0 act as catalysts for this reaction.

Process Controls

The inspection of the incoming colloidal catalyst is critical. Once the colloidal solution is made, the particle size and potential activity are set. Only minor adjustments can be made to maintain the desired density. The variables determining the composition, particle size, and agglomeration or particle separation are pH, $PdCl_2$, $SnCl_2$, and NaCl concentrations, and heating and mixing rates.

In addition to chemical analysis, cyclic voltammetry [21] can be used to measure chemical activity. During the potential cycle, hydrogen is adsorbed and released by the seed, establishing a current peak which may be calibrated with functional test results. Functional tests such as immersion of a coupon in the seed solution, followed by determination of the time for plating continuity, are useful. The test coupons may be measured for total surface Pd/Sn content. These approaches are influenced by coupon surface variables and copper-plating activity.

It should be understood that the seeding is an additive process—a catalyst is added onto the surface. It is obvious that when a substance is taken out, it should be put back, otherwise the possibility of depletion increases by day—or even by hour. Additions to a seeder bath should be made according to the analysis of the seeding process.

SUMMARY

Drilling, hole cleaning, and catalyzation of PTHs should not be looked upon as separate processes, but as one interconnected process. Therefore, any trouble-shooting should consider the interaction of the four processes—drilling, hole cleaning, catalyzation, and plating—until otherwise concluded.

As for the seeding, where the catalyst coverage dictates the coverage of electroless copper, the following measures are suggested for good processing:

The entire process line should be cleaned on a scheduled basis.

All solutions should be clear in appearance and filtered.

The concentrations of the colloidal Pd/Sn should be monitored daily.

The frequency of solution changes (refurnishing) should be established.

Activity measurement of the catalyst should be performed daily.

There should be functional testing of the electroless Cu/seed lines.

Aging/size of the colloidal particles should be monitored.

REFERENCES

1. Wrenner, W. R., "Large Multi-layer Panel Drilling System," *IBM Journal of Research and Development* 27, pp. 285–291 (1983).
2. Schaefer, R. T., and R. J. Houseman, "Carbide Metallurgy and Engineering Applied to PCB Drill Bits," *Printed Circuit Fabrication*, pp. 45–55 (July 1983); pp. 14–18 (October 1983).
3. Nelling, J. M., R. W. Saxton, and W. R. Hewitt, "Micro Drill Technology Update," *Printed Circuit Fabrication* (March 1984).
4. Hughes Aircraft Company, "Drill Parameter Study," Research and Development Report CORADCOM-772640-F, 1979.
5. Arai, M., N. Takano, and I. Hoshi, "Smear Generating Mechanism in Drilling of Multilayer Printed Wiring Boards," *Proceedings from Printed Circuit World Convention IV* WCIV-14 (1987).
6. Kanaya, Y., and K. Arai, "Small Hole Drilling in PWB Fabrication," *Proceedings from Printed Circuit World Convention IV* WCIV-15 (1987).
7. Tuck, J., "Small Hole Drilling," *Circuit Manufacturing* 23, no. 2, pp. 24–32 (1983).

8. Weiss, R. E., "The Effect of Drilling Temperature and Multilayer Board Hole Quality," *Circuit World* 3, no. 3, p. 8 (1977).

9. Weiss, R. E., "Evaluation of Drilled Hole Quality as a Function of Speed and Feed in Multilayer Boards," *Circuit World* 7, no. 1, p. 16 (1980).

10. Dietz, R., "Tune Up Your Drill Operation," *Printed Circuit Fabrication*, 6(4) pp. 33–37 (April 1983).

11. Deckert, Cheryl A., "Quality and Productivity Advances in Plated Through-Hole Technology," presented at Printed Circuit World Convention IV, June 1987.

12. Johnson, J., "Tetra-Etch Etchant for Teflon Board Processing," *Printed Circuit Fabrication* (June 1986).

13. Rust, R. D., R. J. Rhodes, and A. A. Parker, "The Road to Uniform Plasma Etching of Printed Circuit Boards," Proceedings from Printed Circuit World Convention III, Washington, D.C., May 22–25, 1984.

14. Horkans, T., C. Sambucetti, and V. Markovich, "Initiation of Electroless Cu Plating on Non-Metallic Surfaces," *IBM J. Res. Develop.* 28, no. 6, pp. 690–696 (1984).

15. Horkans, J., *J. Electroanal. Chem.* 106, p. 245 (1980).

16. Matijevic, E., A. M. Poskanzer, and P. Zuman, *Plating* 62, p. 958 (1975).

17. Cohen, R. L., and R. L. Meek, *J. Colloid Interface Sci.* 55, p. 156 (1976).

18. Cohen, R. L., R. L. Meek, and K. W. West, "Sensitization with Pd-Sn Colloids I. Role of Rinse and Accelerator Steps," *Plating* 63, p. 5255 (1976).

19. Cohen, R. L., and K. W. West, *Chem. Phys. Letters* 16, p. 128 (1972).

20. Kim, J., S. H. Wen, D.-Y. Jung, and R. W. Johnson, "Microstructure Evolution during Electroless Copper Deposition," *IBM Res. Develop.* 28, no. 6, p. 697 (1984).

21. Horkans, J., "A Cyclic Voltammetric Study of Pd-Sn Colloidal Catalysts for Electroless Depositions," *J. Electrochem. Soc.* 130 (1983).

ACKNOWLEDGMENTS

For their help in producing graphic materials:

R. Chamberlain, IBM Corp.

D. Houser, IBM Corp.

S. Travis, IBM Corp.

PROBLEMS

17.1. What is the advance rate of a drill bit?

17.2. Which two product specifications are directly proportional to drill wander?

17.3. What material requires the use of sodium naphthalene to clean through-holes?

17.4. What is used to remove/etch glass fibers?

17.5. Besides removing debris from the drilled through-holes, what is the other important role of the hole clean processes?

17.6. Which metals are used to catalyze electroless copper?

17.7. What is the colloid Pd/Sn particle charge?

17.8. What are the sensitizer and activator in the two-stage catalyzation process?

CHAPTER 18

ETCHING OF METALS BY WET PROCESSES IN ELECTRONIC PACKAGING

OSCAR A. MORENO
RON C. McHATTON
RALPH S. PAONESSA
ROY H. MAGNUSON

INTRODUCTION

Etching of metal is the removal of unwanted metal by chemical means. In electronic packaging, etching is often used to produce circuit lines on a dielectric (insulating) substrate by etching a pattern in a metal film bonded to the substrate. In contrast to metal plating, which is an additive process, etching is a subtractive technique.

This chapter is concerned with wet metal etching, in which the metal is removed by exposure to a reactive solution, or etchant. Dry etching techniques, such as chemical plasma etching, reactive ion etching, sputter etching, and ion milling, involve attack on metals or other solids by reactive gases. Although these gas phase processes were first utilized mainly in integrated circuit and magnetic memory manufacturing, they are increasingly being applied to circuit packaging. These dry processes are discussed in chapter 13. Many of the parameters that control wet etching are encountered in dry processing as well.

539

The first commercial applications of metal etching to electronic packaging began in the 1940s. A thin sheet of copper was laminated to a flat organic insulating base, or "board." A protective coating, or "resist," was then applied ("printed") onto the copper in those areas that were to remain, and the exposed copper was then removed by exposure to ferric chloride solution. After removal of the resist, there remained a pattern of conductive copper lines on an insulating support. These were the first examples of printed circuit boards or PCBs. Circuit components such as resistors, capacitors, etc., could then be connected in appropriate configurations by soldering them to the copper lines. This process was a significant advance in mass production of electronic circuits for two reasons. First, it replaced much of the time-consuming wiring procedure with an assembly line process utilizing prefabricated boards as connectors. Second, it allowed the rapid production of a wide variety of printed circuits simply by changing the patterns of the etch resist. The lines in these early boards were on the order of a millimeter in thickness and five to ten millimeters wide.

In modern electronic packaging, wet chemical etching is still an important process [1]. Figure 18-1 shows a schematic representation of subtractive fabrication of a circuit pattern. Substrate materials include thin film polymers such as polyimides, board materials such as epoxy-impregnated fiberglass, ceramics, and

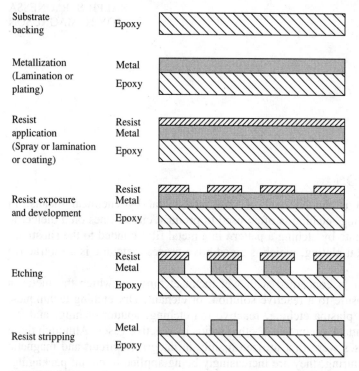

FIGURE 18-1
Steps involved in the fabrication of a printed circuit board by the subtractive process.

other insulating materials. The blanket metal film is produced by techniques such as electroless plating, electroplating, vapor deposition, and lamination of metal foil. Moreover, these metal films are often composed of several layers of different metals or other materials. Etching of multilayer films presents special challenges that are discussed below.

The protective barrier or resist can take several forms. The resist can be sprayed on in a screening process. A blanket coating of a photosensitive resist, or *photoresist*, can be applied and a pattern developed by the methods described in chapter 12. The properties of the resist material often impose requirements on the etching chemistry. For example, certain photoresists are unstable in alkaline solutions and therefore cannot be used with alkaline etchants.

As circuit requirements become more complex and circuit components tend increasingly in the direction of miniaturization, electronic packages require greater numbers of circuit lines compressed into smaller areas. As a result, the demands on metal etching have become more stringent. Second-level circuit boards can be as large as several feet on a side. On such a board lines can range from 10 to 100 μm in thickness. These circuit lines might typically range in width from 50 to 250 μm with corresponding spacings of 100 to 500 μm. For first-level chip carrier modules, where attachment to several hundred input/output (I/O) connections may be required in a small area, circuit lines are as narrow as 25 μm and in the future may become even narrower. The corresponding thickness of these narrow lines is in the range of 4 to 10 μm. One of the challenges facing metal etching and packaging in general is the production of progressively thinner circuit lines and denser circuit patterns to make connections to increasingly smaller chips with greater numbers of I/O connections.

Metal etching is a mixture of science and empiricism. This chapter will discuss the scientific rudiments that underlie metal etching. The purpose is to give the reader a scientific framework to begin to understand etching.

An etching system consists of at least three components. These components are the metal to be etched, an etchant solution that will remove the metal, and a machine or container where the metal and the etchant will come in close contact. Table 18.1 gives examples of etching technology.

In the process, a suitable etchant is placed in contact with the metal, for example by spraying or immersion. The metal is etched or dissolved away by chemical reaction with the etchant. The recession of the metal-etchant boundary is controlled by several phenomena such as kinetics, mass transfer, surface adsorption or passivation, and fluid mechanics. Also, the chemical and physical thermodynamics will dictate the mechanism or mechanisms by which the etching takes place. All these effects will be discussed in the following sections. After the proper shape and dimensions are obtained, the photoresist can be removed. The substrate is then ready for any further processing such as mounting electronic components or combining several substrates together to fabricate multilayer PCBs. The reader is referred to the literature for schematics of etching processes and equipment [1].

TABLE 18.1
Etching systems and operating conditions

Metal	Solution	Temperature	Tank
Al	NaOH(20% weight)	$60°-90°C$	dip
Al	HCl(25% weight)	$80°C$	dip
Al	$FeCl_3(12°-36°Be)$	$25°-50°C$	spray
Cr	$KMnO_4/NaOH$, 1:1	$45°C$	spray
Cr	$K_3Fe(CN)_6/NaOH$, 1:3	$25°C$	spray
Cr	$CS(NH_2)/H_2SO_4$, 1:1	$30°-50°C$	dip
Cr	$Ce(SO_4) \cdot 2(NH_4)_2SO_4/HClO_4$, 1:2	$25°-50°C$	spray
Cu	$CuCl_2/HCl$, 2:1	$25°-50°C$	spray
Cu	$Na_2S_2O_8/H_2SO_4$, 1:1	$25°-50°C$	dip
Cu	$FeCl_3(42°Be)$	$25°-50°C$	spray
Ni	$FeCl_3(42°Be)$	$25°-50°C$	spray
Ni	$HNO_3/H_2SO_4/H_3PO_4/CH_3COOH$, 3:1:1:5 vol.	$80°-95°C$	dip
Pd	$HCl/HNO_3/CH_3COOH$, 1:10:10 vol.	$25°-50°C$	dip

THERMODYNAMIC CONSIDERATIONS

From a chemical perspective, metal etching is the conversion, through a chemical reaction or series of chemical reactions, of a solid, insoluble metal to a soluble form. The extended lattice of metal atoms in the solid state must be broken down so that these atoms can enter the solution as soluble compounds. As a practical matter, most etching reactions take place in aqueous solutions. In this chemical environment, metal compounds are generally stable when the metal atom has a formal positive charge, or is in a positive state. Metal atoms in the metallic lattice have a formal charge, or *oxidation state* or *valence*, of zero. Thus the metal etching reaction must involve the removal of electrons from the metal. This process is termed *oxidation* of the metal. A fundamental requirement of any electron loss reaction is that it be exactly balanced by a corresponding electron gain reaction, or *reduction*. In other words, the electrons that are removed from the metal must go somewhere. Where they go is to the etchant, which functions as an electron acceptor or oxidizing agent. For example, etching of copper by ferric chloride proceeds by the following overall reaction:

$$Cu^0 + 2Fe^{3+} \rightarrow Cu^{2+} + 2Fe^{2+} \qquad (18.1)$$

The ions Fe^{3+}, Fe^{2+}, and Cu^{2+} are complexed in solution by varying numbers of chloride ions and water molecules to give species such as $FeCl_4^{2-}$, $FeCl_3(H_2O)_3$, $FeCl_2(H_2O)_4^+$, $FeCl(H_2O)_3^{2+}$, etc. The etchant is thus seen to be an oxidizing agent, which is reduced in the etching reaction. In the complete metal etching reaction, a reduction reaction is required to maintain charge balance. Metal etching places no restrictions on the reduction that may be employed. In electrochemical etching, electric current is used to carry electrons from the site of metal oxidation to the site of reduction and ionic current completes the circuit. In most cases of chemical etching, reduction occurs at the site of metal oxidation, in which case no electronic or ionic currents are involved.

The thermodynamic tendency for a metal etching reaction to proceed depends on the thermodynamic tendency of its component reactions: electron transfer from and solvation of the metal, and reduction of the etchant. In general, etching reactions will be favored when there is a favorable balance among the free energy required to remove electrons from the metal, the free energy gain when the metal ion forms a solvated complex, and the free energy gain when the etchant is reduced. To determine whether the reaction between a given oxidizing agent and a particular metal is thermodynamically allowed, one can make use of tabulated half-cell potentials for the reaction. Thus, the reaction is broken down into its component half-cells; for example,

$$
\begin{array}{ll}
Cu^0 \rightarrow Cu^{2+} + 2e^- & E_1 = -0.34 \text{ V} \\
2Fe^{3+} + 2e^- \rightarrow 2Fe^{2+} & E_2 = 0.77 \text{ V} \\
\hline
Cu^0 + 2Fe^{3+} \rightarrow Cu^{2+} + 2\,Fe^{2+} & E_t = E_1 + E_2 = 0.43 \text{ V}
\end{array}
\qquad (18.2)
$$

The overall change for E_t for the reaction is the sum of the potentials for the half-cell reactions in the directions written. Tabulated values[1] of E^0 are for half-cell reactions written as reduction reactions. For oxidation reactions, the sign of the potential must be reversed. When the half-cell reactions are added together, the electrons cancel out, and charge is conserved. A positive value of E_t indicates a thermodynamically favorable reaction. These potentials can be altered by concentration changes and addition of species that form complexes with the ions involved. A partial listing of half-cell potentials is given in Table 18.2. The reader can thus verify that, for example, the dissolution of aluminum by ferric ion under standard conditions is thermodynamically favorable, whereas the attack of ferric ion on platinum is not.

It is in theory possible to dissolve a metal by attack of species that will form a zero valent complex with the metal. An example of such a process is the formation of volatile $Cr(CO)_6$ from Cr metal in a CO atmosphere. Such species are well known for many of the metals used in circuit applications, e.g., copper, silver, gold, platinum, and chromium. The thermodynamics of these reactions are generally unfavorable, however, so that these complexes tend to decompose and deposit back to the bulk metal. In addition, the rates of formation of such complexes from bulk metals are quite slow, so that such reactions have not found practical applications to metal etching.

Thermodynamics and Pourbaix Diagrams

As has been noted above, the tendency for a metal to dissolve depends in part on the oxidizing strength of the medium in which it is immersed. This tendency

---

[1] Standard potentials, E^0, are defined for unit activity, 1 atmosphere pressure, 25°C, and referenced against the normal hydrogen electrode (NHE), whose standard potential is defined as 0.000 V.

TABLE 18.2
Standard reduction potentials
Potential is referenced against the normal hydrogen electrode (NHE)

Half cell	Potential
$Al^{3+} + 3e^- = Al^0(0.1M\ NaOH)$	-1.71
$Au^+ + e^- = Au^0$	1.68
$Ce^{4+} + e^- = Ce^{3+}(1M\ H_2SO_4)$	1.44
$Cl_2 + 2e^- = 2Cl^-$	1.36
$Cr^{2+} + 2e^- = Cr^0$	-0.56
$Cr^{3+} + e^- = Cr^{2+}$	-0.41
$Cr_2O_7^{2-} + 14H^+ + 6e^- = 2Cr^{3+} + 7H_2O$	1.33
$Cu^+ + e^- = Cu^0$	0.52
$Cu^{2+} + e^- = Cu^+$	0.16
$Cu^{2+} + 2e^- = Cu^0$	0.34
$Fe^{2+} + 2e^- = Fe^0$	-0.41
$Fe^{3+} + e^- = Fe^{2+}(1M\ HCl)$	0.77
$2H^+ + 2e^- = H_2$	0.00
$MnO_4^- + 8H^+ + 5e^- = Mn^{2+} + 4H_2O$	1.49
$Ni^{2+} + 2e^- = Ni^0$	-0.23
$O_2 + 4H^+ + 4e^- = 2H_2O$	1.23
$Pd^{2+} + 2e^- = Pd^0$	0.83
$Zn^{2+} + 2e^- = Zn^0$	-0.76
$Pt^{2+} + 2e^- = Pt^0$	1.19

varies from metal to metal. Furthermore, most metals can exist in several stable oxidation states, and sometimes in several forms for a given metal and oxidation state. These stabilities depend on the pH and oxidizing strength of the solution. Up to this point, we have tacitly assumed that when a metal is oxidized by an etchant in solution, soluble products form. This is not always the case. Under certain circumstances, insoluble oxidation products may form and adhere to the metal surface. These *passive* films can prevent further oxidation by protecting the underlying metal. The stability of these films is determined in part by the oxidizing strength of the environment and of the solution, and by the concentration of species in solution that might contribute to the films' breakdown.

These considerations are crucial to an understanding of metal etching and for understanding any processes that involve oxidation-reduction reactions of metals. Large volumes of thermodynamic data are available to quantify these phenomena. Particularly useful summaries of such thermodynamic data for metals can be found in graphical constructs known as Pourbaix diagrams [2]. We therefore devote this section to an introduction to these useful tools. Some of the processes for which Pourbaix diagrams can provide understanding are:

1. Etching: $M^0 \rightarrow M^{n+} + ne^-$
   Example: Persulfate etching of copper.

$$Cu^0 + S_2O_8^{2-} \rightarrow Cu^{2+} + 2SO_4^{2-}$$

2. Plating: $M^{n+} + ne^- \rightarrow M^0$
  Example: Nickel electroplating.

$$Ni^{2+} + 2e^- \rightarrow Ni^0$$

3. Displacement reactions: $mM^{n+} + nM'^0 \rightarrow mM^0 + nM'^{m+}$
  Example: Immersion tin coating.

$$Sn^{2+} + 2Cu^0 \rightarrow Sn^0 + 2Cu^+$$

4. Formation of oxide coatings: $xM^0 + yH_2O \rightarrow M_xO_y + 2yH^+ + 2ye^-$
  Example: Chloriting of copper ("black oxide" process).

$$ClO_3^- + 3Cu^0 \rightarrow Cl^- + 3CuO$$

In addition, corrosion is typically an aqueous redox process, and is of considerable interest for the assurance of reliability of circuit packages in the field.

As with any process, reaction conditions for a viable aqueous or "wet" process must be set up so that the desired transformation is thermodynamically favored. Moreover, the reaction should proceed at a rapid rate; in other words, it should be kinetically facile. Process conditions, such as pH, temperature, concentration, and addition of metal complexing agents, are controlled to attain these goals.

Pourbaix diagrams are potential-pH diagrams for aqueous processes. These diagrams have been very useful in their application to corrosion chemistry. A Pourbaix diagram depicts areas of thermodynamic stability as a function of speciation mapped onto a potential (ordinate) versus pH (abscissa) plot. More positive values of the potential are in the oxidizing direction (international convention for sign of the voltage), implying that higher oxidation states are found at the top of the chart. The pH is plotted in the normal fashion, with more basic solutions found at the right hand side. Knowledge of the pH of the solution in which a metal is immersed and the electrochemical potential the metal experiences (whether applied or chemically driven) allows one to determine the thermodynamically spontaneous direction for a reaction.

Dramatic changes can occur in the presence of metal ion complexing agents in solution (see appendix). It is the intent of this section to present an approach to the construction of the potential-pH diagrams, illustrated with specific examples drawn from the thermodynamics of nickel chemistry. The nickel system is chosen because of its simplicity. Thermodynamic information is available from a number of compiled sources, as well as the primary literature [2–5].

Thermodynamic Relationships

The type of process considered here is normally run at constant temperature and pressure, so that the appropriate thermodynamic indicator is the free energy G. For a system written as

$$\text{Reactants} \rightarrow \text{Products}$$

the free energy change is defined by

$$\Delta G = G(\text{products}) - G(\text{reactants})$$

The following possibilities exist: $\Delta G < 0$, the reaction is spontaneous left to right as written; $\Delta G > 0$, the reaction is spontaneous right to left; $\Delta G = 0$, the system is at dynamic equilibrium, no net change. In the potential-pH diagram, boundaries are set up at $G = 0$, and at a glance one can determine the spontaneous direction, since the favored species are mapped into areas of the chart.

Several common thermodynamic relationships will be valuable in the construction of the Pourbaix boundaries. The standard enthalpy change for a reaction is given by [6]:

$$\Delta H^0 = \sum_i n_i \Delta H_f^0(\text{products}) - \sum_i n_i \Delta H_f^0(\text{reactants})$$

where ΔH_f^0 indicates the standard heat of formation, and n the number of moles involved in the reaction. The standard entropy change for the system is given by

$$\Delta S^0 = \sum_i n_i S_i^0(\text{products}) - \sum_i n_i S_i^0(\text{reactants})$$

where S_i^0 is the absolute entropy of a mole of compound i at 298.15 K.

The temperature dependence of the standard free energy is given by the equation

$$\Delta G_T^0 = \Delta H^0 - T\Delta S^0$$

which holds exactly at 25°C, and approximately for temperatures not far different from 25°C. The concentration dependence of the free energy is given by

$$\Delta G_T = \Delta G_T^0 + RT \ln Q$$

where Q is the reaction quotient. For the reaction in condensed phase: $a\text{A} + b\text{B} = c\text{C} + d\text{D}$,

$$Q = \frac{[\text{C}]^c[\text{D}]^d}{[\text{A}]^a[\text{B}]^b}$$

To be precise, the concentrations should be expressed in terms of activities, and referred to standard conditions of 1 molal.[2] For convenience, the following discussion will use molar concentrations, but the reader is cautioned that activity effects can be substantial, especially in the high concentration ranges of many processes. For example, the mean activity coefficient for a 0.1 molal copper sulfate solution is 0.16, leading to chemical activities substantially lower than the molar concentrations would lead one to believe [7].

At equilibrium, $\Delta G_T = 0$ and the activities of species are at equilibrium values, so that $Q = K$, the equilibrium constant, and one may write $\Delta G_T^0 =$

---

[2] One molal = 1 mole of compound per kilogram of total solution.

$-RT \ln K$. The connection to electrochemical potentials is provided by

$$\Delta G_T = -nF\Delta E_T \text{ and } \Delta G_T^0 = -nF\Delta E_T^0$$

which leads to the equation

$$\Delta E_T = \Delta E_T^0 - \frac{2.303RT}{nF} \log Q$$

where n is the number of electrons transferred in the process. Since the cell potential E_T is directly related to the free energy change except for a reverse in sign, the spontaneous direction of a redox reaction can also be determined from the cell potential. In aqueous solution, cell potentials are referenced to the NHE, and solution thermodynamic properties are referenced to the hydrogen ion at unit activity.

The Nickel System—An Example

In this section, a simplified consideration is given to the calculation of a potential-pH diagram for nickel. The species Ni_{aq}^{2+}; nickel metal; and nickel(II) hydroxide, $Ni(OH)_2$, will be incorporated. For inclusion of higher oxidation states of nickel and the nickelate anion, Pourbaix's *Atlas* may be referenced [2]. First, the diagram will be determined at standard conditions, then at nonstandard concentration. The effect of temperature will be examined as well as the influence of a strong complexing agent. Thermodynamic parameters of the nickel system can be found in Table 18.3.

Standard Conditions

For the three nickel forms listed above, three equilibrium boundaries may be drawn. They correspond to the processes of reduction of nickel ion in solution to nickel metal, reduction of nickel hydroxide to nickel metal, and precipitation of nickel ion as the hydroxide.

TABLE 18.3
Thermodynamic parameters for the nickel system

Species	ΔH_f^0, kcal/mole	ΔS_f^0, cal/deg · mole
H_2O	− 68.32	16.72
H^+	0	0
OH^-	− 54.957	− 2.519
H_2	0	31.21
O_2	0	49.003
Ni^0	0	7.2
Ni_{aq}^{2+}	− 15.3	−38.1
$Ni(OH)_2$	−128.6	19

From the CRC Handbook of Chemistry and Physics [3]

1. $Ni_{aq}^{2+} + 2e^- \rightleftharpoons Ni^0$

2. $Ni(OH)_2 + 2H^+ + 2e^- \rightleftharpoons Ni^0 + 2H_2O$

3. $Ni_{aq}^{2+} + 2OH^- \rightleftharpoons Ni(OH)_2$

The boundary for the reduction of nickel(II) ion will be a horizontal line in the Pourbaix diagram, since it involves a change of oxidation state, but no proton or hydroxide ion. Since we are referencing to the normal hydrogen electrode, most often a value of the standard potential of the reaction can be found in a table of standard half-cell potentials, and the Nernst equation applied directly. However, one must be sure that the same set of thermodynamic data is used consistently. For calculation of thermodynamic parameters one must use the net reaction:

$$Ni^{2+} + H_2 \rightleftharpoons Ni^0 + 2H^+$$

For this reaction, $\Delta H^0 = 15.3$ kcal/mole, $\Delta S^0 = 14.09$ e.u., and at 298.15 K, $\Delta G^0 = 11.10$ kcal/mole. One then finds $E^0 = -0.241$ V, independent of the pH.[3] (Note that the potential reference is considered a separate half-cell at unit activity of hydrogen ion.) The Nernst equation then has the following form (with the activity of solid Ni^0 taken as 1):

$$E = E^0 - \left(\frac{0.0592}{2}\right) \times \log \frac{1}{[Ni^{2+}]}$$

where the Δ notation has been dropped, and potentials are assumed to be against NHE. At standard conditions, nickel ion carries unit activity so that the logarithmic term is dropped, and one starts the diagram with a horizontal line at -0.241 V as shown in Figure 18-2.

For determination of the boundary between nickel hydroxide and nickel metal referenced to NHE, the net reaction ($n = 2$) is

$$Ni(OH)_2 + H_2 \ (+2H^+) \rightleftharpoons Ni^0 + 2H_2O \ (+2H^+)$$

where the extra hydrogen ions are included as a reminder that the NHE is independent of the nickel reaction. From the thermodynamic data, one can show that $\Delta H^0 = -8.04$ kcal/mole, $\Delta S^0 = -9.57$ e.u., $\Delta G_{298}^0 = -5.187$ kcal/mole, and $E^0 = +0.112$ V. The Nernst equation describing the boundary is

$$E = E^0 - \left(\frac{0.0592}{2}\right) \times \log \frac{1}{[H^+]^2}$$

or

$$E = E^0 - 0.0592 \times pH$$

---

[3] Since the constant is defined in coulombs per equivalent, ΔG must be expressed in joules per mole to obtain the potential in volts. To convert kilocalories per mole to joules per mole multiply kilocalories per mole by 4.184.

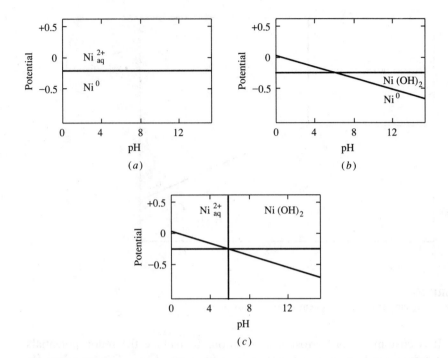

FIGURE 18-2
Construction of the pH-potential diagram.

This boundary will be a diagonal line since it involves both an oxidation state change and a pH dependence. This line has been added in Figure 18-2(b).

The precipitation reaction does not change the oxidation state of the nickel, but does include a dependence on hydroxide ion, linked through the dissociation constant of water, K_w, to the pH. Thus the boundary will appear as a vertical line at the pH at which precipitation begins to occur. The calculation is handled by determination of the standard free energy change for the process, $\Delta G^0 = 21.91$ kcal/mole, then evaluation of the equilibrium constant via $\Delta G^0 = -RT \ln K$.

$$K_{pptn} = \frac{1}{[Ni^{2+}][OH^-]^2} = 1.09 \times 10^{16}$$

At $[Ni^{2+}] = 1$, the equilibrium equation yields $[OH^-] = 9.59 \times 10^{-9}$ and the pH of the boundary is found from $pH + pOH = p(K_w) = 14.00$ at 25°C. In Figure 18-2(c), a vertical line has been added at pH = 5.98.

At this stage, the virtual lines are removed and the stable species indicated, as shown in Figure 18-3. For example, a boundary between $Ni(OH)_2$ and Ni^0 cannot exist at the far left of the diagram, since Ni^{2+} is the stable nickel(II) species at acidic pH. In most cases the regions of stability are readily determined by inspection. In recent years computer programs for the generation of potential-pH diagrams have been written, and algorithms for determination of the stable species developed [8,9].

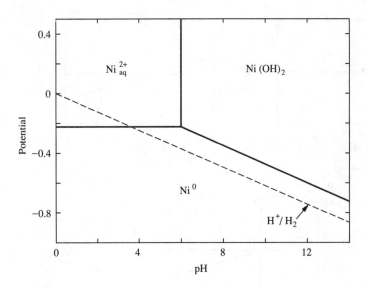

FIGURE 18-3
Pourbaix diagram of the nickel system in aqueous solution.

It is customary for corrosion applications to include the redox potentials for the reduction of protons and oxygen as a function of pH. The boundary for reduction of protons has been indicated as a dotted line in Figure 18-3. It can be seen that nickel metal is susceptible to attack by protons in the pH region from 0 to 4.5. The oxygen couple lies primarily above the range of the diagram, so that nickel would be predicted to be susceptible to oxidation by air. However, to the right of the precipitation boundary, the dissolution is inhibited by the formation of a hydroxide coating, a so-called passivation region.

Ni^{2+} at Nonstandard Concentrations

For most applications it is rare for a solution species to be at unit activity. The diagram boundaries are altered in a straightforward manner and overlaid to show the effect of changes. For the reduction of nickel(II) ion to nickel metal, the full Nernst equation involving Ni^{2+} is used. For example, at an activity of 10^{-2}, the horizontal line appears at

$$E = E^0 - \frac{0.0592}{2} \times \log \frac{1}{0.01} = -0.300 \text{ V}$$

The reduction of nickel(II) hydroxide does not involve Ni^{2+} ion, and thus is unaltered. The vertical precipitation boundary is determined from the equilibrium expression by way of

$$[OH^-] = \sqrt{\frac{K_{pptn}}{[Ni^{2+}]}}$$

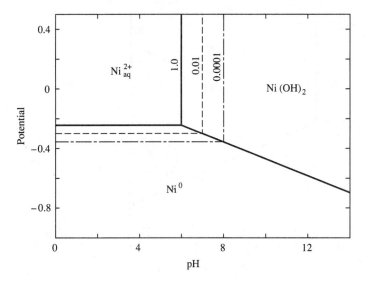

FIGURE 18-4
Concentration dependence of the Pourbaix diagram phase boundaries for the nickel system in aqueous solution.

to give a pH of 6.98. Boundaries for several concentrations of Ni^{2+} have been overlaid in Figure 18-4.

Temperature Dependence

Many processes are carried out at elevated temperatures, so that an adjustment may be necessary. This requires the availability of both enthalpy and entropy changes for the reaction of interest. The free energy changes are then computed using the Gibbs-Helmholtz equation. Equilibrium constants and standard potentials are then recalculated. Note that the temperature coefficient in the Nernst equation does not account for the change in the standard potential with temperature. The nickel system has been worked out in the appendix for $T = 328$ K, and is shown in Figure 18-5, referenced to NHE at 328 K. In practice, potentials are typically measured against more convenient electrodes, such as the saturated calomel electrode, and one should refer to manufacturer's specifications for the potential of the electrode at various temperatures. It should be noted in these calculations that the dissociation constant for water, K_w, changes as well with temperature.

The Effect of Complexing Agents

More substantial changes in the potential-pH diagram can occur when a complexing agent for the metal is present in the solution. Metal ions form a variety of complexes, often in a series of successive equilibria. Nickel(II) ion, for example, forms a 1:1 complex with triethylenetetramine,

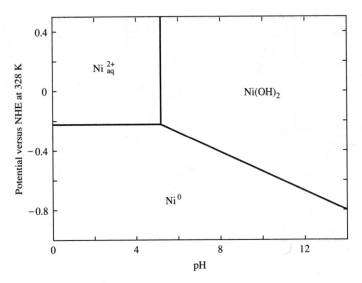

FIGURE 18-5
Pourbaix diagram of the nickel system calculated at 328.2°C.

$H_2N(CH_2)_2NH(CH_2)_2NH(CH_2)_2NH_2$, abbreviated *trien*. The reaction may be written as

$$Ni^{2+} + trien = Ni(trien)^{2+}$$

for which a formation or complexation equilibrium constant, K_f, can be written ([10], p. 566):

$$K_f = \frac{[Ni(trien)^{2+}]}{[Ni^{2+}][trien]} = 1.0 \times 10^{14} \text{ at } 25°C$$

The net effect of the complexing agent will be to reduce the free nickel(II) ion in solution with concomitant changes in the redox and precipitation boundaries. Very many complexing agents have protonation/deprotonation equilibria, so that the effect varies as well with pH, as the nickel ion competes with protons in the system. The trien ligand is an example, having four successive protonation equilibria. Thus the concentrations of $H_4(trien)^{4+}$, $H_3(trien)^{3+}$, $H_2(trien)^{2+}$, and $H(trien)^+$ must also be determined. These successive equilibria are related mathematically by the acid dissociation constants K_1, K_2, K_3, and K_4 ([10], p. 578). In addition to the five trien species (four cations and the neutral ligand) and the two nickel species, we consider also hydrogen ion and hydroxide ion, for a total of nine concentrations to be determined. If the pH and the total amount of nickel and trien in solution are known, one arrives at nine unknowns and nine equations. Although the arithmetic can be complex, the amount of free nickel(II) ion can be found at any pH (see appendix). Figure 18-6 shows the impact of 1.00 M trien on the diagram at 0.01 M total nickel. The boundaries in the absence of trien have been included as dotted lines for reference.

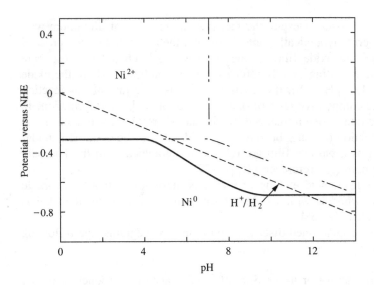

FIGURE 18-6
Calculated phase boundary for nickel (0.01M) in the presence of the trien ligand (1.00M) at 25°C. Ni^{2+}_{aq} in this diagram is a combination of the aqua ion and Ni(trien)$^{2+}$. Dotted lines refer to the nickel system boundaries in the absence of trien.

There are several things to note. First, nickel hydroxide is virtually eliminated from normal pH ranges, and becomes important only at pH 14. This implies that the corrosion of nickel (dissolution into solution) in alkaline solution will be much more likely in the presence of trien, because a protective hydroxide coating formed without the complexing agent will simply dissolve as Ni(trien)$^{2+}$. In addition, the Ni^{2+}/Ni^{0} couple has become more negative, implying greater susceptibility of nickel metal to attack by hydrogen ion in an etching process (provided no kinetic complications are present). This can be seen by examination of the H^{+}/H_{2} boundary—the reduction of hydrogen ion is favored thermodynamically over the pH range 0 to 12, since the Ni^{2+}/Ni^{0} boundary lies below it.

Application of Pourbaix Diagrams to Metal Etching-Passivity

As stated above, the products of metal oxidation in aqueous solution are not always soluble. If these products form and adhere to the metal surface, a protective film forms which prevents further corrosion (or etching) of the metal. These protective coatings are referred to as passive films, and the phenomenon is known as *passivation* or *passivity*. The formation of such films, and the conditions under which they can be removed, have important consequences for metal etching. Pourbaix diagrams are quite useful in understanding these phenomena. Consider the following examples.

Oxide films on metal surfaces are a prime example of passive films. Consider the case of aluminum, which in air or neutral water rapidly forms a protective coating of oxide. Despite the fact that the reaction of aluminum with water at pH 7 is thermodynamically quite favorable, the reaction does not proceed in the presence of the oxide film. If the surface is scratched to expose bare metal, the protective coating rapidly reforms. Under conditions where the oxide is soluble (pH < 4 or pH > 8), the reaction proceeds. The rate of this reaction is governed by the competitive rates of oxide dissolution and oxide formation. In acidic solution the rate is also influenced by the rate of proton reduction, while in basic solution reduction of water becomes a consideration. The conditions under which a metal forms a passive film are represented graphically in the Pourbaix diagram for aluminum (see reference [2], p. 172).

From Figure 18-3, acid etching of nickel is thermodynamically favorable only at pHs < 4, whereas with 1.0 M trien present, acid etching of nickel is favorable at pHs well above 4.

From the above mentioned discussion on Pourbaix diagrams, the following rules can be drawn:

1. If the system lies above or to the right of a boundary line, product is formed (or etching occurs.)

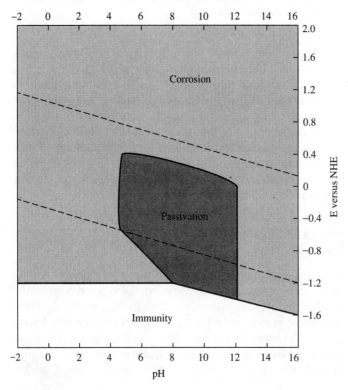

FIGURE 18-7
Simplified diagram for chromium metal in aqueous solution.

2. If the system lies below or to the left of a boundary line, reactants are formed (or no etching occurs).

3. If the system lies on a boundary line, the system is at equilibrium (no net change).

Consider the simplified Pourbaix diagram for the corrosion of chromium metal in aqueous solution in Figure 18-7. Applying the discussion above, it is clear that chromium metal is unstable toward etching if the pH-potential intersection of the etching system lies in the cross-hatched region of the Pourbaix diagram. If the pH-potential intersection lies in the shaded region, the foregoing discussion indicates that chromium metal should be unstable with respect to oxidation; however, the product oxide is insoluble, resulting in formation of a passivating coating. Only in the unshaded region is chromium metal thermodynamically stable as unoxidized or unpassivated.

KINETIC CONSIDERATIONS

Up to this point, the discussion has been based purely on thermodynamics. An additional complication arises from chemical kinetic considerations. While detailed discussion of specific etch mechanisms is beyond the scope of this work, the subject deserves a general exposition. Thermodynamics describes only the equilibrium state and the direction a given system will proceed to attain that state. Kinetics describes the evolution of mass flow throughout the system, including the *approach* to equilibrium and the *dynamic maintenance* of that state. In addition, it describes the detailed mechanism by which a given etch process functions. A system cannot even be crudely understood unless the kinetic and thermodynamic views are to some extent understood and are internally consistent.

Processes that are thermodynamically favorable take place at a finite rate. From a practical viewpoint, it is normally desirable to have the operation proceed at as high a rate as possible, while still maintaining control. Examples of important factors related to kinetics of the system include the geometry (such as etch profile or uniformity of a deposit) and the characteristics of a deposit or coating (such as ductility, hardness, porosity, and composition).

Let us now consider the factors that govern the rate at which a metal dissolution reaction occurs. Of profound importance is the heterogeneous nature of the reaction: the reaction takes place between dissolved (homogeneous) species and the solid metal at the heterogeneous metal-solution interface. The reaction can be generalized by considering a solution of oxidizing agent Ox in contact with metal M. The species Ox must be transported to the surface of the metal, where it extracts electrons from the metal via a succession of one or more molecular processes, and is converted to the reduced form Red. This species then diffuses away from the metal (assuming it remains in solution), and the oxidized metal ion M^{n+} leaves the metal surface and enters the solution. This is illustrated in Figure 18-8. In this regard it might be assisted by solvent molecules or other species to which it can bind. This series of events may be quite complicated

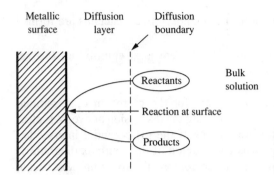

FIGURE 18-8
Pictorial representation of the solution/metal interface

on the molecular scale. Each elementary step has its own rate parameters. The slowest step is the rate-determining step, or bottleneck, in the process. Factors that affect this step, such as concentration of solution species and temperature, can influence the overall rate.

In addition, the reaction rate may be influenced by the rate of mass transfer of reactants and products to and from the surface, respectively. As mentioned earlier, these two extremes are often referred to as activation control and mass transfer control of the reaction rate. In many practical etching systems, the rates are controlled by mass transfer (i.e., diffusion), and hence are strongly dependent on agitation, fluid flow, viscosity of the solution, and geometry at the metal-photoresist surface. This phenomenon is analyzed for the case of electroplating in Chapter 16, and much of that discussion is applicable here.

Mass Transport Control

Clearly the reaction at the surface can take place no faster than the reactants arrive at the surface. If this is the slowest step in the sequence of events, the process is said to be under mass transport control. In the region close to the reacting surface, identified as the diffusion layer in Figure 18-8, there will be a concentration gradient set up. In the limit, the concentration of reactants will be zero at the surface and increase as the distance away from the surface increases, until the concentration of the bulk solution is reached. The distance from the surface to the point where the bulk concentration is reached is dependent on agitation. The rate of flux of material across this layer depends on the concentration gradient and the diffusion coefficient of the species, and is described by mathematical laws (see Fluid Dynamics section in this chapter). In practice, for processes under mass transport control, one can control the concentration gradient, and thence the mass flux and rate, in two ways. First, a steeper slope is obtained at higher bulk concentration. Second, the hydrodynamics of the system can be changed. At higher flow to the surface, the size of the diffusion boundary will be diminished and the gradient and mass transport increased. This will be described in more detail below.

In some cases transfer of material away from the surface can influence the rate of reaction. A common example is the generation of hydrogen gas as a reaction product. Hydrogen bubbles can physically block the surface if they adhere. Normally, this is overcome by the addition of a surfactant.

Kinetic Control

Some processes occur at a rate limited by the rate of reaction at the surface. This is most often called kinetic control, and is intrinsic to the chemistry of the system. One commonly encountered limitation is the rate of electron transfer at the surface. Under some conditions an etching process can be controlled by the rate of electron transfer across the surface when an oxidant such as Fe(III) or Cu(II) is employed. Under electron transfer limitation, the rate is controlled by the rate at which the oxidant can accept an electron from the surface being etched. In other cases, the slow step may be a rearrangement reaction of a species adsorbed directly on the surface. The reaction rate may differ drastically on different metal surfaces. For example, hydrogen reduction is very facile on palladium and very slow on mercury. Etching reactions typically involve a mechanism consisting of several sequential reactions [11,12], an example of which is given below. The observed kinetics depends on the nature of the rate-limiting step.

Kinetically controlled processes are susceptible to catalysis. For instance, it is common for chloride ion to catalyze electron transfer reactions of cations, so that in some cases, chloride ion can be used to increase the speed of an electron transfer limited process.

For a process under kinetic control a variety of responses to changes in concentration can be observed. For a reaction occurring on the heterogeneous surface, an increase in bulk concentration can yield a higher rate at sufficiently low concentrations. However, at higher concentrations the active sites on the metal surface may be fully occupied with reactant and no further increase of rate with concentration can be expected.

Etching processes need not be diffusionally or kinetically controlled over the entire range of etchant concentration encountered. Thus, some etching processes are diffusion controlled at low etchant concentrations, while becoming kinetically controlled at some higher concentrations. An increase in etching temperature may also result in a change in the nature of the rate-controlling process [13].

A necessary property of diffusion-controlled etching is that the observed rate is invariant with respect to the identity of the material to be etched [12]. The presence of catalytic species in the etchant can have profound effects on the observed etch rates. For diffusion-controlled etching processes, agitation of the solution becomes a significant contribution to the etching kinetics. In pattern etching, the slope of the pattern edges depends on the type of kinetics involved [14].

In some cases, an etching reaction may proceed by formation of an adherent surface film, followed by dissolution or further chemical reaction to remove the film. This situation is distinguished from passivation by the rate at which the surface film is removed; the distinction is not always clear-cut, and depends on rate of removal of the surface film. If the rate of removal is very much less than that of formation, the surface will be effectively passivated, and etching will slow to an unacceptable rate.

As stated above, slow kinetic steps in the mechanism can completely override the thermodynamic considerations. Such is the case involving sodium persulfate etching of copper, according to the following reaction:

$$S_2O_8^{2-} + Cu^0 \rightarrow Cu^{2+} + 2SO_4^{2-} \tag{18.3}$$

From thermodynamic considerations alone, the above reaction should proceed to the right ($\log K = 8.4 \times 10^{11}$ at room temperature and neutral pH). The mechanism of persulfate oxidations has been shown to involve a rate-limiting unimolecular homolysis[4] of the oxygen-oxygen bond [15,16] to give two high-energy sulfate radical anions, which subsequently react at extremely high rates to complete the etching reaction:

$$S_2O_8^{2-} \rightarrow 2SO_4\cdot \tag{18.4}$$

$$SO_4^-\cdot + Cu^0 \rightarrow Cu^+ + SO_4^{2-} \tag{18.5}$$

$$SO_4^-\cdot + Cu^+ \rightarrow Cu^{2+} + SO_4^{2-} \tag{18.6}$$

Although the overall etching reaction has a very great thermodynamic driving force, the rate-limiting step requires a sizeable activation energy. Thus, while copper etching by persulfate solutions does proceed, the rate is far lower than it would be if controlled by diffusion.

It is an experimental fact that most reactions in solution behave in common fashion as a function of temperature. Specifically, for rate constants determined by experiment, $\ln k$ is a linear function in $1/T$. This relationship is expressed as the Arrhenius equation:

$$k = Ae^{-E_a/RT} \tag{18.7}$$

where A is an empirical constant known as the *frequency factor* and E_a is the activation energy for the process. In practice, the Arrhenius relationship can be used to predict the changes in the etching rate as a function of temperature. Very rarely is a reaction encountered with a negative activation energy. Thus, as temperature increases, the rate of etching increases.

The foregoing discussion has illustrated the general features of the kinetic concerns that control metal etching. For a more thorough treatment of these subjects the reader is encouraged to consult the current literature and more in-depth treatments on kinetics [6].

ADSORPTION EFFECTS. Other considerations involved in the etching process are adsorption or desorption effects and the formation of passive films. Adsorption

---

[4] Homolysis refers to the cleavage of a two-electron bond in such a way that each of the fragments retains one electron.

and desorption processes must be considered in conjunction with the kinetics of etching reactions. Adsorption of reactant from the etchant solution onto the substrate may produce surface complexes that facilitate etching; however, etching can be drastically slowed or even prevented by the adsorption of nonreactive species or formation of passivating surface films [11].

Certain types of impurities in etchant solutions, although present at very low concentration, may adsorb onto the etching metal surface and hinder further etching [12]. The rate of desorption of gaseous reaction products sometimes limits the rates of etching processes [11]. This is particularly true of acid dissolution of metals, where hydrogen desorption can easily become the rate-limiting process. Normally, this is overcome by the addition of a surfactant.

ANISOTROPY EFFECTS. The directional aspects of etching [12,16] should be mentioned briefly. Most chemical etchants exhibit no noticeable tendency to etch preferentially in one direction. This is especially true in the etching of amorphous materials. These etchants are referred to as isotropic etchants. Since horizontal etching proceeds at a rate equal to vertical etching, the etch boundary recedes at a 45-degree angle relative to the surface. *Anisotropic*, or *directional*, *etching* refers mainly to the tendency of various crystallographic planes to etch at different rates. Various orientations of single-crystal substrates may etch very differently in a given etchant, and substrates of varying roughness may also exhibit large differences in etch rates. Anisotropic etching of single crystal substrates can be most pronounced in the presence of adsorbates that show preferential adsorption to specific crystal planes. In addition, due to variations in current density as a function of substrate shape and cell geometry, electrochemical etching can be highly anisotropic.

ELECTROCHEMICAL CELLS IN ETCHING. The heterogeneous oxidation of metals is further influenced by the ability of metals to conduct electric current. This presents an interesting situation, since it allows the sites of oxidation and reduction to be physically separated on the metal surface. The metal acts as a conductor connecting the two; the situation is analogous to a shorted electrochemical cell. Although it may not at first be obvious that potential difference can exist along the surface of a metal immersed in solution, this can in fact happen. The situation is brought about by local variations in concentration, mass transfer, surface impurities, etc., that alter the environment, and hence the potential, of different regions of the surface. The result is the creation of *local action cells* (similar to differential aeration in corrosion phenomena), whereby the metal corrodes at anodic sites and the etchant is reduced at cathodic sites. Furthermore, initiation of this process may lead to further environmental differences at the anodic and cathodic sites, as metal ions accumulate in solution and oxidant is depleted near the surface. One manifestation is the phenomenon of pitting, illustrated in Figure 18-9. Initially, a surface imperfection or a transient solution inhomogeneity creates a local action cell, and metal dissolution at the anode begins to create a pit. The metal around the pit is protected by its cathodic potential (cathodic

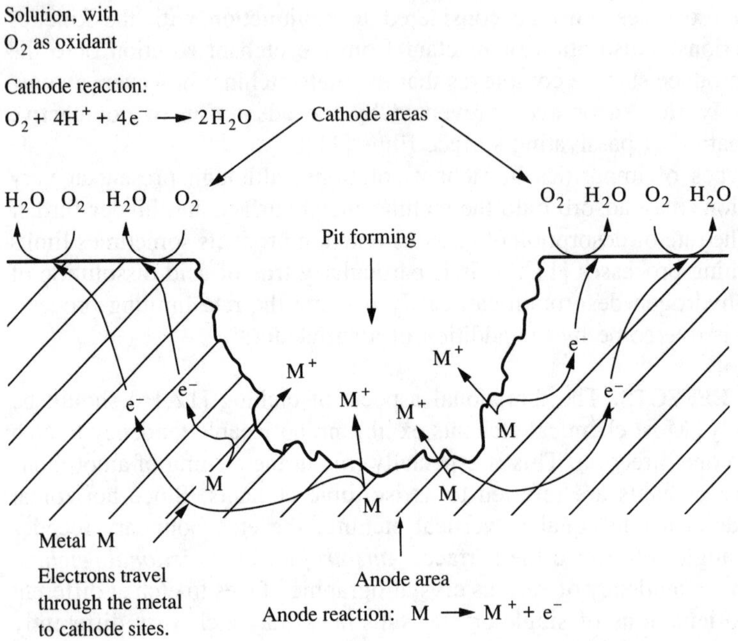

Solution, with
O_2 as oxidant

Cathode reaction:
$$O_2 + 4H^+ + 4e^- \longrightarrow 2H_2O$$

FIGURE 18-9
Schematic representation of the pitting process in an aqueous solution in the presence of chloride ions.

protection). This imbalance continues, with the result that the pit may become quite deep. This type of uneven etching can lead to rough surfaces and uneven line profiles.

ELECTROCHEMICAL KINETICS. The phenomena described above have been observed empirically. To have a grasp of the fundamental laws governing the kinetics of etching one must rely on a theoretical kinetics expression known as the Butler-Volmer (B-V) equation:

$$i_{kin} = i_0[e^{-\alpha nF\eta/RT} - e^{(1-\alpha)nF\eta/RT}] \qquad (18.8)$$

where i_{kin} is the kinetic current density, i_0 is the experimentally obtained exchange current density, α is the transfer coefficient for the reaction, η is the overpotential, and the remaining symbols have the usual meanings. The Butler-Volmer equation relates the electrical current to the amount of energy needed to drive the electron transfer at the interface between a metal and the electrolytic solution. The electrolytic solution in this case is the etchant. For a simple linear electron transfer, the rate estimated from the B-V equation represents the maximum attainable etch rate. In practice, a given etching system may not strictly follow the B-V boundary conditions; however, the equation still allows one to estimate the etching rate.

FLUID DYNAMICS
AND MASS TRANSFER CONSIDERATIONS

In previous sections, the effects of diffusion (or transfer) control have been discussed in the context of other phenomena affecting the etching process. It was understood that there is a boundary in the energy plane where an etching process could be controlled either by the kinetics of the dissolving metal or by the mass transfer of etchant components from and to the metal surface. This section is concerned with ways to estimate mass transfer rates during the etching process when mass transfer is the only effect affecting the metal dissolution. For convenience and consistency with the general literature the term *mass transfer control* will be used instead of *mass transport control*.

The mass transfer of the components in the etchant solution and the motion of the etchant during the etching process can be rate-limiting. The etchant motion is understood by means of fluid dynamics and it can be determined, theoretically, by solving the proper equations that model the motion. Because mathematical models created for real-life situations are quite complicated, empirical relationships (or equations) have been developed to explain mass transfer processes. These equations are statistical correlations of data obtained experimentally under very well controlled conditions. One can use either theoretical or empirical relationships to design a proper etchant flow.

The components in the etchant must be transported to the surface of the metal, and the etching products must be removed to avoid excessive accumulation. This transportation of matter is described by the laws of mass transfer. The first and second laws of diffusion are examples of these equations (see Chapters 25 and 26).

In order to describe a flow motion, it is necessary to describe the geometry where the flow is to occur. A typical geometry found in etching involves the cross section of a developed photoresist. The term *cavity* will be used to describe this cross section. The cavity is formed by the exposed metal surface as the bottom and the photoresist as the side walls. Figure 18-10 shows this typical geometry.

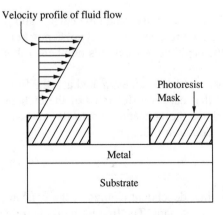

Velocity profile of fluid flow

Photoresist
Mask

Metal

Substrate

FIGURE 18-10
Cavity formed by a photoresist layer in contact with a metal.

In the electronic industry, the most common ways to achieve etchant motion are board dipping, or immersion, and etchant spraying. In the case of board dipping, the board is usually submerged vertically in a tank or container where etchant solution is held. The board is displaced horizontally, while held vertically, in a regular, repetitive manner. The displaced etchant produces a shearing flow on the board's surface, inducing agitation inside the cavities. This type of agitation is slow and inefficient, allowing the etching process to become diffusion limited. A disadvantage of this type of agitation is the long time required to etch a given metal thickness.

In spray etching, the board is placed under an etchant spray within a chamber. Several nozzles are mounted on an array of pipes or manifold. This array is arranged in such a way that the spray generated by adjacent nozzles overlaps slightly to provide spray uniformity. This type of fluid motion is more turbulent and transfers more momentum to the etchant, resulting in an increased convection rate inside the cavities. Thus the mass transfer rates are enhanced, and limitations by diffusion are avoided. For very deep cavities, such as thick photoresist and thick copper, extreme turbulence may be necessary to prevent diffusion control. This turbulence may be achieved by focused ultrasonic agitation.

In the description of flow patterns, it is convenient to differentiate flows by position on the board. Flow inside the cavities is referred as *internal* flow, while flow outside the cavities is referred as *external*. In the typical geometry, the internal flows are induced by external flows, and the velocities of these flows have different orders of magnitude. In order to generate agitation inside the pattern lines, one must have a strong external flow. This flow is induced by either method described above. For details of the fundamentals of fluid dynamics and mass transfer, the reader is encouraged to consult references [17] and [18].

One measurement of etchant flow rate is given by the dimensionless Reynolds number (Re). The Reynolds number defines the ratio of inertial forces to viscous forces in the moving etchant. Hence a low Reynolds number means that the viscous forces are great as compared to the inertial forces.

$$\text{Re} = \frac{LV\rho}{\mu} \tag{18.9}$$

In this relationship L is the cavity width, V is the velocity inside the cavity, and ρ and μ are the density and viscosity of the etchant respectively. Reynolds numbers on the order of 10 to 100 develop inside the cavities. A Reynolds number of this magnitude indicates that flow is laminar.

Once the flow velocity on the metal surface has been estimated and the flow type is known, it is possible to determine the mass transfer rates of the etchant components and etching products. One relates the concentration to the velocity field through a mass balance according to Equation (18.10):

$$\frac{\partial C_i}{\partial t} = -\nabla N_i + R_i \tag{18.10}$$

In this equation the term on the left is the time dependent change in the concentration field of any of the species present in the etchant. The first term on the right is the flux gradient of the electroactive species to and from the metal surface. The second term is the generation or consumption rate of any homogeneous (nonelectroactive) species, such as a complexing agent, in the etchant. This last term is given by a kinetic type of expression that relates the concentration of a given species to the kinetic constant to give a reaction rate. Thus, a basic knowledge of the reaction mechanism is useful.

The equation that gives the second term is

$$N_i = -z_i u_i F C_i \nabla \Phi - D_i \nabla C_i + v C_i \tag{18.11}$$

This equation represents the flux of each component going in or coming out of a differential volume element in the cavity. The first term on the right side represents migration induced by the electric field in the solution. The second term arises from diffusion induced by a concentration gradient. The last term arises from convection induced by etchant flow.

By making the proper substitutions, it can be seen that this mass balance contains terms that represent the mass transport of species by migration, diffusion, and the production of homogeneous species. To solve the system of $n-1$ equations with n species (or unknown concentrations), an extra relationship is required. The following condition is used to complete our system of equations:

$$\sum_1^n z_i C_i = 0$$

This system of equations can be solved practically only for the simplest cases. Owing to the fact that processes discussed in this chapter are quite complex, an empirical approach will be used from now on. The purpose of outlining a rigorous analysis of mass transfer phenomena was to give the reader a feeling of the procedure involved in understanding of the etching process. References [17] and [18] should be consulted for details on the significance and other applicable forms of this procedure.

In the absence of etchant velocity, homogeneous reactions, and electric fields, the mass balance simplifies to Fick's second law of diffusion, which is a very simple and useful form for analysis of systems under mass transfer control. This law governs the mass transfer of etchant components during etching by diffusion only. For steady-state conditions, given the flux constant at the etchant-metal interface, and the bulk concentration at infinity, one can estimate an etching rate limited by diffusion as

$$i_{\text{dif}} = -nFD_i \frac{C_b - C_s}{\delta} \tag{18.12}$$

In this equation i_{dif} is the diffusion-limited current density being transferred in the etching process for the species carrying the electrical charge in solution, n is the total number of electrons transferred in the reaction, F is Faraday's constant, D_i is the diffusion coefficient for the metal ion in a dilute solution, C_b is the concentration of metal ions in the bulk of the etchant, C_s is the concentration of metal ions right at the surface of the metal being etched, and δ is the physical dimension that defines the thickness of a layer where diffusion is the only form of mass transport. In the case under discussion this δ could be replaced by H, the thickness of the photoresist, which defines the depth of the cavity. To convert i_{dif} in current density units to an actual etching rate in units of mils per hour, one uses Faraday's law in the following form:

$$\text{Etch rate (mils per hour)} = 1.42 \times 10^6 \frac{i_{\text{dif}} MW}{nF \rho_{\text{metal}}} \qquad (18.13)$$

where MW and ρ_{metal} are the molecular weight (in grams per mole) and the density (in grams per cubic centimeter) of the metal being etched, respectively. The proper units must be used for dimensional consistency.

For cases where diffusion indeed slows the etching process, agitation of the etchant must be provided. Convection is the type of agitation most desirable during etching. There are two modes of convection: natural and forced. Natural convection arises because of changes in the physical properties of the etchant, in the present case changes in density. Changes in density produce buoyancy effects, which induce the etchant to move upwards, thus producing a flow. Empirical relationships that describe this flow are of the following form:

$$\text{Sh} = a(\text{Gr Sc})^b \qquad (18.14)$$

where Sh is the Sherwood number. This number is a dimensionless mass transfer (or transport) coefficient averaged over the mass transfer surface. Gr is the Grashof number. This number is a dimensionless measure of the buoyancy produced by the difference in densities between the solution close to the surface being etched and the bulk of the solution. Sc is the Schmidt number, which gives the thickness ratio between the fluid motion and the mass transfer boundary layer. The numbers a and b are constants determined by experimentation. The exponent b is 1/3 for horizontal and 1/4 for vertical configurations.

The following relationships define the numbers used in Equation (18.14):

$$\text{Sh} = \frac{k_m L}{D_i}$$

$$\text{Gr} = \frac{g(\rho_\infty - \rho_0)l^3}{\rho_\infty \nu^2} \qquad (18.15)$$

$$\text{Sc} = \frac{\nu}{D_i}$$

where k_m is the mass transfer coefficient, defined as the ratio of the diffusion coefficient to the thickness of the mass transfer boundary layer, D_i/δ. This coefficient is given in units of length per unit time.[5] L is the cavity width, l is the

---

[5] In some references the mass transfer coefficient is defined as the amount of mass in moles transported per unit time per unit area.

line length, C_i is the bulk concentration of any component in the etchant, D_i is the diffusion coefficient of a particular component, ρ_∞ and ρ_0 are the densities for the solution in the bulk and of the solution close to the metal surface, respectively, and ν is the kinematic viscosity of the etchant. By simple inspection of Equation (18.12) one can see that the mass transfer coefficient is directly proportional to the current density and thus to the etching rate. Hence a decrease in thickness in the mass transfer boundary layer will result in an increase in the mass transfer coefficient and an increase in the etching rate.

Forced convection is produced by an external force. Forced flows may be produced by pumps, agitators, or any other devices. Mass transfer by forced convection is represented by empirical relationships of the following form:

$$Sh = ARe^a \tag{18.16}$$

where Sh and Re are the Sherwood and Reynolds numbers as defined earlier.

For the geometry under consideration here, Equation (18.16) takes the form,[6]

$$Sh = 13.04 \left(\frac{L}{H}\right)^{0.37} \left(\frac{Z}{H}\right)^{-0.16} Re_{nozzle}^{0.43} \tag{18.17}$$

where the ratios L/H and Z/H are the ratios of the line width to the photoresist thickness and of the nozzle-to-board distance to the photoresist thickness, respectively. These ratios are used as correction factors because of the geometry of the cavity.

In spray etching it is most desirable to have a flow on top of the pattern lines than to have an impinging flow. The reason is that, as shown in Figure 18-11 [19,20], a jet impinging directly on a cavity does not produce as much stirring as when the jet is slightly offset from the center of the line. It must be emphasized that the shearing flow must be uniform all across the substrate to have uniform stirring inside the lines.

Up to this point, the discussion has been concerned with the local effects of flow inside cavities. However, to determine a proper etching machine one must prescribe the operating conditions, including the manifold pressure necessary to obtain a uniform flow on the board. A process should be designed to give a uniform spray, and a drop size that achieves proper wetting for the process should be determined.

To determine the proper nozzle type and size one must consult handbooks [21] and catalogs of nozzle manufacturers [22]. Information about nozzle geometry and pressure requirements must be determined from estimates for the desired flow. Nozzle-to-board distance is determined from data obtained by the nozzle manufacturer for a given flow rate. Once the proper flow pattern has been determined, one proceeds to lay out the manifold and then to select a proper pump.

---

[6] In this case the Reynolds number is based on the flow of etchant in the nozzle.

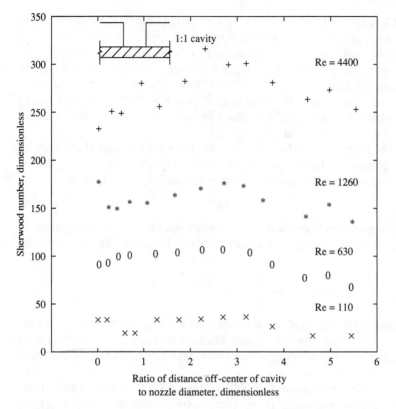

FIGURE 18-11
Effects of mass transfer by offsetting a jet impinging on a PCB.

The previous discussion addresses only the simple aspects of transport phenomena during etching. Other effects such as receding boundaries during etching and eddy flows because of turbulence are beyond the scope of this chapter. The reader is referred to the literature for a more thorough discussion of the subject.

A WORKING EXAMPLE

The discussion that follows will analyze an etchant system using the principles presented in previous sections. The example is that of copper in 0.1 molar sulfuric acid. This system is often used commercially. The overall reaction is

$$Cu^0 + S_2O_8^{2-} \rightarrow Cu^{2+} + 2SO_4^{2-} \tag{18.18}$$

The reaction can be broken down into the component half-cell reactions:

$$
\begin{aligned}
Cu^0 &\rightarrow Cu^{2+} + 2e^- & E_1^0 &= -0.34 \text{ V} \\
S_2O_8^{2-} + 2e^- &\rightarrow 2SO_4^{2-} & E_2^0 &= 2.05 \text{ V}
\end{aligned}
\tag{18.19}
$$

$$E_1^0 + E_2^0 = 1.71 \text{ V}$$

From the half-cell potentials given in Table 18.2, we can see that the overall E^0 for the reaction is 1.71 V, so that it is quite favorable thermodynamically. The high reduction potential of persulfate indicates that it is a very aggressive oxidizing agent. Since the concentration of sulfuric acid is generally on the order of 0.1 molar, the pH is approximately 1. The Pourbaix diagram for copper in aqueous solution ([2], p. 388) shows that at pH $=$ 1 and $E^0 = 2.05$ V, copper dissolves to yield Cu^{2+} ions.

Persulfate oxidations are characterized by a rate-limiting unimolecular homolysis of the persulfate oxygen-oxygen bond to give two high-energy sulfate anion intermediates.

$$S_2O_8^{2-} \rightarrow 2SO_4^-\cdot \qquad (18.20)$$

It is these radical intermediates that attack the metal surface, initially giving rise to copper(I) and sulfate ions.

$$Cu^0 + SO_4^-\cdot \rightarrow Cu^+ + SO_4^{2-} \qquad (18.21)$$

The radical chain is then propagated by reaction of the intermediate Cu(I) ion with persulfate ion:

$$Cu^+ + S_2O_8^2 \rightarrow Cu^{2+} + SO_4^{2-} + SO_4^-\cdot \qquad (18.22)$$

The effect of this reaction is to regenerate the reactive persulfate radical anion. The chain reaction is moderated or terminated by reactions that scavenge Cu(I) and/or the sulfate radical. Examples of these reactions are the reaction of Cu(I) with oxygen to produce Cu(II), and disproportionation of Cu(I) to produce Cu(II) and copper metal.

$$Cu^+ + O_2 \rightarrow Cu^{2+} + O_2^-\cdot \qquad (18.23)$$

$$2Cu^+ \rightarrow Cu^0 + Cu^{2+} \qquad (18.24)$$

The kinetics of radical reactions are complicated by periods of time during which the concentration of radical species builds up to a level sufficiently high to sustain the chain reaction. The induction period can be drastically shortened and the etching rate markedly increased by the addition of catalysts for the formation of the sulfate radical anion. An example of this effect is the experimentally observed catalysis of persulfate etching of copper by the addition of traces of mercurous chloride.

Etching reactions that depend on radical chain mechanisms can prove difficult to control due to the fact that the form of the rate law depends on the chain termination step. Variations in chain termination step from etching to etching due to concentration changes, contamination, or other factors may drastically alter the form of the rate law. The effects of a slow mechanistic step can, to a certain extent, be overcome by elevation of the temperature at which the etching is performed. As discussed earlier, the Arrhenius equation, (18.7), can be used to predict the etching rate at an elevated temperature if the activation energy for the process is known.

In regard to diffusion limitations, it was pointed out that Fick's second law would apply. The analysis of the situation follows Equations (18.12) and (18.13).

Assuming a saturation concentration for $CuSO_4$, a diffusion coefficient for Cu ions of 1×10^{-6} cm^2/sec, a line width of 4 mils, and a photoresist thickness of 2 mils, the calculated etching rate is 67 mils per hour. The actual measured rate is approximately 0.3 mils per hour. Therefore, diffusion does not limit the process, and regular convection would keep the concentration of etching products at a working level.

If the system were limited by diffusion, then vigorous agitation would be prescribed. Again, an estimate of an etching rate with forced convection can be obtained by using Equations (18.17) and (18.13). Assuming a Reynolds number at the nozzle of 2000 and a nozzle-to-board distance of 6 inches, the maximum etching rate as limited by convective mass transfer would be 574 mils per hour. Thus one would require a spray to achieve a good stirring of the etchant and avoid diffusion limitations.

PRACTICAL CONSIDERATIONS

One of the most common practical considerations involves etching of one metal in direct contact with another metal. Such is the case encountered in subtractive circuitization of metallization consisting of layers of Cr and Cu. One of the prime considerations in this type of etching involves etching one of the metals without significant attack on the other metal. Many etchants that function well for one of the metals alone may not etch the sandwich metallization due to passivation of one metal by the other. This can occur by galvanic interaction such as cathodic protection.

Consideration must be given to metal etching in the presence of an organic photoresist. Only rarely is a case of metal etching encountered where a photoresist is not used to define the circuitry. Another concern in practical metal etching involves health and safety. Highly toxic or flammable etchants are usually avoided owing to the difficulties in handling the etchant, ensuring compatibility with equipment materials, and providing a safe working environment. The following discussion elaborates on these issues.

Health and safety issues are important considerations. From the Pourbaix diagram for chromium, it is seen that the metal could be etched with concentrated inorganic acids such as hydrochloric, perchloric, and nitric acids. The products of nitric acid reduction are highly toxic gases or fumes and as such constitute an environmental hazard. In contrast, perchloric acid etching produces stable metal chlorides. The difficulty with perchloric acid etching lies in the explosive potential of the concentrated acid, while etching with dilute acid is usually impractical owing to the slow rate of etching. Concentrated hydrochloric acid represents a handling problem due to its high vapor pressure and corrosive nature. Very few materials will withstand hydrochloric acid in either liquid or vapor state without deterioration. Finally, situations may be encountered where metal etching processes would involve poisonous or carcinogenic additives, such as cyanide ion or phenol, as complexing agents. As described in detail earlier, these complexing agents serve the purpose of shifting the region of thermodynamic stability to lower potentials, thereby enhancing etching. These additives pose a serious health and safety concern and therefore must be carefully controlled.

Compatibility with photoresist is crucial to defect-free etching. Subtractive patterning depends upon the ability to cleanly etch while at the same time protecting the areas of the circuitry under the photoresist. Etchants that attack the photoresist result in resist embrittlement, lifting, and crack formation. Any of these effects will lead to the formation of severe circuit defects such as undercutting, open lines, and etchant entrapment. Therefore, etchants must be selected with an eye toward their compatibility with a particular photoresist. It is usually much easier to tailor an etchant to a photoresist than vice versa. The problem of photoresist-etchant incompatibility can be illustrated by the case of chromium etching with ceric salts. Thermodynamically the cerium(IV-III) couple would etch chromium very easily. However, cerium(IV) is such a potent oxidant that few organic materials can withstand it. It has been observed experimentally that these salts attack most photoresists.

Galvanic effects can be utilized, for example, in the selective etching of chromium in contact with copper. It has been experimentally determined that solutions of sodium chloride with moderate amounts of hydrochloric acid will not etch chromium by itself. But they will, however, rather rapidly etch chromium in electrical contact with copper. This phenomenon is called cathodic protection and results from a potential induced on the surface of the chromium by copper metal in the presence of chloride ion.

Etchant cost can be an important factor in choosing an etchant. This cost not only includes the purchase price but must include the regeneration and disposal of the etchant. If inexpensive regeneration is possible, it can allow important savings in purchasing and disposal. For example, cupric chloride regeneration is accomplished by bubbling chlorine gas or bottled oxygen through the spent etchant. Oxygen regeneration is preferred since it alleviates the hazard and handling difficulties of chlorine gas. There are also electrochemical regeneration techniques. An example is the electrochemical regeneration of cupric or ferric chloride in an electrolytic cell where a platinum basket is introduced and electrical current is passed until no cuprous or ferrous ion remains in the system.

The previous discussion has dealt with the chemical and chemical engineering factors in etching, and earlier chapters have discussed plating and photo technology.

SUMMARY

Photo patterning of resist is currently not a significant limitation on tolerances. Precision artwork, along with parabolic and point source light projection systems, is capable of creating patterns in modern resist materials much more precisely than it is possible to etch them, when dealing with an interconnection thickness of 1.4 thousandths of an inch, which is typical for printed circuit boards. The major factors controlling etch precision are copper thickness, agitation uniformity, and residual resist. A lack of copper thickness uniformity allows penetration of the etchant to the substrate in the thin areas first. When this happens, the sideways etching rate increases in the penetrated area.

To decrease the limitations of diffusion, shear spray patterns are used. The actual agitation patterns are determined by a variety of geometrical design factors, including the density of the interconnections, the pressure of the spray, the pattern of nozzles, the distance from the nozzles to the board, and the depth of the photoresist channels. The added depth of the copper being etched adds to the critical aspect ratio. The depth of the fluid on the surface of the board also becomes a predominant factor under these conditions. Therefore, the system must be set up to prevent puddling on the surface of the board. Board surface curvature becomes significant, since it can easily be equal to or greater than the combined resist-copper thickness.

Residues in the resist channels are equally important since they prevent etching. The best way to remove these is by spray development where agitation is enhanced by the momentum of the droplets.

The key to precision etching is chemical engineering control of agitation patterns on a complex surface during plating, photoforming, and etching. The role of chemistry is to minimize the effects of diffusion by enhancing and controlling the chemical etch rate variables.

REFERENCES

1. J. A. Scarlett, editor, *The Multilayer Printed Circuit Board Handbook*, Electrochemical Publications Ltd., Ayr, UK, 1985.
2. M. Pourbaix, *Atlas of Electrochemical Equilibria in Aqueous Solutions,* Houston, Texas: NACE-Cebelcor, 1974.
3. R. C. Weast, *CRC Handbook of Chemistry and Physics*, 56th ed., Cleveland: CRC Press, 1975.
4. W. M. Latimer, *The Oxidation States of the Elements and their Potentials in Aqueous Solutions*, 2d ed., New York: Prentice-Hall, 1952.
5. A. E. Martell and R. M. Smith, *Critical Stability Constants*, vols. 1–4, New York: Plenum Press, 1974–1982.
6. For detailed accounts of chemical thermodynamics, the reader is referred to standard textbooks:
   W. J. Moore, *Physical Chemistry*, 3d ed., Englewood Cliffs, N.J.: Prentice-Hall, 1962.
   G. W. Castellan, *Physical Chemistry*, 2d ed., Reading, Mass.: Addison-Wesley, 1971.
7. W. J. Moore, *Physical Chemistry*, 3d ed., Englewood Cliffs, N.J.: Prentice-Hall, 1962, p. 350.
8. Ellis D. Verink, Jr., "Improving the Ability to Predict Corrosion Behavior through Corrosion Research," *Corrosion* (Houston) vol. 38, no. 6, pp. 336–47 (1982).
9. C. M. Chen and K. Aral, "A Computer Program for Constructing Stability Diagrams in Aqueous Solutions at Elevated Temperatures," *Corrosion* (Houston) vol. 38, no. 4, pp. 183–190 (1982).
10. W. J. Blaedel and V. W. Meloche, *Elementary Quantitative Analysis*, 2d ed., New York: Harper and Row, 1963.
11. N. Hackerman, in *The Surface Chemistry of Metals and Semiconductors*, edited by H. C. Gatos, New York: John Wiley & Sons, 1960.
12. H. C. Gatos and M. C. Lavine, "Chemical Behavior of Semiconductors: Etching Characteristics," *Prog. in Semiconductors* (GB), vol. 9, pp. 1–45 (1965).
13. Paul R. Camp, "Etching Rate of Single-Crystal Germanium," *J. Electrochem. Soc.*, vol. 102, pp. 586–93 (1955).
14. W. Kern, J. L. Vossen, and G. L. Schnable, "Improved Reliability of Electron Devices through Optimized Coverage of Surface Topography," *Annu. Rel. Phys. (Symp.)*, vol. 11, pp. 214–23 (1973).

15. T. A. Turnay, *Oxidation Mechanisms*, London: Butterworths, 1965.
16. B. A. Irving, in *The Electrochemistry of Semiconductors*, edited by P. J. Holmes, New York: Academic Press, 1962.
17. R. B. Bird, W. E. Stewart, and E. N. Lightfoot, *Transport Phenomena*, New York: John Wiley & Sons, 1960.
18. J. S. Newman, *Electrochemical Systems*, Englewood Cliffs, N.J.: Prentice-Hall, 1973.
19. M. L. Greenlaw, "Mass Transfer in Square Trenches under the Influence of Submerged Impinging Slot Jets," M.S. thesis, University of Illinois, Urbana, 1985.
20. M. L. Greenlaw, Ph.D. thesis in progress, University of Illinois, Urbana, 1986.
21. R. H. Perry, and D. W. Green, editors, *Perry's Chemical Engineers' Handbook*, 6th ed., New York: McGraw-Hill, 1984, pp. 18–48.
22. Catalogs of the following companies are useful. Bete Fog Nozzle, Inc., Greenfield, Mass.; Spraying Systems Co., Wheaton, Il.; Spraco, Inc., Nashua, N.H.

APPENDIX

THE EFFECT OF TEMPERATURE

Very many operations are carried out at elevated temperatures to speed the kinetics, and in such a case it is important to adjust the pH-potential diagram to the higher temperature. This requires knowledge of the enthalpy and entropy changes associated with the chemical transformation, and the assumption that ΔH^0 and ΔS^0 do not change dramatically with temperature (a fairly good approximation for condensed phase reactions since the temperature range accessible in the aqueous solution at one atmosphere pressure is relatively small).

Consider the nickel system at 55°C as an example. For the reaction

$$Ni_{aq}^{2+} + 2OH^- = Ni(OH)_2$$

$$\Delta H^0 = -3386 \text{ cal/mole}$$

$$\Delta S^0 = 62.14 \text{ e.u.}$$

so that

$$\Delta G_{328.2}^0 = \Delta H^0 - (328.2) \times \Delta S^0$$

$$= -23,780 \text{ (cal/mole)}$$

$$\log K_{328.2} = \frac{-(-23,780)}{(2.303)(1.99)(328.2)}$$

$$K_{328.2} = 6.46 \times 10^{15} = \frac{1}{a_{Ni_{aq}^{2+}}(a_{OH^-})^2}$$

Solving the equation for the activity of OH^- at which precipitation occurs with $a_{Ni_{aq}^{2+}} = 1$ we obtain $a_{OH^-} = \sqrt{1/6.46 \times 10^{15}} = 1.24 \times 10^{-8}$, that is, pOH = 7.91.

In determining the pH it must be kept in mind that K_w, the constant for water, also changes with temperature. Measured values for K_w are available in the literature [3], or it may be calculated from the data in Table 18.3, following the procedures in this section.

$$H_2O = H^+ + OH^- \tag{18.25}$$

$$pK_w(55°C) = 13.14 = pH + pOH$$

$$pH = 13.14 - 7.91$$

$$pH = 5.23$$

The Ni_{aq}^{2+}/Ni^0 redox couple potential at 55°C is calculated by using the virtual reaction $Ni_{aq}^{2+} + H_2 = Ni^0 + 2H^+$:

$$\Delta G_{328.2}^0 = \Delta H^0 - 328.2\Delta S^0$$

$$= 15,300 - 328.2(14.09)$$

$$= 10,677 \text{ cal/mole}$$

Since the virtual reaction is referenced to the normal hydrogen electrode (NHE),

$$\Delta E_{328.2}^0 = -\frac{\Delta G^0}{nF}$$

$$= 15,300 - 382.2(14.09) = 10,677 \text{ cal/mole}$$

$$\Delta E_{328.2}^0 = \frac{-(10,677)(4.184)}{(2)(96,487)} = -0.231 \text{ V}$$

At 328.2 K the Nernst equation is

$$E_{328.2} = E_{328.2}^0 - \frac{R(328.2)(2.303)}{2F} \log\left(\frac{1}{a_{Ni_{aq}^{2+}}}\right)$$

$$E_{328.2} = 0.231 - 0.0326 \log\left(\frac{1}{a_{Ni_{aq}^{2+}}}\right)$$

It is important to note that both the intercept and the slope of the concentration dependence shift with temperature.

For the $Ni(OH)_2/Ni^0$ boundary, $\Delta H^0 = -8040$ cal/mole and $\Delta S^0 = -9.57$ e.u., so that at 55°C,

$$\Delta G_{328.2}^0 = -8040 - (328.2)(-9.27)$$

$$= -4,899 \text{ cal/mole}$$

$$\Delta E_{328.2}^0 = -\frac{\Delta G_{328.2}^0(4.184)}{(2)(96,487)} = +0.106 \text{ V}$$

and

$$E = E^0 - \frac{RT(2.303)}{nF}(2)(pH)$$

$$E = 0.106 - \frac{(4.184)(328.2)(2.303)}{(2)(96,487)}pH$$

$$E = 0.106 - 0.0326pH$$

THE EFFECT OF TRIEN COMPLEX

Nickel ion in solution can be complexed with a number of ligands, among them the trien molecule. A formation equilibrium for this may be written $Ni^{2+} + trien = Ni(trien)^{2+}$. For this reaction as written, the formation equilibrium constant is

$$K_f = \frac{[Ni(trien)^{2+}]}{[Ni^{2+}][trien]} = 10^{14} \tag{18.26}$$

where [trien] and $[Ni^{2+}]$ refer to the free unprotonated ligand and free Ni^{2+}, respectively.

The total nickel concentration, $[Ni^{2+}]_{total}$, is given by

$$[Ni^{2+}]_{total} = [Ni^{2+}] + [Ni(trien)^{2+}] \tag{18.27}$$

The trien has a series of protonation equilibria and associated equilibrium constant expressions:

$$H_4(trien)^{4+} = H_3(trien)^{3+} + H^+ \quad K_1 = \frac{[H_3(trien)^{3+}][H^+]}{[H_4(trien)^{4+}]} = 4.8 \times 10^{-4}M \tag{18.28}$$

$$H_3(trien)^{3+} = H_2(trien)^{2+} + H^+ \quad K_2 = \frac{[H_2(trien)^{2+}][H^+]}{[H_3(trien)^{3+}]} = 2.1 \times 10^{-7}M \tag{18.29}$$

$$H_2(trien)^{2+} = H(trien)^+ + H^+ \quad K_3 = \frac{[H(trien)^+][H^+]}{[H_2(trien)^{2+}]} = 6.3 \times 10^{-10}M \tag{18.30}$$

$$H(trien)^+ = trien + H^+ \quad K_4 = \frac{[trien][H^+]}{[H_2(trien)^+]} = 1.2 \times 10^{-10}M \tag{18.31}$$

A mass balance equation may also be written for the trien species.

$$[trien]_{total} = [trien] + [H(trien)^+] + [H_2(trien)^{2+}]$$
$$+ [H_3(trien)^{3+}] + [H_4(trien)^{4+}] + [Ni(trien)^{2+}] \tag{18.32}$$

The concentrations to be determined are those of the five trien species, the two nickel species, H^+, and OH^-. Hydrogen ion and hydroxide ion are coupled through the ion product of water.

$$[H^+][OH^-] = K_w = 1.00 \times 10^{-14} \tag{18.33}$$

If $[H^+]$ is fixed, one has solved the problem in principle: eight equations and eight unknowns. Thus the concentration of free Ni^{2+} ion can be determined at any pH, and the resultant phase boundaries determined as described earlier.

If one can make the assumption that $[trien]_{total} \gg [Ni^{2+}]_{total}$, a direct expression can be readily obtained. This permits elimination of the $[Ni(trien)^{2+}]$ term in equation (18.32). Successive substitutions of equations (18.28)–(18.31) into Equation (18.32) gives Equation (18.34):

$$[trien]_{free} = \frac{[trien]_{total}}{1 + \dfrac{[H^+]}{K_4} + \dfrac{[H^+]^2}{K_3 K_4} + \dfrac{[H^+]^3}{K_2 K_3 K_4} + \dfrac{[H^+]^4}{K_1 K_2 K_3 K_4}} \tag{18.34}$$

Substitution of Equation (18.27) into Equation (18.26) and rearrangement leads to an expression for $[Ni^{2+}]_{free}$ in terms of $[trien]_{free}$:

$$[Ni^{2+}] = \frac{[Ni^{2+}]_{total}}{K_f[trien]_{free} + 1)} \tag{18.35}$$

Substitution of Equation (18.34) into (18.35) gives the overall expression for $[Ni^{2+}]$ in terms of $[H^+]$:

$$[Ni^{2+}] = \frac{[Ni^{2+}]_{total}}{[trien]_{total}\left(1 + K_f + \dfrac{K_f[H^+]}{K_4} + \dfrac{K_f[H^+]^2}{K_3 K_4} + \dfrac{K_f[H^+]^3}{K_2 K_3 K_4} + \dfrac{K_f[H^+]^4}{K_1 K_2 K_3 K_4}\right)} \tag{18.36}$$

The expression for the potential boundary between Ni^{2+} in solution and Ni^0 is the Nernst equation.

$$E = E^0_{Ni^{2+}/Ni^0} - \frac{0.0592}{2} \log \frac{1}{[Ni^{2+}]} \tag{18.37}$$

Substitution of (18.36) into (18.37) yields the general dependence of the phase boundary potential on the hydrogen ion concentration:

$$E = -0.241$$

$$- \frac{0.0592}{2} \log \frac{[trien_{total}]\left(\dfrac{K_f[H^+]}{K_4} + \dfrac{K_f[H^+]^2}{K_3 K_4} + \dfrac{K_f[H^+]^3}{K_2 K_3 K_4} + \dfrac{K_f[H^+]^4}{K_1 K_2 K_3 K_4}\right)}{Ni^{2+}_{total}} \tag{18.38}$$

This equation holds provided $[trien]_{total} \gg [Ni^{2+}]_{total}$.

The curve in Figure 18-6 in this chapter was calculated for $[trien]_{total} = 1.00$ M and $[Ni^{2+}]_{total} = 0.010$ M. The flat portion on the left of this figure is the limiting case where all the trien ligand exists in protonated forms, so that the nickel ion exists almost entirely as the Ni^{2+} aqua ion, and the potential represents that found at $[Ni^{2+}] = 0.01$ M. On the far right-hand side (basic), the trien is virtually all in the form of the deprotonated trien species, and the nickel ion exists as virtually all $Ni(trien)^{2+}$, with 10^{-16} M free Ni^{2+} as the limiting concentration. Since $K_{pptn} = 10^{16} = 1/[Ni^{2+}][OH^-]^2$, the $[OH^-]$ at which precipitation occurs is 1.0 M or pH $= 14$. This is the intersection point between the $Ni(OH)_2/Ni^0$ boundary (diagonal dotted line in Figure 18-6) and the potential boundary calculated in Equation (18.38).

PROBLEMS

18.1. Consider the metal etching reaction in which ferric chloride ($FeCl_3$) is used to etch zinc metal (Zn). (Refer to the table of standard reduction potentials.)

    (*a*) Write out the oxidation half-cell, reduction half-cell, and complete balanced chemical equation for the etching process.

    (*b*) Calculate the potential E^0 and the free energy change ΔG^0 for the reaction (assume 1 mole/liter concentrations of all species). Is the reaction thermodynamically favored?

18.2. The rate of copper etching in ferric chloride ($FeCl_3$) etchant is found to be much greater in a spray etcher than in an immersion etcher with moderate agitation. Explain this observation. (Hint: The rate of transport of etchant to the metal surface is greater in a spray etcher than in the immersion etcher.)

    Is the etching reaction mass-transport- or activation-controlled?

18.3. An engineer etches copper in a beaker of ferric chloride under conditions where the concentration is 2 moles/liter and the pH is 1, and finds it is completely etched in 100 seconds. In an attempt to speed up the process, the engineer increases the ferric chloride concentration to 3.5 moles/liter, pH $= 1$, but now finds that the etching process takes 140 seconds to completion. Both solutions are stirred at the same speed. Suggest an explanation for this result. (Hint: The viscosity of the etchant solution increases with increasing ferric chloride concentration.)

    How does this conclusion compare with that reached in Problem 2?

18.4. An engineer determines the amount of time needed to etch a given thickness of copper in his spray-etching equipment by running tests on blanket metal circuit boards (i.e., boards with no photoresist). He then discovers that it takes longer to etch the copper away from the narrow line channels (spaces between circuit lines) on his circuit boards covered with a photoresist pattern. The line channels are 25 μm wide, and the photoresist is 15 μm thick. Areas with very wide channels (e.g., 500 μm wide) etch at the same rate as found for the blanket metal board. Suggest an explanation for these results.

18.5. A certain etch process produces copper lines with a 45-degree sloping sidewall, because of undercut. A circuit designer wishes to produce a circuit using this process with copper lines separated at the bottom by 50 μm. The width of the lines on the top surface of the coper must be 50 μm. How many lines will fit side by side in a 1-cm space if the copper is 20 μm thick? 5 μm thick?

18.6. For which of the metals in Table 18.2, "Standard Reduction Potentials," is the reaction with hydrogen ion thermodynamically allowed?

18.7. It is possible to use an electric current to oxidize (etch) a metal if the metal is made the anode in an electrochemical cell. The reduction reaction then takes place at the cathode of the cell, which can be a metal such as platinum. The electrons removed from the metal travel through the external circuit to the cathode, where they reduce some appropriate species present in solution. The oxidizing power or potential of the metal being etched can be set to a desired value through the use of a device called a potentiostat. This potential corresponds to the ordinate of the Pourbaix diagram for the anode.

An engineer attempts to etch chromium in this way in a pH-6 solution. She begins with a potential of -2.2 V but finds that the metal doesn't etch. She then increases the potential to -0.9 V and finds that the metal begins to dissolve. As she makes the potential increasingly positive, she finds that the rate of dissolution continues to increase until suddenly, at -0.6 V, the etching stops. Undeterred, she continues to make the potential more oxidizing until, finally, at $+0.8$ V, the metal begins to dissolve again.

Explain the behavior of chromium at each of these potentials.

18.8. What would you expect to happen at each of the potentials in Problem 7 if the metal being etched were (a) aluminum at pH 6? (b) aluminum at pH 12?

CHAPTER 19

JOINING MATERIALS AND PROCESSES IN ELECTRONIC PACKAGING

CHARLES G. WOYCHIK
RICHARD C. SENGER

INTRODUCTION

In electronic packaging an interconnection is an electrical path or bridge between two different levels of circuitry. The silicon chip is attached to a metallized substrate (chip carrier). This assembly is then sealed either by encapsulation in plastic or by brazing or soldering a cap to the substrate. This complete assembly is referred to as the component, which is then attached to a printed circuit card. The card is then attached to the next level of circuitry, which is the printed circuit board. The junction between the chip and the substrate is referred to as a first-level interconnection, and the junction between the component leads and the card is referred to as a second-level interconnection. Current packaging designs are focusing on attaching the chip directly to the board, eliminating one level of packaging and increasing the speed of the system. The processes used to make these junctures are referred to as assembly processes. The main focus of this chapter will be on the different types of metallurgical joining methods (assembly processes) that are used to establish an electrical interconnection. There are many requirements that determine the use of a particular joining method to produce an interconnection. Following are a few of the more important material and process variables to be considered.

Electrical properties. The material that is used to form the interconnection should have a high electrical conductivity.

Thermal properties. The substrate and the heat sink should be of a material having a high thermal conductivity. In the case of the chip carrier, it is necessary to have good electrical insulating properties along with a high thermal conductivity.

Phase stability. No part of the electrical package may exhibit any phase change during cyclic heating and cooling.

Mechanical properties. The interconnections must be designed to withstand stresses imparted as a result of handling during assembly and also the stresses generated as a result of the thermal expansion mismatch between the two levels of circuitry during heating and cooling.

Corrosion resistance. All of the materials comprising the package must be impervious to corrosion. In many cases it is required that the first-level interconnections and the metallization on the substrate operate in a hermetic environment.

Processability. The specified package design must be adaptable to a high-speed joining process in order to achieve high-volume production.

Cost. In addition to the above engineering conditions, a particular package must be cost competitive with other packaging schemes.

Microelectronic packaging requires the selection of suitable materials and assembly processes to build a specific design. A discussion of some conventional packages and metallurgical joining processes used with these packages is given below.

Dual in-line package (DIP). A schematic diagram of this type of package is shown in Figure 19-1. Electrical connections between chip and lead frame are

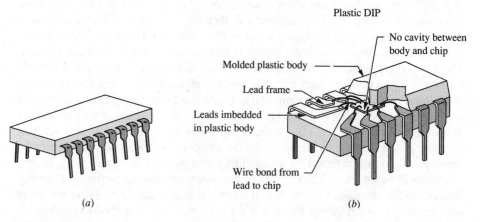

FIGURE 19-1
(*a*) A standard type of DIP package suitable for attach using pin-in-hole technology. (*b*) A detailed illustration of the internal structure of a DIP package.

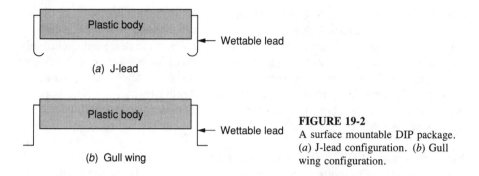

(a) J-lead

(b) Gull wing

FIGURE 19-2
A surface mountable DIP package.
(a) J-lead configuration. (b) Gull
wing configuration.

made by wire bonding, which is discussed later in this chapter. The back side of
the chip is mounted to a substrate, as shown in Figure 19-1(b), and the input/output
(I/O) pads on the chip are attached to the lead frame by wires. Lead frames are
made of kovar, nickel, copper, and aluminum. Encapsulation of the DIP is made
either by plastic or ceramic materials. The former is used mainly for nonhermetic
packages, the latter for hermetic packages for high-performance applications. The
leads are connected to the wall of plated through-holes (PTHs) on the printed
circuit board (PCB) by soldering. Since DIPs have leads along only two edges,
the application of DIPs is limited to a low number of I/O counts. By reforming
pins, DIPs can be easily converted to surface mountable packages, an alternate
method for interconnecting a component to a printed circuit board. Pins can be
bent inward [J lead, Figure 19-2(a)] or outward [gull wing lead, Figure 19-2(b)].

Quadpacks. This type of package has interconnections on all four sides of the
component. There are many packages with different sizes and lead configurations
in this category. Figure 19-3 shows a standard leaded quadpack carrier attached

FIGURE 19-3
An illustration of a standard leaded quadpack package.

to a printed circuit card. The internal circuitry of a quadpack is similar to that of a DIP package. Leaded chip carriers can be mounted either by pin in hole or surface mount lead configurations. Leadless chip carriers are designed exclusively for surface mount applications. The fan-out circuit patterns for these packages are made on the substrate by screen printing metal or alloy paste, which is subsequently fired, or by photoprinting evaporated thin film metals. Wire bonding is used to connect the chip to the carrier (substrate) circuitry. The perimeter leads are brazed or soldered to the substrate.

Pin grid array (PGA). This type of carrier is suitable for a package which requires higher I/O counts than a DIP or quadpack. An illustration of this type of package is shown in Figure 19-4. The most common chip carrier material is sintered alumina. Bonding of the chip circuit to its carrier can be achieved by wire bonding or flip chip C4 (controlled collapse chip connection). The process for the C4 type of soldering will be discussed later. The pins can be locked into the holes of the ceramic substrate by mechanically deforming the top portion of the pin around the bottom side of the ceramic hole (this is a standard pinning process). The head of the pin on the top side of the substrate is then soldered to the connecting metallization on the substrate. In the case of multilayered ceramics (MLC) substrates, the pins are brazed to pads on the bottom side of the substrate. As in the previous case, these PGA packages are attached to the plated through-holes in the card soldering. Figure 19-5 shows a PGA-type of package used in an intermediate range processor. Refer to other parts of the book for thermal and electrical performance of this type of package.

Tape-automated bonding. Tape-automated bonding (TAB) is a concept for an interconnection in which the metallurgical bond is similar to a wire bond. Numerous papers have been written on TAB first- and second-level assembly processes [1–8]. In the past, TAB was used for simple packages having low I/O counts. However, it is gaining rapid momentum for applications in more complicated high-performance packages, due to its automation of the chip assembly

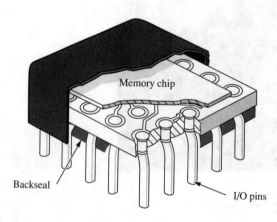

Backseal

Memory chip

I/O pins

FIGURE 19-4
An illustration of a pin-grid-array package.

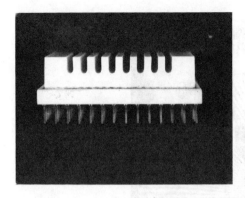

FIGURE 19-5
A photograph of a 35-mm MLC (multilayered ceramic) module used in the IBM 9370 processor. This module has a 14 × 14 area array matrix of pins brazed to the multilayered ceramic substrate.

process. The assembly process for attaching the TAB package to the second-level carrier is not yet well defined. The flexible leads, typically copper, are made by photoprinting an evaporated or a plated layer of metal on a polyimide carrier. The fact that the leads are photoprinted and held in place by the polyimide carrier allows for the attainment of a high density of I/O counts that cannot be achieved in either the DIP or quadpack packages. The polyimide film carrier is strong enough to support the thin lead pattern, but flexible enough to minimize undesirable stresses to the interconnection joints or leads during handling or system operation. The chip circuitry is bonded to inner pads of the lead by thermal compression bonding (this bonding method is discussed later in the chapter) or by gold-tin liquid phase bonding. The outer pads are attached to the card, or next layer of circuitry, by soldering. Figure 19-6 is a photograph of a section of a 35-mm tape having 124 I/O pads.

Although the basic metallurgical principles are the same, joining for electronic packaging applications is in many aspects different from that for bulk structural components. An electronic interconnection is very small in size; for example, a bonding wire is typically of the order of 0.001 in. in diameter. In general a very high density of these interconnections is required on a substrate (first-level interconnection). The small size of these joints requires high precision in handling to ensure good registration of one joining pad to another pad. In the case of a solder interconnection between two copper pads, it is very important to ensure that the pad is not totally dissolved by the tin in the solder (which is usually a Sn-Pb alloy). Also, the rate of solidification will control the grain size. Even for fast cooling rates, the relative grain size of a solder interconnection is large with respect to the overall size of the interconnection. Three methods of joining metallic or alloy bodies are

1. *Fusion welding.* Joining is achieved by melting a portion of the component base that is to be joined. Fusion welding includes a number of processes in which heat is generated by electric arc, laser, electron beam, or by the heat generated due to friction at the surfaces of mating leads. The joining temperatures achieved by this type of attach process are typically very high.

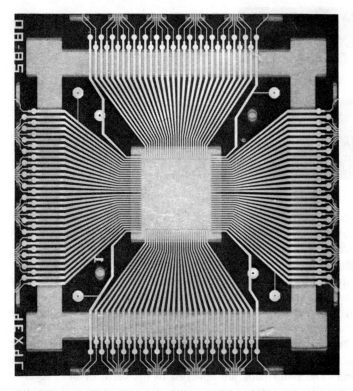

FIGURE 19-6
An illustration of a strip of 35-mm TAB tape.

For example, the temperatures generated during electron beam microsurface welding can be as high as 7000°F.

2. *Soldering and brazing.* Soldering and brazing are processes in which metallurgical joining is achieved by the wetting and flow of an alloy with a lower melting point than either of the metallic leads being attached. Brazing, or *hard* soldering, is distinguished from *soft* soldering merely by the temperature range in which the fillet alloy melts. Alloys used for brazing normally require temperatures above 400°C and rely on wetting as well as diffusion for the bond formation, while soft solders melt at temperatures less than 400°C [9].

3. *Solid state bonding.* The mechanism of this type of bonding is based on the interdiffusion of atoms across the mating interface without involving any molten state of materials. Solid state bonding requires heat to increase the rate of atom diffusion. A certain amount of pressure is also required to facilitate plastic deformation of the surfaces of the joining components and to ensure contact of the two surfaces. Thermal compression, ultrasonic, and thermosonic bonding methods fall into this category.

Since fusion welding requires undesirable melting of joining components, it is not used for the chip-to-chip carrier, or component-to-PCB interconnections. Soldering has long been used in many areas in electronic packaging. Important applications among them are flip-chip bonding and second-level packaging assembly. Thermal compression, ultrasonic, and thermosonic techniques in conjunction with wire bonding or TAB are heavily used for the chip-to-fan-out circuitry interconnection.

In this chapter, we will first discuss the fundamental aspects of soldering and its prime applications in electronic packaging. Brazing will be discussed briefly in this section in relation to pin attachment on multilayered ceramic (MLC) packages. Some of the important design factors for materials and processes will also be discussed. Due to their close tie with wire bond and TAB interconnection schemes, thermal compression, ultrasonic, and thermosonic bonding will be discussed.

SOLDERING

Soldering is, in general, a low-temperature joining process that is reversible, making it especially attractive since repair and rework are necessary considerations in the production of electronic assemblies. Advantages of soldering over other competitive joining techniques such as welding, brazing, or adhesive bonding (commonly used to attach heat sinks to the package, or for backside mounting of a chip to the substrate) are as follows:

Many of the electronic packages consist of materials such as polymers that may be degraded in high-temperature assembly operation. Since solder alloys melt at a relatively low temperature, soldering can accommodate these materials by a proper selection of solder alloys. Low-temperature joining processes are desirable also because unwanted diffusion or structural transformation of materials in the package will be limited. This is one of the major reasons why low-melting eutectic soldering is used in PCB assembly processes.

The solder joints are formed by the nature of the wetting process, even when the heat and solder are not directed precisely to the places to be joined. Molten solder wets selectively; it does not wet polymers and some metals, such as Cr, Al, or Ti, when using a conventional rosin-based flux to clean the base metal during the soldering operation. The wetting characteristics allow molten solder to be confined to desired places. One example of the application of selective wetting characteristics is flip-chip bonding, and another is protective coating (solder mask) on printed circuit boards.

The physical reversibility of the soldering process allows nondestructive removal of the soldered components. This implies that repair is easy compared to other joining techniques.

The soldering process can be accomplished as quickly as the wetting takes place (typically on the order of a few seconds). Thus the process offers high-speed joining and automation for high-volume production purposes.

Materials and Process Considerations for Soldering

SOLDER ALLOYS. Some of the common electronic component leads or pad metallizations to be joined by soldering are Cu, Ni, Au, Pd, or alloys of these containing small amounts of alloying elements. Unsolderable (or difficult to solder) metal parts, such as Ti or Al, are often coated with one of the above metals by electroplating or vacuum deposition. Unless there is contamination or an oxide layer, Cu, Ni, and Au provide a good surface for wetting with common Sn-Pb solders. In addition, they will form an intermetallic phase layer with the tin from the molten solder. While Sn-Pb alloys account for a large portion of the solder used in the electronics industry, a number of other alloy systems are available. Each alloy is unique with regard to its properties and must be chosen to meet the requirements of the assembly, such as ductility, resistance to low cycle fatigue, tensile and shear strength, and rate of base metal consumption (particularly during rework operations). Alloys of Sn-Bi, Pb-Ag-Sn, Pb-In, Sn-Ag, Sn-In, and Sn-Pb-In, and others are often used to meet specific requirements of soldering temperature, corrosion resistance, or increased joint strength. Almost all the soft solders are binary or ternary alloys of Sn.

It is well known that small amounts of certain impurities in the solder alloy influence the mechanical properties of an interconnection, and the degree of wetting of the molten solder. Contamination with small additions of Al, Cd, Cu, Fe, Mg, Ni, S, and Zn are undesirable because they decrease the fluidity of the molten solder and usually cause a gritty appearance of the solder joints [10]. Small amounts of Au do not decrease the wettability of the solder, and have shown to actually improve the ductility and toughness of solder for additions up to 2.5 weight percent [11]. Slow cooling of Au-containing solders often results in solder joints having a frosty appearance. Silver is intentionally added to Sn-Pb solders to prevent precious metal dissolution by the molten solder. The addition of 2 percent Ag to Sn-Pb alloys was found to substantially increase the creep strength of a solder joint [12].

Table 19.1 lists commonly used solders with their compositions, melting points, and some other properties. The operation temperature of soldering is one of the major considerations in selection of the solder alloy since the composition determines the melting point. To supplement this information, equilibrium phase diagrams are shown in Figure 19-7 for the Sn-Pb, Sn-Bi, and Sn-In binary systems.

It is apparent from the phase diagram shown for the Sn-Pb system that the melting temperature changes substantially with composition. A pure metal always melts at a single temperature (points A and B on the binary phase diagram shown in Figure 19-8). The freezing point of a liquid alloy is determined by the liquidus line while the melting point of a solid alloy is determined by the solidus line. In the region between these two curves both a liquid and a solid phase can coexist. At the eutectic point an alloy will transform from the liquid state to a solid state containing two separate phases. This eutectic point is denoted by the letter C on the phase diagram shown in Figure 19-8, and corresponds to a composition of 63 percent Sn and 37 percent Pb, by weight, on the Sn-Pb phase diagram. Any

Common solder alloys: compositions, melting temperatures, and mechanical properties

Sn	Pb	Bi	In	Sb	Cd	Ag	Cu	Solidus	Liquidus	Tensile strength, lb/in²	Density, g/cm³
8.3	22.6	44.7	19.1		5.3				47	5400	8.86
12.0	18.0	49.0	21.0						58	6300	8.58
12.5	25.0	50.0			12.5			70	74	4550	9.50
50.0			50.0						117	1720	7.74
37.5	37.5		25.0						138	5260	8.97
42.0		58.0							138	8000	8.72
	15.0		80.0	5.0					157	2550	8.20
			100.0					150	157	515	7.44
70.0	18.0		12.0						174	5320	7.96
62.0	36.0					2.0			179	N.A.	8.50
63.0	37.0							183	183	7700	8.46
63.0	36.7			0.3				183	183	7800	8.40
60.0	40.0							183	188	7600	8.50
60.0	39.7			0.3				183	188	7600	8.50
50.0	50.0							183	212	6000	8.90
50.0	49.7			0.3				183	215	6300	8.90
50.0	48.5						1.5	183	215	7000	8.90
	50.0		50.0						215	4700	9.14
96.0						4.0		221	221	N.A.	7.50
45.0	55.0							183	224	5600	9.10
			90.0			10.0			231	1650	8.10
	75.0		25.0						231	5450	9.80
100								232	232	1400	7.30
40.0	59.7			0.3				183	234	5400	9.30
40.0	60.0							183	234	5400	9.30
95.0				5.0				236	242	5600	7.30
35.0	65.0							183	245	5200	9.50
28.0	70.5			1.5				185	254	N.A.	N.A.
30.0	70.0							183	255	5000	9.70
20.0	80.0							183	275	4800	10.0
	90.0		5.0			5.0			290	5730	N.A.
5.0	93.5					1.5		296	301	4600	11.10
5.0	95.0								315	4300	11.35

The above table contains some of the solder alloys available in a variety of different solder forms (in order of increasing melting temperature). Melting temperatures are not recommended soldering temperatures; typically 21 to 65°C above the melting point insure good alloy fluidity and wetting characteristics. (Reprinted with permission: *Ref. 9.*)

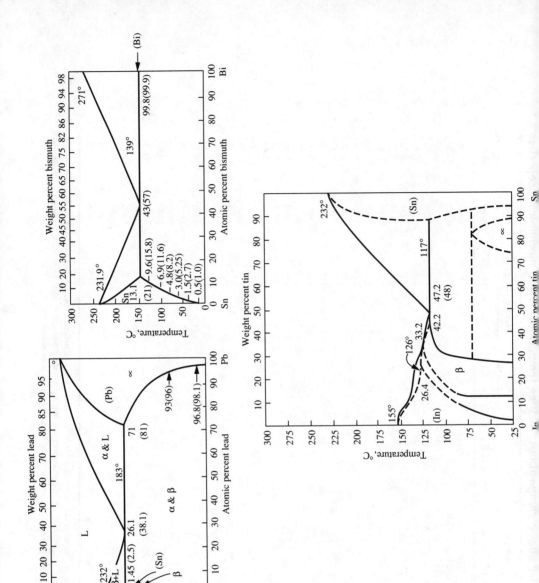

FIGURE 19-7
The phase diagrams for the Sn-Pb, Sn-Bi, and Sn-In systems. (Reprinted with permission: P. Hansen, Constitution of Binary Alloys.)

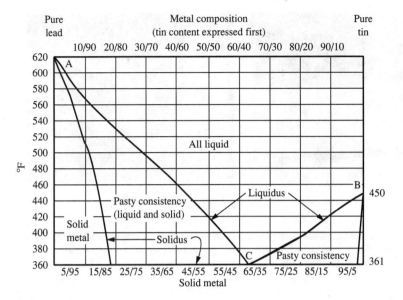

FIGURE 19-8
Another version of the Sn-Pb phase diagram, which illustrates the liquidus and solidus curves. (Reprinted with permission: LIA Solder Manufacturer Committee, "Solders and Soldering: A Primer.")

composition change from the eutectic composition results in an increase of the liquidus temperature of the alloy. Three single-phase regions are present in the Sn-Pb eutectic phase diagram shown in Figure 19-7: They are the Sn-rich phase, denoted by β, the Pb-rich phase, denoted by α, and the liquid phase region, which is denoted by L. In addition there are three two-phase regions, α +L, β +L, and α + β. The horizontal phase boundary line separating the α + β phase region from both the α +L and β +L phase regions corresponds to the eutectic temperature, which is 183°C. At the eutectic composition, liquid will transform to a duplex phase mixture of α and β upon cooling the alloy below the eutectic temperature.

Joining by soldering occurs by a reaction of the liquid solder with a solid surface. An example of the resulting microstructure when using molten eutectic Sn-Pb solder to wet a solid copper surface that is subsequently air cooled and solidified is shown in Figure 19-9. From this micrograph, four distinct phase regions can be identified. The phases identified in the cross section at the solder/copper interface are:

1. An isolated Cu-Sn intermetallic phase in the bulk microstructure.
2. A proeutectic Pb-rich phase that forms in dendrites from the liquid phase.
3. A solidified eutectic solder, which consists of a two-phased lamellar structure. One phase is Sn-rich (β) and the other phase is Pb-rich (α).

(a)   (b)

FIGURE 19-9
(a) A low magnification of a solder interconnection, 240×. (b) A high magnification of the region represented by the arrow in (a), which clearly shows the phases comprising the microstructure of the solder (Sn-Pb alloy) of eutectic composition at the copper interface, 860×.

4. A continuous Cu-Sn intermetallic phase region that forms adjacent to the copper surface.

The microstructure (size and distribution of dispersed phases) of an interconnection will strongly influence its mechanical properties. Since the resulting microstructure is determined by the cooling rate, the soldering process must be carefully controlled. Also, the solder composition must be chosen to meet the requirements of being able to withstand stresses due to handling during assembly, and also possess good thermal fatigue properties during operation of the device. Most often Sn-Pb alloys are used to form an interconnection, and the composition is chosen based upon the sequential joining hierarchy used in manufacturing, with the chip carrier components using high-temperature solder compositions such as 5 Sn/95 Pb or 10 Sn/90 Pb, while the component-to-printed circuit board (PCB) joining uses the lowest-temperature (63 Sn/37 Pb) alloy. In some cases an alloy that melts even lower than the Sn-Pb eutectic alloy may be required, such as the Sn-Bi eutectic alloy.

INTERMETALLIC PHASE FORMATION. To produce a reliable joint, a reaction must occur between the molten solder and the joining surface in order to form an intermetallic phase region at the interface. Intermetallic phases produced by soldering in electronic packaging are stoichiometric binary compounds containing Sn. The most common intermetallic compounds are from the Cu-Sn, Au-Sn, Ni-Sn binary systems, and in certain applications Pd-Sn intermetallic phases will form when a Pd base metal is being soldered. Elemental Pb or Bi does not react with the Cu, Au, Ni, or Pd that are commonly used as base metals. Thus, in soldering with Sn-Pb or Sn-Bi solder, only Sn will react with a base material (most often Cu) to form an intermetallic phase. For example, the occurrence of the Cu-Sn phase upon soldering becomes apparent when viewing the Cu-Sn phase diagram (Figure 19-10), which reveals the presence of two stoichiometric compounds corresponding to compositions of Cu_3Sn and Cu_6Sn_5. Phase diagrams of the Ni-Sn and Au-Sn systems also exhibit similar types of intermetallic phases.

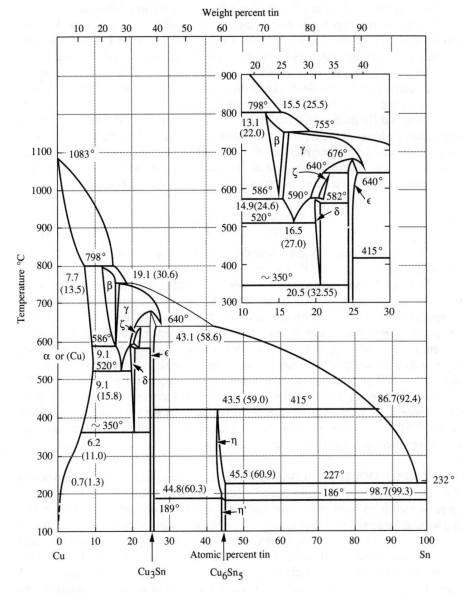

FIGURE 19-10
The Cu-Sn phase diagram. (Reprinted with permission: P. Hansen, *Constitution of Binary Alloys*, Genium Publishers.)

Both of these phase diagrams exhibit stable intermetallic phases similar to the phase regions in the Cu-Sn phase diagram.

When using a particular solder to join certain metals, the phases that will form can be predicted from corresponding phase diagrams. For example, when a Sn-Pb solder is used on a copper surface, the reaction between the solder and solid copper involves the respective Sn, Pb, and Cu elements. The three binary phase diagrams, Sn-Pb, Sn-Cu, and Cu-Pb, can help predict the microstructures that will form at the joining interface. For example, the Cu-Pb phase diagram does not contain any type of intermetallic phase. This explains why Pb is not responsible for the reaction of the solder with the Cu base metal. Similar reactions can occur for other metal surfaces, such as Au or Ag. The number of intermetallic phases will be determined by the Sn-X phase diagram, where X is the other component comprising the binary Sn-X phase diagram.

Kang and Ramachandran [13] have investigated the growth kinetics of Ni-Sn intermetallic phases at the liquid Sn and solid Ni interface. Similar kinetics govern the formation of Cu-Sn intermetallic phases that form at the liquid solder and solid copper interface. The thickness δ of an intermetallic layer will increase with time t and temperature T, and can be expressed by the following equation:

$$\delta = k_0 \exp\left(-\frac{Q}{RT}\right) t^n$$

where k_0 is a constant, n is the parabolic growth rate constant, which is 0.5 for the initial stages of growth, R is the gas constant, and Q is the activation energy.

Almost all the intermetallic phases are brittle. The brittle nature of an intermetallic phase is in part a result of the ordered crystal structure of the compound phase. It is therefore not desirable to have a thick layer of intermetallic compound at the joining interface because of its tendency to fracture. In order to prevent the excessive growth of thick intermetallic layers a barrier layer is often deposited on the copper base metal contacting the molten solder alloy. Nickel is most often used as a barrier layer between solder or pure tin and copper [14,15]. The nickel will react with the Sn to produce Ni-Sn intermetallic phases that grow much more slowly than Cu-Sn intermetallic phases at high temperatures. Iron is more effective as a barrier metal than nickel, but is hardly ever used since the Fe-Sn intermetallic phase easily corrodes [16].

WETTING OF MOLTEN SOLDER. During the joining process, the molten solder must wet the surfaces to form a sound joint. The liquid solder is constrained from spreading away by surrounding the surface around the confines of the joint with a nonwettable material (a solder mask). The ability of a liquid to wet a solid surface is measured by the contact angle, which is the angle that the liquid solder makes with the solid base metal surface. This is illustrated in Figure 19-11, showing the relationship between the dihedral angle θ and the degree of wetting [17]. A contact angle of less than 90 degrees indicates wetting of the molten solder to the surface, while an angle greater than 90 degrees indicates lack of wetting affinity. To ensure good wetting, the molten solder should wet the connecting surface

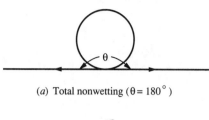

(a) Total nonwetting ($\theta = 180°$)

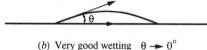

(b) Very good wetting $\theta \rightarrow 0°$

FIGURE 19-11
An illustration of the contact angle (θ), which represents the degree of wetting of a liquid on a solid surface.

with a contact angle of less than 10°. In solder connections, the molten solder normally wets the connecting surface up to the boundary of the solder mask.

The minimum temperature for wetting does not always correspond with the melting temperature of a bulk solder alloy. In the case of Sn-Pb solder temperature 100°F over the melting temperature is recommended for good wetting and solder flow. However, as noted earlier, this will accelerate the growth of the intermetallic phase region. Thus the joining reaction is a compromise in accelerated wetting and joint weakening because of the growth of a larger amount of the brittle Cu-Sn intermetallic phase region.

The temperature dependence of the viscosity of the solder requires a sufficient amount of superheating to ensure that the fluidity of the solder is high enough for the solder to flow into all the narrow spaces by capillary action. Nonuniform surfaces can be caused by inclusions and occlusions (metal oxides, sulfides, etc.) of either the solder or base metal. Wetting is also affected by the presence of nonmetallic foreign particles and absorbed vapors on the surface that are not displaced during the wetting process, as well as variations in intermetallic compound formation [18–20]. Impurities in the solder, such as Zn, Al, Fe, Sb, and Sn-Pb oxides, along with nonmetallic contaminants such as S and P, also decrease the wetting rate [21,22].

The wettability of a solder can be measured using a meniscograph or a wetting balance [10]. Figure 19-12 schematically shows a wetting balance recorder trace to illustrate a good wetting and a bad wetting condition, along with intermediate wetting possibilities. Although a large amount of work has been done on determining the wetting properties of solder, particularly Sn-Pb solders, it has been very difficult to find agreement among data published by different investigators. These discrepancies can be attributed in part to variation in the degree of alloy purity and the accuracy of the testing equipment, as well as a lack of standardization of the variables being reported.

FLUXING IN SOLDERING. The word *flux* comes from the Latin word *fluxus* which means "to flow." The flux in a soldering process is similar to a catalyst in a chemical reaction, which triggers and promotes the process without entering into the end product. A clean, oxide-free surface is imperative to ensure that solder joints are of a uniform quality and good mechanical reliability. Oil, grease,

(a) Non wetting

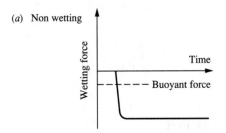

(b) Rapid wetting

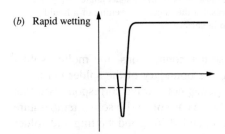

(c) Slow wetting

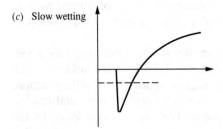

(d) Rapid wetting
followed by
dewetting

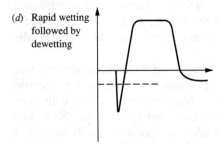

FIGURE 19-12
Wetting balance traces illustrating different degrees
of wetting of solder on a copper surface.

tarnish, fingerprints, pencil markings, tape residues, machine lubricants, and general environmental dirt can all contribute to solderability problems. The importance of cleanliness should be stressed at all stages of manufacturing of components and cards, including storage, handling, and cleaning cycles for the elements to be joined. A flux is used to remove any oxides and other surface compounds that may be present on the base metal in order to promote wetting of the solid metal surfaces with the liquid solder at the soldering temperature. In addition to removing any tarnish or oxides from the metal surfaces, the flux prevents the reoxidation of the surfaces when heated prior to soldering. The flux will lower the surface tension between the molten solder and the metal surface so that solder will readily flow and wet the solid surface. During this wetting process, the flux will be automatically displaced from the solid metal/liquid solder interface to ensure that a good bond is formed between the metal and solidified solder.

The choice of a flux will depend on how difficult it is to remove the tarnish films, the temperature required for soldering, and how susceptible the parts being soldered are to corrosion by ionic residues. The ideal flux should have a melting point slightly below the soldering temperature, along with the following features: It should possess the ability to remove oxides and other surface compounds; protect the surface from further oxidation during soldering; leave no electrically conducting or ionic residue after soldering (so as to be noncorrosive); and be safe to personnel and the manufacturing plant, and not harmful to the environment.

The many fluxes available are normally classified into three categories: inorganic, organic, and rosin fluxes. Inorganic fluxes are the most active, containing inorganic salts such as zinc chloride or ammonium chloride. They are also the most corrosive and conductive of all the fluxes and are effective in promoting wetting on most metals. Because they have both highly corrosive and conductive residues, they are not generally used in soldering of electronic components. Also, it should be mentioned that these fluxes are soluble in water.

Organic fluxes are moderately active and contain mostly organic acids, such as tartaric acid or stearic acid. These fluxes are also soluble in water. They are less active than the inorganic fluxes, but more active than the rosin fluxes. Organic fluxes are used extensively in electronic packaging assemblies in which closely spaced regions that can entrap flux residue are absent, and when water cleaning is sufficient to remove the flux.

The mildest fluxes are the rosin fluxes, which are generally used in electronic component soldering because they are noncorrosive and nonconductive. Rosin has the unique property of dissolving copper oxide when it is heated to its molten state (127°C). The active component in rosin flux is abietic acid, which at the soldering temperature is chemically active and attacks copper oxide, converting it to copper abietate. Initially the flux displaces the atmospheric gas layer on the metallic surface, then, upon further heating, any tarnish film is removed. There are three classifications of rosin fluxes:

Nonactivated rosin (R), frequently referred to as water white rosin, which contains no additional activating agents.

Mildly activated rosin (RMA), the basic rosin with mild activating agents to improve fluxing action.

Activated rosin (RA), the basic rosin with small amounts of strong activating agents added to improve the fluxing ability.

Table 19.2 gives a general guide to flux selection for various types of contacting metals.

TABLE 19.2
Metal solderability chart and flux selector guide

Metals	Solderability	Rosin fluxes			Organic fluxes—water soluble	Inorganic fluxes—water soluble	Special flux and/or solder
		Nonactivated	Mildly activated	Activated			
Platinum Gold Copper Silver Cadmium plate Tin (hot dipped) Tin plate Solder plate	Easy to solder	✓	✓	✓	✓	Not recommended for electrical soldering	
Lead Nickel plate Brass Bronze Rhodium Beryllium copper	Less easy to solder	Not suitable		✓	✓	✓	
Galvanized iron Tin-nickel Nickel-iron Mild steel	Difficult to solder	Not suitable			✓	✓	
Chromium Nickel-chromium Nickel-copper Stainless steel	Very difficult to solder	Not suitable			Not suitable	✓	
Aluminum Aluminum-bronze	Most difficult to solder	Not suitable			Not suitable		✓
Beryllium Titanium	Not solderable						

(Reprinted with permission: LIA Solder Manufacturers Committee, "Solders and Soldering: A Primer.")

Soldering for Flip-Chip Bonding

Wire bonding and TAB interconnection schemes require a face-up configuration of the circuitized side of a chip, whereas the flip-chip bonding concept adopts a face-down configuration. It has been developed as a means of improving the bonding operation from a reliability and cost standpoint. It also increases the I/O count on a given area of the chip. There are two types of flip-chip bonding, beam lead bonding and controlled collapse chip connection (C4). The beam lead concept is shown in Figure 19-13(a). The pad is produced by plating or photoprinting techniques during processing of the undiced wafer. The leads can be bonded by thermal compression or ultrasonic bonding methods. Beam lead bonding is used for very specific applications and will not be discussed further. Flip-chip bonding was developed by IBM for solid logic technology (SLT) in the early 1960s. A thorough discussion of the evolution of first-level attach using flip-chip bonding is given by Gow and Leach [23]. The original concept of flip chip employed a small solder-coated copper ball placed between the chip terminal pad and the corresponding terminal pad on the chip carrier substrate. The solder interconnection is made by exposing the assembly to a temperature above the melting temperature of the solder coating of the ball, which was a eutectic Sn-Pb alloy. This early technology made use of a ball approximately 0.005 in. in diameter. In order to use this approach to interconnect high-density chips to the chip carrier, smaller ball diameters were required. This attach approach worked well for low I/O chips used in SLT; however, for higher I/O chips a better method

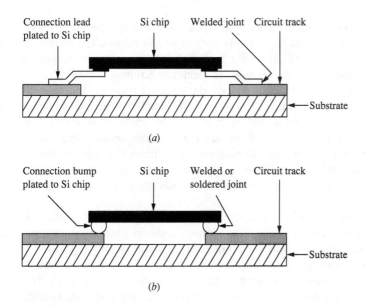

(a)

(b)

FIGURE 19-13
(a) A schematic illustration of a beam lead type of interconnection. (b) A schematic illustration of a C4 type of interconnection.

was needed to eliminate the need for solder ball placement and to improve the reliability of the joint. In C4 chip bonding, a Pb-rich alloy pad is deposited on the chip. Surface tension of the molten solder joint during reflow aligns the chip to the chip carrier and creates a good mechanical interconnection upon solidification of the solder. Chip joining using C4 bonding paved the way for future generations of high-density chips. Systems employing C4 attach that were built in the late 60s using millions of chips are still operating without failure in field installations. The Pb-rich interconnection between the chip and the substrate has a diameter in the range of 0.005 in. and a standoff of 0.002 to 0.003 in. Figure 19-13(*b*) schematically illustrates a C4 joint.

There are many advantages of the C4 technology compared to other interconnection schemes. The most significant of them is the fact that the whole area under the chip can be used for I/O interconnection, whereas only the perimeter of the chip can be used for I/O with wire bonding or TAB. Since a C4 interconnection is made with a small volume of solder, the interconnection is formed without any leads or wires, and provides a good thermal path for heat dissipation. The small volume of the solder interconnection also gives better inductance characteristics than other kinds of connections.

PROCESSING AND DESIGN FACTORS OF FLIP-CHIP BONDING. A schematic illustration of the formation of a wetback on the chip and the subsequent attach of the bumped chip to the substrate is shown in Figure 19-14. This illustration shows the two basic assembly steps that make it possible to form a flip-chip C4 type of interconnection.

Since the trend in the electronics industry is toward producing chips with a greater number of circuits and accordingly increased I/O counts, the feature of an area array capability for I/O, allowing high densities of interconnection, is expected to increase the use of the C4 type of interconnection.

In the joining process, solder is placed on terminal pads on the chip by vacuum deposition through a mask. The deposition results in a cone-frustum-shaped piece of lead, covered by a thin layer of tin. The solder is then reflowed through a hydrogen belt furnace to homogenize the lead and tin and form solder bumps for chip connection. The chip joining process utilizes the surface tension of molten solder to limit the collapse of the reflowed solder bumps on the chip (wetback) to provide a connection to mating pads on a metallized ceramic substrate.

The maximum height of solder that can be evaporated through a mask is limited by the "bridging" of the solder between two adjacent pads during reflow. In order to obtain the required solder volume, the diameter of the deposition is increased beyond the diameter of the chip pad. To obtain the desired solder bump size during reflow, a bump-size-limiting metallurgy technique is required. This process utilizes the nonwetting characteristic of glass. The chip pad metallization [a cross-sectional view of the chip pad metallization is shown in Figure 19-14(*a*)] consists of a thin film of chromium to form an adhesive layer, so that a thick

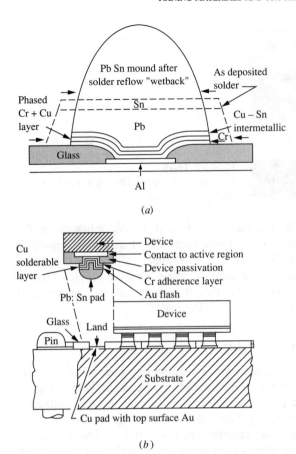

FIGURE 19-14
Schematic illustration of flip-chip C4 attach process steps. (*a*) Illustration of the deposited layers of Pb and Sn before reflow, and the resulting shape of the bumped chip (wetback) after reflow. (P. A. Torta and R. P. Sopher, "STL Device Metallurgy and Its Monolithic Extension," *IBM J. Res. Dev.*, Vol. 13, No. 3, May 1969, p. 226. Copyright 1969, International Business Machines Corporation; reprinted with permission.) (*b*) A simple representation of the flip-chip attach procedure. (L. F. Miller, "Continued Collapse Chip Joining," *IBM J. Res Dev.*, Vol. 13, No. 3, May 1969, p. 239. Copyright 1969, International Business Machines Corporation; reprinted with permission.)

layer of Cu can be evaporated to form the circuitization pattern. The C4 pad sites and also any outer lead sites that will be soldered to pins extending through the chip carrier are masked off, and a thin top layer of chromium is deposited. The remaining pads are usually coated with a layer of gold to protect the copper surface from oxidation and improve the solderability during C4 attach. Solder is deposited on this metallized pad as well as on an annular surrounding area of the chip pad to obtain the required volume of solder. During reflow, the chromium acts as a solder dam and the molten solder is drawn back by the surface tension to the area of wettable metal to form the desired shape of the solder ball (wetback) on the chip.

The ceramic substrate metallization utilizes the nonwetting characteristic of solder on ceramic and chromium to limit the flow of solder to the substrate joining pads. Films of chrome-copper-chrome are photoetched on the ceramic surface to form the connecting circuit lines and pads on the chip carrier substrate. Then the top layer of chrome is etched to expose the copper at those spots on the substrate where the chip solder bumps will make contact. The controlled collapse

100μm   ×430   30kV

FIGURE 19-15
Micrograph of the equilibrium shape of a C4
joint.

chip connection (C4) is made by locating the chip and its solder bumps on the surface of the substrate with flux, and reflowing the solder assembly through a temperature-controlled furnace to form the C4s. The solder volume and the solder limiting pad areas on the substrate and chip combine with the surface tension forces of the molten solder to support the device at a controlled spacing and keep it aligned until the joint solidifies (Figure 19-15).

Soldering for Second-Level Packaging

Thus far much emphasis has been placed on fabrication of components, or first-level packaging (attaching the chip to the chip carrier). Second-level packaging is attaching the component to a printed circuit card. The card in turn is attached to a board; this is third-level packaging. This process of joining parts of one level to form a higher level of package can continue for many stages. The number of levels is a function of the sophistication of the system being designed. Discussion will now move to second-level interconnections.

The purpose of the second-level package is to provide a carrier for the printed circuit conductors, which electrically interconnect the various components, in order to allow communication between the devices. The common card technology consists of multilayered metallization separated by layers of insulating material (impregnated fiber glass cloth). These layers are connected by *vias* and *plated through-holes* (PTHs). A PTH is a hole drilled completely through the card and then plated with copper; it will electrically connect to an internal power or signal metallized layer. The PTH brings this circuitry to the outside surface of the card. A via connects only internal signal layers to one another to allow wiring direction changes and wire design flexibility. A schematic illustration of a multilayer printed circuit package in cross section is shown in Figure 19-16. PTHs and vias can be seen in this illustration. On one side of the card is a surface metallization consisting of round doughnut shaped *lands* around each PTH. The other side of the card has a land and pad pattern, which is defined by openings in a dielectric coating that acts as a solder barrier or mask.

In the assembly of the components to the cards, two technologies are common: (1) pin-in-hole (PIH) technology, which uses components with leads

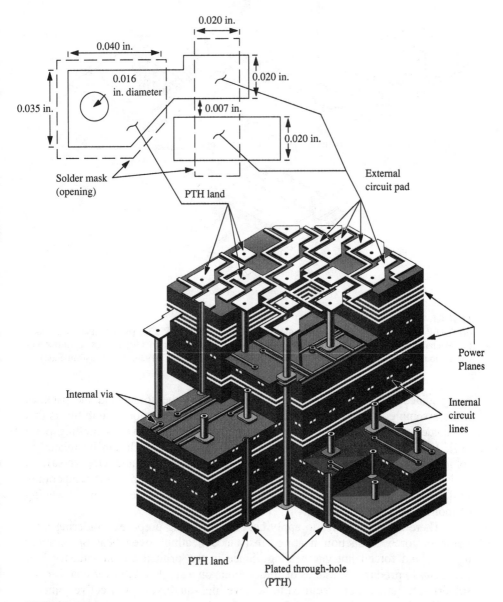

FIGURE 19-16
Isometric view of multilayered printed circuit board cross section.

that are inserted into the PTH of the card and soldered in place, and (2) surface
mount technology (SMT), where component leads are designed to be soldered to
the surface lands or pads on the cards (see Figure 19-17).

SOLDERING PROCESS FOR SECOND-LEVEL PACKAGING. In order to perform soldering it is necessary to have molten solder in contact with the metals

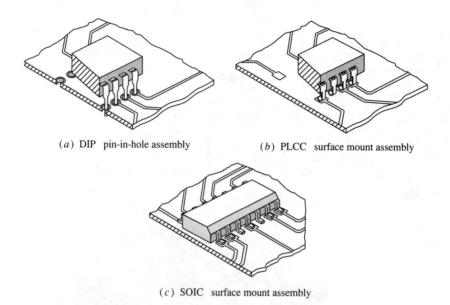

(a) DIP pin-in-hole assembly   (b) PLCC surface mount assembly

(c) SOIC surface mount assembly

FIGURE 19-17
(a) A DIP pin-in-hole assembly. (b) A plastic-leaded chip carrier (PLCC) surface mount assembly.
(c) A small-outline integrated circuit (SOIC) surface mount assembly. (All figures reprinted with
permission: Jerry Mullen, "How to Use Surface Mount Technology," 1984, Texas Instruments.)

being joined. This requires a heat source for both the solder and the package
to be assembled. A general rule of thumb for achieving good wettability is that
the solder temperature should be approximately 100°F above its melting point
[24]. Figure 19-18 shows the effect of wetting times, which can be related to
solderability (short wetting times imply high solderability and vice versa), of
various metallurgies with Sn-Bi eutectic solder, as a function of the temperature
[25]. These curves were generated using a meniscograph (see section on wetting
of molten solder).

There are many ways to apply heat for soldering purposes: soldering iron,
flame or torch, induction heating, resistance heating, oven heating, infrared
heating, and forced hot gas heating. Since in the printed circuit industry it is
necessary to produce thousands of solder joints on a single multilayer board, mass
soldering techniques are required to achieve high-quality, cost-effective joints.

Two basic processing techniques are predominantly used by the packaging
industry: wave soldering and vapor phase reflow soldering.

WAVE SOLDERING. Wave soldering is performed by passing a printed circuit
board, which has been populated with the components (normally PIH) required
for joining, across the crest of a molten, standing solder wave. Only the bottom
of the board is exposed to the molten solder which is the source of heat to the
board and components, as well as the solder source for the solder joint.

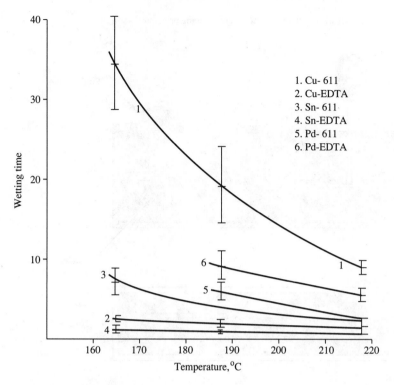

FIGURE 19-18

Wetting time versus temperature of 42% Sn/58% Bi alloy contacting copper, immersion tin, or a palladium surface. The two types of fluxes used were either α 611 or an EDTA flux.

In the case of soldering leads into plated through-holes, capillary action in addition to the compressive force of the wave against the board fills the PTH with solder, giving a reliable solder joint.

The solder wave in this process is formed by the continuous circulation of molten solder driven by a rotating impellor, which forces the solder through a nozzle to form a wave which then drains back to the solder pot reservoir (see Figure 19-19). The movement of the wave continually purges the solder of dross (oxides of solder) and flux residues and assures homogeneity of the solder being used. This flow action also allows faster heat transfer to the board, and the motion of the solder imparts a mechanical wiping action which enhances solderability. Additionally, because the solder is pumped from the bottom of the solder pot, the temperature of the solder wave remains uniform and is well controlled. Several and varied solder wave shapes have been designed, from unidirectional to bidirectional flow, from single wave to multiple cascade waves, or using inclined waves with a progressive separation of component leads simulating a vertical withdrawal from the solder wave [see Figures 19-20(a) and (b)]. Solder waves can be from several inches to several feet long, depending on the product being processed.

Solder wave generation – dual tank pumping principle

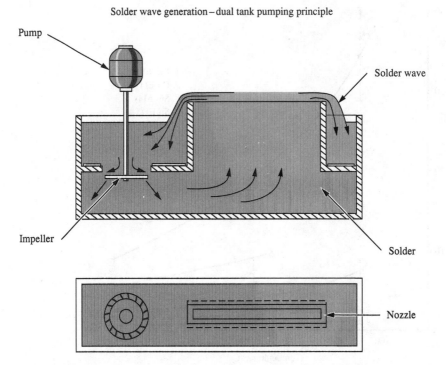

FIGURE 19-19
A schematic illustration of the solder wave generation used during wave soldering. (Reprinted with permission: ASM Committee on Soldering, "Welding, Brazing, and Soldering," *Metals Handbook*, 9th Edition, Vol. 6, 1983.)

The type of wave used for assembly depends in large part on what result is required. For instance, component leads protruding through the bottom of a PTH will require different wave dynamics than will a board that requires depositing solder on a land that does not contain a PTH (surface soldering).

In addition to the solder wave, wave soldering systems contain two other key operational stages—fluxing and preheating. These stages are normally contained in one unit, with a conveyor system carrying assembly work to each station automatically.

Several techniques are used for fluxing in a wave solder system. Wave fluxing is similar to wave soldering, with the board bottom passing through a wave of liquid flux. Flux is applied to the bottom surface with capillary forces carrying it up the PTHs, covering the wall of the PTH and the component leads. Foam fluxing utilizes a liquid flux and air passed through a porous diffuser, which creates fine bubbles of foamed flux. A nozzle guides the foam up to the board surface to achieve the fluxing action. Other less common techniques utilize rotating brushes or sprays to carry the flux to the board.

Preheating of the board is an important factor in the solder assembly operation. If the board travels through a solder wave immediately after fluxing, the flux will be wiped from the base metal before it has had a chance to perform

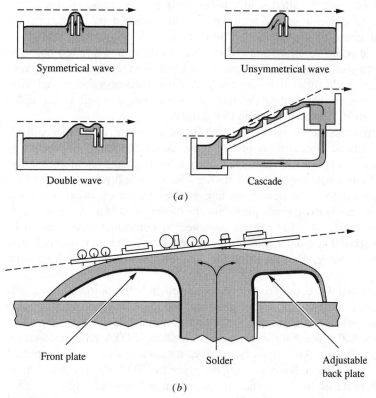

Symmetrical wave

Unsymmetrical wave

Double wave

Cascade

(*a*)

Front plate

Solder

Adjustable
back plate

(*b*)

FIGURE 19-20
(*a*) Illustration of the various types of wave shapes that can be generated. (Reprinted with permission: *Ref. 9*.) (*b*) Controlled planar wave showing adjustable backplate to control wave configuration and flow pattern. (Reprinted with permission: *Ref. 15*.)

its intended cleaning function. Additionally the solvents used as the carrier of the active ingredients of the flux will evaporate violently and cause the solder to splatter. By preheating, the solvent carriers are driven off and the flux is given enough temperature and time to clean and remove the surface compounds from the metal to be joined. Caution must be exercised in not heating the flux to a point whereby the fluxing action is destroyed. The board temperature is also raised to allow a reduced time in the solder wave. Preheating also helps prevent thermal distortions that may occur in the board as a result of high temperatures first being applied to one end of the board as it passes over the hot solder wave. The preheat temperature must not be raised to the point that damage to the printed circuit board results, such as delamination of the materials used to fabricate multilayer structures or cracking of the PTH walls due to thermal expansion. With tin-lead eutectic solder, a preheat temperature of approximately 200°F is common.

IMMERSION WAVE SOLDERING. As printed circuit boards become thicker and more complex, the need to minimize thermal stresses during processing and the need to obtain completely filled solder holes become conflicting requirements. In general, more difficult solder requirements are satisfied by increasing the soldering temperature of the board. Unfortunately, such an increase also raises the occurrence of board delamination and cracking of the plated through-holes. In addition, remelting of the solder on printed circuit boards to repair defects requires temperatures and cycle times that increase with board thickness and complexity. Dissolution of the copper plating by molten solder during repair is an added undesirable factor in board survival and reliability.

Two approaches are available to reduce or eliminate these mechanisms of board degradation. The usual approach is to reduce the peak temperature to which the board is exposed by changing from a high-speed system (generally, wave soldering) to a batch system (oil immersion, oven reflow, or vapor phase reflow) that permits lower temperatures but requires longer exposure times and expensive solder preforms (solder preforms are pieces of solder in the shape of rings, washers, balls, or other shapes designed to conform to the area to be soldered). The second approach is to use a solder with a lower melting point than the normal eutectic tin-lead solder (63% Sn, 37% Pb), which has a melting point of 183°C.

The immersion wave soldering process combines both approaches in a way that avoids the use of high-cost solder preforms and the low throughput of batch processing. The process consists of a fairly conventional solder wave submerged in a heated flux bath. The bath is glycerin with added EDTA (ethylenediamine tetraacetic acid). The mixture serves as a preheat medium, a flux, and a protective blanket for the solder. In addition to using the glycerin/EDTA mixture, the normal Sn-Pb solder is replaced by a eutectic mixture of tin and bismuth (42% Sn, 58% Bi), with a melting point of 138°C. With this soldering process, the populated board is first immersed and preheated in the hot glycerin/EDTA flux. While still immersed, it is passed over a Sn-Bi solder wave. The board is then removed from the flux bath, quenched, and cleaned.

The motivation for using a lower-melting-point solder in place of the eutectic Sn-Pb is the differential thermal expansion of the component parts of the printed circuit board. The difference between the vertical expansion of the base epoxy glass laminate and the copper plated through-hole is the principal cause of board delamination and cracking of the plated through-holes (see Figure 19-21). Reducing the peak temperature of the board to 152–163°C, the temperature range used for Sn-Bi soldering, results in a strain of 15×10^{-3} mm/mm, rather than the 38×10^{-3} mm/mm produced when using conventional Sn-Pb solder.

One reason that Sn-Bi eutectic solder is used for low-temperature soldering is that its mechanical properties closely approximate those of Sn-Pb [26]. In addition, the copper dissolution rate in this alloy is six to seven times lower than that of eutectic Sn-Pb solder.

Immersion wave soldering avoids some of the problems involved in conventional wave soldering. In conventional wave soldering, the combination of wave compression, preheating of the board, and superheating of the solder are used to achieve hole fill by the solder. However, wave compression is relatively ineffec-

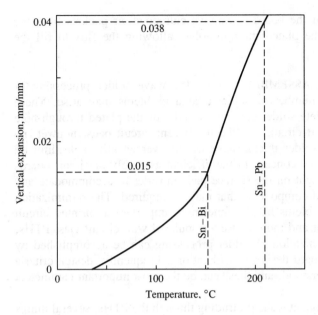

FIGURE 19-21
Vertical expansion of the epoxy-glass printed circuit board as a function of temperature.

tive with small-diameter holes, and capillary forces control hole fill. Preheating of the board is limited by oxidation of the card surface, which results in degradation of solderability. With the use of a Sn-Bi solder during conventional wave soldering it is not feasible to superheat the solder, due to the formation of excessive dross. A process which excludes air from both the board surface and the wave is required. This is achieved by immersion soldering in the glycerin/EDTA.

A schematic of an immersion solder machine is shown in Figure 19-22. The machine consists of a large titanium tank with heating coils and a conveyor. A conventional solder wave is located at the bottom of the tank. A molten solder seal, which is dependent on the density differential between the solder and glycerin, is used to contain the working fluid and permit the use of an external solder wave impeller. The preheat station contains a nozzle that directs

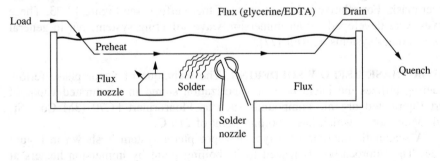

FIGURE 19-22
A schematic illustration of an immersion wave soldering operation.

flux against the bottom of the board to remove entrapped air pockets from the board bottom and from the plated through-holes, allowing the flux to fill the holes.

DESIGN FACTORS FOR ASSEMBLY. During the wave solder processing of printed circuit boards, a number of solder-related problems may arise. These problems include incomplete solder coverage (voids) of the plated through-hole walls, solder bridges (short circuits) between adjacent circuit pads, nonwet (no solder) circuit pads, and solder thickness or height over an allowable limit on leads or connectors. Also of concern to the efficient assembly of circuit boards is the control of solder height on the surface pads in order to accommodate any additional surface-soldered components that may be required. The minimization of these solder-related problems has become more important as printed circuit boards have become larger and more densely populated with circuit vias, PTHs, lines, and pads. The optimization of solder processing can be accomplished by both solder process and board design considerations. Frequently, design criteria are not recognized or understood, and these can be the most important parameters to be considered.

For cards with component leads protruding through the PTHs, several things should be considered in the design in order to minimize solder defects in this type of interconnection. Small-diameter component leads in large-diameter plated through-holes frequently result in solder joint depletion. This occurs because the surface tension of the molten solder in large-diameter clearances fails to support sufficient solder to keep the hole properly filled. Large land areas should be avoided, and an excessive number of solder joints in one area can promote bridging. The thickness of copper circuit pads can contribute to solder bridging, due to solder buildup on the edge of the pad. A square pad configuration will bridge more frequently than a round one of the same diameter and center-to-center spacing [27].

For boards having component leads that terminate within the board thickness, or boards that will experience surface solder only, significant board design parameters include circuit pad geometry, circuit pad surface area, signal-to-ground circuit pad spacing, land width around PTHs, land to PTH diameter ratio, component lead length, solder mask thickness, and the size of the opening in the solder mask. For relative values of some of these effects see Figure 19-23. These curves were developed for an immersion wave soldering system, but in general apply to other systems as well [27].

VAPOR PHASE REFLOW SOLDERING. The technique of vapor phase reflow soldering utilizes the latent heat of vaporization stored in a saturated vapor of inert fluorinated organic compounds such as Flourorinert FC70 (3M Co., St. Paul, Minnesota), which has a boiling point of 215°C.

A schematic diagram of a typical vapor phase system is shown in Figure 19-24. The fluorocarbon is heated to its boiling point by immersion heaters at the bottom of the system. Vapors then rise and fill the chamber above. Because

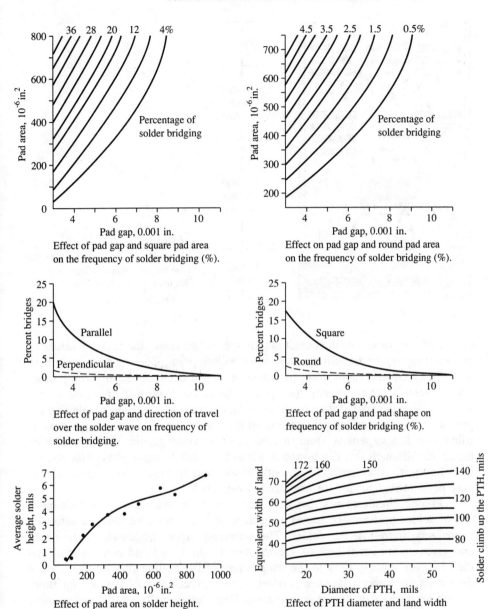

FIGURE 19-23
Printed circuit board wave solder design considerations.

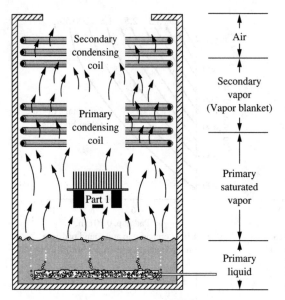

FIGURE 19-24
A schematic illustration of a vapor phase reflow furnace. (Reprinted with permission: "Soldering Part 2," *Circuits Mfg.*, July 1980.)

the liquid and vapor are in contact at atmospheric pressure, the vapor is saturated and its temperature remains constant at its boiling point. Cooling coils are placed above the primary vapor zone to condense and recirculate the diffusing vapor and prevent it from escaping the system. Since the fluorocarbons used in this system are expensive materials, a secondary vapor area is used to blanket the primary zone and further prevent loss of the fluid. The secondary vapor zone is filled with less expensive fluid that has a lower boiling point than the primary material. Although this discussion is given for a batch vapor phase process, in-line conveyorized systems are currently available that make it possible to achieve very high throughput for surface mount assembly.

Printed circuit boards that have been precoated with solder, or that make use of solder preforms or pastes at each joint, are immersed into the saturated vapor zone to reflow the solder. The saturated vapor condenses on the board and gives up its latent heat of vaporization to the board and components, and the assembly quickly rises to the temperature of the boiling fluorocarbon. Since the maximum vapor phase temperature cannot exceed that of the boiling fluid, the temperature of the board is well controlled. Additionally, since the assembly is surrounded totally by the vapor, thermal stresses in the board such as are caused by wave soldering are not experienced. However, preheating is generally required to prevent thermal shock to the components being mounted to the board. An added benefit to this approach of soldering components to cards and boards is that there is no geometry dependence, nor is there any limitation to mixed components on a board such as surface mount and through-hole-mounted devices. Also, oxidation during soldering is minimized since the vapor envelops all parts of the assembly and excludes air from coming in contact with the metallization at the higher temperatures. This permits the use of mildly activated rosin fluxes with parts that have good solderability. The fluorocarbon is inert and does not pose any corrosion concern during the reflow cycle.

SURFACE MOUNT TECHNOLOGY. The discussion in the preceding sections is fundamental to most solder assembly operations whether they be for plated through-hole mounted components or surface mount components (SMCs). Pin-in-hole (PIH) technology uses components with I/O leads that are inserted into the PTHs of the board and soldered in place to effect the electrical interconnection. Surface mount technology (SMT) uses components with I/O leads that are soldered to lands on the surface of the board and eliminates the need for large diameter PTH sites in the card to accept the component leads.

SMT is the newest technology in the advancement of high-density electronic packaging assemblies. One of the main advantages of SMT is that higher circuit densities can be achieved on boards since the space required for a PTH can be reduced from about 0.040 in. to approximately 0.010 in. (this plated through-hole is only for electrical connection between adjacent signal or power layers in the card). This means that circuit line routing can be much more efficient and spacing can be reduced between component (chip carrier) leads with a resultant decrease in package size. Additionally, the components can now be mounted on both sides of the board, whereas with PIH technology component location is normally limited to one side of the board. With a reduction in chip carrier size, the physical attributes such as mechanical strength and thermal expansion mismatch become more compatible with the circuit board. The reduced mass of the chip carrier has better resistance to vibration and mechanical shock. Electrical performance is also enhanced due to lower impedance of the assembly, reduced noise, and improved frequency response.

There are several types of SMCs presently in use. These are resistors, capacitors, inductors, transistors, and integrated circuit (IC) packages. There are a variety of carriers used for integrated circuit chips. They are the SOIC (small-outline integrated circuit), the PLCC (plastic-leaded chip carrier), and the LCCC (leadless ceramic chip carrier). Figure 19-17 shows schematic pictures of the PLCC and SOIC compared to a DIP (dual in-line package), which is used for PIH assemblies.

The assembly process for surface mount components consists of the following steps:

1. *Surface preparation of the circuit board.* As previously described the cleanliness of the circuit boards is of paramount importance. Procedures should be performed to insure the board is ready for soldering.

2. *Paste application.* A solder paste is applied in the proper volume on the board land pattern, which replicates the footprint of the SMC leads. The most frequently used technique is a screening process with either screens or stencils to form the proper pattern. The solder paste is a mixture of solder alloy particles suspended in a flux binder. The paste performs three functions: it supplies the

necessary solder for joining, it supplies the flux for removing metal oxides, and it frequently acts as an adhesive to hold the SMC in place during the reflow process (for single-side placement).

3. *Component placement.* After the paste is applied to the board, the SMCs are placed by automatic pick-and-place equipment. Alignment of the leads to lands is very critical to form a proper solder joint. Normally no more than one-half the lead width is allowed for misregistration.

4. *Dry paste.* When all components are placed on the board, the assembly is baked in order to evaporate some of the moisture from the solder paste to minimize flux and solvent bubbling during reflow soldering. This is required to minimize solder voids and poor wetting of the joints.

5. *Solder reflow.* After the paste has been dried, the prepared board is assembled by reflowing the solder paste in a soldering system. The most frequently used is the vapor phase system. Since the temperature in a vapor phase system is controlled by the boiling point of the primary fluid, and heating of the board is very rapid, time in the vapor becomes the only controlling parameter for the system. Time in the vapor zone, which is considerably longer than time on a solder wave, has been found to affect both the appearance and reliability of the solder joints. Remember that excessive time and temperature cause the formation of detrimental thick intermetallic layers.

6. *Flux removal.* The final step in surface mount assembly is the final cleaning of the board assembly to remove flux residues. Because of the lower temperature of vapor phase soldering, rosin fluxes do not polymerize as readily as in a wave soldering system and as a consequence flux removal is easier. However, the longer times necessary with vapor phase processing can make it harder to clean the flux residue. A large variety of solvents are available for cleaning rosin flux residue. Trichloroethane, or a blend of it with alcohol, has been found to be very effective for most applications.

PIN BRAZING
FOR MULTILAYER CERAMICS

Multilayered ceramics (MLCs) have wide applications for use as chip carriers in electronic packaging. MLC substrates are monolithic structures of personalized and interconnected composites of ceramic and metal layers. Since these substrates have thicknesses exceeding 0.100 in. in many cases, it becomes necessary to braze the pins to lands on the back side of the substrate (a mechanical pinning process cannot be used). The lands on the back side of the MLC substrate consist of a screened and fired thick film of molybdenum, 5 μm of electroless Ni-B, with an outer layer of immersion gold approximately 1000 Å in thickness. The lands on the substrate are screened with a Au-Sn paste of eutectic composition. The flux in the paste is rosin based. The pins are held on the screened pads of the substrate by a fixture. The fixture and substrate are then subjected to a furnace reflow cycle reaching a peak temperature of about 390°C. The atmosphere in the furnace is hydrogen gas. Essentially this brazing operation is very similar to

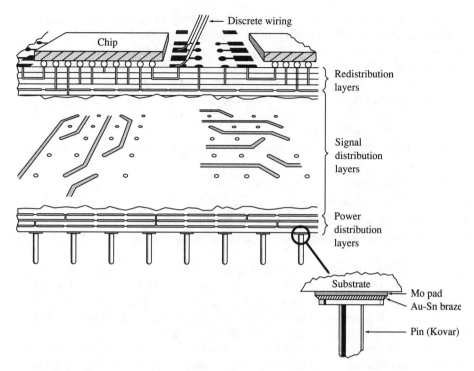

FIGURE 19-25
Cross-sectional view of an MLC substrate. (D. P. Seraphim and I. Feinberg, "Electronic Packaging Evolution in IBM," *IBM J. Res. Dev.*, Vol. 25, No. 5, Sept. 1981, pp. 617–629. Copyright 1981, International Business Machines Corporation; reprinted with permission.)

soldering; however, higher reflow temperatures are used since the eutectic Au-Sn alloy melts at 280°C. Figure 19-25 shows a cross section of an MLC substrate. The pins on this substrate are kovar (Fe-Ni-Co alloy, which has a thermal coefficient of expansion that very closely matches that of the ceramic substrate) plated with nickel and an outer layer of gold, which is brazed to molybdenum pads on the ceramic substrate using a Au-Sn eutectic alloy.

SOLID STATE DIFFUSIONAL BONDING

Materials and Design Considerations for Wire Bonding

Wire bonding is a technique to provide electrical interconnections from the terminals on a chip circuit to those on a chip carrier circuit by using thin wire, typically 25 to 75 μm (0.001 to 0.003 in.) in diameter. It is the most commonly used interconnection scheme in various microelectronic packages. After separation and sorting of the wafer into individual chips, the back side of the chip is attached to an appropriate medium by using polymer adhesive, solder, or eutectic alloy.

This process is commonly referred to as die bonding. Selection of optimum die-bonding material is determined by the hermeticity requirements of the package, thermal dissipation, heat current through the die bond and the thermal expansion coefficient of chip carrier media. Au-Si and Au-Sn eutectic alloys [28,29] and silver-filled epoxy adhesive are widely used for die bonding [30].

General requirements for wire material are:

- High electrical conductivity
- Bondability to other metals by thermal compression, ultrasonic or thermosonic techniques
- Corrosion resistance
- Drawability into thin wire
- Ductility for plastic deformation during bonding operation
- Strength for wire handling during discrete bonding operation

The most commonly used wire materials are aluminum, aluminum alloys (Al-Mg-Si, Al-Si, Al-Mg, Al-Cu) [31,32], gold, and gold-beryllium [33,34]. Copper alloys and palladium are being researched as possible replacements for expensive gold [35,36]. Among various wire materials, pure gold, beryllium-doped gold alloy, pure aluminum and Al/1% Si are most widely used due to their high electrical conductivity, mechanical properties, and bondability. Common terminal metallizations on which the wire is bonded are aluminum, aluminum alloys (containing small amount of Si, Cu, or Mg), gold, palladium-silver, gold-palladium, nickel, and platinum. The terminal metallization is prepared by a thin-film process (vacuum evaporation, sputtering), plating, or a thick-film process (screen printing of metal or alloy paste and then firing). Gold or beryllium-doped gold wire can be bonded by thermal compression, ultrasonic, or thermosonic techniques. Aluminum and aluminum alloy wire can be bonded by ultrasonic techniques. This section will discuss the most commonly used wire and terminal metallization systems, their processes and important technical considerations.

An SEM image of a wire bond is shown in Figure 19-26(a); a high magnification of the bond at the chip and at the chip carrier is shown in Figures 19-26(b) and (c), respectively. The wire bond experiences a combination of stresses from various sources during testing, storage, and operation of the device. The response of the wire bond material and design under these stress conditions can determine the durability of the package [37]. Even though wire bonds are much more flexible than a rigid type of bond such as a C4 joint, mechanical stresses can be generated within the wire bond during thermal excursions due to mismatch of the coefficients of thermal expansion of the device and the chip carrier. Figure 19-27 shows a wire failure after 200 thermal cycles between 0 and 125°C on a 0.001-in.-diameter gold wire. In addition, the molding of plastic around the chip/top portion of the leadframe to form the final encapsulated package will impart residual mechanical stresses in the wire bond. The metal in the bond region is highly deformed and often contains a high density of defects, as well

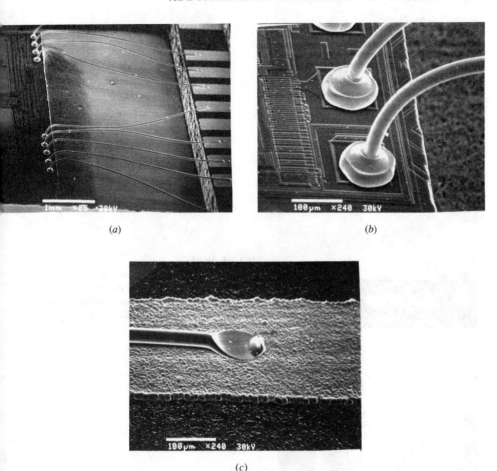

FIGURE 19-26
(*a*) An SEM image of a ball-wedge type of wire bond. (*b*) A higher magnification of the ball bond at the chip pad. (*c*) A higher magnification of the wedge bond at the chip carrier pad.

as impurities that most often are present on the wire and pad surfaces. The geometrical configuration of the heel region can be viewed as a microscopic stress concentrator. Movement of the wire loop due to thermal expansion and contraction can induce great stresses in the metal of the wire bond near the heel region. Mechanically induced bond failures generally occur at the heel. If the loop is too long, prolonged use of the device will anneal and thus soften the wire material. This results in creep, and the wire would begin to sag and eventually could short with adjacent circuitry. On the contrary, if the loop is too short, a larger stress would be applied on the bond interface or heel region. The effects of using rigid or soft wire material correspond to the effects of long or short loop designs respectively. For these reasons optimum design of the geometrical features of the

FIGURE 19-27
Wire bond failure on a gold wire after 200
thermal cycles, 0–125°C. The diameter of the
gold wire is 0.001 in. (Reprinted with permission:
J. O. Honeyecut, IBM Technical Article
TR-51.0157.)

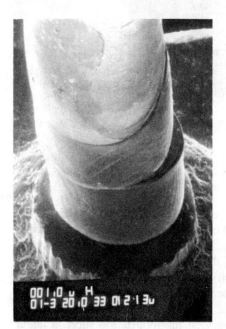

FIGURE 19-28
Cleavage type failure in a gold wire bond after
only ten thermal cycles. Test conditions are −65 to
125°C, and the wire diameter is 0.002 in.
(Reprinted with permission: J. O. Honeycut, IBM
Technical Article TR-51.0157.)

wire bond and proper selection of the wire material are as important as the bonding interfaces in controlling the reliability of wire bond interconnections. Figure 19-28 shows a cleavage type of failure in a wire having a 0.002-in. diameter, due to excessively large grains produced by slow cooling of the wire, directly adjacent to the newly formed ball produced after melting of a small portion of the wire (referred to as flame-off). The test conditions were −65 to 150°C, for ten cycles. Bond pull tests and shear tests have been adopted as screening tests to ensure good interface bonding. Overall wire bonding reliability is assessed by using accelerated thermal cycling testing, which best simulates actual device operation.

Although TAB packaging design features are different from that of wire bonding schemes, the interconnection method is almost identical. Metallurgical aspects of inner lead beam and bumps at the chip have been discussed in the reference by Lieu and Fraenkel [38].

Thermal Compression Bonding

The traditional form of thermal compression (TC) bonding is known as chisel or wedge bonding. This technique requires two separate alignments. A more advanced technique for higher production rate is ball/wedge bonding (Figure 19-29). Since ball bonding requires only one alignment operation, it allows faster interconnection operation of the wire to lead in any direction after the first bond. As illustrated in Figure 19-29, chip-terminal-to-wire interconnection is made by ball bonding, while interconnection of wire to chip carrier circuitry is made by

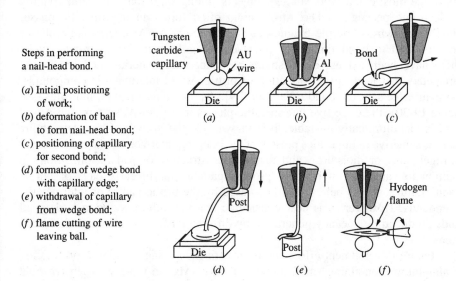

Steps in performing a nail-head bond.

(a) Initial positioning of work;
(b) deformation of ball to form nail-head bond;
(c) positioning of capillary for second bond;
(d) formation of wedge bond with capillary edge;
(e) withdrawal of capillary from wedge bond;
(f) flame cutting of wire leaving ball.

FIGURE 19-29
An illustration of the ball wire-bonding process flow. (Adapted from A. B. Glaser and G. E. Subak-Sharpe, *Integrated Circuit Engineering, Design, Fabrication and Applications*, published by Addison-Wesley.)

wedge bonding. After each segment of wire interconnection, the wire is cut off by breaking it at the wedge bond. A new ball at the wire tip is produced by passing the tip of the wire over a hydrogen flame. The wire is fed through a collar which allows the ball to form by capillarity. A variety of materials can be used for the collar, such as glass, WC, Si_3N_4, BN, Al_2O_3, quartz, and ruby. The temperature of the collar during TC bonding is in the range of 200–250°C. A modern automatic TC bonding machine can interconnect approximately 60 wires in 15 seconds.

Thermal compression bonding consists of heating and applying mechanical pressure to two joining bodies so that plastic flow can be facilitated. The plastic flow produces new clean metal surfaces at the interface as deformation by slip mechanism proceeds. The enhanced mobility of dislocations at elevated temperature allows extensive plastic flow even with mild mechanical pressure. The creation of fresh surfaces by plastic flow enhances interdiffusion of two joining surfaces. Therefore, the ductility of wire material is one of the essential requirements in TC bonding. Also, the ductility contributes to the dispersion of pressure over a wide area and thus reduces stress concentration and potential risk of damaging brittle chip material. However, as pointed out previously, too ductile (soft) a wire will suffer from sagging. Pure gold wire is too ductile to be easily used for wire bonding; gold wire alloyed with a small amount of beryllium (1–2 percent) is more widely used for high-performance IC packages. Beryllium in gold inhibits grain growth at low temperature and thus provides stability of wire properties, and also increases yield strength by precipitation hardening.

Thermal compression bonding of a gold wire to aluminum metallization is one of the most extensively studied subjects among solid state diffusion bonding in electronic packaging. The Au-Al bond often forms an intermetallic phase, which is known as "purple plague" due to its color. Au-Al intermetallic phases have resulted in bond failures at interfaces of gold wires on aluminum pads. There have been extensive investigations concerning materials and processing parameters that lead to the formation of brittle Au-Al intermetallic compounds, and their effects on the bond reliability [39]. As the phase diagram indicates (see Figure 19-30), there are five intermetallic phases in the Au-Al system: the Au-Al bond is metallurgically unstable. It is known that the growth of an intermetallic phase is sensitive to impurities present at the bonding interface, to the environment (nitrogen, air, or moisture), and to the microstructure of pad metallization, in addition to variations in bonding process parameters. In early days of purple plague failure, it was believed to be caused by the brittle nature of intermetallic compounds. However, it is now considered that the failure is due to Kirkendall voids produced by vacancy generation by differential flux of gold and aluminum atoms.

Future development effort is concentrating on using aluminum or an alloy of aluminum (containing small amount of Si or Mg) to replace expensive gold wire and to achieve high reliability and allow high-temperature applications [40]. The process in this case is aluminum ball bonding by a thermosonic technique, a combination of heat (420 to 440°C) with ultrasonic energy. The major problem

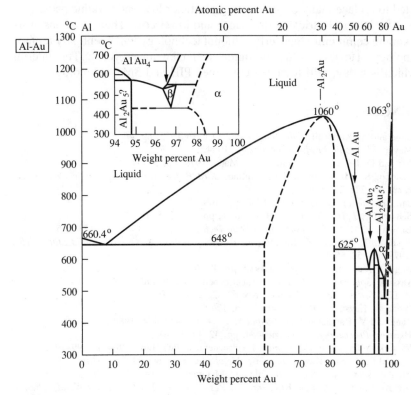

FIGURE 19-30
The Au-Al phase diagram. (Reprinted with permission: E. A. Brandes, ed., *Smithells Reference Book*, 6th ed., Butterworth, 1983.)

encountered in early attempts was the difficulty of forming good quality balls comparable to those produced with gold. Gold gave good uniformity of ball size, ball shape, and surface conditions. Recent work uses a capacitive discharge method along with gas shielding to produce a good ball on aluminum wire.

CONCLUSION

The common methods of first-level assembly are wire bonding and flip-chip soldering. As a result of higher I/O densities in future chip designs, flip-chip soldering will be the appropriate method for first-level assembly. In second-level assembly, surface mount attach is steadily becoming more suitable than PIH technology to accommodate the high component densities demanded for today's technology. In the future, even higher densities will be required, necessitating the development of totally new packaging schemes. One approach to achieve very high chip densities on cards would be direct chip attach to the card. This is a very desirable packaging approach; however, to make it a viable technology will require much development effort in low-thermal-expansion cards. TAB packages are expected to replace many of the existing plastic or low-end ceramic packages, since they offer better electrical performance and lower cost. There is a consensus in the packaging community that surface mount technology will eventually replace PIH technology. However, this migration will take many years. The interim period, which is today, will focus on combined PIH and SMT technologies.

REFERENCES

1. G. Dehaine and K. Kurzweil, *Solid State Tech.*, October 1975, pp. 46–52.
2. D. E. Meyer, *Electronic Products*, October 1, 1984, pp. 71–75.
3. A. Keizer and D. Brown, *Solid State Tech.*, March 1978, pp. 59–64.
4. T. Kawanobe, K. Miyamoto, and M. Hirano, *33rd Proceedings of Electronic Components Conference*, pp. 221–226 (1983).
5. J. Lyman, *Electronics*, December 18, 1980, pp. 100–105.
6. T. A. Scharr, *Int. J. Hybrid Elec.*, vol. 6, no. 1, pp. 561–565, 1983.
7. D. B. Brown and M. G. Freedman, *Solid State Tech.*, September 1985, pp. 173–175.
8. M. Hayakawa, T. Maeda, M. Kumura, R. H. Holly and T. A. Grielow, *Solid State Tech.*, March 1979, pp. 52–55.
9. "Soldering," *Circuits Manuf.*, January 1980, pp. 17–30.
10. R. J. Wassink, *Soldering in Electronics*, Electrochemical Pub. Ltd, 1984.
11. R. Wild, IBM Owego Technical Report No. 67-825-2157, 1967.
12. R. Wild and R. Hagstrom, *Proc. Internepcon*, 1969.
13. S. K. Kang and V. Ramachandran, *Scripta Met.* 14, pp. 421–424 (1980).
14. A. C. Harman, *Proceedings of Internepcon*, pp. 42–49 (1978).
15. R. J. Wassink, *Soldering in Electronics*, Electrochemical Pub. Ltd., 1984, pp. 99–101.
16. P. J. Kay and C. A. Mac Kay, *Trans. Inst. Metal Finish* 54, pp. 68–74 (1976).
17. H. H. Manko, *Solders and Soldering*, 2nd ed., McGraw-Hill, New York (1979).
18. M. L. Ackroyd and C. A. Mac Kay, *Int. Tin Res. Inst.*, Publ. 529, *Circuit World* 3, no. 2, pp. 6–12 (1977).
19. M. L. Ackroyd, C. A. Mac Kay, and C. J. Thwaites, *Metals Tech.*, 2, no. 2, pp. 73–85 (1975).
20. G. Becker and B. M. Allen, *Proc. Internepcon*, pp. II.52–II.66 (1970).
21. K. G. Schmitt-Thomas and H. M. Zahel, *Metall.* 37, no. 1, pp. 43–49 (1983).
22. C. A. Mac Kay, *Electronics* 29, no. 3, pp. 44–48; no. 4, 41–44; no. 5, 51–53 (1983).
23. J. Gow and A. R. Leach, *Circuits Manuf.*, May 1983, pp. 72–74.
24. H. H. Manko, "Solders and Soldering," McGraw-Hill, 1964.
25. C. A. McKay, "An Investigation of the Soldering Properties of Sn-Bi Alloys," Alpha Metals Project M-80-15 for IBM, 1982.
26. J. R. Getten and R. C. Senger, *IBM J. Res. Devel.* 26, no. 3 (May 1982).
27. R. C. Senger, C. S. Jaw, and H. J. Healey, IBM Endicott Technical Report No. TR01.3034, 1985.
28. R. K. Shukla and N. P. Mencinger, *Solid State Technology*, July 1985, pp. 67–74.
29. T. D. Hund and S. N. Burchett, *Int. J. Hybrids and Microelectronics*, Vol. 6, no. 1, pp. 243–250, Oct., 1983.
30. F. K. Moghadam, *Solid State Technology*, January 1984, pp. 149–157.
31. T. J. Matcovich, *31st Proceedings of Electronic Components Conference*, pp. 24–30 (1981).
32. J. Onuki, M. Suwa, and T. Iizuka, *34th Proceedings of Electronic Components Conference*, pp. 7–12 (1984).
33. S. P. Hannula, J. Wanagel, and C. Y. Li, *33rd Proceedings of Electronics Components Conference*, pp. 181–188 (1983).
34. M. Poonawala, *33rd Proceedings of Electronic Components and Conference*, pp. 189–192 (1983).
35. A. Bischoff and F. Aldinger, *34th Proceedings of Electronics Components Conference*, pp. 411–417 (1984).

36. J. Kurtz, D. Cousens, and M. Dufour, *34th Proceedings of Electronics Components Conference*, pp. 1–6 (1984).
37. S. P. Hannula, J. Wanagel, and C. Y. Li, *34th Proceedings of Electronic Components Conference*, pp. 31–35 (1984).
38. T. S. Liu and H. S. Fraenkel, *Int. J. Hybrid Microele.*, pp. 69–76.
39. E. Philofsky, *Solid-State Electronics* 13, pp. 1391–1399 (1970).
40. V. A. Pitt, C. R. Needes, and R. W. Johnson, *31st Proceedings of Electronic Components Conference*, pp. 18–23 (1981).

PROBLEMS

19.1. Why is it necessary to have the chip attached to a substrate instead of being directly attached to the epoxy printed circuit card? In other words, discuss the role of the substrate in electronic packaging.

19.2. What are two desirable features of chip attach using C4 joints?

19.3. The following figure shows a hypothetical module having five C4 joints. Determine the maximum amount of displacement produced at the outer diagonal joints and also the center joint during thermal cycling. T varies from 0 to 100°C. The module is alumina and the chip is silicon. Assume the solder is fully compliant; $\alpha_{Al_2O_3} = 6$ ppm/°C; $\alpha_{Si} = 2$ ppm/°C.

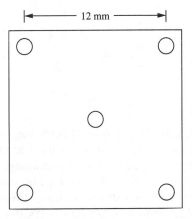

19.4. Even though pure Pb has the optimum creep/fatigue properties for both first- and second-level packaging, small additions of Sn are made. Why?

CHAPTER
20

MECHANICAL PROCESSES

WARREN R. WRENNER
JOHN D. LARNERD

INTRODUCTION

The drive to mount and interconnect more circuitry on fewer boards has placed increased demands on the performance of mechanical equipment used in manufacturing by the computer industry. For example, in first-level packaging, very accurate and highly automated equipment is required for the pinning and standoff processes used to put input/output pins on ceramic substrates. In many cases, the equipment used to produce the holes in large printed circuit boards (PCBs) must operate with a positional tolerance that is a fraction of the diameter of a human hair. The equipment that is used to measure the location of these holes must be even more accurate than the drilling machines.

Note that Chapter 10 discusses packaging components and ceramic materials processes, and Chapter 17 covers plated through-hole processes. This chapter focuses on examples of critical mechanical processes. Complementary chapters are "Robotics and Automation" and "Factory of the Future," Chapters 28 and 31, respectively. The increasing complexity of interconnect designs has driven the need for improvements in accuracy, throughput, and reliability of the mechanical positioning, pinning, drilling, photoprinting (Chapter 12), and inspection technologies. The major technologies developed include special pin-swaging and standoff tools; the use of air bearings for support and guidance on rigid, lapped-granite surfaces; closed-loop positioning systems with feedback transducers, air bearing spindles, and programmable drilling heads; and vision systems with computer-controlled image acquisition and analysis.

PINNING AND STANDOFF OPERATIONS

The pin grid array–type first level ceramic package consists of an integrated circuit (IC) chip, a circuitized ceramic substrate, and a protective cap (see chapter 10). The ceramic substrate provides electrical connections between the IC chip and the printed circuit (PC) card or board. The ceramic substrate must go through pinning and standoff process steps to provide the required electrical connections. Pinning is a process that connects copper pins, called input/output (I/O) pins, to the circuitized ceramic substrate.

A special pin-swaging tool is used to attach all of the I/O pins to the ceramic substrate simultaneously. Figure 20-1 illustrates the sequence of operations. First, the pins are inserted into a holding block, shown in Figure 20-1(a), which very accurately locates the pins in the desired pin matrix. The ceramic substrate is then placed over the pins, and a ram is moved downward to a fixed stop, providing the force necessary to form the pin head. A pin knockout block provides the necessary support to the bottom of the pins. The ram then moves upward to its home position. The shape of the pin after the first stroke of the ram is shown in Figure 20-1(b). Next, the ceramic substrate assembly is lifted a predetermined distance away from the surface of the pin block by raising the knockout block as shown in Figure 20-1(c). The ram descends again to form the pin bulge on the bottom side of the substrate. The shape of the head and pin bulge after the second stroke of the ram is shown in Figure 20-1(d). The knockout block then moves upward to eject the finished assembly from the pin block.

The standoff operation pinches the four corner leads to force the material to expand outward and form a projection on both sides of the pin. A stiffening rib is also formed to help strengthen the lead in the direction perpendicular to the projections.

When the completed module is assembled to the PC card or board, the leads go through plated through holes (PTHs). The module will drop down until the projections on the corner leads interfere with the top of the PTH, thereby holding it away from the surface of the PC card or board. This space is required to provide room for the cleaning solvent to remove the solder flux from under the module after it is soldered to the PC card or board. The space also allows air to freely circulate around the module to provide cooling during normal operation.

The standoff projections must be wide enough to interfere with the PTH and thick enough to withstand the maximum downward force that a component insertion tool would exert on the module during assembly to a PC card or board. This force is usually less than 10 pounds.

The standoff tool uses eight sliding pinch jaws, which are controlled by a ring cam, to pinch the four corner leads, causing them to expand outward and form the projections and the stiffening web at the same time.

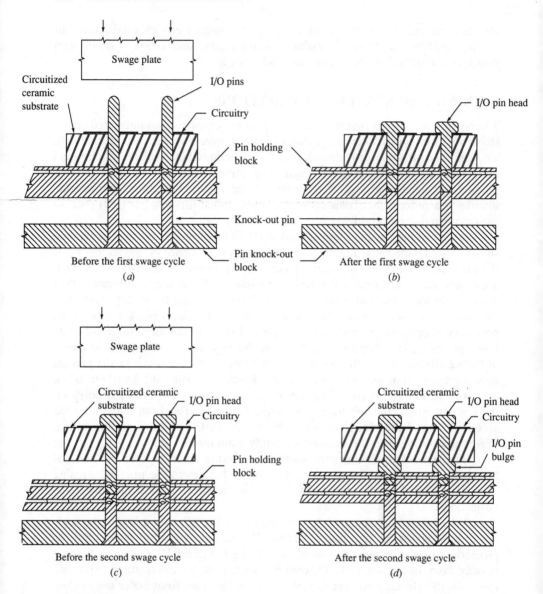

FIGURE 20-1
Sequence of swaging operations.

X-Y POSITIONING SYSTEMS

Precise planar *x-y* motion is required for a variety of operations in the fabrication of printed circuit boards. Often with programmed control and accurate fixturing, *x* and *y* coordinate systems are used in artwork generation, drilling, inspection, test, and rework applications.

Motion Control and Feedback

There are several methods of converting rotary motion into controlled linear motion. The worm and pinion, rack and pinion, and rail and capstan are three examples. Another example is the ball screw actuator, which functions in the same manner as a simple screw thread and nut. If the nut is prevented from rotating as the screw is turned, the nut will travel linearly along the screw. The ball screw is different from the screw thread and nut in that the operating surfaces rest on several circuits of ball bearings. Ball screw assemblies are used extensively in numerically controlled machine tools, precision measuring instruments, and in other equipment requiring accurate positioning and repeatability.

The most common method of motion control used in *x-y* positioning systems is that of attaching the ball nut to a table or carriage, and driving (rotating) one end of the ball screw with a servomotor. The ball screw assembly, shown in Figure 20-2, consists of a screw shaft, a ball nut, ball bearings, and ball tubes for recirculation of the balls. The ball nut has a gothic arch ball groove [1] to minimize friction caused by the tendency of the balls to slip rather than roll.

Control systems are classified either as open-loop or as closed-loop systems. A closed-loop system is also called a feedback control system. In such a system, a controlled variable is fed back and compared with a reference input. A block diagram of a closed-loop positioning system that uses a transducer as a feedback device is shown in Figure 20-3.

Transducers used as feedback devices in machine tools [2] include ball screw resolvers, inductive scales, metal scales, glass scales, and laser interferometers.

FIGURE 20-2
Ball screw assembly (courtesy of NSK Corporation).

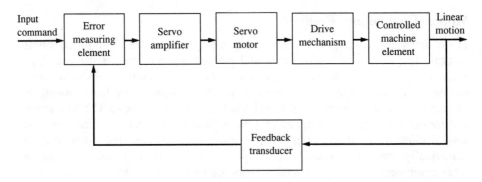

FIGURE 20-3
Block diagram of one axis of a positioning system.

The most accurate is the laser interferometer, with which motion control with 0.1-μm accuracy is routine. An incremental linear transducer consisting of a glass scale and a scanning head is shown in Figure 20-4. The glass scale is made up of identical alternating transparent and opaque increments. The scanning head consists of a miniature lamp, a scanning reticle, and solar cells. Light from the lamp is transmitted through the scale grating and through the transparent spaces of the reticle on the solar cells. Motion of the scanning head relative to the scale causes variations in light intensity on the solar cells, which transform these variations of light into electrical signals. These signals are amplified, subdivided, and compared to the controller command position, and the error signal is used to move the axis until the command and feedback signals are equal. A datum (axis zero) is provided by using another set of solar cells and a reference mark on the scale.

There are a variety of techniques used for support and guidance in x-y positioning systems. One technique, illustrated in Figure 20-5(a), uses ball bushings on hardened and ground shafts. Another method (b) uses crossed roller bearings; the closely spaced rollers are held between two V surfaces. A third method, shown in (c), uses air bearings on lapped granite surfaces. Two guidance mechanisms are used to make an x and y coordinate system, either configured independently as a split-axis system or one supporting the other in a piggyback arrangement. The disadvantage of the piggyback design is off-axis twisting forces that cause errors in positioning.

The split-axis design using air bearings on lapped granite surfaces is the best choice for state-of-the-art positioning systems. In one configuration of this design, the table, which moves in the y direction, is supported and guided with air bearings on a rigid granite base. The base also supports, on steel box castings, a granite beam on which a carriage moves orthogonal to the direction of table travel.

The accuracy of a positioning system depends on more than the feedback transducer and bearings. The stiffness of the drive and the straightness of the

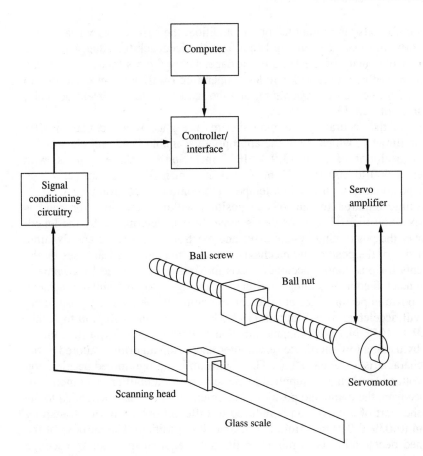

FIGURE 20-4
Precision positioning system using an incremental linear transducer for feedback.

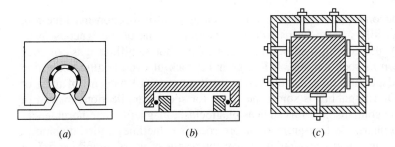

FIGURE 20-5
Schematic illustration of techniques used for support and guidance in *x-y* positioning systems: (a) ball bushing on continuous support shaft, (b) crossed roller vee groove, and (c) air bearings on lapped granite.

guiding ways are also important factors. In addition, the best accuracy is achieved when the driving force, the guiding beam, and the feedback transducer are close to the center of gravity of the table or carriage. Other factors that are important include the use of materials with similar temperature coefficients of expansion to minimize the influence of temperature, and the reduction, to the extent possible, of all sources of heat [3].

Many of the accurate positioning systems use glass scales containing 100 lines per millimeter, which, with the appropriate scanning head and electronics, provides a resolution of 1 μm (3.9×10^{-5} in.). Some of these systems, with axis travel of 600 to 800 mm (23.6 to 31.5 in.), maintain $\pm$ 2 μm ($\pm$ 7.9 $\times$ 10^{-5} in.) positioning accuracy in a temperature-controlled environment.

The time required for an axis to position within a certain error band is called in-position time. The x- or y-axis move time is determined by the length of the move, the positioning system drive and electronics, the static and dynamic characteristics of the positioning mechanism, and in the most critical cases by the requirements for positioning accuracy. Axis move time is affected by command and tachometer feedback scaling, and can be changed within limits to achieve similar in-position performance at different velocities. Some large x-y positioning systems will accelerate to 5–8 m/min (197–315 in./min) and will stop to within 2 μm (7.9×10^{-5} in.) of command position in 0.15 second. Typically, this is achieved by using a low-inertia dc servomotor and a stop algorithm tailored to the physical characteristics of each axis. The algorithm is a programmed deceleration or list of voltage levels that is supplied to the servomotor amplifier as a function of the distance from the command position. Maximum drive-voltage is applied to the motor at the start of a move and is reduced to different optimum levels, starting at 1–2 mm (0.039–0.079 in.) from the desired stop position. The purpose of the programmed deceleration is to minimize the time required to make an accurate stop.

Granite and Granite Components

Granite has become popular with drilling, routing, and inspection machine-tool builders, especially during the past decade. Granite is one of the igneous rocks [4] formed by the cooling and solidification of the hot semifluid mass of rock material known as rock magma. It is a granular rock of even texture consisting mainly of feldspar and quartz and often a small amount (10 percent) of mica or hornblende. Diorite-gabbro is sometimes used for special applications instead of granite. Diorite contains minerals such as plagioclase, magnetite, and hornblende, but almost no quartz. A fine-grained gabbro rock is sometimes called diabase.

Naturally aged and stress-relieved over thousands of years, granite is free of internal stresses. Granite has a low coefficient of expansion, of the same order of magnitude as those of glass and cast iron. Bulk granite takes a long time to reach thermal equilibrium due to its poor thermal conductivity; for the same reason, the influence of small temperature excursions over short periods of time is negligible.

The density (mass per unit volume) of granite is the same as for aluminum and slightly more than one third the density of cast iron.

Air Bearings

The notion that air might be used as a lubricant was first mentioned [5] during the middle of the nineteenth century. In 1897 Albert Kingsbury built an air-lubricated bearing and published some of the earliest experimental data. Prior to that, only oil and water were considered to be acceptable lubricants. Today, because of low frictional losses, gas bearings are used to support loads and maintain position in precision instruments, dental drills, high-speed electrical motors, and machine tools. Gas bearings are used in both high-temperature and low-temperature (cryogenic) applications because viscosity remains within acceptable values over wide temperature ranges. A film of air is used to support magnetic heads used in disk memory and for the rotating shafts in high-speed spindles. Even heavy objects, such as the machine tools used in industry, can be moved with relative ease by using large air bearing pads with a portable compressor.

There are two classes of gas bearings: in the self-acting gas bearing, the supporting film is produced by relative tangential movement of the two surfaces, while in the externally pressurized bearing, the film is produced by an external supply of compressed gas. Self-acting bearings are not used in machine tools because they cannot provide as much support as externally pressurized bearings and are only suitable for applications where surfaces are always moving when under load [6].

Figure 20-6 shows two types of air bearings commonly used in drilling equipment. One type, used for both support and guiding in linear motion, is the sliding bearing (a) in which there is relative tangential motion between two flat surfaces. Filtered compressed air is supplied to the bearing through flexible tubing connected to the end of one air channel. The other channel openings are sealed with threaded plugs. Depending on the design parameters, this type of bearing will typically provide a lift of 5 to 25 μm (0.0002 to 0.001 in.) off of the reference surface. A second bearing configuration is the journal (b), in which the two surfaces are concentric cylinders. The radius of the journal is r and the clearance between the journal and the bearing is c. The clearance ratio (c/r) has been made large in the illustration for clarity. For this type of bearing, air is supplied to peripheral rows of orifices or jets into the clearance space between the journal and the bearing.

Positioning Accuracy and Repeatability

Each axis of a numerically controlled (NC) positioning system is evaluated mathematically using positioning error data measured with a laser interferometer (LI). This technique, used in dimensional metrology, measures physical displacement

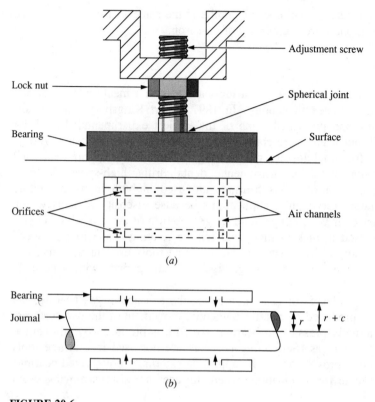

FIGURE 20-6
Two types of air bearings commonly used in drilling equipment: (a) sliding bearing and (b) journal bearing.

in units of the wavelength of light [7] from a helium-neon (He-Ne) laser source. The basic resolution of the LI is one quarter wavelength of the laser light (about 0.16 μm or 6.3 microinches), but can be extended several orders of magnitude using suitable electronics. Figure 20-7 shows how a portable laser interferometer is used to measure one axis of an x-y positioning system. The laser transducer system measures the change in relative position between the interferometer and the cube-corner.

The cube-corner consists of three mutually perpendicular mirrors, a shape similar to the corner of a room. This device reflects a beam of light parallel to the incident beam. In this example, the interferometer is secured to the moving y-axis table, and the cube-corner is mounted to the stationary overhead carriage. A block diagram of the Hewlett-Packard laser measurement system is shown in Figure 20-8. The laser produces light of two different frequencies, f_1 and f_2, with opposite circular polarizations. One of the two frequencies, f_1, is optically separated and directed to the cube-corner mounted on the carriage. The other frequency, f_2, is optically separated and directed to the reference cube-corner

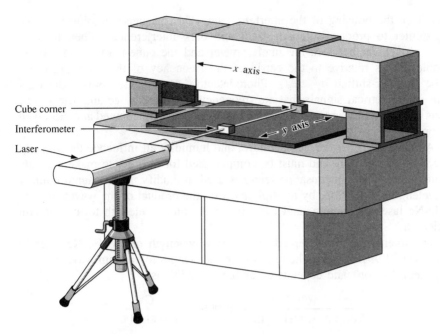

FIGURE 20-7
Measurement of accuracy and repeatability using a laser interferometer.

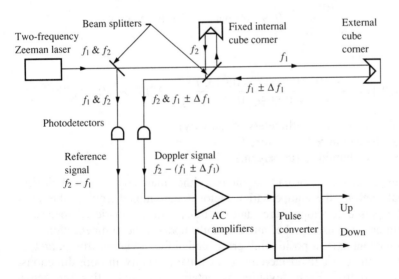

FIGURE 20-8
Hewlett-Packard laser measurement system.

mounted on the housing of the interferometer. This frequency rejoins f_1 at the beam-splitter to produce alternate light and dark interference. When there is relative movement between the inteferometer and the cube-corner, e.g., when the table moves relative to the carriage, the frequency of the returning beam will be Doppler-shifted by $\pm\Delta f_1$. Photodetectors are used to convert the light signals into electrical signals that are frequency separated by ac amplifiers. The outputs of these amplifiers are converted into pulses that are counted and displayed electronically.

Atmospheric pressure, temperature, and humidity all influence the velocity of the He-Ne laser light and must be compensated for in order to make accurate measurements. The compensation factor is used to modify the measured number of wavelengths of motion, by multiplying it by the product of the wavelength of the He-Ne laser in vacuum, and the ratio of the air wavelength to the vacuum wavelength.

The internationally accepted value for wavelength of the He-Ne laser in vacuum is 632.991399 nm. The ratio of air wavelength to vacuum wavelength is always less than one and depends on ambient conditions:

$$\frac{\text{Air wavelength}}{\text{Vacuum wavelength}} = 0.999 + \frac{C}{1,000,000}$$

The compensation factor C relates the index of refraction of air to pressure, temperature, and humidity [8].

$$C = \frac{10^{12}}{N + 10^6} - 999,000 \text{ parts per million (ppm)}$$

In the metric system,

$$N = 0.3836391P\left[\frac{1 + 10^{-6}P(0.817 - 0.0133T)}{1 + 0.0036610T}\right] - 3.033 \times 10^{-3}Re^{0.057627}T$$

where P = air pressure (in millimeters of mercury)
  T = air temperature (in degrees Celsius)
  R = relative humidity (in percent)

The compensation factor can be entered either manually or automatically. In the manual method, the compensation factor is determined from a table after measuring the pressure, temperature, and humidity. An automatic compensator, equipped with an air sensor probe, automatically updates the compensation.

To evaluate an axis, a positioning error, called the deviation from target, is measured at a number of fixed increments of travel, usually in both directions, under program control. Each deviation is obtained by subtracting the actual position (measured with the LI) from the target position. The test is repeated to obtain the spread (dispersion), shown in Figure 20-9, which normally occurs in a number of trials. In practice, positioning accuracy is measured on both sides of a carriage or table in order to prevent a false indication due to a curved guiding

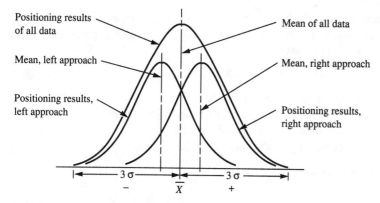

FIGURE 20-9
Positioning results shown as deviation from mean.

beam or rail. Table 20.1 shows the deviations from target measured for 20 points at 25-mm increments of axis travel. Each target position was measured three times, using a bidirectional approach, for a total of six measurements.

According to the National Machine Tool Builders Association [9], numerical control system accuracy at a point is defined as being the sum of the signed value

TABLE 20.1
Deviation from target data

Point no.	Target position, mm	Deviation from target, μm					
		↓	↑	↓	↑	↓	↑
1	25	0.7	0.3	0.9	0.5	0.9	0.6
2	50	0.9	0.6	1.3	1.0	1.4	1.0
3	75	0.7	0.4	1.0	0.7	1.0	0.7
4	100	0.7	0.6	1.0	0.9	1.0	0.8
5	125	0.7	0.5	1.0	0.8	1.1	0.8
6	150	0.3	0.1	0.6	0.3	0.5	0.4
7	175	−0.5	−0.5	−0.1	−0.3	−0.1	−0.3
8	200	−0.7	−0.8	−0.5	−0.6	−0.3	−0.5
9	225	−0.6	−0.7	−0.3	−0.5	−0.2	−0.4
10	250	−0.3	−0.6	−0.1	−0.3	0.1	−0.2
11	275	−0.3	−0.4	0	−0.2	0.1	−0.1
12	300	−0.4	−0.5	−0.1	−0.2	0	−0.2
13	325	−0.6	−0.7	−0.4	−0.5	−0.1	−0.3
14	350	−0.3	−0.5	−0.1	−0.2	0.1	−0.1
15	375	0.2	0	0.3	0.2	0.6	0.4
16	400	0.7	0.4	0.8	0.7	1.0	0.8
17	425	−0.2	−0.4	−0.1	−0.2	0.1	−0.1
18	450	−1.5	−1.6	−1.2	−1.4	−1.0	−1.2
19	475	−1.4	−1.8	−1.7	−1.9	−1.5	−1.6
20	500	−1.3	−1.5	−1.7	−1.3	−0.9	−1.1

of the difference between the mean and the target at any point, plus or minus the value of the dispersion at that same point, using the sign that gives the largest absolute sum. This is expressed as

$$A = \Delta X \pm 3\sigma \qquad (20.1)$$

in which A = accuracy at any point
ΔX = difference between mean and target (a perfect master measure); a signed number
3σ = expected dispersion on each side of mean

The expression for the mean is

$$\overline{X} = \frac{\sum X}{N} \qquad (20.2)$$

The dispersion is calculated using the following:

$$s = \left[\frac{\sum (X - \overline{X})^2}{N} \right]^{1/2}$$

$$3\sigma = 3s \left[\frac{N}{N-1} \right]^{1/2}$$

$$= 3 \left[\frac{\sum (X^2)}{N-1} - \frac{(\sum X)^2}{N(N-1)} \right]^{1/2} \qquad (20.3)$$

where X = data value
$\sum X$ = sum of all data values
N = sample size
s = sample standard deviation

Example. Determine the axis positioning accuracy using the deviation from target data shown in Table 20.1.

Solution. Starting with the first target position,

$$\sum X = 0.7 + 0.3 + 0.9 + 0.5 + 0.9 + 0.6 = 3.9 \ \mu m$$

$$\overline{X} = \frac{\sum X}{N} = \frac{3.9}{6} = 0.65 \ \mu m$$

$$3\sigma = 3 \left[\frac{\sum (X^2)}{N-1} - \frac{(\sum X)^2}{N(N-1)} \right]^{1/2}$$

$$= 3 \left[\frac{0.7^2 + 0.3^2 + 0.9^2 + 0.5^2 + 0.9^2 + 0.6^2}{6-1} - \frac{3.9^2}{6(6-1)} \right]^{1/2}$$

$$= 0.70 \ \mu\text{m}$$

$$X + 3\sigma = 0.65 + 0.70 = 1.35 \ \mu\text{m}$$

$$X - 3\sigma = 0.65 - 0.70 = -0.05 \ \mu\text{m}$$

The calculations, repeated for each target position, are listed under *Before zero shift* in Table 20.2. The next step is to determine the maximum and minimum values in the $\Delta x + 3\sigma$ and $\Delta x - 3\sigma$ columns.

$$\text{Maximum} = 1.89 \ \mu\text{m (point 2)}$$

$$\text{Minimum} = -2.21 \ \mu\text{m (point 19)}$$

The zero shift is

$$\frac{1.89 - 2.21}{2} = -0.16 \ \mu\text{m}$$

Finally, the *After zero shift* column is completed by adding $0.16 \ \mu\text{m}$ to each value in the *Before zero shift* column. The result is that the measured positioning accuracy for one side of the axis is $\pm 2.05 \ \mu\text{m}$. The performance of this axis is plotted in Figure 20-10.

TABLE 20.2
Accuracy and repeatability data

				Results in micrometers				
			Before zero shift			After zero shift		
Point no.	Σx	3σ	Δx	$\Delta x - 3\sigma$	$\Delta x + 3\sigma$	Δx	$\Delta x - 3\sigma$	$\Delta x + 3\sigma$
1	3.9	0.70	0.65	−0.05	1.35	0.81	0.11	1.51
2	6.2	0.86	1.03	0.17	1.89	1.19	0.33	2.05
3	4.5	0.68	0.75	0.07	1.43	0.91	0.23	1.59
4	5.0	0.49	0.83	0.34	1.32	0.99	0.50	1.48
5	4.9	0.64	0.82	0.18	1.46	0.98	0.34	1.62
6	2.2	0.53	0.37	−0.16	0.90	0.53	0	1.06
7	−1.8	0.54	−0.30	−0.84	0.24	−0.14	−0.68	0.40
8	−3.4	0.53	−0.57	−1.10	−0.04	−0.41	−0.94	0.12
9	−2.7	0.56	−0.45	−1.01	0.11	−0.29	−0.85	0.27
10	−1.4	0.70	−0.23	−0.93	0.47	−0.07	−0.77	0.63
11	−0.9	0.56	−0.15	−0.71	0.41	0.01	−0.55	0.57
12	−1.4	0.56	−0.23	−0.79	0.33	−0.07	−0.63	0.49
13	−2.6	0.65	−0.43	−1.08	0.22	−0.27	−0.92	0.38
14	−1.1	0.61	−0.18	−0.79	0.43	−0.02	−0.63	0.59
15	1.7	0.61	0.28	−0.33	0.89	0.44	−0.17	1.05
16	4.4	0.59	0.73	0.14	1.32	0.89	0.30	1.48
17	−0.9	0.49	−0.15	−0.64	0.34	0.01	−0.48	0.50
18	−7.9	0.67	−1.32	−1.99	−0.65	−1.16	−1.83	−0.49
19	−9.9	0.56	−1.65	−2.21	−1.09	−1.49	−2.05	−0.93
20	−7.8	0.85	−1.30	−2.15	−0.45	−1.14	−1.99	−0.29

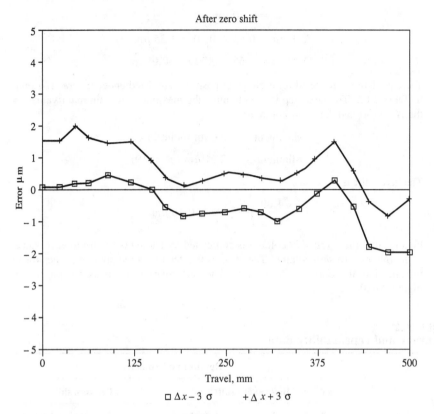

FIGURE 20-10
Plot of axis accuracy and repeatability data.

MECHANICAL DRILLING

The strong need to increase computer operating speed led to a higher level of integration, first on the chip and module and finally on the board itself, where the size and path length of the printed circuit lines have been reduced and the holes have been moved closer together. A dramatic increase in wiring requirements for second-level packaging (as described in Chapters 1 and 2) has resulted in more lines per channel and a demand for a larger number of smaller-diameter holes in a thicker board—the multilayer board. At the same time, the requirement for drilled hole location accuracy was increased to guarantee adequate insulation between circuits. In one design (Figure 20-11), six or nine multichip modules are plugged into a board that contains thousands of via holes and tens of thousands of through-holes. The via holes, when copper plated, allow for internal plane-to-plane signal connections, and the through-holes are used for the logic service terminals, module terminals, and cable termination. The biggest challenge to mechanical drilling was producing the through-holes through 20 layers of copper and 19 layers of epoxy glass.

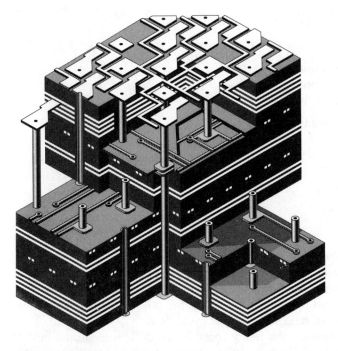

FIGURE 20-11
Section of multilayer board showing high-aspect-ratio through-holes.

Drilling Parameters

Drilling parameters can differ for a variety of reasons, including diameter of the drill, aspect ratio of the hole, composition and thickness of the board, and performance of the drilling equipment. As an example, Figure 20-12 shows the parameters that are related to creating a hole from the start to the end of one drill cycle. The linear displacement of the drill, from the top-of-stroke position (start) to the bottom-of-stroke position (stop), is called the in-feed. The return from the bottom-of-stroke position to the top-of-stroke position is called the out-feed. The

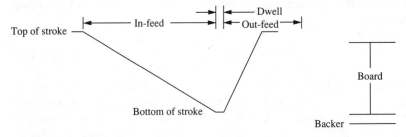

FIGURE 20-12
Linear displacement of drill during one drill cycle.

time for one drill cycle, t_z, is

$$t_z = t_i + t_d + t_o$$

where t_i = z axis infeed time
t_d = z axis dwell time
t_o = z axis outfeed time

and since $s = vt$, the infeed rate is

$$v_i = \frac{s}{t_i}$$

and the outfeed rate is

$$v_o = \frac{s}{t_o}$$

where v_i and v_o are expressed in centimeters per minute.

In general, a smooth z axis infeed is important not only for good hole wall quality, but also for minimizing drill wander and drill breakage. A fast outfeed reduces the heat generated by the drill bit while it is drilling the hole and offers an additional advantage of reducing machine cycle time.

An additional parameter, the rotary speed of the drill, is measured in drill revolutions per minute (rpm). Carbide drills, used for drilling printed circuit (PC) boards, operate most effectively at surface speeds of 183 to 213 surface meters per minute (SMM), which is 600 to 700 surface feet per minute. The surface speed is calculated from spindle rpm and the diameter of the drill:

$$\text{SMM} = \frac{\pi d}{1000} \times \text{rpm} \tag{20.4}$$

where d = drill diameter in millimeters.

The relation of rpm to surface feet per minute (SFM), for the range of drill diameters used in the PC board industry, is shown in Table 20.3. The English system of units is used because it remains in widespread use by drill bit manufacturers in the United States.

Another expression commonly used as a guide in selecting parameters is chip load (also called advance per revolution or bite), which relates the infeed rate to spindle rpm.

$$\text{Chip load} = \frac{v_i}{\text{rpm}} \tag{20.5}$$

Drill chip loads typically range from less than 25 μm/rev (0.001 in./rev) [10] for the smaller-diameter drills used to produce the high-aspect-ratio holes in multilayer boards, to four or five times that amount for the less demanding applications of drilling. The smaller diameter drills are constrained by both the reduced strength of the drill and by physical limitations of the equipment.

TABLE 20.3
Relation of rpm to surface feet per minute for the range of drill diameters used in the printed circuit board industry

Fraction	Wire gage	Drill diameter	Surface feet per minute									
			300	350	400	450	500	550	600	650	700	750
			Revolutions per minute									
	80	0.0135	84,882	99,030	113,177	127,324	141,471	155,618	169,765	183,912	198,059	212,206
	79	0.0145	79,028	92,200	105,371	118,543	131,714	144,886	158,057	171,228	184,400	197,571
1/64		0.0156	73,456	85,699	97,941	110,184	122,427	134,669	146,912	159,155	171,397	183,640
	78	0.016	71,620	83,556	95,493	107,429	119,366	131,303	143,239	155,176	167,112	179,049
	77	0.018	63,662	74,272	84,882	95,493	106,103	116,713	127,324	137,934	148,544	159,155
	76	0.020	57,296	66,845	76,394	85,943	95,493	105,042	114,591	124,141	133,690	143,239
	75	0.021	54,567	63,662	72,756	81,851	90,945	100,040	109,135	118,229	127,324	136,418
	74	0.0225	50,929	59,418	67,906	76,394	84,882	93,371	101,859	110,347	118,835	127,324
	73	0.024	47,746	55,704	63,662	71,620	79,577	87,535	95,493	103,450	111,408	119,366
	72	0.025	45,837	53,476	61,115	68,755	76,394	84,034	91,673	99,312	106,952	114,591
	71	0.026	44,074	51,419	58,765	66,110	73,456	80,802	88,147	95,493	102,838	110,184
	70	0.028	40,925	47,746	54,567	61,388	68,209	75,030	81,851	88,672	95,493	102,314
	69	0.0293	39,110	45,628	52,146	58,664	65,183	71,701	78,219	84,738	91,256	97,774
	68	0.031	36,965	43,126	49,287	55,447	61,608	67,769	73,930	80,091	86,252	92,412
1/32		0.0312	36,728	42,849	48,971	55,092	61,213	67,335	73,456	79,577	85,699	91,820
	67	0.032	35,810	41,778	47,746	53,715	59,683	65,651	71,620	77,588	83,556	89,524
	66	0.033	34,725	40,512	46,300	52,087	57,874	63,662	69,449	75,237	81,024	86,812
	65	0.035	32,740	38,197	43,654	49,111	54,567	60,024	65,481	70,937	76,394	81,851
	64	0.036	31,831	37,136	42,441	47,746	53,052	58,357	63,662	68,967	74,272	79,577
	63	0.037	30,971	36,132	41,294	46,456	51,618	56,779	61,941	67,103	72,265	77,427
	62	0.038	30,156	35,182	40,207	45,233	50,259	55,285	60,311	65,337	70,363	75,389
	61	0.039	29,382	34,279	39,177	44,074	48,971	53,868	58,765	63,662	68,559	73,456
	60	0.040	28,648	33,422	38,197	42,972	47,746	52,521	57,296	62,070	66,845	71,620
	59	0.041	27,949	32,607	37,265	41,924	46,582	51,240	55,898	60,556	65,215	69,873
	58	0.042	27,284	31,831	36,378	40,925	45,473	50,020	54,567	59,115	63,662	68,209
	57	0.043	26,649	31,091	35,532	39,974	44,415	48,857	53,298	57,740	62,181	66,623
	56	0.0465	24,643	28,751	32,858	36,965	41,072	45,179	49,287	53,394	57,501	61,608
3/64		0.0469	24,433	28,505	32,577	36,650	40,722	44,794	48,866	52,938	57,011	61,083
	55	0.052	22,037	25,710	29,382	33,055	36,728	40,401	44,074	47,746	51,419	55,092
	54	0.055	20,835	24,307	27,780	31,252	34,725	38,197	41,670	45,142	48,614	52,087
	53	0.059	19,422	22,659	25,896	29,133	32,370	35,607	38,845	42,082	45,319	48,556
		0.060	19,099	22,282	25,465	28,648	31,831	35,014	38,197	41,380	44,563	47,746
1/16		0.0625	18,335	21,390	24,446	27,502	30,558	33,613	36,669	39,725	42,781	45,837

An additional factor is heat. The frictional force between the drill and the wall of the hole produces heat that can melt the epoxy and form an insulation layer, called smear, between the plated through-hole and an internal plane. The amount of heat generated increases with rpm and decreases as the chip load is increased. Weiss [11] measured the drill temperature of 1.18-mm (0.0465-in.) drills in a multilayer board using nonfunctional lands as calorimeters. He reported that the drilling temperature decreases as the chip load is increased, and the decrease of temperature with chip load is independent of rpm above 75 μm/rev (0.003 in./rev).

The problem of hole wall temperature is aggravated in high-aspect-ratio hole drilling because it is more difficult to extract chips, causing additional heat to be generated in the hole. A second problem caused by high-aspect-ratio hole drilling is drill wander. Experience has shown that with optimized drill bit and process parameters, drill wander for the most part is due to the influence of the glass fibers on the path of the drill as it travels through the board.

One approach that aids the extraction of chips and reduces drill wander is to use repeated advances and withdrawals (peck feeding) of the drill as it creates the hole. Another method is the use of a two-step drilling process. This technique consists of using a drill with a short flute to start the direction of the hole and then using a drill with a longer flute to complete the drilling process.

A highly important consideration in the two-step drilling approach is the need for a very accurate and repeatable drilling machine. Even a small positioning error between the center (pilot) and through-hole drilling operations will increase drill breakage and drill wander and cancel the advantage of the two-step drilling process. A second important consideration in both approaches is the increase in drilling time, which can add to the manufacturing cost of the product.

In practice, drilling parameters are determined from an experiment designed using a matrix so that the effect of changing a single variable can be isolated. Variables such as entry and backup materials, the drill manufacturer, and the number of drill hits are just as important as the in-feed rate, the out-feed rate, and the drill speed. Hole quality can be quantified by applying a scaled value (for example, 1 to 10) to epoxy smear, nailheading (smearing of the copper, which has the appearance of the head of a nail in cross section), sidewall roughness, fiber fracture, and burr height. This can be accomplished by carefully examining scanning electron microscope photographs of axial sections of holes.

Drilling Heads

A drilling head could consist of a single spindle with a self-contained z axis with unique drive, or a cluster of spindles with a common drive and select/deselect capability. Mechanisms used to actuate the z axis drill cycle could be pneumatic, or hydraulic, or could use a motor-driven cam, rocker arm, or ball screw. In some applications where there is no heavy load, a linear dc motor could be used. An illustrative example of a drilling head consisting of a dc servomotor-driven ball screw arrangement [12] that is completely programmable is shown in Figure

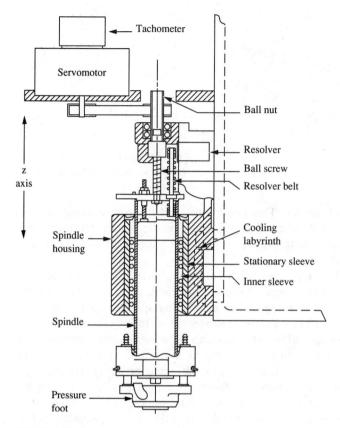

z
axis

FIGURE 20-13
Drilling head consisting of a dc servomotor-driven ball screw arrangement.

20-13. The spindle is held in a sleeve guided inside a second sleeve by a ball cage. The dual sleeve assembly is contained in a spindle housing secured to the overhead axis of the drilling machine. One end of a small ball screw is attached to the inner sleeve containing the spindle. During a drilling cycle, vertical motion of the spindle results when the servomotor rotates the ball nut. Depth control of the drill is provided by feedback from a resolver coupled to the spindle.

Spindle Considerations

Spindles are used throughout industry for drilling, boring, grinding, milling, and routing. Applications include PC board drilling, silicon chip slicing, and lens grinding, to mention a few. Both ball bearing and air bearing spindles are used for drilling PC boards; however, air bearing spindles are often favored in multilayer board drilling because of requirements for higher rpm and lower runout.

When the spindle rotor shaft is operated at high speed, small inaccuracies in machining cause a deviation of the axis of rotation from the position of the

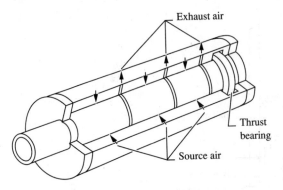

FIGURE 20-14
Diagram of an air-brearing spindle.

principal central axis of inertia. This condition, called dynamic unbalance, is the cause of runout, which is speed dependent. Spindle manufacturers reduce the unbalance, to a considerable extent, by selective removal of material at key locations on the rotor shaft.

Air bearing spindles, in principle, have almost infinite life provided they are supplied with air that is free of contamination, and are never subjected to metal-to-metal contact due to overloading. Some air bearing spindles contain integral high-frequency electric drive motors, while others are air turbine driven. The heat generated by electric motor driven spindles can be a significant contributor to hole location error, especially in multiple-spindle machines. Spindle movement due to heat can be minimized by using cooling labyrinths or liquid cooling. Air-driven spindles can essentially eliminate thermally induced spindle movement; however, early experiments with suitable air turbine spindles proved that it was not possible to maintain a constant chip load when drilling high-aspect-ratio holes, especially at a variable hit rate. A diagram of an air bearing spindle is shown in Figure 20-14. The rotating shaft is supported within a housing by compressed air supplied through rows of very small diameter—approximately 100 μm (0.004 in.)—orifices. The thrust bearing is basically two circular, flat plates. Under axial load, the distance between those plates will tend to lessen, causing an increase in pressure, resulting in automatic compensation within its load capacity. Factors that should be considered when selecting spindles include static and dynamic runout, rpm, temperature rise, size, and power output.

Spindle Packaging

The number and configuration of drilling heads chosen for a drilling machine will depend on many factors, including base capacity requirements for throughput (related to the type of product and product mix), physical weight or hole location accuracy (depending on whether a moving or fixed head is being considered), the type of z axis drive chosen, and thermal considerations. One factor can interfere with another. For example, a motor-driven ball screw arrangement chosen for its smooth in-feed and programmability could make it impossible to achieve optimum spindle-to-spindle spacing because of its physical size.

The influence of the product can be significant. A board containing tens of thousands of holes located uniformly on a fixed grid could be drilled very efficiently with a stationary head containing a large number of spindles (perhaps 40–60) in a single station. In a situation where a large percentage of the holes are located in module sites that occupy a small percentage of the board area, a configuration of small clusters or small groups of individual spindles spaced according to the distance between module sites could be used. Within some practical framework, the number of spindles in a station and the number of stations are chosen to meet the base capacity requirements for throughput.

Machine Hit Rate and Throughput

Cycle time for drilling machines is defined as the sum of the x-y move time, the z axis in-feed time, the z axis dwell time, and the z axis withdrawal time.

$$t_c = t_{x,y} + t_i + t_d + t_o$$

where t_c = machine cycle time
$t_{x,y}$ = x-y move time
t_i = z axis in-feed time
t_d = z axis dwell time
t_o = z axis out-feed time

The machine hit rate (MHR) for a single spindle is

$$\text{MHR} = \frac{1}{t_c}$$

and, since $s = vt$, it follows that

$$\text{MHR} = \frac{1}{t_{x,\,y} + (s/v_i) + t_d + (s/v_o)} \tag{20.6}$$

in which s = z axis stroke
v_i = in-feed rate
v_o = out-feed rate

A combination of z axis in-feed rate and spindle rpm is chosen to produce acceptable hole quality. The x and y axis move time is determined by the length of the move, the positioning system electronics, the static and dynamic characteristics of the positioning mechanisms, and in the most critical cases by the requirements for drilled hole location accuracy. Throughput is related to the machine hit rate, the number of drills drilling during each machine cycle (N_D), the number of holes in the panel (N_C), and stacking of the product. The base capacity (BC), expressed in boards per hour, is

$$\text{BC} = \frac{N_D}{N_C} \times \text{MHR} \times \text{SF} \times 3600 \tag{20.7}$$

where SF is the stacking factor, and in the case of a multiple-station machine,

N_D is the product of spindle utilization, the number of spindles in a station, and the number of stations in the machine.

> **Example.** Assume that a drilling machine had to produce 4000 0.76-mm-diameter holes in 254-by-381-mm panels that are 1.52 mm thick, at an in-feed rate of 457 cm/min and an out-feed rate of 1016 cm/min. The machine is configured with ten stations (five at the front and five at the back), spaced so that one spindle is over each station. Stack the panels three high to maximize throughput. The z axis stroke necessary to drill the stack must include 1.27 mm for top-of-stroke clearance and 0.51 mm for additional drill travel into a backer board. Find the base capacity using 0.3 seconds for the average x-y move time and 0.02 seconds for the z axis dwell time.

Solution

$$\text{Stroke} = (3 \times 1.52) + 1.27 + 0.51 = 6.33 \text{ mm} = 0.633 \text{ cm}$$

$$\text{MHR} =$$

$$\frac{1}{0.3 \text{ s} + \dfrac{0.633 \text{ cm}}{457 \text{ cm/min} \times 1 \text{ min/60 s}} + 0.02 \text{ s} + \dfrac{0.633 \text{ cm}}{1016 \text{ cm/min} \times 1 \text{ min/60 s}}}$$

$$= \frac{1}{0.3 + 0.0831 + 0.02 + 0.0374}$$

$$= 2.27 \text{ cycles per second}$$

$$\text{BC} = \frac{(1)\,(1)\,(10)}{4000} \times 2.27 \times 3 \times 3600 = 61.29 \text{ boards per hour}$$

System Configuration

Today, computer-controlled drilling systems are common and, in some larger installations, a combination of computer and microprocessors is used with three or more drilling machines operated by a single computer. The configuration of the drilling system, shown in Figure 20-15, consists of a computer, a controller interface, and the drilling machine. Communication with a host computer is through a sensor-based communications adapter (SBCA). A display terminal is used to select the product-dependent part numbers, to display operator-requested information, e.g., parameter checks, and to facilitate the use of special maintenance routines to debug and calibrate various portions of the drilling machine. By entering a part number, the file of hole location data is sent from the host to the system computer diskette for storage and drilling use. The printer is used to record data on machine performance.

The controller interface contains the functional controllers for the x, y, and z axes, and handles direct digital input and output (DI/DO) information from the operator control panel, the drilling machine, and the computer. Direct DI/DO includes control switches, nonhardwired indicators, x, y, and z axis limit switches,

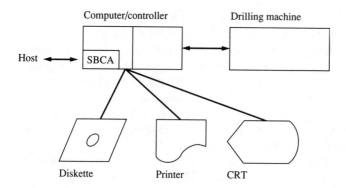

FIGURE 20-15
Drilling system configuration. The host computer supplies the *x-y* hole location data through the sensor-based communications adaptor (SBCA) to the computer. The computer saves the data on its own disk and controls the drilling machine through the controller.

and the x and y axis display feedback. The computer manages the functional controllers, handles interrupts, and communicates with the peripheral equipment attached to the system.

Process Control

The use of x-y measurement of drilled hole location is becoming more widespread in the computer industry, especially in the manufacture of multilayer printed circuit boards. These measurements provide a simple yet effective means for process control of the drilling equipment used to produce the thousands, and in some cases tens of thousands, of holes in each board. Straightforward statistical treatment of measured errors in drilled hole location enables the manufacturing engineer to monitor the x and y axis positioning accuracy and repeatability as well as tooling and spindle location. Gradual changes can be observed and corrective action taken before hole location errors exceed the process window. For example, the control specification S_C for the location of a drilled hole is given by

$$|\overline{X}_F| + 3\sigma_F \leq S_C \tag{20.8}$$

where $|\overline{X}_F|$ is the absolute value of the mean error for drilled holes measured on the top side of the panel and σ_F is the standard deviation for the location data for these holes. In the case of high-aspect-ratio holes, drill wander becomes an additional concern. In this case we could address the specification for a drilled hole in the product, S_P, as

$$|\overline{X}_F| + 3[\sigma_F^2 + \sigma_w^2]^{1/2} \leq S_P \tag{20.9}$$

where σ_w is the standard deviation of the drill wander, obtained by measuring the same holes on the back of the panel.

Another area where process controls can be effective in yield management is computer-selected drilling parameters. Historically, in multiple-product environments, adjustments were necessary to the feed rate, the spindle rotation rate, and in some cases, the position of the drilling heads, for the drill bits to clear both the product and the locating pins at the top-of-stroke position. These adjustments were made manually by each operator, thereby increasing the risk of improper adjustment and, as a result, possible panel damage or reduced quality of the drilled holes. This problem can virtually be eliminated by using a system where drilling parameters such as in-feed rate, dwell time, out-feed rate, top-of-stroke and bottom-of-stroke positions, and spindle speed are all controlled by the computer and selected via a product-dependent part number.

Computer-controlled in-process checks can be used to monitor environmental and machine-related parameters that affect yield, such as air temperature and x and y axis positioning accuracy, and then stop the machine and alert the operator if any operating limits are exceeded. The importance of monitoring environmental conditions cannot be overstated for many drilling applications. Small changes in room temperature will affect the performance of incremental linear transducers used for position feedback and therefore influence the positioning accuracy of both the x and y axis. Equally important for accurate drilling systems is the temperature of the incoming air to the machine. An increase in temperature, for example due to the failure of a chiller-dryer in the compressed-air line, can influence spindle temperature and result in thermally induced spindle movement.

Spindle performance can be monitored by periodically measuring rpm with a portable electronic stroboscope. A better method is to continually monitor the rpm of each spindle with the computer. In this case, each spindle has a built-in sensor that provides an output signal whose frequency is proportional to rpm. For communication with the computer, the frequency-dependent signal would have to be modified with suitable signal-conditioning circuitry.

Other in-process checks often used in drilling systems monitor the performance of the x, y, and z axis positioning mechanisms. In-position tests can be made at each x and y location by electronically comparing the computer counters with the feedback counters. If the position error exceeds a pre-established limit, the computer stops the machine and a message is displayed on the terminal telling the operator to call maintenance. This type of control can prevent the loss of product by responding to a failing servomotor, a loose coupling to the ball screw, or a faulty servomotor amplifier. The z axis in-feed rate and stroke can be monitored using an attached incremental linear transducer, or in the case of a dc motor drive, with an integral high-resolution rotary encoder.

INSPECTION SYSTEMS

Various inspection methods are used to increase printed circuit board quality and yields. Automatic systems are used to inspect glass masters, drilled hole location, top and/or bottom side surface-mounted devices, through-hole components, pad center-to-center distances, and circuit integrity. They can detect missing, mislocated, and even misoriented parts. As mentioned earlier, the statistical treatment of measured *x-y* hole location data provides the manufacturing engineer with a means of monitoring the performance of drilling systems. A second and more important reason for measuring location exists in the case of multilayer boards. The measurement of drilled holes is one of several methods (x-ray of copper reference dots is another) used to gather data that, when treated statistically, is used to control the dimensional changes [13] that accumulate from artwork generation through etching, plating, lamination, and drilling on each board.

Example of Automatic Measurement

One configuration of an automatic inspection system designed for the *x-y* measurement of drilled hole location is shown in Figure 20-16. The system consists of a controller, an inspection machine, and a vision system. A display terminal and keyboard are used to aid the operator in running menu-driven programs. The printer is used to generate a report on each board that includes all data points, measured errors, and a statistical analysis of the measured errors, including the mean, standard deviation, and regression analysis. Hole location data and measurement data are stored on a diskette by board serial number.

The inspection machine is a four-axis positioning system that is capable of handling boards up to 710 by 1220 mm. The board to be inspected is held by a vacuum chuck mounted on the *y* axis table. The table is supported and guided with air bearings on a large granite base. The base also supports a granite bridge on which the *x* axis slide moves orthogonally to the direction of table travel. A

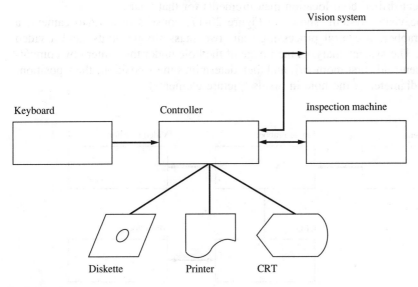

FIGURE 20-16
X-Y inspection system configuration.

third (z axis) positioning mechanism is attached to the x axis slide, which also is supported and guided with air bearings. Both the x and y axes have a closed-loop positioning system that uses a dc servomotor-driven leadscrew and a laser measurement system for feedback. A compensator is used with the interferometer to automatically correct for the effects of atmospheric conditions on the velocity of the laser light.

The inspection head is mounted on a z axis slide that moves in the vertical direction and provides automatic focusing capability for the objective lens. The z axis is driven by a closed-loop positioning system that uses a linear incremental transducer, consisting of a glass scale and an optical scanning head, for feedback. The automatic focus system uses a small, randomly polarized helium-neon laser to produce light that is reflected off of the part to be held in focus.

The inspection head contains a fourth computer-controlled axis used for selecting lens magnification. It consists of a rotary turret with four lenses to provide an objective magnification of 1.2, 2.5, 10, or 40 times. Illumination, provided by a tungsten-halogen light source, is transmitted through fiber-optic cables to the inspection head.

Three panel locating pins are secured by "keepers" mounted on the table. Each pin has a calibration hole in its geometric center. The calibration holes in two of the three pins are used to define panel zero at the beginning of each board measurement cycle. At the start of the measurement cycle, both the x and y axes move so that the inspection head is over the first calibration hole. The vision system measures the center of the hole, and then the process is repeated at the second calibration hole. Board zero is then calculated to be the datum for the subsequent drilled hole location measurements for that board.

The vision system, shown in Figure 20-17, consists of a video camera, a frame grabber, a central processing unit, two mass storage units, and a video monitor. The system analyzes the image of the hole under the camera by computing the centroid (first moment), and then determines the x position, the y position, and the diameter of the hole in pixels (picture elements).

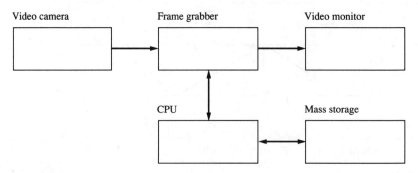

FIGURE 20-17
Vision system configuration.

System Operation

After first loading the machine control and vision system programs from the system diskette, the operator homes each axis to establish a datum point. The home program also interrogates the weather station to determine the compensation factor for the effects of barometric pressure, temperature, and humidity on the accuracy of the laser feedback system. Next, the system diskette is replaced with a diskette that contains the hole measurement programs and utilities used to set the threshold of the vision system, generate printed reports, and create or modify hole location data.

The operator selects a program by typing the appropriate serial number, commands the system to start the measurement routine, and then observes the analog and digitized images of each hole that is measured.

CONCLUSION

Higher levels of circuit board integration will continue to place further demands on positioning systems, mechanical drilling, and inspection. Faster and more powerful computers will require multilayer boards with more layers and higher circuit densities resulting in more vias, smaller-diameter high-aspect-ratio holes, and increasing demands for accuracy.

Today, we know how to build highly productive drilling systems and very accurate drilling systems. The challenge of the future will be to develop drilling systems that are both productive and accurate. Innovations in spindle design and z axis drives, together with more efficient packaging, will drive the need for automatic handling and hole location measurement, as well as for improvements in diagnostics, automatic drill changers, and process control.

REFERENCES

1. M. Ninomiya, "Maintaining Ball Screw Precision," *Machine Design*, April 1979, pp. 105-107.
2. H. N. Norton, *Handbook of Transducers for Electronic Measuring Systems*, New York: Van Nostrand Reinhold, 1972.
3. W. R. Wrenner, "Large Multi-Layer Panel-Drilling System," *IBM J. Res. Develop.* 27, pp. 285–291 (1983).
4. E. S. Dana and W. E. Ford (revised by C. S. Hurlbut, Jr.), *Dana's Manual of Mineralogy*, New York: John Wiley & Sons, 1950.
5. W. A. Gross, *Gas Film Lubrication*, New York: John Wiley & Sons, 1962.
6. N. S. Grassam and J. W. Powell, *Gas Lubricated Bearings*, London: Butterworth, 1964.
7. J. Koch, "Laser Interferometers: Opto-electronic Yardsticks with Microinch Resolution," *Machine Design*, February 1975, pp. 92–97.
8. Hewlett-Packard Model 5501A Laser Transducer System Operating and Service Manual, section 2, "Laser and Optics," 1982.
9. *NMTBA Definition and Evaluation of Accuracy and Repeatability for Numerically-Controlled Machine Tools*, 2d ed., McLean, VA: National Machine Tool Builders Association, 1972.
10. H. Aoyama, "Drilling Automation and Small/High Aspect Hole Drilling," *PC Fab.*, April 1986, pp. 64–74.
11. R. E. Weiss, "The Effect of Drilling Temperature on Multilayer Board Hole Quality," *Circuit World* 3, no. 3, April 1977, pp. 8–14.

12. W. E. Kosmowski, "Method and Apparatus for High Speed Precision Drilling and Machining," U.S. Patent 4,088,417, 1978.
13. D. P. Seraphim, "A New Set of Printed-Circuit Technologies for the IBM 3081 Processor Unit," *IBM J. Res. Develop.* 26, pp. 37–44 (1982).

PROBLEMS

20.1. What spindle rpm would you use at a feed rate of 228.6 cm/min to obtain a chip load of 25.4 μm/rev?

20.2. What feed rate would you use to obtain a chip load of 50 μm/rev with a drill rotating at 80,000 rpm?

20.3. What rpm is required for a #68 drill to operate at a surface speed of 650 SFM?

20.4. Determine the hit rate (in cycles per second) for a drilling machine operating with the following parameters:

Total z-axis stroke	3.4 mm
Infeed rate	229 cm/min
Outfeed rate	1016 cm/min
Average x-y move time	0.25 s
z-axis dwell time	0.02 s

20.5. Calculate the base capacity (in boards per hour) of a four-station drilling machine with three spindles over each work station. This machine is capable of 2.9 drilling cycles per second with a spindle utilization of 90 percent. The board pattern contains 4698 holes.

20.6. Determine the axis positioning accuracy ($\Delta X + 3\sigma$ and $\Delta X - 3\sigma$) from the deviation-from-target data shown below.

Point no.	Target position, mm	Deviation from target, μm					
		↓	↑	↓	↑	↓	↑
1	25	−0.4	−1.3	−0.1	−1.1	−0.3	−1.0
2	50	−0.8	−1.5	−0.6	−1.3	−0.5	−1.1
3	75	−0.7	−1.5	−0.5	−1.3	−0.6	−1.2
4	100	0	−0.7	+0.2	−0.6	+0.2	−0.5
5	125	+0.1	−0.5	+0.4	−0.3	+0.3	−0.2
6	150	+0.2	−0.5	+0.5	−0.3	+0.3	−0.2
7	175	−0.5	−1.1	−0.2	−1.0	−0.4	−1.0
8	200	−1.4	−1.9	−1.2	−1.7	−1.1	−1.8
9	225	−1.8	−2.4	−1.6	−2.2	−1.5	−2.2
10	250	−1.7	−2.2	−1.4	−2.0	−1.4	−2.1
11	275	−1.7	−2.2	−1.5	−2.0	−1.5	−2.1
12	300	−1.6	−2.2	−1.2	−1.9	−1.3	−2.0

BULK
ANALYSIS
IN ELECTRONIC
PACKAGING

JAMES SPALIK

INTRODUCTION

This section deals with a number of bulk analytical methods that are very useful in electronic packaging. In Chapters 22, 23, and 24, electron spectroscopy, ion beam methods, and electron beam microscopy are described. These methods primarily treat the surface regions of analytes; in contrast, bulk analysis, which is the subject of this chapter, considers the sample as a whole.

Components of electronic packaging have been described in other chapters. For analytical purposes, it is better to consider these as types of materials, such as organic/inorganic, compounds/elements, solids/liquids/gases, or composites/mixtures/pure substances. Since the properties of these materials can be greatly affected by their composition (sometimes by trace impurities), analysis plays an important role in assuring their functional integrity.

The term *bulk analysis* refers to the application of an analytical technique to a segment, an area, or a whole component. It also applies to a portion of a process solution even when that portion is meant to represent the whole. Both trace and major components are to be considered in this category. Sample size may range from micrograms to grams, or even larger in some cases.

Some examples of situations requiring bulk analysis will help in understanding its use. Consider, for example, cleaning baths containing either deionized water or organic solvents. Trace contaminants can render them unusable. On the other hand, for a plating bath, as discussed in Chapter 16, all major components must be maintained at concentrations defined by the process window (described in Chapter 9). The range of concentrations to analyze can be from below parts per billion (ppb) to percent.

Next, consider solids, which may consist of laminates, composites (such as laminated circuit boards), ceramic substrates, thin films, or coatings. The analyte may not be homogeneous. Instead, it may be distributed in layers, zones, cavities, or interstices. The concentration range requirements will be the same as for liquid solutions; but, it is now necessary to define the type of distribution.

Analytical techniques include a variety of methods, instruments, specific variations, and a number of classical wet procedures that can make up complete methods or simply parts of a sample preparation step. In selecting methods for discussion in this chapter, the focus is on techniques commonly used in the electronic packaging industry. Even though problems in bulk analysis can often be solved using classical wet chemistry, including titrations, filtrations, and extractions, these have been omitted to allow room for a basic coverage of the principles related to three broad areas of analysis: chromatography, spectroscopy, and thermal analysis.

Among chromatographic methods, size exclusion chromatography or gel permeation chromatography (SEC/GPC), which is a procedure of critical importance in characterizing polymers used in electronic packaging, receives substantial coverage. Similarly, thermal analysis receives broad coverage because it is widely used to determine the state of cure advancement of resins and photoresists, either by differential scanning calorimetry (DSC), or in the case of photoresists, by photodifferential scanning calorimetry.

Electronic packaging places greater demands on analysis of trace components than almost any other field. This is due to the critical nature of chemical processes used in electroless plating, photolithography, plasma reactions, and in other operations affected by the continued reduction in scale, and the drive toward higher transmission rate.

This chapter starts with a brief discussion of some principles that should be observed in selecting and taking a sample for analysis.

PREPARATION FOR ANALYSIS

There is a false impression among the inexperienced that with expensive, sophisticated instrumentation, analysis is simply a matter of inserting a probe, or placing a bit of the sample in the instrument, and reading the results from a computer display. In fact, as the sensitivity of the instrument increases, proper sampling and handling become even more critical. Some principles that will aid in obtaining analytical results more rapidly will be presented.

History

In general, the more that is known about a sample, and the more specific the analytical request, the easier it is to obtain an answer. Trying to identify and quantify a trace contaminant can take months, whereas proving the presence of a specific constituent may be done in a matter of minutes. A history of the sample should be prepared, including a list of other materials that may have contributed to the problem either by contact or transfer through the air or solution. Obviously, a recurring problem has a history that can be used to quickly determine if the problem is originating from a single source.

Sampling

There are at least two principle considerations in obtaining a proper sample for analysis. The first requires that the sample be representative of the material under investigation. It is essential that the sample be taken at the right time and from the appropriate source. The second consideration is to understand the effect of aging in order to ensure that evaporation, corrosion, or surrounding contamination do not adversely alter the results.

Handling

It is very important to avoid extraneous contamination of the sample. For liquids, the proper choice of container material and proper rinsing with the sample are essential. There are no universal generalizations in regard to the choice of container material; however, there are some specific things to avoid. Glass, while generally inert to most solvents, is attacked by strong base and fluoride solutions. Plastic caps are another source of extraneous contamination due to the leaching of plasticizer by organic solvents. This contamination is enhanced when organic samples are allowed to stand in plastic containers or glass containers with plastic caps. Generally, teflon liners or containers are an excellent, but expensive, choice.

ANALYTICAL TECHNIQUES

Chromatography

Chromatography (Figure 21-1) is a term used to signify techniques utilized to separate complex mixtures of substances with very similar properties. Literally it means "color writing," because in its earliest applications, it was used to separate plant pigments. The separated pigments produced a multicolored pattern called a chromatogram (color graph). The appearance and intensity of the visible colors were used to identify and quantify the components of the mixture.

Because of the near impossibility of determining the presence and quantities of individual species in mixtures without tedious separation procedures, chromatography has developed into a major component of many analytical procedures. Modern instruments utilize combinations of techniques in what are referred to as hyphenated instruments. One familiar combination couples a gas chromatograph (GC) with a mass spectrometer (MS), giving a GC-MS.

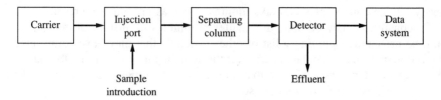

FIGURE 21-1

Schematic of a chromatographic system showing five basic components: (*a*) a carrier fluid system that can deliver gas or liquid at a constant, non-pulsating rate, (*b*) a sample introduction port consisting of a septum or valve that will not interrupt the carrier flow; (*c*) a separating column that may be kept in a temperature programmable oven, (*d*) a detector that can monitor the presence of species as they elute from the column, and (*e*) an output device such as a chart recorder.

Strictly speaking, chromatography only provides a means of separating species in a mixture. To obtain an analysis, a detector that responds to the presence of the analyte is needed. For the colored pigments described previously, the eye will serve, but for most individuals quantitative precision based on color intensity is very poor. Therefore, a large variety of detectors giving high specificity and sensitivity have been developed. Also, as a result of the ever-increasing need to separate more complex mixtures, and the practicality of expanding the variety of solid, liquid, and gas mixtures into the analytical scheme, the number of variations of chromatographic techniques has also increased. Among the more frequently used techniques are (1) gas-liquid chromatography (GLC), usually called gas chromatography (GC), (2) high-performance liquid chromatography (HPLC), (3) ion chromatography (IC), and (4) size exclusion chromatography (SEC).

A description of the general chromatographic principles will aid understanding of the four commonly used techniques (see Figure 21-2). Basically, a sample mixture is introduced at one end of a tubular column, which may be packed with a porous material, or have only the wall coated with a similar, nonmobile material. The sample mixture is carried through the column by an inert gas or liquid (mobile phase). As the sample moves through the column, the individual species will move through the column at different rates depending on their affinity for the packing or wall coating, which has been selected for its ability to provide differential retention. The packing generally consists of a support composed of small, inert, spherical particles a few microns in diameter, coated with viscous liquid (the *stationary phase*) having solvent properties compatible with the species in the mixture. Wall-coated columns can have either a support coated by liquid, or only the liquid. These are designated support-coated open tubular (SCOT) and wall-coated open tubular (WCOT) columns, respectively. With the proper selection of stationary phase, each species will elute from the exit end of the column sequentially rather than as a mixture. In most applications, the practice is to chemically bond the stationary phase so that it will not be moved by the mobile phase.

Gas chromatography is suitable for species that are volatile, and can be carried through the column without decomposing. Many organic processing solutions

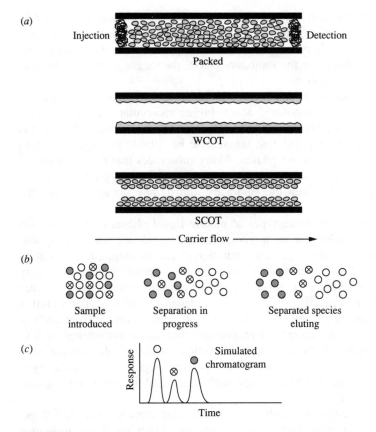

FIGURE 21-2
(a) Types of separating columns: Packed-Typical packings consist of an inert support coated with a low volatility liquid, an inert porous material (molecular sieve), or ion exchange resins. WCOT-Wall Coated Open Tubular columns have the liquid phase placed directly on the wall. SCOT-Support Coated Open Tubular columns have the liquid coated inert support attached to the wall. Open tubular types have less back pressure, allowing much longer columns for greater separation. (b) The separation process: (1) A sample is introduced into a carrier as a slug of material, (2) separation begins as the slug is carried into the column, (3) separated species sequentially exit the column directly into the detector, (c) detection of the individual species is represented graphically as a response (concentration) versus time (retention) plot.

fall into this category. To analyze a sample by GC, a small amount (≈ 1.0 μL) is injected with a syringe, through a silicone rubber septum, into a low-volume chamber (injection port) kept at a temperature sufficient to instantly vaporize all of the sample. A steady flow of inert carrier gas, usually helium, is maintained, flowing through the injection port–column–detector assembly. The column is placed in a temperature-programmable oven, which can be operated isothermally or programmed to operate over an appropriate temperature range. The range is selected to achieve a fast separation of species. For isothermal runs, the optimum temperature is usually below the boiling point of the highest boiling species.

This provides an opportunity for the vapor phase of each component to partition between the mobile gas phase and the stationary liquid phase, and hence achieve separation based on the relative solubilities of each species in the stationary phase. The more soluble a species in the stationary phase, the longer it will take to elute from the exit end of the column.

High-performance liquid chromatography is suitable for species that cannot be vaporized without decomposing. Many higher molecular weight organics, organic salts, and surfactants fall into this category. Because the mobile phase is a liquid, the only requirement is that the analyte be soluble to varying degrees in both the stationary and mobile phases. Many substances that can be analyzed by GC, with the notable exception of gases, can be analyzed by HPLC. Again, partitioning of species between the stationary and mobile phases provides the separation.

Since there are many more types of mobile liquid phases than gas phases (all gases are equally soluble in each other, whereas liquids are not mutually soluble), it is possible to achieve greater separation in certain complex mixtures by selecting mobile and stationary phases based on the different degrees of polarity of the liquid phases. In fact, it is possible to achieve continuously varying degrees of polarity using mixtures of solvents, or by a technique called gradient elution programming. Using this technique, one may start with a polar solvent such as water, gradually introduce increasing proportions of a less polar solvent such as methanol, and continue with increasing proportions of a nonpolar solvent such as heptane until the mobile phase is 100 percent heptane. The polar type molecules, such as salts, would elute first with water, and the non-polar organics would elute last with the heptane.

Size exclusion chromatography (SEC) in earlier literature was called gel permeation chromatography (GPC). As the new name implies, it is a technique for separating species according to their size. Polymers used in packaging are generally received as high-molecular-weight resins, and it is important to be able to characterize the size distribution in the uncured resin. In SEC the analytical column is packed with a material consisting of small spherical particles having small holes or craters in their surfaces. As the dissolved resin is carried past the spherical particles, molecules of the resin polymer stray into the holes and are delayed in traversing the column. Small molecules stay within the holes longer than the large molecules, which may not fit deeply into the craters. Therefore, the larger molecules elute before smaller molecules, resulting in a separation distributed according to decreasing molecular weight. The column packing can be selected with crater sizes appropriate to a particular molecular weight range.

This is a particularly helpful technique to use in understanding the range of molecular size in epoxies. The photoreacting behavior appears to be very sensitive to the tails of the molecular distribution. Screened coatings also are in this category since they tend to "bleed" and lose their live acuity if the molecular distribution is not well controlled. Some of these considerations are reviewed in Chapters 11 and 12.

Samples of chromatograms using each technique are shown in Figure 21-3.

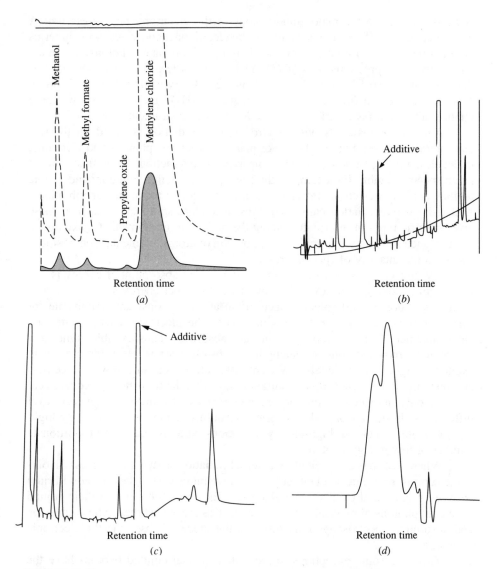

FIGURE 21-3
Typical chromatograms: (*a*) GC-MS of a photoresist stripping agent showing it to be a four-component mixture; (*b*) HPLC of typical additive fraction in photoresist; (*c*) GC-FID of the same additive fraction; (*d*) GPC of the binder in photoresist. The bimodal molecular weight distribution suggests a blend of two grades of polymethyl methacrylate.

Spectroscopy

Spectroscopy covers a broad category of analytical techniques based on the emission or absorption of radiation.

Electromagnetic radiation is a form of energy that travels through space at the speed of light. Gamma, x-ray, ultraviolet, visible light, infrared, microwave, and radio wave are the major subdivisions of the electromagnetic spectrum. A more precise way of defining a region of the electromagnetic spectrum is in terms of either the energy, frequency, or wavelength of the radiated photons. These are related by the expressions $E = hC/\lambda$ and $E = hv$, where E is in joules (J), h is Plank's constant (6.625×10^{-34} J·s), C is the velocity of light (3.0×10^8 m/s), λ is the wavelength (m), and v is the frequency (Hz). Table 21.1 shows these relationships, and Figure 21-4 shows schematics of instrumentation.

Analysis of materials with electromagnetic radiation is based on the fact that interactions (see Figure 21-5) take place between the photons and electrons, molecules, nuclei, or ions. Usually one type of interaction predominates for a given spectral region. For example, the principle interaction in the uv and visible regions is photon-electron. Infrared radiation gives rise to molecular vibrations. And radio wave radiation causes changes in nuclear spin states. Each of these gives specific types of information about the molecules or atoms that make up the sample, such as structure of molecules, position of atoms, elemental composition, and even the quantity of species present.

UV and visible spectroscopy are similar in that absorption of energy is due to electrons moving from lower (ground state) to higher energy states. Electrons in chemical bonds, and valence electrons in atoms or ions, are responsible for the absorptions observed in this region. When the electronic energy states are close together, less energetic photons are absorbed. Thus, visible light (low energy/low frequency/long wavelength) is absorbed when electrons are easily excited to higher energy levels. UV light, alternately, is absorbed when electrons are more highly restricted. It is important to note that the transitions are quantized; they are only activated by photon energies that exactly match the energy level differences. In other words, low-energy transitions cannot be activated by high-energy photons, nor can high-energy transitions be activated by a proportionate number of low energy photons.

When electrons are excited by external radiation, a dynamic process involving electrons returning to their ground state rapidly establishes an equilibrium between ground and excited state electrons. Electrons returning to their ground state emit photons of precisely the same energy as those absorbed. Thus, emission and absorption spectroscopy only differ in the mode of observing the electronic transition.

There are some exceptions to the statement that emitted photons have the same energy as those absorbed, and such exceptions form the basis for important analytical instruments. Electrons excited to certain high-energy states may return

TABLE 21.1
Major subdivisions of the electromagnetic spectrum

Region name	Energy, J	Wavelength	Frequency, Hz
Ultraviolet	9.9×10^{-19} to 5×10^{-19}	200–400 nm	1.5×10^{15} to 7.5×10^{14}
Visible	5×10^{-19} to 2.5×10^{-19}	400–800 nm	7.5×10^{14} to 3.8×10^{14}
Infrared	6.6×10^{-20} to 4×10^{-21}	2.5–50 μm	1×10^{14} to 6×10^{12}
Radio wave	4×10^{-25} to 6.6×10^{-28}	0.5–300 m	6×10^8 to 1×10^6

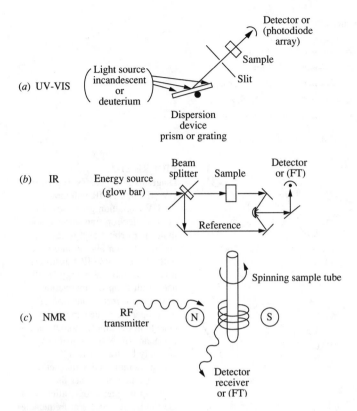

FIGURE 21-4

Schematics of instruments utilizing electromagnetic radiation sources and detectors. Each operates in a frequency/wavelength range that causes specific types of interactions in species of interest, and results in a characteristic spectrum. (a) UV-VIS spectrophotometer—the optics and sample holder must be transparent to ultraviolet and visible light. Two basic types differ in the way a spectrum is obtained. One uses a rotating prism or diffraction grating to scan the spectrum while a photodetector monitors changes in intensity of monochromatic radiation as a function of wavelength. The other uses a photodiode array consisting of hundreds of detectors positioned to detect small intervals in the spectrum. Scanning instruments can give excellent spectral resolution; but require more time to scan the spectrum. Photodiode array instruments can collect a total spectrum in a second or less. (b) IR spectrophotometer—optically an infrared spectrophotometer is similar to a UV-VIS spectrophotometer. Primary differences consist of a heated glower for an IR radiation source, a fast response thermocouple for a detector, and IR transparent optics. Similarly with UV-VIS, scanning instruments exist; but, designs that utilize fourier transformation (FT) of interferograms can acquire spectra in one second intervals. Thus, FT and photodiode array instruments are useful in conjunction with chromatographic systems and in monitoring reaction kinetics. (c) NMR spectrometer— these instruments use radio frequency transmitters as energy sources, and radio frequency receivers as detectors. The sample is placed in a homogeneous, strong magnet field. Magnets are either permanent, electro, or super-conducting types. Scanning types exist that scan either the magnetic field or the radio frequency; but, FT instruments offer significant advantages in sensitivity as well as time.

(a) UV-VIS electronic transitions

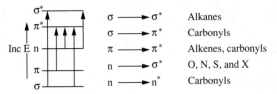

$\sigma \longrightarrow \sigma^*$	Alkanes
$\sigma \longrightarrow \pi^*$	Carbonyls
$\pi \longrightarrow \pi^*$	Alkenes, carbonyls
$n \longrightarrow \sigma^*$	O, N, S, and X
$n \longrightarrow n^*$	Carbonyls

(b) IR molecular vibrations

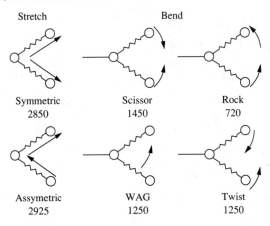

Stretch		Bend	
Symmetric 2850	Scissor 1450	Rock 720	
Assymetric 2925	WAG 1250	Twist 1250	

(c) NMR nuclear spins (precession)

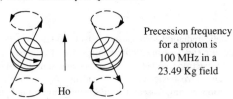

Ho

Precession frequency for a proton is 100 MHz in a 23.49 Kg field

FIGURE 21-5
Diagrams of processes excited by UV, IR, and NMR radiation. (a) UV excitation processes involve electron transitions among molecular orbital energy levels, and result from photon interaction with electrons. (b) IR transitions involve vibrational energy levels, and result from the interaction of electromagnetic radiation of the proper frequency causing a change in the dipole moment about the bond. (c) NMR transitions can only be studied in a strong, homogeneous magnetic field. Certain nuclei act like spinning magnetic tops, align with the field, and precess at frequencies that depend on the strength of the field. Electromagnetic radiation of the same frequency as the precession causes the nuclei to flip opposed to the field with the resulting absorption of energy.

to ground state by different paths, whereby they emit photons from intermediate states with correspondingly lower energies. Three principle modes of emission result from such transitions: fluorescent, phosphorescent, and nonradiative (thermal heating). While each of these have important applications, they are beyond the scope of this chapter.

Infrared spectroscopy utilizes the spectral absorption characteristics of materials for their chemical identification. Infrared radiation is in the portion of the electromagnetic spectrum between the visible and microwave regions. Of greatest practical use is the portion between 4000 and 600 cm^{-1}, known as the mid-IR region, though the near infrared (14,290–4000 cm^{-1}) and far infrared (700–50 cm^{-1}) have become more utilized in recent years. The values given in units of cm^{-1} (reciprocal centimeters) are simply reciprocals of the corresponding wavelengths.

When infrared radiation is absorbed by a molecule, it is converted to energy of molecular vibration. The wavelength absorbed corresponds to the energy of vibration, which in turn depends on the force constant of bonds, the mass of atoms connected by the bonds, and the type of vibration (stretching, bending, scissoring, or wagging). In general, smaller masses and higher bond strengths give rise to vibrations at higher frequencies, as predicted by Hooke's law. Since infrared spectroscopy involves primarily vibrational properties of molecules, a vibrational designator is most commonly used to identify a characteristic absorption.

It has been found that certain structural groups give rise to characteristic bands at or near the same frequencies, regardless of the structure of the rest of the molecule. These characteristic bands allow the analyst to obtain useful structural information by inspecting the infrared absorption spectrum of an unknown compound.

The most useful vibrations of organic compounds occur in the mid-IR region of the spectrum, which has been very well characterized. Some common vibrations used for identification are the N–H and O–H stretches ($3700-3100$ cm^{-1}), C–H stretches ($3100-2700$ cm^{-1}), C=O stretches ($1900-1550$ cm^{-1}), C–H bends ($1500-1250$ cm^{-1}), and C–O stretches ($1300-1000$ cm^{-1}). The region below 1600 cm^{-1} is known as the fingerprint region of the infrared spectrum for organic compounds. Absorptions in this region can be used for positive identification of pure compounds, in the same way a fingerprint can be used to identify a person. In practice, a very pure sample must be prepared before it can be identified. The characteristic bands are used to narrow the identity to certain classes of compounds, and other techniques such as mass spectrometry (MS) and nuclear magnetic resonance spectroscopy (NMR) can be used to assist in the identification. Once the identity is determined, comparison with a reference spectrum of a known compound completes the confirmation.

The analysis of inorganic compounds by infrared spectroscopy is much more difficult. Some compounds, such as sulfates, silicates, metal carbonates, and phosphates, have vibrations in the mid-IR region that have also been characterized, and are easily identified. Usually, the infrared absorbances of other inorganic species are weak, and occur in the far infrared region of the spectrum. For these species, infrared spectroscopy is normally used only after the major components of the unknown inorganic compound have been identified, and a verification is needed for the major functional group associated with the compound.

Many of the samples brought for analysis are mixtures. If the components are known, it is possible to quantify species if they have concentrations in the percent range, and if they have absorption bands different from other species present. Generally, infrared is not used for quantitative analysis, and is incapable of trace analysis.

Samples for infrared analysis can be liquids, solids, or gases. They can be examined as thin films, supported or mixed with a transparent medium such as KBr, NaCl, CsI, Nujol, or even HDPE. The only requirements are that the supporting medium be transparent to infrared in the region of interest, and that the sample be of the proper thickness to give a good spectrum without overabsorbing. Special cells are available to give thin, constant thickness optical paths for liquids, and long paths for gases (ten-meter gas cells). Films and coatings on nontransparent materials can be analyzed by reflectance techniques.

Dispersive type infrared spectrophotometers, employing prisms or diffraction gratings with scanning monochrometers, have been standard equipment in analytical laboratories for years. Advances in computers have made Fourier transformation (FT) of frequency data rapid and relatively inexpensive. This allows the sample to be exposed to the full spectral region simultaneously instead of having to be scanned slowly. The absorbed frequencies are recorded as a free induction decay (FID), and the mathematical transformation converts the result to a frequency spectrum. As a result, FT instruments can obtain high-resolution spectra in seconds instead of minutes. By accumulating multiple spectra, higher sensitivity is achieved. Because of the increased sensitivity, many accessories that typically have low energy throughput (and therefore high spectral noise) can be used to characterize difficult samples. A recent example is the use of infrared microscopes to resolve and identify particles as small as 20 μm in diameter.

Fourier transform infrared analysis (FTIR) has been used extensively in electronic packaging to characterize organic materials and the reactions that occur during processing. A recent application is the use of its rapid scanning capabilities to follow the curing reactions of a photosensitive epoxy resin used as a photoresist (Figures 21-6 and 21-7). Figure 21-7(a) shows infrared absorbance spectra of the photoresist before and after curing. The band near 910 cm^{-1} is due to epoxy groups in the material, and its loss can be correlated with the extent of cure. Using an in situ exposure unit and a heated cell, the process conditions can be modeled in the instrument. The loss of epoxide groups at various stages of the process can be measured, and the process conditions can be rapidly adjusted to obtain optimum cure levels for the required chemical to obtain the needed physical properties.

Thermal Analysis

Thermal analysis (equipment schematics shown in Figure 21-8) involves the measurement of some change in a physical or chemical property of a material as a function of temperature. A change in the latent heat of a material can be used to determine specific heats, heats of phase transition, glass transition temperatures of polymers, and melting points. The technique for this type of analysis is called differential scanning calorimetry (DSC). Loss of weight as a function of temperature is termed thermal gravimetric analysis (TGA). Expansion or contraction, or softening or melting, as measured by a probe touching the sample as the temperature is changed, can give accurate measurement of coefficients of expansion, and is called thermal mechanical analysis (TMA). Even changes in tensile strength, flexibility, or springiness as a function of temperature can be measured by attaching a probe to flex the sample at specified frequencies as the temperature is ramped, and observing how the material recovers from the oscillatory stress. This method is called dynamic mechanical analysis (DMA). These are but a few of the more common techniques of thermal analysis.

Since there are so many polymer materials to choose from for a variety of component applications, a fast screening technique such as TGA is required. For example, if subsequent processing requires the material to be stable for extended

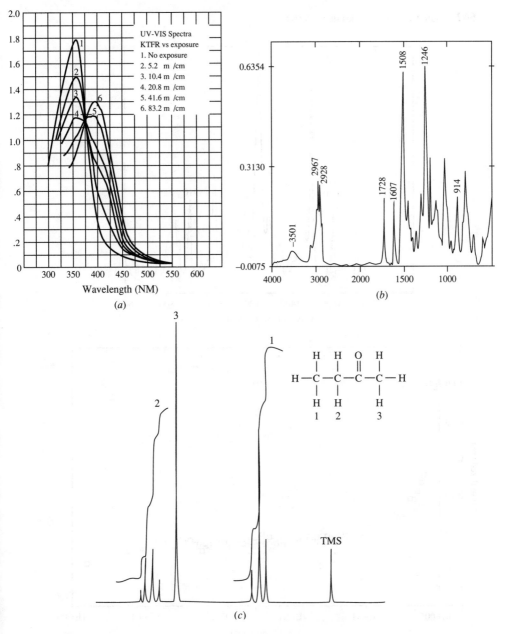

FIGURE 21-6

Typical UV-Vis, IR, and NMR spectra. (*a*) Multiple plot UV-Vis spectra of a photoresist showing the decrease in initiator and the increase in another species as a function of exposure. (*b*) IR of a photoresist showing O—H at 3501 cm^{-1}, C—H at 2967 cm^{-1}, C—O at 1728 cm^{-1}, aromatic ring C—C stretch at 1607 cm^{-1}, and the epoxy C—O at 914 cm^{-1}. (*c*) Proton NMR of MEK showing the three groups of equivalent protons. The position left of TMS (tetramethyl silane standard) at 0 ppm is called the chemical shift, and helps to define the relative position of the C—H groups. The number of peaks in each multiplet indicates the number of protons on the adjacent carbon by the formula (n − 1). The area integrations give the number of equivalent protons causing the multiplet.

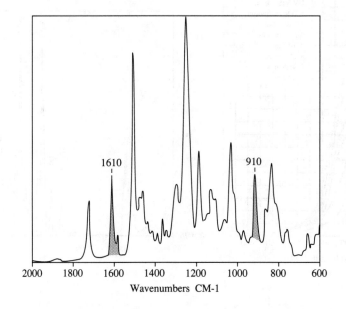

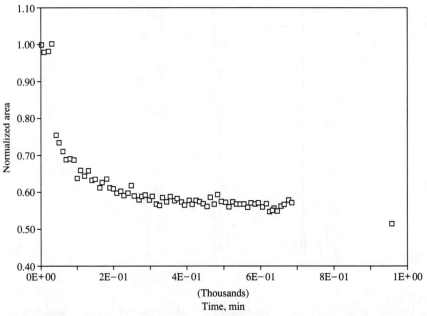

FIGURE 21-7

Application of Infrared Spectroscopy to the study of photoresists. (*a*) Infrared absorbance spectra of an epoxy based photoresist showing the epoxy band at 910 cm^{-1}, and the aromatic ring breathing vibration at 1610 cm^{-1}. (*b*) Plot of the area ratio of the 610 cm^{-1} band to the 1610 cm^{-1} band showing the decrease in epoxide groups with time at room temperature after initial exposure to UV light.

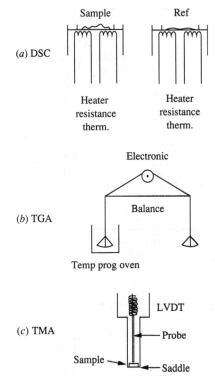

FIGURE 21-8

Schematics of DSC, TGA, and TMA instruments. (*a*) DSC consisting of matched calorimeters. Each has a platinum resistance heater, and a platinum resistance thermometer. When the temperature is ramped, a feedback circuit provides current to the cell lagging in temperature. Differential current: (Energy = I2R) is plotted as a function of temperature, giving a thermalgram. (*b*) TGA consisting of an electronic balance with a sensitivity of 0.1 microgram, and a temperature programmable sample heater. Weight is plotted as a function of temperature. (*c*) TMA consisting of a quartz probe mounted in a LVDT that monitors expansion, or softening of the sample as a function of temperature.

time at elevated temperatures (such as soldering), one can make a rapid choice of those that pass the TGA scan. Weight loss due to solvent evaporation, water evaporation, and loss of plasticizer can also be studied and modeled by TGA.

TMA of materials used in electronic packaging is important in finding or customizing materials that have similar coefficients of expansion so that thermal cycling will not cause rapid fatigue failure. With the growth of surface mount assembly techniques (see Chapters 2 and 19), a large number of composite material combinations are being investigated to obtain a thermal coefficient of expansion match to ceramics and silicon. Some of these combinations are noted in Chapter 11.

Since polymers exhibit unique calorimetric responses associated with their probabilistic origin, DSC can be used to detect differences in the way the basic repeat units are arranged. These differences are virtually impossible to find by chemical analysis or by any other instrument technique.

Typical thermal analysis plots are shown in Figure 21-9. Since more detail on thermal analysis of polymers is given in Chapter 11, only DSC principles and applications will be elaborated here.

DSC instrumentation consists of a pair of microcalorimeters, each having matched platinum resistance thermometers and heaters. The heaters are controlled by a feedback circuit that monitors the temperature in each as the temperature is ramped either up or down. If the temperature lags in the sample cell, the

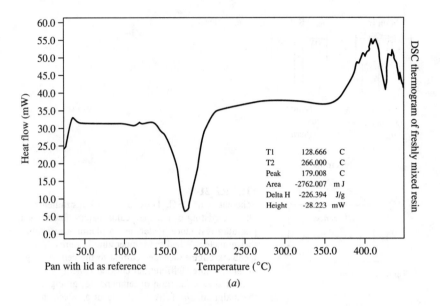

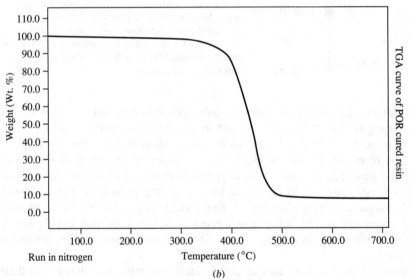

FIGURE 21-9
Typical DSC, TGA, and TMA plots. (*a*) DSC of a polymeric material showing the exothermic reaction as a result of curing. (*b*) TGA of the same material showing the loss of the anhydride component of the hardener. (*c*) TMA showing the softening point, and the coefficients of expansion before and after softening.

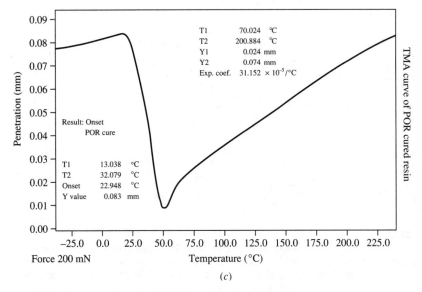

Force 200 mN

(c)

FIGURE 21-9
(continued)

feedback circuit provides just enough current to catch up with the reference cell. The differential electrical current is precisely monitored, and the differential heat required by the sample in the sample cell is given as a function of the Joulean heating expression $E = I^2/R$. A plot of the differential heating rate in millijoules per second, or millicalories per second, versus temperature shows characteristic thermal properties of the material.

Three useful phenomena in characterizing polymers are easily observed: (1) glass transition, (2) melting, and (3) crystallization. Polymers do not crystallize in precise steric order; consequently, the majority of polymers assume only a glassy solid state. At the glass transition temperature T_g the glassy polymer begins to soften and turns into a liquid. Even crystalline polymers exhibit glass transitions; i.e., they contain domains of amorphous material. These domains arise because it is not possible to incorporate an entire polymer molecule into a crystal. End groups, folds protruding from the crystalline region, and molecules bridging from crystal to crystal prevent complete crystallization. This "impurity" in the crystals and unequal distribution in crystal sizes leads to the rather broad melting and crystallization peaks typically observed for synthetic polymers (see Figure 21-10).

Melting points obtained by DSC have been used to measure tin content of tin solders with excellent precision, as shown in Figures 21-11 and 21-12. The advantage of using melting points is in the simplicity of obtaining an analysis and the versatility of using very small samples in situ on their components. The use of various solder compositions is reviewed in Chapter 19. Samples containing a few milligrams of solder are placed in a disposable sample pan. The temperature is ramped at 20 degrees per minute through the melting point. Percent tin is

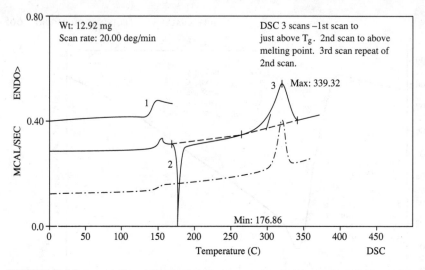

FIGURE 21-10

Triple DSC scan showing: (1) the glass transition temperature, T_g, (2) the exothermic curing reaction, and the endothermic melting reaction. T_g characterizes where the glassy polymer softens and turns into a liquid. Even crystalline polymers exhibit T_gs due to the domains of amorphous material. These domains arise because it is not possible to incorporate an entire molecule into a crystal.

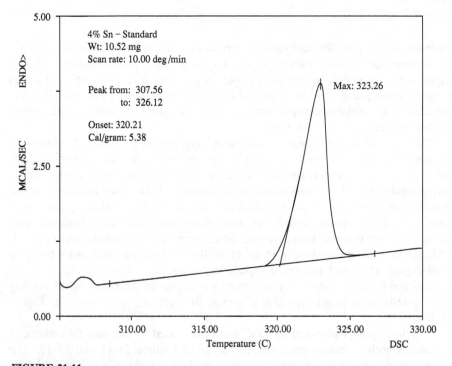

FIGURE 21-11

DSC thermalgram of a 4/96 tin/lead solder. The melting point is determined at the intercept of the baseline by the tangent of the inflection point of the onset slope. Area under the curve is proportional to the heat of fusion.

666

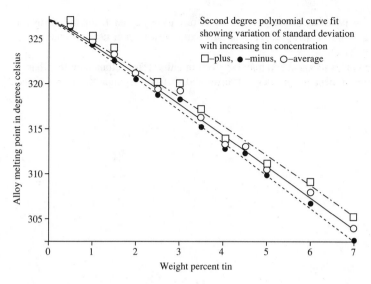

FIGURE 21-12

Plot of melting points versus percent tin in low tin solders. Melting points shown correspond to the solidus line in the tin-lead phase diagram. This method does not require a sample weight, can be used even if the solder is still attached to a part of the joint, will give good precision with sub-milligram samples, and can achieve $=/-$ 0.5% precision with only a few milligrams of sample.

determined from a calibration curve derived from standards. This technique is applicable to any alloy system where there is a significant enthalpy release or absorption at a phase boundary.

PROBLEMS

21.1. What type of information is usually gained using infrared spectroscopy? Specifically, discuss whether the information relates to organic vs. inorganic, elemental vs. molecular, and quantitative vs. qualitative.

21.2. The accepted range for the analytical infrared spectrum is from 4000 cm^{-1} to 200 cm^{-1}. Calculate the wavelengths corresponding to these values, in microns.

21.3. Describe and compare the phenomena resulting from excitation by radiation in the UV-VIS, IR, and NMR regions of the electromagnetic spectrum.

21.4. The carbonyl group exhibits an absorption in the 1600 cm^{-1} to 1700 cm^{-1} region and is a distinctive feature when present in the IR spectrum. Assuming stretching vibrations, determine the force constant of the C$=$O bond.

21.5. The abscissas in UV-Vis, IR, and NMR spectra are usually presented in nanometers (nm), reciprocal centimeters (cm^{-1}), and megahertz or ppm chemical shift from TMS, respectively. Each are related through the velocity of light and the energy per photon ($E = hC/y = h\nu$). Calculate the wavelengths in meters corresponding to a) 100 MHz, b) 1690 cm^{-1}, and c) 520 nm.

21.6. Describe at least three types of thermodynamic data that can be obtained using a DSC.

21.7. What are the two fundamental types of information gained from the chromatographic analysis of a sample? How is this same information used to characterize polymers by SEC?

21.8. The ordinate in an IR spectrum may be given in either transmittance or absorbance. Calculate the precentage of energy transmitted that corresponds to an absorbance of 0.95.

CHAPTER 22

SCANNING ELECTRON MICROSCOPY IN ELECTRONIC PACKAGING

DAVID E. KING

INTRODUCTION

As packaging technology advances, the demands increase for finer circuit lines, with smaller spacing between them in all three directions. Meeting product specifications with ever-increasing packaging density has become an increasingly difficult task. This trend requires paying attention to product reliability or process yield issues that require a variety of materials analysis capabilities. Chapter 9 addresses reliability and Chapter 31 proposes future trends in manufacturing.

Effective failure analysis and materials characterization require the use of a variety of complementary analytical techniques. Among the most important tools is the scanning electron microscope (SEM).

The edge shape of photoresist line channels for additive plating, as shown in Figure 22-1, is an example of structures where small features found with the SEM are important. (Chapters 12 and 16 provide more detail on additive circuit manufacturing.) If severe undercutting occurs at the base of the resist, reduced line-to-line spacing results, with no significant increase in current carrying capacity. A nearly vertical wall is best. This example shows a slight undercutting, and some residual resist in the line channel. The SEM is effective at inspecting for these conditions.

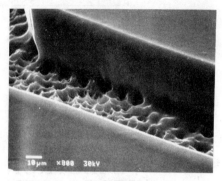

SE image
65° tilt

SE image
65° tilt

FIGURE 22-1
Photoresist line channel profile.

When developing new manufacturing processes and supporting existing ones, the integrity of the materials and the interfaces between them must be assured. Reliability tests can reveal problems, as discussed in Chapter 9. These conditions as well as those found by quality control can be further understood with the aid of the SEM.

Energy-dispersive x-ray spectroscopy (EDS) is a common attachment that supplements the imaging ability of the SEM, providing a means to determine what elements are present in the sample.

There are many variations in the way that the SEM can be applied to problems in electronic packaging. Selecting the correct SEM setup and sample preparation technique can make the critical difference in an analysis. A thorough understanding of the principles of the SEM is necessary to effectively make these choices. This chapter provides an overview of these principles. Readers are encouraged to seek additional information from the references listed.

An important ingredient in successful SEM analysis is the interaction between the person requesting the analysis and the person who performs the analysis. Both the requestor and the SEM analyst must share the need to solve the problem. Without proper background information and cooperative planning, the result is often just a pile of SEM micrographs. Collaborative efforts are also required during the analysis, and when interpreting the results.

THE PRINCIPLES OF SCANNING ELECTRON MICROSCOPY

History

Development of the SEM started in the 1930s. The transmission electron microscope (TEM) was developed before the SEM, and has about ten times better resolution. But the faster sample preparation and ease of image interpretation of the SEM has made it a more widely used tool in electronic packaging technology.

Optical examination of samples can often provide useful information. Although light microscopy is easily used, it is limited in application. Compared to visible light, high-speed electrons have a very short wavelength, which gives the SEM better resolution. Furthermore, the perception of depth that SEM images provide is similar to what we are used to in normal vision. This is due to the shadowing effect created with the usual detector. In both techniques, when the magnification is increased, the depth of focus is proportionally reduced. However, depth of focus in the SEM is about 500 times greater than that of light microscopy.

The Basic Instrument

Generating an image by using electrons is far from a trivial task. Early investigators began with the TEM because it is a natural extension of light microscopy. Preparation of the thin samples needed for transmission of electrons (≈ 1000 Å) is, however, very tedious, artistic, and time-consuming. These difficulties led to interest in imaging bulk specimens. The scanning approach, as opposed to transmission, provided the required capabilities.

Figure 22-2 shows a simple SEM apparatus and the associated peripherals. The electron gun produces a stream of electrons and accelerates them down the column. Magnetic lenses focus this stream into a small beam, commonly referred

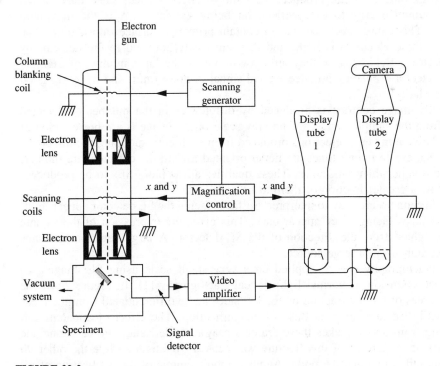

FIGURE 22-2
Schematic SEM.

to as the probe, which is rastered across the sample. Several different detectors are positioned near the sample to sense the electrons emitted from the interactions of the incident beam with the sample. The signal from a detector is displayed on a cathode ray tube (CRT). The electron beam rastering on the sample is coordinated with the display raster.

Detectors and Signals

A variety of particles and radiation are produced when a beam of electrons impinges on a solid. Electrons and x-rays are the most useful. The variation of either one of these, as a function of position, provides the needed image contrast. Detection has been the subject of much of the development in the SEM field.

Efficiency of collection is very important. A fine beam is required for useful resolution, but it produces very low fluxes of electrons and x-rays. If a given signal must be greatly amplified to produce an image, a large amount of electronic noise also results. High detector efficiency is therefore required to suppress noise.

SECONDARY ELECTRONS. Emitted low-energy electrons (< 50 eV) are referred to as secondary electrons (SE). They are produced by interactions of beam electrons with the electrons in the target material, on their way in and on their way back out. They are produced near the material surface, since they do not have enough energy to escape from far below (about 30 Å is the maximum depth). This causes an SE image to contain primarily surface information. The flux of these electrons is high, and they are easily attracted to the detector by an electrical field because they are slow moving. The large number of SEs and high detector efficiency produce a good signal-to-noise ratio, and therefore good image clarity.

SE image contrast is produced by differences in the number of detected SEs from the sample surfaces. The gray scale of the image is a representation of the relative number of electrons produced from each corresponding picture point. Parts that create lighter tones are better oriented toward the detector, are thinner, or are rougher than dark areas. These qualities allow more SEs to be produced, or allow a greater fraction of them to be collected.

SE images are easy to interpret. Surfaces facing the SE detector appear light, and surfaces facing away appear dark. This gives the impression of the sample being lighted from the direction of the SE detector. A large detector simulates diffuse light from a large source.

Fractography can be applied on a very small scale using SE imaging as a direct extension of normal light optical fractography [1]. In Figure 22-3, the small areas of fatigue fracture on the left side can be differentiated from the areas of ductile fracture, even at this low magnification. The effect of apparent soft lighting from above makes these features easy to understand. We can conclude that about 10 percent of this fracture was caused by fatigue, while the other 90 percent failed in a ductile mode. Another good example of the application of SE

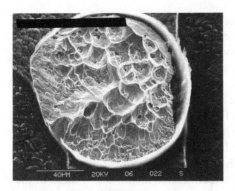

SE image
35° tilt

FIGURE 22-3
Fractography of a solder joint.

imaging is, as mentioned previously, in the evaluation of photoresist (see Figure 22-1).

The view in Figure 22-4 is taken at the outside of a right angle bend in a circuit line produced by a subtractive process, viewed at about 87 degrees of tilt. The SE image provides a clear definition of the circuit line profile after etching, together with the overhanging Cr layer. The high tilt angle adds to the impact of such an image, but the highlights and shadows also provide understanding of the circuit features.

BACKSCATTERED ELECTRONS. Backscattered electrons (BSE) have much higher energies than SE. The yield of BSEs is directly related to the average atomic number of the sample. Their high energy (i.e., high velocity) makes them very directional, so low-angle detectors provide high topographic contrast, like low-angle spot lights. By using an array of high-angle detectors, topographic contrast can be minimized, and sensitivity to average atomic number can be maximized.

SE image
87° tilt

FIGURE 22-4
Subtractively produced circuit line.

Higher average atomic number materials have larger nuclei, giving rise to a higher flux of reflected electrons. This creates atomic number contrast, which is very helpful when imaging a sample containing unknown materials, although topographic contrast is also present and must be dealt with. BSE imaging does not provide a complete elemental analysis, but when combined with a knowledge of the sample, or x-ray analysis, it can be a powerful tool.

The fractured circuit board shown in Figure 22-5 illustrates how SE imaging can be greatly supplemented by BSE. The bromine fire retardant in the epoxy (see Chapter 11) raises the average atomic number of the resin so that it is higher than that of the organic contaminant. Notice the even higher average atomic number of the glass fibers compared to the resin, which makes them more visible in the BSE image also. SE only shows the fine surface structure, which in this case is only confusing.

Metallographically polished samples are short on topographic detail, so they are prime candidates for the BSE mode. Cracks are shown more clearly, and with more contrast, using the BSE mode (Figure 22-6). The edges of cracks produce a high flux of SEs. While the crack is naturally dark (i.e., produces no SEs), the bright edges blend into it, creating a masking effect. This is not a problem with BSE imaging. The location of the copper (Cu) is also made much more obvious by BSE, due to atomic number contrast. The void into which the Cu has been plated is shown by the arrow. In a reliability test, this plated void would have a high probability of causing a short circuit. If only SE imaging were used, such a potential risk site could easily be missed.

Subsurface high-atomic-number materials embedded in low-atomic-number materials can normally be seen best by BSE imaging. SE images are more diffuse in such a case. Figure 22-7 shows an extraneous piece of Cu below the surface of a circuit board. A comparison of a light image to the SE and BSE images is given. More detail is shown in the BSE image. Because the BSEs produce many

SE image
0° tilt

BSE image
0° tilt

FIGURE 22-5
SE versus BSE image of a PC board fracture section.

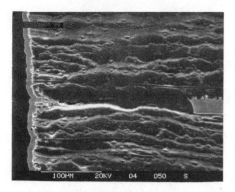
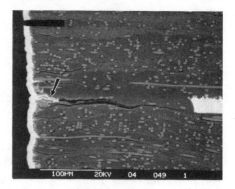

SE image
35°tilt

BSE image
35°tilt

FIGURE 22-6
Cracks and atomic number contrast by BSE.

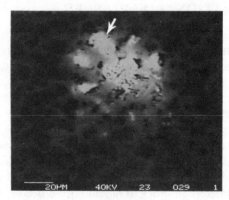

Light image

BSE image
35°tilt

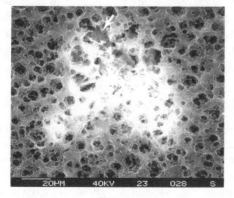

SE image
35°tilt

FIGURE 22-7
Light versus SE and BSE—subsurface Cu.

SEs on their way out of the rough epoxy, a broadening effect is produced in the SE image of the Cu. Higher beam energy is needed to find deeper metals.

X-RAYS. X-rays, the energies of which are characteristic of the elements in the sample, are also emitted in the SEM. Computer assisted x-ray spectroscopy has been developed to produce a list of the elements present. By using x-ray analysis, micron-sized materials in the sample can be identified.

Two types of x-ray analysis systems are common SEM accessories: energy-dispersive spectroscopy (EDS), and wavelength-dispersive spectroscopy (WDS). The difference between the two is in the detectors used. While WDS provides much higher sensitivity, there are many operational advantages to EDS. Only EDS will be discussed in this chapter, but the main principles of EDS also apply to WDS.

SEM OPERATIONAL VARIABLES

The principles of beam formation and detectors are not complex. The aspect that makes the art and science of SEM so challenging is that not all signals (i.e., information sources) can be adequately detected with the same selection of variables. This is especially important if the examination is attempting to determine the presence of a condition or material near the limit of detection.

Fine surface detail is an example. Just because fine texture cannot be seen in a micrograph doesn't mean it is not there. If the beam is too large, details will be missed. But when a fine beam is used, there may not be enough BSE intensity to produce adequate image contrast, and x-ray production may be too weak for spectral analysis. Samples must often be investigated using a number of different machine parameters in order to obtain all the needed information.

Many tradeoffs must be considered when an investigation is undertaken. Since changes in machine setup must often be followed by extensive readjustments in operational variables, it is advantageous to be able to estimate the proper setting, without having to rely upon a trial-and-error approach. Beam current, final aperture size, accelerating voltage, and chamber geometry are the primary variables in SEM operation. There are a host of sample preparation variables also; they will be addressed later.

Beam Current (Spot Size)

The SE resolution of an SEM can be no better than about the diameter of the beam (or probe). The amount of current in the beam determines its minimum size and directly affects the amount of signal produced. A strong signal requires a large beam, and hence reduced resolution.

A fine beam creates problems, as well as potentially high resolution. The beam current is usually low (normally in the micro- to nanoamp range), and the signal is only a fraction of that amount. Detectors are inherently inefficient, due to the difficulty of collecting the image-producing electrons; a considerable

Small beam   Large beam

15-second
photo

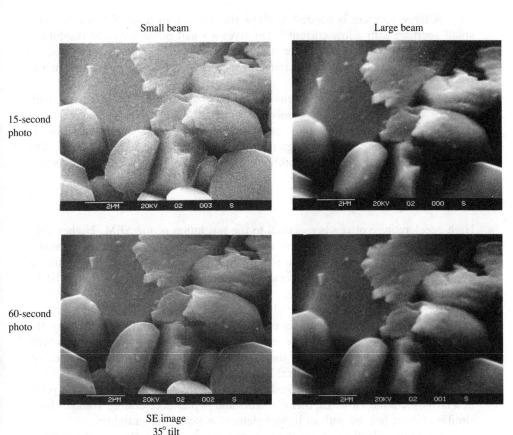

60-second
photo

SE image
35° tilt

FIGURE 22-8
Beam size effects on noise and resolution.

amount of signal amplification is necessary. In the process of raising the signal to a usable level, noise is amplified also. This noise produces random darkening and lightening of image points, thus creating visual confusion, commonly referred to as "snow."

A photograph can be taken at a slow rate, in order to collect more signal, and average the noise out of the image. However, it is the difficulty of adjusting the instrument and focusing the beam under these conditions that makes noise such a problem. Figure 22-8 shows the effect of spot size and photo time. The smaller beam shows more detail, but the noise can mask it. Using a long photo record time with a large spot cannot improve resolution.

Final Aperture Size

The final aperture limits the probe diameter by physically masking out stray electrons. Since the beam formation process creates a near-gaussian distribution of electron density versus diameter, this approach is very effective.

A large aperture is needed to allow the passage of a large beam. Using a small aperture with a low current beam gives a small probe for good resolution. Aperture size also has a significant effect on depth of field. As with light optics, a larger aperture results in a larger range of incident electron angles, which makes focus more critical.

Most instruments have a choice of three apertures. The largest is suitable for x-ray spectroscopy and noise-free imaging up to several thousand times magnification. Many applications can use this choice. The smallest aperture is needed for only the highest magnification, and cannot be used for EDS. The middle size is for in-between conditions; some restriction in EDS work exists.

Accelerating Voltage

High accelerating voltage is essential to the resolution of the SEM. Higher voltage produces higher beam current density, allowing more signal in a smaller probe. There are reasons why the maximum voltage is not the setting to use in every analysis. The effect of interaction volume and sample charging must be considered.

Interaction volume is the volume of material from which the signal is produced. The incoming beam is elastically and inelastically scattered by the electrons and nuclei of the sample material. The depth of penetration varies with the beam voltage and material of the sample. Figure 22-9 shows the penetration depth that can be expected at several beam voltages, in several common materials. The Monte Carlo simulation method is useful in showing typical electron paths in a cross-sectional view. Lighter elements and higher accelerating voltages have similar effects, but the path in lighter-element materials is straighter.

Secondary electrons come from approximately the top 50 Å of the sample. They are produced by the incoming as well as the exiting electrons. Most BSEs are produced within the top third of the interaction volume. Deeper beam penetration broadens the region where SEs are produced, reducing resolution and affecting the image by subsurface features. When looking for surface detail on low-density materials, it is often better to use somewhat lower voltage to keep the beam penetration and broadening to a minimum.

Figure 22-10 shows an example of how a different beam voltage can provide a different amount of surface detail. This circuit pad contained a Cr layer that needed to be removed. Normal etch solutions were ineffective at removing all of it. By noting the difference in visible details between 20 and 5 kV, the presence of a film is shown in the central area of this pad; this film kept the etching solution from reaching the metal surface. The film was later proven to be residual photo resist.

Low Beam Voltage

Low beam voltage (below 2 kV) is important because it reduces sample charging, which means avoiding conductive coating of nonconductive samples. Normal

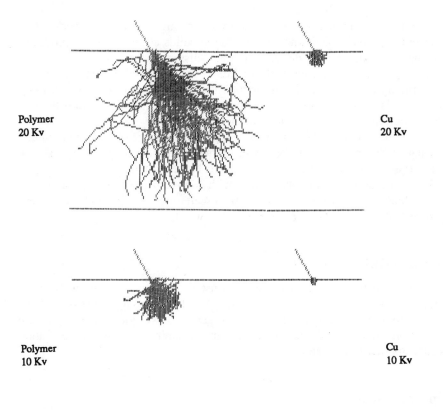

Polymer
20 Kv

Cu
20 Kv

Polymer
10 Kv

Cu
10 Kv

FIGURE 22-9
Monte Carlo models of beam penetration.

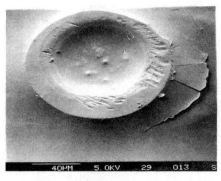 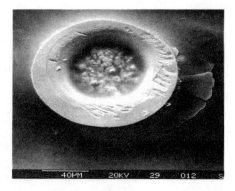

SE image
35°tilt

SE image
35°tilt

FIGURE 22-10
Effect of kV on depth of image information.

SEM sample preparation employs a conductive coating to allow the excess electrons from the probe to be drained off. If charge builds up, the voltages produced on the specimen surface will deflect the beam, distorting the image.

At low beam voltage, the yield of SEs is considerably less than at the kV levels normally used in an SEM. At about 1 kV, the number of electrons leaving the specimen is about the same as the number arriving [2,3]. This means there is no charge buildup, and no need to coat the sample.

Low-voltage work is of value in at least three situations: (1) when continued processing of the material is to be done, (2) when further electrical testing may be needed, and (3) when surface analysis is to be performed.

There are shortcomings associated with low-voltage work. Resolution is greatly reduced. Contrast is very low. Backscattering is very slight also, and solid state BSE detectors will not respond below 5 to 10 kV.

SAMPLE PREPARATION

The condition of the sample, and the steps used to prepare it, can easily be as important as the choice of machine settings. Light microscopy must be applied first. Metallographic sectioning and microsurgery are two good ways to prepare samples. Sometimes, the as-received condition is examined. A conductive coating is usually applied to insulators.

Sample Selection

The selection of a representative sample is extremely important in SEM work because of the amount of time involved in preparing them, and the possibility of generating misleading results. Light microscopy and a knowledge of the manufacturing processes are useful here.

Light Microscopy

Light microscopy provides information that may not be available by SEM. It also yields information on what preparation steps must be taken. Qualities of color and transparency must be investigated, marked, and photographed before SEM examination.

Figure 22-11 shows an example of a condition related to poor solderability. Areas of different colors were found by light microscopy, but when imaged by SE, only a thick overcoating material could be seen on the left side of the pad. Other colored areas were examined by EDS; the area that was copper-colored was found to be normal, but the other colored areas were found to be created by a layer of residual processing material, which degraded solderability. Without the light microscopy, this SEM analysis would have been ineffective, due to the lack of color and transparency variation in the SEM. Only the thick overcoating

Dark field light image
0° tilt

SE image
35° tilt

Color and transparency.

would have been found, addressing only one of the causes of poor solderability.

Metallographic Sections

Metallographic sectioning is often needed for circuit boards or other multilayered electronic components. This technique was developed for use with light microscopy to investigate material microstructures, and has been extended for use on manufactured parts. Subsurface areas can be examined with minimal deformation of the structure. A metallographically prepared sample also provides the flatness needed for accurate quantitative x-ray elemental analysis.

The resolution of metallography is inherently limited. Even the most skilled polisher produces slightly smeared and deformed material, in the tenths of micrometer range. This size is about the best that light microscopy can resolve, but an SEM can resolve much finer details.

The flat surface of a metallographic section is difficult to image in SE mode, since roughness and orientation differences are the conditions that produce SE image contrast. Only polishing relief, cracks, and the BSE component of the SE image are available for imaging. BSE imaging provides lower resolution, but metallographic specimens are weak in available resolution anyway. This makes a reasonable match. Only a two-dimensional view is possible; the great depth of field of the SEM is not utilized here.

Transparency makes subsurface features visible with light optics, but for the SEM, true transparency exists only in the tenths of micrometers, depending on the beam voltage used. This means that some conditions visible in light optics cannot be found at all in the SEM.

A technique called *microsurgery* is often a better way to reach the area of interest for SEM analysis.

Microsurgery

The practice of microsurgery involves cutting, tearing, peeling, or otherwise separating the parts of a sample. We call it microsurgery because it often utilizes microscopes and scalpels.

The precursor to microsurgery was fracture sectioning, which is in itself of value in packaging materials analysis. Figure 22-12 illustrates the complementary information a metallographic section (left) and a fracture section (right) can provide. This sample is composed of a polymer film with copper (Cu) on both sides. The metallographic section shows a good representation of the cross section, so that dimensional measurements can be taken. Grain structure of the Cu is also shown, since an appropriate etchant was used.

This fracture section shows the ductility of the Cu foils by the extent of their deformation. Both Cu foils show nearly 100 percent reduction in area, which is very good. The columnar type of grain structure found metallographically in the thinner Cu was of concern, since some cases of columnar structure are brittle and weak. Noting the degree of deformation, which is a result of good ductility, there should be no concern here. A glassy fracture of the dielectric between the Cu layers can also be seen. This is an indication of complete polymerization.

The size and composite nature of electronic packages often necessitate a more delicate and selective approach than typical fracture sectioning. Careful peeling is often used when a strong enough top layer is involved. Cutting can also be used to reach the region of interest, or to guide the path of fracture. Cleaving and low-temperature brittle fracture are also useful. The key is to know

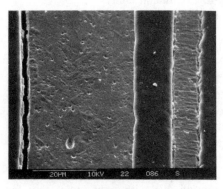

SE image
35° tilt

SE image
35° tilt

FIGURE 22-12
Metallographic and fracture sections.

what was done, so that it can be differentiated from the actual condition under investigation. Careful observation is important. Normally, cutting all the way to the area of interest is unacceptable. Natural fracturing on a plane of weakness is where microsurgery really pays off.

The concept of cohesive and adhesive failure (discussed in Chapter 27) is basic to understanding where the weakness is. When the strength of bonds between polymers, metals, and ceramics is questioned, it is of prime importance to know which of these failure modes occurred. BSE imaging and x-ray spectroscopy after separation will usually be adequate to determine if one material remains on the other or has diffused into the other. Diffusion of polymers and metals at interfaces is an important part of composite technology. The fundamentals of these processes are discussed in Chapters 25 and 26.

When the failure is cohesive, the bond was stronger than the materials involved, and the bulk properties of the fractured material itself have to be improved to create a stronger joint. If an adhesive failure is found (separation at an interface), the bond is weaker than the materials.

Figure 22-13 shows one case where areas of cohesive fracture and areas of adhesive fracture were found on the same part. It was suspected that this copper pad in the interior of the circuit board had broken loose from the epoxy during processing. Whether the fracture was caused by excessive stress or an abnormally weak bond had to be determined. The piece of defective circuit board was metallographically sectioned from the top surface of the board down to the top of this loose copper pad. Microsurgery was used to cut around the remaining epoxy, and the pad was then gently pried out, revealing the failed interface. The SE image shows the cavity on the right, and the land on the left. BSE shows significant areas of cohesive fracture, where epoxy remains on the Cu, as well as areas of adhesive fracture, where the Cu is clean. "Good" lands showed only cohesive fracture.

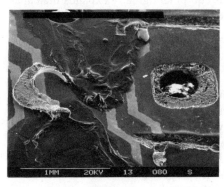
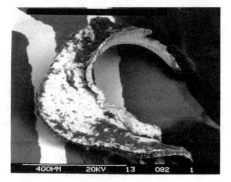

SE image
35° tilt

BSE image
35° tilt

FIGURE 22-13
Regions of cohesive and adhesive fracture.

The SEM may fail to differentiate cohesive from adhesive failures if the remaining amount of material is extremely thin. The thickness where this begins to be a problem is dependent upon the interaction volume. Lower beam voltages will reduce this limit. If the layer is thin enough, surface analysis techniques must be employed.

Coating of Samples

Samples must usually be covered with a thin, conformal, conductive coating to avoid electrical charging. Thermal damage from the electron beam can also be reduced, and resolution in very light materials can be improved. The selection of the correct coating material is important. Interfering x-ray lines and coating artifacts must be avoided.

A variety of coating materials and techniques are available. Some of the most common materials used are carbon (C), gold (Au), gold-palladium (Au-Pd), platinum (Pt), and aluminum (Al). The metals are usually sputter deposited because it is faster and provides a more continuous coating on rough samples than other methods. Carbon is most often thermally evaporated because it does not sputter efficiently. The major reason for using C is the freedom from x-ray line interferences, but C does not dissipate heat very well. Al is a fine heat conductor. Au-Pd is the finest grained, but Pt combines a favorable balance of all these qualities. A thorough discussion of coating techniques can be found in references [2] and [3].

ENERGY-DISPERSIVE X-RAY SPECTROSCOPY

The most common attachment for an SEM is an energy-dispersive x-ray spectroscopy (EDS) system. The addition of this equipment greatly expands the capability of an SEM. Valuable elemental information can be obtained from the x-rays created by the electron beam.

EDS Theory

The interaction of an electron beam with any material produces two types of x-rays: characteristic and continuum. Characteristic x-rays are produced as the result of the collisions of beam electrons with inner shell electrons. In addition, electrons coming into close proximity with target atoms produce a range of x-ray energy, from zero to the energy of the incident beam, called continuum or braking radiation. This forms the background in the EDS spectrum. The characteristic x-rays form peaks, superimposed upon the background.

Emitted x-rays are measured with a detector that differentiates between the energies of the x-rays, hence the name *energy dispersive*. Detailed descriptions of the EDS detector are given in references 2 and 3. The normal detector is a lithium-doped silicon chip. X-rays induce a current in the detector, proportional to their energy. The current is amplified and a pulse is formed, which is fed to a multichannel analyzer. The analyzer accumulates the number of pulses in each energy channel and displays them on a CRT to produce the EDS spectrum. The presence of a given element results in a series of peaks at specific energies in the spectrum. EDS can detect elements as light as carbon. The size of a peak is directly proportional to the amount of that element in the sample. Using standards makes it possible to quantify the composition of the sample.

Spectral interpretation is far from trivial. Overlap of peaks from different elements requires the use of at least two peaks per element, whenever possible. There are several artifacts that can create peaks where they should not be. Detection electronics sometimes cannot differentiate two x-rays that arrive at very near the same time. In this case only one x-ray is counted, and the energies of the two are added together. This is a sum peak. Another commonly misinterpreted peak is created when a silicon x-ray escapes from the detector. This is an escape peak. Enough counts must be obtained to smooth out random variations also.

Not only is qualitative elemental information available, but it is also possible to determine the spatial distribution of elements. Careful selection of operating parameters can provide both spatial and depth information on the various elements detected in a sample.

X-ray production and detection is a statistical process; therefore, to obtain adequate information it is necessary to achieve high count rates and to count for a significant length of time. Spectral acquisition times of about five minutes are normal. Large beams provide adequate x-ray generation, at the expense of image resolution. An accelerating voltage of at least twice the energy of the x-ray of interest needs to be used. All the interaction volume, spot size, and aperture considerations mentioned earlier apply here. Figure 22-14 shows how beam voltage affects relative peak heights, because of interaction volume effects. A pair of EDS spectra were recorded from the marked area to show the amount of chromium on this circuit line. With a high voltage (40 kV), a small chrome (Cr) peak is noted. At 10 kV, the Cr content is strongly detected. Both spectra were smoothed, and the 10 kV spectrum was normalized with respect to the background level at the 40 kV Cr peak. The greater peak-to-background ratio with 10 kV makes quantitative determination more accurate at this voltage.

With the development of computers and interfaces, EDS systems have become increasingly sophisticated. They are now capable of rapidly performing complicated calculations and involved digital image processing. These systems can control the electron beam so that it is possible to do digital elemental mapping and line scans as well as record digital video images. These routines have made it possible to determine spatial distribution of elements and to do particle sizing with elemental quantification. Data from these operations can be manipulated, stored and graphically presented.

EDS Use on Contaminants

Packaging processes are often sensitive to contaminants. Polarized light microscopy (PLM) is very useful for identifying particulates, particularly minerals and

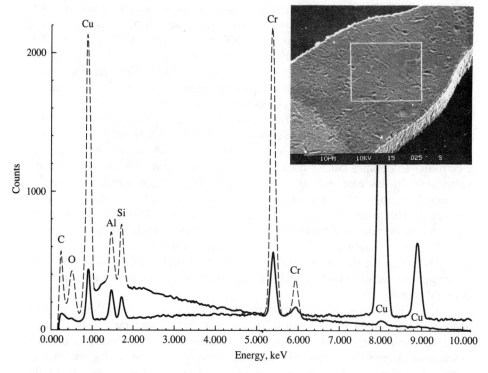

FIGURE 22-14
Effect of beam kV on Cr peak height.

fibers [4]. X-ray diffraction and infrared spectroscopy are powerful tools for identifying corrosion products and organics, respectively. SEM with EDS can provide shape, size, and compositional information useful for contaminant identification.

Figure 22-15 is an example of dendritic growth between circuit lines, which is known to be produced by the presence of ionic contaminants, a migratable metal, moisture, and an applied voltage. SE imaging was no help in finding the cause of this short circuit. BSE easily showed the dendrites that had grown between the two leads. EDS was used to identify the dendrite material (at arrow) as lead (Pb) with some indium (In). These elements are present in the solder joints, at the left and right of the photo. Conformal coating of the space between these solder joints eliminated this failure mechanism.

Figures 22-16, 22-17, and 22-18 show the combined use of SEM and EDS, with two different types of sample preparation. Drill smear is shown in Figure 22-16 at the arrow, as revealed by metallographic sectioning. This is the traditional approach to determining if drill smear has been adequately removed. Drill smear is the circuit board material redeposited in the drilled hole due to the heat and scrubbing in the through-hole drilling operation. EDS (Figure 22-17) identified the material between the copper layers, at the circles, to be organic with bromine,

BSE image
35°tilt

FIGURE 22-15
Dendritic Pb growth.

which is known to be an epoxy fire retardant. In addition, there was some silicon and calcium found, which was difficult to understand. Figure 22-18 shows drilled holes that were cut in half (dry sectioned) after smear removal and examined in the SEM. A relatively clean hole (left) compares to a very poorly cleaned hole (right). BSE shows the epoxy and glass fibers in sharp contrast to the copper metal (bright band across the center of both photos). EDS was used to confirm the elemental composition of the brominated epoxy and the glass fibers. The shapes of the glass fragments are easily compared to those in the glass fiber bundles, above and below the copper.

Fingerprint residue has been an especially troublesome contaminant in some processes. Figure 22-19 is an example of a baked-on skin flake. The dark material was found on both the substrate, between the solder pads (left), and on the solder ball (right). The solder joint was poor because of the physical barrier created by this material. Figure 22-20 is the EDS spectrum taken from the area at the

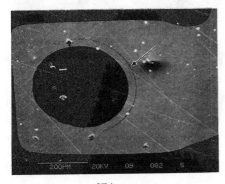

SE image
35° tilt

SE image
35° tilt

FIGURE 22-16
Drill smear, by metallographic section.

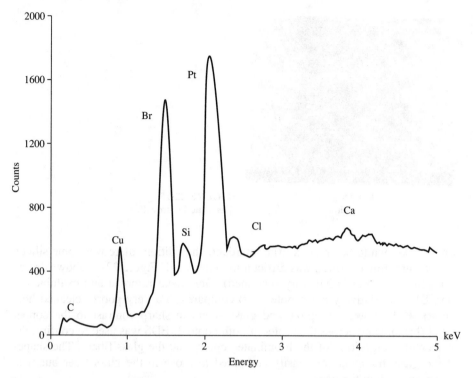

FIGURE 22-17
Drill smear EDS spectra.

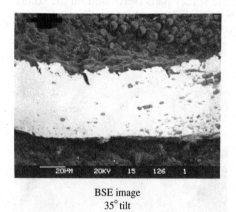

BSE image
35° tilt

BSE image
35° tilt

FIGURE 22-18
Drill smear, by dry section.

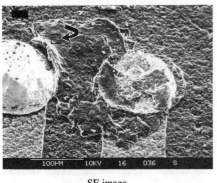 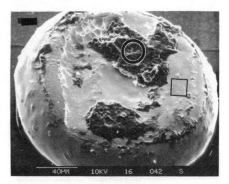

SE image
35° tilt

SE image
35° tilt

FIGURE 22-19
Skin flake residue.

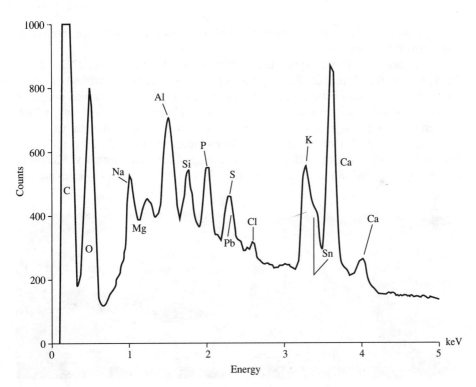

FIGURE 22-20
Skin flake residue EDS spectra.

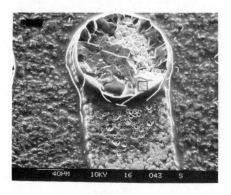
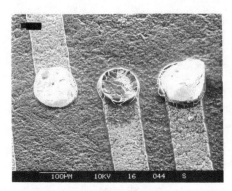

SE image
35° tilt

SE image
35° tilt

FIGURE 22-21
Rosin flux residue.

arrow, which identified it as a probable skin flake. Comparisons with references confirmed this conclusion.

Residual rosin flux is a potential problem in some areas. Figure 22-21 shows the typical brittle fracture of flux residue. A ring of flux is shown around this pad. Light microscopy is of little value here, because the material is clear and colorless in thin films and small amounts. EDS shows some tin content in used flux, which is often used as a "fingerprint," along with the brittle fracture characteristics.

The charred fiber on the array of solder spheres in Figure 22-22 could interfere with the soldering operation at the next higher level of assembly. Titanium content was an indication of coloring dye. This fiber must have been present

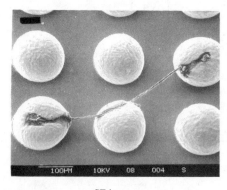
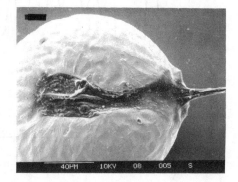

SE image
35° tilt

SE image
35° tilt

FIGURE 22-22
Charred fiber on solder balls.

before the solder balls were reflowed, since the degraded fiber affected the shape of the solder balls.

SUMMARY

The SEM with EDS is a valuable tool in the analysis of electronic packaging components. Processing variations, physical and electrical test failures, and yield detractors can be understood by proper selection of the sample, its preparation, and the SEM operational variables.

No single set of SEM operational conditions will provide the needed information to solve all problems. While many decisions must be made in the use of the equipment, a similar number of decisions must be made in the area of sample preparation.

REFERENCES

1. American Society for Metals, *Fractography and Atlas of Fractographs*, Metals Handbook, vol. 9, 8th ed., 1874.
2. Goldstein, Newbury, Echlin, Joy, Fiori, and Lifshin, *Scanning Electron Microscopy and X-ray Microanalysis*, New York: Plenum Press, 1981.
3. Oliver C. Wells, *Scanning Electron Microscopy*, New York: McGraw-Hill, 1974.
4. Walter McCrone, *The Particle Atlas*, 2d ed., vols. 1–6, Ann Arbor Science Publishers, 1987.

PROBLEMS

22.1. What sample preparation and SEM setup should be used to determine the cause of fracture for the following wire? About an inch from a connector termination, a round, solid conductor Cu wire is often found to be electrically open, at the end of the manufacturing process line. There are 15 of these fluorocarbon-polymer-insulated wires on each connector. What approach to preparing the fracture, and what instrument parameters will be best for a first-pass look? You are given the connector with the cable attached. Consider the following possible contributors: flexing fatigue, corrosion, defects in the Cu metal, nicks or other damage created in assembly, excess tension, etc.

22.2. Would 10 or 30 kV be best for a first look at a metallic part with a suspected thin film of surface contaminant causing poor adhesion? Why?

22.3. What can be said about the fracture of a polymer-to-conductor bond (no separate adhesive used) when SEM and EDS analysis shows that some of the polymer remained on both the surfaces? What could be done to further strengthen the joint?

22.4. If an SE image shows fracture of a normally ductile material to be without significant deformation, what else can be done to add information to help the materials engineer?

22.5. What kV is best for looking at heavy-element materials?

22.6. What qualities of SEM images are different from those of light optics images?

22.7. What specific information could be missed if the following setups were used?
(*a*) Coarse beam on a fatigue fracture.
(*b*) Fine beam on a dielectric-to-metal bond fracture interface.

      (*c*) Low kV on a metal with the potential for thin (several tenths of microns thick) organic films.

      (*d*) High kV beam on a polymer specimen.

      (*e*) Metallographic cross section of a fracture.

22.8. Which detector provides the most information when examining a fatigue fracture, and why: SE, BSE, or x-ray?

22.9. Which aperture is best for rapid survey viewing at a tilt of 35 degrees—small, medium, or large? Why?

22.10. Which aperture is best for photographing a rough fracture surface at 75 degrees—small, medium, or large? Why?

22.11. If a metallic surface known to produce a poor bond looks the same as one known to produce a good bond, which action is most likely to help differentiate them? Assume 20 kV, medium aperture, medium spot size, and SE imaging were used on that first look.

CHAPTER 23

ION BEAM ANALYSIS

JAMES W. MAYER

INTRODUCTION

The integrated circuit is based on processing control of the outer micron-implanted dopant distributions, diffusion profiles, oxide thickness, silicide formation, intermetallic diffusion barriers, and interconnects. As microelectronic packaging progresses to thinner deposits and finer structures, many of the techniques developed for analysis become applicable to packaging. In many cases, for example, processing control is based on the ability to measure the depth profile of the elements within this outer micron.

Previous chapters in this book, and the following chapter, have been based on the use of electron spectroscopies in conjunction with sputter sectioning to establish depth profiles (Figure 23-1). Chapter 26, "Case II Diffusion," provides a clear illustration of the use of ion beam analysis to understand processes at interfaces. This chapter concentrates on atomic scattering—both Rutherford backscattering (RBS) and forward recoil spectrometry (FRES)—in which the depth scale is determined by the energy loss of the analyzing particles, typically megaelectron-volt (MeV) helium ions or protons. These techniques are nondestructive in that sample erosion is not required and they are quantitative, as the scattering and stopping cross sections have been established over decades of research in atomic and nuclear physics. A brief description is given of secondary ion mass spectroscopy (SIMS), which is the most sensitive of the analysis techniques, but one that relies on sputter sectioning for depth profiles. The chapter

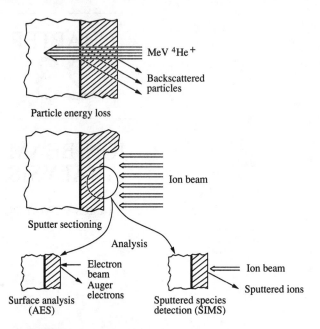

FIGURE 23-1
Schematic diagram of two approaches to obtain depth profiles in thin films. With particle energy loss techniques, the thickness of the layer is determined from the energy loss of the energetic particles. With sputter sectioning techniques, the amount of material probed is determined by the sputtering yield. The surface composition can be directly analyzed either by surface analysis or sputtered species detection.

concludes with a discussion of the application of ion beam analysis, RBS and FRES, to interdiffusion in polymers.

Rutherford backscattering analysis of complex thin-film systems (see Chapters 10 and 13)—for example the Si/PtSi/TiW/Al structure used on silicon integrated circuits and shown in Figure 23-2—is straightforward and permits an evaluation of film thickness and composition in 10 to 15 minutes. In the remaining sections of this chapter, we will outline the physical concepts that are used to evaluate such data. More detailed discussions are given in *Backscattering Spectrometry* (Chu et al., 1978), *Fundamentals of Surface and Thin Film Analyses* (Feldman and Mayer, 1986), *Materials Analysis by Ion Channeling* (Feldman et al., 1982), and *Handbook for Ion Beam Analysis* (Mayer and Rimini, 1978).

Aside from the accelerator that provides a collimated beam of MeV particles (usually ^{4}He ions), the instrumentation for RBS and FRES is simple (Figure 23-3). Semiconductor nuclear particle detectors are used, which have an output voltage pulse proportional to the energy of the particles scattered from the sample into the detector. The technique is also quantitative as MeV He ions undergo close-impact scattering collisions governed by the well-known Coulomb repulsion between the positively charged nuclei of the projectile and the target atom. The kinematics of the collision and the scattering cross section are independent of chemical bonding,

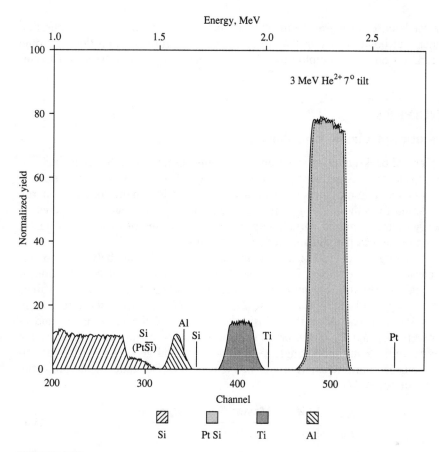

FIGURE 23-2
Rutherford backscattering spectrum of a multilayer sample used in evaluation of metallization schemes on silicon devices. The sample has been annealed so that Pt has reacted with Si to form PtSi.

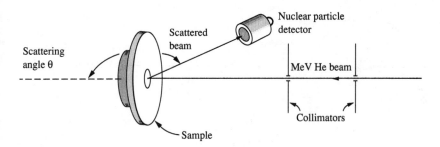

FIGURE 23-3
Schematic of the experimental setup for Rutherford backscattering with collimated beam of He ions incident on a sample. Particles scattered to an angle θ are detected by a nuclear particle detector.

and hence backscattering measurements are insensitive to electronic configuration and chemical bonding within the target. To obtain information on the electronic configuration one must employ the electron spectroscopies discussed in Chapter 22.

PRINCIPLES

Kinematics of Elastic Collisions

In Rutherford backscattering spectrometry, monoenergetic particles in the incident beam collide with target atoms and are scattered backwards into the detector-analysis system that measures the energies of the particles. In the collision, energy is transferred from the moving particle to the stationary target atom; the reduction in energy of the scattered particle depends on the masses of incident and target atoms and provides the signature of the target atoms.

The energy transfers, or kinematics, in elastic collisions between two isolated particles can be solved fully by applying the principles of conservation of energy and momentum. For an incident energetic particle of mass M_1, the values of the velocity and energy are v and $E_0(E_0 = \frac{1}{2}M_1 v^2)$, while the target atom of mass M_2 is at rest. After the collision, the values of the velocities v_1 and v_2 and energies E_1 and E_2 of the projectile and target atoms are determined by the scattering angle θ and recoil angle ϕ. The notation and geometry for the laboratory system of coordinates are given in Figure 23-4. The ratio of the projectile energies for $M_1 < M_2$ is

$$\frac{E_1}{E_0} = \left[\frac{(M_2^2 - M_1^2 \sin^2 \theta)^{1/2} + M_1 \cos \theta}{M_2 + M_1} \right]^2 \tag{23.1}$$

This energy ratio, called the kinematic factor K, determined only by the masses of the particle and target atom and the scattering angle. A subscript is usually added to K, e.g., K_{M2}, to indicate the target atom mass.

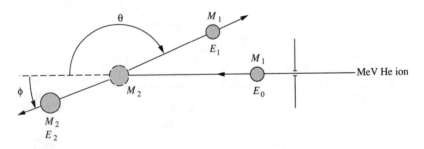

FIGURE 23-4
Schematic representation of an elastic collision between a projectile of mass M_1, velocity v, and energy E_0 and a target mass M_2 that is initially at rest. After the collision, the projectile and the target mass have velocities and energies v_1, E_1 and v_2, E_2, respectively. The angles θ and ϕ are positive as shown.

For direct backscattering through 180 degrees, the energy ratio has its lowest value, given by

$$\frac{E_1}{E_0} = \left(\frac{M_2 - M_1}{M_2 + M_1}\right)^2 \tag{23.2}$$

and at 90°,

$$\frac{E_1}{E_0} = \frac{M_2 - M_1}{M_2 + M_1} \tag{23.3}$$

For $\theta = 180$ degrees, the energy E_2 transferred to the target atom has its maximum value, given by

$$\frac{E_2}{E_0} = \frac{4M_1M_2}{(M_1 + M_2)^2} \tag{23.4}$$

with the general relation given by

$$\frac{E_2}{E_0} = \frac{4M_1M_2}{(M_1 + M_2)^2} \cos^2 \phi \tag{23.5}$$

The ability to distinguish between two types of target atoms that differ in their masses by a small amount ΔM_2 is determined by the ability of the experimental energy measurement system to resolve small differences ΔE_1 in the energies of backscattered particles. Energy resolution values of 10–20 keV, full width at half maximum (FWHM) for MeV ^{4}He ions, can be obtained with conventional electronic systems. For example, backscattering analysis with 2.0-MeV ^{4}He particles can resolve isotopes up to about mass 40 (the chlorine isotopes, for example). Around target masses close to 200, the mass resolution is about 20, which means that one cannot distinguish among target atoms ^{181}Ta and ^{201}Hg.

Figure 23-5 shows a backscattering spectrum from a sample with approximately one monolayer of Cu, Ag, and Au. The various elements are well separated in the spectrum and easily identified. Absolute coverages can be determined from a knowledge of the absolute cross section discussed in the following section.

In forward recoil spectrometry, used primarily in hydrogen and deuterium analysis, the mass of the incident projectile is greater than that of the target atom. The incident energy is transferred primarily to the lighter target atom in a recoil collision (Equation 23.5). The energy of the recoils can be measured by placing the target at a glancing angle (typically 15 degrees) with respect to the beam direction and by moving the detector to a forward angle ($\phi = 30$ degrees) as shown in the inset of Figure 23-6. This scattering geometry allows detection of hydrogen and deuterium at concentration levels of 0.1 atomic percent and surface coverages of less than a monolayer.

The spectrum for ^{1}H and ^{2}H (deuteron) recoils from a thin polystyrene target are shown in Figure 23-6. The recoil energy from 3.0-MeV ^{4}He irradiation at a recoil angle ϕ of 30 degrees can be calculated from Equation 23.5 to be 1.44 MeV and 2.00 MeV for ^{1}H and ^{2}H, respectively. Since ^{2}H nuclei recoiling from

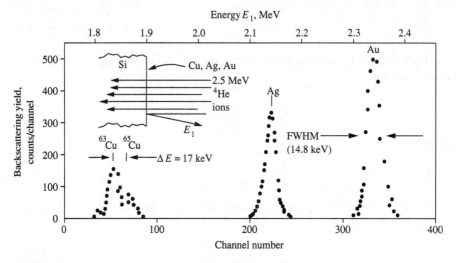

FIGURE 23-5
Backscattering spectrum (θ = 170 degrees) for 2.5 MeV He ions incident on a target with approximately one monolayer coverage of Cu, Ag, and Au. The spectrum is displayed as raw data from a multichannel analyzer, i.e., in counts per channel and channel number. The energy scale of the backscattered particles is shown at the top of the figure.

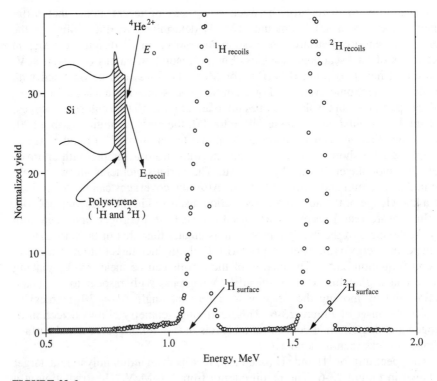

FIGURE 23-6
The forward recoil spectrum of ^1H and ^2H (deuterium) from 3.0 MeV ^4He ions incident on a thin ($\approx$ 200 Å) polystyrene film on silicon. The detector is placed so that the recoil angle ϕ = 30 degrees, and a 10-μm-thick mylar film is mounted in front of the detector.

the surface receive a higher fraction ($\approx 2/3$) of the incident energy E_0 than do ^{1}H nuclei ($\approx 1/2$), the peaks in the spectrum are well separated in energy. The energies of the detected recoils are shifted to lower values than the calculated position due to the energy loss in the mylar film placed in front of the detector to block out He ions scattered from the substrate.

Scattering Cross Section

As shown in Figure 23-5, the identity of target atoms is established by the energy of the backscattered particle after an elastic collision. The number N_s as of target atoms per square centimeter is related to the yield Y or the number Q_D of detected particles by

$$Y = Q_D = \sigma(\theta)\Omega Q N_s \tag{23.6}$$

where Ω is the detector solid angle, Q is the total number of incident particles in the beam, and $\sigma(\theta)$ is the scattering cross section. The value of Q is determined by the time integration of the current of charged particles incident on the target.

The scattering cross section can be calculated from the force that acts during the collision between the projectile and target atom. For most cases in backscattering spectrometry, the distance of closest approach during the collision is well within the electron orbits so that the force can be described as an unscreened Coulomb repulsion of two positively charged nuclei with charges given by the atomic numbers Z_1 and Z_2 of the projectile and target atoms.

The scattering cross section is

$$\sigma(\theta) = \left(\frac{Z_1 Z_2 e^2}{4E}\right)^2 \frac{4}{\sin^4 \theta} \frac{\left\{\left[1 - \left(\frac{M_1 \sin \theta}{M_2}\right)^2\right]^{1/2} + \cos \theta\right\}^2}{\left[1 - \left(\frac{M_1 \sin \theta}{M_2}\right)^2\right]^{1/2}} \tag{23.7}$$

which can be expanded for $M_1 \ll M_2$ in a power series to give

$$\sigma(\theta) = \left(\frac{Z_1 Z_2 e^2}{4E}\right)^2 \left[\sin^{-4}\frac{\theta}{2} - 2\left(\frac{M_1}{M_2}\right)^2 + \dots\right] \tag{23.8}$$

where the first term omitted is of the order of $(M_1/M_2)^4$. For He ($M_1 = 4$) incident on Si ($M_2 = 28$), $2(M_1/M_2)^2 \approx 4$ percent, so that even the second term is small. For a rule of thumb, one can use the leading term, which gives the scattering cross section originally derived by Rutherford

$$\sigma(\theta) = \left(\frac{Z_1 Z_2 e^2}{4E}\right)^2 \frac{1}{\sin^4(\theta/2)} \tag{23.9}$$

To give an example of orders of magnitude, the scattering cross section for 1-MeV ^{4}He ions ($Z_1 = 2$) incident on Si ($Z_2 = 14$) and scattered through 180 degrees has a value of about 10^{-24} cm^2 (a *barn*) using Equation 23.9 and the relationship $e^2 = 14.4$ eV $\cdot$ Å.

When evaluating the relative amounts of the constituents of components of an intermetallic compound, a quick evaluation can be made of the cross section ratio by using the square of the atomic number: $\sigma_1/\sigma_2 \approx (Z_1/Z_2)^2$. This approach could be used in Figure 23-5, for example, to find the ratio of Au to Aq atoms/cm^2 (using Equation 23.6)

$$\frac{N_{Au}}{N_{Aq}} = \frac{Y_{Au}}{Y_{Aq}} \cdot \left(\frac{Z_{Ag}}{Z_{Au}}\right)^2 \qquad (23.10)$$

It also serves as a rough estimate when evaluating thicker films, as shown in Figure 23-7, for 2000 Å of Co-Ta on Si. The ratio of the shaded areas (easily accessible from the data analysis system) gives the composition—in this case Co/Ta = 1.

The scattering cross section for ^{4}He ions on medium- to heavy-mass elements has a smooth dependence on scattering angle θ and incident energy E_0. There are deviations at high energies where resonances in the cross section occur and at low energies where screening effects of the orbital electrons influence the effective charge in the scattering process. In forward scattering of hydrogen and

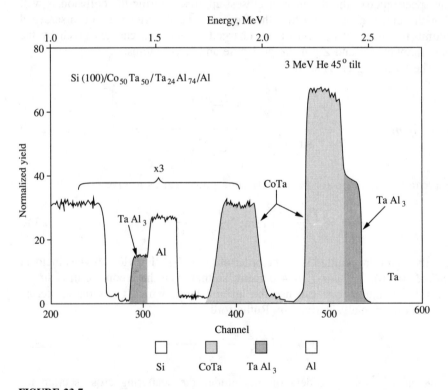

FIGURE 23-7
Rutherford backscattering spectrum of an amorphous diffusion barrier, CoTa, with a layer of TaAl$_3$ deposited between the CoTa and Al layers.

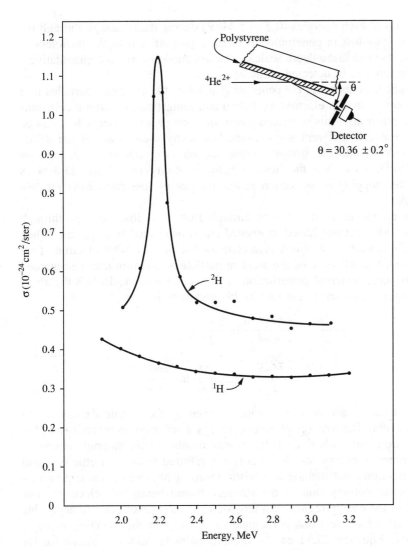

FIGURE 23-8

The scattering cross-section for the ^{4}He-^{1}H (proton) and ^{2}H (deuterium) recoil reaction at a scattering angle of 30.36 ± 0.20 degrees obtained from polystyrene targets.

deuterium, the deviations are sufficiently great that experimentally determined cross sections, such as those shown in Figure 23-8, are used.

DEPTH PROFILES

Energy Loss of He and H Ions in Solids

In the previous section, it was tacitly assumed that the atoms to be identified were at the surface of the materials. Here, we consider composition depth profiles in which the depth scale is established by the energy loss dE/dx of light (H$^+$, d$^+$, and He$^+$) ions at high energies (0.5 to 5 MeV) during their passage through the solid. The energy lost in penetration is directly proportional to the thickness of material traversed so that a depth scale can be assigned directly and quantitatively to the energy spectra of detected particles.

For light ions such as ^{4}He penetrating a solid, the energetic particles lose energy primarily through electron excitation and ionization in inelastic collisions with atomic electrons—this is termed *electronic-energy loss*. Energy loss can be expressed in several different ways. Some frequently used measures are dE/dx in units of eV/Å, and the stopping cross section $\epsilon = (1/N)dE/dx$ in units of eV/(atoms/cm^2), where N is the atomic density in atoms/cm^3. Figure 23-8 gives values for the stopping cross section ϵ, and the energy loss rate dE/dx for ^{4}He and ^{1}H in Al.

When an He or H ion moves through matter, it loses energy through interactions with electrons raised to excited states or ejected from atoms. When the projectile velocity v is much greater than that of an orbital electron (fast-collision case), the influence of the incident particle on an atom may be regarded as a sudden, small external perturbation. This picture leads to Bohr's theory of stopping power. The energy loss rate in this treatment is

$$-\frac{dE}{dx} = \frac{4\pi Z_1^2 e^4 n}{mv^2} \ln \frac{2mv^2}{I}$$
$$= \frac{2\pi Z_1^2 e^4}{E} \cdot NZ_2 \cdot \frac{M_1}{m} \ln \frac{2mv^2}{I} \tag{23.11}$$

where $E = 1/2 M_1 v^2$ and $n = NZ_2$ with N given by the atomic density in the stopping medium. The average excitation energy I for most elements is roughly $10Z_2$ in electron volts, where Z_2 is the atomic number of the stopping atoms.

The complete energy loss formula (often referred to as the Bethe formula) contains corrections that include relativistic terms at high velocities and corrections for the nonparticipation of the strongly bound inner-shell electrons. For helium ions in the energy regime of a few MeV, relativistic effects are negligible and nearly all target electrons participate ($n = NZ_2$) in the stopping process. Consequently, Equation 23.11 can be used to estimate values of dE/dx for He ions at energies above 1.0 MeV. For example, the electronic energy loss of 2-MeV ^{4}He ions in Al has a value (calculated from Equation 23.11) of 31.5 eV/Å using values of $n = NZ_2 = 0.78/Å^3$ and $I = 10Z_2 = 130$ eV. This is close to the more exact values shown in Figure 23-9.

Energy Width in Backscattering

As MeV He ions traverse the solid, they lose energy along their path at a rate dE/dx between 30 and 60 eV/Å. In thin-film analysis, to a good approximation, the total energy loss ΔE to a depth t is proportional to t. That is,

$$\Delta E_{\text{in}} = \int^t \frac{dE}{dx} dx \approx \left.\frac{dE}{dx}\right|_{\text{in}} \cdot t \tag{23.12}$$

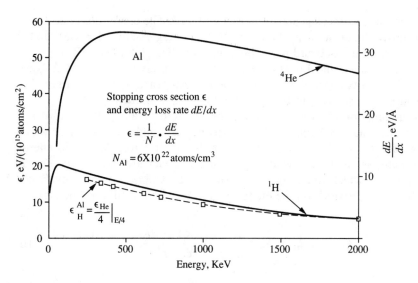

FIGURE 23-9
Stopping cross-section, ϵ, and energy loss rate, dE/dx for ^{4}He and ^{1}H in Al. The open squares for the hydrogen data are scaled from the He data by evaluating the stopping powers at the same velocity ($E/4$) and scaling by 4 for the Z_1 dependence.

where $dE/dx|_{in}$ is evaluated at some average energy between the incident energy E_0 and E_0.

The total energy difference ΔE between particles scattered at the surface and from a depth t is also linear with thickness to a good approximation. The energy of a particle at depth t is

$$E(t) = E_0 - t \cdot dE/dx|_{in} \tag{23.13}$$

After scattering, the particle energy is $KE(t)$; the particle loses energy along the outward path and emerges with an energy

$$E_1(t) = KE(t) - \frac{t}{|\cos\theta|}\frac{dE}{dx}\bigg|_{out}$$

$$= -t\left(K\frac{dE}{dx}\bigg|_{in} + \frac{1}{|\cos\theta|}\frac{dE}{dx}\bigg|_{out}\right) + KE_0 \tag{23.14}$$

The energy width ΔE of the signal from a film of thickness Δt is

$$\Delta E = \Delta t\left(K\frac{dE}{dx}\bigg|_{in} + \frac{1}{|\cos\theta|}\frac{dE}{dx}\bigg|_{out}\right) = \Delta t S \tag{23.15}$$

The subscripts *in* and *out* refer to the energies at which dE/dx is evaluated, and S is often referred to as the backscattering energy loss factor.

The backscattering spectrum at $\theta = 170$ degrees for 3-MeV ^{4}He incident on a 4000-Å Al film with thin Au markers (about three monolayers of Au) on the front and back surfaces is shown in Figure 23-10. The energy loss rate dE/dx along the inward path in Al is about 22 eV/Å at energies around 3 MeV and is about 29 eV/Å on the outward path at energies of about 1.5 MeV ($K_{Al} \approx 0.55$). Inserting these values into Equation 23.15, we obtain an energy width ΔE_{Al} of 165 keV. The energy separation between the two Au peaks is slightly larger, 175 keV, because one uses K_{Au} in Equation 23.15 along with dE/dx (Al).

The assumption of constant values for dE/dx or ϵ along the inward and outward tracks leads to a linear relation between ΔE and the depth t at which scattering occurs. For thin films, $\Delta t < 1000$ Å, the relative change in energy along the paths is small. In evaluating dE/dx, one can use the *surface energy approximation* in which $(dE/dx)_{in}$ is evaluated at E_0 and $(dE/dx)_{out}$ is evaluated at KE_0. In this approximation the energy width ΔE_0 from a film of thickness Δt is

$$\Delta E_0 = \Delta t S_0 = \Delta t \left(K \frac{dE}{dx} \bigg|_{E_0} + \frac{1}{|\cos \theta|} \frac{dE}{dx} \bigg|_{KE_0} \right) \qquad (23.16)$$

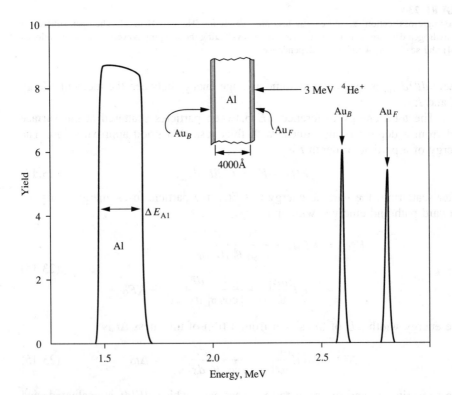

FIGURE 23-10
Backscattering spectrum ($\Theta = 170°$) for 3.0 MeV He ions incident on a 4000 Å Al film with thin Au markers on the front and back surfaces.

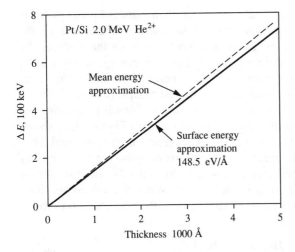

FIGURE 23-11
Rutherford backscattering energy widths of Pt films on a silicon substrate. The energy width ΔE versus Pt film thickness is calculated from the surface energy and mean energy approximations.

where the subscripts denote the surface energy approximation. A comparison between the surface energy and the mean energy approximations is shown in Figure 23-11 for 2.0-MeV He scattering from a Pt film. In the surface energy approximation, the conversion between energy width ΔE and thickness Δt is 148.5 eV/Å. In the mean energy approximation the ΔE-versus-Δt relation deviates from a straight line, and the value of ΔE for a 5000-Å film exceeds by about 3 percent the value from the surface energy approximation. The comparison between the mean energy and surface approximations serves as a quick estimate of the probable error introduced by using the surface energy approximation. A backscattering spectrum can be viewed as a *linear* depth profile of the elements within the sample.

Depth Profiles with Rutherford Scattering

The energy loss of light ions follows a well-behaved pattern in the MeV energy range. The values of dE/dx or ϵ can be used to obtain composition depth profiles from the energy spectra of backscattered particles. Let us consider backscattering analysis of a 1000-Å Ni film on Si. Nearly all the incident ^{4}He beam penetrates microns into the target before it is stopped. Particles scattered from the front surface of the Ni have an energy given by the kinematic equation (Equation 23.1), $E_1 = E_0 K$, where the kinematic factor K for 2-MeV ^{4}He backscattered at a laboratory angle of 170 degrees is 0.76 for Ni and 0.57 for Si.

As the particles traverse the solid, they lose energy along their incident path at a rate of about 64 eV/Å (assuming a bulk density for Ni of 8.9 g cm^{-3}). In thin-film analysis, to a good approximation, energy loss is linear with thickness. Thus, a 2-MeV particle will lose 64 keV penetrating to the Ni-Si interface. Immediately after scattering from the interface, particles scattered from Ni will have an energy of 1477 keV, as derived from $K_{Ni} \times (E_o - 64)$. On their outward path, particles will have slightly different energy loss due to the energy dependence of the energy loss process, in this case 69 eV/Å. On emerging from the surface, the ^{4}He ions scattered from Ni at the interface will have an energy of 1408 keV. The total energy difference ΔE between particles scattered at the surface and near the interface is 118 keV, a value which can be derived from Equation 23.16.

In general, one is interested in reaction products or interdiffusion profiles (see Chapters 25 and 26). For example the energy spectrum of Figure 23-12 shows schematically an Ni film reacted to form Ni_2Si. After reaction, the Ni signal ΔE_{Ni} has spread slightly, owing to the presence of Si atoms contributing to the energy loss. The Si signal exhibits a step corresponding to Si in the Ni_2Si. It should be noted that the ratio of heights H_{Ni}/H_{Si} in the silicide layer gives the composition of the layer. To a first approximation the expression of the concentration ratio is given by

$$\frac{N_{Ni}}{N_{Si}} \approx \frac{H_{Ni}}{H_{Si}} \frac{\sigma_{Si}}{\sigma_{Ni}} \approx \frac{H_{Ni}}{H_{Si}} \left(\frac{Z_{Si}}{Z_{Ni}} \right)^2 \qquad (23.17)$$

where we have ignored the difference in stopping cross sections along the outward path for particles scattered from Ni and Si atoms. The yield from the Ni or Si in the silicide is given closely by the product of signal height and energy width

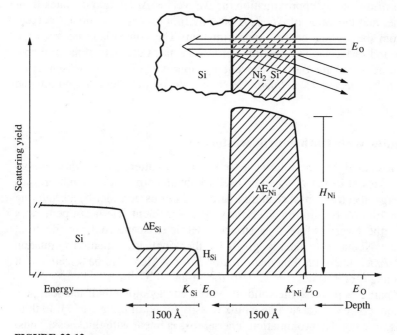

FIGURE 23-12
Schematic backscattering spectra for MeV ^{4}He ions incident on 1000 Å Ni film on Si after reaction to form Ni_2Si. Depth scales are indicated below the energy axes.

ΔE. Therefore, a better approximation to the concentration ratio of two elements A and B uniformly distributed within a film is

$$\frac{N_A}{N_B} = \frac{H_A \Delta E_A \sigma_B}{H_B \Delta E_B \sigma_A} \tag{23.18}$$

In this case of Ni_2Si the difference between application of Equations 23.17 and 23.18 corresponds to a 5 percent difference in the determination of the stoichiometry of the silicide.

Depth Resolution and Energy Loss Straggling

With backscattering spectrometry one can determine composition changes with depth. In this section we consider the limits to depth resolution δt in backscattering spectrometry. For a thin marker layer at a depth t below the surface, the energy resolution δE_1 is given by the full width at half maximum (FWHM) of the signal. The energy resolution is translated to a depth resolution by the use of the S factor of Equation 23.15:

$$\delta t = \frac{\delta E_1}{S} \tag{23.19}$$

The energy resolution normally includes two components: detector resolution, δE_d, and energy straggling, δE_s. Assuming both contributions are gaussian,

$$(\delta E_1)^2 = (\delta E_d)^2 + (\delta E_s)^2 \tag{23.20}$$

Although the detector resolution depends on particle energy and mass, as a first approximation we consider it as a fixed value with $\delta E_d = 15$ keV.

An energetic particle that moves through a medium loses energy via many individual encounters. Such a discrete process is subject to statistical fluctuations. As a result, identical energetic particles that all have the same initial velocity do not have exactly the same energy after passing through a thickness Δt of a homogeneous medium. The energy loss ΔE is subject to fluctuations. The phenomenon is called energy straggling. Straggling sets a fundamental limit to the depth resolution obtainable with ion beam energy loss techniques. For He ions incident on layers thinner than 1000 Å, the straggling is small compared to the resolution of a solid state detector, and hence plays no role in the obtainable depth resolution. For depths greater than a few thousand angstroms, energy loss straggling sets the limit in depth resolution. In Si, for example, films approximately 5000 Å thick can be analyzed before straggling becomes comparable to normal detector resolution of 15 to 20 keV.

Computer-Based Data Analysis

Backscattering and forward recoil scattering have now been established as laboratory analytical tools. In order to facilitate data analysis, computer programs such as RUMP, developed at Cornell University, are used routinely. Such programs account for scattering and stopping cross sections as well as detector resolution and energy straggling. These programs allow for evaluation of the composition and thickness of thin-film and layered structures as well as impurity distributions. One can simulate spectra as a guide to design of experiments (Doolittle, 1986).

HYDROGEN AND DEUTERIUM DEPTH PROFILES

Forward recoil spectrometry is a method for nondestructively obtaining depth profiles of light elements in solids. For the geometry shown in Figure 23-13, the technique can be used to determine hydrogen and deuterium concentration profiles in solid materials to depths of a few microns by using ^{4}He ions at energies of a few MeV. The forward recoil technique is similar to backscattering analysis, but instead of measuring the energy of the scattered helium ion, the energies of the recoiling ^{1}H or ^{2}H nuclei are measured. Hydrogen is lighter than helium, and both particles are emitted in the forward direction. A mylar foil ($\approx 10 \ \mu m$ thick) is placed in front of the detector to block the penetration of the helium ions while permitting the passage of the ^{1}H ions (protons). The stopping power of ^{1}H ions is sufficiently low compared to that of He ions (Figure 23-8) that a 1.6-MeV ^{1}H ion only loses 300 keV in penetrating a film that completely stops 3-MeV ^{4}He ions. The mylar absorber does introduce energy straggling, which, combined with the energy resolution of the detector, results in an energy resolution at the sample surface of about 40 keV.

Depth profiles are determined by the energy loss of the incident He ion along the inward path and the energy loss of the recoil ^{1}H or ^{2}H ion along the outward path. The diffusion of deuterium (^{2}H) in polystyrene can be determined from spectra such as shown in Figure 23-13. In that case a ^{2}H ion detected at an energy of 1.4 MeV corresponds to a collision that originated a ^{2}H recoil from a depth about 4000 Å below the surface. The use of forward recoil spectrometry allows the determination of hydrogen and deuterium diffusion coefficients in the range of 10^{-12} to 10^{-14} cm^2/s, which are difficult to determine by conventional techniques.

As an example of depth measurement, consider a layer of hydrocarbon on both sides of a self-supported 4000-Å aluminum film. We will use a symmetrical scattering geometry with the sample inclined at an angle α to the beam and the detector at an angle 2α, so that the path length to the back surface (sample thickness t) is the same ($t/\sin \alpha$) for both the inward He ion and the outward proton. Hydrogen recoils originating from the front surface will have an energy $E_1 = K^1 E_0$ where $K^1 = 0.480$ for $2\alpha = 30$ degrees (Equation 23.5) and K^1 indicates the recoil kinematic factor. Hydrogen recoils originating from the back surface, at depth t, will have an energy $E_1(t)$ given by

$$E_1(t) = K^1 E_0 - K^1 \Delta E_{\text{He}} - \Delta E_{\text{H}} \qquad (23.21)$$

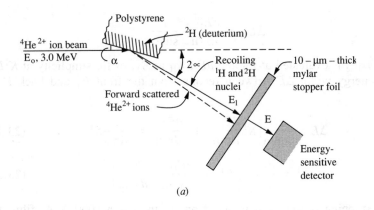

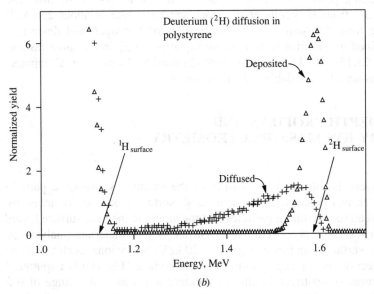

FIGURE 23-13
(a) Experimental geometry for forward recoil spectrometry experiments to determine depth profiles of ^1H and ^2H in solids. (b) Recoil spectrum of ^2H diffused in a sample of polystyrene for 1 hour at 170°C. The sample consisted of a bilayer film consisting of 120 Å of deuterated polystyrene on a large molecular weight ($M_w = 2 \times 10^7$) film of polystyrene (*from Mills et al.*, Appl. Phys. Lett. 45).

where ΔE_{He} is the energy loss of the He on the inward path and ΔE_H is the energy loss of the hydrogen on the outward path.

$$\Delta E_{He} = \left. \frac{dE}{dX} \right|_{He} \cdot \frac{t}{\sin \alpha}$$

and

$$\Delta E_H = \frac{dE}{dx}\bigg|_H \cdot \frac{t}{\sin \alpha}$$

where $dE/dx|_{He}$ is evaluated at E_0 and $dE/dx|_H$ at E_1, or for simplicity at $K^1 E_0$.

The energy width ΔE between recoils from the front E_1 and back $E_1(t)$ surfaces is

$$\Delta E = \frac{t}{\sin \alpha}\left(K^1 \frac{dE}{dx}\bigg|_{He} + \frac{dE}{dx}\bigg|_H\right) = \frac{t}{\sin \alpha} S^1 \qquad (23.22)$$

$$S^1 = \frac{dE}{dx}\bigg|_{He}\left\{K^1 + \frac{dE}{dx}\bigg|_H \bigg/ \frac{dE}{dx}\bigg|_{He}\right\} \qquad (23.22)$$

where the stopping power ratio is about one-sixth. For a 4000-Å Al film, the energy width ΔE from H recoils at the front and back side is about 250 keV for 2-MeV He ions. Consequently, depth profiles of hydrogen and deuterium can be determined in a routine fashion using Equation 23.22 in the same fashion that Equations 23.15 and 23.16 are used in Rutherford backscattering. Computer-based programs are also available for data analysis.

SPUTTER DEPTH PROFILES AND SECONDARY ION MASS SPECTROMETRY

Sputtering

Depth profiles can be obtained by erosion of the sample by energetic particle bombardment. In this process, called sputtering, surface atoms are removed by collisions between the incoming particles and the atoms in the near surface layers of a solid. This surface layer removal is carried out in materials analysis by bombarding the surface with low-energy (0.5–20 keV) heavy ions, such as O^+ or Ar^+, which eject or sputter target atoms from the surface. The yield of sputtered atoms, the number of sputtered atoms per incident ion, lies in the range of 0.5 to 20 depending upon ion species, ion energy, and target material. Surface-sensitive electron spectroscopy techniques can be used after each layer is removed to determine the composition of the new surface, and hence deduce the depth profile of the atomic composition. It is also possible to analyze the sputtered atoms, generally the ionized species, to determine the composition of the sputter-removed materials. This technique of secondary-ion mass spectroscopy (SIMS) has been used extensively in depth profiling.

The ion bombardment–induced erosion rates are characterized by the sputtering yield Y, which is defined as

$$Y = \text{sputtering yield} = \frac{\text{mean number of emitted atoms}}{\text{incident particle}} \qquad (23.23)$$

The sputtering yield depends in general on the structure and composition of the target material, the parameters of the incident ion beam, and the experimental geometry. Measured values of Y cover a range of over seven decades; however, for the medium-mass ion species and keV energies of general interest in depth profiles, the values of Y lie between 0.5 and 20. Sputtering yields can be accurately predicted by theory for single-element materials. Figure 23-14 shows the energy and incident particle dependence of the sputtering yield of Si. The experimental values are in good agreement with calculations (solid line) based on nuclear energy loss mechanisms. A number of review articles and books on the topic of sputtering are listed in the bibliography at the end of this chapter.

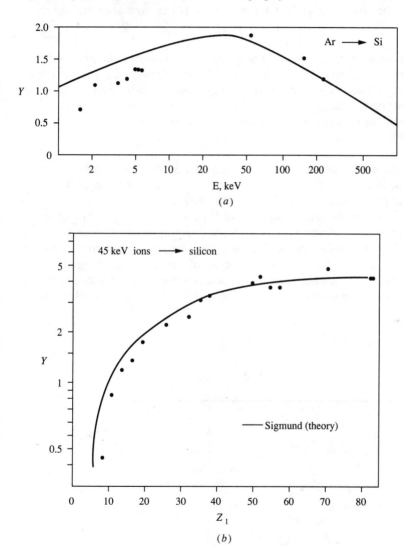

FIGURE 23-14
(a) Energy dependence of the Ar sputtering yield of Si, and (b) Incident ion dependence of the Si sputtering yield. The solid line represents the calculations of Sigmund and the data is from Andersen and Bay in Behrisch, 1981.

In the sputtering process, atoms are ejected from the outer surface layers. The bombarding ion transfers energy in elastic collisions to target atoms that recoil with sufficient energy to generate other recoils (Figure 23-15). Some of these backward recoils (about one to two atoms for a 20-keV Ar ion incident on Si), will approach the surface with enough energy to escape from the solid. It is these secondary recoils that make up most of the sputtering yield. The most important parameter in the process is the energy deposited at the surface. We can then express the sputtering yield Y for particles incident normal to the surface as

$$Y = \Lambda F_D(E_0) \tag{23.24}$$

where Λ accounts for all the material properties such as surface binding energies, and has values of 0.2 to 0.4 eV/Å; and $E_d(E_0)$ is the density of deposited energy at the surface and depends on the type, energy, and direction of the incident ion and the target parameters Z_2, M_2, and N.

The deposited energy at the surface can be expressed as

$$F_D(E_o) = \alpha N S_n(E_0) \tag{23.25}$$

where N is the atomic density of target atoms and $S_n(E)$ is the nuclear stopping cross section. In this equation, α is a correction factor with a value of 0.25 for a normal angle of incidence of the beam to the surface.

A charged particle penetrating a solid loses energy through two processes: (1) energy transfer to electrons, or electronic stopping, and (2) energy transfer to the atoms of the solid, or nuclear stopping. In both cases the interaction is basically of a Coulomb type; for the case of electronic stopping it is pure Coulomb, while in the nuclear case it is a form of screened Coulomb potential. If a screened Coulomb potential with a $1/r^2$ dependence is used, the nuclear energy

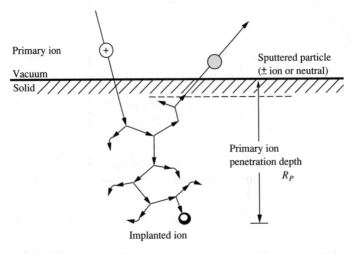

Primary ion

Vacuum

Solid

Sputtered particle
($\pm$ ion or neutral)

Primary ion
penetration depth
R_P

Implanted ion

FIGURE 23-15
Schematic of the ion-solid interactions and the sputtering process.

loss, $dE/dx|_n$ is independent of energy; i.e.,

$$\frac{dE}{dx}\Big|_n = NS_n = N\frac{\pi^2}{2}Z_1Z_2e^2a\frac{M_1}{M_1 + M_2} \qquad (23.26)$$

The yield Y of sputtered particles from single-element amorphous targets was expressed in Equation 23.24 as the product of two terms: one, Λ, encompassing material parameters and the other, F_D, the deposited energy. The derivation of Λ involves a description of the number of recoil atoms that can overcome the surface barrier and escape from the solid. The details of the derivation for the linear cascade regime are given by Sigmund (Behrisch, 1981). The result is

$$\Lambda \approx \frac{0.042}{NU_0}(\text{Å/eV}) \qquad (23.27)$$

where N is the atomic density (in Å^{-3}), and U_0 is the surface binding energy (in eV). The value of U_0 can be estimated from the heat of sublimation ($\approx$ heat of vaporization) and typically has values between 2 and 4 eV. For Ar incident on Cu, the value of $NS_n = 110$ eV/Å; the surface binding energy U_0 is about 3 eV based on a heat of vaporization of about 3 eV. The sputtering yield with $N = 8.5 \times 10^{-2}$ atoms / Å^3 is

$$Y = \frac{0.042 \times 0.25 \times 110 \text{ eV/Å}}{8.45 \times 10^{-2}/\text{Å}^3 \times 3 \text{ eV}} = 4.5 \qquad (23.28)$$

which is in reasonable agreement with measured values of about 6. More detailed treatments of sputtering yield are given in Behrisch (1983) and Feldman and Mayer (1986).

These estimates hold for the ideal case of an amorphous single-element target. The sputtering yields from single crystal, polycrystalline, or alloy targets may deviate significantly from the simple estimates above. With polyatomic targets, preferential sputtering of one of the elements can lead to changes in composition of the surface layer. These changes will be reflected in the Auger yields, which give the composition of the altered, not original, layer. Another complication is ion beam mixing (redistribution within the collision cascade), which can lead to broadening of the interface when profiling layered targets. It is possible in many of these cases to use Rutherford backscattering to establish layer thicknesses and the concentration of the major constituents. This provides a calibration for the sputter profile.

Secondary-Ion Mass Spectrometry (SIMS)

Surface layers are eroded by the sputtering process, and hence the relative abundances of the sputtered species provide a direct measure of the composition of the layer that has been removed. One of the most commonly used analysis techniques is the collection and analysis of the ionized species—the secondary ions. As shown in Figure 23-16, the secondary ions enter an energy filter, usually

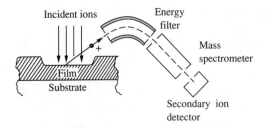

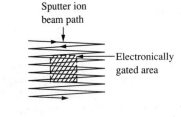

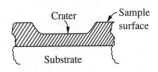

FIGURE 23-16
Schematic of a SIMS apparatus. An incident ion beam results in sputtered ionic species that are passed through an electrostatic energy filter and a mass spectrometer and finally into an ion detector. The beam is usually swept across a large area of the sample and the signal detected from the central portion of the sweep. This avoids crater edge effects.

an electrostatic analyzer, and are collected in a mass spectrometer. All SIMS instruments possess a capability for surface and elemental depth concentration analysis. In one mode of operation the sputter ion beam is rastered across the sample where it erodes a crater in the surface. To insure that ions from the crater walls are not monitored, the detection system is electronically gated for ions from the central portion of the crater. There are also direct imaging instruments — ion microscopes — in which the secondary ions from a defined microarea of the sample are detected so that an image of the surface composition can be displayed.

SIMS is often used in analysis of dopant profiles in semiconductors because the detection sensitivity extends down to 10^{15}–10^{16} atoms/cm^3. The data for As in Si shown in Figure 23-17 is a typical example of depth profile techniques. Another strong feature of SIMS analysis is the ability to analyze hydrogen over a wide range of concentrations.

In spite of the strength of SIMS in analysis of impurity levels, the yield of secondary ions is extremely sensitive to surface conditions — the variation in ionization yield is one of the major difficulties encountered in quantitative analysis. This problem can often be overcome by use of suitable standards.

INTERDIFFUSION IN POLYMERS

Ion beam analysis techniques have been used extensively to investigate interdiffusion and reactions in ion-implanted silicon, in thin-film layers such as silicide

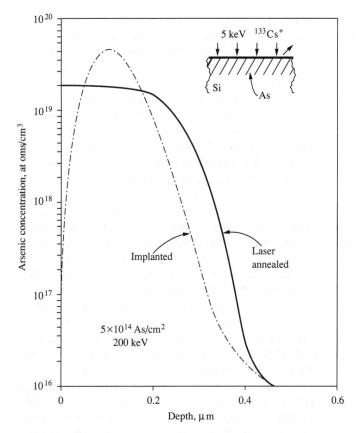

FIGURE 23-17
SIMS concentration profile of As implanted in Si and redistributed by pulsed laser melting of the outer Si layer. The measured concentration profile extends to levels of $5 \times 10^{15}/cm^3$. (From C. Magee, RCA Laboratories.)

and aluminide layers, and more generally in determining concentration profiles in surfaces modified by lasers, ion or electron beams, and short-duration heat pulses. More recently, ion beam techniques have been applied to interdiffusion in polymers. In these cases, as outlined below, Rutherford backscattering and forward recoil spectrometry have been used to measure diffusion coefficients in the range of 10^{-12} to 10^{-15} cm^2/s in polymer layers a few microns thick—the range of interest in packaging applications. Conventional techniques for measuring diffusion in polymers are difficult to use in this range because they lack the required depth resolution or they do not measure directly the concentration profile of the diffusing species.

Diffusion of entangled polymer chains determines properties such as adhesion of two polymer surfaces or phase separation of polymer blends. Diffusion of polymer chains is envisaged as a reptation process in which a given chain molecule crawls within a virtual tube defined by the constraints imposed by the other molecules. The diffusion of polymer A into B can be studied in a bilayer configuration, shown in Figure 23-18, by either introducing a diffusion marker (Au particles as shown in Figure 23-18) or by tagging the diffusing molecules by using deuterated polymer A.

In the case where markers are used, Rutherford backscattering determines the depth location of the A/B interface. During diffusion, the shift of the marker in depth reflects the different diffusion coefficients of the polymers because of the different molecular weights of the polymers. The slower-moving species has the higher molecular weight. For a lower-molecular-weight species A on a higher-molecular-weight polymer B, the marker in Figure 23-18 will shift toward the surface as A diffuses in B.

The diffusion coefficients of polystyrene (PS) molecules into high-molecular-mass ($M_w = 2 \times 10^7$) polystyrene is shown in Figure 23-19; the data marked by crosses was obtained with the marker displacement technique (Green et al., 1985). Shown in the same figure by circles are the self-diffusion coefficients of deuterated polystyrene (d-PS) molecules diffusing into the same high-molecular-weight substrate. For these determinations of the diffusion coefficient (Mills et al., 1984), the concentration profiles of d-PS were measured by forward recoil spectrometry as shown earlier in Figure 23-13. The agreement between the two techniques—marker displacement and concentration profiles—is excellent over nearly four orders of magnitude of the diffusion coefficient. These measurements have been used to gain insight into the reptation process (Green et al., 1984).

The combination of backscattering and forward recoil measurements has also been used to measure mutual diffusion of chemically dissimilar yet compatible polymer blends by means of diffusion couples consisting of two polymer blends with a slight composition difference. In the present case, the miscible polymer blend is polystyrene:poly(xylenyl ether), or PS:PXE, with the PS molecules being the faster diffusing species. Two techniques were used: in one (forward scattering)

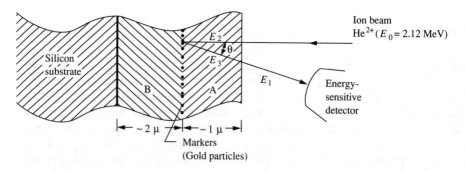

FIGURE 23-18
Schematic drawing of a bilayer, A/B, polymer interdiffusion couple representing the geometry of a Rutherford backscattering experiment with a thin Au marker at the interface between a polystyrene layer B with high molecular weight and a layer A with low molecular weight.

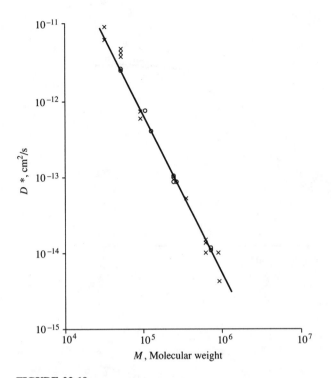

FIGURE 23-19
Self-diffusion coefficients of d-PS, deuterated polystyrene (open circles), in high-molecular-weight PS ($M_w = 2 \times 10^7$) as a function of molecular weight M of PS. Data obtained by forward recoil. Marker displacement measurements (crosses) of diffusion coefficients were determined by Rutherford backscattering from Au markers (Figure 23-18).

the PS molecules were deuterated (d-PS); in the other (backscattering) bromine was used to stain only the PXE component in the miscible blend after diffusion.

The mutual diffusion couples consist of two films of d-PS:PXE. The top film contains a lower volume fraction (typically about 0.5) of deuterated polystyrene than the bottom film (about 0.6). Forward recoil measurements give the depth distribution of deuterium in the diffusion couple, which consists of two d-PS:PXE blends that differ by 10 percent in concentration. Figure 23-20 shows the volume fraction profile of d-PS versus depth for such a couple before (a) and after (b) diffusion for 1800 seconds at 206°C. The solid line for the as-deposited depth profile (Figure 23-20(a)) corresponds to a step function convoluted with the instrumental depth resolution function. After diffusion, the composition profile becomes significantly broadened and the solid line represents a diffusion depth profile with erf(x/w) terms where $w = 2(Dt)^{1/2}$ and the diffusion coefficient had a value of 1.1×10^{-13} cm²/s.

Mutual diffusion in polymer blends of PS:PXE was also studied by tracing the depth profiles of PXE molecules after diffusion (Composto, 1987). To reveal

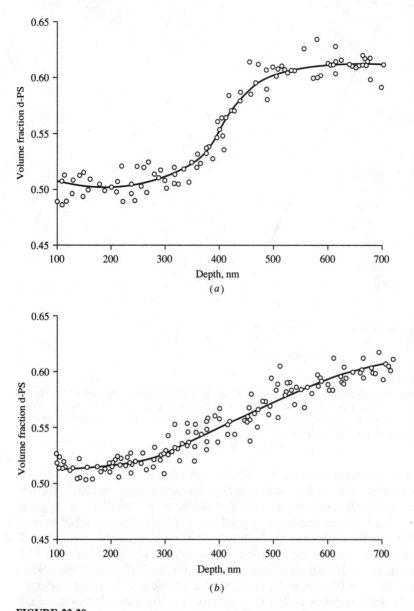

FIGURE 23-20
Volume fraction of deuterated polystyrene in a d-PS:PXE mutual diffusion couple: (*a*) as deposited and (*b*) after diffusion for 1800 s at 206°C. The solid lines represent a best fit to the data using an instrumental resolution of 80 nm in (*a*) and an erf diffusion profile with a D of 1.1×10^{-13} cm²/s in (*b*).

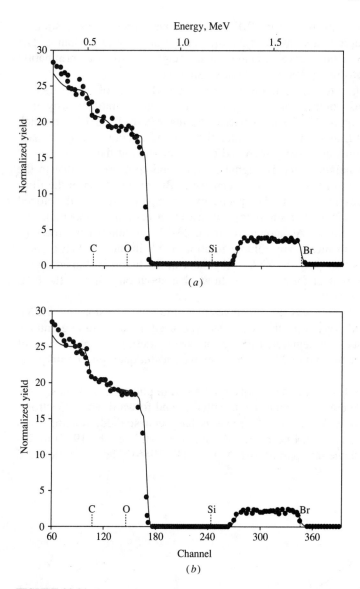

FIGURE 23-21

RBS spectra of 2.20 MeV He ions backscattered from (a) pure PXE ($\phi = 0.0$) and a PS:PXE blend ($\phi = 0.5$). Both samples were stained in a bromine and methanol solution for 24 hours. The solid lines are simulated spectra where the thickness and mer unit of PXE are 915 nm and $C_8 H_{7.70} OBr_{0.14}$. The energies at which the He ions would be backscattered by carbon, oxygen, silicon, and bromine nuclei at the surface are marked.

the depth profile after diffusion, the PXE molecules were preferentially strained by exposing the couple to a solution of 2 mole percent bromine in methanol. The covalently-bound Br atoms serve as heavy nuclear tags that allow determination of the PXE depth profile by Rutherford backscattering.

To calibrate the bromine staining of PXE, PS:PXE films of uniform composition and various volume fractions of PS were stained simultaneously in a bromine solution. Figure 23-21 shows backscattering spectra from 100-nm-thick films of (*a*) pure PXE and (*b*) a blend with a 0.5 volume fraction of PS. The signals from the Br atoms are clearly seen and extend throughout the films. The 50 percent decrease in height of the Br signal in the blend compared to that in the pure sample is a result of the 50 percent decrease in Br concentration in the film. The solid lines represent the fit to the spectra by a computer program developed by Doolittle (1985, 1986) that simulates spectra for a certain composition and depth profile in the sample. As shown in Figure 23-22, the number of Br atoms decreases linearly with increasing volume fraction of PS in the PS:PXE blends. This result indicates that only the PXE component in the blend is stained by the bromine. The measurement for pure PXE shows that about one-third of the PXE molecules are brominated.

The staining of PXE was used to investigate the interdiffusion of PS:PXE blends. The methodology was the same as was used in determining mutual diffusion coefficients with deuterated PS by forward scattering. As in the case of diffusion in polystyrene (Figure 23-19), both analysis techniques gave comparable results.

The two examples, diffusion in polystyrene and in polystyrene blends, were chosen to illustrate the application of backscattering and forward recoil spectrometry to polymers. A wide variety of problems has been studied, ranging from diffusion of polymer rings in linear polymer matrices (Mills et al., 1987) to diffusion of trichloroethane into photoresist (Mills et al., 1986). The straightforward

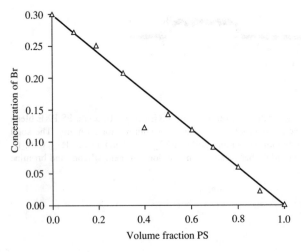

FIGURE 23-22
Number n of bromine atoms per average mer versus the volume fraction ϕ of PS in a PS:PXE blend. The solid line is a linear fit to the data given by $n = 0.294(1 - \phi_{PS})$.

application of these ion beam analysis techniques has also stimulated joint university-industry research programs.

REFERENCES

Behrisch, R., editor, *Sputtering by Particle Bombardment*, vols. 1 and 2, Berlin: Springer-Verlag, 1981 and 1983.

Chu, W. K., J. W. Mayer, and M.-A. Nicolet, *Backscattering Spectrometry*, New York: Academic Press, 1978.

Composto, R. J., Ph.D. thesis, Cornell University, 1987.

Composto, R. J., J. W. Mayer, E. J. Kramer, and D. M. White, *Phys. Rev. Lett.* 57, p. 1312 (1986).

Doolittle, L. R., *Nucl. Instr. and Meth.* B9, p. 344 (1985) and B15, p. 227 (1986).

Feldman, L. C., and J. W. Mayer, *Fundamentals of Surface and Thin Film Analysis*, New York: North-Holland, 1986.

Feldman, L. C., J. W. Mayer, and S. T. Picraux, *Materials Analysis by Ion Channeling*, New York: Academic Press, 1982.

Green, P. F., P. J. Mills, C. J. Palmstrom, J. W. Mayer, and E. J. Kramer, *Phys. Rev. Lett.* 53, p. 2145 (1984).

Green, P. F., C. J. Palmstrom, J. W. Mayer, and E. J. Kramer, *Macromolecules* 18 p. 501 (1985).

Mayer, J. W., and E. Rimini, *Ion Beam Handbook for Material Analysis*, New York: Academic Press, 1977.

McCrea, J. M., "Mass Spectrometry" in *Characterization of Solid Surfaces*, edited by P. F. Kane and G. B. Larrabee, New York: Plenum Press, 1974.

McHugh, J. A., "Secondary Ion Mass Spectrometry," in *Methods of Surface Analysis*, edited by A. W. Czanderna, New York: Elsevier Publishing, 1975.

Mills, P. J., P. F. Green, C. J. Palmstrom, J. W. Mayer, and E. J. Kramer, *Appl. Phys. Lett.* 45, p. 957 (1984).

Mills, P. J., J. W. Mayer, E. J. Kramer, G. Hadziioannou, P. Lutz, C. Strazielle, P. Rempp, and A. J. Kovacs, *Macromolecules* 20, p. 513 (1987).

Mills, P. J., C. J. Palmstrom, and E. J. Kramer, *J. Materials Sci.* 21, p. 1479 (1986).

Oechsner, H., editor, *Thin Film and Depth Profile Analysis*, New York: Springer-Verlag, 1984.

Werner, H. W., "Introduction to Secondary Ion Mass Spectrometry (SIMS)" in *Electron and Ion Spectroscopy of Solids*, edited by L. Fiermans et al., New York: Plenum, 1978.

Ziegler, J. F., *Helium, Stopping Powers and Ranges in all Elements*, New York: Pergamon, 1977.

PROBLEMS

A monolayer (10^{15} atoms/cm^2) of Au is deposited on the front and back surfaces of a 4000 Å (4×10^{-5} cm)–thick Al film. The Au-Al structure is deposited on a carbon substrate. You carry out Rutherford backscattering analysis at a 170-degree scattering angle (assume 180 degrees for calculation) with 2.0-MeV ^{4}He ions.

23.1. What is the energy of the ^{4}He ions scattered from Au and Al at the surface? What is the energy of the recoiling Al atoms? What is the ratio of Au to Al scattering cross sections? Indicate what the value of these quantities would be if the scattering angle were 120 degrees.

23.2. How much energy would the He ion lose in penetrating the Al film? Assume a constant energy loss at the value for 2-MeV ions. If the He is backscattered from the Au layer at the back surface and emerges from the film, what is the energy difference between detected He ions scattered from Au atoms at the front and rear surfaces?

23.3. Calculate the energy width for the Al signal assuming an inward energy loss at energy $E_0 = 2$ MeV and an outward energy loss at an energy $K_{Al}E_0$.

23.4. Assume there is a monolayer of hydrogen on the surfaces of the Al film instead of the Au layers. If the sample is inclined at an angle $\alpha = 15$ degrees to the beam, what is the energy of detected hydrogen recoils that originate at the front and the back surfaces?

23.5. Assume that you want to sputter the Al film with 1-keV argon ions. Assume a Thomas-Fermi screening radius of 0.15 Å. What is the nuclear energy loss value? If the surface binding energy U_0 is 3 eV, what is the sputtering yield?

CHAPTER 24

ELECTRON SPECTROSCOPY

LUIS J. MATIENZO
FRANCIS EMMI
ROBERT W. JOHNSON

OVERVIEW

In this chapter three specific techniques widely used in surface analysis (complementary to the analysis techniques described in Chapters 21, 22, and 23) will be considered. These techniques provide information on elemental and structural compositions of solid surfaces. They are x-ray photoelectron spectroscopy (XPS) or electron spectroscopy for chemical analysis (ESCA), Auger electron spectroscopy (AES), and ultraviolet photoelectron spectroscopy (UPS). Other surface-sensitive techniques, such as low-energy ion scattering spectroscopy (ISS), secondary-ion mass spectrometry (SIMS), appearance potential spectroscopy (APS), inelastic tunneling spectroscopy (IETS), and low-energy electron diffraction spectroscopy (LEED), have also found applications in electronic packaging but are not considered here. In specific cases, a combination of more than one technique will yield complementary information in problem solving, but these capabilities may not be readily available in less-equipped laboratories.

Historically, the origin of electron spectroscopies goes back to the early 1900s. In 1905, Einstein [1] described the photoelectric effect, which is the excitation of electrons from a material by photon stimulation. In general, excitation from various orbital levels for photoemission depends on the energy source available. Photoemission of inner-shell electrons is commonly obtained by using soft x-rays while excitation from outer-shell valence levels is obtained by using ultraviolet light.

Doubly ionized atomic processes, such as those found in the de-excitation of atoms that have lost a negative charge during ionization and lose a second electron in a subsequent step, take their name from the first observations reported by Pierre Auger on the interactions of gases with x-rays [2]. In subsequent years, other researchers confirmed and demonstrated that the mechanisms of de-excitation involving x-ray beams could proceed by either the above described transitions or by x-ray fluorescence, that is, the emission of x-ray radiation following the movement of an electron from an outer orbital into an inner hole. The probabilities of these two processes depend on the atomic number of the element under study. For low atomic numbers, i.e., $Z < 20$, the double ionization process has a maximum probability and the fluorescence yield has a low probability. For atomic numbers $Z > 20$, fluorescence as a decay process is more probable.

Due to the limits of high-vacuum systems available until the 1940s, detection of the two de-excitation processes was hampered by the additional collisions introduced by residual gases in the system. In addition, the observation of these photoelectrons was not possible until appropriate controls on magnetic fields were designed by Siegbahn and his coworkers at Uppsala University, Sweden. In the United States, an x-ray photoelectron spectrometer was developed by Steinhardt in the research for his doctoral dissertation at Lehigh University [3]. This work was not followed by others because it was believed its commercialization was not feasible. In 1958, the Uppsala group published their first report on copper oxides [4]. The Uppsala group over the years compiled a list of binding energy shifts for a variety of chemical elements in different types of chemical compounds [5]. The term ESCA was coined to describe this analytical method. In more recent years, this technique became known as x-ray photoelectron spectroscopy (XPS) to denote the electron spectroscopic technique using x-ray excitation.

Ultraviolet photoelectron spectroscopy (UPS) was pioneered by Turner and al-Joboury [6] in England by using uv radiation from a helium discharge lamp. This technique was originally developed to study the valence levels of gases, which correspond to narrow bands with additional fine structure originating from vibrational modes.

Auger electron spectroscopy (AES), initially limited to the study of metals, was advanced in the 1970s. This technique gained a lot of popularity in the microelectronics industry because it allows better dimensional resolution than either XPS or UPS on conducting substrates. In addition, the elemental sensitivity of AES is excellent and information on chemical environments for some elements in molecular bonds can be obtained by analyzing AES line shapes. When used in conjunction with sputter etching, it is possible to determine composition profiles as a function of depth. Developments in data processing and storage allowed mapping of elements, which had only been available with microprobe instruments. The limitations of AES were (*a*) the necessity to have conductive materials for analysis, (*b*) electron beam damage such as carbonization of organic layers and redeposition of carbon on unwanted areas, and (*c*) the need to differentiate the signal to obtain elemental resolution from the background signal. Further developments in electron beam diameter and overall instrumentation allowed the commercialization of scanning Auger systems that offered better dimensional resolution and continued to outweigh the disadvantages of the earlier AES instruments.

Present day XPS instruments are able to focus x-ray beams to diameters of about 100 μm by means of crystal monochromators or slits; therefore, chemical analysis of small insulator surfaces is possible. AES instrumentation with beam resolution of about of 250–300 Å is also possible and extends the analytical range of work for conductors and semiconductors. The present instrumental limitations in sensitivity will need to be surmounted to keep pace with additional miniaturization of components and lines of future electronic systems.

This chapter introduces the reader to the application of these techniques in electronic packaging by first discussing the basic principles and limitations of each technique. Then, real examples of surface analysis in characterizing surface interactions, and the use of these techniques in process development, failure analysis, and reliability studies, will be given.

PRINCIPLES

In this section, basic processes involved in electron spectroscopies, such as the origin of the electrons, will be discussed, as well as surface sensitivity and the types of information that can be obtained.

Electron spectroscopic techniques use either photons (x-rays or uv light) or electrons to probe the sample; the resulting excitation ejects electrons with well-defined energies. The kinetic energies of these electrons are measured, giving elemental information and sometimes yielding information on the chemical environment of an atom or a molecule. The depth of analysis using these techniques is generally on the order of 5–100 Å.

The depth information depends upon the energy of the emitted electrons. The distance an electron travels before undergoing inelastic collisions is known as the mean free path and is more dependent on the kinetic energy of the emitted electrons than on the material from which they are emitted. The universal curve for electron mean free paths versus electron energies is shown in Figure 24-1. Normally, electrons escape from depths of 20 Å or less, but on some occasions, they can originate from several mean free paths below the surface.

X-Ray Photoelectron Spectroscopy (XPS)

In XPS, core electrons are ejected from atoms as a result of a photoelectric effect. Bound electrons are emitted by exposing a solid to "soft" (low energy) x-rays as depicted in Figure 24-2, showing a K-shell electron ejected to a free state. The kinetic energies of electrons are accurately measured and are characteristic of the particular atoms from which they originate. By the law of conservation of energy, binding energies of photoelectrons can be determined from the difference between

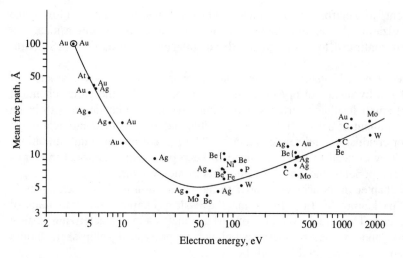

FIGURE 24-1
Universal curve for electron mean free paths. (From G. Somorjai, *Chemistry in Two Dimensions: Surfaces*, Ithaca, N.Y.: Cornell University Press, 1981. Reprinted with permission.)

the x-ray energy and the measured kinetic energy, plus corrections for the work function of the sample and the spectrometer. The latter is determined specifically for each instrument and accounted for during calibration procedures. The law of conservation of energy is represented by

$$hv = E_b + KE + (\Phi_s + \Phi_{spec}) \tag{24.1}$$

where hv is the energy of the monoenergetic x-ray photons, KE is the kinetic energy of the photoelectrons, E_b is the binding energy, and Φ_s and Φ_{spec} are the work functions for the sample and spectrometer.

In a typical XPS spectrum, the number of electrons detected is plotted as a function of binding energy. In addition to peaks due to core level excitation,

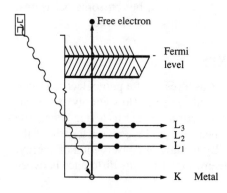

FIGURE 24-2
Excitation process in x-ray photoelectron spectroscopy.

there are peaks caused by x-ray induced Auger transitions, shake-up satellites, shake-off satellites, x-ray satellites, energy loss lines, plasmons, and ghost peaks. Additional information on these processes can be found in the suggested reading list.

Ultraviolet Photoelectron Spectroscopy (UPS)

Ultraviolet photoelectron spectroscopy uses lower-energy photons to excite the sample. Typically, the source is a gas discharge lamp that produces high-intensity radiation with small natural linewidth; He I (21.2 eV) and He II (40.8 eV) are frequently used. These energy sources are not sufficient to excite core electrons; only electrons from the valence bands are involved. Information on the electron band structure and molecular orbitals is obtained by this technique, but it is not useful for elemental identification.

Auger Electron Spectroscopy (AES)

In Auger electron spectroscopy, electrons are emitted as the result of excitation of the sample with an electron beam. As described in Figure 24-3, the Auger electron results from a secondary process. The first step in the process involves an electron filling a vacancy created in a lower energy level by the electron beam used for excitation (for example, L_1–K in Figure 24-3). In the next step, enough energy is transferred to another electron ($L_{2,3}$) for it to escape the atom.

Kinetic energies of the Auger electrons are characteristic for each element. In a typical analysis, data is collected and usually displayed in the differentiated form, which allows small peaks to be more easily distinguished in a large background signal. Figure 24-4 shows AES spectra in the undifferentiated and differentiated modes.

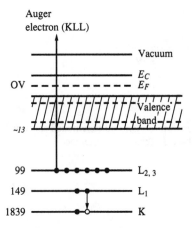

FIGURE 24-3
Schematic representation of the $KL_1L_{2,3}$ Auger de-excitation process for a silicon surface.

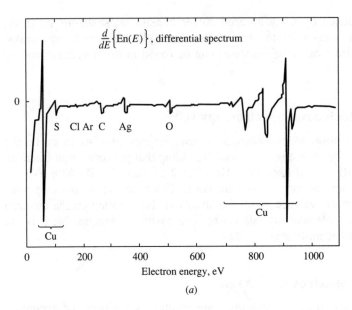

(a)

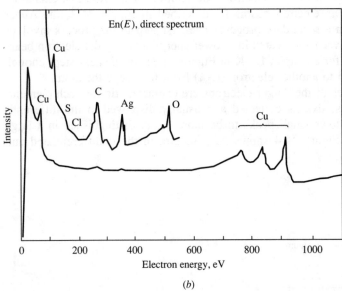

(b)

FIGURE 24-4

Differentiated and direct Auger electron spectra for a copper surface. The undifferentiated spectrum shows the region between 100 and 600 eV on an expanded scale. Taken from reference [13].

INSTRUMENTATION

Basic instrumentation needed to perform XPS, UPS, and AES includes an excitation source, an energy analyzer, and a detector enclosed in a high-vacuum chamber. Following sample irradiation, ejected electrons are separated according to their kinetic energies in an analyzer, and counted by a detecting system. Figure 24-5 shows a schematic representation of the apparatus used for photoelectron spectroscopy. Each of these items will be discussed briefly in the following sections.

Excitation Sources

XPS. Various x-ray excitation sources are available to irradiate samples for XPS. The most common source used is $K_{\alpha1,2}$ x-rays produced from electron bombardment of Al or Mg targets. The bandwidth for these $K_{\alpha1,2}$ x-rays is about 1 eV; an increase in x-ray bandwidth contributes to the overall peak broadening and affects spectral resolution. Certain measures can be taken to improve resolution. A monochromator or a different energy source with a narrower bandwidth may be used. Monochromatization is limited by the natural linewidth of the excitation source, but other energy sources such as synchrotron radiation may be employed for greater resolution. Several x-ray sources are given in Table 24.1.

The spot size, and hence sample area, from conventional sources is about several square millimeters. However, during the early 1980s, a new approach to obtain XPS spectra with improved spatial resolution was developed. It uses the principle of x-ray monochromatization developed in the early 1970s, and a focusing monochromator crystal that meets the requirements of Bragg's law for a particular anode source and is capable of producing a beam of reduced size. A reduced spot size reduces contributions from background and x-ray satellite peaks.

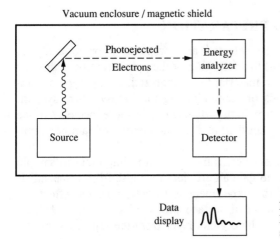

FIGURE 24-5
Basic components of a photoelectron spectrometer.

AES. The excitation source for AES is an electron beam, generated from a heated filament and focused on the sample by a lens system. The diameter of the electron beam, which defines the sampling area, is small, typically from 100 to 0.2 μm in diameter. Further discussion of excitation sources can be obtained from the suggested reading list.

Detection

Electrons emitted from sample surfaces are separated according to energy (i.e., analyzed) in a device known as an energy analyzer, and then counted. The most common types of analyzers used for XPS are the hemispherical mirror analyzer (HMA) and the cylindrical mirror analyzer (CMA); the CMA is typically used in AES. Spectral resolution is determined by the pass-energy window in the analyzer. For survey scans, the normal pass energy is on the order of 50–100 eV. High-resolution spectra are usually collected with pass-energy windows of 25 eV. Electrons are commonly counted by a channel electron multiplier. In more recent designs, improvements in the detecting system, such as parallel detectors that permit a more efficient collection of signals over a wide range of energies, are also used. Principles associated with the operations of analyzers and detectors are presented in the references.

Vacuum System

For surface analysis, a requirement is a clean vacuum system, preferably a base vacuum on the order of 10^{-10} torr. Following the introduction of a sample into the analysis chamber, residual gases must be pumped out. The importance of ultrahigh vacuum (UHV) is twofold: (*a*) to minimize chamber and sample surface contamination, and (*b*) to reduce unwanted collisions, thereby reducing photoelectron count rates. Usually, AES and XPS analyses are performed in a vacuum between 10^{-7} and 10^{-11} torr.

SAMPLE SELECTION AND PREPARATION

A representative specimen must be selected and prepared prior to analysis. Important considerations include (*a*) the definition of the problem to be addressed by the analysis, (*b*) the origin of the material to be analyzed, (*c*) the type of handling required, (*d*) the possibilities of contaminating the high-vacuum system, (*e*) the sensitivity of the sample to radiation damage, (*f*) the expected surface characteristics of the sample during service, and (*g*) the method of sample preparation.

The objectives of an experiment determine the factors important to sample selection and preparation, in order to gain an efficient response from a specialist. If, for example, the intention of the experiment is to determine the effects of temperature on the properties of a metal surface protected by a thin lubricant film, it may be necessary to remove the organic layer from the top of the metal.

TABLE 24.1
Photon sources used for XPS

Source	Energy, eV	Depth, Å	Linewidth, eV
Mg Kα	1254	40	0.7
Al Kα	1487	70	0.85
Ti Kα	4510	300	2.0
Cr Kα	5417	600	2.1
Y Mλ	132	—	0.47
Si Kα	1740	—	1.0
Cr Lα	573	—	3.0

On the other hand, if one intends to determine the surface composition of the organic film, direct XPS analysis may be possible, provided that the film has a low vapor pressure and does not contaminate the vacuum system. Solids may also vaporize or decompose in the analysis chamber and contaminate the system; in microelectronic processing, two solids that potentially introduce these problems are ammonium fluoroborate and benzotriazole. The former is used in the thermal treatment of aluminum parts and also as a component of fluxes. Benzotriazole is commonly used as a corrosion inhibitor for copper surfaces.

Electron or x-ray beams used for analysis may chemically alter the sample composition and structure. Undeveloped photoresist films will be quickly altered and may not be easily handled by these techniques. Other organic films have been shown to suffer degradation. For example, ion beam damage effects in poly(methyl methacrylate) have been shown to occur through the reduction of –C–O– and –COO– groups by the elimination of oxygen [7]. In studying the reaction of epoxy resins with metallic films, some epoxy resin components may degrade under x-ray irradiation and produce unexpected results. Boulouri et al. have reported that the diglycidylether of bisphenol A can undergo ring opening as a result of x-ray exposure, yielding a homopolymer rather than the expected epoxy resin, even in the absence of an amine curing agent [8]. The authors illustrated these changes by using XPS and AES techniques.

Another point that is often ignored is the change to a surface caused by sample handling and storage. For example, chromate conversion coatings are altered considerably during storage after coating deposition. Degradation of the adhesion of a polymer film to the protected metal surface may be due to water loss from the coating.

The most common type of surface contaminant, fingerprints, is introduced by sample handling and may lead to erroneous conclusions. An unavoidable type of contamination is the rapid deposition of adventitious hydrocarbon under normal atmospheric conditions and in vacuum systems. Although this type of contamination is almost always present in high-vacuum systems, depending on the nature of the sample, this contribution may be minimized by proper handling.

In many laboratories, EFFA dusters are used to blow off particles from the samples prior to analysis. Normally, this treatment does not introduce any contaminants, but in same cases the propelling solvent may attack organic films and destroy the sample. A typical example of this problem is the destruction of polyoxyethylene terephthaloyl polymer (Mylar or PET) by this treatment.

Most of the samples of interest in microelectronic packaging are solids, and they can be handled in several ways. First of all, samples for analysis must be prepared in a clean environment with properly cleaned tools and handled with gloves to minimize contamination. Solid samples are usually mounted on flat holders by using either conductive paste, methanol-based carbon dag, clips, or other fixtures to insure proper electrical contact. If powders are to be studied, these can be mounted by carefully placing the powder on a foil of a soft metal such as indium or over adhesive tape. In both cases, the material must cover the underlying surface to eliminate unwanted spectral contributions. Another approach to the proper handling of powder involves the compression of the powder into a pellet and mounting to the sample holder. If a fresh surface is required, appropriate cleavage of the pellet must be performed.

Analytical systems may utilize an introduction chamber for sample transfer. This small-volume chamber permits fast pumping to achieve pressures near those existing in the main chamber, thus allowing rapid introduction to the main system. An added advantage of the introduction chamber is that it allows for a close control on unnecessary outgassing in the analytical system and can reduce the chances of contamination. In addition, carrousels that can handle a considerable number of samples are also available in some commercial systems to satisfy the need for repetitive, routine samples such as for quality control of critical processes.

Solid samples inside the analytical chamber may also be treated to expose the surfaces of interest, and various approaches have been successfully used. Among these: (1) ion gun sputtering, (2) heating in ultrahigh vacuum, (3) fracturing materials, and (4) scraping. In some cases, combined cycles of ion bombardment and sample heating may be necessary to produce atomically clean surfaces for calibration purposes. The methods listed above must be used judiciously, and in some cases they may be limited to very specific applications. Recent work by Lazarus and Sham has shown that mechanical scraping of tantalum foil and titanium dioxide single crystals efficiently removes the oxide layer of Ta_2O_5 in the case of the foil and the water present in the case of the single crystal; this technique was shown to be more efficient than argon ion beam cleaning [9].

DATA ANALYSIS

This section describes information that can be obtained from XPS and AES spectra. The problems or uncertainties associated with data analysis are also discussed.

Survey and High-Resolution Spectra

In practical surface analysis, the type of chemical information needed, such as composition or molecular structure, determines the requirements on spectral resolution. For example, if knowledge of the elemental surface composition of an alloy is required rather than the chemical state of the components, only fast survey spectra over a wide range of kinetic energies need be collected. These spectra are limited by the capabilities of the electron gun in AES or by the energy range covered by the anode material used to produce x-rays in XPS. If the composition of the surface is needed, or if the environment around an atom needs to be determined, limiting the pass-energy of the analyzer will allow for data collection over the energy window of interest. High-resolution spectra will take longer to collect, and curve-fitting routines may be needed to assign binding energies and determine various chemical environments for an atom.

Two examples are given below to illustrate the type of spectra obtained in routine XPS and AES analysis. Figure 24-6 presents the survey and high-resolution scans obtained by XPS for a sample of a polyimide film deposited on a silicon wafer and stored in a laboratory chamber. Polyimides are thermally stable insulators used in packaging, particularly in chip carriers (other applications are discussed in Chapters 10 and 11). The most commonly used polyimide is the reaction product of pyromellitic anhydride and oxydianiline, normally represented as PMDA-ODA [10]. The polymeric material contains imide groups, aromatic groups, and nitrogen- or oxygen-carbon linkages. The XPS spectrum of this polymer was recorded using a magnesium $K\alpha$ anode. The survey scan shows the expected elements for the polyimide film, namely, carbon, oxygen, and nitrogen. Other peaks are also present, mainly those arising from x-ray–generated Auger transitions, a ghost peak originating from impurities present in the anode material (in this case, aluminum contamination of the magnesium source results in a carbon ghost peak), and unexpected contaminants such as fluorine (F 1s) signal observed at about 700 eV, and the small copper (Cu $2p_{3/2}$) signal at about 950 eV. If the pass-energy window is reduced to 25 eV, high-resolution spectra of the elements found in the survey scan can be obtained. For the same sample, the high-resolution spectrum of the C 1s level is shown. Three peaks are expected for an aromatic polyimide. One peak is attributed to the carbonyl groups in the imide ring, one to C–O, C–N, and some contribution from C in the PMDA aromatic ring, and another peak to the C–C bonds from the remaining aromatic rings. These are identified on the spectrum. Peaks at the higher binding energies are due to a combination of the C shake-up peaks and CF_2 and CF_2–CFH contributions.

The second example represents the type of information revealed by survey and high-resolution AES scans. Figure 24-7 presents AES spectra obtained for an aluminum alloy chemically etched with an alkaline cleaner. The surface contains a layer of oxidized aluminum over which sulfur and chlorine, initially trapped during the preparation of the alloy, are exposed; adsorbed hydrocarbon is also present. Due to the etching action of the cleaner, other impurities appear on the surface. The high-resolution spectrum of the same sample in the energy range of 550–950 eV is given in Figure 24-7. Analysis of the peak energies and shapes reveals that the concentrated surface impurities are associated with manganese,

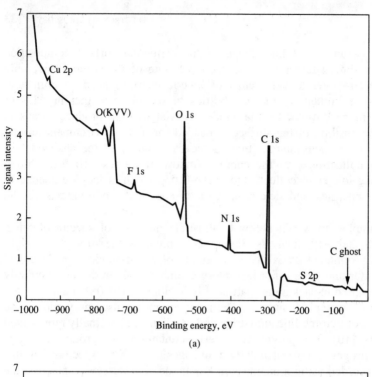

(a)

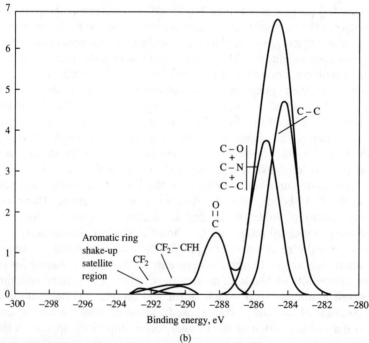

(b)

FIGURE 24-6
(a) XPS survey spectrum of a polyimide film on a silicon wafer. (b) High-resolution XPS spectrum of the C 1s region of the same film showing fluoropolymer contaminant groups in addition to the expected signals for PMDA-ODA polyimide.

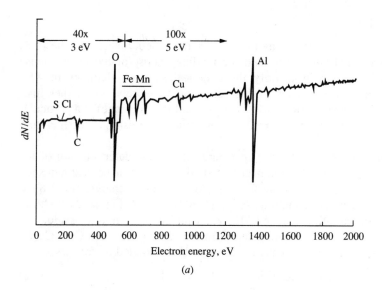

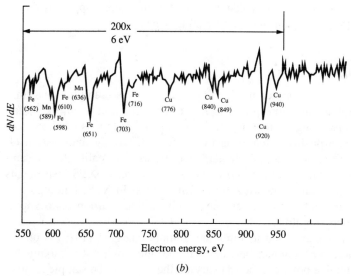

FIGURE 24-7

(a) AES survey spectrum of a piece of aluminum alloy 3003 after alkaline cleaning. (b) High-resolution AES spectrum of the same sample in the 550–950 eV region showing the presence of copper, iron and manganese. After reference [11].

iron, and copper. A more precise interpretation of the etching process with the assistance of other analytical techniques demonstrated that the impurities were segregated in islands and corresponded to alloy intermetallics with aluminum. These intermetallic species were insoluble in the pH range of the cleaning process [11].

Auger Parameter

The Auger parameter α, defined as the sum of the kinetic energy of the most intense x-ray–generated Auger line and the binding energy of the most intense photoelectron line, can help identify the chemical state of elements in XPS [12]. This sum of energies can be measured more accurately than either line. In addition, α is independent of sample charging and is a constant for a particular chemical compound; values of α have been determined and listed in the literature [13].

An example of the use of the Auger parameter is the determination of the oxidation states of copper. Although cupric oxide (CuO) can be distinguished from metallic copper (Cu^o) based on their core level binding energies (933.8 eV and 932.6 eV, respectively), cuprous oxide (Cu_2O) cannot. Figure 24-8 shows the XPS 2p spectra and XPS Auger LMM lines for copper, cuprous oxide, and cupric oxide. Use of the Auger parameter allows distinction between Cu^o, Cu_2O, and CuO. In this case, the Auger parameters can be calculated as follows:

$$\alpha = KE(LMM) + BE(2p_{3/2}) \tag{24.2}$$

Table 24.2 lists the values for α, calculated for the different copper oxidation states, and shows that these can indeed be distinguished.

Sample Charging

A potential problem in photoemission spectroscopy is sample charging. Nonconducting samples are more prone to this effect. Electron beam excitation in AES leaves the insulating sample negatively charged, while loss of photoelectrons in XPS leaves a positive charge on the surface. Charging manifests itself in kinetic energy shifts. An internal standard can be used to calibrate for sample charging, whereby kinetic energy corrections can be made. XPS instruments contain flood guns for charge neutralization of insulators. In XPS, binding energies are frequently corrected by referencing to adventitious carbon or to vapor-deposited metals such as gold.

Several techniques are used to minimize sample charging [14]: (a) photo-conduction, (b) increasing contact between sample and holder, (c) mixing the sample with a conductive powder, and (d) varying the angle of the sample relative to either the source or the analyzer.

Depth Profiling

Often, chemical information obtained by analysis of the surface is enhanced by information gained from below the surface. Several techniques can be used to yield elemental or chemical information as a function of depth. These techniques include surface analysis combined with sputter depth profiling, angle-resolved XPS, and XPS with higher-energy x-ray sources [15].

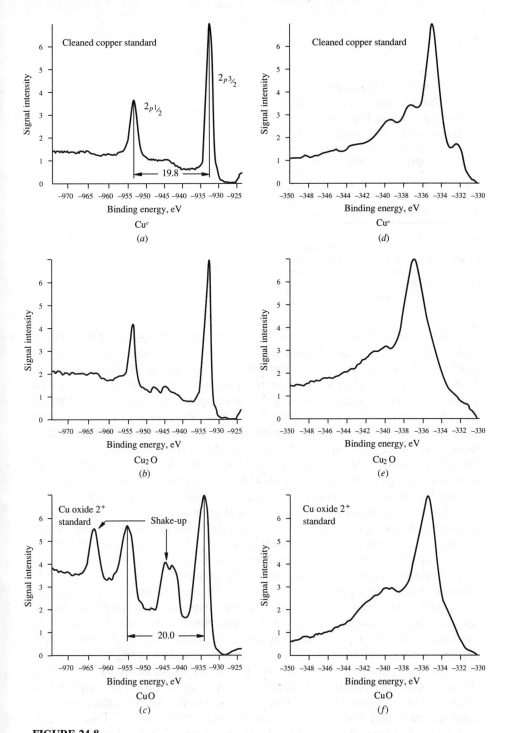

FIGURE 24-8

Cu 2p photoelectron lines and x-ray–generated Auger LVV lines for copper metal, cuprous oxide, and cupric oxide. The Auger parameter α is obtained by adding the energies of the Auger transition and the Cu $2p_{3/2}$ line.

737

TABLE 24.2
Core level binding energies, auger line kinetic energies, and auger parameters for copper, cuprous oxide, and cupric oxide

	$2p_{3/2}$	LMM	α
Cu	932.6	918.4	1851.0
Cu_2O	932.6	916.6	1849.2
CuO	933.8	917.9	1851.7

Sources: C. D. Wagner, W. M. Riggs, L. E. Davis, J. F. Moulder, and G. E. Muilenberg, *Handbook of X-ray Photoelectron Spectroscopy*, Perkin-Elmer Corporation, Physical Electronics Division, Eden Prairie, Minn., 1979; and C. D. Wagner, unpublished data.

Both XPS and AES can be combined with ion beam sputtering techniques. The sample is usually bombarded with ion beams of energies in the range of 1–15 keV. Ions from inert gases (e.g., Ar) are commonly used. To obtain useful profiling information, the analysis probe (x-ray or electron beam) must be well contained within the crater formed by ion beam sputtering to avoid contributions from the sidewalls of the crater. Since the electron beam diameter is usually much smaller than the ion beam, this is not normally a concern for AES (unless spatial information is required, such as is the case when performing elemental maps). For XPS, however, the ion beam may need to be rastered, i.e., scanned over an area large enough to ensure that the area of the crater is larger than the excitation or sampling area; this is less of a concern when using small-spot XPS.

In order to obtain a depth profile, a depth scale must be determined by (a) measuring the sputtering rate for a particular material and ion beam current density, (b) measuring the time to sputter through a layer of known thickness, or (c) measuring the depth of the crater formed as the result of sputtering.

Although chemical information can be obtained as a function of depth, artifacts in data interpretation may arise due to changes that may occur as a result of the interaction between the ion beam and the sample material. Depth resolution, for example, is a function of the sputter depth. Some oxides may be reduced by ion beams [16,17], or preferential sputtering may occur [18,19]. A typical depth profile for the components of an aluminum alloy is shown in Figure 24-9.

Nondestructive techniques can also be used to obtain information regarding the elemental distribution as a function of depth up to the escape depth of an electron. For example, electrons emitted perpendicular to the surface originate from deeper depths than electrons emitted from oblique angles. This principle is depicted in Figure 24-10. Figure 24-11 shows an example of angle-resolved XPS spectra for an Al_2O_3 film on Al; variations in effective sampling depth and relative surface sensitivities as a function of angle are shown in Table 24.3. To increase the mean free path of electrons, a higher-energy x-ray source can be

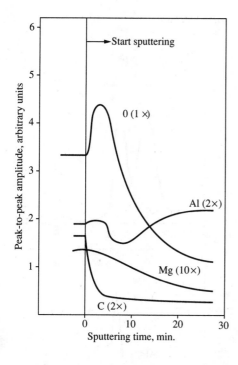

FIGURE 24-9
Auger beam depth profiles for some 3003 aluminum alloy components. The scans were done by following the peak-to-peak amplitudes of the most intense lines for each element. Taken from reference [11].

used for excitation. The table also includes a partial listing of some x-ray sources and their corresponding analysis depths.

Other approaches make use of the differences in electron kinetic energy to differentiate between signals arising from various depths of the sample. In the energy range between ≈ 100 eV and ≈ 1000 eV, mean free path lengths are between 5 and 20 Å. These differences may be useful in determining whether a material is located at the surface or in the bulk. For example, since Cu has peaks present in the AES spectrum at kinetic energies of 76 eV and 920 eV, surface (top 5 Å) versus bulk (≈ 20 Å) contributions can be determined. Figure 24-12 presents AES spectra for a clean Cu surface and a contaminated Cu surface, respectively. The thickness of the contaminate layer can be estimated by the attenuation of either Cu signal using treatments beyond the scope of this chapter. Additional information on this subject can be found in the suggested reading material.

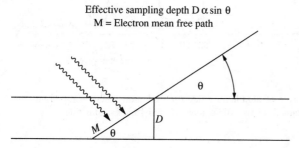

FIGURE 24-10
Basis of angular resolution in XPS.

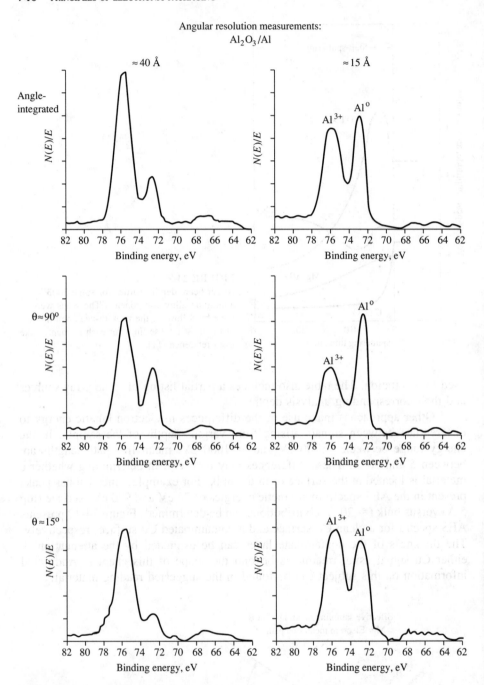

FIGURE 24-11
Angle-resolved XPS spectra for a layer of aluminum oxide on aluminum metal; θ represents the take-off angle. The spectra were obtained for two films about 40 and 15 angstroms thick.

TABLE 24.3
Effective sample depth as a function of electron take-off angle θ for XPS measurements

θ^1	Effective sampling depth (relative)[2]	Relative surface sensitivity
90°	1.0	1.0
30°	0.5	2.0
11°32'	0.2	5.0
5°44'	0.1	10.0

[1] θ is defined as the electron take-off angle as shown in Figure 24-10.
[2] Effective sample depth relative to normal value.

Quantitation/Sensitivity

XPS. Quantitative analysis can be accomplished since the intensities of the peaks are proportional to the elemental concentration. Among other factors, intensities (I_x) are proportional to the atomic photoelectric cross sections (σ_x), the escape depth of the electron (L_x), the efficiency of the photoelectric process (y_x), and the spectrometer transmission function (T_x). If, for example, the relative concentrations for elements A and B, n_a/n_b, needs to be determined, Equation 24.3 must be used:

$$\frac{n_a}{n_b} = \frac{I_a}{I_b} \times \frac{\sigma_b \times L_b \times y_b T_b}{\sigma_a \times L_a \times y_a T_a} \tag{24.3}$$

If $y_a \approx y_b$, and the peaks have similar energies, then $L_a \approx L_b$ and $T_a \approx T_b$. The ratio then simplifies to

$$\frac{n_a}{n_b} = \frac{I_a}{I_b} \times \frac{\sigma_b}{\sigma_a} \tag{24.4}$$

Photoelectric cross sections for different subshells have been calculated [20], and they show a strong Z dependence. As an alternative to the above calculation, primary standards may be used to determine sensitivity factors, which describe the response of an element in a particular chemical environment to photoemission. Sensitivities for most elements have been measured by Wagner [21]; however, errors as large as 50 percent can be introduced as the result of matrix effects.

AES. Although the use of AES is primarily qualitative analysis, quantitative analysis is also possible [22–24]. Feldman and Mayer [22] give a good summary of quantitative analysis by using AES to determine the absolute concentration of

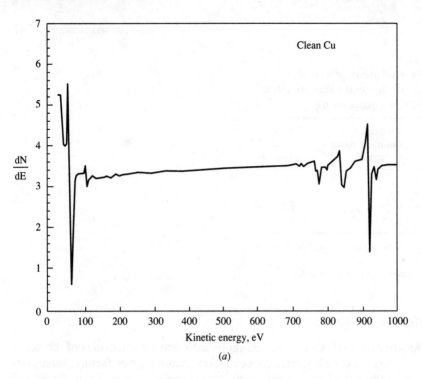

(a)

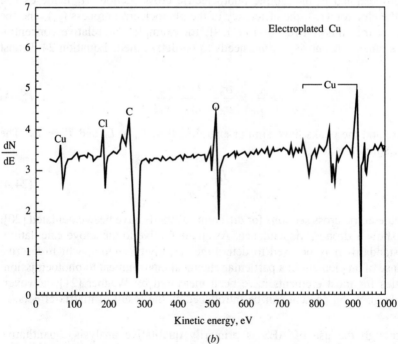

(b)

FIGURE 24-12
(a) Derivative AES survey spectrum of clean copper surface. (b) Derivative AES survey spectrum of the same film after contamination. The thickness of the contamination layer can be derived using the attenuation of the copper signal at about 60 eV.

an element x in a matrix:

$$Y_A(t) = N_x \Delta t \sigma_e(t)(1 - w_x)e^{-t\cos\theta/\lambda}I(t)Td\Omega\frac{1}{4\pi} \tag{24.5}$$

where N_x = number of x atoms per unit volume

$\sigma_e(t)$ = ionization cross section at depth t

$Y_A(t)$ = yield of KLL Auger electrons

w_x = fluorescence yield

λ = escape depth

θ = analysis depth

T = analyzer transmission

$d\Omega$ = analyzed solid angle

$I(t)$ = electron excitation flux at depth t

Or, when using external standards with known concentrations:

$$\frac{N_x^S}{N_x^T} = \left(\frac{Y_x^S \lambda^T}{Y_x^T \lambda^S}\right)\left(\frac{1 + R_B^T}{1 + R_B^S}\right) \tag{24.6}$$

where N_x^S = known concentration of element x

N_x^T = concentration to be found in the test sample

R_B = backscattering factor

Elemental sensitivity factors, although less accurate, are probably most frequently used for determining atomic concentrations of elements, by the relationship:

$$C_x = \frac{I_x/S_x}{\sum_y(I_y/S_y)} \tag{24.7}$$

where C_x is the concentration of element x, S_x and S_y are the relative sensitivities for elements x and y, and I_x and I_y are the signals for elements x and y, respectively [25–27].

APPLICATIONS
TO ELECTRONIC PACKAGING

Some mention of the applications of these techniques in the electronic packaging industry has already been made. Because of the complexity of electronic packages, the analysis of a single, localized area may not be representative. In an industrial environment, surface analysis is generally used in four different ways: (1) to characterize surface and interfacial reactions, (2) to guide process development, (3) to find production defects, and (4) to determine how reliable the product is under environmental exposure. Each of these applications will be considered separately in the following pages.

Surface and Interfacial Reactions

Surface and interface properties are important concerns in electronic packages. The durability of an adhesive bond between similar or dissimilar materials is a good example of what is important to maintaining the mechanical integrity of a variety of components. The oxidation of metallic films during deposition or plating can strongly influence the quality and performance of the system. Selected examples will be mentioned below for the following cases:

1. Metal/inorganic layer, metal/polymer, and polymer/polymer interactions.
2. Surface modification
3. Reactions on metal surfaces
4. Fluxes for soldering
5. Corrosion inhibition

METAL/INORGANIC LAYER, METAL/POLYMER, AND POLYMER/POLYMER INTERACTIONS. In recent years, a considerable amount of attention has been given to the problems of adhesion between similar and dissimilar layers involved in packaging. Chapter 27 expands on the subject of adhesion and is complementary to the discussion here.

Surface analysis has taken an important role in separating the chemical contributions from the mechanical aspects of adhesion. In general, adhesion is the result of chemical, mechanical, and electrostatistic interactions at an interface. On a microscopic scale, the roughness of the interfacial layers contributes to the mechanical strength of the joint, whereas surface reactions at specific sites contribute to the chemical aspect of adhesion. Surface charges contribute to the electrostatic component of adhesion. Issues of adsorption and desorption of solvents, and contamination, and the effect of these on mechanical integrity and failure analysis have been addressed by a variety of surface analysis techniques.

Processing parameters, deposition conditions, and environment in general influence strongly the properties of metallic films. An example is the behavior of copper films deposited on top of existing copper films, such as those found in electrical connections. Lefakis and Ho [28] have shown, using AES and depth profiling, that when a 200-Å-thick copper film is evaporated onto copper, the expected interface oxide cannot be found. In films with a thick layer of oxide prior to the deposition of the 200 Å Cu, the oxygen is shown to be distributed throughout the film. The conclusions of these experiments demonstrate that copper oxide decomposes and loses oxygen through diffusion into the copper matrix (see Chapters 25 and 26 on Case I and Case II diffusion). The practical importance of these results is that if thin copper oxides are present on real samples, they can dissociate in some temperature ranges and so would have little effect on copper-to-copper contacts.

Bonding of metals to SiO_2 films has been studied by using surface analysis techniques. There is general agreement that adhesion of a metallic film to SiO_2 is dependent on the metal deposited. Two typical behaviors are found. Good adhesion to SiO_2 is always present for metals such as Ti, V, Nb, Hf, and Zr. The metals Ni, Pd, Pt, Cu, and Ag do not adhere well to SiO_2. Chromium has a behavior that does not fit either of the two groups. Cros et al. [29], have recently shown that when ultrathin layers of chromium are evaporated on SiO_2, they actually react at room temperature by forming a thin region of chromium silicide in a matrix containing chromium oxide and segregated silicon. The use of angle-resolved XPS spectroscopy and AES analysis of the samples following evaporation and in situ annealing suggest that a similar room temperature reaction could be expected for low-temperature-grown SiO_2 films.

Hybrid packages contain a combination of metal films and organic polymeric films used as insulators. These films must show excellent mechanical integrity during their expected life span. Adhesion of metals to polymeric materials has been studied recently to determine the contribution of mechanical, chemical, and electrostatic factors to adhesion using surface analysis. Several routes can be taken to study the interaction of a polymer with a metallic film: (a) a metallic film is deposited onto a fully cured polymeric film, (b) a metallic film is formed prior to the deposition of freshly synthesized organic film, or (c) a metallic film is allowed to interact with a precursor of the polymer. Case (a) may be used to study the interaction of the metal on the cured film, while cases (b) and (c) may be used to attempt to understand the affect of the polymeric film as it forms on the metallic substrate. The distinctions between these two cases (i.e., metal on polymer versus polymer precursor(s) on metal) are subtle but may be used to clarify individual interactions of the initial precursor of the polymer with the metal, as opposed to the single interaction of an intermediate compound with the metal film.

The third route involves the interaction of the polyamic acid with the metal surface coated with a thin layer of oxide. In this instance, metal selectivity and interaction with the organic material seem to influence the stability of the metal-polymer joint.

Ho et al. [30] have shown that UPS and XPS give considerable information on the interaction of metals with cured polyimide (PMDA-ODA) films, and that UPS can also be used to follow precisely the curing reaction of the polyamic acid to polyimide. Since water is eliminated by ring closure and can adsorb on the cured film, an additional anneal step was necessary to produce a dry film. After anneal, the experimental UPS bandshapes were in agreement with the calculated band spectrum of the polymer. Figure 24-13 presents the UPS spectra and the calculated band shape for PMDA-ODA polyimide to show the similarity between experimental and theoretical results. Following deposition of a few monolayers of copper (Figure 24-14), the valence band spectrum changes due to the appearance of the Cu 3d level superimposed on the valence band region of PMDA-ODA. After anneal the lineshape more resembles the initial polyimide spectrum, indicating the weak interaction of copper with the polymer and the possibility of intermixing with the polymer layers below the surface. XPS spectra of the C 1s region of these samples show a gradual decrease of the carbonyl signal in PMDA-ODA, with copper coverage represented by the number of monolayers

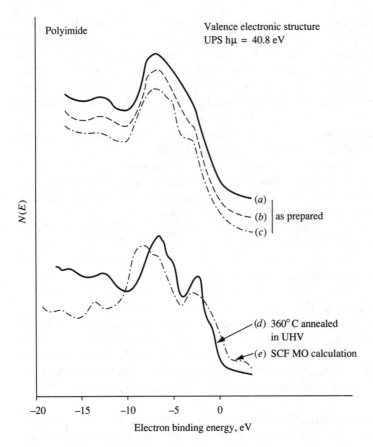

FIGURE 24-13
UPS spectra (a)–(c) of three PMDA-ODA films in the as-prepared state. UPS spectra (d)–(e) of the same film after anneal at 360°C in UHV and simulated using SCF MO calculations. After reference [30].

(ml) deposited. However, after anneal the carbonyl intensity recovers, indicating little or negligible interaction with the polymer.

If the metallic film is aluminum, UPS is not as useful because the valence states of aluminum have weak intensities. However, high-resolution XPS spectra of the C 1s, Al 2p, N 1s, and O 1s regions show that at low monolayer coverages, aluminum interacts with the carbonyl groups, and as the coverage increases, a new peak corresponding to the formation of the Al–O–C bond appears.

The investigation of interactions of the above type has been expanded to other polymeric systems of practical importance [31]. Thin aluminum films deposited onto polyacrylic acid and low-density polyethylene have been studied by using conventional and angle-resolved XPS analysis. Experimental results revealed that in the polyacrylic system, the metal interacts with the polymer through the formation of Al–O–C bonds, and with polyethylene through the

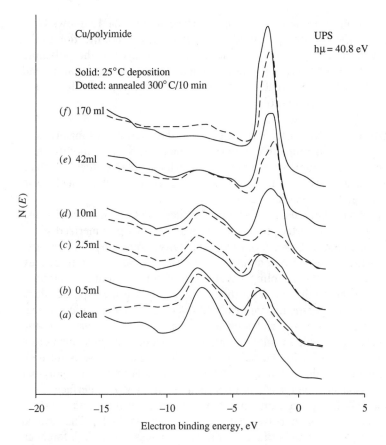

FIGURE 24-14
UPS spectra measured during in-situ formation of the Cu-polyimide interface as a function of Cu monolayer coverage (ml). After reference [30].

formation of Al–C groups. Extension of this work to copper led to the conclusion that copper does not form this type of bond with these polymers.

When the polyamic acid of PMDA-ODA is formed in the vapor phase and allowed to interact with a polycrystalline silver surface, the individual reactants behave differently; namely, ODA adsorbs associatively while PMDA decomposes. Vapor phase–formed polyamic acid reacts with the metal at about 120°C and fully imidizes, but the structure contains silver-carbon bonds. Partially decomposed PMDA is also capable of forming these bonds [32].

Sometimes, it is important to bond a layer of a polymer to an already cured layer of polymer material and produce a good adhesive joint. For a cured PMDA-ODA polyimide film, surface modification of the initial layer may be required. For instance, the cured film can be hydrolyzed by bases, which results in an opening of the imide rings and regeneration of carboxylate groups. Addition of a layer of polyamic acid on the modified surface layer followed by a thermal treatment results in the formation of a relatively strong interface. Figure 24-15 shows C 1s high-resolution XPS spectra for a polyimide film after surface modification to the carboxylate salt and after deposition and curing of a thin layer of the polyamic acid. Using the methods of curve resolution, an additional C 1s signal at around 288 eV is detected for the base-treated film. This signal indicates the formation of a carboxylate group during hydrolysis.

SURFACE MODIFICATION. Surface modification is the alteration of materials to make them compatible or incompatible with other materials. Surface modification can be achieved by preferentially reacting a material at active sites. Surface modification approaches may include some energetic treatment such as an electron beam, plasma source, or corona discharge.

Alkoxysilanes have been used for several decades due to their reactivity and ability to react with specific surface hydroxyl groups to form a polymerized siloxane network on the material being modified. Typically, some metallic surfaces that contain thin oxide layers also contain these hydroxyl groups; therefore, they can potentially react with the coupling agent. Aluminum, titanium, chromium, and silicon films can be modified. With the ability to choose the type of silane coupling agent, one can specifically tailor a hydrophobic or hydrophilic surface. In addition, one can select a material with a tail that contains a reactive group and can effectively cross-link to another organic molecule. In this manner, a transitional layer compatible with the bulk of the added organic phase is formed, and an enhancement of adhesion may also be obtained. Polymer-into-polymer diffusion, as discussed in Chapter 25, is important at this type of interface.

Surface analysis has played an important role in determining the surface changes that enhance adhesion between two dissimilar materials. The critical requirements on the amount of surface modification necessary to reinforce an unsaturated polymer for use in microelectronic applications can be specified. Inorganic fillers, such as micas, can react with coupling agents and produce materials dispersible in organic media. If an azido-terminated silane is used, one can couple effectively the surface-modified mica to a double bond in an unsaturated polymer [33]. Precise evaluation of the surface modification level and the mechanical strength of the composite, as measured by XPS determination of pre-reaction surface concentrations of azido groups (N_3), is valuable in controlling the performance of the composite. Figure 24-16 illustrates the XPS survey spectrum for mica after treatment with a solution of an azido-terminated silane. The small amount of nitrogen in this spectrum reveals the incorporation of the coupling agent on the mica. The N 1s high-resolution spectrum also shown in this figure indicates the two different nitrogen environments expected for an azido group. Quantitative XPS analysis of the total nitrogen level can be correlated with the concentration of the silane solution.

Sometimes, surface modification is introduced by plasma polymerization followed by an additional reaction on a newly generated surface. Inagaki et al. [34] have shown that this combination of reactions can produce novel moisture-sensitive devices. These are used to monitor the exposure of components and

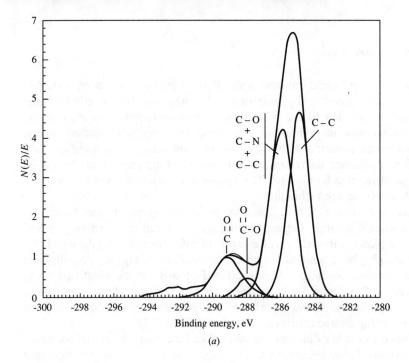

(a)

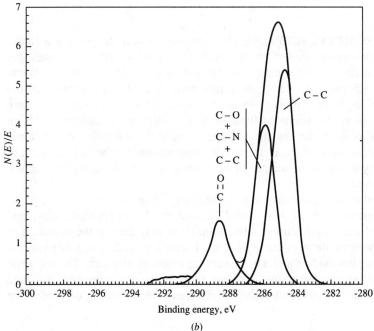

(b)

FIGURE 24-15
(a) XPS spectrum of the C 1s region in a partially hydrolyzed film of PMDA-ODA showing the formation of carboxylate groups at around 288 eV. (b) XPS spectrum of the C 1s region of the film shown above after reaction and curing with a thin film of polyamic acid. The curve-resolved signals are added for clarification.

boards to moisture in humid environments. Plasma polymerization of trimethyl sylildimethylamine, bis(dimethylamino)methylsilane, and bis(dimethylamino)-methylvinylsilane has led to the formation of polymeric networks on ceramic substrates treated with methyl bromide to yield ammonium-terminated layers. These films contain quaternary nitrogen ions that are sensitive to moisture, and their electrical resistance decreases by four orders of magnitude in the 20–90 percent range of relative humidities. The sequence of reactions was followed by XPS and IR spectroscopic techniques.

XPS has also been used to follow changes occurring on surfaces treated by plasmas or corona discharge. Organosilicone polymers can be effectively etched in pure oxygen plasmas to yield surface layers of SiO_2 for thin-film devices [35]. Gerenser et al. [36] have pointed out special techniques of tagging polyethylene surfaces after corona discharge. Newly created groups can be identified using angle-resolved XPS. Corona discharge treatment can segregate some materials to the surface. Polyethylene shows a preferential segregation of stabilizers that can decrease the wetting characteristics of the treated films [37].

The above examples illustrate the role of surface analysis on surface modification processes. Other references to this topic may be found in the suggested reading list.

REACTIONS ON METAL SURFACES. The reactions of a metal surface with the surrounding environment can affect the performance of an electronic package. For example, in epoxy resins, many brominated compounds are added as flame retardants (see Chapter 11). These materials must be carefully selected: Torrisi and Pignataro [38] have presented evidence, using XPS, or migration of these components from epoxy resins to their interface with copper laminates. Using XPS, it was possible to follow the kinetics of bromine migration on fractured surfaces subjected to a thermal treatment. Preparation of the copper surface appeared to play a role in the migration of bromine as well as the type of surface reactions observed.

High-reliability circuits may use hermetically sealed microelectronic packages with materials such as gold-plated Kovar, an alloy of nickel, iron, and cobalt. Thermal excursions during fabrication steps may lead to the diffusion of the alloy components through the gold film. Following surface segregation, the metallic components oxidize and adsorb varying amounts of water. The sequence of reactions leading to these results has been analyzed by angle-resolved XPS and applied to the analysis of encapsulated parts after various stages of production [39].

FLUXES FOR SOLDERING. Parts or components of electronic packages are often connected by specific solders, as discussed in Chapter 19. As a surface preparation procedure, the base metal may need to be cleaned with a chemical compound that removes oxides, sulfides, carbonates, or corrosion products and permits wetting of the solder. The preparatory step is accomplished by applying organic or inorganic compounds as fluxes. The flux may react with the base

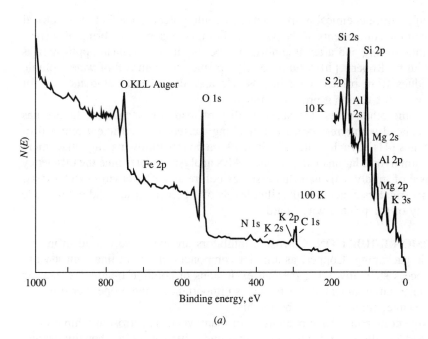

(a)

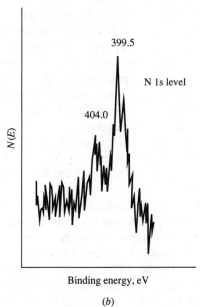

Binding energy, eV

(b)

FIGURE 24-16

(a) Survey XPS spectrum of a surface-modified mica with an azido-terminated silane. (b) High-resolution XPS spectrum of the N 1s region of the same sample showing the two expected nitrogen environments for an azido group. After reference [33].

metal and produce chemical by-products that could potentially affect the electrical response of the components of the package. For this reason, monitoring of surface composition during and after assembly may be required. A typical application is pointed out by Roberts [40] who used XPS to study the removal of water-soluble flux residues from printed wire boards. He also describes a rapid method for evaluation of the cleaning process efficiency.

In some other cases, the susceptibility of leads to mechanical failure has been associated with stress corrosion cracking caused by insufficient removal of flux residues [41]. Lynch et al. [42], in work on boards following chip attachment to thick-film hybrid laminates, have used AES analysis to determine the efficiency of removal of mildly activated flux residues before and after environmental tests. These investigators concluded that flux levels of 1 percent or less did not degrade the quality of the printed wire board.

CORROSION INHIBITION. Corrosion inhibitors are extensively used in microelectronic packaging. Copper, as a major component of circuit lines, must withstand process steps involving thermal excursions or solution treatments. During field service, environmental effects are also important for the proper performance of the package; corrosion must be inhibited.

Azole compounds have been used for many years as corrosion inhibitors for copper and its alloys. Among such compounds, benzotriazole, benzimidazole, and imidazole are used in their pure forms or in proprietary formulations. The efficiency of these compounds for corrosion inhibition is related to the reactivity of the ligand molecule with surface oxides of copper. Although there is controversy on the definitive modes of complexation, it is fairly well established that the interaction of benzotriazole with cuprous oxide may lead to the passivation of the copper surface. Because there are many ways to prepare a surface for processing, there are different degrees of interaction between the organic molecule and the metal. Concentration, pH, degree of oxidation, application temperature, and time of contact between the inhibitor and the metal are important in selecting process parameters. Monolayers or multilayers may be formed on the protected surface. Since packaging is a dynamic and sequential process, corrosion inhibition requirements may vary from step to step.

Surface analysis is a key tool for determining the behavior of the metal surface through simulated or real production lines. In printed wire boards that require a determination of surface composition on the board itself, conventional XPS is applicable. On the other hand, if the inner surface of a plated through-hole needs to be examined, small-spot XPS may meet the requirements. Where there are spatial limitations, such as in restricted or particular areas on lands of plated through-holes, AES can be used to obtain information on inhibitor surface coverage.

Several publications have reported the applications of surface analysis to the above problems. Miller et al. [43] have reported AES studies on the interactions of HCl-etched copper with benzotriazole and benzimidazole as a function of ligand concentration, temperature, and time of immersion at neutral pH. Both ligands form corresponding cuprous-ligand complexes, but the thickness of the resulting layers varies; benzimidazole forms layers ten times thicker than those formed by the benzotriazole complex. XPS has been used to study the surface composition of copper protected by azole corrosion inhibitors after long storage periods prior to fluxing and soldering. The performance after soldering has also been classified by using the XPS results [44].

Only a few examples of applications of surface analysis in microelectronic packaging have been mentioned here. Many more exist, and frequently it is necessary to couple several techniques to achieve a complete understanding of a simulated or real process.

The Role of Surface Analysis in Process Development

This section will describe situations in which surface analysis plays an integral role in process development. Surface analysis is used to test and compare various alternatives for a process step, to characterize surface and interface reactions, and to examine products subjected to simulated or real environmental exposures in order to define acceptable materials.

Wire leads of similar or dissimilar materials joined by a variety of methods are used in packaging. Aluminum-copper wire bonding presents some economic advantages; however, its application to working systems has been hampered by the difficulty in producing a reliable bond. Olsen and James [45] have reported that microstructure, test conditions, and prebonding and postbonding conditions will affect the reliability of the bond. Ultrasonic bonding of aluminum-copper joints has been shown by Pignataro et al. to produce good bonds provided that the surface composition of the copper prior to bonding is properly controlled [46]. These researchers followed the copper surface composition as a function of surface preparation by means of XPS. After characterization, the bonds were subjected to high-tensile-strength (HTS) measurements and pull tests after pressure cooker tests (PC). Figure 24-17 gives the results obtained for a control and three different surface preparations. The x-ray–generated Cu LMM transitions are shown for the control (a), sample heated to 500°C in nitrogen (b), sample heated to 500°C in hydrogen (c), and sample etched in 5 N HCl and rinsed before bonding (d). HTS results indicate that preparation method (b) loses bond strength in about 50 hours. On the other hand, the three other methods give relatively good results with over 300 hours of exposure. In the pull tests following PC exposure, samples from (c) yielded the best bond strength retention.

In the attachment of a silicon wafer to a package, a gold-silicon preform is often used. A gold layer is normally deposited on the back side of the wafer for protection against contamination or oxidation prior to sawing. Several methods are possible for attachment of the wafer to a metallic layer on the board. Among these, rapid thermal excursions are standard processes. A major concern in defining a process sequence is to determine whether the gold film must be evaporated or sputtered to yield acceptable performance. Also, if the plant has limited capacity,

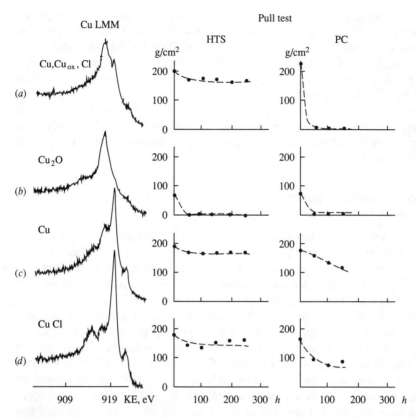

FIGURE 24-17
Cu LMM XPS signals of four different copper surfaces, with bond strength data, and pressure cooker test data. See text for description of samples. After reference [46].

it may be important to know at what stage of production the process can be disrupted without affecting yields.

Linn and Bajor [47] have described an approach to optimize these processes. Following different methods of surface preparation for silicon wafers, gold films were evaporated or sputtered. Samples were subjected to a humid environment prior to peel testing. During the last step, it was noticed that the samples that were evaporated failed rather easily. On the other hand, the sputtered samples had high peel strength. XPS analysis of the evaporated interface revealed that the silicon wafer contained high levels of SiO_2. SEM micrographs demonstrated that the two different deposition methods resulted in different morphologies for the gold layer. Due to the differences observed, it was proposed that water diffusion to the interface was the cause of failure in the evaporated layer. Rapid thermal anneal of either type of gold deposition technique yielded excellent bonds to the package. Based on these observations, it would be necessary to introduce rapid thermal anneal in the process to produce reliable parts if the sequence of fabrication were expected to be disrupted.

An example of process optimization was reported by Lea and Howie [48], who determined the effects of electroless copper on the outgassing of boards in automated soldering of plated through-holes in printed wire boards. Outgassing defects may occur during final copper plating and will induce discontinuities in the wall of plated through-holes. The performance of 14 production lines was evaluated keeping constant all processing steps with the exception of the electroless deposition of copper. Several boards made of FR-4 epoxy glass laminate were evaluated before and after the individual process steps, and characterized by AES, XPS, and SEM techniques. The remaining parts of the experiment were allowed to continue through photoresist exposure, personalization, and production of plated through-holes under identical conditions. The parts were copper plated by a single source under conditions to produce sets of samples containing 10, 20, or 30 μm of deposited copper. Finally, the boards were soldered and inspected visually for dicontinuities in the copper film. The results in plating quality were related to the thickness of the copper-plated barrel, and considerable variations in each thickness was found.

Opacity measurements of the electroless coatings were made. Coatings were separated into different groups, from those with virtually perfect opacity to those containing large voids. XPS analyses of these coatings for the relative concentration of the palladium activator correlated with the arbitrary scale of opacity, as shown in Figure 24-18. (See also the discussion of seeding in preparation for copper deposition in Chapter 17.)

Since in the electroless deposition process tin is used as a sensitizing agent and palladium as the activator, less tin and more palladium would be expected in a good electroless film. Quantitative XPS analysis of several samples demonstrated that after some time, the palladium-to-tin ratio remains constant and cannot be correlated to copper quality measured by the opacity scale. Further work to clarify

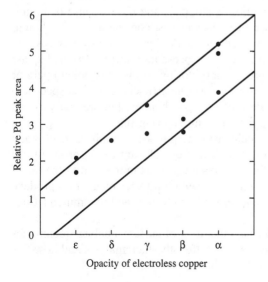

FIGURE 24-18
Palladium coverage of the wall of plated through-hole surface as measured by XPS and the degree of opacity of the electroless copper coating. After reference [48].

the composition of these films was pursued by using AES combined with argon ion depth profiling. This approach revealed that, initially, the surface was rich in tin, and it was only after removal from the accelerator bath that the tin at the surface decreased in concentration. The shape of the depth profiles after removal from the bath gave an indication of the sequence of reactions leading to the electroless film formation, in agreement with the proposed reaction mechanism of electroless deposition. The authors concluded that XPS and AES alone could not provide the information needed to solve the problem.

SEM micrographs were used to determine the morphology of the various films during deposition. Film quality seemed to be related to the number of small-sized copper nuclei and their growth rate. It was also shown that the metal catalyst coverage on the epoxy matrix and glass fibers played an important role in overall performance.

The conclusions of this detailed study were that defects present in the electroless copper film, and an inadequate sealing of these defects by the electrodeposited copper, are the main cause of outgassing during a soldering step. Only with proper evaluation and strict control of process parameters can the percentage of defects as a function of copper barrel thickness be adequately reduced and understood.

Failure Analysis

Reliability testing exposes products or parts to simulated operating and accelerated test conditions, such as higher temperature and humidity, increased voltages, mechanical shock, and thermal and electrical cycling. The methodology used in predicting product durability under various conditions is described in Chapter 9, on reliability and testing. Failure analysis is required when a part or product does not meet the expected life requirement.

Characterization of product defects found in manufacturing line-down situations is of a different nature. Production is stopped, and the response of failure analysis must be immediate, because of yield and cost considerations. Although manufacturing lines may be well controlled for standard processing conditions, the following can influence yields: (a) inadequate process monitoring, (b) inadequate process adjustments in response to process monitoring, (c) unintentional operator error, (d) unauthorized process modification, (e) insufficient materials specification, and (f) inadequate documentation of process changes. Failure analysis can contribute also in improving the yield of a manufacturing line.

An understanding of failure mechanisms requires a combination of good technical skills and the ability to obtain information relevant to the problem. To determine the failure mechanisms, the samples must be examined "after the fact." After information on the failure is obtained, simulation experiments to reproduce the failures are desirable. A model is developed and tested, and recommendations are made to correct the problem.

In this section, several practical applications of surface analysis techniques are summarized to illustrate the importance of after-the-fact analysis and how this information leads to solutions of costly problems.

Optical microscopy, SEM, and Auger depth profiling techniques combined have often been used to examine via holes interconnecting two metal layers separated by an insulator. A typical problem began with electrical testing of some vias showing higher-than-usual resistance readings. Analysis revealed that residual polymer restricted the contact between the two metal layers in the via. As a result, moisture was able to penetrate between the metal layers to the via interface, and oxidize the metal. Based on failure analysis results, recommendations were made to correct the erratic behavior of the polymer etching process and include an additional step to insure removal of residual polymer from the via.

In a lamination process, a copper foil available from a manufacturer was initially evaluated and shown to have the desired tensile strengths and levels of ductility. This copper foil was used for production lamination to an epoxy prepreg. Weeks later, several parts failed minimum requirements for adhesion strength between the copper and the epoxy. Characterization of the epoxy polymer indicated that the degrees of cure were the same on both the good and bad parts. Additional discussions with production engineers revealed that the manufacturer had delivered two different lots of the same copper foil. Mechanical tests of both copper lots showed equivalent mechanical properties and textures; therefore, it was concluded that the two lots were identical. Surface analysis of the as-received foils from the supplies was conducted. XPS survey scans for a sample of the first lot used in preproduction runs showed a surface treatment of oxidized chromium. The second lot contained only oxidized copper. Corrections were made and additional specifications to include a surface treatment as well as the desired mechanical properties were requested from each copper foil supplier.

During fabrication of glass-reinforced epoxy-to-epoxy bonds to produce structural laminates, some specimens failed shear strength testing. SEM analysis indicated possible adhesive failure at the interface. Once the location of failure had been established, elemental compositions of both sides of the failed joint were determined by surface analysis. XPS survey scans indicated C, O, N, and Si were present on both sides. C, O, and N are known components of the epoxy. Possible sources for Si were glass fibers used as the reinforcement in the laminate, and a silicone-based polymer used as a mold release agent in a lamination step. Because XPS could not differentiate well between these two Si chemical environments, absorption-reflectance infrared spectroscopy was used [49]. IR spectra gave positive identification for a silicone-based polymer.

The detection of a suspected mold release material is not necessarily an indication of failure. Depending on the desired mechanical requirements of the joint, the level at which failure is induced by the contaminant must be determined [50]. In order to ensure that excessive levels of mold release agents are not transferred, process modifications must be made. Two approaches to control the transfer to the surface are mechanical abrasion prior to bonding and the use of a physical barrier, such as a peel ply, that separates the laminate and mold surfaces [51].

One interesting problem involved delamination of a metal line from a substrate. In this particular instance, soldering was used to join a chip to a carrier.

Instead of separation at the solder/chip interface, or within the solder, separation occurred between the metal lines and the substrate. Optical microscopy showed that separation occurred between the Cu layer and the Cr underlayer used for improving adhesion to the substrate.

Since in situ peeling could not be done in the analysis chamber, a different approach was used. After peeling the line, with the interface preserved, AES sputter depth profiles were used to reach the interface. Results showed high levels of oxygen present at the Cr/Cu interface. This indicated that vacuum problems occurred during metal deposition. On-line residual gas analysis records demonstrated that high levels of oxygen were present during this time frame. Measures were taken to correct the problem and additional monitors were put in place.

In another case, delamination of metal to a polyimide layer was noticed at the end of a process. If such delamination were left uncorrected, it could cause early failure of the package. Samples with low and high adhesion, as measured by a peel strength test, were selected for surface analysis. XPS surveys on the delaminated samples showed lower than normal concentrations of oxygen and nitrogen for a polyimide surface. In addition, the nitrogen region of the spectrum contained a nitrogen signal with a lower binding energy than expected for polyimide. These results led to an investigation to determine the reason for the change in polyimide character.

The thermal history of these polyimides was suspect. Simulation experiments were performed to test the model by depositing metal on polyimide at various temperatures. XPS results showed that there was a relationship between the relative intensity of the low-binding-energy peak and peel strength as a function of temperature. The conclusions of this work suggested that polyimide had been exposed to higher than normal temperatures during metal deposition. Recommendations were made to check the temperature-sensing devices and heaters in the system.

The next example illustrates how surface analysis may provide an answer to a question, but may not determine the underlying causes and permanent solutions to sporadic problems. Failure of solder to wet a copper interconnection pad was observed during inspection. AES survey scans showed that copper, carbon, and oxygen were present. The distribution of these elements was determined by AES elemental mapping. Figure 24-19 shows the maps for copper and carbon on the pad; the oxygen distribution, not shown, was similar to that of copper. To determine if there was another contaminant under the carbon layer, energy-dispersive x-ray analysis (EDS) with greater penetration depth ($\approx 1\,\mu$m) was used. This demonstrated that Cr was also present on the copper surface.

A good solder joint must have a good metallurgical bond. The tin from the solder must interact with the copper to form an intermetallic compound at the interface. For this to occur, a clean copper surface is needed. Based on the experimental results, the observed nonwet was caused by a carbon overlayer on chromium. Although the approach answered the initial question properly, it did not provide information as to the source of the problem or a final solution.

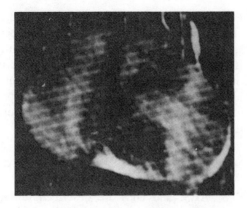

(*a*) Electron image

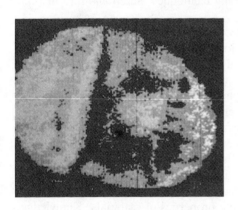

(*b*) Copper distribution

(*c*) Carbon distribution

FIGURE 24-19
(*a*) Electron image for a interconnection pad exhibiting solder nonwet. (*b*) Auger elemental map for copper distribution. (*c*) Auger elemental map for carbon distribution.

It is important to determine the nature of the carbon because it may indicate the step or steps in the process that deposit, or are deficient in removing, the organic layer. For example, photoresist is used to define circuitry and to protect specific features. After development, the photoresist layer is removed by a solvent to expose Cr for subsequent etching. Following removal of the chromium layer, the copper surface is exposed for soldering.

Several techniques can be used to determine the composition of the organic layer. Although XPS can be used to obtain some information regarding the chemical nature of the film, complementary analysis by infrared spectroscopy and mass spectrometry are needed to establish if the organic layer is residual photoresist or other materials.

Positive identification of the film as photoresist would make the process parameters for solvent removal suspect. Assuming the process parameters are within specifications, the quality of the photoresist film should be investigated. If the film is not photoresist, the spectral fingerprint of the contaminant layer can be compared with those available for other organic materials used in the process. Failure to identify the contaminant layer after this procedure will lead to a conclusion that the contamination is spurious. This information should be properly documented and kept for reference.

SUMMARY

The principles of XPS, AES, and UPS are described in this chapter. These techniques can provide qualitative or quantitative chemical information on the top layers (≈ 100 Å) of a variety of materials. When used in conjunction with nondestructive and destructive depth-profiling techniques, additional information can be obtained.

Applications of electron spectroscopy to electronic packaging are found in characterizing surface and interface reactions, in process development, and in failure analysis. Often, the use of other analytical methods, as well as completely detailed knowledge of process steps, is required to a solve a problem.

A purpose of this chapter is to emphasize the role played by surface and interface phenomena in controlling the package's performance and its reliability in service environments. It is clear that surface analysis capabilities have become essential in product and process development and other applications.

In the future, electronic packaging will require smaller components and circuit lines, which place additional demands on surface-sensitive techniques. It is possible that the importance of surface and interface effects in electronic packaging may overwhelm many of the bulk properties of materials. Knowledge of surface sciences and the development of new surface analysis capabilities will become increasingly important in industry.

REFERENCES

1. A. Einstein, "On the Electrodynamics of Moving Bodies," *Ann. Physik* 17, p. 132 (1905).
2. P. Auger, "Secondary Rays Produced in a Gas by X-rays," *Compt. Rend.* 177, p. 169 (1923).
3. L. Steinhardt, "An X-ray Photoelectron Spectrometer for Chemical Analysis," Ph.D. dissertation, Lehigh University, 1950.
4. C. Nordling, E. Sokolowski, and K. Siegbahn, "Shifts of Photo- and Auger Electron Lines," *Ark. Fys.* 13, p. 483 (1958).
5. K. Siegbahn, C. Nordling, A. Fahlman, R. Nordberg, K. Hamrin, J. Hedman, G. Johansson, T. Bergmark, S. Karlsson, I. Lindgren, and B. Lindberg, *ESCA, Atomic Molecular and Solid State Structure Studied by Means of Electron Spectroscopy*, Uppsala, Sweden: Almquist & Wicksells, 1967.
6. D. W. Turner, and M. L. al-Joboury, "Molecular Photoelectron Spectroscopy: The H_2 and N_2 Molecules," *J. Chem. Soc.*, pp. 5141–5147 (1963).
7. M. J. Hearn and D. Briggs, "Ion Beam Damage in Poly(Methyl Methacrylate)," *Surf. Interface Anal.* 9, p. 411 (1986).

8. H. Bolouri, J. M. R. MacAllister, R. A. Pethrick, and S. Affrossman, "Study of Epoxy Resins: Sensitivity of a Diglycidyl Ether to X-ray or Electron Irradiation," *Appl. Surf. Sci.* 24, pp. 18–24 (1986).

9. M. S. Lazarus and T. K. Sham, "The Performance of Mechanical Scraping Tools in Ultrahigh Vacuum: An XPS Study," *J. Electr. Spectrosc. Rel. Phenom.* 31, pp. 91–96 (1983).

10. K. L. Mittal, *Polyimides: Synthesis, Characterization, and Applications,* New York: Plenum Press, 1985.

11. L. J. Matienzo and K. J. Holub, "Surface Studies of Corrosion Preventing Coatings for Aluminum Alloys," *Appl. Surf. Sci.* 9, pp. 47–73 (1981).

12. C. D. Wagner, "Auger Parameter in Electron Spectroscopy for Identification of Chemical Species," *Anal. Chem.* 47, pp. 1201–1203 (1975).

13. D. Briggs and M. P. Seah, *Practical Surface Analysis,* pp. 181–189, Chichester: John Wiley & Sons, 1983.

14. H. Gonska, H. J. Freund, and G. Hohlneicher, "On the Importance of Photoconduction in ESCA Experiments," *J. Electron Spectrosc. Rel. Phenom.* 12, p. 435 (1977).

15. K. L. Smith and J. S. Hammond, "Destructive and Non-Destructive Depth Profiling Using ESCA," *Appl. Surf. Sci.* 22, pp. 288–298 (1985).

16. R. Kelly, "An Attempt to Understand Preferential Sputtering," *Nucl. Instr. Methods* 149, pp. 553–558 (1978).

17. S. Storp and R. Holm, "ESCA Studies of Chemical Shifts for Metal Oxides," *J. Electron Spectrosc. Rel. Phenom.* 16, pp. 183–189 (1979).

18. L. I. Yin, S. Ghose, and I. Adler, "Investigation of Possible Solar Wind Darkening on the Lunar Surface by Electron Spectroscopy," *J. Geophys. Res.* 77, pp. 1360–1367 (1972).

19. A. Turos, W. F. Van der Weg, D. Sigurd, and J. W. Mayer, "Change of Surface Composition of Sulfur Dioxide Layers During Sputtering," *J. Appl. Phys.* 45, pp. 2777–2779 (1974).

20. J. H. Scofield, "Hartree-Slater Subshell Photoionization Cross-Sections at 1254 and 1487 eV," *J. Electron Spectrosc. Rel. Phenom.* 8, pp. 129–137 (1976).

21. C. D. Wagner, "Sensitivity of Detection of the Elements by Photoelectron Spectroscopy," *Anal. Chem.* 44, pp. 81A–82A (1972).

22. L. C. Feldman and J. W. Mayer, *Fundamentals of Surface and Thin Film Analysis,* New York: Elsevier Science Publishing Co., 1986.

23. M. P. Seah, "Quantitative Auger Electron Spectroscopy and Electron Ranges," *Surf. Sci.* 32, pp. 703–728 (1972).

24. F. Meyer and J. J. Vrakking, "Quantitative Aspects of Auger Electron Spectroscopy," *Surf. Sci.* 33, pp. 271–294 (1972).

25. A. W. Czanderna, editor, *Methods of Surface Analysis,* pp. 180–184, Amsterdam: Elsevier Scientific Publishing Co., 1975.

26. P. W. Palmberg, G. E. Riach, R. E. Weber, and N. C. Mac Donald, *Handbook of Auger Electron Spectroscopy,* Edina, Minn.: Physical Electronics Industries, 1972.

27. P. W. Palmberg, "Quantitative Auger Electron Spectroscopy Using Elemental Sensitivity Factors," *J. Vac. Sci. Technol.* 13, pp. 214–218 (1976).

28. H. Lefakis and P. S. Ho, "Oxide Characterization and the Behavior at Cu-Cu and Cu-CuO$_x$ Interfaces," *Thin Sol. Films* 136, pp. L31–L34 (1986).

29. A. Cros, A. G. Schrott, R. D. Thompson, and K. N. Tu, "Influence of Sputtering Damage on Chemical Interactions at Cr-SiO$_2$ Interfaces," *Appl. Phys. Lett.* 48, pp. 1547–1549 (1986).

30. P. S. Ho, P. O. Hahn, J. W. Bartha, G. W. Rubloff, F. K. LeGoues, and B. D. Silverman, "Chemical Bonding and Reaction at Metal/Polymer Interfaces," *J. Vac Sci. Technol. A* 3, pp. 739–745 (1985).

31. B. M. DeKoven and P. L. Hagans, "XPS Studies of Metal/Polymer Interfaces-Thin Films of Al on Polyacrylic Acid and Polyethylene," *Appl. Surf. Sci.* 27, pp. 199–213 (1986).

32. M. Grunze and R. Lamb, "Characterization of Ultra-Thin Polyimide Films ($d < 10\text{Å}$) Formed by Vapor Deposition of 4, 4′ Oxidianiline and 1, 2, 3, 5 Benzenetetracarboxylic Anhydride," *J. Vac. Sci. Technol.* A5, p. 1685 (1987).

33. L. J. Matienzo and T. K. Shah, "Surface Treatment of Phlogopite Particles for Composite Applications: XPS and DRIFT Studies," *Surf. Interface Anal.* 8, pp. 53–59 (1986).

34. K. Inagaki, K. Suzuki, and K. Oh-Ishi, "Moisture Sensor Devices Composed of Thin Films Plasma-Polymerized from Amino-Groups-Containing Silanes," *Appl. Surf. Sci.* 24, pp. 163–172 (1985).

35. N. J. Chou, C. H. Tang, J. Paraszczak, and E. Babich, "Mechanism of Oxygen Plasma Etching of Polydimethyl Siloxane Films," *Appl. Phys. Lett.* 46, pp. 31–33 (1985).

36. L. J. Gerenser, J. F. Elman, M. G. Mason, and J. M. Pochan, "ESCA Studies of Corona-Discharge-Treated Polyethylene Surfaces by Use of Gas-Phase Derivatization," *Polymer* 26, pp. 1162–1166 (1985).

37. F. C. Schhwab and M. A. Kadash, "Effect of Resin Additives on Corona Treatment of Polyethylene," *Plastic Film and Sheeting* 2, pp. 119–126 (1986).

38. A. Torrisi and S. Pignataro, "Kinetics of Bromine Migration and Copper Reactions at Copper-Epoxy Interfaces," *Surf. Interface Anal.* 9, pp. 441–444 (1986).

39. J. H. Linn and W. E. Swartz, Jr., "An XPS Study of the Water Adsorption/Desorption Characteristics of Transition Metal Oxide Surfaces: Microelectronic Implications," *Appl. Surf. Sci.* 19, pp. 154–166 (1984).

40. R. F. Roberts, "Detection of Water Soluble Flux Residues on Printed Wiring Boards by ESCA," *Surf. and Interface Anal.* 7, pp. 88–92 (1985).

41. B. D. Dunn and C. Chandler, "The Corrosive Effect of Soldering Fluxes and Handling of Some Electronic Materials," *Welding J.* 59, pp. 289–307 (1980).

42. J. T. Lynch, R. T. Bilson, N. R. Matthews, and A. Boetti, "The Measurement of Flux Residues from Chip Carrier Attachment and Their Effect on Other Thick Film Hybrid Components," *IEEE Trans. on Components, Hybrids, and Manufacturing Technol.* CHMT-7, pp. 336–348 (1984).

43. D. C. Miller, D. J. Auerbach, and C. R. Brundle, "Auger Electron Spectroscopic Studies of the Thickness of Films Formed on Copper by Benzotriazole," private communication, 1987.

44. M. Bakszt, "Providing Solderability Retention by Means of Chemical Inhibitors," *Printed Circuit World Convention III Proceedings* WCIII-04, Washington D.C., 1984.

45. D. R. Olsen and K. I. James, "Effect of Ambient Atmosphere on Aluminum-Copper Wire Bond Reliability," *IEEE Trans. Compon. Hybrid Manuf. Technol. Proceedings*, vol. CHMT-7 n4, pp. 357–362, New Orleans, 1984.

46. S. Pignataro, A. Torrisi, O. Puglisi, A. Cavallaro, A. Perniciaro, and G. Ferla, "Influence of Surface Chemical Composition on the Reliability of Al/Cu Bond in Electronic Devices," *Appl. Surf. Anal.* 25, pp. 127–136 (1986).

47. J. H. Linn and G. Bajor, "Analysis of Gold Films Used in Microcircuit Chip-Attachment," *Appl. Surf. Anal.* 26, pp. 211–218 (1986).

48. C. Lea and F. H. Howie, "Blowholing in PTH Solder Fillets, Part 5: The Role of the Electroless Copper," *Circuit World* 113, pp. 28–42 (1986).

49. L. J. Matienzo and T. K. Shah, "Detection and Transfer of Mold Release Agents in Bonding Processes," *15th National SAMPE Technical Conference Proceedings*, pp. 604–616, Cincinnati, Ohio, 1983.

50. D. L. Messick, D. J. Progar, and J. P. Wightman, "Surface Analysis of Graphite Fiber Reinforced Polyimide Composites," *15th National SAMPE Technical Conference Proceedings*, pp. 170–189, Cincinnati, Ohio, 1983.

51. L. J. Matienzo, J. D. Venables, J. D. Fudge, and J. J. Velten, "Surface Preparation for Bonding Advanced Composites. Part 1: Effect of Peel Ply Materials and Mold Release Agents on Bond Strengths," *30th National SAMPE Meeting Proceedings*, pp. 302–317, Anaheim, Calif., 1985.

SUGGESTED READING LIST

Principles of techniques

1. A. W. Czanderna, editor, *Methods of Surface Analysis*, vol. 1, Amsterdam: Elsevier Publishing Co., 1975.

2. E. Kuppers, *Low Energy Electrons and Surface Chemistry*, Weinheim: Verlag Chemie, D-694, 1974.

3. D. Briggs and M. P. Seah, editors, *Practical Surface Analysis*, Chichester: John Wiley & Sons, 1983.
4. J. H. D. Eland, *Photoelectron Spectroscopy*, 2d. ed., London: Butterworths, 1984.
5. L. Ley and M. Cardona, *Topics in Applied Physics*, vol. 27, *Photoemission in Solids*, Berlin: Springer Verlag, 1979.
6. T. A. Carlson, *Photoelectron and Auger Spectroscopy*, New York: Plenum Press, 1975.
7. L. C. Feldman and J. W. Mayer, *Fundamentals of Surface and Thin Film Analysis*, New York: North-Holland, 1986.
8. C. R. Brundle and A. D. Baker, editors, *Electron Spectroscopy*, London: Academic Press, 1977.

Surface modification

1. E. P. Plueddemann, *Silane Coupling Agents*, New York: Plenum Press, 1982.
2. D. E. Leyden and W. T. Collins, *Silylated Surfaces*, New York: Gordon and Breach Science Publishers, 1980.
3. M. Shen and A. T. Bell, *Plasma Polymerization*, ACS Symposium Series No. 108, Washington D.C.: 1979.
4. H. V. Boenig, *Plasma Science and Technology*, Ithaca, N.Y.: Cornell University Press, 1982.
5. N. G. Einspruch and D. M. Brown, editors, *VLSI Microelectronics Microstructure Science*, vol. 7, New York: Academic Press, 1984.

Adhesion and adhesive joints

1. L. H. Lee, editor, *Characterization of Metal and Polymer Surfaces*, vol. 2, New York: Academic Press, 1977.
2. G. P. Anderson, S. J. Bennett, and K. L. DeVries, *Analysis and Testing of Adhesive Bonds*, New York: Academic Press, 1977.
3. D. H. Buckley, *Surface Effects in Adhesion, Friction, Wear, and Lubrication*, Tribology Series, vol. 5, Amsterdam: Elsevier, 1981.
4. D. H. Kaeble, *Physical Chemistry of Adhesion*, New York: Wiley-Interscience, 1980.

Polymers and polymer surfaces

1. D. T. Clark and W. J. Feast, *Polymer Surfaces*, Chichester: John Wiley & Sons, 1978.
2. W. Klopffer, *Introduction to Polymer Spectroscopy*, Berlin: Springer Verlag, 1984.
3. D. W. Dwight, T. J. Fabish, and H. R. Thomas, editors, *Photon, Electron, and Ion Probes of Polymer Structure and Properties*, Washington, D. C.: American Chemical Society, 1981.

Thin films

1. J. M. Poate, K. N. Tu, and J. W. Mayer, *Thin Film—Interdiffusion and Reactions*, New York: John Wiley & Sons, 1977.

Fluxes and soldering

1. R. W. Woodgate, *Handbook of Machine Soldering*, New York: John Wiley & Sons, 1983.

Applications

1. N. Winograd and S. W. Gaarenstroom, *Physical Methods in Modern Chemical Analysis*, vol. 2, New York: Academic Press, 1980.
2. D. A. Shirley, editor, *Electron Spectroscopy*, Amsterdam: North Holland, 1972.
3. N. G. Einspruch and D. M. Brown, *VLSI Microelectronics Microstructure Science*, vol. 7, pp. 497–523, New York: Academic Press, 1983.
4. T. L. Barr and L. E. Davis, *Applied Surface Analysis*, Philadelphia: American Society for the Testing of Materials, 1980.

5. K. L. Mittal, editor, *Surface Contamination*, vols. 1 and 2, New York: Plenum Press, 1979.
6. H. Leidheiser, Jr., *Corrosion Control by Coatings*, Princeton, N.J.: Science Press, 1979.

PROBLEMS

24.1. In a friction experiment, a flat piece of iron was treated with methyl chloride and its AES spectrum recorded. The survey spectrum revealed the presence of carbon, chlorine, and iron on the surface of the metal. After sliding a piece of silver 250 times over the iron, the friction coefficient decreased dramatically. An AES survey spectrum of the iron surface after the experiment showed the presence of carbon, chlorine, iron, and silver. How can you explain the reduction of the friction coefficient based on the surface analysis results?

24.2. A sample of copper foil from a processing line is said to be creating some manufacturing problems because it contains a mixture of metal and cuprous and cupric oxides. If you are asked to determine what is present on the surface of the copper foil, how would you distinguish the following:

(a) A mixture of copper (II) oxide and metal
(b) A mixture of copper (I) oxide and copper (II) oxide
(c) A mixture of copper (I) oxide and copper metal
(d) A mixture of copper (II) oxide, copper (I) oxide and some copper metal

24.3. In an adhesion experiment, a bond containing a metal film and a polymer film is made to fail. Either polyvinyl chloride or polyethylacrylate, with monomeric repeating units of $[CH(Cl)CH_2]$ and $[CHCOOC_2H_5]$, respectively, was used in the experiment.

(a) What kind of high-resolution XPS spectra would you expect to find in each case?
(b) XPS spectra of both sides of a joint show comparable levels of carbon after fracture. What is the most likely mode of failure?

24.4. Epoxy/graphite laminates are used in some microelectronic applications. Is it possible to characterize these surfaces by AES?

24.5. A metallic film is etched in a CF_4/O_2 gas plasma chamber. The AES and XPS spectra of the etched surface reveal the presence of carbon, fluorine, metal, and oxygen signals. How would you distinguish if:

(a) A fluoropolymer layer is deposited onto the metal?
(b) The metal film became fluorinated?
(c) You had a combination of (a) and (b)?

24.6. In a new process development, Al_2O_3 ceramic is proposed as the substrate material but it must contain a layer of gold of thickness about 250 Å. In order to bond the gold to the ceramic, two 50-Å layers of titanium and chromium must be applied sequentially onto the ceramic. The calibrated sputtering rate for the argon ion beam is about 12.5 Å/min.

(a) What is the expected profile for the film on the ceramic?
(b) An additional step requires heating of the device to 300°C. If chromium diffusion into gold is present and diffusion of titanium into chromium is expected as well, what would a depth profile look like at this point?

24.7. XPS scans of a piece of pure aluminum nitride show aluminum, nitrogen, and small amounts of oxygen and carbon. If surface oxidation of the nitride to oxynitride is possible:

(a) How many nitrogen signals are expected and how would you assign them?

(b) How would you determine what is concentrated in the top 5–10 Å of the film?

24.8. The fabrication of a 16 K RAM circuit uses a device that is implanted with phosphorus. The final cleaning of the device involves washing with nitric acid to remove organic contaminants and a rinse with deionized water. Phosphorus can diffuse out during electron beam treatment. After inspection of the device, scanning AES analysis identifies small spots containing C, O, and N. Following a stage involving glass reflow, pinholes are seen. AES spectra of these areas show they contain phosphorus, carbon, nitrogen, and enhanced levels of oxygen. These pinholes can introduce serious problems during metallization. How do you explain these results and how can you correct them?

24.9. At the end of a process, delamination occurred in a composite film containing two dissimilar polymer layers with dimensions of 100 by 300 μm. How would you determine:

(a) The mode of failure?

(b) If a contaminant is present at the delamination site?

(c) In a flow diagram, outline the approach to be used for a complete failure analysis.

24.10. In the processing of an experimental electrical device, two metal layers are deposited sequentially and then subjected to an oxidation step in a chamber kept at room temperature. The optimum electrical response is obtained when the second metal layer contains 5 Å of its oxide as determined by ellipsometry.

(a) How would the surface composition be determined?

(b) If the oxidation process started earlier than expected, how would you determine if the first metal layer was oxidized?

(c) If the metals were codeposited instead of being deposited sequentially, what would you expect the elemental depth profiles to show?

CHAPTER
25

INTRODUCTION TO SOLID STATE DIFFUSION

BRIAN S. BERRY

INTRODUCTION

The ability of atoms to move around or diffuse in solids is of profound importance in many areas of technology, but nowhere is this more the case than in the field of microelectronics. At the chip level, the diffusion of doping elements into silicon is an essential step in the fabrication of transistors. The formation of an insulation layer of SiO_2 on the surface of a silicon wafer by high-temperature oxidation is another example of an important diffusion-controlled reaction. At much lower temperatures, the choice of metallic interconnections and contacts is also frequently influenced by diffusion-controlled phenomena. For instance, the degradation of polycrystalline aluminum stripes by electromigration involves preferential diffusion at grain boundaries. Care must be taken in the choice of metals that are allowed to come into contact with each other to safeguard against interdiffusion and the formation of undesirable intermetallic compounds.

Apart from diffusion in relatively well-defined monoatomic crystalline solids such as silicon or metals, we are also increasingly concerned in packaging technology with the use of polymers such as the polyimides or epoxies. In these high-molecular-weight and frequently amorphous materials, diffusion phenomena can become extremely complicated and are often poorly understood. The simplest cases are those that involve the diffusion of relatively small molecules such as those naturally or unintentionally present in the atmosphere. Small-molecule diffusion can be important if the polymer is required to provide hermeticity or protection to a package, or if the swelling associated with the sorption of such species causes problems with dimensional stability or the generation of internal stresses.

The purpose of this chapter is to provide a short introductory account of diffusion in solids, with a bias towards packaging materials that will be continued in the next chapter. The reader should be cautioned that the coverage given here is far from complete and is only intended to provide an initial exposure to a large and diverse field [1–8,19].

SOME COMMON DENOMINATORS IN SOLID STATE DIFFUSION

We have already alluded to the great variety of host materials in which diffusion may occur. When combined with a similar diversity in the various species that may diffuse in any one host material, it can be readily appreciated that no single diffusion mechanism can be expected to explain every situation. Nevertheless, certain basic features are common to most cases of solid state diffusion, and these will be considered in this section.

The notion that atoms (or, sometimes, molecular groups) can and do diffuse in solids may at first seem somewhat curious. While we can readily understand the diffusive motion of molecules in a fluid, we are confronted with the fact that the characteristic rigidity of solids is caused by atoms freezing into a set of fixed equilibrium positions, about which the atoms may vibrate more or less vigorously depending on the temperature, but to which they are confined. If this simple picture were exactly true, solid state diffusion could not occur. To account for diffusion, we must recognize that each excursion of a vibrating atom represents an attempt to escape from its present site, and that on rare occasions an atom may by chance have enough vibrational (thermal) energy to carry it over the potential barrier separating it from an adjacent site. This sudden *thermally activated* transition or *jump* over an energy barrier to a neighboring site is the basic event responsible for solid state diffusion. The schematic diagram of Figure 25-1 shows an atom (or other diffusing entity) occupying an equilibrium site labeled 1, about which it oscillates with a mean frequency ν_0. We may write the

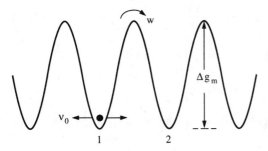

FIGURE 25-1
A schematic representation of solid state diffusion in terms of thermally activated jumps over a potential barrier.

jump rate w for transitions to the specific neighboring site 2 as the product of ν_0 and p_{12}, the probability of success per excursion along the path $1 \rightarrow 2$. If site 1 is surrounded in the three-dimensional solid by a total of z equivalent sites to which jumps may occur, the total jump rate Γ at which an atom leaves a given site is zw. We thus have

$$\Gamma = zw = z\nu_0 p_{12} \tag{25.1}$$

Whereas z and ν_0 can be regarded as essentially temperature-independent factors, the probability term is strongly temperature dependent through an exponential relationship of the form

$$p_{12} = \exp\left(\frac{-\Delta g}{kT}\right) = \exp\left(\frac{-\Delta G}{RT}\right) \tag{25.2}$$

where Δg is the free energy of activation per diffusing particle, ΔG is the free energy of activation for a mole of diffusing particles, T is the absolute temperature, and k and R are, respectively, Boltzmann's constant and the gas constant.

It should be noted that the exponents $\Delta g/kT$ or $\Delta G/RT$ in Equation (25.2) are identical dimensionless ratios in which the barrier height is expressed as a multiple of the mean thermal energy. Exponential terms of the type shown in Equation (25.2) are frequently referred to as Boltzmann factors, because they can be traced to the Boltzmann distribution for the sharing of energy among a collection of identical particles. Combining Equations (25.1) and (25.2), we obtain the jump rate Γ in the form

$$\Gamma = z\nu_0 \exp\left(-\frac{\Delta g}{kT}\right) \tag{25.3}$$

Although temperature appears explicitly in Equation (25.3), it should be recalled that Δg is also a function of temperature. From basic thermodynamic definitions, we have

$$\Delta g = \Delta h - T\Delta s \tag{25.4}$$

where Δh and Δs denote the enthalpy and entropy differences introduced by conveying the diffusing particle from the equilibrium to the activated or saddle-point configuration at the top of the barrier. We thus may write Equation (25.3) in the form

$$\Gamma = \Gamma_0 \exp\left(-\frac{\Delta h}{kT}\right) \tag{25.5}$$

where the temperature-independent prefactor Γ_0 is given by

$$\Gamma_0 \equiv z\nu_0 \exp\left(\frac{\Delta s}{k}\right) \tag{25.6}$$

and the enthalpy term Δh is commonly (and somewhat loosely) referred to as the activation energy for diffusion.

In the absence of driving forces, the sequence of jumps made by any particular atom produces a completely random path known as a random walk. An important property of such random motion is that n^2 jumps are required to produce a root-mean-square displacement equal to n jumps laid end to end in the same direction. We may state this result in the form

$$(\overline{R^2})^{1/2} = \lambda N_j^{1/2} \tag{25.7}$$

where the quantity on the left-hand side is the rms displacement after N_j jumps each of distance λ. Equation (25.7) can be used to demonstrate the prodigious number of jumps an atom must make to travel even the microscopic distances with which we are typically concerned in solid state diffusion. For example, if the jump distance is 0.1 nm, the number of jumps required to produce an rms displacement of a hair's breadth (50 μm) is 2.5×10^{11}. Although the rms displacement is useful as an indication of a mean diffusion distance, we must not lose sight of the fact that it is an average based on a large number of imaginary trials. In such trials of N_j jumps each, some atoms will have travelled less, and others more, than the rms displacement. This distribution is the basic reason why diffusion from a concentrated source produces a smooth diffusion profile rather than the propagation of an abrupt front.

It should be apparent from the foregoing remarks that diffusion is a manifestation of an intrinsic atomic mobility, which, in principle at least, is possessed by all solids at temperatures above absolute zero. In this sense diffusion is always taking place in solids, whether or not they are in thermodynamic equilibrium. In some cases, a readily observable manifestation of diffusion may be lacking, as with self-diffusion in a pure element. To demonstrate diffusion in such a case, rather special techniques are required. The most direct and important of these involves the use of radioactive isotopes as a source of tracer atoms. Radiotracer techniques are of relatively modern origin, and became available long after the phenomenon of diffusion was recognized through simple experiments in which a concentration gradient produced an easily observable flux, as in the diffusive spreading of a drop of dye in a still liquid. Historically, then, it is not surprising that the earliest formulation of the "laws" of diffusion involved what we now call chemical diffusion, or diffusion in a concentration gradient.

FORMAL CONSIDERATIONS

Fick's Laws

The relationship now known as Fick's first law deals with the flux J in a steady state system containing a gradient in the diffusant concentration, C. For one-dimensional diffusion along the x axis, the relation is

$$J = -D\frac{\partial C}{\partial x} \tag{25.8}$$

where D is known as a diffusion coefficient. Historically, the intent of the law was to identify D as a simple proportionality constant uniquely relating J and $\partial C/\partial x$. While this is true in certain simple situations, it is now recognized that this is not always the case. To retain the law in its original form, we must be prepared to allow for a reinterpretation of D, as we shall see later. Returning to Equation (25.8), we note that the negative sign is introduced to satisfy the physical requirement that flow occurs in the direction in which the concentration C decreases. If C is expressed as the mass of the diffusing species per unit volume, the flux J is expressed as the mass crossing unit area per unit time. The dimensions of the diffusion coefficient are thus $(\text{length})^2(\text{time})^{-1}$. In the cgs system of units D is expressed in units of cm^2/s. Nowadays, there is an increasing tendency to use SI units and to express D in units of m^2/s. It should be noted that Equation (25.8) does not contain time as a variable; the phrase *steady state* implies consideration of a system which has settled down to a condition where the *same* flux J is being transported across every cross section. Under these conditions there can be no accumulation or depletion of the diffusing species in a slab dx and the concentration $C(x)$ does not change with time.

If D is independent of the concentration, Equation (25.8) can be integrated to the form

$$J = -D\frac{C_2 - C_1}{x_2 - x_1} \tag{25.9}$$

and the steady state *diffusion profile* is linear, as drawn in Figure 25-2. On the other hand, if D were a function of concentration, the steady state diffusion profile would be nonlinear. A special case of Equation (25.9) concerns diffusion-controlled permeation of a gas (or vapor) through a membrane, when the emergent gas is so effectively removed that the surface concentration C_2 on the exit face becomes negligibly small in relation to the concentration C_1 on the entrance face. Equation (25.9) then reduces to

$$J = \frac{DC_1}{a} \tag{25.10}$$

where a is the membrane thickness. Alternatively stated, Equation (25.10) shows

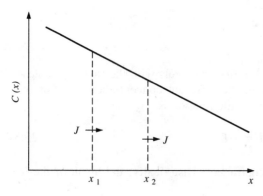

FIGURE 25-2
Illustration of one-dimensional steady state diffusion.

that the permeation rate per unit area for a unit thickness is simply the product of the diffusion coefficient and the concentration in solution at the entrance face of the membrane.

Examples of steady state diffusion are relatively rare, and are mostly contrived in the laboratory for the purpose of measuring D. Most often, we are confronted with the situation of transient or non–steady state diffusion, where both the flux J and concentration C are functions of both position and time t. As illustrated for one-dimensional diffusion in Figure 25-3, a difference between the entrance and exit fluxes for the slab dx produces a change in the concentration of the diffusing species within the slab. Application of the conservation of mass gives the continuity equation

$$\frac{\partial C}{\partial t} = -\frac{\partial J}{\partial x} \tag{25.11}$$

Since the concentration $C(x, t)$ can only change by an infinitesimal amount in the interval dt, we may combine Fick's first law [Equation (25.8)] with Equation (25.11) to obtain a differential equation governing transient behavior. This is commonly referred to as Fick's second law, and has the form

$$\frac{\partial C}{\partial t} = \frac{\partial}{\partial x}\left(D\frac{\partial C}{\partial x}\right) \tag{25.12}$$

In an isothermal system, if D is independent of position by virtue of being independent of concentration, Equation (25.12) simplifies to

$$\frac{\partial C}{\partial t} = D\frac{\partial^2 C}{\partial x^2} \tag{25.13}$$

which is also the basic differential equation for the diffusion of heat. (In that case, concentration is replaced by temperature and D becomes the thermal diffusivity of the material.) Substantial books have been written on the solutions of Equations (25.12) and (25.13) for a variety of initial and boundary conditions [9]. Here, we can do no more than catalog the results for a few illustrative cases of one-dimensional diffusion.

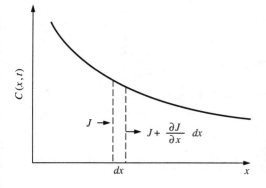

FIGURE 25-3
Illustration of one-dimensional transient diffusion.

Some Solutions
of the Second Diffusion Equation

We shall restrict attention to concentration-independent diffusion, and focus on solutions of Equation (25.13) for three different situations of experimental importance. The first of these deals with the diffusion of a thin radioactive film into a specimen of effectively infinite thickness. (In practical terms, the actual thickness need be only a few times the penetration distance $\bar{x}$, defined below.) In this situation, diffusion into the specimen depletes the surface supply of tracer atoms, and the concentration at the surface ($x = 0$) continually decreases. The concentration profile $C(x, t)$ for x and $t > 0$ is given by

$$C(x, t) = \frac{m}{(\pi Dt)^{1/2}} \exp\left(-\frac{x^2}{4Dt}\right) \tag{25.14}$$

where m is the mass per unit area of the radiotracer atoms present on the surface of the sample at $t = 0$. Illustrative profiles for various values of Dt are given in Figure 25-4. Using Equation (25.14), it is evident that the concentration always equals $1/e$ of its current surface value at a distance $\bar{x}$ that satisfies the condition

$$\bar{x} = 2(Dt)^{1/2} \tag{25.15}$$

As a consequence, $\bar{x}$ is a convenient measure of a diffusion length, and Equation (25.15) is frequently used for rough estimates of penetration distances.

A second situation of interest is obtained by changing the boundary conditions in the arrangement considered above. The surface concentration $C(0, t)$

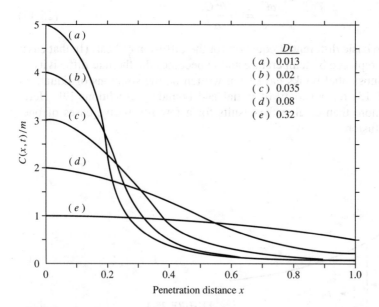

	Dt
(a)	0.013
(b)	0.02
(c)	0.035
(d)	0.08
(e)	0.32

FIGURE 25-4
Diffusion profiles for the thin-film solution represented by Equation (25.14).

is now not allowed to decrease, but rather is held constant at a value C_0. This situation can, for example, sometimes be obtained by feeding diffusant to the sample from the gas phase. The concentration profile now has the form shown by the universal curve in Figure 25-5, and is given by

$$C(x, t) = C_0(1 - \text{erf}(u)) \tag{25.16}$$

where

$$u \equiv \frac{x}{2(Dt)^{1/2}}$$

and erf (u) is the *error function*, available from standard mathematical tables.

For our third and final example, we consider a situation where the sample is no longer thick compared with the diffusion length. A particular case involves diffusion-controlled desorption from a slab of thickness a, subject to the conditions $C = C_0$ for $0 < x < a$ at $t = 0$, and $C = 0$ for $x = a$ and $x = 0$ at $t > 0$. The solution to the diffusion equation is now given by the infinite series

$$C(x, t) = \frac{4C_0}{\pi} \sum_{n=1}^{\infty} \left[\frac{1}{2n + 1} \sin \frac{(2n + 1)\pi x}{a} \right] \exp \left[-\frac{(2n + 1)^2 \pi^2 Dt}{a^2} \right] \tag{25.17}$$

An illustrative set of diffusion profiles corresponding to Equation (25.17) is shown in Figure 25-6. By integration of Equation (25.17) over the thickness of the membrane, we find that the relative mass of diffusant remaining in the membrane

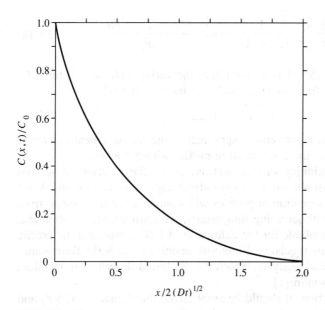

FIGURE 25-5
Master curve for the diffusion profile corresponding to Equation (25.16).

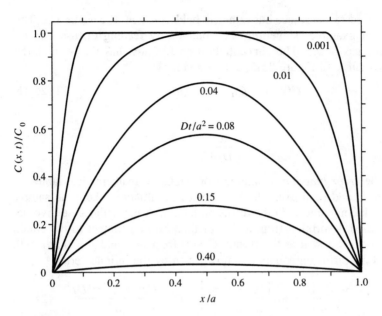

FIGURE 25-6
Diffusion profiles for desorption from a rectangular slab, as given by Equation (25.18).

after time t is given by

$$\frac{m(t)}{m(0)} = \frac{8}{\pi^2} \sum_{n=0}^{\infty} \frac{1}{(2n + 1)^2} \exp\left[-\frac{(2n + 1)^2 \pi^2 Dt}{a^2} \right] \qquad (25.18)$$

From a plot of Equation (25.18) as a function of the variable $(Dt/a^2)^{1/2}$, the ratio $m(t)/m(0)$ is found to fall from its initial value of unity to 0.5 when

$$(Dt/a^2)^{1/2} = 0.22 \qquad (25.19)$$

Equation (25.19) provides a convenient approach to the measurement of D that avoids a need to determine the concentration profile within the slab.

As noted at the beginning of this section, all of the solutions described above are based on the assumption that D is independent of concentration. When this is not the case, the concentration profiles will assume a different shape from those discussed above. Without going into details, we shall simply mention that analysis procedures are available for the extraction of $D(C)$, provided the profile is determined with adequate precision. Foremost amongst these is the Boltzmann-Matano method of graphical analysis, which can be found described in standard texts such as that of Shewmon [1].

To conclude this section, it should be emphasized that Equations (25.8) and (25.12) are restricted to a host medium that is both homogeneous and isotropic.

Homogeneity is required so that diffusion from a uniform source in the yz plane will produce a concentration profile that is solely dependent on x. On the microscopic scale characteristic of solid state diffusion, many materials are inhomogeneous and consist of an aggregate of different phases for which D may have distinctly different values. Even in single-phase materials structural inhomogeneities such as grain boundaries or dislocations may provide localized paths through which diffusion can occur at a greatly accelerated rate (as discussed later in this chapter). Complications of this sort can be of great practical significance but go beyond the scope of Fick's laws as given in Equations (25.8) and (25.12). Finally, the assumption of isotropy is involved whenever the diffusion coefficient is written as a scalar quantity, so that the flux (a vector) is always considered to be parallel to the concentration gradient (another vector). This condition is satisfied in random (nonoriented) amorphous materials, and in crystalline solids of cubic symmetry. It is not satisfied in materials of lower than cubic symmetry. For example, the diffusion tensor for tetragonal or hexagonal crystals contains the nonzero components $D_{11} = D_{22}$, and D_{33}. These define two independent coefficients for diffusion perpendicular and parallel to the unique crystal axis [1].

An Alternative Formalism for Fick's First Law

Despite its intuitive appeal, the interpretation to be placed on Fick's first law [Equation (25.8)] is not as straightforward as it may appear at first sight. One difficulty lies in the implication that a flux can only be produced by a concentration gradient, and the corollary that a concentration gradient cannot exist in a system in thermodynamic equilibrium. Since neither of these conclusions is true, we are faced with the prospect that Equation (25.8) is either inadequate or has a more subtle interpretation than we have considered so far. In this section we consider an alternative formalism for the flux equation which clarifies some of these difficulties, and which at the same time gives some insight into the meaning of the Fickian diffusion coefficient appearing in Equation (25.8).

In thermodynamics, we learn that equilibrium of a solute (diffusant) species corresponds to a spatial equalization of its chemical potential μ. This quantity is related to the solute concentration by the expression

$$\mu = \mu^0 + kT \ln \gamma c \qquad (25.20)$$

where μ^0 is the chemical potential of the solute in a reference state, γ is the activity coefficient for the particular concentration considered, and c is the concentration expressed in dimensionless terms as a mole fraction. It is to be emphasized that, in general, γ is not simply a constant, but is a function of c. However, if solution of the diffusant occurs without chemical interactions, the behavior is said to be ideal (from the viewpoint of simplicity), and $\gamma = 1$.

The simplest one-dimensional flux equation that can be written down in terms of the chemical potential is the simple proportionality

$$J = -M\frac{\partial \mu}{\partial x} \tag{25.21}$$

Although this equation has an obvious resemblance to Equation (25.8), it is now formulated in terms consistent with the equilibrium requirement that the flux vanish when $\partial \mu / \partial x$ vanishes and μ is spatially uniform. We may also observe that $\partial \mu / \partial x$ has the dimensions of force [note from Equation (25.20) that μ has the dimensions of energy], so that the proportionality coefficient M does not have the same dimensions as the diffusion coefficient D. Whereas Fick's first law is written in terms that emphasize the appearance of a flux as a consequence of a concentration gradient, Equation (25.21) emphasizes the more general concept of a flux as the response to an energy gradient or effective force.

To find the relation between M and D, we first use Equation (25.20) to obtain

$$\frac{\partial \mu}{\partial x} = kT\left(\frac{1}{C}\frac{\partial C}{\partial x} + \frac{\partial \ln \gamma}{\partial x}\right)$$

$$= \frac{kT}{C}\Phi\frac{\partial C}{\partial x} \tag{25.22}$$

where Φ is a dimensionless *thermodynamic factor* given by

$$\Phi \equiv 1 + \frac{\partial \ln \gamma}{\partial \ln c} \tag{25.23}$$

Inserting Equation (25.22) into (25.21), and comparing the result with Equation (25.8), we find

$$D = \frac{MkT}{C}\Phi \tag{25.24}$$

Since $\Phi = 1$ for an ideal solution, it is seen from Equation (25.24) that MkT/C has the significance of the diffusion coefficient that would be exhibited if the solution were ideal and diffusion were caused by concentration differences only. To recognize this, we may write

$$D^* = \frac{MkT}{C} \tag{25.25}$$

and think of D^* as an ideal diffusion coefficient. (We will see later why D^* is more commonly called a *tracer* diffusion coefficient.) Combining Equations (25.24) and (25.25), we have

$$D = D^*\Phi \tag{25.26}$$

which concisely expresses the result that, in general, the Fickian diffusion coefficient D can be thought of as a hybrid or composite quantity involving an ideal diffusion coefficient and a correction factor for nonideality. The correction factor is needed because, in nonideal solutions, the concentration gradient $\partial C/\partial x$ appearing in Fick's law is only partially responsible for the flux J. Another contribution comes from the effective driving force originating from nonideality of the solution, which adds a second contribution to J. To see this, we need only insert Equation (25.22) into Equation (25.21) to obtain

$$
\begin{aligned}
J &= -MkT\left(\frac{1}{C}\frac{\partial C}{\partial x} + \frac{\partial \ln \gamma}{\partial x}\right) \\
&= -D^*\left(\frac{\partial C}{\partial x} + C\frac{\partial \ln \gamma}{\partial x}\right)
\end{aligned}
\tag{25.27}
$$

This is an equivalent but more understandable and useful form of Fick's first law, where D has been replaced by D^* and $\partial C/\partial x$ has been replaced by $(\partial C/\partial x) + C(\partial \ln \gamma/\partial x)$. Use of Equation (25.27) instead of Equation (25.8), for example, enables us to understand some occasional cases of "uphill diffusion," where the flux is directed up the concentration gradient. In terms of Equation (25.8), such a state of affairs corresponds to the seemingly incomprehensible situation that D is negative. On the other hand, we readily see from Equation (25.27) that this can occur when the nonideal term $\partial \ln \gamma/\partial x$ is of opposite sign to $\partial C/\partial x$, and is large enough (due to a strong attractive interaction between the diffusant atoms) to override the effect of the concentration gradient.

We may now inquire into the conditions under which a measured diffusion coefficient D actually corresponds to D^*. Since thermodynamic ideality holds for self-diffusion in pure elements and for solute diffusion at sufficient dilution, the D measured in these cases by chemical profiling, radiotracers, or any other technique is D^*. Less obviously, the D for self-diffusion of a component species in a nonideal concentrated solution is also D^*, provided the sample is chemically homogeneous. To see this, we return to Equation (25.27) and observe that, since γ is only a function of C, the term $\partial \ln \gamma/\partial x$ vanishes for isoconcentration conditions. Since this means that the term $\partial C/\partial x$ also vanishes, it may appear at first sight that this result is of little consequence, since it apparently precludes the possibility of observing a diffusion flux under isoconcentration conditions. Fortunately, however, self-diffusion of a small quantity of a radioisotope into a homogeneous solid-solution effectively satisfies the isoconcentration requirement, since the radioisotope is *chemically* indistinguishable from the inert isotope of the same species present at the finite uniform concentration C in the test sample. Fick's first law for tracer diffusion may therefore be written

$$
J^* = -D^*\left(\frac{\partial C^*}{\partial x}\right)_C
\tag{25.28}
$$

where the starred quantities refer to tracer atoms only. Since the condition of constant C implies that $\partial \ln \gamma/\partial x$ vanishes from Equation (25.27), it is clear that the isoconcentration tracer diffusion coefficient D^* defined by Equation (25.28) is the same as the ideal diffusion coefficient introduced earlier. Indeed, it is now customary to refer to the ideal diffusion coefficient simply as the tracer diffusion coefficient.

We are now ready to pursue the interpretation of the formal theory given above in atomistic terms. In the next section, we consider a simple model that links D^* to the jump rate Γ introduced earlier, and also provides a physical picture for the meaning of the nonideality term appearing in Equation (25.27).

ATOMISTIC CONSIDERATIONS

Derivation of Fick's First Law

We start by considering the net flux produced by atomic jumps between a pair of adjacent planes separated by a distance λ, as shown in Figure 25-7. We denote the concentration of diffusant atoms on plane 1 as n_1 per unit area, and the average rate at which an atom jumps from plane 1 to plane 2 as Γ_{12}. With a similar notation for plane 2, the flux $J = J_1 - J_2$ is given by

$$J = n_1\Gamma_{12} - n_2\Gamma_{21} \qquad (25.29)$$

Using the relations $n_1 = C_1\lambda$, $n_2 = C_2\lambda$, and $C \equiv (C_1 + C_2)/2$, Equation (25.29) can be rearranged into the form

$$J = \lambda\left[\left(C - \frac{\lambda}{2}\frac{\partial C}{\partial x}\right)\Gamma_{12} - \left(C + \frac{\lambda}{2}\frac{\partial C}{\partial x}\right)\Gamma_{21}\right] \qquad (25.30)$$

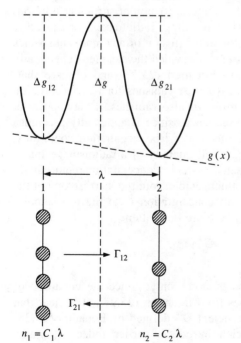

FIGURE 25-7
A simple model for diffusion by thermally activated jumps between adjacent planes. The envelope $g(x)$ of the site occupation energy is tilted to represent the effect of non-ideal solution behavior.

Gathering terms and writing $\overline{\Gamma} = (\Gamma_{12} + \Gamma_{21})/2$, we obtain

$$J = -\lambda^2\overline{\Gamma}\left(\frac{\partial C}{\partial x} - \frac{C}{\lambda}\frac{\Gamma_{12} - \Gamma_{21}}{\overline{\Gamma}}\right) \tag{25.31}$$

and by comparison with Equation (25.27) we find

$$D^* = \lambda^2\overline{\Gamma} \tag{25.32}$$

and

$$-\frac{\partial \ln \gamma}{\partial x} = \frac{\Gamma_{12} - \Gamma_{21}}{\lambda\overline{\Gamma}} \tag{25.33}$$

Equations (25.32) and (25.33) are both important results. Equation (25.32) provides a link between D^* of the formal theory and the atomic jump rate, whereas Equation (25.33) shows that the atomistic significance of nonideality is that $\Gamma_{12} \neq \Gamma_{21}$ or, in other words, that the jump rate is a function of concentration. Specific expressions for the jump rates Γ_{12} and Γ_{21} may be obtained using analogs of Equation (25.3) in conjunction with the energy diagram shown at the top of Figure 25-7. This is a tilted version of that shown in Figure 25-1, so as to include the energy gradient or driving force created by nonideal solution behavior. The activation energies Δg_{12} and Δg_{21} are given by

$$\Delta g_{12} = \Delta g + \frac{-\lambda}{2}F \tag{25.34}$$

and

$$\Delta g_{21} = \Delta g + \frac{\lambda}{2}F \tag{25.35}$$

where F, the effective force on an atom, is given by $F \equiv -\partial g/\partial x$ and has a positive sign when directed down the energy gradient. Writing the jump rates in terms of a Boltzmann expression of the type represented by Equation (25.3), we find

$$\frac{\Gamma_{12} - \Gamma_{21}}{\overline{\Gamma}} = \exp\left(\frac{F\lambda}{2kT}\right) - \exp\left(-\frac{F\lambda}{2kT}\right) \tag{25.36}$$

Since $F\lambda/kT$ is almost always much less than unity, Equation (25.36) can be further simplified by a first-order expansion of the exponential terms, to obtain

$$\frac{\Gamma_{12} - \Gamma_{21}}{\overline{\Gamma}} = \frac{F\lambda}{kT} \tag{25.37}$$

Equation (25.37) shows that the difference in the jump frequencies, $\Gamma_{12} - \Gamma_{21}$, is directly proportional to the effective chemical force F acting on an atom. By comparison with Equation (25.33) it is also evident that, in thermodynamic terms, this force is given by

$$F = -\frac{kT\partial \ln \gamma}{\partial x} \tag{25.38}$$

We also note for completeness that if Equations (25.32) and (25.37) are inserted into Equation (25.31), Fick's law becomes

$$J = -D^*\frac{\partial C}{\partial x} + \frac{CD^*F}{kT} \tag{25.39}$$

This form is also a useful starting point for the consideration of diffusion under the influence of *external* forces, such as those produced by an electric field or a stress gradient. Each of these cases calls for an appropriate redefinition of the force F, as has been discussed by Manning [2,3].

In preparation for a discussion of specific mechanisms, we consider now some particular forms of Equation (25.32) for ideal diffusion in cubic crystal structures. In the model of Figure 25-7, we presumed that the jump distance was the same as the interplanar spacing λ. This is true only for the simple cubic structure where an atom may jump in six different directions and where $\Gamma_{12} = \Gamma/6$. For this case, Equation (25.32) becomes

$$D^* = \frac{\Gamma\lambda^2}{6} \tag{25.40}$$

More detailed considerations show that Equation (25.40) has wider applicability to cubic structures if λ is simply redefined as the jump distance, rather than the interplanar spacing. Thus for diffusion amongst nearest neighbor lattice positions in the simple cubic, face-centered cubic, body-centered cubic, and diamond cubic structures, λ takes the values a, $a/\sqrt{2}$, $a\sqrt{3}/2$ and $a\sqrt{3}/4$, respectively, where a is the lattice parameter. Correspondingly, Equation (25.40) takes the form

$$D^* = \beta\Gamma a^2 \tag{25.41}$$

where β has the respective values 1/6, 1/12, 1/8, and 1/32. Alternatively, we may relate D^* to the rate of a specific atom jump, w, by replacing Γ by zw [Equation (25.1)], where z is the number of directions in which jumps may occur. For the sequence of structures given above, $z = 6, 12, 8,$ and 4 respectively, with the result that we may also write

$$D^* = \alpha w a^2 \tag{25.42}$$

where α takes the corresponding values 1, 1, 1, and 1/8.

The Interstitial and Vacancy Mechanisms of Diffusion

To proceed further with atomistic considerations, we now need to be concerned with the specific details of the jump process, i.e., with the particular diffusion mechanism involved. Two basic mechanisms, which in some respects are oppo-

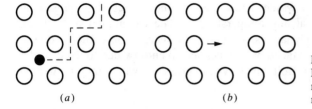

FIGURE 25-8
Illustrations of (a) the interstitial mechanism and (b) the vacancy mechanism of diffusion.

sites of each other, are shown in Figure 25-8. The simplest situation, (a) in the figure, involves a dilute concentration of solute atoms which are small enough to dissolve interstitially in the host material. Such interstitial solute atoms occupy their own set of sites, which is distinct from that occupied by the host atoms. Because of this, and the assumption of a dilute solution, an interstitial atom is surrounded by neighboring sites which are virtually always empty. In this case, the free energy of activation Δg is simply the height of the barrier to migration, labeled Δg_m in Figure 25-1, and Equation (25.3) becomes

$$\Gamma = z \nu_0 \exp\left(-\frac{\Delta g_m}{kT}\right) = z \nu_0 \exp\left(\frac{\Delta s_m}{k}\right) \exp\left(-\frac{\Delta h_m}{kT}\right) \tag{25.43}$$

where the last step involves the separation of Δg_m into terms involving a migration entropy Δs_m and a migration enthalpy Δh_m. Inserting this result into Equation (25.41), we obtain D^* for interstitial diffusion in cubic materials in the form

$$D^* = D_0^* \exp\left(-\frac{\Delta h_m}{kT}\right) \tag{25.44}$$

where the pre-exponential constant D_0^* is given by

$$D_0^* = \beta_i z \nu_0 a^2 \exp\left(\frac{\Delta s_m}{k}\right) \tag{25.45}$$

and β_i is a geometrical factor which depends on the location of the interstitial sites in the particular structure considered.

We now turn to the problem of *self-diffusion* in a simple crystalline solid. Here we must confront a problem that may have already presented itself to the reader, namely that jumps to neighboring sites are blocked by the presence of other atoms. Initially, it was thought that self-diffusion must therefore involve a simultaneous place exchange of two atoms. However, in one of the most interesting episodes in the history of diffusion, it gradually became clear that some form of *defect-mediated* mechanism provided an energetically-preferable alternative to direct place exchange. In most cases the defect of principal interest in promoting self- (or substitutional solute) diffusion is the vacant lattice site or vacancy [Figure 25-8(b)], which is present in real crystalline solids as an intrinsic defect species. We mean by the use of the term intrinsic that vacancies are present not merely as accidents of growth, but that they possess a characteristic free energy of formation Δg_f which determines the mole fraction $\bar{c}_v$ present in thermodynamic equilibrium at any temperature T. The calculation of $\bar{c}_v$ is an exercise in statistical thermodynamics, in which the increase in enthalpy due to the creation of vacancies is balanced by the increase in entropy due to the disorder they introduce. The result is a Boltzmann-type expression, which may be written in the alternative forms

$$\bar{c}_v = \exp\left(-\frac{\Delta g_f}{kT}\right) = \exp\left(\frac{\Delta s_f}{k}\right)\exp\left(-\frac{\Delta h_f}{kT}\right) \tag{25.46}$$

where Δs_f and Δh_f represent the entropy and enthalpy terms for vacancy formation. It is of interest to note that even at the melting point of most solids, $\bar{c}_v$ does not exceed 10^{-4} mole fraction, and can be orders of magnitude smaller at temperatures where diffusion processing steps are carried out. As indicated in Figure 25-8(b), the significance of the vacancy in self-diffusion is that it provides a site into which any one of the neighboring atoms may jump. To obtain an expression for the atomic jump rate via the vacancy mechanism, we may return to Equation (25.1) and the relation $w = \nu_0 p_{12}$ for the jump rate of a chosen atom 1 into a specific neighboring site 2. For the vacancy mechanism, the probability p_{12} involves the product of the probability for the atom jump out of site 1 and the probability that the chosen site 2 is vacant. Since the vacancy population is randomly distributed, the latter probability is simply $\bar{c}_v$. Hence we have

$$\Gamma/z = w = \nu_0 p_{12} = \nu_0 \bar{c}_v \exp\left(-\frac{\Delta g_m}{kT}\right)$$

$$= \nu_0 \exp\left(-\frac{\Delta g_f + \Delta g_m}{kT}\right) \tag{25.47}$$

where the last step uses Equation (25.46).

To complete the calculation of D^*, we must recognize that the introduction of the vacancy mechanism necessitates a modification to Equations (25.40), (25.41), and (25.42). These equations were based on the assumption that successive jumps occurred in a completely random manner. For a defect mechanism, however, this is no longer true. An atom which has just completed one jump has a much higher than average probability of making the return jump, simply because the vacancy is still adjacent to it. Clearly, a *correlation effect* of this nature tends to retard diffusion, and can be accounted for by the introduction of a correlation factor $f \leq 1$ into the right-hand side of Equations (25.40), (25.41), and (25.42). Specifically, Equation (25.42) then becomes

$$D^* = f\alpha a^2 w \tag{25.48}$$

and by the use of Equations (25.47) and (25.48) we finally obtain

$$D^* = D_0^* \exp\left(-\frac{\Delta h_f + \Delta h_m}{kT}\right) \tag{25.49}$$

where

$$D_0^* = f\alpha a^2 \exp\left(\frac{\Delta s_f + \Delta s_m}{k}\right) \qquad (25.50)$$

The correlation factor for self-diffusion by the vacancy mechanism in fcc and bcc metals has the approximate values 0.78 and 0.72 respectively. Although correlation is thus a minor correction in these cases, the reader should bear in mind that correlation effects become considerably more important (and complicated) when extended to the case of substitutional solute diffusion by the vacancy mechanism.

Whether diffusion occurs by the interstitial or vacancy mechanism, we see from Equations (25.44) and (25.49) that the diffusion coefficient can be expected to follow an empirical relationship of the type

$$D = D_0 \exp\left(-\frac{Q}{kT}\right) \qquad (25.51)$$

where the activation energy Q corresponds to Δh_m for interstitial diffusion, and to $\Delta h_f + \Delta h_m$ for diffusion by the vacancy mechanism. Equation (25.51) has been verified experimentally in a wide variety of cases and, despite some exceptions, can be regarded as the normal behavior expected of the diffusion coefficient. A short collection of data for both interstitial and vacancy mechanism diffusion is given in Table 25.1, from which it can be seen that activation energies of 0.8 to 2.5 eV (77 to 240 kJ/mol) are fairly typical for metals. The very strong temperature dependence that energies of this magnitude impose on the diffusion coefficient is conveniently summarized in Figure 25-9.

TABLE 25.1
Parameters for interstitial solute diffusion and self-diffusion (by the vacancy mechanism) in some metals

System	Structure	D_0, cm^2/s	Q, eV
Fe-C (ferrite)	bcc	0.004	0.83
Fe-C (austenite)	fcc	0.25	1.5
Nb-H	bcc	0.0005	0.11
Nb-O	bcc	0.009	1.17
Nb-N	bcc	0.021	1.51
Cu	fcc	0.16	2.07
Au	fcc	0.04	1.76
Ni	fcc	0.92	2.88
Al	fcc	0.05	1.28
Pb	fcc	1.37	1.13

A comprehensive tabulation of diffusion data is given in Reference [8].

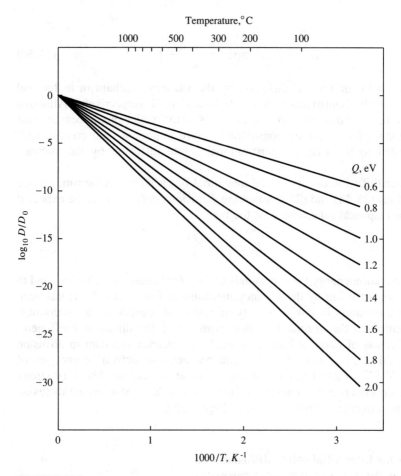

FIGURE 25-9
Diffusion data are often presented as Arrhenius (or rate) plots of log D versus $1/T$. These calculated curves illustrate the power of the Boltzmann factor for a range of activation energies typically encountered in solid state diffusion. To convert to other units in common usage, note that 1 eV = 23.0 kcal/mol = 96.2 kJ/mol.

OTHER CATEGORIES OF DIFFUSION

The preceding section described two basic mechanisms of diffusion and introduced the important concept of defect-mediated diffusion. We now proceed to branch out from these considerations in a number of different directions, so as to obtain a wider perspective of the field as a whole.

Substitutional Solute Diffusion and the Kirkendall Effect

Most metallic solid solutions are of a substitutional character where both solvent and solute atoms share a common set of lattice sites. As a result of this common-

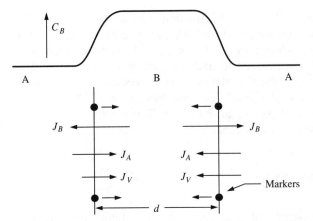

FIGURE 25-10
A configuration used to detect marker shift in the chemical interdiffusion of two species A and B that form a continuous range of substitutional solid solutions. The markers (e.g., fine Mo wires), inserted at the original interfaces of the ABA sandwich, move together if $J_B > J_A$ and apart if $J_B < J_A$.

ality, and the vacancy mechanism described above, the diffusion of one species becomes linked to that of the other. For example, in the dilute regime where we may think of a single solute atom B surrounded entirely by solvent (A) atoms, it is clear that the diffusivity of B cannot depend solely on the rate at which B exchanges with a vacancy, for without any motion of A atoms, B can do no more than switch back and forth repeatedly between the same pair of sites. In fcc alloys, the basic model describing solute diffusion involves consideration of no less than five different jumps [2,3,7]. This model has met with considerable success, including an explanation of the effect of solute additions on the self-diffusion of the solvent.

The fact that A and B atoms exchange with a vacancy at different rates has an important consequence when diffusion occurs in a chemical concentration gradient (e.g., at the junction of a diffusion couple made from solid solutions of two different compositions). Here, the imbalance in the atomic fluxes J_A and J_B gives rise to a vacancy flux moving in the direction of the slower species. If there are enough local sources and sinks for vacancies to preserve the vacancy concentration near its equilibrium value, the vacancy flux corresponds to a shifting of layers of atoms from one side of a reference plane to the other. As a consequence, the reference plane shifts or migrates with respect to the ends of the sample. One type of experiment in which such an effect can be demonstrated is shown in Figure 25-10. Marker motion induced by a net vacancy flux is known as the Kirkendall effect. If was first demonstrated in a definitive manner in 1947, and represented one of the earliest and most compelling observations in favor of the vacancy mechanism of diffusion.

Partially Interstitial or Amphoteric Solutes

Although most binary solid solutions are either exclusively of the interstitial or substitutional type, there are some cases of unexpectedly rapid diffusion where the explanation is believed to involve a solute with a mixed or amphoteric solution behavior. These situations involve a solute which is predominantly substitutional, but which also possesses a small interstitial fraction with a very much higher diffusivity. Examples of this type are found for both metals in metals (e.g., Au in Pb), and for metals in semiconductors (e.g., Cu in Ge and Au in Si). However, the details of the mechanisms involved in each of these cases are believed to be different [4,5]. For example, the Cu interstitial in Ge becomes substitutional by combining with a vacancy (the Frank-Turnbull mechanism). On the other hand, recent work indicates that the Au interstitial in Si becomes substitutional by ejecting a host atom (the kick-out mechanism).

Diffusion Involving Other Types of Defect Disorder

The vacancy is but one example of a general class of mobile point defects produced by the local addition, removal, or substitution of one or more host atoms. The opposite of the vacancy is the extra host atom squeezed between its fellows, and known as the self-interstitial. In metals, it appears that the jump rate of a self-interstitial may be very much higher than that of a vacancy. However, because the formation energy of the self-interstitial is also much larger than that of the vacancy, the equilibrium concentration of self-interstitials is usually so small that their contribution to diffusion under equilibrium conditions is negligible. Interestingly, an exception to this statement appears to exist for the case of self-diffusion in silicon at high temperatures [5]. Also, it should be borne in mind that the role of self-interstitials may become much more important under conditions of high-energy particle irradiation, where the numbers of vacancies and interstitials are not dictated by equilibrium conditions and tend to be of a similar magnitude.

Another point defect of occasional interest is the divacancy. Again, the formation energy of this defect is relatively high compared with that of the single vacancy, and the equilibrium concentration is correspondingly smaller. Nevertheless, because of evidence for a substantially increased mobility, it has been argued that divacancies may contribute sensibly to equilibrium diffusion in some metals at high temperatures.

Ordered compounds, and especially those of a strongly ionic character, possess a number of interesting possibilities with respect to the defects that mediate diffusion. In compounds where the ions have a fixed valence, stoichiometry must be preserved to maintain charge neutrality in the pure compound. In the pure alkali halides, the defect population is of the *Schottky* type, i.e., containing equal numbers of A^+ and B^- vacancies. In silver bromide, on the other hand, the defects are of the *Frenkel* type, with disorder on one sublattice only, consisting of equal numbers of silver vacancies and interstitials. In ceramic oxides containing transition metal ions, the ability of the metal ion to assume a higher valence state permits a deviation from stoichiometry while charge neutrality is maintained. For example, the conversion of two Fe^{2+} ions to Fe^{3+} ions is associated with the introduction of a vacancy on the metal sublattice of FeO, thereby promoting the mobility of the remaining metal ions. Analogous considerations are involved when ionic compounds are doped with solutes whose valence is different from that of the host species. A classic example is provided by the introduction of divalent ions such as Ca^{2+} into the alkali halides. In this case charge neutrality is maintained by the introduction of an equal number of alkali ion (cation) vacancies, which can serve to enhance the cation mobility and resulting ionic conductivity at temperatures below the regime dominated by intrinsic defects.

High-Diffusivity Paths and Short-Circuit Diffusion

So far in this chapter, we have considered diffusion as a process that occurs throughout the volume of a homogenous solid. In doing so, we have ignored the fact that real materials often possess free surfaces, grain boundaries and dislocations, all of which represent localized paths along which diffusion may occur at greatly enhanced rates. Because these planar or line defects offer a lower resistance to diffusion, this type of accelerated diffusion is frequently referred to as *short-circuit* diffusion. Although the mechanistic details are usually not as well understood as for volume diffusion, the central experimental fact in short-circuit diffusion is that the activation energies involved are substantially smaller than for competing bulk mechanisms, with the result that short-circuit diffusion can take over as the dominant transport process at low temperatures. As an example, we may consider the case of grain boundary diffusion where, as a rule of thumb, the activation energy Q_b for boundary diffusion is only about half of that for lattice diffusion, Q_l. A rough estimate of the ratio of the fluxes transported down the boundary and through the interior of the grain is then given by

$$\frac{J_b}{J_l} \approx \frac{b}{g} \exp\left(\frac{Q_l - Q_b}{kT}\right) \approx \frac{b}{g} \exp\left(\frac{Q_l}{2kT}\right) \tag{25.52}$$

where b is the effective grain boundary thickness and g is the grain size. If we make use of the empirical correlation for fcc metals that $Q_l/kT_m \approx 18$, where T_m is the melting point and $k = 8.61735 \times 10^{-5}$ eV/K, Equation (25.52) becomes

$$\frac{J_b}{J_l} \approx \frac{b}{g} \exp\left(\frac{9T_m}{T}\right) \tag{25.53}$$

Since b can be taken as about 5 nm, a grain size of 0.5 mm corresponds to a b/g ratio of 10^{-6}. In this case, we find from Equation (25.53) that short-circuiting diffusion becomes dominant ($J_b/J_l > 1$) when $T < 0.65\ T_m$. Diffusion experiments conducted at such relatively low temperatures reveal the presence of short-circuiting paths by the appearance of an extended tail on the diffusion profile. In addition, the Arrhenius plot of log D against $1/T$ becomes less steep, corresponding to a smaller effective activation energy. The importance of short-circuiting diffusion in thin-film applications has been discussed in a number of books and review articles [10,11].

Noncrystalline Inorganic Solids and Polymers

A large number of interesting and useful materials lack the long-range periodic structure that is the essential characteristic of a crystalline solid. Noncrystalline materials may be referred to either as amorphous or glassy, the latter designation being particularly appropriate if the amorphous structure is the result of solidification of a liquid without crystallization. The two largest and oldest families of amorphous materials are the oxide glasses and organic polymers. In recent years two other groups, amorphous semiconductors and metallic glasses, have undergone rapid development and exploitation.

On a local or short-range scale, amorphous materials frequently contain atomic arrangements or structural units that are reminiscent of those found in comparable crystalline solids. However, whereas these units are precisely defined in crystals, this is not the case in amorphous solids. Here, local disorder is manifested by the existence of a distribution of interatomic spacings, bond angles, etc., with the result that site occupation energies and saddle-point energies each become distributed variables exhibiting dispersion about a mean value. As a result, and in contrast to Figure 25-1, a schematic illustration of the free energy diagram for diffusion in an amorphous solid has the form shown in Figure 25-11. One approach to modeling diffusion in such a disordered potential is to calculate and analyze a large number of particle trajectories, developed according to an appropriate set of rules that serve to define the atomic disorder. As an alternative to this Monte Carlo approach, Kirchheim has given an analytic treatment based on the simplifying assumptions of a Gaussian distribution of site energies and a constant saddle-point energy. This treatment has had considerable success in explaining certain characteristic features of the diffusion of hydrogen in metallic glasses, and of alkali ions in the oxide glasses [12,13].

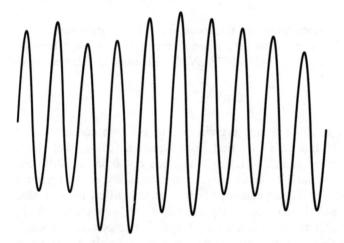

FIGURE 25-11
A schematic illustration of the free energy potential for diffusion in an amorphous solid.

Due to the structural complexity of amorphous polymers, it is not surprising that treatments of small-molecule diffusion in these materials are more tenuous than those developed for simpler crystalline solids. Nevertheless, it is interesting that most of the models proposed appear to fall into one of two categories. One group contains models that are basically similar to the interstitial mechanism discussed earlier. Models of this type have been used to explain the observation that in some cases the activation energy and prefactor vary systematically and strongly with the size of the diffusing molecule. Other models approach diffusion in polymers with the concept of holes or free volume, employed originally to explain diffusion in liquids. It should be noted that whereas a vacancy in a crystalline solid has a continuous existence during its lifetime between source and sink, the idea of the hole defect (which should not be confused with a missing electron) is quite different. A hole is pictured as a transient defect or instantaneous microcavity formed and dissipated by random thermal energy fluctuations. Although free volume models of diffusion have some appealing features, they tend to be formulated in phenomenological terms and do not lead easily to explicit experimental predictions concerning, for example, the temperature dependence of the diffusion coefficient. A useful review of these and other models proposed for small-molecule diffusion in polymers has been given by Frisch and Stern [14].

To conclude this discussion, we touch briefly on the subject of volume changes associated with diffusion. In inorganic materials, such changes are generally quite small, but in certain circumstances (e.g., for layered or modulated thin-film structures), they may cause significant self-stresses because of the high elastic stiffness of the materials involved [15]. In polymers, modest degrees of swelling ($\approx$ 1 percent in volume) due to the limited sorption of small molecules is a well-known phenomenon, particularly for the case of moisture absorption [16]. On the other hand, penetrants with a large solubility in certain polymers can lead to much larger amounts of swelling, and may induce or trigger structural relaxation changes within the polymer host itself, which proceed on a time scale comparable to or longer than that involved in diffusion. Those more complicated situations have been categorized as Case II diffusion, and are discussed in the following chapter.

A CASE HISTORY: SULFUR PERMEATION IN BACKSEAL POLYMERS

To illustrate the practical importance of a knowledge of diffusion in packaging technology, we conclude the present chapter with a case history in which diffusion played a key role in the occurrence, and later prevention, of a painful series of hardware failures in the field. In the early and middle 70s, large numbers of circuit modules were produced by solder-joining silicon chips to half-inch-square ceramic substrates carrying a pattern of metallic lands applied by screening and firing a silver-palladium paste. Due to a tendency for the lead-tin solder used in chip joining to deplete palladium from the lands, small regions could be formed near the chip sites where the lands had been converted to essentially pure silver. These silver-rich regions became the trouble spots for an unexpected failure mechanism that turned out to be exacerbated by the use of two types of silicone polymers in the module package. One of these was a silicone gel, applied at the chip sites as a precaution against the phenomenon of silver migration between adjacent lands. The other was the silicone rubber sealant used to secure a protective aluminum cap over the substrate. This backseal covered the underside of the substrate and penetrated any crevices between the edges of the substrate and the cap. When these modules were placed in service in locations where the atmosphere contained high trace levels of elemental sulfur, the palladium-depleted silver-rich regions corroded with unexpected rapidity to form the sulfide Ag_2S, which led to failures by the formation of an open circuit.

To understand and avoid this problem, it was necessary to investigate the kinetics of sulfur permeation both in the silicones mentioned above and in various candidate replacements for the backseal, of which the most promising proved to be an epoxy of the dgeba (diglycidyletherbisphenol-A) type. As we saw in Equation (25.10), permeation involves the product of two factors, namely a solubility term and a diffusion coefficient. Measurements of both types were carried out [17] with the aid of a sulfur source containing a known fraction of the radioisotope S^{35}. This isotope, which is a β-emitter with a convenient half-life of 87 days, provides nanogram sensitivity for the detection of sulfur by standard liquid-scintillation counting techniques. Solubility measurements conducted on both the silicone rubber and gel showed that the sorption of sulfur conformed to a reversible solution process obeying Henry's law:

$$X = H(T)P \tag{25.54}$$

where X is the equilibrium weight percentage of sulfur in solution, P is the sulfur vapor pressure, and $H(T)$ is the Henry's law coefficient at temperature T. This result implies, through simple mass action considerations, that the solution process does not involve dissociation of the S_8 molecule known from vapor density measurements to exist in the vapor phase. Some of the solution results obtained for the silicone rubber are shown in Figure 25-12, where the equilibrium solubility is plotted on a logarithmic scale against $1/T$. The data fall on two straight lines of opposite slope, corresponding to the use of two different sets of experimental conditions. The lower line represents the solubility under the isobaric conditions obtained by varying the temperature of the sample while holding the sulfur source at a constant lower temperature so as to fix the sulfur vapor pressure at a constant value. This line is represented by the equation

$$X = AP \exp\left(-\frac{\Delta h_s}{kT}\right) \tag{25.55}$$

where A is a constant and Δh_s is the enthalpy or heat of solution. By contrast, the upper line of Figure 25-12 represents the maximum or saturated solubility X_{sat} obtained by allowing the temperature of the sulfur source to increase along with that of the sample. The equation of this line is

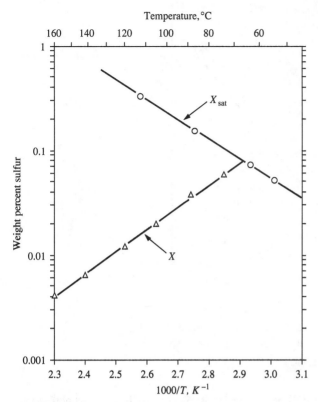

FIGURE 25-12
Solubility data for sulfur in silicone rubber. The isobaric data (lower branch) were obtained with the sulfur source at 341 K. From reference [17].

$$X_{\text{sat}} = B \exp\left(-\frac{\Delta h_{\text{vap}} + \Delta h_s}{kT}\right) \qquad (25.56)$$

where B is a constant and Δh_{vap} is the heat of vaporization of sulfur (≈ 1 eV). The opposite slopes of the two lines shown in Figure 25-12 reflect the situation that Δh_{vap} is larger and positive, while Δh_s is smaller and negative. This means that the increase in solubility with temperature under saturated conditions is due to the rapid rise in the vapor pressure of sulfur, which is sufficient to override the decrease in solubility observed under isobaric conditions. From the slope of the isobar in Figure 25-12, Δh_s is found to be -0.43 eV, where the negative sign corresponds to exothermic solution behavior. The magnitude of the heat of solution is substantial, and leads to a Henry's law coefficient H_{25} (i.e., for 25°C) orders of magnitude larger than previously encountered for permeants in silicone rubber. This is illustrated in Figure 25-13, where the result for sulfur has been added to a previous compilation demonstrating an empirical correlation

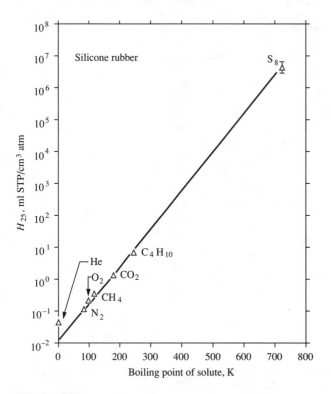

FIGURE 25-13

Correlation of the Henry's law coefficient for solubility at 25°C with the boiling point of the solute species. From reference [17].

between H_{25} and the boiling point of the solute species [18]. Whereas the previous data were restricted to solutes with boiling points below room temperature, the addition of the result for sulfur more than doubles the temperature range of the plot and extends the ordinate by six decades. In practical terms, the significance of this finding is the realization that silicone rubber is an extraordinary getter for sulfur vapor. When coupled with rapid diffusion, we thereby obtain a high-permeability backseal that can feed sulfur to the internal reactive sites. Attempts to measure the diffusion coefficient of sulfur in membranes of silicone rubber by conventional permeation measurements quickly revealed that the diffusion rate was indeed large. In fact, diffusion was so fast that the available flux of sulfur atoms evaporated from the radioactive source was transmitted through the membrane under steady-state conditions without the development of a discernible concentration gradient. As a consequence, this method of measurement had to be abandoned in favor of a vacuum desorption technique based on the use of Equation (25.19). As shown in Figure 25-14, the diffusivities so obtained fell typically in the range of 10^{-7} to 10^{-6} cm^2/s. They are represented by Equation (25.51) with a D_0 of 4×10^5 cm^2/s and a Q of 0.74 eV. Figure 25-14 also shows the much

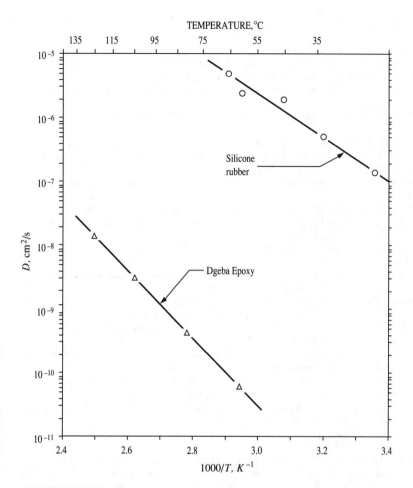

FIGURE 25-14
Arrhenius plots for the diffusion of sulfur in a silicone rubber and a dgeba epoxy. From reference [17].

slower diffusion behavior observed in the epoxy material by membrane permeation experiments. In this case the parameters D_0 and Q are 3×10^5 cm^2/s and 1.06 eV, respectively. Since D_0 is about the same for both materials, the observed reduction of a comfortable four orders of magnitude in the diffusion coefficient is due entirely to the larger activation energy, and illustrates once again the power of the Boltzmann factor in influencing the rate of diffusion. Interestingly, equilibrium solubility measurements performed on the epoxy showed no major change compared with the levels found in the silicone rubber. As a consequence, the success of the epoxy as a replacement backseal that eliminated the corrosion failures can be attributed wholly to its effectiveness as a diffusion barrier to the ingress of sulfur.

REFERENCES

1. P. G. Shewmon, *Diffusion in Solids*, New York: McGraw-Hill, 1963.
2. *Diffusion*, Metals Park, Ohio: American Society for Metals, 1973.
3. J. R. Manning, *Diffusion Kinetics for Atoms in Crystals*, Princeton: Van Nostrand, 1968.
4. A. S. Nowick and J. J. Burton, editors, *Diffusion in Solids: Recent Developments*, New York: Academic Press, 1975.
5. G. E. Murch and A. S. Nowick, editors, *Diffusion in Crystalline Solids*, New York: Academic Press, 1984.
6. J. Crank and G. S. Park, editors, *Diffusion in Polymers*, New York: Academic Press, 1968.
7. M. B. Bever, editor, *Encyclopedia of Materials Science and Engineering*, vol. 2, pp. 1175, 1180, 1186, Oxford: Pergamon Press, 1986.
8. E. A. Brandes, editor, *Smithells Metals Reference Book*, 6th ed., London: Butterworths, 1983, Chapter 13.
9. J. Crank, *The Mathematics of Diffusion*, Oxford: Oxford University Press, 1946.
10. J. M. Poate, K. N. Tu, and J. W. Mayer, editors, *Thin Films—Interdiffusion and Reactions*, New York: Wiley Interscience, 1978.
11. D. Gupta and P. S. Ho, editors, *Diffusion Phenomena in Thin Films*, Park Ridge, N.J.: Noyes Publications, 1988.
12. R. Kirchheim, "Solubility, Diffusivity and Trapping of Hydrogen in Dilute Alloys, Deformed and Amorphous Metals—II," *Acta Metall.* 30, p. 1069 (1982).
13. R. Kirchheim, "Influence of Disorder on the Diffusion of Alkali Ions in SiO_2 — and GeO_2 — Glasses," *J. Non-Crystalline Solids* 55, p. 243 (1983). See also R. Kirchheim and U. Stolz, "Modelling Tracer Diffusion and Mobility of Interstitials in Disordered Materials," *J. Non-Crystalline Solids* 70, p. 323 (1985).
14. H. L. Frisch and S. A. Stern, "Diffusion of Small Molecules in Polymers," *CRC Critical Reviews in Solid-State and Materials Sciences* 11, p. 123 (1983).
15. A. L. Greer and F. Spaepen, "Diffusion" Chapter 11 in *Synthetic Modulated Structures*, edited by L. L. Chang and B. C. Giessen, New York: Academic Press, 1985.
16. B. S. Berry and W. C. Pritchet, "Bending-Cantilever Method for the Study of Moisture Swelling in Polymers," *IBM J. Res. Dev.* 28, p. 662 (1984).
17. B. S. Berry and J. R. Susko, "Solubility and Diffusion of Sulfur in Polymeric Materials," *IBM J. Res. Dev.* 21, p. 176 (1977).
18. W. L. Robb, "Thin Silicone Membranes—Their Permeation Properties and Some Applications," *Ann. N.Y. Acad. Sci.* 146, p. 119 (1968). See also G. J. Van Amerongen, "Diffusion in Elastomers," *Rubb. Chem. Technol.* 37, p. 1065 (1964) and V. Stannett in Chapter 2 of reference 6.
19. R. J. Borg and G. J. Dienes, *An Introduction to Solid State Diffusion*, Boston: Academic Press, 1988.

PROBLEMS

25.1. Look up the word *diffusion* in a dictionary. Discuss whether the definition that applies to atomic migration is technically correct. Can you improve the definition?

25.2. Make a crude estimate of the atomic vibration frequency ν_0 of a monoatomic solid of atomic weight 50 and density 8 g/cm^3 whose Young's modulus of elasticity is 200 GPa. Use the relation that the vibration frequency of a mass m on a spring of stiffness constant K is $(1/2\pi)(K/m)^{1/2}$.

25.3. What is the total path length traveled by an atom if the rms displacement executed by a random walk is 0.1 mm and the jump distance is 0.1nm?

25.4. At room temperature (22°C) the diffusion coefficient for moisture in a circuit board epoxy has been measured as 2.5×10^{-9}cm^2/s. Is it realistic to consider vacuum drying circuit boards at room temperature if the path length to a free surface is 1 mm? How much faster would drying occur at 90°C if the activation energy were 0.4 eV?

25.5. Interstitial carbon atoms in bcc iron occupy the octahedral sites located midway along the cube edges and at the center of each cube face. What is the appropriate value of β_i to be inserted into Equation (25.45)?

25.6. By inspection of Eq. (25.39), describe conditions under which a steady state flux is given by the relation $J = CD^*F/kT$, where F now refers to an externally imposed force acting on the diffusant. Show that D^*F/kT corresponds to a drift velocity v and thus that v/F, a quantity called the drift mobility, is given by D^*/kT.

CHAPTER
26

CASE II
DIFFUSION

RONALD C. LASKY
THOMAS P. GALL
EDWARD J. KRAMER

INTRODUCTION

While there are many examples of diffusion phenomena in the literature which do not obey Fick's laws [1], such non-Fickian behavior is most importantly observed for diffusion of relatively small organic molecules into polymer glasses. A certain limiting case of such behavior is commonly called Case II diffusion. Case II diffusion is particularly important in packaging because many of the processes that are in widespread use now, or will be in the future, require exposure of polymer glasses (e.g., photoresists or dielectric layers) to such organic molecules.

Non-Fickian behavior can be most easily observed with a simple gravimetric experiment where the weight gain of a polymer sheet, which is exposed to a liquid or vapor penetrant, is monitored as a function of time. The initial weight gain in such an experiment can often be expressed as

$$w = bt^n \tag{26.1}$$

where w is the weight gain, t is time, and b and n are constants. If Fick's laws are obeyed, n will be equal to 0.5. The resulting diffusion is often referred to as Case I diffusion. Cases where n differs from 0.5 are non-Fickian. Under many conditions the weight gain is linear with time, i.e., $n = 1$. This subset of non-Fickian diffusion is now commonly called Case II diffusion.

Microscopic analyses of cross sections of polymer sheets in Case II diffusion experiments have revealed sharp penetrant concentration profiles that advance with constant velocity [2–5]. The leading edge of these profiles is often referred to as the *front*. Since front position and w are linear functions of time, the term *linear kinetics* is used to describe the Case II process. Linear kinetics and a sharp diffusion profile are the observable, distinguishing characteristics of Case II diffusion.

Early attempts to explain Case II diffusion considered the effects of changing surface concentration, time-dependent diffusion coefficients, internal stresses, and surface skins [6–8]. None of these approaches achieved success. By the mid 1950s Bagley and Long [9], Newns [10], and Long and Richmond [11] had suggested that Case II diffusion was a two-stage process. The diffusion-controlled first stage involved elastic expansion of the polymer network; whereas the slower second stage was not diffusion related but was a rearrangement of the polymer chains.

Alfrey, in 1965 [12], was the first to use the term *Case II* to describe linear sorption kinetics. At the same time he also proposed three criteria for Case II diffusion:

1. A sharp advancing boundary separates a glassy core from an outer swollen and rubbery shell.
2. Behind the advancing penetrant front, the swollen gel is nearly in equilibrium.
3. The boundary between the swollen gel and glassy core advances at a constant velocity.

These criteria consolidated earlier work with Alfrey's and formed the foundation for the current theories of Case II diffusion.

Figure 26-1(a) shows how the Case II front develops with time. Curves 1, 2, and 3 show the induction period, during which the solvent concentration decreases smoothly with depth. After the critical volume fraction, ϕ_C, is reached at the surface (curve 4), the solvent concentration remains constant for some depth, then falls off abruptly. The point where the solvent concentration decreases abruptly is called the Case II front. The front advances into the glass at a constant velocity with a Fickian precursor ahead of it.

THEORIES OF CASE II DIFFUSION

Modified Fickian Theories

Following qualitative suggestions by Scott [13] and Hartley [5], Crank, in 1953, was first to propose a quantitative model for Case II diffusion [7]. He used Fick's equations with a diffusion coefficient that depended on both the sample history and penetrant concentration. He also introduced the important concept of differential swelling stress and its effects on the diffusion process. Using this approach and selecting the right parameters, he was able to reproduce the sharp front of Case II diffusion but not the linear kinetics. While not in use today, Crank's approach is the foundation for some of the current accepted theories.

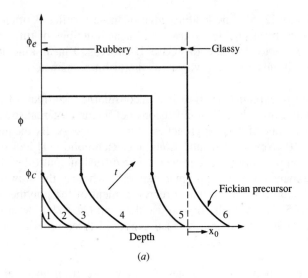

(a)

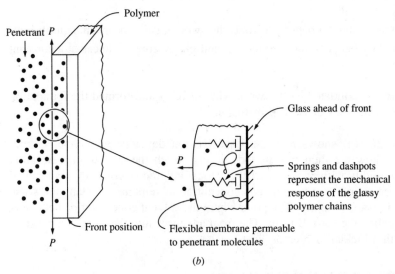

(b)

FIGURE 26-1
(a) A family of penetrant volume fraction depth profiles. (b) The diffusion of a penetrant into a polymer. The effect of the resulting osmotic pressure on the polymer is modeled by a system of springs and dashpots.

Thomas and Windle [14] assumed that the diffusion in the glass ahead of the front is important to the kinetics. They then developed two equations that coupled the diffusion of penetrant down a gradient in its activity with the time-dependent mechanical deformation in response to the osmotic stress. Their approach appears, with proper parameter selection, to predict qualitatively all the essential aspects of Case II sorption [4,14–18].

The Thomas and Windle Model
of Case II Diffusion

The swelling of a polymer glass by penetrant molecules must occur by these molecules occupying "interstitial" sites between polymer chains. Since the glass is a roughly random close-packed arrangement of chain segments to begin with (i.e., it has little free volume) there are only a limited number of such sites available without concurrent motion of the polymer chains. If the equilibrium volume fraction of penetrant is much larger than the volume fraction of free volume, there is a kinetic problem associated with sorption. In a rubber or melt far above its glass transition temperature T_g, the chains can move rapidly apart by processes involving bond rotation, and equilibrium can be achieved almost instantaneously. In the glass such processes are extremely slow, so that the initial penetrant volume fraction ϕ is less than the equilibrium volume fraction ϕ_e.

The sorption process can be represented conceptually by the simple mechanical analog shown in Figure 26-1(b). One can consider the differences between the actual penetrant volume fraction ϕ and the equilibrium volume fraction ϕ_e to result from an osmotic pressure of the penetrant molecules in the swollen polymer. The osmotic pressure is resisted (mechanical equilibrium is maintained) by stress exerted by the glassy polymer chains, which are represented in the model by springs and dashpots. As bond rotation and motion of the polymer chains slowly occurs, this stress is relaxed and new sites for the penetrant molecules are created, allowing an increase in ϕ and a decrease in osmotic pressure. In a semi-infinite solid, as shown in Figure 26-1(b), the osmotic pressure tends to swell the polymer in a direction normal to the surface due to the lateral constraint of the underlying glass. The polymer molecules in the swollen outer layer are observed to be oriented perpendicular to the surface.

Thomas and Windle showed that the osmotic pressure at any point in the polymer can be approximated by [14]

$$P = \frac{k_B T}{\Omega} \ln \frac{\phi_e}{\phi} \qquad (26.2)$$

where P is the osmotic pressure, k_B is Boltzmann's constant, T is the absolute temperature, and Ω is the penetrant molecular volume. In the mechanical model of Figure 26-1(b), the springs represent the elastic response of the polymer, whereas the dashpots represent the viscous response which Thomas and Windle assumed to be linear in pressure. The penetrant volume fraction is

$$\phi = \frac{P}{E} + \int \frac{P}{\eta} dt + \phi_{fv} \qquad (26.3)$$

where E is an effective elastic modulus of the polymer, η is the viscosity, and ϕ_{fv} is any initial "free volume." The first and second terms in the above equation are the elastic and viscous contributions respectively. Differentiation of Equation (26.3) results in

$$\frac{\partial \phi}{\partial t} = \frac{P}{\eta} + \frac{(\partial P / \partial t)}{E} \tag{26.4}$$

In a typical application the osmotic pressure is less than 3×10^7 N/m^2, and the polymer elastic modulus E is about 3×10^9 N/m^2. Hence the elastic contribution to ϕ will be less than 0.01, wheras ϕ_e is typically greater than 0.1. The small elastic contribution to ϕ is usually negligible, resulting in

$$\frac{\partial \phi}{\partial t} = \frac{P}{\eta} \tag{26.5}$$

Combining Equations (26.2) and (26.5) and assuming that the viscosity decreases exponentially with penetrant concentration as

$$\eta = \eta_0 \exp(-m\phi) \tag{26.6}$$

where m and η_0 are material constants, results in

$$\frac{\partial \phi}{\partial t} = \frac{k_B T}{\eta_0 \Omega} \exp(m\phi) \ln \frac{\phi_e}{\phi} \tag{26.7}$$

The swelling described by Equation (26.7) is occurring at every point in the polymer as is diffusion. Recognizing this relationship between swelling and diffusion, Thomas and Windle coupled Equation (26.7) to Fick's first law written in terms of activity:

$$J = -D \frac{\phi}{a} \frac{\partial a}{\partial x} \tag{26.8}$$

Thomas and Windle also assumed that D increased exponentially with ϕ. By comparing the numerical solution of Equations (26.7) and (26.8) to gravimetric data and optical measurements of the front position, Thomas and Windle showed that this theory predicted qualitatively the central aspects of Case II diffusion [14].

A straightforward extension of the Thomas and Windle model [19–23] leads to a simple expression for the Case II front velocity V:

$$V = \sqrt{\frac{D(\phi_m)}{\phi_m} \left(\frac{\partial \phi}{\partial t} \right)_{\phi_m}} \tag{26.9}$$

where $D(\phi_m)$ is the diffusivity of the penetrant in the glass and ϕ_m is the penetrant volume fraction at the maximum in the osmotic pressure, which occurs ahead of the front. This expression for V has been shown to accurately predict the Case II velocities for various temperatures and activities of iodohexane diffusing into polystyrene [21,22].

Several limitations to this model should be expected. Equation (26.2) represents the osmotic pressure as being proportional to the logarithm of a ratio of penetrant volume fractions. The osmotic pressure P is actually proportional to the logarithm of a ratio of activities [24], i.e., $P = (k_B T / \Omega) \ln (a_e / a)$. The former representation is accurate for small values of ϕ, where the activity is given by

$$a = \gamma\phi \qquad (26.10)$$

where γ is a constant. However, a more accurate representation of a is given by [25]

$$a = \phi \exp [(1 - \phi) + c_1(1 - \phi)^2] \qquad (26.11)$$

where c_1 is the polymer-solvent interaction parameter. This more nearly correct expression for activity could be used in deriving an alternate set of coupled diffusion/relaxation equations. While these equations can still be solved numerically, the analytical methods of Hui et al. [19] cannot be applied to find asymptotic solutions.

In most applications it has not been possible to represent the mechanical response of polymer glasses by simple linear viscoelastic models such as the Maxwell model of Figure 26-1(b) [26]. Typically the mechanical response of a polymer glass can only be modeled by a system of nonlinear springs and dashpots. When the model is restricted to a very low stress regime, linear viscoelasticity may be used; however, at high stresses an extremely broad distribution of relaxation times is needed to represent the data. Hence the assumption of linear viscoelasticity and a single relaxation time (one dashpot) used in deriving Equation (26.4) is probably much too simplified a representation of the mechanical response of the glassy chain.

Attempts to improve the prediction of the swelling rate by substituting a non-Newtonian, i.e., P-dependent, viscosity in Equation (26.6) improves the description of the swelling at low penetrant activities markedly [21], but even this modification fails for high activities, where the polymer glass appears to undergo a yield phenomenon similar to mechanical yielding [20]. Nevertheless, the expression for the front velocity [Equation (26.9)] is still accurate if the empirical swelling rate is determined by Rutherford backscattering (RBS) measurements of surface swelling [20].

EXAMPLES OF CASE II DIFFUSION IN PACKAGING

Photoresist Debonding Caused by Case II Diffusion

An important example of Case II diffusion in glassy polymers is developing images in photoresist. The photoresist is exposed to radiation to create a pattern, then the developer, a solvent, is used to remove the unwanted photoresist. For example, 1, 1, 1-trichloroethane (TCE) is often used to develop photoresists made primarily of polymethylmethacrylate (PMMA).

Mills et al. [24,27,28] showed that the debonding of photoresist from a copper substrate is driven by stresses occurring when the photoresist swells due to absorption of a solvent. Figure 26-2 shows the apparatus used to measure

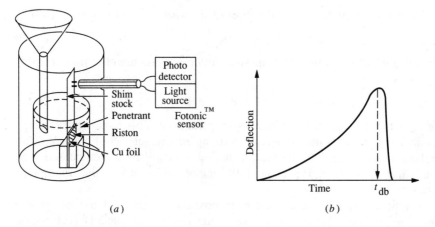

(a)   *(b)*

FIGURE 26-2
(a) Drawing of the apparatus for the debonding experiment. (b) Schematic of a typical deflection-versus-time curve.

debonding time and a typical experimental result. The bilayer film bends because the photoresist swells as it absorbs solvent. When the film debonds the elastic strain that caused the sample to bend is released and the copper snaps back elastically.

For copper circuits coated with photoresist to provide protection from an etchant, premature debonding of the photoresist would cause a circuit defect. The photoresist is generally removed later in the process by swelling with a different solvent, and residual photoresist is also considered a defect. So control of the photoresist debonding is critical to reducing defects in electronic packaging.

The concentration profile of the solvent as measured by RBS shows the characteristics of Case II diffusion, i.e., a uniformly swollen layer formed behind a sharp front which propagates into the polymer at a constant velocity. Comparison of debonding time to measurements of the solvent front penetration for a number of different solvents shows that the debonding occurred when the front had penetrated about one-fifth of the photoresist layer thickness.

The correlation of the debonding time to the Case II front velocity shows that the debonding is controlled by the propagation of the Case II diffusion front. Since the debonding occurs when the solvent front is only one-fifth of the way through the film, the mechanism of debonding is not due to an attack of the solvent on the interfacial bond between the polymer film and the copper substrate, but has to do with the stress produced by the swelling of the top layer.

When the elastic (swelling) strain energy G released by the crack is greater than the fracture toughness G_{Ic} of the interface, the film will debond. The strain energy release rate when the crack propagates in the interface between the photoresist and the substrate depends quadratically on both the depth x_f of the swelling layer and on the volume swelling strain ϵ_s in the material behind the front, i.e.,

$$G = A\,\epsilon_s^2\,x_f^2 \qquad (26.12)$$

where A is a constant that depends on the elastic properties and geometry of the swollen film and substrate. Debonding is predicted to occur when G exceeds the critical strain energy release rate G_{lc} of the interface, that is, when

$$x_f \geq \left(\frac{G_{lc}}{A \epsilon_s^2} \right)^{0.5} \tag{26.13}$$

Equation (26.13) thus predicts that x_f at debonding will be approximately constant if G_{lc}, A, and ϵ_s are approximately constant, a prediction that is in good agreement with the observation that the photoresist film debonds when the Case II front has reached approximately one-fifth of the way through the film for all of the solvents tested.

Mills et al. have also shown that the diffusion of solvent varies markedly with the batch of developer solvent used. The batches differed mainly in the amount of inhibitors present. In a manufacturing situation the concentration of inhibitor in the developer should be controlled to maintain consistent photoresist performance.

Dissolution Rate of Polymers

The sharply defined lines required for a dense package are made possible by the high resolution of photolithography. Developing away photoresist to open up areas on a sample for etching or depositing metal is a crucial step in achieving fine lines. In order for the photoresist to be useful, exposure to radiation must change the rate at which it dissolves in the developing solution. A photoresist dissolves either faster or slower when it has been exposed to radiation, depending on whether it is a positive or negative resist.

The photoresist dissolves by the following process: First, the resist absorbs the developer, then the developer molecules cause the polymer chains that make up the photoresist to swell, and finally the swollen chains disentangle and diffuse away from the surface. The rate of dissolution has been measured using an interferometer to track the thickness of the polymer layer [29]. The rate of dissolution is linear with time, which suggests that the diffusion is Case II.

Surface Modification

Manufacturing processes involve many steps that change the surface of a polymer. Simply exposing parts to air can cause oxidation and swelling due to absorption of moisture. Diffusion kinetics are very sensitive to the conditions at the surface. A small amount of water will preplasticize some polymers, causing a large increase in the solvent diffusion rate [29].

Plasma or reactive ion etching (RIE) steps are a part of many packaging processes. As an example, oxygen plasma etching is commonly used to remove photoresists. The changes caused by a plasma or reactive ion treatment depend on the detailed nature of the bonding of the polymer. As ions penetrate the surface they randomly break chemical bonds, forming free radicals. The way that the free radicals recombine is a statistical process that depends on the chemical species present. The reactive species can form new bonds between chains (cross-linking) or break bonds (chain scission). Both process will occur, but the relative rate varies widely.

Cross-linking will increase the polymer molecular weight, making it stiffer and more resistant to solvent attack. Chain scission will reduce the glass transition temperature, making the polymer softer, and more likely to dissolve. The net change in the average molecular weight due to a reactive ion etch process depends on the polymer involved. If a polymer is heavily cross-linked, the extent of swelling can be limited and Case II diffusion might not occur. In the case of chain scission the smaller fragments will plasticize the polymer, increasing the rate of diffusion.

Figure 26-3 compares the absorption of iodohexane by a control sample of polystyrene, and by one that has been reactive ion etched with oxygen for 45 seconds. The control sample shows the sharp concentration front characteristic of Case II diffusion, while the oxygen etching has reduced the solvent absorption, inhibiting the formation of a Case II front.

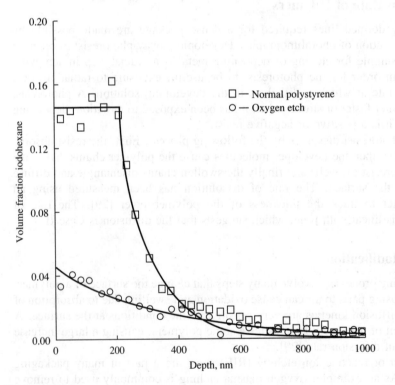

FIGURE 26-3
The effect of reactive ion etching with oxygen on diffusion of iodohexane in polystyrene.

SUMMARY

Case II diffusion is only now being quantitatively understood. The Thomas and Windle model of Case II diffusion is an important step in predicting the interactions of polymers and solvents. In order to apply the model to practical situations, material parameters such as viscosity will have to be measured experimentally. Future work in this field will aim to improve predictability and control of many of the processes involved in manufacturing electronic packages.

APPENDIX

MECHANISMS OF POLYMER DIFFUSION IN POLYMER MELTS

THOMAS P. GALL
RONALD C. LASKY
EDWARD J. KRAMER

Considerable progress has been made in the past decade in understanding the mechanisms by which a long polymer chain can move through a melt of entangled chains above the glass transition temperature. The theory of reptation proposed by de Gennes [30] describes the motion of a single chain along its own contour through a set of fixed constraints (see Figure 26-4). For a melt of low-molecular-weight chains, the constraints imposed by entanglements with neighboring chains are not fixed, and so the polymer molecule gains an extra degree of freedom of motion due to constraint release [31].

Diffusion in intermediate-length polymers, or in a polydisperse sample with a wide range of molecular weights, can be described by a mixture of reptation and constraint release. The reptation contribution to the diffusion coefficient depends strongly on the molecular weight of the diffusing chain and not at all on the molecular weight of the matrix, whereas constraint release depends only weakly on the molecular weight of the diffusing chain but strongly on the molecular weight of the matrix. The change in the molecular weight dependence can be seen in the changes in the slope of the curve [32] in Figure 26-4(b). For a more detailed review of reptation and constraint release see Kramer [33] or Klein [34]. The experimental technique used to collect the data in the figure is discussed in Chapter 23.

The relevance of interdiffusion to electronic packaging comes in the self-adhesion of polymer layers. Diffusion of chains across the interface is generally accepted as the main mechanism for the adhesion of uncross-linked polymers [35]. An example of the importance of interdiffusion can be seen in the adhesion of polyimide layers used as insulators in ceramic substrates. Brown et al. [36] have shown that the peel strength of a bilayer sample correlates with the interdiffusion distance.

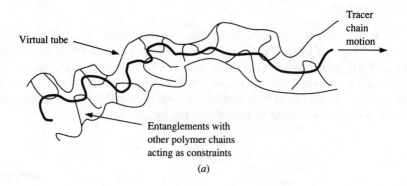

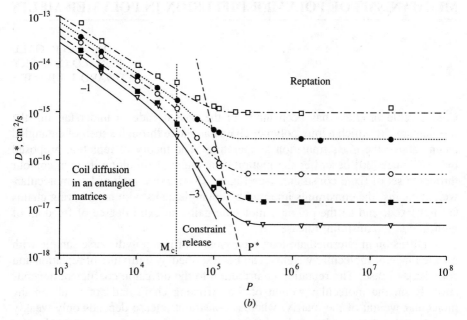

FIGURE 26-4

(*a*) Entanglements with other polymer chains constrain the motion of a polymer molecule. The only two ways the molecule can move is either by crawling along its contour (reptation) or by one of the neighboring chains acting as a constraint diffusing away (constraint release). (*b*) The dependence of the tracer diffusion coefficient D^* for a series of polystyrene chains with molecular weights, M, (M = 55,000 (□), M = 110,000 (●), M = 255,000 (○), M = 520,000 (■), M = 915,000 (▽)) on the matrix molecular weight, P. The weak dependence of the reptation contribution to D^* on P and strong dependence on M produces flat lines widely spaced, while the opposite dependencies gives steeply sloped lines closely spaced in the region of the plot dominated by constraint release.

REFERENCES

1. J. Crank, *The Mathematics of Diffusion*, 2d ed., Oxford: Oxford University Press, 1975, p. 254.
2. G. S. Hartley, *Trans. Faraday Soc.* 42B, p. 6 (1946).
3. C. Robinson, *Trans. Faraday Soc.* 42B, p. 12 (1946).

4. N. L. Thomas and A. H. Windle, *Polymer* 19, p. 255 (1978).
5. G. S. Hartley, *Trans. Faraday Soc.* 45, p. 820 (1949).
6. J. Crank and G. S. Park, *Trans. Faraday Soc.* 47, p. 1072 (1951).
7. J. Crank, *J. Poly. Sci.* 11, p. 151 (1953).
8. G. S. Park, *J. Poly. Sci.* 11, p. 97 (1953).
9. E. Bagley and F. A. Long, *J. Am. Chem. Soc.* 77, p. 2172 (1955).
10. A. C. Newns, *Trans. Faraday Soc.* 52, p. 1533 (1956).
11. F. A. Long and D. Richmond, *J. Am. Chem. Soc.* 82, p. 513 (1960).
12. T. Alfrey, *Chem. Eng. News* 43, (Oct. 11), p. 64 (1965).
13. J. R. Scott, *Trans. Inst. Rubber Industries* 13, p. 109 (1937).
14. N. L. Thomas and A. H. Windle, *Polymer* 23, p. 529 (1982).
15. N. L. Thomas and A. H. Windle, *Polymer* 18, p. 1195 (1977).
16. N. L. Thomas and A. H. Windle, *J. Membrane Sci.* 3, p. 337 (1978).
17. N. L. Thomas and A. H. Windle, *Polymer* 21, p. 619 (1980).
18. N. L. Thomas and A. H. Windle, *Polymer* 22, p. 627 (1981).
19. C. Y. Hui, R. C. Lasky, E. J. Kramer, and K. C. Wu, *J. Appl. Phys.* 61, p. 5129 (1987).
20. C. Y. Hui, R. C. Lasky, E. J. Kramer, and K. C. Wu, *J. Appl. Phys.* 61, p. 5137 (1987).
21. R. C. Lasky, E. J. Kramer, and C. Y. Hui, *Polymer* (MSC Report Number 6184), in press.
22. R. C. Lasky, Ph.D. thesis, Cornell University, 1986.
23. R. C. Lasky, E. J. Kramer, and C. Y. Hui, *Polymer* (MSC Report Number 6141), in press.
24. P. J. Mills and E. J. Kramer, *J. Materials Sci.* 21, p. 4151 (1986).
25. P. J. Flory, *Principles of Polymer Chemistry*, Ithaca, N.Y.: Cornell University Press, 1953, Chapter 13.
26. R. B. Bird, R. C. Armstrong, and O. Hassager, *Dynamics of Polymeric Liquids*, New York: John Wiley & Sons, 1977, Chapters 7–9.
27. P. J. Mills and E. J. Kramer, *J. Materials Sci.*, in press.
28. P. J. Mills, C. J. Palmstrom, and E. J. Kramer, *J. Materials Sci.* 21, p. 1479 (1986).
29. F. Rodriquez, P. D. Krasicky, and R. J. Greole, *Solid State Tech.*, May 1985, p. 126.
30. P. G. de Gennes, *J. Chem. Phys.* 55, p. 572 (1971).
31. W. W. Graessley, *Adv. Polym. Sci.* 47, p. 67 (1982).
32. P. F. Green and E. J. Kramer, *Macromol.* 19, p. 1108 (1968).
33. E. J. Kramer, *Mat. Res. Soc. Symp. Proc.* 40, p. 227 (1985).
34. J. Klein, *Macromol.* 19, p. 105 (1986).
35. K. Jud, H. H. Kausch, and J. G. Williams, *J. Mater. Sci.* 16, p. 204 (1981).
36. H. R. Brown, A. C. M. Yang, T. P. Russell, W. Volksen, and E. J. Kramer, *Polymer*, to be published (1988).

PROBLEMS

26.1. For a photoresist stripper that works by swelling the resist to cause it to debond,

(a) Estimate the depth of the swollen layer, x_f, required to cause a 50-μm thick layer of photoresist to debond.

(b) For thick layers the induction time t_0 for Case II diffusion to begin is small compared to the time for debonding, so can often be ignored. Assuming that $t_0 = 0$, and given that the debonding time for the resist in part (a) is 300 s, calculate the time needed to strip a 75-μm thick layer of photoresist.

(c) If the photoresist in part (b) did not debond, but had to be dissolved, calculate the time it would take to dissolve. Assume that diffusion is the rate-determining step in the dissolution, and use the same properties as in part (b).

26.2. Exposure to light increases the rate at which a certain photoresist dissolves when it is developed. (Ideally the resist remaining after development would be completely insoluble, but in a real system it can partially dissolve.)

(a) Assuming that diffusion is the rate-limiting step in dissolution, calculate the minimum time to develop the photoresist whose properties are listed below.

Property	Exposed to light	Unexposed
Volume fraction solvent ϕ_m	0.1	0.1
Swelling rate $(d\phi/dt)\vert_{\phi_m}$	0.1	0.003
Diffusion coefficient $D(\phi_m)$, cm^2/s	1×10^{-9}	1×10^{-9}
Initial thickness, x_0, μm	50	50

(b) Calculate the thickness of the photoresist layer remaining after developing.

CHAPTER 27

CHEMICAL BONDING AND STRESS RELAXATION AT METAL-POLYMER INTERFACES

PAUL S. HO

INTRODUCTION

In microelectronics, the large-scale integration of devices necessitates the use of multilayered metallization structures on the chip level and for packaging. Such structures are usually fabricated using alternate metal and insulating layers. This incorporates metal-insulator interfaces throughout the structure, which, depending on the configuration and dimension of the devices, can have varying degree of complexity. These interfaces have to be designed with certain functional characteristics, particularly good chemical and adhesion properties. For this purpose, it is important to understand these basic properties of the metal-insulator interface.

Among the insulators, ceramics and polymers are the most commonly used. Comparing these two classes of materials, the ceramics have the advantage of high mechanical strength and thermal stability. For advanced packaging applications, the ceramic materials have certain undesirable characteristics, including high

809

dielectric constants (about 6–8) and difficulty in forming line patterns with small dimensions, e.g. less than about 10 μm. These factors restrict the speed and density of the packaging structure. In contrast, polymers generally have low dielectric constants of about 3 to 4 and can be processed by lithographic techniques to yield patterns with dimensions in the micron range, making them well suited for packaging applications.

The current trend in microelectronic packaging is to incorporate polymer layers into ceramic structures to make optimum use of the properties of these two classes of materials. The use of polymer layers provides additional wiring levels, which, with proper design, can enhance substantially the packaging performance by improving the interconnect routing and wiring distribution. One fundamental concern of such a structure is its mechanical properties, particularly its stress relaxation behavior during thermal cycling or heat treatment. The stress arises from the differential thermal expansions of metal, polymer, and ceramic substrate (see Table 27.1). Depending on the material combination, the stress can reach a very high level, exceeding the yield stress of metal in some cases. To maintain the integrity of the structure under such circumstances, strong adhesion at the interface is needed.

The adhesion strength of the solid-solid interface is largely determined by the bonding characteristics and structure of the interface [1]. In this regard, the metal-polymer interface is distinct and interesting. While the metal is characterized by an ordered atomic structure with close-packing density, the polymer has a disordered, loose, and interwound molecular structure. The bonding characteristics of such a structural combination is difficult to predict a priori. In spite of its wide range of applications, metal-polymer bonding has not been investigated until recently [2–5].

This chapter reviews the results of the recent investigations on properties pertinent to adhesion at metal-polymer interfaces formed on pyromellitic dian-

TABLE 27.1
Properties of some common packaging materials between 0 and 100°C

	Thermal expansion coefficient, $10^{-6}/°C$	Thermal conductivity, W/m°C	Young's modulus, 10^{11} dyne/cm^2
Cr	6.5	91.3	27.9
Cu	17.0	397	13.0
Ti	8.9	21.6	12.0
Ni	13.3	88.5	20.0
Si	3.0	83.5	11.3
Quartz	0.57	2.0	7.3
Polyimide	~ 35	0.20	0.32
$3Al_2O_3 \cdot 2SiO_2$	3.8	7	25

hydride-oxydianiline (PMDA-ODA) polyimide. The discussion is focused on the chemical and mechanical properties of the interface with the aim to understand the microscopic nature of interfacial adhesion. One key finding in these studies is the reactive nature of the metal-polymer interface, particularly for a high-temperature polymer such as polyimide [4–7]. This was manifested by material reactions at the interface, such as intermixing and cluster formation. The extent of these reactions can affect significantly the structure and morphology of the interface. This material aspect of the interface and its correlation to chemistry are important for understanding the mechanical behavior of the interface.

This chapter first reviews briefly interfacial adhesion and its dependence on chemical bonding and interfacial morphology. This is followed by three sections addressing chemical bonding, microstructure, and stress relaxation of metal-polymer interfaces formed on PMDA-ODA polyimide. This polymer is chosen in the discussion because it is commonly used in microelectronics and packaging, and the information about its interfaces is the most complete at this time. For other polymer interfaces, one can refer to other review articles and conference proceedings [8–12]. Due to the emerging interest and the increasing activities in studying metal-polymer interfaces, this chapter is not intended to be a comprehensive review; instead its materials are selected to describe the current understanding of the microscopic nature of the metal-polyimide interface.

INTERFACIAL ADHESION

When an interface, or surface, area is created in a solid, work is required. The reversible work required to create a unit interfacial, or surface, area is the interfacial, or surface, tension. It can be expressed as

$$\gamma = \left(\frac{\partial G}{\partial A} \right)_{T,P,n} \tag{27.1}$$

where γ is the interfacial, or surface, tension, G is the Gibbs free energy, and $\partial G/\partial A$ is evaluated under the conditions of constant temperature T, pressure P, and number of moles n in the system.

The work required to separate reversibly the interface between two bulk phases α and β to infinity per unit area is the work of adhesion [1]:

$$W_a = \gamma_\alpha + \gamma_\beta - \gamma_{\alpha\beta} \tag{27.2}$$

where W_a is the work of adhesion, γ_α and γ_β the surface tensions of α and β phases respectively, and $\gamma_{\alpha\beta}$ the interfacial tension between α and β. According to Equation (27.2) the work of adhesion comes from the decrease of Gibbs free energy per unit area when an interface is formed.

The interfacial tension is determined by the bonding characteristics and structure of the interface. Bonding can be classified into the two broad categories of molecular and chemical bonds. The molecular bond originates from intermolecular forces commonly known as van der Waals forces. The chemical bond comes from interatomic forces. Chemical bonds can be further classified as covalent, ionic, and metallic bonds [13]. In general, the energy of a chemical bond is higher than that of a molecular bond. The bonding at the metal-polymer interface is chemical in nature, and its formation is a result of charge transfer from the metal to the polymer. The details will be discussed later in this chapter, under "Chemical Bonding and Electronic Structure."

The structure of an interface is controlled by the extent of material reaction that occurs when two materials are brought into intimate contact. A common form of reaction is interdiffusion driven by a concentration gradient. Such reactions can affect significantly the adhesive strength of the interface. For example, two polymers with relatively weak molecular interaction can form a strong interface by interdiffusion, i.e., interpenetration of polymer segments, at the interface. For metal-polymer interfaces, chemical bonding plays an important role in controlling the extent of interfacial reaction.

Adhesion is a property measuring the maximum amount of mechanical work or energy that can be transferred across an interface before separation. When an interface is pulled apart, the work required is the work of adhesion plus the work required to deform, elastically and/or plastically, the two bulk materials. The total work is called the fracture energy G, which can be expressed as

$$G = W_a + W_p \tag{27.3}$$

where W_p is the irreversible work of deformation. Thus the adhesive strength of a joint depends on the mechanical properties of both the interface and the two bulk phases. To design a joint with high adhesive strength, it is not sufficient to have only a strong interface; the mechanical properties of the bulk phases have to be optimized with respect to the interface in order to provide the overall strength.

Many methods have been developed to test the mechanical strength of an adhesive joint [14]. These tests generally involve making a joint of a certain configuration, then separating it by applying a mechanical stress. Depending on the joint configuration, the test measures the fracture stress or the fracture energy of the joint. Three of the commonly used tests—tensile test of the butt joint, shear test of the lap joint, and peeling test—are shown schematically in Figure 27-1. All these tests measure the fracture stress of the joint, from which one can derive the fracture energy by analyzing the deformation behavior of the test structure. It should be emphasized that these tests do not measure directly the work of adhesion of the interface. For polymeric materials, which can be deformed extensively before fracture, the contribution of the plastic work to the fracture energy is often large compared with the work of adhesion. Thus the results of these tests have to be corrected for the deformation energy in order to derive the work of adhesion.

In this chapter, emphasis will be placed on studies of the nature of the chemical bonding and its effect on the microstructure of the interface. These are two aspects essential for understanding adhesion at the metal-polymer interface. Discussions on other aspects of adhesion, such as interfacial thermodynamics and mechanistic analysis of adhesive bonds can be found in reference [11].

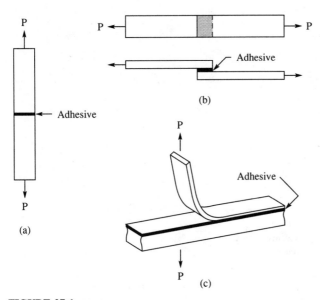

FIGURE 27-1
Schematic diagram showing three common methods for measuring the fracture stress of adhesive joints: (*a*) tensile test of a butt joint, (*b*) shear test of a lap joint, and (*c*) peeling test.

CHEMICAL BONDING AND ELECTRONIC STRUCTURE

Electronic Structure of Polyimide Surface

Uv photoemission (UPS) and x-ray photoemission (XPS) spectroscopies have been used to probe the electronic structure of interfaces. Figure 27-2 shows schematics of these photoemission techniques. In UPS, uv photons generated by gas excitation or from a synchrotron source, with energy of about 100 eV or less, are used to excite the valence electrons. The electrons with sufficient energy to escape from the surface can be detected and their energy analyzed by a spectrometer. The results give the distribution of the valence states. Since the valence states are directly involved in the formation of the chemical bond, UPS seems to be ideally suited for probing the bonding characteristics. For metal-polymer interface studies, this is complicated by the fact that the polymer molecule contains various atomic elements. The energy levels of the valence states of these atoms or their complexes are close together and usually overlap in the energy range of the measurement. This makes it difficult to identify the specific complex or bonding sites responsible for bond formation. In XPS, electron excitations are made by x-ray photons of energy usually exceeding 1000 eV. Thus, XPS measures the core states lying several hundred electron-volts below the valence states. Unlike UPS, the core levels detected by XPS are well separated for the various atomic elements in the polymer molecule. Although

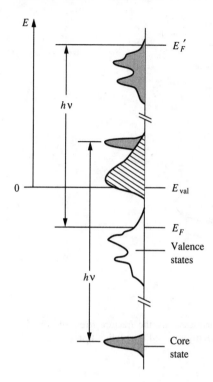

FIGURE 27-2
Schematic diagram showing the basic principle of
UPS and XPS spectroscopies.

those core states are not directly involved in the formation of chemical bonds, their energy levels and spectral features are affected by bond formation. The atomic specificity of XPS is useful for determining the role of different atomic species in bond formation at the interface.

Both of these spectroscopy techniques have very high surface sensitivity, since they measure only those electrons within the escape distance of the surface, which is of the order of 10–50 Å. This restricts the probing depth of an interface to about the escape distance. As a result, most spectroscopy studies have been carried out on interfaces with only a few monoatomic layers of metal coverage. The surface sensitivity of these studies enables one to deduce the morphology of the metal coverage by measuring the intensity variations of the metal and the substrate as a function of metal coverage. Spectroscopy experiments have been carried out during metal deposition on the polymer. By observing the changes of the spectral features during deposition, one can monitor the evolution of the valence and core states during the course of interface formation. Molecular orbital (MO) calculations have been carried out to derive the electronic structure of various metal-polymer complexes. By combining the spectroscopy results with MO calculations, the nature of the chemical bonds at the interface can be deduced.

These experiments were carried out in a multitechnique ultrahigh-vacuum chamber equipped with a hemispherical electron spectrometer for UPS and XPS measurements. The chamber was designed to allow a clean in-situ evaporation of metal while carrying out spectroscopy measurements without changing the sample position. The samples were fabricated using Dupont 5878 polyimide. To insure a smooth substrate surface for the polyimide overlayer, a polished Si wafer was used with thin Al layers deposited on both sides. The polyimide layer of 100–300 Å was first spun onto the Al layer, then heated to 85°C to drive off most of the n-methyl pyrolidone solvent. This was followed by curing at 350°C for 30 minutes in inert gas to complete the imidization reaction for formation of the polyimide molecular structure. This sample configuration provided a conducting Al layer under the polyimide for reducing electrical charging, and a separate Al layer on the bottom for sample heating. All x-ray core level spectra were collected at room temperature in a grazing emission of 10 degrees with respect to the surface in order to maximize the surface sensitivity.

First, the bonding characteristics of the polyimide surface were investigated [15]. Since polymer surfaces are susceptible to damage under the common cleaning techniques, such as ion sputtering, considerable effort has been spent to establish a procedure to prepare a clean and intrinsic surface. Annealing was found to be most effective in obtaining a reproducible surface. Figure 27-3 shows subtle differences in surface chemistry for three prepared samples, as detected by UPS. For these UPS measurements, 21.2-eV photons generated by a helium resonance lamp were used. Upon annealing at 360°C for about a half-hour, the differences disappeared and the spectra converged to one typical of the polyimide surface, as represented by curve (d). The nature of this annealing treatment was investigated by carrying out a series of annealing experiments, on as-cured polyimide as a function of temperature. Evolution of water was detected and the corresponding changes in the spectral features showed that water desorption from the polyimide surface accounts for most of the changes observed during annealing.

To verify that the UPS spectrum in curve (d) of Figure 27-3 indeed represents the valence density of states (DOS) of polyimide, theoretical calculations of the molecular orbital structure of polyimide were carried out. Quantum chemical programs are readily available to carry out self-consistent-field molecular orbital (SCF MO) calculations of organic molecules of limited size. Such a program, GAUSSIAN-80 [16], was chosen to study the molecular orbitals of polyimide [6,17].

The fundamental unit (building block) of the polyimide chain is shown in the bottom of Figure 27-3. It consists of two parts: (1) a PMDA part, which contains four C=O bonds, a benzene ring, and two N atoms; and (2) an ODA part, which consists of two benzene rings connected through an ether oxygen. The calculations yield molecular orbitals whose energy distribution can be used to deduce the DOS of the molecule, including both valence and core states. The theoretical results can be compared to the observed spectra by integrating the energy distribution of the DOS with a proper instrumental broadening of the spectrometer. The agreement between the measured [curve (d)] and calculated [curve (e)] valence DOS's in Figure 27-3 is remarkably good. The main peaks near −10 and −14 eV are well reproduced by the calculation, and the shoulder near −18 eV also appears.

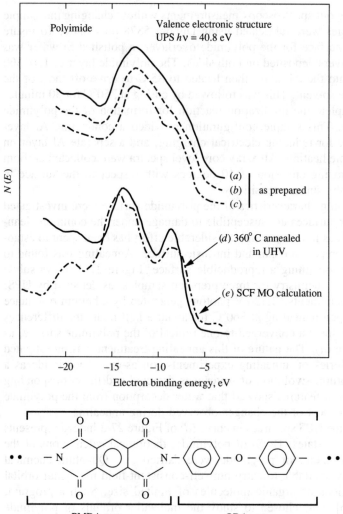

FIGURE 27-3
Above: (a)–(c) UPS spectra from three different polyimide surfaces in the as-prepared state, (d) UPS spectrum after 360°C in situ annealing in UHV; and (e) valence density of states from SCF MO calculations, broadened with 2.0 eV FWHM for comparison with experiment. Below: Basic unit of the molecular structure of polyimide.

This agreement indicates that the 360°C in-situ annealing yields a UPS spectrum characteristic of intrinsic polyimide. It also permits us to identify the molecular orbital origin of the spectral peaks. The -10-eV peak consists of three types of orbitals: those associated with the benzene rings, i.e., the C 2p orbitals lying perpendicular to the molecule; the N 2p lone pair orbitals (also perpendicular to the molecule); and the O 2p lone pair orbitals lying perpendicular to the molecule for the ODA part and along the molecular axis for the PMDA part. The peak near -14 eV is composed of the C–C σ bonds, C–O orbitals perpendicular to the molecular plane, and C–H π bonds. The peak structure between 17 and 20 eV is C(2s). In essence, the -14 eV peak represents bounded valence states that form the backbone structure of the polymer, while the -10 eV peak contains the unbounded valence states, which are available to interact with the electronic states of the deposited metal atoms. As seen from subsequent discussions on interfaces, these valence states are primarily responsible for forming the chemical bond at the interface.

Among the core spectra observed by XPS, the C 1s spectrum provides the most useful information, so it is relied upon in probing the effect of metal interaction on the polymer chemistry. As shown in Figure 27-4, it consists of three main peaks reflecting the different chemical environments of the carbon atoms in the PMDA-ODA polymer repeat unit. These include one peak at -288 eV for the carbonyl carbon atoms, one at -285 eV for the aromatic carbons of the six-

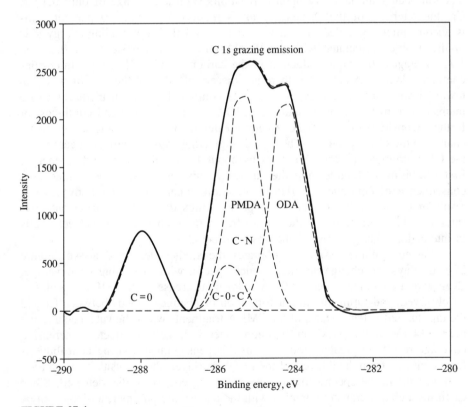

FIGURE 27-4

XPS C 1s spectrum calculated for the clean PMDA-ODA surface. The deconvoluted contributions reflect the different chemical environments of the density of states.

member rings in PMDA as well as carbon atoms bonded to oxygen and nitrogen atoms, and one at -284 eV for certain aromatic carbon atoms in the phenyl rings of ODA. The core level spectrum of oxygen exhibits only two peaks, derived from the carbonyl and ether oxygens. The spectrum of nitrogen is even simpler, consisting of only one peak, reflecting its single chemical environment in the repeat unit.

The electronic structure of the interface has been investigated for a number of metals, including Cu [3,5,18,19,20], Ni [3,21], Al [22,23], Ti [24,25], Cr [3,18,20,26], Co [27], Ag [3], Au [28], and Ge [29]. In the following sections, results are discussed for Cr and Cu to illustrate strong and weak bonding characteristics.

Chromium-Polyimide Interface

The spectral series in Figure 27-5 illustrates the evolution of the C 1s core states upon Cr deposition. The result indicates that for this strongly interacting metal, reaction occurs immediately upon deposition. With a coverage of only 0.12 Å Cr, the intensity of the carbonyl peak is reduced by about 50 percent. This is accompanied by a decrease in the intensity on the high-binding-energy side of the doublet associated with the PMDA unit. After depositions of 0.2- and 1-Å coverages, the rapid reduction in the carbonyl and PMDA peak intensities continues. With 1-Å coverage, there is a discernible shift of the overall spectrum toward lower binding energy as well as a continued reduction in the emission intensities from the carbonyl and the PMDA fragment. This trend continues for further deposition of Cr. At coverages greater than 2 Å we note a significant change in the C 1s spectrum. The signal originating from the aromatic carbons in the ODA fragment (-284 eV) shifts by about 1.8 eV to lower binding energy, first exhibiting a shoulder and then a peak. This result can be explained as a consequence of continued overlayer formation, where coverages above 2–3 Å result in Cr atoms located atop the phenyl rings of the ODA fragment after saturating the interaction with the PMDA fragment. Such a matrixing effect results in interactions that give rise to the observed peaks.

Valence photoemission corroborates the studies described above. Figure 27-6 displays the changes in the valence spectra with increasing Cr coverage. Even with a coverage of only 0.08 Å a drastic decrease in the -10 eV peak can be observed, indicating the initial formation of chemical bonds between Cr and the unsaturated valence states of the PMDA fragment, while the valence structure near -14 eV of the backbone bonding states is largely unaffected. Increasing coverage results in greater reductions of this peak with increasing contributions from emission from the d levels of the metal overlayer (the intensity near -5 eV). A rigid shift in the spectra toward lower binding energy is also detected. Since Cr forms a charge transfer complex with the underlying polymer, a dipole can be expected to form which shifts the kinetic energy of the photoemitted electrons. This effect is observed in core level as well as valence spectra with a shift toward lower binding energy of 1.2 eV, saturating above 3–4 Å coverage.

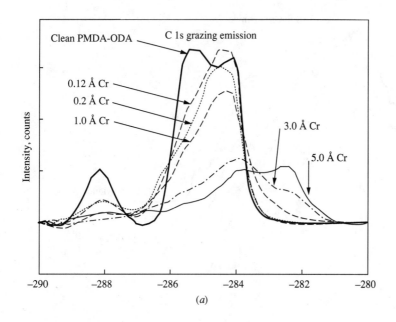

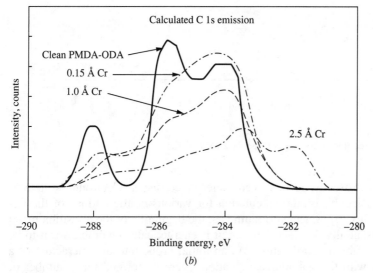

FIGURE 27-5
(a) C 1s series with increasing Cr coverage; (b) Calculated C 1s spectra with increasing Cr coverage.
For 0.15 and 1.0 Å coverages, Cr atoms were distributed over the 5- and 6-member rings of PMDA.
For the 2.5 Å coverage additional Cr has been placed atop the phenyl rings of ODA.

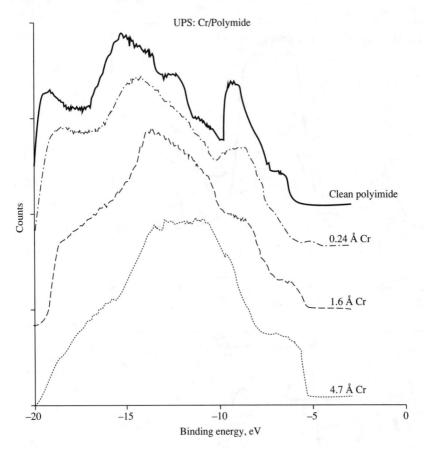

FIGURE 27-6
UPS spectra for increasing Cr coverages using 21.2 eV radiation.

These spectroscopy results were interpreted using electronic structures derived from molecular orbital calculation for various configurations of the Cr-polyimide complex [6]. Since chromium is at the beginning of the transition metal series and is essentially 3d electron poor, it is energetically favorable for it to be located at sites of high coordination. As an initial probe into the interaction of a single Cr atom with the polymer, calculations were performed for a number of different complexes involving one chromium atom interacting with the PMDA or the ODA unit. The most stable complex formed was the one with the Cr atom sitting above the central six-member ring, that is, at the position of the highest coordination. Another stable complex with PMDA was found with Cr above the center of the five-member ring. The ODA part yields only one such type of site for complex formation with Cr, namely, above either of the two phenyl rings.

For the most stable Cr-PMDA complex, the lowest unoccupied $a_2\pi$ molecular orbital (LUMO) interacts with the $3d_{xy}$ Cr orbital, which is filled and lying in a plane parallel to the plane of the molecule [6]. These orbitals are shown at the top of Figure 27-7. In the calculation, the distance of the Cr atom above the molecular plane was varied to determine the minimum energy configuration. The resultant bonding orbital is shown at the bottom of Figure 27-7. In this complex, the charge transferred from the Cr atom resides mainly at four of the C atoms composing the central benzene ring; two of the C atoms on this ring receive no charge since these C atoms had essentially zero LUMO amplitude. There is also some small amount of charge transfer to the carbonyl C atoms. As a result of these interactions, the carbonyl C 1s levels were found to shift to lower binding energies while the intensity of the -288 eV peak was reduced. These changes in electronic states can explain the spectral features of the C 1s electrons observed with the initial coverage of Cr, i.e. a rapid reduction in the intensities and a shift in energy of the carbonyl and PMDA peaks.

Similarly, when the chromium atom is placed above one of the five-member rings, there is charge transfer with a resultant shift of the C 1s core levels to lower binding energy. Since the chromium atom is not equidistant from all of the carbonyl groups, the C 1s carbonyl levels split into inequivalent pairs, causing an overall broadening of the spectrum. Such an asymmetry in the charge transfer results in shifts of all the C 1s levels associated with the carbon atoms on the central ring.

Results of the molecular orbital calculations can be further evaluated by calculating the C 1s spectra as a function of Cr coverage. Coverages can be

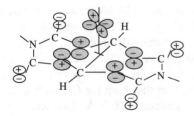

FIGURE 27-7

(*a*) Lowest unoccupied level of the PMDA fragment and Cr $3d_{xy}$ orbital (occupied). The orbitals are shown at a large separation (noninteracting). (*b*) Occupied orbital of the complex that occurs as a result of the interaction of the unoccupied a_2 orbital and filled Cr $3d_{xy}$ orbital. For all orbitals, the opposite phases are indicated by $+$ or $-$ and the filled and empty orbitals are shaded and unshaded, respectively.

simulated with appropriate combinations of clean polyimide and Cr-PMDA and Cr-ODA complexes. As shown in Figure 27-5(b), the calculated spectra provide a satisfactory account of the observed results, not only for the peak shifts but also for the overall spectral features.

A model of the interaction of Cr with PMDA-ODA can now be constructed. Chromium initially bonds in highly coordinated sites on the PMDA fragment, forming a Cr-polymer complex. Increased coverage results in saturation of such energetically attractive sites and forces Cr to interact with the phenyl rings of the ODA fragment. This occurs as the overlayer thickness approaches one monolayer. Central to the formation of the Cr-polymer complex is the charge transfer from the deposited Cr atom to the lowest unoccupied molecular orbital of the repeat unit of the monomer. The relative stability of these complexes depends on the site coordination. For Cr, the most stable complex is the one with the highest coordination on PMDA. The manner in which charge is transferred from Cr to the various atoms in the polymer repeat unit indicates that the chemical bond between Cr and the polyimide is delocalized in nature. A similar mechanism of delocalized bonding can be applied to interpret the spectroscopy results of other metals; copper-polyimide bonding is discussed next.

Copper-Polyimide Interface

With respect to chemical reactivity, Cu represents a contrast to Cr, since its filled d orbitals lead to significantly weaker interaction with the polymer. In core level studies [18,20], increasing Cu coverage produced minimal changes in the observed spectra (Figure 27-8(a)). Even at coverages of up to 3 Å, the carbonyl signal persists, being reduced only 40 percent relative to the clean spectrum. The -285 eV peak diminishes more rapidly than the one at -284 eV, suggesting that Cu interacts primarily with the PMDA fragment. This is similar to the case with Cr, except that the reduction in peak intensity is significantly smaller at comparable coverages, indicating a weaker chemical interaction at the interface. Since Cu is more electron-rich than Cr, it is not as energetically favored as Cr to occupy high-coordination sites. Such chemistry weakens the interaction with the underlying polymer, and consequently incomplete coverage due to island formation was observed [4]. Such an effect of chemical reactivity on the interfacial structure has been found to be important for all the metal-polyimide interfaces studied so far. This subject will be discussed further in the next section.

These observations are consistent with the valence spectra of the Cu-covered polymer (Figure 27-8(b)). Although initial Cu interaction with PMDA was observed, as evidenced by the decrease of the -10 eV peak, this effect saturates. Even above 2 Å coverage, the -10 eV peak is still observable, while at this coverage in the Cr case this peak had completely disappeared. This result suggests that patches of the polymer remained uncovered, consistent with island formation and weak Cu chemistry. Interestingly, a dipole shift is observed for this system, although it is considerably smaller than that seen for Cr, with a shift to lower binding energy of only 0.4 eV at a coverage of 3 Å. Copper, then,

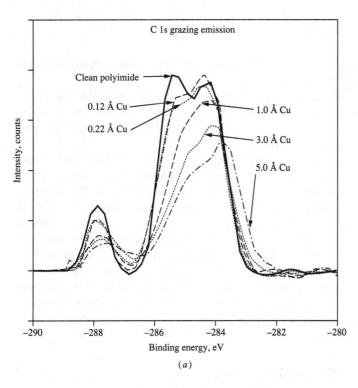

(a)

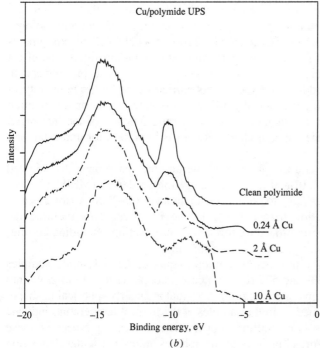

(b)

FIGURE 27-8
(*a*) C 1s spectra for increasing Cu coverage; (*b*) UPS spectra for increasing CU coverage.

presents a case with chemical reactivity significantly weaker than that of Cr when deposited onto the PMDA-ODA.

Although the relative energies of various complexes are difficult to determine accurately due to the weak interaction of Cu, the observed spectroscopy results are consistent with the formation of weak charge transfer complexes, predominantly with the PMDA fragment.

Polyimide bonding characteristics have been investigated for three other 3d metals, Ni [3,21], Ti [24,25], and Co [27]. Together these studies reveal a consistent chemical trend in which the bonding strength and the chemical reactivity depend upon the number of the d electrons in the metal. The overall bonding characteristics are well explained by the formation of metal-polymer complexes. On this basis, Ti is similar to Cr, as it interacts strongly with the polyimide, while Ni is similar to Cu, although its reactivity is somewhat higher. The trend in the chemical reactivity is also reflected in the interfacial morphology obtained with initial metal coverages. Again Ti is similar to Cr, with uniform coverages, while Ni is close to Cu, with island formation. Aluminum, a metal with p valence states, has also been investigated [22,23]. Its bonding strength and chemical reactivity are intermediate between those of Cr and Cu.

INTERFACIAL MICROSTRUCTURE

Results of the spectroscopy studies on chemical bonding indicate that the chemical reactivity between the metal and polymer can change the interfacial morphology, e.g., from a uniform coverage of Cr to an island formation of Cu. The experiments discussed in the previous section were performed at room temperature; the effect of chemistry on interfacial structure was expected to be more pronounced at elevated temperatures. Additional experiments carried out at higher temperatures for Cu [4] and Al [22,23] have indeed confirmed the nature of thermal activation of the process. Since the interfacial tension is a strong function of the interfacial structure, this aspect of the material characteristics is important for understanding adhesion.

The microstructure and morphology of the interface have been investigated by cross-sectional TEM [30] and medium-energy ion scattering (MEIS) [7]. For TEM studies, the samples have to be thinned to about 1000–2000 Å (for electron transparency) in a direction perpendicular to the interface. This required the development of an elaborate procedure for sample preparation, including cutting, gluing, mechanical polishing, and final ion milling.

Figure 27-9 shows cross-sectional micrographs of Cu/polyimide samples prepared at 293, 423, 523, and 527 K. In all cases, copper is seen to migrate into the polyimide films to form copper agglomerates that are nearly spherical in shape. Diffraction patterns obtained from the samples at the highest temperature indicate that each of the spheres is crystalline copper. One interesting aspect of these observations is that the spread in sizes of the spheres at any particular deposition temperature is very small. At high temperature, some of the Cu aggregates rest at the polyimide/substrate interface. Apparently, the diffusion length of either the

CROSS-SECTION TEM
Cu ON POLYIMIDE

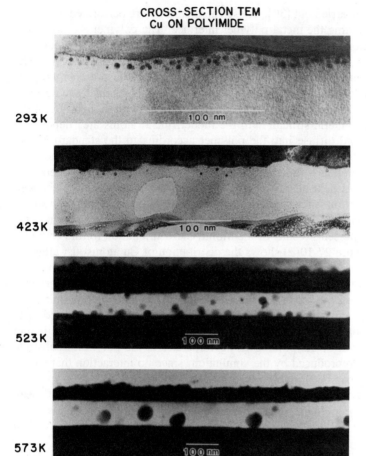

FIGURE 27-9
Cross-sectional TEM micrographs showing morphologies of Cu/polyimide interfaces formed at 293, 423, 523, and 527 K. Deposition rate is about one monolayer per minute.

individual Cu atoms or the agglomerates is sufficiently large so that the aggregates become pinned at this interface. At lower temperature, the diffusion length is shorter and the aggregates are imbedded nearer the surface of the polyimide film. The observed temperature dependence of Cu morphology leads one to infer that Cu atoms have a high diffusivity in polyimide, and upon diffusing inside, they aggregate, probably because the binding energy between Cu atoms is higher than that of Cu to polyimide. These results are a consequence of the weak chemical interaction between Cu and the polyimide and are consistent with the spectroscopy studies.

To elucidate the role of chemical interactions in the formation of the metal-polymer interface, related Monte Carlo calculations on a model metal-polymer interface have been carried out [30]. These calculations were performed on a two-phase system simulating the metal-polymer interface. The polymer was idealized as a set of interaction sites with the metal. For the case of Cr, such sites may well be the high-coordination sites associated with the PMDA-ODA polymer [6]. The interaction sites were allowed to move about, consistent with the energetics of the model. The calculations were performed on a two-dimensional square grid. The number of interaction sites introduced in the model provides a measure of the openness of the polymeric structure. In the model, metal atoms are "deposited" at the top surface and diffuse into the polymer according to the standard Monte Carlo algorithm. All single-metal-atom jumps and single-polymer-interaction-site jumps were considered as a function of the interaction energies and temperature. Nearest-neighbor and next-nearest-neighbor interactions were assumed between the polymer interaction sites and between the metal atoms. Only a nearest-neighbor interaction was assumed between the metal atoms and interaction sites.

Figure 27-10 presents results of the Monte Carlo simulation for the Cu/polyimide interface. Figure 27-10(a) shows the distribution of Cu atoms just after the deposition of six monolayers of Cu. One notes that the Cu agglomerates are near the surface with, however, a number of free single Cu atoms distributed throughout the bulk. Figures 27-10(b) and (c) show that migration and agglomeration can occur as a result of annealing even after the deposition has been terminated. The extent of migration and agglomeration depends on the annealing temperature and time. The observation of the pinning of agglomerates at the polyimide/substrate interface cannot be reproduced by the simulation, since no interaction that would yield such pinning has been incorporated into the model.

The morphology for deposition at lower temperature has also been simulated. It has been found that the reduced thermal activation yields complexes closer to the polyimide surface, consistent with TEM observation (Figure 27-9).

Formation of the metal-polymer interface is also a function of the deposition rate of Cu. The micrographs shown in Figure 27-9 were obtained from samples prepared with extremely slow evaporation rates (about one monolayer per minute). When this rate is increased, the details of the interfacial morphology change significantly. For evaporation at room temperature, a slight increase of the evaporation rate (a few monolayers per minute) causes the Cu to form a uniform film at the surface on the polyimide instead of penetrating into the film. This is shown in Figure 27-11(a). Annealing of the sample after the rapid room-temperature evaporation will result in the formation not of Cu clusters in the polyimide, but of islands on top of the polyimide [Figure 27-11(b)]. This is apparently the regime for which single metal atoms deposited at the surface have a good chance of interacting with other metal atoms at the surface before they can diffuse into the bulk. When the evaporation rate is increased to a few monolayers per second, even high-temperature evaporation does not produce Cu clusters inside the polyimide film.

Figure 27-12 shows the microstructures obtained when Cr is evaporated at slow rates onto hot and cold polyimides, respectively. These micrographs show

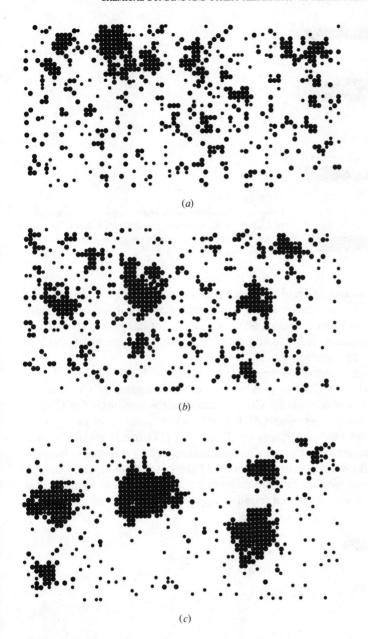

(a)

(b)

(c)

FIGURE 27-10
Results of Monte Carlo simulation for the Cu/polyimide interface: (a) as-deposited; (b and c) showing migration and agglomeration of Cu atoms after deposition as a function of time.

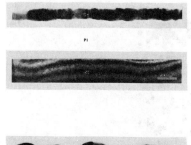

FIGURE 27-11
Cross-sectional TEM micrographs showing morphologies
of Cu/polyimide interfaces formed with a deposition rate
of several monolayers per minute: (*a*) as-deposited, and
(*b*) upon annealing at 300°C.

clearly that there is no formation of Cr precipitates inside polyimide and negligible
intermixing due to Cr penetration. Monte Carlo simulation for this more strongly
interacting metal has shown that most of the Cr atoms get hung up at the surface
and are unable to penetrate into the film [25], in qualitative agreement with the
observed Cr/polyimide morphology.

The composition distribution measured by MEIS on TEM samples [7] confirmed the intermixing of Cu into polyimide and the formation of Cu clusters,
while no penetration was detected for Cr. The extent of the intermixing of Cu was
again found to be strongly dependent on the rate and temperature of deposition.

Observations on other interfaces of Al and Ni [17] reveal morphological
features intermediate between Cr and Cu. Aluminum is like Cr in that it tends to
form a uniform film on the polyimide in most of the cases studied, even though
with slow evaporation rates (a few monolayers per minute) and at temperatures
of about 300°C, Al atoms have been found to agglomerate to form surface islands

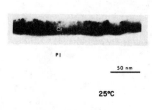

FIGURE 27-12
Cross-sectional TEM micrographs showing morphologies of
Cr/polyimide interfaces formed with deposition temperatures at
(*a*) 25°C and (*b*) 300°C.

[22]. In comparison, Ni is similar to Cu, as it tends to diffuse into the polyimide to form Ni clusters near the polyimide surface, particularly under the conditions of a low evaporation rate and a high deposition temperature.

The observed effects of evaporation rate and temperature on the interfacial structure reveal the kinetic nature of the formation process of the metal-polymer interface. For example, in the case of Cu, the high binding energy between Cu atoms favors the formation of Cu clusters in the polyimide or surface islands. However, the actual morphology of the interface depends on the rate of Cu diffusion into the polyimide relative to diffusion along the surface, and the outcome is controlled by the deposition rate and temperature. The bonding between copper and the polyimide is weak, so one can promote the indiffusion by a high temperature and a slow rate of deposition. Under such conditions, an interface can be formed with considerable intermixing and cluster formation of Cu in the polyimide. For strongly bonded metals, such as Cr, the rate of diffusion into the polyimide is very low, so that in most of the cases studied, only a uniform interface was obtained. Such an interplay of the energetics of chemical bonding and the kinetics of the formation process plays an important role in controlling the structure and morphology of the interface.

The observed variation in the interfacial structure and morphology can influence significantly the mechanical properties of the interface, such as the adhesion strength. For example, experimental results suggest that the weakly bonded Cu-polyimide interface can be strengthened by promoting the extent of the intermixing at the interface.

It is interesting to estimate the macroscopic adhesion of the Cr-polyimide interface by assuming a sharp interface with bonding primarily through the Cr-PMDA complex. The binding energy of a single Cr atom to the six-member or the five-member ring in the PMDA and the ODA fragments is about 4–5 eV. Since we estimate a surface repeat unit density of 2.5×10^{14} per cm^2 and there are three Cr atoms bonded to one PMDA unit (one to the six-member ring and two to the five-member rings), and two to one ODA unit, this gives an interfacial bonding energy of about 6×10^{19} eV/m^2. This corresponds to about 4 J/m^2. Measurements of interfacial adhesion for this interface yields a fracture energy of about 40 J/m^2 which is an order of magnitude higher than the estimated bonding energy. This suggests that the adhesion strength of an interface even with strong chemical reactivity cannot be accounted for by the bonding energy between the metal and the polymer repeat unit.

STRESS RELAXATION

Method of Stress Measurement

In packaging structures, a high-level stress can be generated during thermal treatments due to the thermal mismatch of the materials. Stress analysis of the metal/polyimide/quartz structure [31] reveals a significant difference in the stress levels of metal and polyimide in a multilayered structure, most of which has to be accommodated at the metal-polyimide interface. The ability of the interface to withstand such thermal stresses will determine to a large extent the reliability of the multilayered structure.

The behavior of stress relaxation during thermal cycling has been investigated in structures formed by depositing metal on a quartz or polyimide-quartz substrate [31,32]. The stress was measured by a cantilever bending beam technique. This is a well-known technique and its fundamentals have been reviewed by Campbell [33]. A schematic diagram of the computer-controlled system used is shown in Figure 27-13. The sample assembly contains a thin quartz cantilever, used as a substrate, and a thick reference quartz beam. The metal or metal-polyimide structure to be studied is deposited on the thin quartz beam. The sample assembly is placed in an environment chamber where measurements can be made during thermal cycling in vacuum or under a gaseous atmosphere.

The film stress is determined from the deflection δ of the beam along its length, measured by a laser deflection technique. In all the studies reported

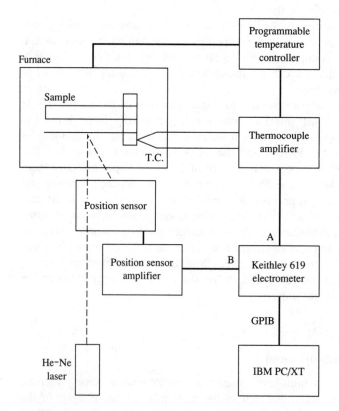

FIGURE 27-13
Schematic diagram of a system used to measure stress generated in thin film structures.

here, the thicknesses of the metal and polyimide films are small (about a few microns) compared with the substrate (about 100 microns in thickness). Under this condition, if the stress is uniform and isotropic, its value can be related to the end deflection δ of the beam as

$$\sigma = \frac{E_b t_b^2}{3(1 - \nu_b)L^2 t_f} \delta \tag{27.4}$$

where E_b = Young's modulus of the beam
$\qquad\qquad$ (7.3×10^{11} dyn/cm^2 for fused quartz)
$\qquad t_b$ = beam thickness
$\qquad \nu_b$ = Poisson's ratio for the beam (0.16 for fused quartz)
$\qquad L$ = unclamped length of the beam
$\qquad t_f$ = sample thickness

The stress σ acts in the direction parallel to the length of the beam. In deriving Equation 27.4, it has been assumed that the beam is infinitely hard, so that its curvature is much greater than its thickness, which in turn is greater than the film thickness. A more general expression for the average stress in each layer of the sample has been derived in reference [31], which takes into account the plastic deformation of the polymer.

$\qquad$ The film stress σ measured using the bending beam technique equals the sum of thermal and intrinsic stresses, i.e.,

$$\sigma = \sigma_{\text{thermal}} + \sigma_{\text{intrinsic}} \tag{27.5}$$

The thermal stress σ_{thermal} is generated by the thermal expansion mismatch between film and beam in going from temperature T_1 to T_2. It can be calculated using

$$\sigma_{\text{thermal}} = \int_{T_1}^{T_2} \frac{E_f(T)}{1 - \nu_f} [\alpha_s - \alpha_f(T)] \, dT \tag{27.6}$$

where $E_f(T)$ = Young's modulus of the film
$\qquad \nu_f$ = Poisson's ratio for the film
$\qquad \alpha_s$ = thermal expansion coefficient of the quartz substrate
$\qquad \alpha_f(T)$ = thermal expansion coefficient of the film

Polyimide/Quartz Structures

Results of stress measurement for a polyimide/quartz structure subject to thermal cycling [34] are shown in Figure 27-14. The polyimide layer was prepared by spinning onto the quartz substrate a dilute DuPont 5878 polyamic acid, then baking at 85°C for ten minutes. The stress in the as-prepared film is tensile (i.e., positive), reaching a level of 2×10^8 dyn/cm^2 at room temperature. This is probably the intrinsic stress generated as a result of the inability of the polymer to relax, at room temperature, the stress developed during spin coating. As temperature increases, thermal stress begins to develop and to compensate for the intrinsic stress since the polymer has a higher thermal expansion coefficient

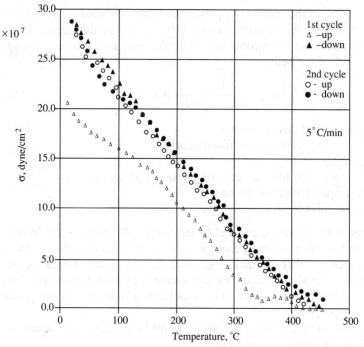

Average film stress of 10.5 μm polyimide on 5 mil fused quartz

FIGURE 27-14
Stress-temperature relationship of polyimide/quartz structure obtained during thermal cycling.

than quartz (see Table 27.1). At the same time, heating initiates the imidation process and drives out solvent from the polymer layer. This complicates the stress measurement since the film thickness and composition change simultaneously. The total stress continues to decrease with increasing temperature and is almost completely relaxed at about 350°C. This is close to the glass transition temperature of the polyimide. Further heating does not seem to increase the level of stress. Cooling from 450°C results in a linear increase of tensile stress with decreasing temperature. The stress level in the first cooling cycle reaches 2.7×10^8 dyn/cm² at room temperature.

As seen from Figure 27-14, repeated thermal cycling does not markedly change the stress-temperature relationship observed. This, plus the linear stress-temperature relationship, indicates that the stress generated in the polyimide layer is mainly thermal in origin and the polyimide is able to sustain the relatively low level of thermal stress with little stress relaxation.

Metal/Quartz Structures

Stress relaxation has been investigated for several metals using a metal/quartz structure, including Cu, Ti, and Cr [31,32]. The behavior was found to be significantly different from that of the polyimide/quartz structure. First, the stress level in the metal film was found to be more than an order of magnitude higher because of the higher modulus of the metal film (see Table 27.1). Second, depending on the mechanical properties of the metal film, the stress generated during thermal cycling can be substantially relaxed. In general, the kinetics of stress relaxation in the metal film are different under compression than under tension, so the stress relaxation behavior observed is different for the heating versus the cooling cycle. Interestingly, stress relaxation depends not only on the metal film but also on the bonding characteristics of the metal/substrate interface, as demonstrated by contrasting behaviors observed for Cu and Cr.

Results of stress measurement during thermal cycling are shown in Figure 27-15 for a Cu/quartz structure. The Cu film was deposited at 80°C, so its initial stress state at room temperature is tensile. Upon heating, the tensile stress changes into compression above 80°C, as expected. During the first heating cycle, significant stress relaxation takes place, as the stress almost vanishes at 250°C. Upon cooling from 250°C, a tensile stress is developed in Cu since it contracts more than the quartz substrate. The maximum stress is 3.3×10^9

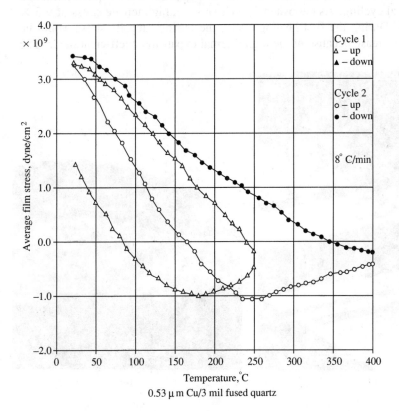

0.53 μm Cu/3 mil fused quartz

FIGURE 27-15
Results of stress management for a Cu/quartz structure during thermal cycling.

dyn/cm^2 at the end of the cooling period. This value is significantly lower than the theoretical thermal stress of 12×10^9 dyn/cm^2 calculated assuming no stress relaxation, suggesting that the thermal stress has been substantially relieved during the cooling cycle. Similar behavior was observed for the second thermal cycle between room temperature and 400°C. It is interesting to observe that the maximum compressive and tensile stresses reached in the second thermal cycle are almost identical to those of the first cycle. Thus, in spite of the higher cycling temperature, the Cu film can relax to about the same final stress level.

The change in the microstructure of the Cu film subject to thermal cycling has been examined by electron microscopy. Scanning electron microscopy has detected hillock growth in the Cu film. This occurs most probably during the heating cycle due to the relaxation of the compressive stress. Cross-sectional transmission electron microscopy observes grain growth and the presence of slip bands [31]. As shown in Figure 27-16(a), the slip bands observed in a Cu film on an Si substrate are oriented on the $\{111\}$ plane with a $\{0\bar{2}2\}$ slip direction, which are the normal slip characteristics for face-centered cubic materials.

Figure 27-17 shows the stress relaxation observed in a Cr/quartz structure during thermal cycling. As deposited, the Cr shows a high tensile stress of 9.3×10^9 dyn/cm^2. During the first heating cycle, the intrinsic tensile stress would be expected to decrease because of the high thermal expansion coefficient of Cr. The

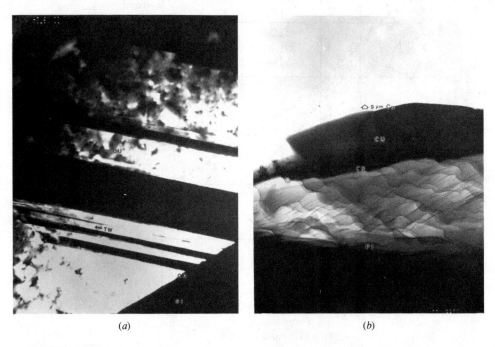

(a)　　　　　　　　　　　　　　　　(b)

FIGURE 27-16
Cross-sectional TEM micrographs of thermal cycling samples showing (a) slip bands at the Cu/quartz interface, and (b) deformation contrast in the polyimide at the Cu/polyimide interface. In both samples, a thin Cr layer of about 500 Å was used as a glue layer at the interface.

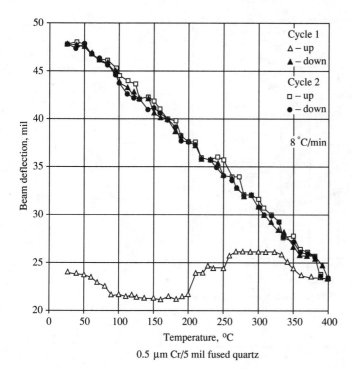

0.5 μm Cr/5 mil fused quartz

FIGURE 27-17
Stress-temperature relationship obtained for a Cr/quartz structure during thermal cycling.

amount of relaxation of the intrinsic stress actually observed is, however, very small in comparison to Cu; indeed, the stress level at the maximum temperature of 400°C is about the same as in the as-deposited film. This reveals a drastically different stress relaxation behavior for a body-centered cubic refractory metal such as Cr, as compared to a face-centered cubic metal such as Cu. Because of the high melting point and high activation energy for mass transport in body-centered cubic refractory metals, their thermal relaxation processes are much slower than those of Cu at the same temperature [35].

Upon cooling, tensile stress in the Cr film continues to accumulate with decreasing temperature, eventually reaching a very high level of 18.4×10^9 dyn/cm^2 at room temperature. The additional stress generated during thermal cycling is reproducible and varies linearly with temperature. Its magnitude can be almost completely accounted for by the thermal mismatching between Cr and quartz. This indicates that there is almost no stress relaxation in the Cr film under the thermal cycling conditions used in this study.

Metal/Polyimide/Quartz Structures

When polyimide is introduced as an intermediate layer into the metal/quartz structure, the stress behavior of the structure under thermal cycling can be significantly

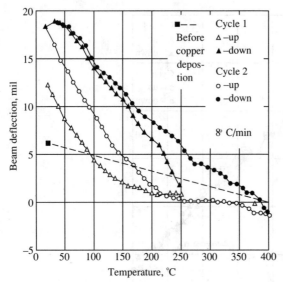

5.3 µm Cu/2.9 µm 5878 polyimide/5 mil fused quartz

(a)

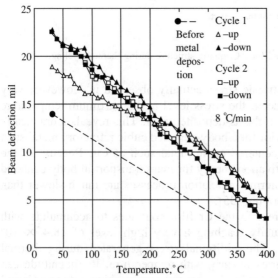

0.5 µm Cr/7.5 µm polyimide/5mil fused quartz

(b)

FIGURE 27-18
Beam deflection observed as a function of temperature during thermal cycling for (a) Cu (0.53 µm)/polyimide (2.9 µm)/quartz (125 µm) structure, and (b) Cr (0.5 µm)/polyimide (7.5 µm)/quartz (125 µm) structure.

altered. The result of a study for a Cu (0.53 μm)/polyimide (2.9 μm)/quartz (125 μm) structure [36] is shown in Figure 27-18(a). The overall behavior of this structure is similar to that of the Cu/quartz structure, except that the final stress at the end of the cooling cycle is almost 25 percent less. Thus the polyimide film serves as a stress-buffering layer, facilitating the release of the thermal stress in Cu. The extent of the stress reduction has been investigated as a function of the thickness of the polyimide layer [36]: The results showed a maximum reduction of about 30 percent, with the effect saturating at about 3 μm.

The effect in Cr was studied by inserting a 7.5-μm polyimide layer into the Cr/quartz structure of Figure 27-17 (this thickness was used to ensure a complete buffering of the stress). The observed beam deflections during thermal cycling for this trilayered structure are shown in Figure 27-18(b), together with the beam deflection from the polyimide/quartz structure measured before metal deposition. Compared with the beam deflections observed in Figure 27-17, it is clear that the polyimide layer is very effective in reducing the stress generated during thermal cycling. During the first heating period, the polyimide enabled the Cr layer to relieve its as-deposited stress by as much as 80 percent at 400°C. During subsequent thermal cycles, the maximum residual stress at the end of the cooling period was reduced by about 50 percent. Since we know from the results in Figure 27-17 that Cr by itself alone is not effective in relieving the thermal stress because of slow kinetics, the stress relaxation observed with polyimide has to occur primarily through the polyimide layer.

The mechanism for the stress relaxation was examined by cross-sectional TEM. Figure 27-16(b) shows the morphology of the Cu/polyimide interface in the multilayered structure after one hour of annealing at 400°C. It can be seen that stress relaxation has taken place near the interface, giving rise to the image contrast observed in the polyimide. The extent of the deformation contrast is about 1 to 1.2 μm, indicating the presence of a sharp stress gradient near the interface in the polyimide. The range of the stress gradient observed probably explains the saturation effect at a polyimide thickness of 3 μm. In contrast to the Cu/Si interface, stress relaxation takes place not only in the Cu film, but also through the Cu/polyimide interface by development of Cu intrusions into the polyimide layer. These intrusions initiate the wavy deformation features in the polyimide. The deformation is to be expected due to the steep stress gradient at the metal/polymer interface. Interestingly, fine fissure lines are present in the polyimide, suggesting that shear deformation parallel to the interface has occurred in the polyimide near the interface.

SUMMARY

In this chapter, interfacial adhesion was first briefly discussed in relation to the bonding and structure of the interface. This was followed by reviews of bonding, microstructure, and stress relaxation of the metal-polyimide interface. The metal-polyimide interface has been studied under well-controlled conditions using a combination of analytical techniques. These include photoemission spectroscopy to monitor the valence and core states, TEM to observe the morphology and microstructure, and the cantilever bending beam technique to measure stress relaxation. These experimental investigations have been complemented by theoretical studies, including molecular orbital calculations to deduce the electronic states of metal-polymer complexes, and Monte Carlo simulation to derive the interfacial morphology. Detailed microscopic information on the chemical and material characteristics of the metal-polyimide interface has been obtained. Material reactions such as intermixing and cluster formation are important phenomena in this class of interfaces. They are important for understanding the mechanical behavior of the interface, particularly for high-temperature polymers, such as polyimide.

Among the metals studied, Cu forms a weak chemical bond to polyimide, so Cu atoms can diffuse readily into polyimide to form clusters. In contrast, Cr bonds strongly to polyimide, giving rise to a high-strength interface with uniform morphology. The difference in the bonding characteristics is reflected in the stress relaxation behaviors of these interfaces.

In summary, the metal/polyimide interface exhibits interesting chemical and mechanical characteristics. Many of the observed characteristics seem to be general for other polymers as well, particularly for those with high-temperature applications. The success of using such metal-polymer interfaces for packaging applications may well depend on the extent of fundamental understanding of such interfacial properties.

ACKNOWLEDGMENTS

Results reported in this article come from collaboration of the author with many coworkers in the IBM T. J. Watson Research Center. He is indebted to their efforts and wishes to thank particularly P. O. Hahn, J. W. Bartha, G. W. Rubloff, F. K. LeGoues, B. D. Silverman, A. Rossi, R. Haight, P. Sanda, R. C. White, C. K. Hu, S. T. Chen, F. Faupel, and C. H. Yang. During the course of this work, the author has also benefited from discussions with many colleagues from the IBM Almaden Research Center, System Technology Division at Endicott, New York, and the General Technology Division at Fishkill, New York.

REFERENCES

1. For a general review on interfacial adhesion, see Chapters 10 and 11 in *Polymer Interface and Adhesion* by S. Wu, New York: Marcel Dekker Inc., 1982.
2. J. M. Burkestrand, *J. Vac. Sci. Technol.*, 16, p. 363 (1979); and *J. Appl. Phys.* 52, p. 4795 (1981).
3. N. J. Chou and C. H. Tang, *J. Vac. Sci. Technol.*, A2, p. 751 (1984).
4. P. S. Ho, P. O. Hahn, J. W. Bartha, G. W. Rubloff, F. K. LeGoues, and B. D. Silverman, *J. Vac. Sci. Technol.*, A3, p. 739 (1985).
5. P. O. Hahn, G. W. Rubloff, J. W. Bartha, F. LeGoues, and P. S. Ho, *Mat. Res. Soc. Symp. Proc.* 40, p. 251 (1985).

6. A. R. Rossi, P. N. Sanda, B. D. Silverman, and P. S. Ho, *Organometallics* 6, p. 580 (1987).
7. R. Tromp, F. K. LeGoues, and P. S. Ho, *J. Vac. Sci. Technol.*, A3, p. 782 (1985).
8. C. B. Duke, *Int. J. Quantum Chem.* 13, p. 267 (1979).
9. T. J. Fabish, *CRC Critical Reviews in Solid State and Materials Science*, Cleveland: Chemical Rubber Co., p. 383 (1979).
10. E. D. Feit and C. W. Williams, editors, *Polymer Materials for Electronic Applications*, ACS Symposium Series 184, Washington, D.C.: American Chemical Society, 1982.
11. E. Geiss, K. N. Tu, and D. R. Uhlmann, editors, *Proceedings on Electronics Packaging and Materials Science*, Materials Research Society, 1985.
12. *Proceedings on Electronic Packaging and Materials Science*, R. Jaccodine, K. A. Jackson, and R. C. Sundahl, editors, Mat. Res. Soc., 1988.
13. See, for example, L. Pauling, *The Nature of Chemical Bond*, 3rd ed., Ithaca, New York, Cornell University Press, 1960.
14. See, for example, Chapter 14 of Ref. 1.
15. P. O. Hahn, G. W. Rubloff, and P. S. Ho, *J. Vac. Sci. Technol.*, A2, p. 756 (1984).
16. J. S. Binkley, R. A. Whiteside, R. Koishnan, R. Seeger, D. J. DeFrees, H. B. Schlegel, S. Topiol, L. R. Kahn, and J. A. Pople, *Quantum Chemistry Program Exchange (QCPE)* 10, p. 416 (1981).
17. B. D. Silverman, P. N. Sanda, P. S. Ho, and A. R. Rossi, *J. Polym. Sci. Poly. Chem. Ed.* 23, p. 2857 (1985).
18. R. C. White, R. Haight, B. D. Silverman, and P. S. Ho, *Appl. Phys. Lett.* 51, p. 487 (1987).
19. M. J. Goldberg, J. S. Clabes, C. A. Kovac, and J. L. Jordan, in *Proceedings on Electronic Packaging Materials Science*, ed. by R. Jaccodine, K. A. Jackson, and R. C. Sundahl, Materials Res. Soc. (1988).
20. R. Haight, R. C. White, B. D. Silverman, and P. S. Ho, to appear in *J. Vac. Sci. Technol.*, A6, p. 2188 (1988).
21. J. W. Bartha, P. O. Hahn, G. W. Rubloff, and P. S. Ho, to be published.
22. J. W. Bartha, P. O. Hahn, F. LeGoues, and P. S. Ho, *J. Vac. Sci. Technol.*, A3, p. 1390, (1985).
23. Lj. Atanasoska, H. M. Meyer III, G. Anderson, and J. H. Weaver, *J. Vac. Sci. Technol.*, A5, p. 3325 (1987).
24. F. S. Ohuchi and S. C. Freilich, *J. Vac. Sci. Technol.*, A4, p. 1039 (1986).
25. J. L. Jordon, P. N. Sanda, J. F. Morar, C. A. Kovac, F. J. Himpsel, and R. A. Pollak, *J. Vac. Sci. Technol.*, A4, p. 1046 (1986).
26. J. L. Jordon, C. A. Kovac, J. F. Morar, and R. A. Pollak, *Phys. Rev.* B36, p. 1369 (1987).
27. S. G. Anderson, H. M. Meyer III, and J. H. Weaver, *J. Vac. Sci. Technol.*, A6, p. 2205 (1988).
28. H. M. Meyer III, S. G. Anderson, Lj. Atanasoska, and J. H. Weaver, *J. Vac. Sci. Technol.*, A6, p. 30 (1988).
29. Lj. Atanasoska, H. M. Meyer III, S. G. Anderson, and J. H. Weaver, *J. Vac. Sci. Technol.*, A6, p. 2175 (1988).
30. F. K. LeGoues, B. D. Silverman, and P. S. Ho, *J. Vac. Sci. Technol.*, A6, p. 2200, (1988).
31. S. T. Chen, C. H. Yang, F. Faupel, and P. S. Ho, to be published in *J. Appl. Phys.* (1988).
32. C. K. Hu, D. Gupta, and P. S. Ho, *Proceedings of VLSI Multilevel Interconnection Conference*, IEEE Publication 85CH2197-2, p. 187 (1985).
33. D. S. Campbell, in *Handbook of Thin Film Technology*, ed. by L. I. Maissel and R. Glang, New York: McGraw-Hill, 1970.
34. H. M. Tong, C. K. Hu, C. Feger, and P. S. Ho, *Polymer Engineering and Science* 26, p. 1213 (1986).
35. H. J. Frost and M. F. Ashby, *Deformation Mechanism Maps,* London: Pergamon Press, 1982.
36. S. T. Chen, C. H. Yang, H. M. Tong, and P. S. Ho, *Mat. Res. Soc. Symp. Proc. 56*, ed. by R. Nemanich, P. S. Ho, and S. S. Lau (1986).

PROBLEMS

27.1. The molecular configuration of the PMDA-ODA repeat unit (shown in Figure 27-3) has been determined by Takahashi et al., *Macromolecules* 17, 2583 (1984). The distances for O–C, C–C and C–N bonds were found to be 1.36 Å, 1.39 Å and 1.42 Å respectively. Assuming, for simplicity, that all the angles in the five-member and the six-member rings are equal, calculate the close-pack molecular density of the PMDA-ODA surface.

27.2. (*a*) During the initial stage of interfacial formation, assume that the Cr atoms interact only with the central six-member ring of the PMDA. Calculate the minimum thickness of Cr required to provide a complete reaction at this stage.

(*b*) Upon further deposition of Cr, all the six-member and the five-member sites in the PMDA and the ODA units will become completely occupied. Assuming that the binding energy for all the Cr complexes is 5 eV each, calculate the cohesive energy in units of joules/m^2 for the Cr-PMDA-ODA interface.

27.3. Refer to the stress measurement for the polyimide/quartz structure shown in Figure 27-14. Given the following properties at 20°C for the polyimide—$E_f = 2.4 \times 10^{10}$ dyne/cm^2, $\alpha = 29.5 \times 10^{-6}$ per °C, and $\nu = 0.33$—calculate the thermal stress generated during the first heating cycle from 20° to 350°C by assuming the polyimide properties to be independent of temperature. Based on the result, calculate the total stress in the polyimide/quartz structure at 350°C and compare it with the observed stress.

27.4. The difficulty of stress calculation in the previous problem comes from the assumption of temperature independence of the polyimide properties. Since all these parameters vary considerably with the temperature, this factor has to be taken into account in order to deduce the correct thermal stress. As it turns out, for the polyimide in the test temperature range, the product of E_f and $\alpha/(1 - \nu)$ is approximately constant and equals 6.9×10^5 dyne/cm^2/°C. Using this value, calculate the total stress of the polyimide during thermal cycling and show that most of the stress observed is thermally induced.

27.5. The material parameters of Cu for stress calculation are less temperature-sensitive than those of the polyimide. Assuming that they are independent of the temperature, calculate the thermal stress to be expected for the first heating cycle in the experiment shown in Figure 27-15. Compare your result with the total stress observed. What can you conclude about the extent of the stress relaxation in this structure? How about the stress relaxation for the first cooling cycle?

CHAPTER 28

ROBOTICS AND AUTOMATION

EDWARD M. COLLINS
JOSEPH P. PAWLETKO

INTRODUCTION

This chapter will focus on robotics applied to packaging applications. The first section will be a brief tutorial on robotics, describing types of robots and discussing drive systems and sensors.

The section "Robotic Applications in Printed Circuit Manufacturing" will describe three applications that have been implemented. Along with a description and pictorials for each robotic application is a set of guidelines to follow when an automation project is being initiated.

ROBOTICS AND AUTOMATION

This section describes the uses for automation and robotics, gives the differences between the two, discusses the requirements for electronic packaging applications, and suggests best approaches to both robotics and automation.

An automated or adaptive system is used to assemble the many elements of an electronic package, be it a PC board or a special module. The basic distinction between robotics and single-purpose automation is programmability and adaptability. Automation is the optimum solution to a large-scale high-volume

841

production line where most of the lines are single-purpose, i.e., designed for one specific product. By breaking down the various assembly steps, special equipment can be designed to perform an operation at a very high speed, thus achieving high production rates. Robotics, on the other hand, allows a product mix and short production runs, because it allows each workstation to be reprogrammed for a new task in a very short time.

The basic requirements of a robot in the semiconductor industry are high precision, a clean room environment, reasonable speed, and adaptability by use of sensory feedback or vision; also, a robot must be reliable and require minimum intervention.

There are many manufacturing operations for electronic products that require specialized robots. Two basic sizes are involved; one is capable of chip placement and wire-bonding applications; the other involves board testing and module placement on printed circuit boards. Each of these applications requires unique attributes. A robot used in chip placement and testing requires high precision, tactile feedback, and/or vision, as well as a reasonable speed of operation. Because it moves very small masses like semiconductor chips, the mechanical structure can be small and light. Precision of such units is derived from appropriate feedback systems, which will be discussed later in this section.

The second type of application involves higher payloads, more insertion force for module placement, and a larger workspace, which is determined by the largest printed circuit board a given system must handle. The latter requirements imply a need for an electromechanical entity using either dc servo or stepping motors as the dynamic elements.

Types of Robots

Two basic types of robots are those using polar and rectangular coordinate systems respectively (Figure 28-1). The polar types are subdivided into cylindrical, polar, and revolute robots. The revolute is also referred to as anthropomorphic, because the construction involves a shoulder, an elbow, and a wrist, allowing humanlike motion.

The polar variety has a wide range of applications in the automobile industry, especially for heavy work like welding and painting. These robots are also used in radioactive environments where remove controlled robots use vision operation in places lethal to humans. The precision of such structures is more than adequate for typical applications. The magnitude of the task determines the size of the robot.

For less demanding applications, pneumatic actuators are used to control simple tasks like pick and place activity where close tolerances are not required. Figure 28-2 shows a pneumatic cylindrical extension boom mounted on a base that allows the boom to rotate as well as elevate. This particular system has grippers as the end effector, which could also be a special adapter or a vacuum cup for a pick-and-place application [1].

Configurations

The physical form of a robot's manipulator is described by the coordinate system to which its axes of motion conform.

Spherical
Robot moves inside a spherical work envelope, along R, θ and φ axes (GE Model GP 66).

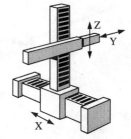

Cartesian
Robot moves along straight line X, Y and Z axes which form a rectangular work envelope (IBM 7565).

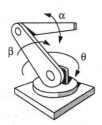

Jointed Arm (Revolute)
Anthropomorphic robots have several rotary joints and, generally, a rotating base. Wrist, elbow and shoulder are terms often applied to the joints (ASEA IRb-6).

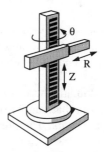

Cylindrical
Robot moves inside a cylindrical work envelope, along Z, R and φ axes (ASEA MHU Senior).

FIGURE 28-1
Classes of robots.

Figure 28-3 shows two other polar robots, a heavy-duty large work envelope for noncritical applications (*a*), and a smaller, very rigid unit for precision applications (*b*) [1].

The smaller robot is frequently used for component placement applications because of its ability to reach the component storage area and position the component over the work space. This type of robot consists of a vertical column that carries the servoed rotating arm. The end effector is mounted on the z axis, which can be either servoed or have vertical motion to a specific stop position.

The z axis end effector is usually job specific and can be a gripper, a suction cup, or a variety of other tools. The actuators are mostly dc servo motors mounted in the joint directly or coupled through a timing belt.

FIGURE 28-2
Pneumatic drive point-to-point (made by Seiko Instruments USA, inc.). The device has a 2.2-lb load capacity, ±0.001-in. repeatability, and 18- to 43-in. reach.

Robotic Actuators

There are many methods of imparting motion to a robotic structure. Actuators range from hydraulic to pneumatic and electrical. For the class of robots under discussion, electrically driven actuators are of main interest. Both rotary and linear actuators are used.

In the electrical domain there are dc servo motors that can be brush or brushless, stepping motors that can be of hybrid or permanent magnet type, and variable reluctance type motors. The prevalent element in many robots is the ser-

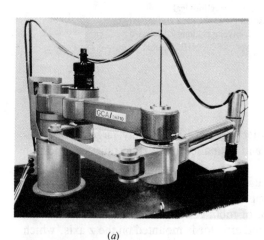

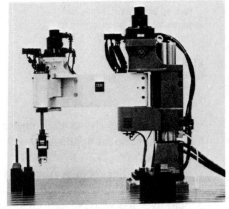

(*a*)   (*b*)

FIGURE 28-3
(*a*) Electric dc servo (pneumatic z axis), point-to-point or controlled path (made by GCA Corporation). It has ±0.02-in. repeatability and 113-in. reach. (*b*) Electric dc servo and stepping motor drive with pneumatic z axis, point-to-point (made by IBM). It has ±0.002-in. repeatability and 40-in. reach.

vo-controlled dc motor. Many robots use stepping motors in similar applications. For a detailed discussion of these motors, refer to the references.

The three-phase variable reluctance type is a new class of motors with a very high torque-to-volume ratio and high angular resolution. A motor of this type consists of six three-phase units forming a circle. The flux passes transversely through the rotating armature ring, as in an IBM-developed linear motor (Figure 28-4) [2]. Such motors can be mounted directly in the joints because of their characteristics.

Rectangular type robots can be cantilever or gantry type as shown in Figures 28-5 and 28-6. Cantilever design provides more open space in the assembly area, but it has several shortcomings. A long settle-out time after arrival at the desired location is one, while another is the variable accuracy dependent on the distance from the pivot point. For more precise positioning a gantry system is desirable. Accuracy and repeatability are functions of the mechanical structure as well as the precision of the positioning system. Dynamics is better controlled in the gantry design, because this design has higher mechanical stiffness and the reflected load is independent of position within the work envelope.

Several approaches are used to achieve linear motion in these robots. The low-end systems use a leadscrew drive, which is actuated by a dc servo or stepping motor. There are several disadvantages in this implementation. The major one is speed of access to a given point. One trades off resolution and repeatability for speed by changing the leadscrew pitch. The size of the work envelope is limited by a practical leadscrew length. A two-arm robot is not possible using this design, because two independent systems cannot be mounted on one leadscrew.

To circumvent these shortcomings, rack-and-pinion designs have been developed. Dc servo motors mounted on the gantry with a pinion shaft engaged in a stationary rack result in translational linear motion of the gantry. Attempts have been made to use hydraulic motors for this type of configuration. However, the hydraulic hose system is unwieldy, thus forcing the use of one motor for the gantry. Gantries driven from one end suffer more problems than the cantilever systems. Hydraulic systems are not compatible with clean room environment standards.

In gantry systems, accuracy is a function of the distance from the control element, and acceleration is a function of the gantry stiffness and position of the load. Binding of the gantry occurs, stalling or slowing the system down, and also causing positional hysteresis. To avoid many of these shortcomings, linear motors are being used. There are linear motors in existence of permanent magnet, hybrid, and variable reluctance types.

Permanent magnet linear motors are not very popular because of their limited performance, heavy armature using alnico magnets, high flux leakage, and poor efficiency. A new magnet material, neodymium-iron-boron, may revolutionize many motor designs. Its high magnetic energy per unit volume allows compact and lightweight design.

Dc commutated motors consist of a stator bar with many coils along its length. The armature is a samarium-cobalt magnet with brushes mounted on it.

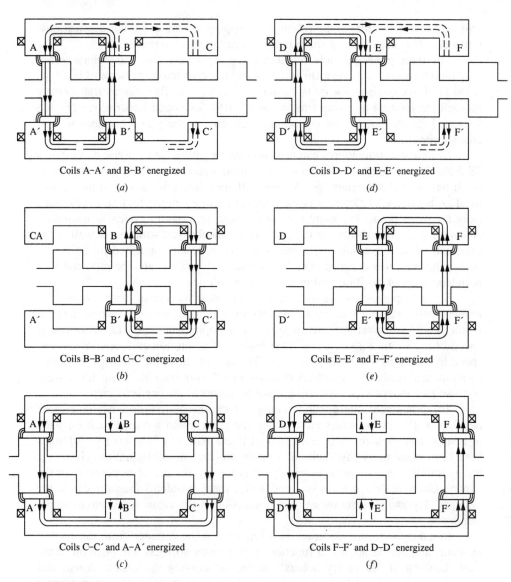

Coils A–A′ and B–B′ energized

(a)

Coils D–D′ and E–E′ energized

(d)

Coils B–B′ and C–C′ energized

(b)

Coils E–E′ and F–F′ energized

(e)

Coils C–C′ and A–A′ energized

(c)

Coils F–F′ and D–D′ energized

(f)

FIGURE 28-4
Linear stepping motor flux distribution—two-phase energization.

By energizing the appropriate coils the desired motion is achieved; the brushes make the motor self-commutating.

A hybrid linear motor, also called a Sawyer motor [3], is a four-pole, two-phase configuration (Figure 28-7). The magnetic field is commutated to the desired poles in the proper sequence by the coils.

Variable reluctance motors offer many advantages. The IBM design can

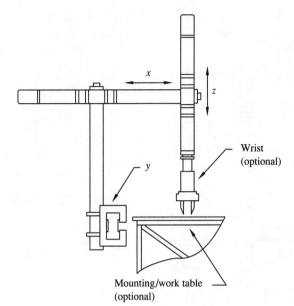

FIGURE 28-5
Cartesian cantilevered robot.

achieve a very high force-to-mass ratio, which results in high dynamic performance (Figure 28-4). It is a three-phase motor with an armature structure positioned on both sides of the stator bar.

Implementing a gantry with variable reluctance motors offers many advantages. The stator bars can be used as a supporting structure and as a transducer grid. Driving the motors in tandem results in a very stiff gantry. Several gantries

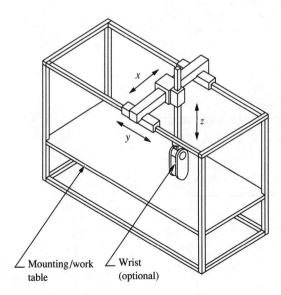

FIGURE 28-6
Cartesian gantry robot.

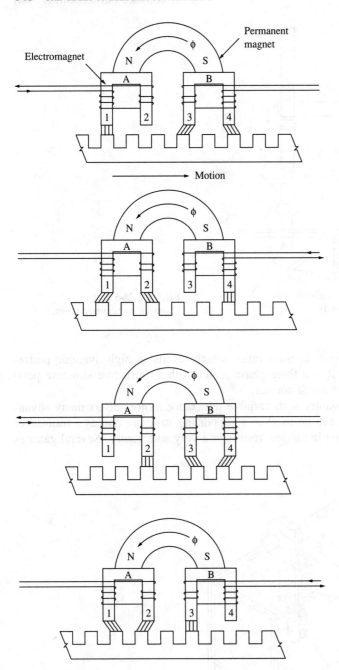

FIGURE 28-7
Principle of the Sawyer linear motor.

can operate on the stationary stator bars independent of each other. The motors are highly reliable, exposed to bearing wear only, and can operate in a clean room environment. High positional accuracy and repeatability have been achieved over a large workspace area.

SENSORY BASE

Vision is a very important adjunct to the robot. The most precise robot requires that the positioning of the workpiece be more accurate than the accuracy of the unit. It is impractical to achieve this in a production environment without adding vision capability. As each workpiece is positioned, a vision system redefines the positional coordinates for the new condition. In tight tolerance cases the component's image is superimposed on the placement site before a component is placed on it.

Position feedback can be derived in many ways, depending mainly on the type of precision required and what dynamic element is to be controlled. In noncritical applications, like the pneumatic pick and place robot, microswitches or simple photocells may be sufficient to sense and stop the robot at a specific location. This approach is rather primitive, but still in use today. In this case, reprogramming means moving the sensors to new positions.

Tactile feedback is also important. Many components are delicate, and a pickup device could easily crush them. By sensing the gripping or insertion force, damage to a component or substrate can be prevented.

Most tactile feedback systems use strain gages to sense the gripping forces on the workpiece. When the grip force can be sensed, it can be controlled to a specific value. Other systems use a magnetoresistor element in a bridge configuration that allows minute displacements of a magnet over the resistor to be translated to a force value. This approach is not as sensitive to overloads, because it is a noncontact system.

Piezoceramics are used in many applications to sense both pressure and accelerating forces. One disadvantage to this approach is that static pressure cannot be sensed reliably because of gradual decay of the piezoelectric charge.

In many applications proximity sensing is sufficient to calibrate a system or establish new coordinates for subsequent operation. Capacitive or eddy current probes are frequently used to locate a reference surface or object.

Dc servo and closed-loop stepping motors require feedback systems from which position and velocity can be derived. Rotary motor applications frequently use incremental resolvers. Leadscrew-driven systems have two options for position sensing: one senses carriage position from a linear grid, and the other uses the known leadscrew pitch and the resolver output.

Many resolvers are available commercially, including optical grid, capacitive disk, and inductively coupled discs. Linear grids follow a similar pattern with the addition of a magnetically recorded wire and a pickup coil for sensing position by sensing the recorded signal.

ROBOTIC APPLICATIONS
IN PRINTED CIRCUIT MANUFACTURING

The development and implementation of robotics and the integration of these applications into a printed circuit manufacturing environment is a rigorous but exciting endeavor. This section describes several installed systems and discusses some of the lessons learned from their implementations. Critical guidelines have been established and will be discussed. Design criteria and the interaction of these criteria with the manufacturing process are also considered.

Issues to be considered in the development of an automation/robotic application are as follows:

Of primary importance is the design of the product. This is perhaps the most important consideration, and can have the most impact upon the success or failure of such an application. Typically, the biggest hindrance in using robots in manufacturing is the product design itself, and not the design of the robotic/automation system. A component or assembly designed for robotic/automation manufacturing processes usually has a minimal number of parts and an ease of manufacture. Such an assembly would also be easily produced by using manual techniques. This is primarily true due to the reduced number of parts involved and decreased fits and adjustments.

Operations that involve special robot tools or fixtures should be avoided. They increase the overall cost of the robotic system and add to the system cycle time. This results in an increased process cost.

All required motion and access points must be easily accessible. Again, any need for special tools and/or special robotic moves and subroutines adds to cycle time and increases operational costs.

Detail parts should be self-locating and easily distinguishable from similar parts. Orientation notches or other features should be identifiable on individual parts.

The required operations should be within the capabilities of the robot. New development should not be required in order to implement a robotic/automation application. The robot should have a large enough reach, or envelope, for it to access all the physical points required by the application.

Automated Core Stamp/
Automated Punch Press System

The automated core stamp system and the automated punch press system are two robotic applications in the laminate machining area at IBM in Endicott, New York. Both utilize IBM 7535 robots (SCARA type) for material handling.

The automated core stamp system (Figure 28-8) stamps a part number on each individual circuit core. The sequence of operations for the system is as follows:

A product cart, loaded with printed circuit panels to be stamped, is loaded onto the product input elevator.

A stack of empty carriers is placed in the empty tub elevator.

Panels are transferred to the stamp head.

Robot places an empty carrier at the product output elevator.

Printed circuit board panels are located under the head and then stamped [Figure 28-8(c)].

Robot picks up stamped panel and places it in a carrier, inserting spacers as required [Figure 28-8(d)].

Completed printed circuit panels, stacked in carriers, are removed by the operator from the product output elevator.

The printed circuit panel is very sensitive to damage at this point in the process since it has just been laminated with fresh copper on both sides. Eliminating product damage was a key requirement to be met in this application. In order to meet this requirement, vacuum grippers were used to handle the product, as shown in Figure 28-8(a). An air bearing–and–vacuum technology was also utilized in handling and locating the product to be stamped. As shown in Figure 28-8(b), the flat plate has a large number of small-diameter holes to which air pressure is applied from underneath. In the center of the figure there is a locator "puck." At the same time that the air pressure is applied from below, the puck applies a vacuum through holes in the puck's underside. This causes the product to be securely held by the puck, which then rotates and moves the panel against the edge locators. Once the panel is successfully located [Figure 28-8(c)], the part number is stamped. Proper adjustment of both the air pressure and the vacuum allows the product to ride on an air "bearing," and thereby minimizes product damage while locating the panel.

Once the product is stamped, it is picked up by the robot gripper and placed into a container. It should be noted that very thin, flat, smooth products are being handled, which have the tendency to adhere to each other when placed in a vertical stack. This "ringing" effect is a problem that has to be overcome in a great number of applications. In this case a technique is used to flex the product gently while it is held by the gripper, in order to insure single-part pickup. The IBM 7535 robot system is used as a material-handling device in this application. It stacks the stamped panels into containers, inserts spacer sheets at the appropriate points, and retrieves and positions an empty container when the previous one is fully loaded.

The automated punch press system is similar to the automated core stamp system, and in fact uses some common subassemblies. Basically, this system is also a materials-handling application.

The system is designed to handle printed circuit panels that are to be trimmed to the correct size and shape in the vertical punch press. Figure 28-9 identifies the major pieces of the system. The sequence of operation for the system is as follows:

The operator loads a stack of printed circuit board panels [Figure 28-9(a)], to be trimmed, at the input shuttle.

Panels are transferred to the panel locator area, where they are positioned prior to insertion into the die [Figure 28-9(c)].

Panels are then transferred into the die of the punch press, where they are trimmed to correct size and shape [Figure 28-9(b)].

Trimmed panels are transferred to the stamper unit, where they are stamped with the appropriate information.

Robot places empty carriers on the product output elevator.

Robot picks up trimmed and stamped panels from the stamper unit, and places them in a carrier.

Carriers are removed by the operator at the product output elevator area.

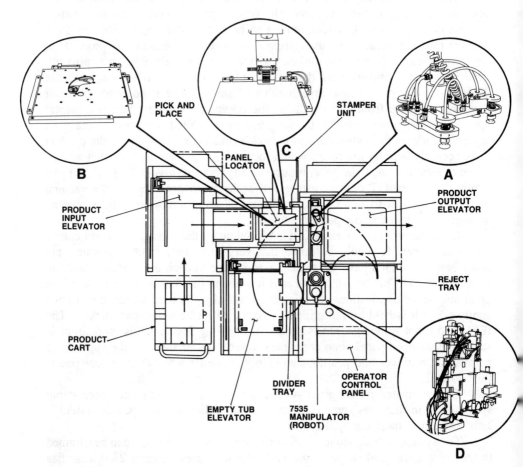

FIGURE 28-8
Automated core stamp system.

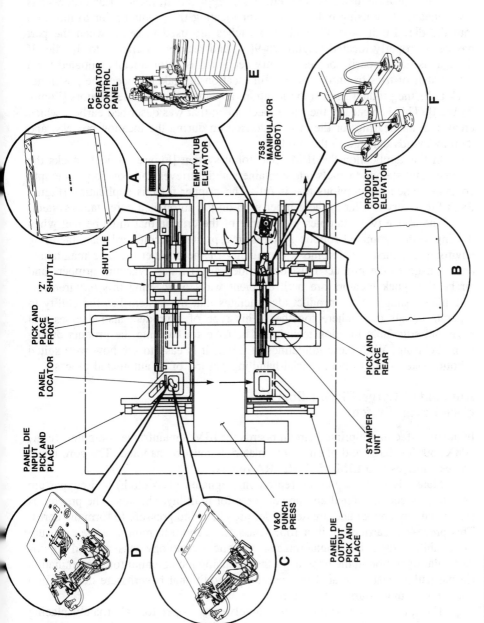

PC
OPERATOR
CONTROL
PANEL

A

SHUTTLE

'Z'
SHUTTLE

PICK AND
PLACE
FRONT

PANEL
LOCATOR

PANEL DIE
INPUT
PICK AND
PLACE

D

EMPTY TUB
ELEVATOR

E

7535
MANIPULATOR
(ROBOT)

F

PRODUCT
OUTPUT
ELEVATOR

B

PICK AND
PLACE
REAR

STAMPER
UNIT

C

V&O
PUNCH
PRESS

PANEL DIE
OUTPUT
PICK AND
PLACE

FIGURE 28-9
Automated punch press system.

853

Edge location methods are used. In this application, the location process has been enhanced by using hole location for locating the product prior to insertion into the die. Limit switches and air switches are used to sense when the part has been properly located [Figure 28-9(c)] and may be inserted into the die. If the part is not properly located, a retry sequence is automatically initiated for a set number of tries or until a successful location has taken place. The positioner puck pulls the product against the location rails and over the hole locators [Figure 28-9(d)]. This is basically the same mechanism that was described earlier, which creates an air bearing for the product to ride on during the location process, and thereby avoids damage to the product.

In this application the IBM 7535 robot system [Figure 28-9(e)] stacks the trimmed and stamped panels into containers and retrieves and positions an empty container when the previous one is full. The gripper for this application [Figure 28-9(f)] is very similar to the gripper used on the automated core stamp system.

The above two examples yield some of the criteria that must be met when dealing with automated robotic systems in a printed circuit card manufacturing environment. First and foremost, the quality of the product must be maintained and damage to the product must be avoided at all times. Vacuum grippers and air bearing puck locators are both efficient ways of meeting this requirement. The planar shape of the product also dictates that the system has the ability to separate panels that adhere together. This type of a product also dictates that edge and edge/hole location techniques be used for locating the product during manufacturing. With these guidelines in mind, it is easy to see how well suited robotic systems are to the material-handling portion of an automated process.

Automated Large Circuit Board Core Layup System

In the manufacture of printed circuit boards for IBM's mainframe computers (e.g., 308X, 309X), large and often fragile materials must be handled. The core layup process utilizes two IBM 7535 SCARA robots.

Materials such as prepreg (epoxy-impregnated glass cloth) and copper are used to produce the signal and power cores for multilayer boards. The prepreg is surrounded by copper and pressed under high heat and pressure to form the cores. This process is carried out in a 100K-class clean room to minimize the exposure to reliability-impacting contaminants. Robotic technology was applied to this particular operation in the hope of further reducing the potential for contamination (in particular, that generated by operators). Additional benefits are derived in a reduction in unit hours and cycle time.

The manual core layup process involves two operators alternately placing the various layers of prepreg, copper, and planishing plates (stainless steel plates used to separate the individual cores) onto a transport plate, which carries the cores into the press. In order to duplicate this process, two IBM 7535 robots were used. Figure 28-10 identifies the major pieces of the system.

These robotic systems have the necessary precision and accuracy to place several sheets of raw materials to construct the cores. The robotic system also successfully meets the requirements for the 100K clean room. An operator is required to supply raw materials and to move the transport plates, but minimal human support is necessary while the cores are layed up. This allows the operator to work at another task while also monitoring and assisting system performance.

Each robot can handle two different materials. Materials are presented to the robots on carts [Figure 28-10(b)], which are precisely aligned to the robot system mounting tables. Each cart can handle only one material; the carts are encoded to prevent the operator from placing materials in the wrong station. A conveyor extension was added to the existing manual station to carry the transport plate between two robots. The materials are transferred from the carts to the transport plate by means of a vacuum gripper end effector [Figure 28-10(c)].

The sequence of operations for the automated large circuit board core layup system is as follows:

The operator puts loaded material carts into place and locks them into position.

The operator places a press pad (blotter paper used to provide uniform heat and pressure) on the transport plate and sends it into the robot station.

The operator selects, from a menu screen, the type of core to be layed up.

The robots perform the layup of the cores.

When the robot cycle is complete, an operator attention light flashes and the operator sends a completed transport plate to the manual layup station.

A top press pad is put in place and the vacuum frame is placed over the entire package. The completed transport plate assembly is sent to the lamination press and a new plate is brought in to begin the cycle again.

As in many robotic system applications, the end effector [Figure 28-10(c)] required the most design effort. This system required the ability to pick up a 24-by-28-inch sheet of copper, prepreg weighing only a few ounces, and planishing plates weighing ten pounds. In addition, the static electricity that tends to make prepreg sheets stick together had to be overcome.

The final gripper design has eight suction cups on retractable cylinders mounted around the perimeter, and one cup located in the center. The outside cups pick up the edge of the material and the retractable cylinders lift while the center cylinder remains down. This action breaks the static forces [Figure 28-10(c)].

The vacuum cups were designed especially for this system. Veined inserts are molded as part of the cup to prevent raw material damage. The cup is clean room compatible and has good durability. A certain amount of compliance was designed into the cup to allow for a flex pickup [Figure 28-10(d)].

Vacuum sensors are used to detect if the material has been picked up, and a reflectivity sensor determines if the correct side of the copper is up (oxide side must go against prepreg). A stack height sensor is used to determine when the z axis is fully down, which aids in reducing the cycle time of the robot.

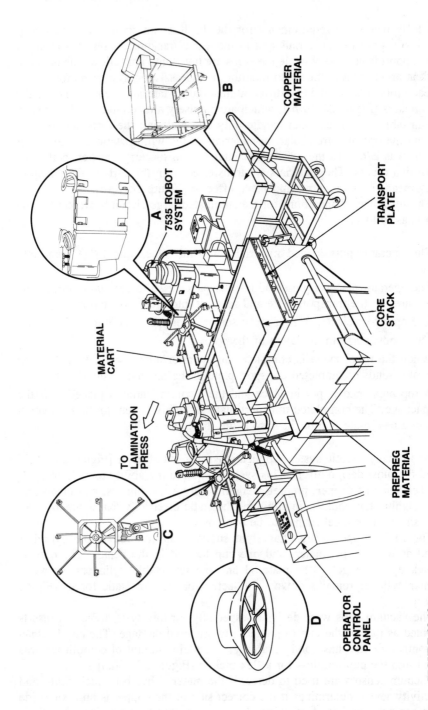

FIGURE 28-10 Automated large circuit board core layup system.

Due to the large size of sheets handled in this application, modifications to the IBM 7535 robot arm were required. In order to reach all of the application points, a three-inch arm extension [Figure 28-10(a)] was added to each robot arm.

The entire system is controlled by an IBM Personal Computer, which acts as a system host. The Personal Computer gives the operator a menu for application selection and downloads the application programs to the robot controllers. Each robot operates independently. One digital input (DI) and one digital output (DO) per robot are used for cross-synchronization, allowing the robots to operate within the same work envelope without the risk of a collision. An error/stop routine provides information to the operator in the event of a system problem. To protect the operator, a cage surrounds the entire work envelope of the robot systems.

As stated in previous examples, a key concern in this system was maintaining product quality and avoiding damage to the product at all times. Once again, vacuum grippers were used and edge location techniques were employed to locate the material at the input station. Reduced contamination levels and increased productivity were achieved.

SUMMARY

The above information only scratches the surface on robotics and robotic applications. The applications shown are "islands" of automation; i.e., they only represent the automation of one part of a process sector within a complete line. When an automation project is started, decisions must be made on what level of automation is required. Typically, four levels can be considered [4], as suggested by the following diagram:

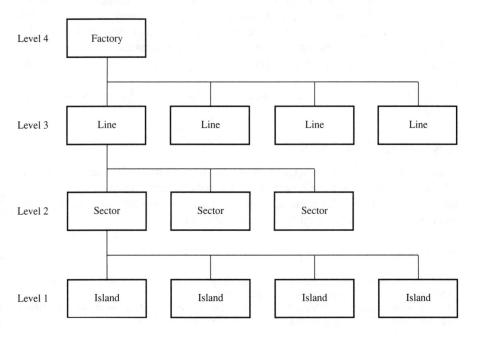

The level of planning and investment increases dramatically as one moves up from automation level 1 toward automation level 4.

In addition to the various considerations discussed earlier, the following issues should also be addressed when robotics or automation is being considered.

1. Flexibility versus capacity
2. Cost (initial and operating)
3. Space
4. Life
5. Expansion
6. Part number mix
7. Setups
8. Reliability
9. Modularity
10. Process flow
11. Material handling

These issues must be analyzed at each level, and a common strategy must be developed, upon which all levels of automation are based. This is an extremely important point and should not be treated too lightly; because if a common strategy is not in place and controlled, there will be as many strategies as there are islands, sectors, lines, and factories. This will lead to difficulty in material handling and information processing between islands, sectors, etc.

One additional point to make is that the automation should be simulated before equipment is purchased and installed. There are many good computer-based simulators available that can quickly identify problem areas such as bottlenecks and perform line balancing. Before committing to hardware, effort should be made to do thorough research, plan carefully, and look at several successful operations.

References 5–7 are excellent sources of information for topics covered in this chapter.

ACKNOWLEDGMENTS

We are indebted to Walter Bojan, Jeffrey Lee, David Mingarelli, Edward Tasillo, and George Chisam for their support of the robotic applications figures; Kevin Opp and Bill Struble for their artwork support; Vince Scotto for editing; and Jennie Sladky for her diligent typing and criticism.

REFERENCES

1. Robotics Technology Advanced Manufacturing Engineering, Dept. 460-087, IBM San Jose, California, "Robotic Equipment Survey," May 1983.
2. J. P. Pawletko and H. D. Chai, "Linear Step Motors," *Proc. Second Symp. on Incremental Motion Control Systems and Devices*, University of Illinois, April 1973.

3. W. E. Hinds and B. Nocito, "Sawyer Linear Motor," *Proc. Second Symp. on Incremental Motion Control Systems and Devices*, University of Illinois, April 1973.
4. John R. Holland, "Factory Area Networks—The Key to Successful Factory Automation Strategies," *Flexible Manufacturing Systems*, pp. 35–61 (1984).
5. J. P. Pawletko and H. D. Chai, "Linear Stepping Motor with Uncoupled Phases," *Proc. Thirteenth Symp. on Incremental Motion Control Systems and Devices*, University of Illinois, May 1984.
6. D. W. Tausz, "The Application of Continuous Flow Manufacturing Techniques, 'The Department 066 Story,'" *Proceedings, IBM Design for Automation Symposium*, pp. 165–167; New Orleans, 1985.
7. J. Andresakis and R. Klein, "Robots Streamline Large Circuit Board Core Layup," *Manufacturing Technology Digest* 3, no. 1, 1985.

PROBLEMS

28.1. Select a robotic application that you are familiar with and provide a list of checkpoints to be considered in developing this application. Discuss each and tell why each is important.

28.2. The major areas of robotic applications are

> Materials handling
> Machine load/unload
> Welding
> Spray coating
> Process operation(s)
> Assembly
> Inspection

Give examples of each. Discuss application-specific design considerations for each type of application and explain why they are important.

28.3. Discuss what types of steps must be taken to insure that there is no product damage incurred during a robotic application. Give examples of preventive steps that can be taken.

28.4. Discuss the importance of robotic application workplace layout and what can be done to optimize it. What are some of the limitations that are involved in this optimization?

28.5. What various types of interlocks and sensors are to be considered during a robotic application installation and why? What type of malfunction can be avoided by the proper selection and application of interlocks and sensors?

28.6. An assembly designed for robotic assembly is also more easily assembled by a person. Is this statement true? Discuss and explain this concept.

28.7. Can robotic application technology be linked to group technology? And if so, how?

28.8. Name the types of robots common in electronic manufacturing. List typical applications.

28.9. What distinguishes robotics from automation?

28.10. Why is it necessary to have feedback systems in the robotics area?

CHAPTER 29

FINANCIAL ASPECTS OF DEVELOPMENT

ALAN J. BIALKOWSKI

INTRODUCTION

In the preceding chapters the focus has been on the technical aspects of electronic packaging. Some of the considerations in designing a high-performance product for the marketplace are discussed in Chapter 1, "Packaging Architecture." In this chapter the financial aspects of engineering will be discussed in more detail, as well as their influence on a product strategy for the design and development of an electronic package. Three major topics will be addressed: engineering in industry, the investment decision, and the management of development.

ENGINEERING IN INDUSTRY

In industry, there are three basic engineering disciplines: research, development, and manufacturing engineering.

The most fundamental scientific thoughts are associated with research both in their basic and applied forms. Basic research deals with advancing the frontiers of scientific thought and may not have immediate impact in engineering applications. The financial return on basic research is often uncertain at best, and carries a fair amount of risk that it may never leave a laboratory environment.

A viable approach is to carry out basic research in a university environment supported jointly by industry and government. This approach is beneficial for two reasons: first, it allows concentration of scientific talent in specific fields, thereby providing efficiency; and second, it is an economically attractive alternative to performing basic research in an industrial environment. The financial risk associated with this type of effort may be reduced by sharing the funding among several companies, the university, and government.

As basic research projects mature, economic choices need to be made to determine whether it is financially attractive to invest further in these projects, and to gauge the potential for developing products that can be marketed. At this point in-house capabilities of the industries are developed because a marketable product is possible. This is the world of applied research, where the feasibility of turning scientific thoughts into an application is studied and evaluated. This transition from basic research to application can be a long drawn-out process. A classic example is Einstein's theory of relativity, which was postulated in 1905 and remained in the realm of basic research until 1942, when Enrico Fermi split the atom at the University of Chicago, which eventually lead to the development of atomic weapons and nuclear power. Today we are witnessing the emergence of superconductivity from a laboratory phenomenon to practical applications in the fields of power transmission, electronics, and transportation. Once potential applications are identified, they belong to the second engineering discipline, development engineering.

Development engineering is responsible for translating scientific knowledge into products and processes that provide a revenue and profit stream to a corporation consistent with the firm's long-term financial or marketing goals. Development represents an investment much in the same way as a building or a piece of equipment. As an investment, it is subject to business procedure, since it must provide an attractive rate of return to the firm. The development process is characterized by prototype products that are often hand crafted. The development process ends when engineering has delivered a functioning product or process that can be scaled up for mass production in a manufacturing environment.

It is the manufacturing engineering, the third discipline, that assumes responsibility for the scale-up of the manufacturing process. This includes process flow, plant layout, manufacturing and quality standards, and the installation and start-up of the equipment. It is the activities of this discipline that remain with the product until the end of its life. In the start-up phase of the production cycle, both development and manufacturing engineering are intimately involved in making the transition from laboratory to plant floor. Once stability is achieved, the development engineer returns to the laboratory while the manufacturing engineer maintains and enhances the production process.

The focus of this chapter will be on development engineering. This is the area that is most closely linked with the business objectives of the firm, and the area where the major investment decisions are made.

THE INVESTMENT DECISION

Development is a business investment, the same as a building or a piece of equipment. The decision to invest in a project is subject to the strategic direction

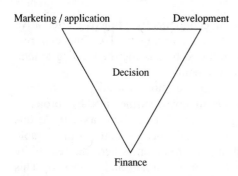

Marketing / application   Development

Decision

Finance

FIGURE 29-1
Interaction process involving marketing,
development, and finance.

of the firm and must support its business objectives or goals. These goals can range from market share to revenue growth to sustained profitability to return on assets or equity to market presence. Development investments are strategic in nature. They are designed to introduce products/processes that will sustain future growth; as a result, they are subject, in the short term, to the impact of a firm's current financial performance. In troubled times, there is a tendency to reduce development and long-term growth to satisfy short-term financial considerations; therefore, once the decisions are made to support specific projects, the quality and financial returns associated with the projects should be of such a magnitude as to insulate them from this type of tactic. Successful firms recognize the need to balance long-term growth with short-term financial considerations. As a result, great care is exercised in supporting development projects and even greater care is exercised in curtailing them.

The investment decision is an interactive process involving marketing, development, and finance (see Figure 29-1). These activities will be discussed separately in this section.

Marketing

The role of marketing in development is unique, in that it defines the direction of the development effort, based on its understanding of the marketplace, the customer's needs, and the competitive pressures in that marketplace. In vertically integrated organizations, application groups exist who provide an internal marketing interface to the operating units of the organization. These application groups coordinate the product or technical requirements of the operating units and work them into a basis for a development program. The functions of marketing and applications groups are identical.

Development projects are driven either by the marketplace ("a market in search of a product") or by technology ("a product in search of a market"). In the first instance, the marketing groups have identified a need for a product or solution to a specific problem. This identification serves as the basis for formulating a development project. Marketing and development now jointly analyze the alternatives available to them. These alternatives are analyzed from a technical as well as financial aspect. Tradeoffs are made between cost and function. Once the alternatives are defined, they are analyzed against competitive offerings, both today's offering and those yet to come.

This competitive analysis is driven by the marketing group. It is their responsibility to define who and what is the competition. The *what* are the products in the marketplace, the *who* are the firms that produce those products. It is essential to understand the nature of the competition, because it is necessary to project their future capabilities. A key factor in this type of analysis is the ability to anticipate the direction in which competition and technology will move in the future, so that the requirements placed on development will satisfy a future need in the marketplace. Throughout this analysis, marketing and development work together in correlating marketplace demands with the future direction of technology.

Once the requirements are defined, it is necessary to establish a target price for the product. This requires marketing to forecast future price developments in the marketplace and to anticipate the reaction of the marketplace to the introduction of the product. Once price targets are established, preliminary estimates are made as to whether development can produce a product that satisfies the targeted price. This is generally an iterative process that involves tradeoffs in cost, performance, and design. Once the preliminary design is established and initial price and cost issues have been resolved, marketing provides the sales forecast, which serves as the basis of the financial analysis that leads to the decision to support the program. In summary, where development projects are designed to address a specific need, the role of marketing or applications engineering is the definition of the environment for the development engineer.

A second route in a new product introduction begins with development postulating a product without first establishing a need in the marketplace. These development-driven projects generally result from scientific breakthroughs in a laboratory environment. The most successful example of this is the home computer, which created an entire market where previously there was none. The work on superconductivity falls into this category, as industrial applications of this technology are only now being defined. In these instances, the marketing force is seen in its traditional role of sales. Their focus is on stimulating a demand for these products. Once these products are established, the follow-on development projects would be treated as marketing requirements.

The second route of development involves a high degree of risk, since it deals with a laboratory phenomenon in search of a market. While industry will support some level of this activity, it clearly prefers to operate in response to market forces and developments. The risk in the first approach is more manageable than depending on technological breakthroughs to stimulate or create a market.

In summary, marketing and applications organizations play a key role in providing a direction to the development effort and defining the boundaries of various projects in terms of performance and price.

Development

The role of development in the investment process is one of concept and design. Design in industry is constrained by financial and marketing considerations. A product design must satisfy marketplace wants, at a price determined by the marketplace or by a price set by competition. As a result, product design may be a compromise between an optimum design and affordable design. Cost targets and prices are provided to the designer by finance. These targets are based on the competitive analysis provided by marketing and their understanding of the marketplace and competition. The designer is then taxed to produce a product that will satisfy these requirements.

Figure 29-2 represents a competitive analysis chart for a proposed development project, Product A. The prices include the cost of manufacture, general and administrative costs, and profit, and are designed to recover the development investment. Competitive analysis provides a price band in which the project must perform. Ideally, the product price should be close to the lower limit, which is the "best of breed." However, there are reasons why a product might be sold at somewhat higher prices. For example, the product may provide added function, or in some instances a firm's reputation for service and reliability may allow it to compete at a higher price. The dotted line in the figure reflects the incremental cost of that product. In this case it reflects the cost to manufacture the product and the unique engineering effort to introduce and support it in a manufacturing

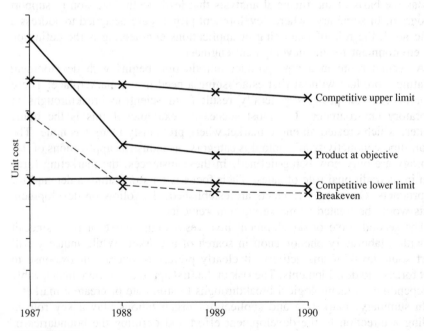

FIGURE 29-2
Competitive analysis, Product A.

process. The financial strategy in the case of Product A would be to announce the product at an average price whereby the firm would absorb a loss in the early years, at the low end of the learning curve associated with a product introduction, and to defer profits until later in the product cycle.

Figure 29-3 represents a project that would require redirection in the development effort, Product B. It is clear that the firm would have problems competing in this market unless there is something so unique about this product that competitive price is not a factor. Instances exist in technology where performance is the overriding concern as opposed to price; however, these instances are few and far between. In addition, performance favorability is a fleeting factor because over time competition will match or exceed this advantage.

When dealing with Product B, the development engineer is at a disadvantage. His training has been toward optimizing solutions without consideration for financial consequences. Now he is faced with constraints of product costs and development funds. As a result, the development process enters a cycle of tradeoffs, which involve not only finance but also marketing. In order to achieve cost reductions, it may be necessary to modify performance or function. Redesign becomes a reiterative effort to find a balance between final design, cost, and marketing requirements.

Throughout this process, the development engineer continues to identify his resource requirements in terms of man-years of effort, capital, and spending.

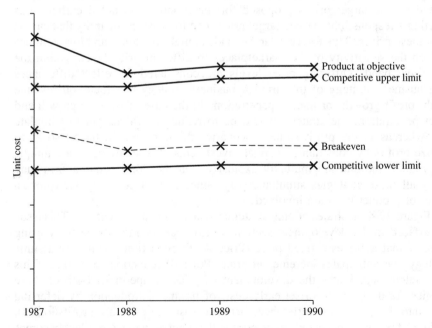

FIGURE 29-3
Competitive analysis, Product B.

Each of these elements is factored into the business case and used to evaluate the investment decision. There is a need to establish a close working relationship with the manufacturing arm of the organization in order to insure that it can support the design in a scale-up and achieve the targeted costs. This involvement should occur relatively early in the product cycle, once the final design has been completed. Manufacturing strategy is discussed in Chapter 31.

Finance

The role of finance in the development cycle is to provide support to marketing and development decisions. The decision to proceed with a project is a line responsibility, while finance is a staff organization. Finance may assess the project and highlight its concerns and recommendations, but line management has the final say in making the decision. This decision can be influenced by factors other than the financial aspects of the proposal, e.g., market presence, a response to a competitive product, or the need to advance technology.

Finance is responsible for developing the investment case and analyzing all the financial aspects of the proposed project. The key concerns are whether the project will be profitable, the timing of that profitability, and the return on the investment.

Throughout the development process, finance assesses the tradeoffs between development and marketing until a final design has been accepted. This analysis is based on the target prices proposed for the product. As noted earlier, it is marketing's responsibility to peg target prices and forecast the quantity that can be sold at these prices. This forecast can be made at high, medium, and low volumes based on the sensitivity of the marketplaces to different price levels within the competitive band. Differing price levels are considered to reflect differences in the business strategy of the firm. A business strategy can focus on revenue growth, profit growth, or market penetration. In the case of revenue growth and market penetration, the strategy would be to reduce profit margin to stimulate sales, whereas in the profit strategy scenario, the price would be set so as to optimize profits by constraining capital investments. These strategies are unique to a product line and are mutually exclusive. On the other hand, a firm may employ all these strategies simultaneously, since from a company viewpoint a number of product lines are involved.

Figure 29-4 illustrates pricing strategies that a company might use. This issue first surfaced in the development section. A firm can average its costs by using a strategy that assumes a fixed price (Price A) through time, or it may assume a strategy that anticipates incremental price (Price B) reductions over time. This latter strategy anticipates the development of price competition. Both of these strategies are designed to offset early costs of product introduction by deferring profits into the future (the differences between cost and price are profits/losses); therefore, it becomes necessary to determine the point at which a product becomes profitable.

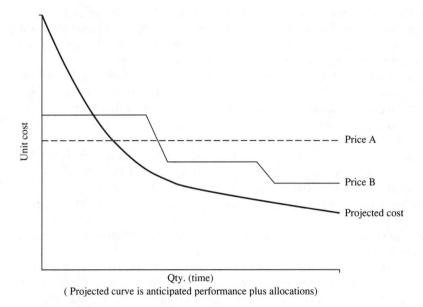

Qty. (time)
(Projected curve is anticipated performance plus allocations)

FIGURE 29-4
Program analysis. Projected curve is anticipated performance plus allocations.

Figure 29-5 is a crossover analysis. The amount of investment is plotted as a cumulative number and includes both development and manufacturing, capital, and inventory. This is offset by the projected profits. In the early phases of a program, the bulk of activity is investment; initial losses accompany a production

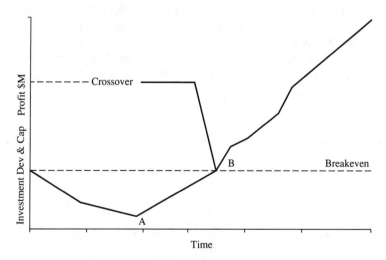

Time

FIGURE 29-5
Investment analysis. The difference between cost and price are profits/losses.

start-up. The period revenues become profitable at point A in the figure. The purpose of this analysis is to determine the magnitude of the investment, and the time it takes to recover the investment (breakeven—point B). The greater the delay to profitability, the less attractive the investment.

The revenue calculation is based on marketing's target prices and the volume forecasts at those prices. Manufacturing costs are provided by finance and reflect the capital investment required to support the product. Development costs are provided by the engineering organizations. In normal business practice, administrative and marketing costs are allocated to a product by finance based on a preset algorithm. The key issue is the margin and whether that margin is acceptable from a financial or a business standpoint. If the project fails these hurdles, it is resubmitted for additional evaluation by marketing and development. The process is reiterative until these criteria are satisfied, with tradeoffs in design-cost-price-profit.

The algorithms used in the analysis of a product are relatively simple and are outlined in Table 29.1.

Once a project satisfies profitability criteria, the issue is one of affordability. A basic postulate in development is that there are always more development projects than there are funds. Although the project has passed its first series of hurdles, it has to be reprioritized relative to other projects. This occurs in the budget process, which is discussed later in the chapter.

In conclusion, the decision to invest in a development project is driven by the interaction of marketing, development, and finance. It is marketing's primary responsibility to identify a market requirement for a product and communicate that requirement to development, whether it is something as ordinary as a washing machine or as complex as a computer. Once the technical and pricing requirements are identified, the development process becomes reiterative in nature, as tradeoffs are made between marketing design and cost. Once the design issues are resolved and the financial analysis is complete, line management is in a position to make the go/no go decision based on the strategic direction of the firm. The strategic direction can encompass competitive requirements, market presence, and financial goals. Business strategies are also subject to the same type of tradeoffs as designs.

TABLE 29.1

Forecast quantities	AA	
Target price	BB	
Revenue		YY
Less:		
Manufacturing cost		(ZZ)
Development cost		(ZZ)
Administrative and marketing costs		(ZZ)
Profit before taxes		XX
Margin		%

Program Management

In the preceding pages, the decision process for approving a development project has been described. In this section, the management of that project will be discussed.

Throughout a product cycle there are specific checkpoints for formally reviewing its progress and determining whether it is meeting the goals that were established at its inception. Many terms are used for these reviews: for example, *phase reviews*, *progress reviews*, or *checkpoints*. The terminology may differ from company to company but the purpose is the same: to evaluate progress and confirm the commitment of funds to the project. (In the succeeding discussion the term *phase reviews* will be used to describe the process.)

The phase review process is diagrammed in Table 29.2. The process provides a road map whereby a project is measured by its ability to meet key technical and financial criteria.

The discussion of the investment decision corresponds to a Phase 0 review. The general outlines of the project have been established and its financial viability has been determined. Once the decision is made to proceed, the project moves to Phase I, wherein resources are committed in terms of capital, manpower, and funding. This is also the point when formal commitments are made to specifications and delivery schedules.

TABLE 29.2

Phase review process	Characteristics
Phase 0 Program proposal	Initial requirements Tradeoffs in specifications and forecast
Phase I Program definition	Concept Investment case Decision to commit funds Establishment of technical and financial checkpoints Establishment of commitments and dependencies
Phase II Design	Design Assessment of performance versus program requirements Prototypes
Phase III Development	Final designs Shippable product Go/no go decision point Rigorous technical and financial criteria Transition from laboratory environment to manufacturing environment
Phase IV/V	Customer shipments Full manufacturing capability Resolution of outstanding technical problems Acceptance by the marketplace
End of life	Withdrawal from the marketplace

One of the most overlooked factors in development is the issue of dependencies. A dependency occurs when one engineering group is dependent on another group to deliver a product or a technology that is an integral part of the project's design. Often these dependencies are assumed to be in place, but then, as the project evolves, they are found to be not in place, resulting in schedule slippages and added expenditures. The Phase I process insures that these dependencies are highlighted and other groups are committed to supporting the project.

At Phase II, designs have stabilized and working models become available. The project can be physically tested and evaluated against the criteria established at Phase I, both from a technical as well as a financial standpoint. If the project meets the criteria, the decision to invest is clearcut. However, if there is shortfall in performance, the decision maker has the following options: terminate the project, rescope the requirement, or redesign the project. Each of these options incurs serious financial consequences.

The termination of a project is the most serious of the three since it implies the loss of a business opportunity. The write-off of the investment, while regrettable, is not catastrophic, for the money has already been spent out of current profits. On the other hand, it would be an extremely serious mistake to pursue and fund a development project that will not meet its objectives, on the basis that an investment has already been made and the incremental investment would be small.

Rescoping a project means accepting it although it fails to satisfy its initial objectives. This occurs when original objectives are overly optimistic; when technology is not available on the scheduled dates or at the scheduled cost; when market requirements change, e.g., more function for the same cost; or when competitive prices are lower than anticipated. The decision to proceed will be based on business considerations.

Redesign is what it implies, a return to the original starting point by writing off the current investment and starting from scratch.

Once a product exits Phase II, the potential for it being released to manufacturing is relatively good. Therefore, the Phase II checkpoint is critical.

In Phase III, development involvement is at its peak. The focus here is delivering a product or project that can be scaled up for mass production. Designs are stable and manufacturing investment is in full swing. The Phase III checkpoint is the final step before a product is revealed to the public. This step consists of a stringent evaluation of the product's technical and financial characteristics. A subtle question at this juncture is whether this is a product that represents the company and its goals. If that question is resolved favorably, the product is announced and price schedules are published.

The phases that follow the announcement are designed to measure the acceptance of the product in the marketplace, the ability of manufacturing to achieve its cost targets, and the ability of development to resolve any outstanding technical issues that remain after the announcement. These subsequent phase reviews also offer the opportunity for formal price reviews to determine whether prices should be adjusted.

In summary, the phase review process provides a road map of technical and financial checkpoints that a project must clear if funding is to continue. It is this vigorous review that insures a firm will have the products when they are needed.

MANAGING DEVELOPMENT

When the needs of development projects exceed available funds, management is faced with the decision of allocating funds among a number of competing projects. Major firms allocate a predetermined percentage of their revenues to support research and development. In this manner, they are trading off current financial results in order to sustain future growth, therefore, competing projects must be prioritized against one another. If the projects were always mutually exclusive, the resolution would be rather simple, but projects may address different aspects of a firm's operations, such that all could be supported if funding were available.

The selection of projects is a judgmental decision based on any or all of the following factors and the weight assigned to each of them by the decision maker: rate of return on investment, marketplace requirements, and new markets.

Rate of return on the investment. The financial analysis described earlier places competing and nonrelated projects on a common base, considering revenue, expense, profitability, and rate of return. If this were the only basis of comparison, the solution would be trivial. Projects would prioritize on the basis of the return, with those projects that have the best returns on the top of the list. The decision maker would then proceed down the list until the available funds were exhausted. However, other factors are involved.

Marketplace requirements. This is generally the key factor. A corporation is expected to produce a range of products in its marketplace that will satisfy customer needs in terms of performance and price. Failure to do so would result in abandoning that market segment to competition. The marketplace requirements approach is a viable and acceptable strategy, and as such it would be the basis of a funding decision, with return on investment becoming a secondary consideration.

New markets. This is a corollary to the above. In this approach, the firm determines whether they should choose to compete in a market segment outside of their traditional marketplace, e.g., an electronics group specializing in consumer products entering the arena of computers. This decision is market-oriented as opposed to financial, and can also serve as the basis for allocating funds.

The factors just described, either individually or in concert, form the basis of making business judgments. It is the decision maker, that is, management, that assigns the appropriate weighting to each of these factors and provides the guidelines for prioritization. In essence, what we have been discussing is the budgetary process of a company.

The budget process is a formalized review and repriorization of development projects. It is an annual exercise in most companies, while in others it is performed annually with a midyear review and reprioritization, if required. The process supplements the phase review process by assessing progress at key checkpoints. This review is far less rigorous than phase review; it can be categorized as an executive overreview of development programs.

Budget review also serves to reconcile the development effort to revenue and profit plans, either by validating them or highlighting potential problems in sufficient time so that corrective action can be taken. This corrective action can run the gamut from extending the life of existing products through enhancements or price actions to terminating development programs that are causing the problem or increasing their investment streams to insure that they meet the revenue plan.

The budget is the final arbiter in supporting developing programs, since it forces choices based on the goals of the corporation. It is the document that literally makes things happen because it commits a firm to a stated course of action. It also forces development management to view their operations within the context of the firm's overall goals, whether they are technical, financial, or market-oriented. The process is extremely effective for building business acumen in the technical community by forcing choices to be made on business-oriented issues rather than on technical issues.

There are three types of budgets commonly found in industry:

Operating budgets, which are the most familiar. These budgets authorize the expenditure of funds by project, and within project by category, e.g., manpower or supplies.

Capital budgets, which identify the equipment and facilities that must be purchased to support a development program. Often the capital required for a development project equals or exceeds the development investment itself since this budget not only identifies the immediate need for capital but also forecasts the capital requirements for when a development project results in a production run. When analyzing the feasibility and costs associated with a development strategy, it is essential to identify the total investment, including the capital required to start up a production run.

Cash budgets are primarily financial documents. They consolidate items from operating and capital budgets that require outlays of cash, and the timing of those outlays. This analysis is then offset by the revenue projections to determine if sufficient funds are available to support all the programs required and still allow the company to meet its profit objectives and pay its dividends. If current plans and projections support the investment stream, the decision is one of implementation. On the other hand, if there are shortfalls, corporate management has the options to borrow funds to support these programs or scale back and reprioritize them.

The budgets interact with each other throughout the year. They are updated to reflect changes as new programs are added, while others are dropped because they no longer satisfy the strategic goals of the corporation or because business conditions limit the funds available. A budget is a living document in this respect.

The budget not only serves as a planning or authorization document but also as a control vehicle. Projects are authorized based on outlays of funds and the expected results. It is the budget that will quickly identify and highlight the potential for overruns, which in turn could expose key schedules. Once overruns are detected in a project, it is reanalyzed to determine if the original commitment can be met and whether the project still satisfies the corporation's strategic direction. The options again are to continue, to terminate, or to scale back.

It is necessary to keep one point in focus when using budgets as control mechanisms. Budgets are spread on a monthly basis and many expenses may be sporadic in nature; as a result, short-term overruns may be just that. Both the finance and the engineering community must remain in close communication so that current financial performance is not only understood, but also can be accurately projected. These projections are critical to an effective control mechanism.

In summary, the budget is the authorization document to proceed with a development program. Its primary use is to prioritize unrelated development programs within the overall objectives of the firm. It deals with the fact that funds are finite and only the most efficient investments can be supported. The phase estimate process, when used in conjunction with the budget process, ensures the efficient expenditure of funds. The budget is a living document; it must reflect the realities of business and is heavily influenced by financial results. When business is good, budgets are expansionary, but as performance becomes lackluster, the pressure to improve profits at the expense of development increases. Whether times are good or not, decisions must be based on the efficient deployment of funds and capital to meet specific business goals; otherwise, many projects will be quickly terminated in a downturn.

SUMMARY

There are two factors that must be kept in mind when proposing a development project or strategy. First, development is an investment in the future, and its outcome must be consistent with goals of the firm; second, it is subject to the short-term constraints of funding.

The investment decision to proceed is based on the ability to identify a market need and specify a competitive solution to solve that need. This is the role of a marketing or applications group. These groups are responsible for postulating the program to the development engineer. The development engineer's responsibility is to produce a design with current or anticipated technology that will provide a solution to the marketing problem. Both organizations then interact with finance to determine whether this solution is affordable and whether it can be introduced into the marketplace at the targeted price and still satisfy the corporation's business goals. This investment decision is an iterative process involving the interplay of specifications, performance, and cost. Once a competitive solution is identified and funds are approved, progress is managed through a phase review process.

The phase review process remains basically unchanged when a project is based on a technological breakthrough, i.e., when development defines the product or application. In this instance, marketing establishes the potential marketplace, but the financial analysis and interplay is just as rigorous, if not more so, since the risk to proceed is far greater. In this instance, it is necessary to develop a new market.

The phase review process provides the technical and financial road map for a program, identifying key checkpoints and specifying the technical and financial goals that must be satisfied before the program can continue. The process culminates at a Phase III review where the final go/no go decision is made. It is at this point that the program is announced to the public.

The first control element in the development arena is the budget process. This is an annual or semiannual review of all development programs that are in process, as well as those that are being prepared. The budget process is an attempt to prioritize unrelated projects against the available funds. In essence it rations available funds based on the efficiency of investments. The more substantial the return on an investment, the more likely it is to be supported even in an economic downturn. Although this process has a formal aspect, it is an area that is continually under review throughout the year as new projects are added and marginal ones deleted.

This is the development process as seen within the context of a business environment. It is probably the most sensitive area of the business, because decisions made here effect the long-term viability of the corporation in markets that have become highly competitive in the last decade.

REFERENCES

Bogen, J. I., editor, *Financial Handbook*, 4th ed., New York: Ronald, 1968.
Helfert, E. A., *Techniques of Financial Analysis*, 5th ed., Homewood, Ill.: Dow Jones–Irwin, 1982.
Maynard, H. B., editor, *Handbook of Business Administration*, New York: McGraw-Hill, 1967.
Vancil, R. F., editor, *Financial Executives Handbook*, Homewood, Ill.: Dow Jones–Irwin, 1975.

CHAPTER 30

THE EVOLUTION
OF LOGIC
CIRCUIT
PACKAGING
TECHNOLOGY

GEORGE G. WERBIZKY
FRANK W. HAINING

INTRODUCTION

The evolution of logic circuit packaging goes back to the last century, when Dr. Herman Hollerith, a distinguished statistician employed by the United States government in the compilation of the tenth census, was wrestling with what seemed an insurmountable problem [1]. From personal experience Dr. Hollerith was fully aware of the enormous difficulty the Census Bureau was having in reducing the mountains of facts gathered by the census takers into usable information. The facts had been collected, as the law provided, in 1880, but five years later the bureau was still struggling to compile them. Considering the rate at which the country was growing, it was not hard to foresee a time when, with existing methods, it would be time to take the next census before the last one could be published. So Dr. Hollerith set to work to find a way by which all this recording, tabulating, and analyzing of facts could be done by machinery.

The 10th census report was finally completed in 1887, with the 11th census only three years away. By that time Dr. Hollerith had worked out the essential features of his mechanical system for recording, compiling, and tabulating census facts, a system which was to prove useful in far wider applications than those of the census, and to introduce unprecedented speed, efficiency, and precision into the organization of large masses of data.

The demand for more powerful data-processing technology has been increasing steadily since Dr. Hollerith's time because of the need for greater productivity in every field of human activity. The computational cost required to meet these needs has steadily decreased, as shown in Table 30.1. In general, the increased performance and reliability of the machines can be attributed to the following:

1. The switching speed inherent in the device. The electromechanical devices start in the millisecond (thousandths of a second) range, the vacuum tubes extended the technology to the microsecond (millionths of a second) area, and the solid state devices are capable of being switched in nanoseconds (billionths of a second).
2. The power required for switching the devices. As the power is reduced, the size and complexity of the power supplies required can be reduced, as well as the cooling systems needed to get rid of the power dissipated. High power and high potentials are a major contributor to the failure rates of the machines.
3. The size of the devices. As the size of the devices can be made smaller, more and more devices can be placed in the same unit of volume, not only reducing the cost, but also improving performance, since the time required to transmit a signal from one device to another can be reduced.

The greatest strides in circuit packaging have been driven by the level of integration in the solid state devices. As more functional circuits can be placed on one chip, the partitioning of function within the machine can be relegated to lower levels of packaging, resulting in more computational power in one module than was once available in an entire machine. For example, there are more functional circuits in one of the thermal conduction modules of the 3081 processor than there were in the System 370 Model 145 computer.

TABLE 30.1
Cost of processing

Year	Cost*	Processing time*
1955	$14.54	375 seconds
1965	0.47	29 seconds
1975	0.20	4 seconds
1985	0.07	1 second

* This chart shows the cost and time for large IBM computers to complete a fixed amount of data processing. The figures in each column are based on an identical mix of 1700 typical data processing operations, involving millions of computer instructions. Included is a cross section of payroll, discount computation, file maintenance, table lookup, and report preparation. Figures show costs of the period, not adjusted for inflation.

RELAYS

Slate Base Relay

The slate base relay was the logic-switching device used in the 034 card sorter and the 080 card sorter [see Figure 30-1(a)]. This relay required a potential of 40 V of direct current, and a power of 400 mw to be activated. The switching speed

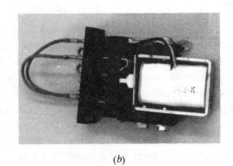

(a)   (b)

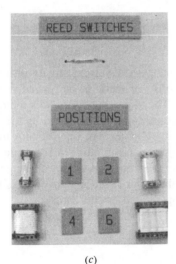

(c)

FIGURE 30-1

(a) Slate base relay. The 2-by-3-by-$\frac{1}{4}$-in. base of the relay was made of slate, a naturally occurring ceramic material. The metal parts were of brass, except for the copper wire used in the electromagnet. The wire was insulated with varnish. A cotton string was used to adjust the stroke of the armature. (b) Wire contact relay. The 1-by-2-by-$\frac{3}{4}$-inch relay was made of manmade polymeric materials, although copper was used for the wires in the electromagnet, and brass for the fittings and connectors. (c) Pluggable Reed relays. Above is the $\frac{3}{4}$-inch glass tube that encloses the two metal wires forming the switch. In the first row below, on the left, is a single relay, which consists of the glass tube, wound with a coil of insulated wire that forms the electromagnet, and the two pluggable connectors on the ends. On the right is a double relay. In the bottom row are a four-relay package (left) and a six-relay package (right).

of this device was 600 ms (0.6 second). The relays were mounted to a frame with screws, and wires were run by hand between them.

The 601 calculating punch [1] used these slate base relays as logic-switching circuits. It was first shipped to customers in 1930. It had a machine cycle time of 3.16 seconds, although it could read and punch 25 cards per minute (1 card in 2.4 seconds).

Wire Contact Relay

The 407 accounting machine, first shipped in 1949, used wire contact relays in the logic-switching circuits [Figure 30-1(b)]. These relays were much more advanced than the slate base relays, as evidenced by a much faster switching speed (0.01 s). Although they were not as fast as vacuum tubes, they were more reliable. The wire contact relay drew only half the power of the slate base relay (200 mw).

The 407 accounting machine also had a feature called a plug board that allowed the operator to program which columns of the card would be read, and the arithmetic functions to be performed. The output of the machine was a printed document, which appeared at 150 lines per minute (0.4 s per line) [2]. Data was stored in print wheels. The machine thus had almost all the features needed for a rudimentary computer.

Reed Relay

The electromechanical switching device was developed to its ultimate point in the Reed relay [Figure 30-1(c)]. The mass of metal to be moved was reduced to a thin wire sealed inside a glass tube. The switching speed was reduced to 1 ms, and the power was only 50 mw, one-fourth of that consumed by the wire contact relay. The Reed relay was much more reliable than the vacuum tube (up to 2 billion operations), and could be plugged (and unplugged for replacement) into printed circuit cards. The Reed relays could be used alone, or ganged into two-, four-, or six-device packages. Reed relays were used in data processing machines well into the 1950s, being mounted on printed circuit cards.

VACUUM TUBES

Vacuum Tube Assembly

In 1947, the 604 electronic calculator [3] made use of vacuum tubes for the logic-switching circuits [see Figure 30-2(a)]. Sixteen vacuum tubes were wired into one unit. The vacuum tubes required 10 mw of power at a 50 V potential to switch. They could switch in 10 μs, much faster than any electromechanical device.

The 604 electronic calculator used 800 vacuum tubes and had a machine cycle time of 100 cards per minute (0.6 s per card), much faster than the IBM 601 calculating punch. The 604 also had a memory feature: the numbers were stored in trigger circuits.

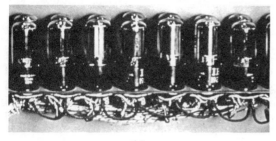

(a)

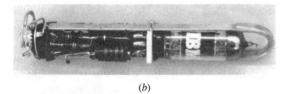

(b)

FIGURE 30-2
(a) Vacuum tube assembly. The vacuum tubes, approximately 3 in. tall and 1 in. in diameter, were plugged into tube sockets mounted on a strip of chassis material. The insulated wire "spaghetti" was connected to the pins on the tube sockets, connecting the whole assembly. (b) Single-tube pluggable unit. The 6-in.-long unit contained a vacuum tube, resistors, capacitors, and a radio tube plug at the bottom. This unit could easily be replaced in the machine, as the components failed.

Single-Tube Pluggable Unit

Vacuum tubes were used in the 650 data processing machine, a true computer. Each vacuum tube was incorporated into a single-tube pluggable unit for easy replacement [Figure 30-2(b)]. This unit became a field-replaceable unit (FRU), which the customer engineer could replace easily in the customer's office. The base of the tube included the passive components (resistors, capacitors, etc.) that made up the functional circuit.

The 650 data processing machine, first shipped to customers in 1954, contained all the elements of a computer, and had a magnetic drum memory, which was capable of random access of magnetically coded information. The machine cycle time (one clock pulse) was 16 μs.

The 650 also used the first solid state (semiconductor) components in an IBM computer (see Figure 30-3). The germanium diodes were plugged into the FRU.

Eight-Tube Pluggable Unit

The FRU concept was extended to an eight-tube pluggable unit used in the 704 data processing machine, a large-scale electronic computer. Thus all the functional circuits associated with eight vacuum tubes could be replaced at one time.

FIGURE 30-3
Diode pluggable unit. This 6-in.-long pluggable unit contained germanium diodes instead of vacuum tubes. The ends of the diodes were pushed into a bifurcated connector.

SOLID STATE DEVICES

Standard Modular Systems (SMS)

The introduction of the 1401 computer in 1959 required a circuit packaging technology that used solid state (transistor) switching devices soldered to printed circuit cards, which were plugged into (wire-wrapped) back panels. The solid state devices avoided all the reliability problems of the vacuum tubes, and had all the advantages of fast switching speed (0.7 μs), low potential (12 V), and low power (10 mw).

The solid state switching devices (Figure 30-4) were hermetically sealed to prevent problems caused by the environment. Several of these transistors, along with other components, were clinched by their leads to holes in the card and then soldered on the back surface of the card (see Figure 30-5).

These SMS cards were plugged into a back panel (Figure 30-6), and a cooling fan blew air over them to remove the power dissipated as heat. The connections between SMS cards were made by wrapping wires (using a machine) on pins in the back panel that were connected to the card edge connectors.

The 1401 computer set the standard for many modern computing systems. It had a machine cycle time of 11 μs, a memory utilizing ferrite cores, and a complete set of peripheral equipment (card reader, card punch, printer, sorter, magnetic tape drives, disc drives, etc.).

Solid Logic Technology (SLT)

The introduction of the System 360 computers in 1964 extended the printed circuit wiring concept to the boards into which the cards were plugged. The solid state switching devices were attached to pinned ceramic substrates, which were soldered into plated through-holes in multilayer cards. The switching speeds of these devices were as high as 30 ns (0.000,000,030 seconds).

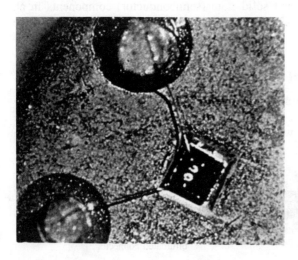

FIGURE 30-4
Transistor. The $\frac{1}{4}$-inch-square transistor on the lower right was made of silicon and had two leads, which functioned as the input and output of the component.

FIGURE 30-5
Standard modular systems (SMS) card.
On this 3-by-5-in. card, made of paper
impregnated with polymer, are various
electronic components: transistors in cans,
Reed relays, resistors, capacitors, etc.,
whose leads are pushed through holes in
the card. On the back side of the card is
the printed circuitry, coated with solder to
form the connections, and at the bottom
of the card are the edge connectors.

The solid logic technology pioneered a new logic circuit packaging architecture (see Figure 30-7).

The integrated circuit chips diced from the silicon wafers were bonded to a
ceramic substrate with the circuit side down [4]. This evolved into a technology
where evaporated solder pads on the chips were reflowed on the substrate to form
a controlled collapse chip connection (C4) joint [5].

These "flip chip" modules had screened circuits on one surface to connect
the active devices (integrated circuit chips) and passive devices (trimmed resistors)
to pins attached to the other side of the substrate.

The sealed module was soldered to a multilayer printed circuit card by
inserting the pins of the module into plated through-holes in the card, and passing
the assembly over a solder wave.

FIGURE 30-6
Standard modular systems (SMS)
back panel with cards. This 18-
by-22 inch back panel can house
up to 126 SMS cards, which are
plugged into a connector system on
the front side. At the bottom are
signal cables, and to the left is a
fan, which blows cooling air over
the cards.

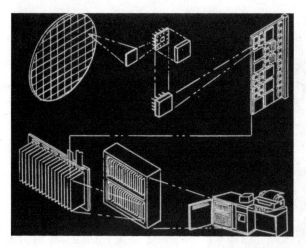

FIGURE 30-7
Solid logic technology (SLT) packaging architecture. Starting from the top left: a silicon wafer, an integrated circuit chip diced from the wafer, the chip mounted on a substrate, and the capped substrate mounted on a card. From the lower left: several cards plugged into a board, two boards attached to a gate, and the gate in a machine.

The printed circuit card, as a field-replaceable unit (FRU), was plugged into a printed circuit board by compressing connector springs on the card to reformed pins soldered into plated through-holes in the multilayer printed circuit board.

Several printed circuit boards with their cards, along with signal cables to communicate between boards and power cables to bring power to the boards, were attached to a frame (gate) in the machine.

The SLT chip was protected from the environment with a glass layer on top of the evaporated (and etched) metallization. One or more of these chips were then joined to the half-inch-square substrate, on which were screened (and fired) circuit lines and resistors. The substrate was then placed inside a protective metal cap, and the back side sealed. These components, and others packaged in different ways, were soldered into various sizes of cards [see Figure 30-8(a)].

The System 360 Model 30 computer had a machine cycle time of 750 ns and an ever-growing list of peripheral equipment.

Advanced Solid Logic Technology (ASLT)

The System 360 Model 90 computer utilized several improvements in the basic SLT to increase the switching speed to 10 ns. The number of circuits on one integrated circuit chip was increased, resulting in more circuits per module, card, and board.

Monolithic System Technology (MST)

Further improvements in SLT resulted in a new technology designation, and a new family of computer systems—the System 370 [6]. The switching devices on the chip increased in number and speed. All the active and passive devices

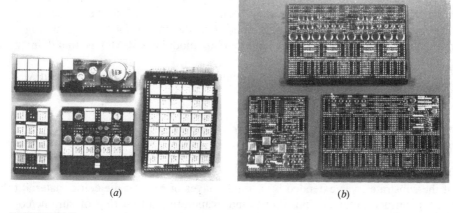

(a)   (b)

FIGURE 30-8
(a) Solid logic technology cards. Top left: a small SLT card containing six SLT modules, with an SLT connector system on the bottom. To its right is a larger SLT card with various components: an SLT module, two transistors with heat sinks, a power transistor, and several resistors and capacitors. On the right is a larger card with 36 SLT modules, several ceramic resistors and capacitors, a top card connector, and an SLT connector on the bottom. (b) Vendor transistor logic cards. These VTL cards look very much like the SLT cards, with one important difference. Plastic dual in-line packages (PDIPs) are mounted on these cards, along with other components purchased from vendors.

required to form a functional circuit were fabricated on the silicon (monolithic). The System 370 Model 145 computer, when delivered in 1971, had a machine cycle time of 200 ns, and its memory was also on integrated circuit chips.

Vendor Transistor Logic (VTL) Technology

In the late 60s and early 70s, the dual in-line package (DIP) was developed by several manufacturers to house integrated circuit chips. Printed circuit cards were developed [Figure 30-8(b)] to use these purchased circuits. Several other approaches to integrated circuit packaging are discussed in the chapters on ceramics in electronic packaging and joining materials and processes.

Metallized Ceramic (MC) Technology

In the early 1970s, a technology for the vapor deposition of metallization on ceramic substrates (thin film) was developed. The metals (chromium and copper) were etched to form the circuits after a suitable photoresist is applied. This new process was an alternative to the screened and fired paste (thick film) process developed for SLT modules [7].

The size of the ceramic substrate was increased to one inch square, to fan out the early large-scale integration (ELSI) chip I/Os to the substrate I/Os (pins). The larger module also had more pins (72) than the half-inch module, and used an improved polymeric backseal material.

With the improvement in logic circuit packaging came an improvement in memory circuit packaging. The half-inch substrates were stacked, one on top of another, and six memory chips bonded to the top surface. A discussion of multichip modules (MCMs) and single-chip modules (SCMs) is found in the chapter on packaging architecture.

The System 370 Model 135 computer used both MC (thin film) and MST (thick film) modules.

Metallized Ceramic Polyimide (MCP) Technology

The wireability of the multichip modules, which was restricted to the two surfaces of the substrate, was extended by adding a layer of organic dielectric material to the top surface (after circuitization), and evaporating a thin film of chrome/copper/chrome on top of the dielectric. The organic dielectric material chosen was polyimide, because it is capable of being etched to provide vias to the buried signal layer, and it can withstand the temperatures required for chip joining. A metallized ground plane buried in the ceramic substrate enhanced the electrical properties of the buried signal layer. These modules were called metallized ceramic polyimide (MCP) modules.

The 3090 processor used 28-mm MCP modules as logic supports on the 16-megabyte memory cards in central storage.

Early Large-Scale Integration (ELSI) Technology

As the processes for making integrated circuit chips have improved, more and more circuits have been able to fit on one chip. Arbitrary definitions have been established to define this level of integration: ELSI, LSI, VLSI, etc. The ELSI (early large-scale integration) chips contained 100 to 500 integrated circuits per chip, and these logic gates could be switched in 6 ns. Another new feature of the ELSI technology was the pluggable module. When plugged into a card, the module became an FRU [8].

The System 370 Model 148 computer used both ELSI and MST technology. The machine cycle time was down to 180 ns. The memory used field effect transistor (FET) chips.

Multilayer Ceramic (MLC) Technology

The ever-increasing performance and circuit/bit densities of the integrated circuit chips required an increasing amount of power and number of interconnections between chips at the substrate level. To fulfill this need a multilayer ceramic, multichip module technology was developed [9].

The substrates for these modules had up to 23 layers and were approximately 4 mm thick and either 35 or 50 mm square. Each layer of the module began as a part of a continuous cast sheet of ceramic material, which was cut into pieces

175 mm (about 7 in.) square, then punched with holes at high speed so that electrical connections could later be made between layers.

Conductive paste was then extruded onto the green sheets through metal masks, forming a wiring pattern unique to a given layer. Stacks of these sheets, with the required configurations of conducting line and insulating layers, layer-to-layer connections, and reference and power planes, were laminated together and trimmed to form individual modules, which were then fired in a furnace to harden the ceramic. Finally, the upper surface of the module was plated with LSI chip sites.

After electrical testing of the ceramic modules and the attachment of the input/output pins, a number of chips, at most nine but more typically six, were joined to their plated sites using C4 joints. Finally, to protect the chips, a cap filled with an inert gas was brazed atop the module.

Up to 10 m (nearly 33 ft) of wiring was concealed within these 4-mm-thick multilayer modules. Two module sizes, 35 and 50 mm square, were used to meet differing input/output needs. The larger had 361 I/O pins, and the smaller, 196. These pins formed a grid pattern on standard 100-mil centers.

Large-Scale Integration (LSI) Technology

In 1979, the 4331 and 4341 computer systems, based on LSI technology, were shipped to customers. The LSI chip consisted of a 704 logic gate array. The chip could switch in 1 ns. Up to nine of these chips could be contained on one multilayer ceramic (MLC) module.

The I/O pins of the ceramic modules were wave-soldered into plated through-holes in the printed circuit cards, to form the next level of packaging [10]. These cards had up to eight layers of circuitry and a grid of plated through-holes with 100-mil center-to-center spacing. Their four voltage planes were made of copper, and were subtractively etched. But the four signal planes (two internal and two external) were made by an additive copper plating process, since the tolerances of the circuit lines had to be held to ± 1 mil to reduce coupled noise and impedance to the levels demanded by the chip technology.

Cards were connected to the board by means of a bifurcated spring connector system [11]. Along one edge of the card was attached a 268-I/O blade carrier, which plugged into the bifurcated springs soldered into the board. This reversed the SLT procedure of having the springs on the card and the pins in the board, and resulted in a much higher connection density. The board held 18 printed circuit cards, and 8 groups of signal cable I/O cards.

The bottom surface of the board accommodated all engineering changes and other discrete (overflow and repair) wires. The engineering changes, overflows, and repairs were made with twisted pairs of 3.5-mil-diameter insulated wires, reflow-soldered to pads on the board's bottom surface.

Also developed for the 4300 processors was a 64,000-bit random access memory (64 K RAM) chip.

The LSI cards used not only the latest technology, but older technologies as well, where appropriate. An LSI card could contain 50-mm MLC modules,

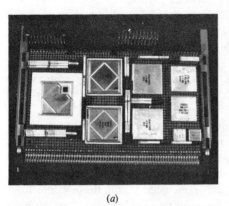

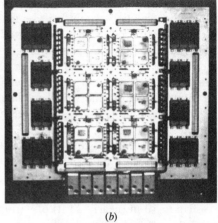

(a)   (b)

FIGURE 30-9
(a) Large-scale integration (LSI) card. This 6-by-4-in. card contains a variety of the modules used in the LSI machines. From left to right are a 50-mm MLC module, two 35-mm MLC modules, two 28-mm MC modules, and two 1-in. modules, below which are two $\frac{1}{2}$-in. (paste) modules. Also on the card are ceramic terminating resistors and capacitors, a 268 I/O connector on the bottom, and top card crossover connectors on the top. (b) Very-large-scale integration (VLSI) board. This 24-by-28 board services six of the thermal conduction modules. On the front side of the board is a $\frac{3}{4}$-in.-thick aluminum stiffener, which serves as a fixture for attaching all the hardware. Surrounding the TCMs is a U-shaped power laminar bus. At both sides of the board are the signal I/O cable housings. At both sides, top, and bottom are openings for the pluggable terminating resistors.

35-mm MLC modules, 28-mm MLC modules, one-inch MC modules, half-inch paste MST modules, terminating resistors, decoupling capacitors, and top card crossover connectors [see Figure 30-9(a)]. The LSI cards were thus an evolutionary development of the earlier card technologies (Table 30.2).

The 4341 processor, when it was shipped in 1979, had a machine cycle time of 100 ns.

TABLE 30.2
Card technologies

Technology	SMS	SLT	MST	ELSI	LSI
Year	1959	1964	1969	1973	1979
Machine	1401	360	370	370	4341
Signal planes	1	2	2	4	4
Power planes	0	1	2	2	4
PTH grid*	0.100	0.125	0.125	0.100	0.100
Through-hole*	0.047	0.043	0.043	0.043	0.033
Via hole*	—	—	—	—	0.014
Lines/channel	1	2	2	2	3
Line width*	0.015	0.013	0.006	0.0055	0.004
Thickness*	0.062	0.040	0.040	0.040	0.052

* All dimensions are in inches.

Thermal Conduction Module (TCM) Technology

The 3081 processor [12], shipped in 1981, used the 704 circuit/chip LSI technology with a larger (90-mm) multilayer ceramic module (MLC) to take advantage of the 1-ns switching speed in a higher-performance computer. Up to nine of the thermal conduction modules (TCMs) could be plugged into a large (24-by-28-inch) board, reducing the chip-to-chip, module-to-module, and board-to-board delays. One level of packaging (the card) was eliminated completely, breaking the tradition established by solid logic technology (SLT).

The 3081 chip set consisted of the 704 logic chip, a 32-by-18 local store chip, and a 256-by-9 cache array chip [13]. One module, incorporating 100 chips, was principally a logic unit. The second, a 118-chip design, was principally an array module. Since each chip's pin-out configuration was identical with regard to signal and power, logic and array chips could be mixed as required by any particular logic partition implementation. As a result of these flexible design features, TCMs characteristically package whole system partitions. Typical modules carried 25,000 logic circuits and 65,000 array bits.

The size of the module had thus evolved from the half-inch paste module to the 90-mm TCM (see Table 30.3). The TCM module cap removed heat by direct contact of a piston to the back of the semiconductor chips. The pistons were backed by springs and placed in machined holes in the inner surface of the module cap. At assembly, the TCM was filled with helium to further aid heat transfer between the crown piston surface and the chips. A water-jacketed cold plate was attached to the module cap to take the heat out of the machine. The TCM had 1800 pins brazed to the MLC substrate, which plugged into the board.

Very-Large-Scale Integration (VLSI) Technology

The 600-by-700-mm (24-by-28-in.) high-performance board package was designed to support and service the TCM [see Figure 30-9(*b*)]. In the module areas of the board, there were groups of bifurcated spring connectors soldered into plated through-holes in the board [14]. The 1800 connectors were arranged in a staggered grid, 2.5 mm (0.0984 in.) in one direction and 1.25 mm (0.0492

TABLE 30.3
Module technologies

Year	Module	Technology	Input/output pins
1964	Half-inch	Thick film (SLT)	12
1973	One-inch	Thin film (MC)	72
1979	28-mm	Thin film (MC)	116
1979	35-mm	Multilayer (MLC)	196
1979	50-mm	Multilayer (MLC)	361
1980	90-mm	Multilayer (TCM)	1800

in.) in the other. Of this number, 1200 were signal connections; 500 were power connections, at −4.25 V dc, −1.25 V dc, and ground; and 100 connections were reserved as spares. The 1800 pins of the TCM were inserted into the connectors with zero insertion force (ZIF), but the pins were actuated into the bifurcated springs with a wiping action by the use of a special cam tool. This wiping action between the surfaces produced a very reliable connection.

Besides the plated through-holes, the 3081 board had programmable vias between adjacent pairs of signal planes to allow changes in printed circuit line direction [15]. The directions of the printed circuit lines on the pairs of signal planes were orthogonal to each other (one vertical, one horizontal). The programmable vias were offset 1.25 mm (0.0492 in.) from the LSTs, and this greatly increased the ability to wire the four-line-per-channel board using design automation algorithms [16].

The machine cycle time of the 3081 processor was 26 ns.

The 3084 processor was essentially two 3081 processors operating in parallel, and cut the effective machine cycle time for some jobs in half. With this computer, the chip capacity of the TCM was extended to 133 and the level of integration was increased (see Table 30.4).

The 4381 processor, shipped in 1984, extended the concept of the air-cooled MLC to a 64-mm module, which could service up to 36 chips, and packaged 22 of these on the large (24-by-28-in.) printed circuit board [17]. When the 4381 Model Group 3 was announced in October of 1984, a significant advance in memory packaging was revealed. A 256 K-bit memory chip, packaged eight to a module, went into a 2-megabyte memory card.

The 3090 processor [18] used emitter-coupled logic (ECL) chips instead of the transistor-transistor logic (TTL) used in the 3081. As a result, the 3090 Model 200, first delivered to customers in 1985, could process commercial jobs up to two times faster than the 3081 Model KX. There was also a vector processor, designed for engineering and scientific tasks, which ran up to three times faster. The 3090 Model 400 was twice as fast as the Model 200.

The printed circuit boards used in these processors drew twice as much current as the 3081, and supplied three potentials and ground to the ECL chips. The VLSI boards used in these latest machines [19] were an evolutionary devel-

TABLE 30.4
Level of integration

Year	Circuits per chip
1964	1
1965	3
1969	5
1973	134
1979	704
1983	1500

TABLE 30.5
Board technologies

Technology	SLT	MST	LSI	VLSI	VLSI	VLSI
Year	1964	1969	1979	1981	1984	1985
Machine	360	370	4341	3081	4381	3090
Signal planes	2	4	6	6	8	6
Power planes	2	3	8	12	10	16
PTH grid*	0.125	0.125	0.049	0.049	0.098	0.098
Through-hole*	0.033	0.033	0.018	0.016	0.018	0.018
Via hole*	—	0.033	0.010	0.006	0.010	0.010
Lines/channel	2	2	2	2	4	4
Line width*	0.010	0.005	0.004	0.003	0.004	0.004
Thickness*	0.059	0.059	0.200	0.180	0.212	0.220
Board width*	10	10	10	24	24	24
Board length*	15	15	15	28	28	28

* All dimensions are in inches.

opment of earlier technology boards (see Table 30.5). The central storage memory used 1-megabit (1000 K) memory chips, a new standard for the industry.

Some of the packaging considerations that went into these machines are discussed in the chapter on packaging architecture.

ACKNOWLEDGMENT

The authors would like to acknowledge the contributions of Norman G. Jones, without whose knowledge of the development of circuit packaging this chapter could not have been written.

REFERENCES

1. International Business Machines Corporation, "Machine Methods of Accounting," AM-6, 1939.
2. C. J. Bashe, L. R. Johnson, J. H. Palmer, and E. W. Pugh, *IBM's Early Computers*, Cambridge, Mass.: The MIT Press, 1986.
3. C. J. Bashe, W. Buchholz, G. V. Hawkins, J. J. Ingram, and N. Rochester, "The Architecture of IBM's Early Computers," *IBM Journal of Research and Development* 25, no. 5, pp. 363–375 (1981).
4. P. A. Totta and R. P. Sopher, "SLT Device Metallurgy and its Monolithic Extension," *IBM Journal of Research and Development* 13, pp. 226–238 (1969).
5. L. F. Miller, "Controlled Collapse Reflow Chip Joining," *IBM Journal of Research and Development* 13, pp. 239–250 (1969).
6. A. Padegs, "System 360 and Beyond," *IBM Journal of Research and Development* 25, pp. 377–390 (1981).
7. D. J. Bendz, R. W. Gedney, and J. Rasile, "Cost/Performance Single Chip Module," *IBM Journal of Research and Development* 26, pp. 279–285 (1982).
8. J. C. Mollen, "IBM Pluggable Modules," *Proceedings of the 30th Electronic Components Conference*, pp. 349–354 (1980).
9. A. J. Blodgett, "A Multilayer Ceramic, Multichip Module," *Proceedings of the 30th Electronic Components Conference*, pp. 283–285 (1980).

10. G. G. Werbizky, P. E. Winkler, and F. W. Haining, "Making 100,000 Circuits Fit Where at Most 6,000 Fit Before," *Electronics* 54, August 1979, pp. 109–114.

11. J. B. Harris, K. M. Hoffman, D. W. Hogan, and V. P. Subik, U.S. Patent 3,915,537, "Universal Electrical Connector," 1975.

12. R. F. Bonner, J. A. Asselta, and F. W. Haining, "Advanced Printed Circuit Board Design for High Performance Computer Applications," *IBM Journal of Research and Development* 26, pp. 297–305 (1982).

13. B. T. Clark, "Designing the Thermal Conduction Module for the IBM 3081 Processor," presented at the 31st Electronic Components Conference, 1981.

14. G. G. Werbizky and F. W. Haining, "New Bifurcated Spring Connector ZIF-System for the IBM 3081 Processor," presented at Productronica, Munich, W. Germany, 1981.

15. G. G. Werbizky and F. W. Haining, "Some Design/Process Considerations Concerning Blind and Buried Vias in Printed Circuit Boards," *Proceedings of the Fourth Annual International Electronics Packaging Conference*, pp. 507–513 (1984).

16. P. W. Case, M. Correia, W. Gianopolous, W. R. Heller, H. Ofek, T. C. Raymond, R. L. Simek, and C. B. Stieglitz, "Design Automation in IBM," *IBM Journal of Research and Development* 25, pp. 631–646 (1981).

17. G. G. Werbizky, J. T. Kolias, R. L. Weiss, and F. W. Haining, "Dense Packaging Puts Processor on One Board," *Computer Systems Equipment Design*, 1984, pp. 21–25.

18. G. G. Werbizky, S. Boyko, and F. W. Haining, "Increased Power Density for High Performance Printed Circuit Boards," *Proceeding of the Fifth Annual International Electronics Packaging Conference*, pp. 224–242 (1985).

19. Hayao Nakahara, "Complex Multilayer Boards Vie for Space in Japanese Computers," *Electronic Packaging and Production*, February 1987, pp. 70–75.

CHAPTER 31

FACTORY
OF THE
FUTURE

K. DALE ROE
PATRICK A. TOOLE

INTRODUCTION

The term "factory of the future" usually evokes the popular image of a revolutionary, dimly lit, windowless, peopleless facility populated by computer-controlled robotic machines tirelessly turning out the products of tomorrow.

A century ago, the phrase might have conjured up images of a cavernous building full of machines powered by a steam-driven engine turning scores of wheels, shafts, and pulleys. Wide, flat belts would transmit power to spin the cutting and shaping tools needed to produce the products of the day—horse-drawn buggies, ice boxes, and kerosene lamps. Large windows would admit the light of day, and electric lights would allow this "factory of the future" to operate around the clock, extracting maximum return on the owner's substantial investment in such a "modern" facility.

This description, while futuristic a hundred years ago, is obviously archaic when viewed in the light of today's technology. Thus, the philosophy of the factory of the future is as dynamic as the technology within—a moving target driven by the intellectual onslaught of creative men and women determined to improve on that which their predecessors viewed as the ultimate.

In fact, the factory of the future is more a philosophy than a building full of robots, more an attitude of design and development than a facility without people, a product of evolution rather than of revolution. The practical factory of the future falls somewhere between the well-lit, people-inhabited factories of today and the conceptual totally robotic, "lights out" factory of tomorrow.

This chapter will deal mainly with the factory of the future as it relates to manufacturing complex printed circuit packages for the electronics industry. The chapter will review some of the manufacturing concepts in place during the infancy of the technology and some of the evolutionary steps that led to today's facility. Then it will briefly look into what the printed circuit factory of the future might be.

Finally, the chapter will discuss people as they relate to work, and the ability and opportunity for each individual to effectively contribute, realize their potential, and be appropriately rewarded in the factory of the future.

OBJECTIVE OF A FACTORY

When purchasing a product, a customer usually wants high quality, high reliability, an attractive price and on-time delivery. Therefore, when deciding how a factory should be designed, it is good to keep in mind the objective of a factory—why such an enterprise exists.

A factory is a *system* designed to deliver high-quality manufactured products, on time, and at a competitive cost to the marketplace. In addition, it has to produce these products so they can be marketed at a reasonable profit.

HISTORY OF PRINTED CIRCUIT MANUFACTURING

A 25-year history of printed circuit packaging technology is discussed in Chapter 30 and in reference [1].

The main driver in advancing printed circuit manufacturing technology has been the need for greater and greater interconnection density (interconnection length per unit area), which has grown exponentially over the years. As higher levels of integration are reached at the component level, the number of component I/O pins increases, requiring higher circuit density at the card or board level.

During the 1950s, when components had only two or three leads, a single layer of printed wiring was adequate to interconnect the circuits. These printed circuits were produced by screening an etch resist material onto a layer of copper foil laminated to an epoxy/paper substrate. Passing the panel through an etching solution removed the unwanted copper, leaving the desired circuit pattern. Holes punched or drilled in the panel allowed the component leads to be inserted and soldered to the circuit pattern, completing the assembly.

The evolution of integrated circuit chips (ICs) and multichip carriers has driven the number of component I/Os from two or three to several hundred, resulting in very complex printed circuits, which require manufacturing facilities such as the one shown in Figure 31-1. In fact, some high-performance multichip carriers have nearly 2000 pins, and the number will undoubtedly grow to several thousand in the future. This will place tremendous demands on the printed circuit factory of the future to produce the high-density, high-performance panels needed to interconnect these advanced components.

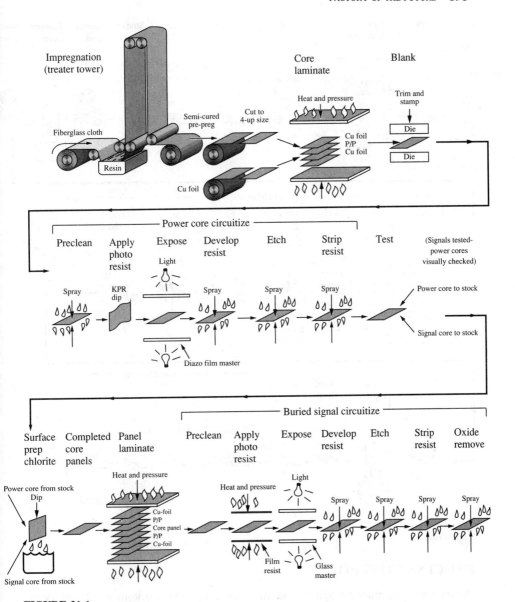

FIGURE 31-1
Process steps and process flow used in a complex printed circuit manufacturing facility.

New materials, finer lines, tighter line spacing, increased power-handling requirements, more signal and power distribution layers, and higher-aspect-ratio holes to interconnect these layers will all have significant impact on the design of future printed circuit factories. High-precision, highly reliable tools and processes orchestrated by extremely complex information-handling and process control systems will be required to maintain high yields and competitive costs.

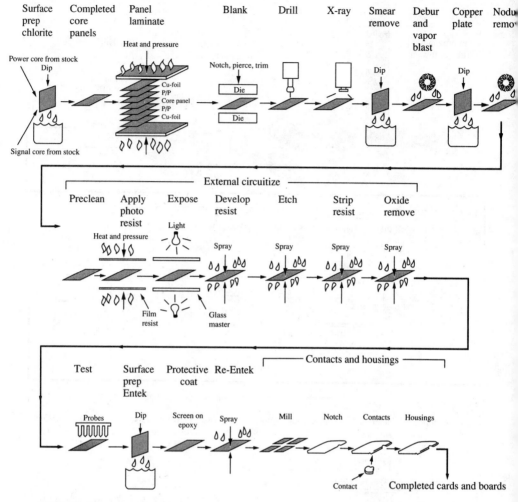

FIGURE 31-1
(*continued*)

PROCESS CONTROL

Developing and making a new manufacturing process work in a laboratory is often quite different than putting that same process into a full-scale, high-volume production facility. What works well in the laboratory often is not transferable to manufacturing without some modifications.

Large production volumes stress hardware. Fragile drills a fraction of a millimeter in diameter that effectively drilled scores of holes in printed circuit materials in the laboratory may self-destruct in a high-speed, high-volume production environment.

Chemicals that were measured in milliliters in the laboratory are measured in hundreds or thousands of gallons to make up production-sized process baths.

As product is processed through the line, chemicals are dragged from one tank to the next, diluting or contaminating subsequent baths, moving the operation outside the effective process window. Adding chemicals to an existing bath can cause deviations in the process. Adding too much of one chemical at once can throw the bath out of specification. Adding chemicals to compensate can result in a bath in oscillation and out of control.

In a process step where large volumes of product are handled rapidly or simultaneously, a malfunction could result in an enormous amount of scrap, especially if the situation went undetected for a significant period of time. Therefore, processes of this type must be extremely reliable to maintain acceptable yields and costs.

In the factory of the future, process baths will require sophisticated, highly accurate sensor-based systems and controls in a closed-loop system (Figure 31-2) to accurately and automatically maintain constant process stability. Mechanical process tools such as drilling machines will have to contain devices to detect tool wear and breakage to increase equipment availability and productivity and reduce

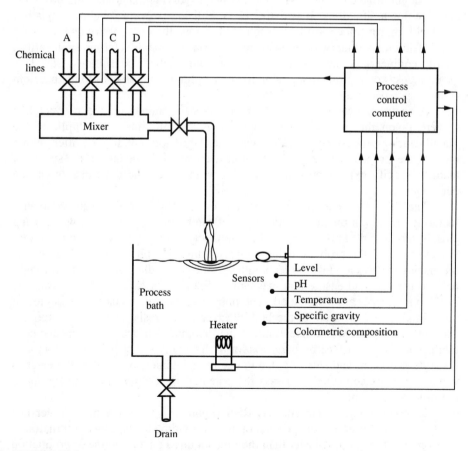

FIGURE 31-2
Closed loop process control system to automatically maintain process stability.

scrap. Virtually every major tool in the manufacturing facility will require a process control computer either imbedded within or associated with the tool.

THE MINILINE CONCEPT

The processes used to make printed circuits include physical and chemical phenomena governed by fundamental laws of material science, physics, and chemistry. At IBM, the tools and techniques of pure science are applied to solve problems amid the hustle and bustle of the production line, using the *miniline* approach.

The miniline concept [2] is useful for scaling up a laboratory-developed manufacturing process to a mass production environment. Scaling the process upwards in two or three increments using miniline techniques can often identify potential problems in the full-scale model. These problems can be fixed in the miniline, effectively reducing the chances of process yield reductions in the final manufacturing line.

The miniline concept can also be applied to solving manufacturing problems and line-down situations. It is extremely useful in resolving reliability, quality, and yield improvement issues. In such applications the concept brings together a team of interdisciplinary skills from manufacturing, manufacturing engineering, quality engineering, development, research, universities, consultants, etc. This group focuses on a problem and identifies the optimum approaches to the problem from all perspectives.

A traditional approach is to run matrix experiments on the product line to zero in on the problem. The approach in the miniline is to duplicate the manufacturing process on a small scale under laboratory conditions, often within the chambers of sophisticated analytical instruments. It is the task of the specialist team to specify experiments and the tools with which the experiments can be performed.

The success of the approach depends on the ability to duplicate the manufacturing process in the laboratory, where it can be tightly controlled while being closely monitored. This is where the interdisciplinary team swings into action. For example, if the problem being addressed involves photoresist adhesion, the mechanics expert can develop techniques to measure adhesion or surface roughness accurately, and ensure that mechanical phenomena (stresses due to swelling of the photoresist) are understood. The photochemist will focus on the degree of cross-linking and ensure that the bottom layer of the resist is properly exposed. The surface analyst will concentrate on the incoming surface. What are the contaminants? Is the surface properly oxidized? The polymer chemist may focus on resist development and swelling due to chemical interaction. Finally, the electrochemist may be consulted to assess the potential for corrosion and blistering in the process environment.

Laboratory equipment such as RBS (Figure 31-3) or scanning Auger can easily reproduce a step in the production line, providing an experiment turnaround time orders of magnitude faster than the time required on a pilot line or production line. Correlations can be achieved by analyzing production line parts.

FIGURE 31-3
Rutherford back-scattering (RBS) system at Cornell University.

Chemicals from the production line can be tested in the miniline to verify their effectiveness and quality. Often, contaminants from exposure to solutions from earlier tanks in the sequence (drag in) have an effect. This can be easily uncovered in miniline experiments by using pure chemicals, precisely characterizing the production line in order to accurately duplicate the conditions, and comparing the results.

The interaction of scientists with manufacturing engineers is synergistic. The versatility of moving the miniline to an industry research or university laboratory expands the application of their expertise and analytical tools.

Minilines can serve both to generate initial data bases in newly established processes and to diagnose and prevent detrimental changes in the manufacturing line. They can also be used to reduce a process to its simplest form without investing in significant capital tooling or a pilot line. With manufacturing processes becoming ever more complex and precise, and quality, yield, and cost becoming more and more important, the miniline concept is an integral element in the factory of the future.

TOOLS OF THE FACTORY

Tools and the space they occupy are only a small part of the total factory system. As manufactured products become more complex and the processes required to produce them become more sophisticated and precise, the amount of data needed to operate and maintain the factory becomes a significant element.

The types of information involved include the following:

Cost data	Maintenance data
Parts tracking/logistics	Setup
Quality/yield/scrap data	Process parameters
Reasons for process variance	Analysis data
Mean time to fail (MTTF)	Measurements
Mean time to repair (MTTR)	Equipment performance data

The real challenge becomes how to automate the flow of this data through the factory and use it effectively to maintain the overall health of the manufacturing system.

The enterprise that controls the tools, space and processes is a data management system called, in this chapter, the computer-integrated enterprise, or CIE. This system integrates all the elements of the facility and becomes the mind of the factory of the future.

The large number of varied circuit patterns that a printed circuit factory may be called upon to produce mandates a high degree of information automation between the circuit designer and the manufacturing floor. This is accomplished through automated design systems that supply logic, test, assembly, and carrier geometry data to the CIE.

A typical system, shown in Figure 31-4, allows the generation of design data both manually and automatically. An engineer in a product development laboratory uses a design automation system to design a printed circuit card or board. The automated system uses logic designs as the input, and the engineering design system (EDS) processes the logic against a set of functions and technology rules within the system.

Once all the rules and logic functions are satisfied, data is transmitted to the manufacturing control systems on a technical data interface (TDI). These systems apply manufacturing rules to the design information, creating data to drive the production control, financial, and manufacturing floor control systems, as well as generating manufacturing routings and numerical control machine data to produce the parts.

In addition to the data pertaining to the physical characteristics of the printed circuit card, such as circuit pattern, hole placement, and land size/placement, data is also generated for the component set, logic, and functional testing operations as well.

The manual system is for unique or special designs that fall outside the parameters of the EDS, allowing the circuit designer to draw a desired circuit configuration on a graphics screen. Once the design is completed, data is transmitted to manufacturing, where the design is reviewed to make sure it can be processed through the factory. The output from both systems goes into the factory's information management system. However, unlike the EDS-generated data, the manual system output contains only physical data and little or none of the logic and test data required to assemble and test the card.

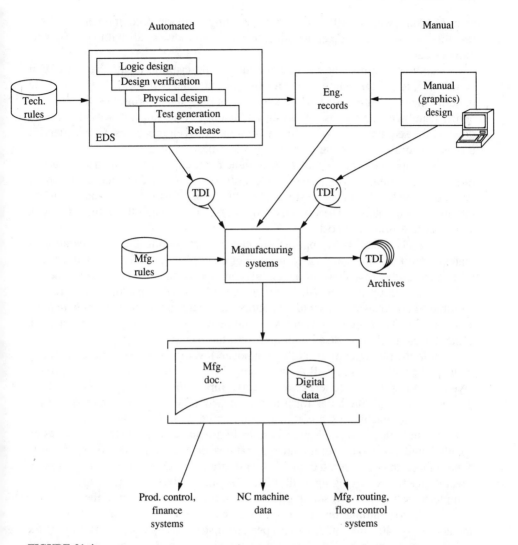

FIGURE 31-4
The engineering design system (EDS) effectively links the product design engineer with the manufacturing facility.

Each *special* item entering the factory requires significant manual intervention for verification and checking, adding to the cost of the product and to turnaround time from order receipt to delivery. Automated design systems for the factory of the future should optimally be flexible and extendible enough that specials will be virtually nonexistent.

In a printed circuit factory, the information management system automatically brings forward from the system product customer an order for printed circuit boards, along with a bill of material including individual components such as resistors, capacitors, and semiconductor chips or modules, all with specific part number data.

Contained within the bill of material is a description of the printed circuit board, including the number of signal, ground, and voltage planes, the hole pattern, the fabrication sequence, and the routing and subroutings through the entire manufacturing process. The routing includes every single process step, even down to the sequence of rinses and drying steps between processes. Consequently, a typical routing may contain several hundred process steps.

Each operation has a unit hour assigned, an expected yield, and a burden rate for the capital tooling, space, and utilities. Thus an expected cost can be derived immediately for the customer and the actual cost can be computed when the order is completed. The objective, of course, is to match the actual cost with a competitive projected cost.

As each operation is approached, the correct routing for that operation is matched with the order and acted upon. For example, the order may require 100 printed circuit boards. Thus, at the circuit core operation, 100 circuit cores having the required circuit pattern will be processed, plus whatever additional cores are required to compensate for yield detractors. The pattern will have been generated automatically on a glass master by a computer-driven, high-precision artwork generator controlled by data generated from the EDS.

Hole-drilling operations at the composite level (a composite is a completely laminated panel containing all the necessary dielectric, signal, power, and ground layers) of the process sequence also receive input from the design system. Here the locations of all the holes for a particular printed circuit board pattern are fed to the drilling machine by a computer timeshared with other operations.

These high-speed, high-precision drilling machines can have as many as 50 spindles and often incorporate several stations on the same positioning table to handle more than one printed circuit board at a time. One operator may monitor several machines, changing drill bits, clearing problems, and inspecting the completed panels prior to sending them on to the next operation in the sequence.

Tools for automation also need to be considered. A wet process line may include 30 or 40 tanks with the proper chemistry in sequence. A trolley at the front end picks up a basket of panels or signal cores and carries them from tank to tank through the sequence of operations. Robots may simply load and unload the panels. This may be absolutely necessary if the parts are too heavy for an operator to handle. In fact, load and unload operations are extremely boring, so they are popular sites for implementing simple, low-cost robots.

Designing a factory to produce state-of-the-art products requires state-of-the-art processes, tools, and materials, especially in the electronics industry. As circuit densities and performance requirements increase, tolerances become tighter, new materials are introduced, and the entire manufacturing process becomes more critical. Often, off-the-shelf equipment does not meet the requirements and must be significantly modified, or new equipment must be custom designed and built to equip the factory.

Capital expenditures for custom equipment can run into the hundreds of thousands or even millions of dollars to develop and build. Therefore, the key to maximizing return on investment lies in designing the required tools and processes for maximum flexibility and extendibility.

YIELD MANAGEMENT

In any factory producing complex products through hundreds of process steps, yields at each step of the process must be optimized for maximum output at the end of the line. Sensors connected directly to a computer can monitor machine conditions, temperatures, air velocities, feed rates, and other process parameters to develop an on-line data base.

By matching orders against the data base on a regular basis, variations in the yields at each process step can be correlated directly to the process parameters at any given point in time. If one step has a 99 percent yield and another has a 70 percent yield, engineering can model both situations to determine the variables that lead to good lots versus bad lots.

It is also possible to determine factors causing scrap in the line by evaluating variations in the process data base during times when an individual process was yielding good parts versus times it was yielding scrap.

The *process window* is the operating range of a process within which acceptable parts are produced. If the process deviates outside the process window, scrap results. Materials, and how the process chemicals interact with them, play an important part in the process windows. Giving operators a 10–15 percent process window tolerance along with the proper training allows them to recognize a problem and take corrective action long before the product goes out of specification.

Occasionally a contaminant in a load of new chemical will go undetected until it shows up as a negative influence in the process. Placing chemical inspection responsibility back to the vendor before the chemical is shipped and analyzing the test data before accepting the shipment can significantly reduce this type of influence on yields.

WHEN TO AUTOMATE

The printed circuit factory of the future is not a darkened room full of robots. It is a facility containing highly skilled people with selective automation for process, precision, health, and safety reasons, or to obtain a logistical advantage over labor-intensive tasks that lend themselves to the benefits of automation. It is a facility where people work in concert with sophisticated information-handling networks linked to state-of-the-art tools to meet the schedule, quality, and cost objectives of the factory.

Sophisticated automation may not be as cost-effective as the manufacturing employee—the person working on the factory floor. However, for a very repetitive operation that will probably be used for several years, automation can be justified.

Even if a part may change in detail, but will remain the same functionally, you may automate certain process steps with a flexible robot, as is done in the automotive industry. As long as cars have doors, they will have to be attached to the body. The shape of the door may change due to styling changes in the overall design of the car, but the fundamental operation of attaching the door will remain basically the same; therefore, it makes good economic sense to automate that particular task.

Tasks involving a product that will be changing its components or form factor within a short period of time and with limited volumes would not be practical to automate, unless it became necessary to do so for process, health, or safety reasons.

AUTOMATION BY OBJECTIVES

To develop a unique, totally integrated automated factory involves a significant economic undertaking and must be approached very cautiously.

IBM developed such a factory in the mid 1980s to manufacture their Pro-Printer for the IBM Personal Computer [3], and the printed circuit card assembly contained within the printer [4]. Development of such a product went hand in hand with development of the factory that would produce it, and the objectives for both were defined early in the program. The objectives were to cut the traditional product development cycle by nearly 50 percent, reduce the number of parts in the printer by 66 percent relative to similar printers on the market, and design the product for full automation, including as far as possible the making of parts. Manufacturing was involved early in the development cycle.

Manufacturing costs were substantially reduced by assembling the product in layers—from the bottom to the top. Parts were self-aligning so that robots could position them. Screws, fasteners, springs, pulleys, and other parts requiring human adjustment were eliminated wherever possible.

The result of all this engineering showed that a product well designed for automated manufacturing could also be easily assembled by hand. Demonstrations have shown the entire printer can be manually assembled in a matter of minutes, merely by snapping most of the parts together.

Cost was reduced by the innovative use of materials. Nearly 70 percent of the printer's parts were plastic—not ordinary plastics, but plastics filled with glass for strength, with Teflon for wear resistance, and with carbon for draining off charges of static electricity. Most of the plant's inventory was stored in-house in the form of plastic pellets, greatly simplifying logistics.

While this futuristic, automated factory is a marvel of modern engineering, it appears that people could perform the actual assembly operations competitively with robots. The real benefit of this highly automated factory is the automation of data that is usually handled by support employees, such as equipment maintenance, measurements and performance data, expediting, quality, and all the other items required to determine if the factory is healthy and operating efficiently.

DETERMINING THE HIERARCHY
OF NEED FOR AUTOMATION

Automation of tools and related operations should be determined by a hierarchy of need, starting with health and safety. If a particular operation is hazardous to humans, there are several ways to approach the problem. One approach is to develop a new process that will significantly reduce or even eliminate the hazard, while still obtaining the desired end result for the product. Another approach is to develop and install into the process safety features that reduce the risk to the operator. However, this can be an extremely expensive approach when working with certain chemical processes if special abatement and handling systems are required. A third approach is to change the product to eliminate the need for the hazardous operation. Finally, if none of the above methods are possible or economically feasible, the potential human operator can be replaced with a robot.

Following health and safety as a reason to automate comes process requirements and/or precision. Some manufacturing processes may require the speed or stamina of a machine, as in a highly repetitive, high-volume operation such as the insertion of electronic components into printed circuit boards. Or a process may require a degree of precision that is economically impractical for a human operator. In cases such as these automation may be the answer.

About on a par with process and precision requirements as a reason to automate is cost—particularly reductions that can be obtained through work-in-process (WIP) inventory. In addition, with certain operations it may be more cost-effective to automate than to use a person with all of the necessary health and safety concerns, salary, and benefits. However, it would be difficult to justify replacing a $40,000- or $50,000-a-year manufacturing employee, including the cost of benefits and overhead, with a $300,000 automated machine tool if the useful life of the operation is only three or four years.

The totally automated factory of the future will not become a reality until the reliability of robotics becomes significantly better and sensor-based process control systems become more reliable and sophisticated. If a robot has to be backed up by two or three maintenance technicians, then an operator is merely being replaced by a technician with little or no benefit to production cost.

A robot is rarely a stand-alone tool, but rather a complement to a larger island of production equipment. Often an off-the-shelf robot is not available with all the features required for highly sophisticated operations. It may require considerable engineering effort to provide sensors, product orientation devices, and other changes to satisfy particular requirements.

Robots must also have the flexibility to change from one operation or part number to another with minimal setup and reprogramming. It is often more cost-effective to use an operator than to try to develop a robot with the required sophistication for some of today's and tomorrow's complex processes, especially in the printed circuit business, where process deviations and chemical interactions between process baths exist.

Many elements affect the decisions for automating a factory. Product life cycle, product design, ease of entering a new product into the factory, process complexity, and cost of automating are only a few of these.

THE FUTURE
MANUFACTURING EMPLOYEE

Manufacturing employees in the past had little involvement in quality control, little say in how jobs were set up and run, and minimal meaningful communication with management. They took instructions with very little feedback.

Since that time, social changes and higher levels of education in the work force have had an effect on the role of the factory worker. Employees' growing dissatisfaction with the status quo, plus new factors appearing in the manufacturing equation, put pressure on management to make changes.

One new factor took form when mechanical and electromechanical products of the past started to give way at the beginning of the electronic era in the 60s. In place of stand-alone machining operations in the factory, complex chemical processes began to appear. Industry had to learn how to put these processes together, control them, train people to operate them, and make them safe for both the operators and the environment.

As a result, morale began to deteriorate year by year, causing productivity and quality to suffer. People were more educated and sophisticated; consequently, they wanted a stronger role in how their jobs were done, more ownership, more involvement.

Teaching employees problem-solving techniques and allowing them time to evaluate their jobs and develop solutions to problems can result in significant paybacks in the amount of time spent directly on the job itself—sometimes as much as a 30–40 percent increase in quality and productivity is achieved, with less time spent directly on the job.

CONTINUOUS FLOW MANUFACTURING
(CFM)

Continuous flow manufacturing (CFM) is a method of running an integrated manufacturing facility with no buffers, resulting in minimal work-in-process (WIP) inventory. Manufacturing employees manage the WIP by starting and stopping specific operations within the line themselves rather than using a dedicated production control group.

With CMF, parts arrive at each operation just in time for processing and leave the operation just in time for processing in the next step in the manufacturing line. Moving in this manner, all parts are in the line only the amount of time required for processing, resulting in minimal cycle time compared to traditional manufacturing methods.

Lines with large variations in process times from operation to operation require fewer operators than operations, allowing operators to perform multiple tasks during the course of a work shift. Operators also have the authority to stop the line if a problem develops, and meet with their peers to solve the problem.

The CFM concept is conducive to developing the manufacturing employee of the future. Manufacturing jobs can be enhanced by adding content through more integration of support jobs and developing ongoing manufacturing education programs to support this environment (Figure 31-5). In addition, a more professional atmosphere can be created by increasing employee involvement in decision making, while minimizing day-to-day measurements and focusing instead on overall department output and quality.

Unit hours, number of pieces produced by an employee per day, and process utilization were the traditional manufacturing measurements. These measurement methods were not always cost-effective and often caused management to make wrong decisions. The focus is shifting to qualitative rather than quantitative measurements, such as are used for professional jobs. More emphasis is placed on team efforts and measuring output, cost and quality on a department or sector basis across three shifts, rather than at each individual process or operation.

It is also necessary to develop manufacturing managers that can manage in this new environment. They must have skills in manufacturing engineering, quality, maintenance, production control, and automation, as well as comprehensive people management skills to manage a broader range of employee levels, from entry level operators to engineering professionals. In order to evaluate individuals for salary increases and promotions, the manager must spend time in the work area becoming intimately familiar with the work employees are doing.

THE PRODUCTION TEAM

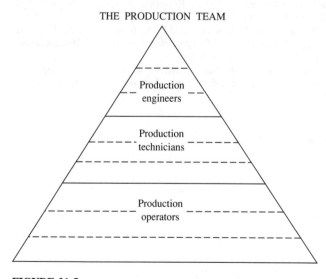

FIGURE 31-5
The modern manufacturing employee can progress from a machine operator to a production engineering level within the manufacturing organization through a combination of experience, education, and performance.

New focus must be placed on hiring and developing more highly skilled manufacturing professionals having the ability to work and grow in a totally integrated, highly participative environment. In the past the manufacturing employee had only 2 or 3 levels of advancement before having to leave manufacturing for career advancement opportunities. The factory of the future provides an environment for career growth, job satisfaction, and recognition within the manufacturing organization by expanding the role of the manufacturing employee. In addition to operating the equipment required to meet production, operations such as process control, diagnostics, quality control, and equipment maintenance are being added to the skills of the manufacturing employee.

SUMMARY

The factory of the future is more than a facility of process machines, robots, and computers. It is the careful integration of machines, processes, information, concern for the environment, and most important of all, people. To design a true factory of the future, a planner's task is as follows: Develop a new production facility not only for today but also for tomorrow. Integrate into it the necessary tools and processes, designed to insure maximum flexibility and extendibility for future generations of products. And when addressing manufacturing problems of today, don't just fix the problem, fix the future.

REFERENCES

1. I. Feinberg and D. P. Seraphim, "Electronic Packaging Evolution in IBM," *IBM Journal of Research and Development* 25, pp. 617–629 (1981).
2. D. E. Barr, W. T. Chen, R. Rosenberg, D. P. Seraphim, and P. A. Toole, "Miniline: Research Applied to Manufacturing," *IEEE Transactions on Components, Hybrids, and Manufacturing Technology* CHMT-9, no. 4 (December 1985), pp. 410–416.
3. G. D. Austrian, "Always Gaining on That Finish Line," *Think Magazine*, 1987, no. 5, pp. 2–4.
4. A. Socolovsky, "IBM Charlotte: Winner of the 1988 Factory Automation Award," *Electronic Business*, February 15, 1988.

CHAPTER 32

FUNDAMENTAL LIMITS FOR ELECTRONIC PACKAGING*

WILLIAM PENCE
J. PETER KRUSIUS

INTRODUCTION

The fundamental limits for electronic packages are explored in this chapter. Here the term electronic package is used in the broadest possible sense to include every packaging level from the lowest (chip package) to the highest (system package). Performance limits for electronic packages arise either from technological constraints or from fundamental physical laws. The former are specific to each packaging technology and the manufacturing methods used to fabricate the packaging structures, while the latter arise from universal laws of physics. Technological limits are often overcome by gradual evolutionary improvements or by adopting new and better technologies, if sufficiently strong driving forces exist. Fundamental limits, however, are hard constraints, which cannot be violated regardless of which technology is adopted. During the early days of semiconductor electronics (1960–1980), electronic packages were exclusively limited

* Work supported by IBM Corporation (Endicott, New York, Dr. W. Chen) and the Semiconductor Research Corporation by the Electronic Packaging Program at Cornell.

by a variety of technological limits. Because of continued exponential evolution, driven by demands for lower cost per circuit function and higher reliability, electronic packages are rapidly approaching several fundamental limits simultaneously. Fundamental limits will soon dictate the future of electronic packaging and electronic systems in general.

The objective of this chapter is to explore the fundamental limits for electronic packages and relate these limits to present and future electronic systems. The discussion starts with a description of past and current evolutionary trends. Next, a common description for all electronic packages is established. Based on this description, the next sections identify the most significant fundamental limits, describe the limits mathematically, and relate them to the present state of the art for electronic packages. Finally, future opportunities for electronic packaging, compatible with the fundamental limits, are discussed.

EVOLUTION OF ELECTRONIC SYSTEMS

Since the invention of the point contact semiconductor transistor at Bell Laboratories in 1949 [1] and the monolithic silicon integrated circuit in 1958 at Texas Instruments, the density of on-chip circuitry has doubled on the average every 1.2 years [2]. Reduction of the size of semiconductor devices (device scaling) has played a key role in this evolution, making it possible to fabricate 32-bit microprocessors and 1-megabit dynamic random access memories (DRAM) on single silicon chips, not larger than 1 cm^2, in 1981 and 1984 respectively [3, 4]. Until recently, however, the importance of the way these chips were connected together in the electronic system was secondary, since chip and package design were generally done independently, and the latter aroused significantly less interest, as it had little effect on the system performance. Today, the package is emerging as a critical part of system design, with communication between chips seen as the crucial factor limiting overall system performance, and with physical and architectural limitations dictating how the chips can and should be interconnected. It is increasingly clear that the traditional hierarchy of relatively independent packaging levels must be replaced with a more global one. Package design must be an integral part of systems design, and it will not only influence the electrical design but also determine the architecture and systems partitioning of forthcoming electronic systems.

While the problems encountered in packaging high-performance systems, such as mainframe and supercomputers, are the most complex, all systems, both those with general-purpose and those with special-purpose architectures, will face limitations on speed, density, and power dissipation. A general-purpose architecture can be differentiated from a special-purpose one by the communication requirements between subelements within the system. A general-purpose system must allow any two subelements to communicate, usually via one or several system busses, whereas a special-purpose system may process information serially in a pipelined or systolic array fashion, requiring only that each subelement communicate with the stage immediately preceding and immediately following it in the information path. We will show that it is possible to examine the physical limitations for electronic systems of both types from a global level using the same approach. This allows us to predict what forms future electronic systems will have to take in order to comply with the fundamental limitations, and where the ultimate limits of solid state electronics lie.

CLASSIFICATION OF ELECTRONIC SYSTEMS

In order to fully characterize the problems to be faced by designers of future electronic packages, we analyze all relevant physical laws. To keep the analysis simple and understandable, we express these physical laws with simple relationships and as few descriptive parameters as possible. The chosen parameters should still describe the essential properties of the electronic system, and be capable of expressing the fundamental limitations imposed on these properties. Using these parameters reduces the complexity of the description of electronic packaging hierarchies and facilitates the understanding of the relationships and relative significances of the physical limitations.

The minimum number of descriptive parameters that can classify all electronic systems and express all relevant fundamental physical limitations is three. The three generalized descriptive parameters are: the spatial separation of subelements in the system (denoted x), the signal propagation delay from one subelement to another one (denoted t), and the spatial separation of the signal lines used to communicate between subelements of the system (denoted s). This is illustrated in Figure 32-1. It is possible to describe a wide variety of electronic systems, from integrated circuits to the global telephone system, with these descriptive parameters, irrespective of the actual physical structures used to implement the systems function.

The subelement spacing x refers to the geometrical center-to-center spacing of subelement blocks, whether they be gates, chips, or entire boards. For most

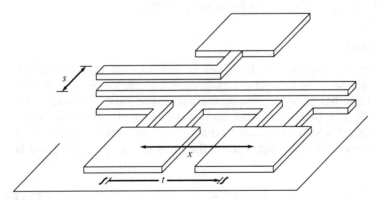

FIGURE 32-1
Definition of generalized descriptive parameters x, s and t.

general-purpose systems this spacing is equal to the geometrical size of the subelement blocks as well. The signal propagation delay t is the time required for a signal to pass from the input to the output of a subelement block, and to the input of the next subelement. Using these two descriptive parameters it is possible to characterize electronic systems across a wide range of dimensions. For example, on a single chip, at the lowest level, the logic gate propagation delay is the time required for a signal to travel through a single gate on a chip, while, at a higher level, the clock rate for a microprocessor gives the time required to effect a register-to-register transfer operation. In the first case the subelement spacing is the geometrical center-to-center spacing of the gate circuits, and in the second it is the spacing of the two most widely separated registers in the microprocessor, which is on the order of the chip size. For general-purpose architectures, the higher the level of the system, the more the subelement spacing approaches the size of the system itself.

These two descriptive parameters alone are not sufficient to completely describe the limitations imposed on electronic systems. Noise constraints as well as wiring space requirements must be represented. This requires another descriptive parameter. The simplest choice is the geometrical spacing s of communication wires connecting subelements. Wire spacing is the most economical and convenient choice, because it adds only one new descriptive parameter, and can describe both geometry effects and noise simultaneously. With the three parameters chosen, a set of models representing the physical limits important for electronic packaging can be constructed.

FUNDAMENTAL LIMITS

In this section the fundamental physical limits for electronic systems are presented and expressed in terms of the generalized descriptive parameters x, s, and t. These three descriptive parameters can be considered to span a three-dimensional space, which is here called the *electronic packaging space*. The following discussion is organized such that first we consider those fundamental limits that can be expressed in the xt plane, followed by those in the st plane.

Light Speed Limit

The first and perhaps most obvious limitation is that imposed by the speed of light. Since it is not possible for two points to communicate via a signal with a velocity greater than that of light, a minimum delay time exists, depending on the separation of the two points. Whether the two points are circuit blocks on the same chip or on chips widely separated in the system, and whether metal lines or optical paths are used for the signal lines, the minimum propagation delay is given by

$$t_{\min} = \frac{nx}{c} \tag{32.1}$$

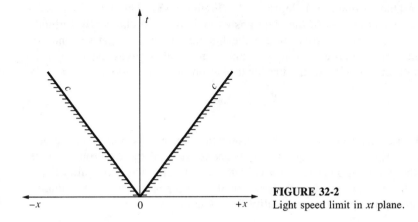

FIGURE 32-2
Light speed limit in xt plane.

where c is the velocity of light in free space and n is the index of refraction in the medium through which the electromagnetic field of the signal must travel. This limit can be shown as an oblique line in the xt plane, as in Figure 32-2. This figure is known as the *Minkowski* diagram of special relativity. The light speed line represents a hard limit and all systems must lie above it. Lossless transmission lines with other insulator materials have a velocity proportional to the inverse of the square root of the permittivity. For subelements fabricated in a semiconductor, the speed restriction is even more severe, since the signal is transmitted by charge carriers (electrons and holes), which travel much slower than an electromagnetic wave propagating in free space.

Charge Carrier Speed Limit

Charge carriers in solids can only travel with the maximum group velocity given by the band structure of that solid, whether these carriers move subject to classical scattering limits or ballistically. Signal speeds, for example in semiconductor devices, are limited to the maximum group velocity, which for all semiconductors is of the order of 1×10^7 cm/s at room temperature. Thus the ratio of charge carrier speed in a semiconductor to the speed of light in a vacuum is 1 to 1000. This explains the drive towards shorter and shorter channel lengths in FET devices and shorter base regions in bipolar devices at the cost of longer interconnect lines. Optical and transmission line lengths can be three to four orders of magnitude longer than semiconductor lines for balanced time-of-flight limits.

Quantum Limits: Tunneling

We next consider limitations imposed by quantum physics. Quantum-mechanical tunneling sets a lower bound for the subelement spacing for isolated subelements. If any two subelements are spaced more closely than this lower bound, electrons will tunnel through any potential barrier separating them and the two subelements will not be able to function independently (Figure 32-3). This tunneling occurs because the wavefunction of the electrons extends into the classically forbidden regions in the potential barrier, where the kinetic energy is negative, and penetrates thin barriers entirely. Simple quantum mechanics gives the ratio of the wavefunction on one side of the barrier to that on the other side as

$$\frac{\psi_2}{\psi_1} = e^{-|k|x} \tag{32.2}$$

where k is the electron wavenumber inside the barrier. Its maximum value is equal to $(2m^* V_0/\hbar^2)^{1/2}$. Here m^* is the effective mass of the electron, V_0 is the height of the potential barrier, and $\hbar$ is Planck's constant. The probability that an electron will be found on the other side of the barrier is given by the square of the wavefunction; hence the probability for tunneling is

$$P \propto e^{-2|k|x} \tag{32.3}$$

For the tunneling probability to be low, $2|k|x$ should be much greater than 1. From this we can find the approximate minimum value for the barrier width x as

$$x \gg \frac{1}{2|k|} \tag{32.4}$$

This limit depends on the effective mass m^* of the materials involved. Assuming a barrier height of 10 eV, the tunneling limit for Si is calculated to be 10 Å and for GaAs 50 Å. Note that this is a very optimistic limit, since for these values of x, the tunneling probability is $1/e$. Any real device technology would require a much lower tunneling probability to be reliable. We must make simplifying assumptions in order to determine a reasonable barrier height for the above calculation. Typical isolation methods used with silicon integrated circuits include p-n junction and dielectric isolation [5]. The latter has replaced junction isolation for both modern MOS and bipolar ICs because of the higher achievable circuit densities. For dielectric isolation structures made of SiO_2, the barrier height for an electron is roughly one-half of the SiO_2 band gap, or 4.5 eV, so the value of 10 eV used in the calculation should give a good estimate of the tunneling limit. Barrier heights for compound semiconductor technologies are always lower than the smallest band gap; in GaAs/AlGaAs the barrier is about 0.3 eV, much less than the 1.3-eV band gap of GaAs.

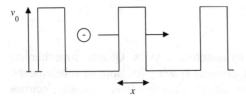

FIGURE 32-3
Quantum mechanical tunneling through a potential barrier with a height of V_0 and width x.

Quantum Limits: Uncertainty Principle

The uncertainty principle in quantum physics fundamentally restricts any physical system, whether it be electrons in the shells of an atom, molecule, or solid, or electrons in a semiconductor with an embedded semiconductor device. The relevant uncertainty principle here, for energy versus time, states that a system cannot have changed its state with certainty until a time greater than $\hbar/\Delta E$ has elapsed:

$$t \geq \frac{\hbar}{\Delta E} \tag{32.5}$$

We take as an example the thermal energy at room temperature, 26 meV, as the absolute minimum energy change from one state to another for irreversible computation. In irreversible computation a sufficient amount of entropy has to be generated to insure that the system cannot spontaneously return to its prior state. For this type of computation we must be sure that ambient noise levels are insufficient to effect such a return, hence the absolute minimum is given by $kT = 26$ meV (at room temperature). The minimum time required for the system to settle into the new state is calculated to be

$$t_{\min} = 1.6 \times 10^{-13} \text{ s} \tag{32.6}$$

Both quantum-mechanical limits are represented in the xt plane in Figure 32-8. Tunneling shows up as a vertical line and the uncertainty principle as a horizontal one. Notice that the tunneling limit presents a real challenge to device scaling in the near future, with minimum feature sizes approaching 1000 Å, but it is unlikely that the uncertainty principle will ever be a real concern. We shall see that ΔE must generally be many orders of magnitude larger than kT for reliable computation, making $t_{\min}$ many orders of magnitude smaller than the number given above. In addition, other restrictions will probably prevent systems from getting very near to the minimum delay time imposed by the uncertainty principle.

Thermal Noise

The minimum energy change associated with a transition from one state to another was taken in the above example to be the thermal energy. In reality, the switching energy must be large in comparison to the thermal energy kT so that device switching can be distinguished from random thermal fluctuations. To calculate exactly how high the switching energy should be, we use the Boltzmann distribution, which gives the probability for finding a noise pulse in a certain energy range. The relative probability that a thermal fluctuation of a magnitude equal to the switching energy E_{sw} occurs can be computed from the Boltzmann distribution, giving

$$P \propto e^{-E_{sw}/kT} \tag{32.7}$$

If we require that the probability for such a noise pulse occurring be less than

0.001, the lowest possible switching energy is 180 meV (at room temperature). This is considerably higher than the absolute minimum (26 meV) assumed above. The switching energy is equal to the power-delay product for a particular device, a useful figure of merit for any family of electronic components. Thus we have established a new minimum for this quantity as well. For large serially organized systems the probability for error with $E_{sw} = 180$ meV is still likely to be too large, in spite of error correction codes and redundant hardware, and hence one expects that the power-delay product for semiconductor device families cannot be indefinitely scaled (reduced) along with the layout dimensions. This is an extremely important fact and will be discussed further in the next section.

Restoring Logic

The switching energy minimum calculated above is based on simple physics and does not depend on the type of logic employed in the system hardware. This section discusses the type of logic that is used almost exclusively in integrated circuits today because it is highly reliable and requires no auxiliary circuitry to implement. Restoring logic gets its name from the fact that any simple gate circuit in this logic family will restore the logic level to its proper value. This means, for example, that an inverter circuit will always generate an output in one of the voltage ranges designated as "high" and "low." If this were not the case, a long chain of gates could each degrade the signal a little bit, until the end result is a signal in the undefined region between high and low. To possess this restorative property, the gate must exhibit gain. Logic families that are not restoring, for example Josephson logic families, have to use separate amplifiers to restore the logic levels to their proper values after a certain number of gates have been traversed [6].

Energy balance requirements for restoring logic set a lower bound on the switching energy, since a minimum stored energy will be dissipated during each switching operation exhibiting gain. For capacitively coupled switching technologies, such as CMOS, the total energy required to effect a change of state can easily be estimated. For a capacitively loaded CMOS inverter, it is given by

$$E_{sw} = CV_{DD}^2 \qquad (32.8)$$

where C is the total output node capacitance and V_{DD} is the total power supply swing [7]. E_{sw} is equal to the electrostatic energy needed to effect the logic transition. The voltage transfer characteristic of such an inverter is shown in Figure 32-4. A hypothetical $1/4$-μm CMOS technology with a 60-Å gate oxide and a $1/4$-by-$1/4$-μm minimum gate area is used here to estimate this switching energy. The total node capacitance is 0.0012 pF and the voltage swing 2 V. With these values the minimum switching energy for this type of restoring logic will be 4.8×10^4 eV. This is two orders of magnitude larger than thermal noise considerations alone would dictate. The requirement that logic devices exhibit gain is thus a very real constraint. We should mention, however, that this minimum is rarely approached. Even though device scaling has made it possible to reduce E_{sw}

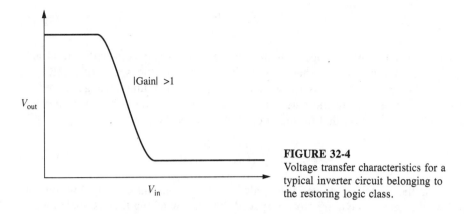

FIGURE 32-4
Voltage transfer characteristics for a
typical inverter circuit belonging to
the restoring logic class.

considerably, the desire for faster switching usually results in a much larger value of E_{sw}. A larger switching energy also increases the reliability of the computation by reducing drastically the probability for an error-inducing noise pulse, and it obviates the need for external amplifier circuitry to generate correct logic levels. While E_{sw} cannot be represented directly in the xt plane, the minimum switching energy will critically determine, in addition to thermal noise limits, heat removal limits for the packaged circuitry as a whole.

Cooling Limits

The limit imposed by cooling is harder to quantify than the other limits discussed so far, but is perhaps more important. System cooling has already become important for high-end computing systems, and will perhaps emerge as the most important limit to reducing the size of future mainframe computers, a trend important from the point of view of time-of-flight signal delays. Already, the IBM 3081 mainframe computer has had to employ an elaborate package technology called the thermal conduction module (TCM), which removes heat from the chips through metal pistons to a liquid-cooled jacket [8]. Even more exotic cooling solutions, such as direct liquid cooling of silicon chips via etched fins, have been proposed [9]. In order to estimate the cooling limits, we have to specify the packing density of the heat-producing elements in addition to the maximum thermal flux that can be passed through the system boundary. The resulting cooling limits will be expressed in the xt plane via a minimum total signal delay time.

A simple model for the total signal delay has been given by Keyes [10]. He has written it as the sum of two parts: a time-of-flight delay and a switching delay. The time-of-flight delay depends only on the separation of the two gates and the signal velocity between them, while the switching delay depends on the switching power and the switching energy:

$$t = t_{tof} + t_{switch} \tag{32.9}$$

If the average distance between logic gates is ma, where m is a positive real number and a is the geometrical gate pitch, then we find

$$t = \frac{ma}{c} + \frac{E_{sw}}{P} \tag{32.10}$$

where P denotes the average switching power for a subsystem with a total area A. This equation assumes that the signal travels with the speed of light, i.e., $n = 1$. The gate pitch a is dictated by the power per unit area generated by dissipative processes in the gates and the maximum tolerable heat flux per unit area, $Q[\text{W/cm}^2]$, provided by the cooling apparatus. With this,

$$t = \frac{m}{c}\sqrt{\frac{P}{Q}} + \frac{E_{sw}}{P} \tag{32.11}$$

As the switching power P is scaled down, a constant cooling environment allows the gates to be more closely spaced, but the switching time is made longer. If the switching power is boosted in an attempt to decrease switching times, the gates must be spaced farther apart, if the generated heat is to be removed with a constant heat flux and the temperature of the circuitry is to remain within the operating range. The optimum power P_{opt} minimizes the total delay. This is easily found to be

$$P_{opt} = \left(\frac{2cE_{sw}}{m}\right)^{2/3} Q^{1/3} \tag{32.12}$$

The optimum power thus increases as the heat removal capability is raised.

Substituting the optimum value of P into the equation for the total delay gives an expression for the minimum delay allowed by cooling and propagation effects:

$$t_{min} = \left(\frac{m}{c}\right)^{2/3}\left(\frac{E_{sw}}{Q}\right)^{1/3}(2^{1/3} + 2^{-2/3}) \tag{32.13}$$

$$t_{min} = 1.89\left(\frac{m}{c}\right)^{2/3}\left(\frac{E_{sw}}{Q}\right)^{1/3} \tag{32.14}$$

The minimum delay time satisfying the cooling requirements can now be calculated under a variety of conditions. A very optimistic limit is obtained if the switching energy is chosen to be at the level of the background noise at $T = 300$ K and a nucleate boiling scheme is employed to cool the circuits. Studies indicate that the largest surface heat flux practically attainable with this method is limited to about 20 W/cm^2 [11]. This is lower than the theoretical maximum capability, mostly due to the fact that it is difficult to ensure that boiling is initiated uniformly across the chip surface; superheating of the coolant can occur before it starts to boil and chip temperatures can be driven to dangerously high levels. At higher heat flux densities film boiling at the chip surface can result in thermal runaway. The delay calculated from Equation (32.14), assuming $E_{sw} = kT$, $Q = 20$ W/cm^2, and $m = 1.5$ (half of the connections to nearest neighbors, half to second-nearest), is 1.4×10^{-14} s. A more realistic minimum switching energy was calculated earlier in the discussion of restoring logic; it was 7.7×10^{-15} J. If this is used in the above equation as E_{sw}, the minimum delay time becomes 1.8×10^{-12} s. This result is easily scaled for different values of Q; results for several cooling schemes can be considered. For example, Tuckerman and Pease have developed a micro heat exchanger that can extract 970 W/cm^2 from a grooved planar Si surface [9]. To date (1989) this represents the highest attainable heat flux from an integrated circuit chip. Together with a switching energy of kT, an "ultimate" cooling limit can be obtained. A system may, from the thermal point of view, be designed to lie in the space above the cooling limit line, but extended operation below it is not possible. From the above model a minimum subelement separation x can be deduced, once the optimum power is calculated, but it is not correct to assume that such a minimum x would represent a fundamental lower bound for that parameter. The cooling model is based on a minimized delay time and a true minimum exists for the total delay as a function of switching power; x alone as a function of P does not have a true minimum, since P could be decreased as much as desired to boost the packing density.

Rise Time Effects

In this section we consider signal line rise time and transmission line effects. Rise time limits in the xt plane are simply represented by horizontal lines; a minimum time is assumed to be necessary for the output of a gate to reach a stable final value following a switching operation (Figure 32-5). Time-of-flight limits on long transmission lines are, as discussed earlier, represented by skewed lines. The rise time limit merges into the transmission-line limit gradually and serves to delineate a *short line* region from a *long line* region [12]. A long line is defined as a line that is longer than the distance a signal propagates during an interval equal to the signal rise time. In this case the line must be treated as a transmission line. A short line is one that is shorter than the distance a signal propagates during an interval equal to the rise time—that is, the signal reaches the end of the line before it has completed its transition from high to low or vice versa. In this case the line can be treated as a lumped element, and the delay calculated as for a simple resistor/capacitor (RC) network [13]. In the xt plane the minimum delay for each value of the rise time can be plotted as a horizontal line for simplicity,

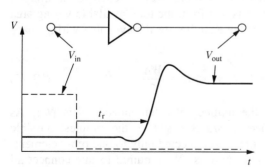

FIGURE 32-5
Voltage rise time at the output of a driver circuit with an ideal step input pulse.

though a more rigorous analysis would of course have to use the lumped element RC time constants, and the delay would thus actually depend on the length of the line. The delay limit would then merge into the transmission line regime at the appropriate value of x.

Wire Limit

The third generalized descriptive parameter we consider is the wire spacing s. No matter what the method used to interconnect different subelements in the system, there will be restrictions on how closely any two signal lines can be placed for a given length of line, because of noise effects. Additionally, the lines cannot be spaced too far apart if all of the connections are to fit into the available wire space. These two considerations restrict the wire spacing to a certain range of values in the xs plane.

First, consider the limit imposed by the fact that there is a finite amount of space available to accommodate the signal lines, called interconnections below, between subelements. For the sake of demonstration the wiring limit will be established for a planar multilayer interconnect scheme typical for board-level wiring. A strictly geometrical model with the following eight definitions will be used:

1. $N =$ number of subelements to be interconnected in a plane
2. $A =$ total area of each wiring plane containing interconnections (cm^2)
3. $k =$ total number of all wiring levels
4. $A' =$ the area of each of the N subelements (cm^2)
5. $A_v =$ cross-sectional area of each via interconnecting wiring levels (cm^2)
6. $N_v =$ the total number of vias required to accommodate all interconnections
7. $<L> =$ average length of each interconnection wire (cm)
8. $W =$ width in centimeters of each of the interconnections, which number $N/2$

The above definitions stem from two assumptions. First, that each "net" (i.e., interconnection) has two terminals only. The fact that we restrict our attention to two-point nets only leads to the $N/2$ term in the 8th definition. Second, we assume that when a net requires one or more vias to reach its destination terminal, that a certain amount of area is lost from the total available wiring area for each via used. With these symbols the equation for the available wiring space becomes

$$kA - NA' - N_vA_v = <L> \frac{WN}{2} \qquad (32.15)$$

Here we have assumed that the number of interconnections is $N/2$. As we stated above, the specific choice of $N/2$ is based on the simplest possible interconnect scheme, where each element is connected only to one other element. We need a way to estimate the number of vias, N_v, required to interconnect all

k layers. Clearly the minimum number of vias required to interconnect k wiring planes is $k - 1$. A reasonable approach to via counting is to partition the excess wire, which does not fit into a single layer, equally among all of the remaining levels. The number of vias is proportional to the number of interconnections, and the partition function gives that fraction of the total number of interconnections that will require vias:

$$N_v = \text{(number of planes)(number of interconnects)(partition function)} \quad (32.16)$$

which, with the present symbols, becomes

$$N_v = k\frac{N}{2}\frac{(<L> WN/2 - A)}{(<L> WN/2)} \quad (32.17)$$

Note that the number of vias is also assumed to be proportional to the number of wiring levels k. This indicates that a via cannot in general penetrate an arbitrarily large number of wiring planes, and that as the number of planes increases so must the number of interconnecting vias between them.

Substituting N_v into Equation (32.15) from Equation (32.17) gives an equation relating N and A in terms of the parameters $<L>$, A', k, A_v, and W. The subelement spacing can be defined to be the square root of the ratio A/N. Solving for this quantity from Equation (32.15) gives

$$x = \sqrt{\frac{A}{N}} \quad (32.18)$$

$$= \sqrt{<L> W\frac{2A' + kA_v + <L> W}{2 <L> Wk + 2kA_v}} \quad (32.19)$$

The limit for $k \to \infty$, which corresponds to an infinite stack of wiring planes, is easily seen to be

$$x(k \to \infty) = \sqrt{\frac{<L> W}{2}}\sqrt{\frac{A_v}{A_v + <L> W}} \quad (32.20)$$

For a single level of wiring ($k = 1$) we find from (32.19) that

$$x(k = 1) = \sqrt{<L> W\frac{2A' + A_v + <L> W}{2 <L> W + 2A_v}} \quad (32.21)$$

These two extreme wiring modes, $k = 1$ and $k \to \infty$, give the same result for the subelement spacing x if the size of the subelements and the width of the interconnects is small compared to the via area. Specifically, if $2A' << A_v$ and $<L> W << A_v$, then the $k = 1$ result collapses to the expression for $k \to \infty$. Under these conditions, x is independent of the subelement and via sizes. Of course, if the via area were vanishingly small, x would approach zero, and the infinite stack of wiring planes would provide an infinite amount of wiring space.

The model therefore predicts that for realistic via sizes ($\approx$ 5 mils), increasing the number of wiring levels in the package tends to compensate for the area lost to the subelements themselves, but it does not provide an infinite wiring area, since the ever-increasing number of vias require more and more space. Scaling via sizes is clearly beneficial from the wireability point of view, and also from the electrical point of view, since each via contributes a series inductance and parasitic capacitance to the signal path.

An ideal layout design minimizes the length of interconnect lines between the subelements. If half of the interconnections are to nearest and half to next-nearest neighbors, i.e., $< L > = 1.5x$, and the wires are butted against each other ($s = W$), then the maximum wireability for any number of wiring levels is the same as that obtained with a single empty plane for wiring. The wiring limit problem is thus reduced to connecting a set of N dimensionless points in a plane of area A with wires of width w. The absolute limit is reached when the entire area A is filled with wiring. Of course a real system could not tolerate signal lines in contact with one another, or even very closely spaced, because of coupled noise, but the above model is helpful in establishing an absolute maximum wireability.

Noise Limits

We now discuss noise in the xs plane. The physical noise mechanisms for coupled signal line and intrinsic line noise depend on the type of wires employed. Coupling and intrinsic line noise for electrical signals are due to impedance coupling (coupled noise) and self-inductance effects (Δ-I noise). Coupling in optical waveguides occurs via evanescent fields with a range of about one wavelength; intrinsic line noise is due to optical nonlinearities and dispersion. In the following we will only deal with electrical signal wires, since they are used almost exclusively in present-day packaging technology, and since this results in a more conservative estimate of the noise limits. Irrespective of the type of wires used, coupled noise is easily representable in the xs plane, while the intrinsic line noise is not.

For simplicity we assume two identical lines run parallel to one another for some distance l (Figure 32-6). A pulse on line 1 (active) will in the general case produce a forward- and reverse-propagating noise pulse on an adjacent line 2 (sense) because of capacitive and inductive coupling of these lines. Capacitive coupling generates two current pulses on the sense line, which propagate in both directions with equal amplitude. Inductive coupling, on the other hand, produces a single current pulse on the sense line, which only propagates into the direction opposite that of the pulse on the active line. This results from Lenz's law, which states that a current induced by the magnetic field of the active line will try to counteract the change in the magnetic field of the active line and thus requires a reverse-propagating current pulse.

The reverse-propagating pulse voltage is determined by the sum of the capacitively and the inductively coupled current components, while for the for-

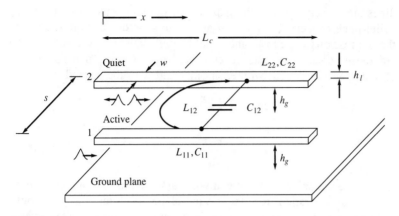

FIGURE 32-6
Coupled noise for parallel signal and sense lines above a common ground plane. Voltage pulses and coupling impedances are described in the text.

ward pulse voltage on the sense line the capacitively and inductively coupled pulses have to be subtracted from each other. As a consequence the reverse-propagating pulse always carries the same sign as the primary pulse on the active line, whereas the forward pulse on the sense line can have the same or the opposite sign, or even add to zero. The last case occurs, if the active and the sense lines are identical and embedded in an infinite homogeneous isotropic dielectric medium.

The mathematical equations describing the coupling can be derived from Maxwell's equations. Two distinct quantitative results emerge from such a treatment of coupled noise, depending on the signal rise time on the active line and the length of the line section where the coupling occurs. Here, as before, a line is considered long if the signal rise time is much shorter than the round trip propagation delay. For a pair of long wires the coupled noise amplitude on the sense line does not depend on the signal risetime:

$$\frac{V_{coupled}}{V_{signal}} = \frac{\gamma}{2} \tag{32.22}$$

where γ denotes the ratio of the self-capacitance (to ground), C_{self}, and the mutual capacitance (between lines), C_{mut}. This result is valid only for identical active and sense lines. If the line is short enough so that reflections from the opposite end arrive before the signal has reached 90 percent of its maximum amplitude, then it can be modeled as a simple resistor-capacitor (RC) network with R and C given by the distributed impedance of the wire. Here both the propagation delay and the signal rise time determine the coupled noise voltage:

$$\frac{V_{coupled}}{V_{signal}} = \gamma \frac{T_{delay}}{T_{rise}} \tag{32.23}$$

If the lines are closely spaced, the gap between them forms, to first order, a simple parallel-plate capacitor, neglecting all fringing fields. Thus in the limit as $s \rightarrow 0$, the expressions for capacitance should give satisfactory results. The magnitudes of these capacitances can be estimated by assuming that the wires form parallel-plate capacitors both with each other and with the ground plane:

$$C_{\text{mut}} = \frac{\epsilon h_l x}{s} \tag{32.24}$$

$$C_{\text{self}} = \frac{\epsilon w x}{h_g} \tag{32.25}$$

Here h_l and h_g denote the height of the metal wire, assumed to have a rectangular cross section, and the separation from the ground plane. With these assumptions γ is $h_l h_g / ws$. Substituting this expression into Equation (32.22) and (32.23) gives, for the long line,

$$\frac{V_{\text{coupled}}}{V_{\text{signal}}} = \frac{h_l h_g}{ws} \tag{32.26}$$

and for the short line,

$$\frac{V_{\text{coupled}}}{V_{\text{signal}}} = \frac{h_l h_g (x/c)}{ws T_{\text{rise}}} \tag{32.27}$$

In Equation (32.27) T_{delay} has been replaced with x/c, where x is the subelement separation and c is the maximum signal propagation speed in the wire (assumed to be the speed of light). If we arbitrarily require that the ratio $V_{\text{coupled}}/V_{\text{signal}}$ be restricted to a value smaller than 0.1, then a family of noise limit curves can be generated in the xs plane for different signal rise times.

A phenomenon known as Δ-I noise occurs in metal wires when a large number N of driver circuits simultaneously switch and pull current through the power supply lines (Figure 32-7). The large sudden change in current di/dt on these N lines produces a noise voltage pulse across the supply line inductance L_{eff}, given by Faraday's law:

$$V_{\Delta-I} = N L_{\text{eff}} \frac{di}{dt} \tag{32.28}$$

The average value of di/dt can be estimated, if we assume that the current switched is just V_d/Z, where V_d is the nominal driver voltage swing and Z is the impedance seen by the power regulator looking into the circuit load. Large capacitors are often distributed liberally throughout the power supply network, usually for each board or chip module, to provide a local store of energy on which the driver circuits can rely. This decreases the length of the power supply path and hence decreases the effective inductance through which current is pulled during a switching operation. Nevertheless, some minimum effective inductance will always exist. If the current is switched in a time t_{switch}, then

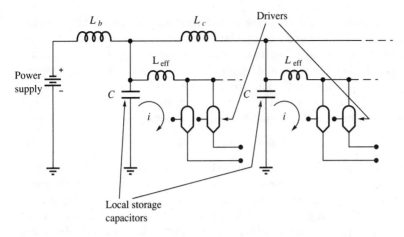

FIGURE 32-7
Equivalent circuit for Δ-I noise arising from power supply line inductances and driver circuits.

$$V_{\Delta-I} = NL_{\text{eff}} \frac{V_d}{Zt_{\text{switch}}} \qquad (32.29)$$

and

$$\frac{V_{\Delta-I}}{V_d} = \frac{NL_{\text{eff}}}{Zt_{\text{switch}}} \qquad (32.30)$$

If we arbitrarily restrict the noise amplitude to be less than 10 percent of the nominal value, then

$$\frac{NL_{\text{eff}}}{Zt_{\text{switch}}} < 0.1 \qquad (32.31)$$

and

$$N < 0.1 \frac{Zt_{\text{switch}}}{L_{\text{eff}}} \qquad (32.32)$$

All that now remains is to estimate the value of L_{eff}. For simplicity we model the power supply network as a set of b parallel strip lines of length l, width w, and dielectric thickness d. The well-known expression for the self-inductance of such a set of lines is

$$L_{\text{eff}} = \frac{\mu_0 d l}{bw} \qquad (32.33)$$

We now make a very optimistic estimate of this quantity. First take d at the tunneling limit of 20 Å, with $w = 10$ mils and $l = 0.5$ cm, smaller than the size of a single chip. L_{eff} is then equal to about $5/b$ pH. The particular parameter values have been selected here to have an "absolute" minimum value for the power line inductance. This result scales linearly with the insulator thickness. Since higher circuit densities will force w and l to smaller values, they will scale together and their ratio will have less of an effect on the wire inductance than will the dielectric thickness d. The thickness d is limited by problems with insulator breakdown and possibly tunneling (Fowler-Nordheim type), so our estimate is quite optimistic. Substituting $Z = 50 \Omega$ as a typical total load impedance and $t_{switch} = 1$ ns sets an upper bound on the number of simultaneously switching drivers as

$$N < 1000b \ (t_{switch} = 1\text{ns}) \tag{32.34}$$

Note that the result scales inversely with the switching time. Thus, faster circuitry will exacerbate the problem. We will leave it to a more careful study to reveal how a limit on the number of simultaneously switching off-chip drivers will limit the overall system speed. While this restriction on N is difficult to represent in the xs plane, it nevertheless is an important result which illustrates well the problems with noise in a switching environment.

The limits in the xs plane have now been completed. All systems must comply with the noise limits and the limits imposed by wiring space requirements. The limits in the xs plane, together with those in the xt plane, form a set of three-dimensional constraints in the xst space, which is the subject of the next section.

ELECTRONIC PACKAGING SPACE

The three generalized descriptive parameters span the three-dimensional electronic packaging space, into which each of the above fundamental limits introduces one boundary plane or surface. Figures 32-8 and 32-9 show the composite set of limits in the xt and xs planes. Since the allowed domain for each limit is one half space, the full set of boundary planes determines a three-dimensional domain, which has to contain all allowed electronic systems. The perspective view of this three-dimensional domain given in Figure 32-10 shows that it has the appearance of an irregular pyramid-like object with its tip pointing toward the quadrant with small element separation, small propagation delay, and small wire separation. This tip is cut off by tunneling and cooling limits; toward the opposite direction the domain is open.

In order to understand the implications of the fundamental limits shown in Figures 32-8 through 32-10, we have plotted some typical systems into the electronic packaging space. For IC technologies, such as CMOS, NMOS, and TTL, simple circuit blocks are characterized by their average linear layout dimension, input-to-output-node propagation delay, and minimum linespace. Larger systems are described by their largest dimension, cycle time, and signal interconnect linespace. An example of a biological system, which is subject to the same fundamental limits, is the human brain, and it has been included at two levels. The neuron, and specifically the synaptic cleft between dendrites of two communicating neurons, is taken as the logic element, although it employs threshold logic instead of binary, and it may act in an inhibitory, as well as an excitatory, mode.

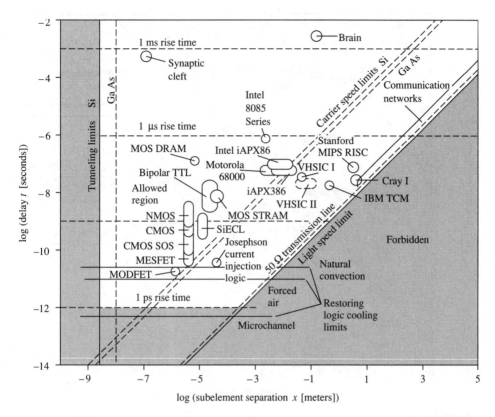

FIGURE 32-8

The *xt* plane of electronic packaging space. Fundamental limits are shown with solid and dashed lines. Hatched areas are beyond ultimate fundamental limits. Sample electronic systems represent both current level of manufacturing as well as leading edge research.

The brain itself can be thought of as a package consisting of about 15 billion neurons, although usually the number of neurons is considered to be less important than the total number of dendritic connections for human intelligence. The latter is much larger than the number of neurons alone, and yet if the neurons were considered by themselves and assumed to be capable of only two excitatory states, the memory capacity of the brain would still be 2^{15} billion bits. Already, many researchers are building *neural network* computers, with architecture mimicking that of the brain, but on a far more simplistic level. Based on conduction rates in the three basic types of neurons, a rough system speed can be estimated from the size of the brain itself. Wire spacing is taken as the average neuron spacing, computed from the volume of the brain and the number of neurons it contains. With these estimates, the biological electronics found in animals, and specifically man, have been plotted together with the other systems mentioned in this and the following sections.

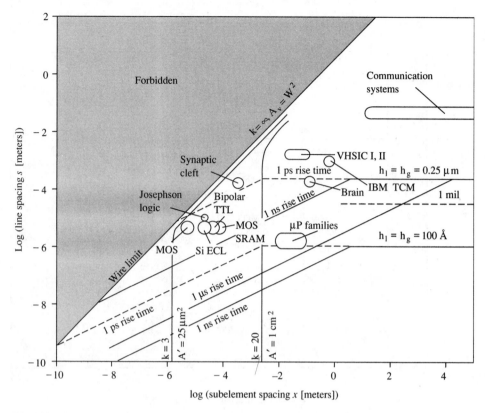

FIGURE 32-9
The *xs* plane of electronic packaging space. For additional information see caption of Figure 32-8.

It is the logarithmic nature of Figures 32-8 through 32-10 which allows such a wide range of systems to be included. As a matter of convention, the *xt* plane is subdivided into the three customary electronic packaging levels. The first level covers systems smaller than about 1 cm, and includes simple gates, memory cells, and dendritic synapses. Technologies represented in Figure 32-8 include MOS, bipolar, and MODFET devices, as well as Josephson interfero-meters and electrochemical neurotransmitters. The range from 1 cm to about 0.5 m is called the second level of packaging, and anything larger belongs to the third level (Figure 32-11). Level 2 can be thought of as a transforma-tion level, since it is the level of the package where the chip pins are routed to the board through some intermediary such as a multilayer ceramic pack-age, or bypass the board altogether in systems such as the IBM 3081, where chip interconnection is nearly all accomplished at the module level. Third-level packaging is the final level in the system, consisting of module, board, and unit interconnections. This may consist of cabling between major subassem-blies and peripheral devices in addition to terminals and power supply and cool-ing equipment. Examples of third-level packages in Figure 32-8 are the Cray-I, a supercomputer, and the global telecommunication network. It is evident

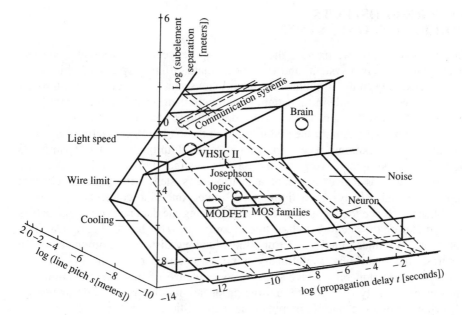

FIGURE 32-10

Three-dimensional view of electronic packaging space (x, s, t). Full lines signify the intersections of either two fundamental limits, or one fundamental limit and a coordinate plane. Dashed lines are either hidden intersections of two fundamental limits or projections of fundamental limits beyond coordinate planes. Only a few sample electronic systems have been included for clarity.

from Figure 32-8 that systems at levels 2 and 3 generally operate an order of magnitude or more slower than the level 1 components from which they were constructed. As long as the entire system cannot be implemented with lowest-level elements, one should expect that bottlenecks will exist at any package level boundary.

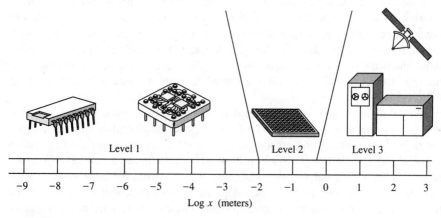

FIGURE 32-11

Hierarchy of packaging levels with representative size scale.

FUTURE PROSPECTS
FOR ELECTRONIC SYSTEMS

In this final section we explore the relationship between the different electronic packaging levels in a system, and attempt to relate different systems architectures to the packaging limits formulated in the previous sections. Figures 32-8 through 32-10 serve as our guide in this attempt, and help us to visualize the complex set of physical limits that govern the evolution of future electronic systems.

We can easily identify the major trends in the evolution of electronic systems for all three packaging levels from Figures 32-8 through 32-10. Distributed and localized systems have gone in their own distinct directions. The telephone network, the classical example of a distributed system with large subelement separations, has always been light-speed limited once the connection has been established. Hence the focus has been on increasing the data rates on signal cables and structuring the network architecture accordingly. This has lead to optical fiber cables and a reduced number of trunk lines for long distance communication. In contrast, the overriding evolutionary trend in localized electronic systems has been size reduction on all levels. As a consequence of the size reduction, the subelement density is increased and the propagation delay reduced dramatically, allowing for ever-enhanced computational capabilities.

We first consider level 1, the chip level, of the package. Both the minimum and maximum characteristic dimensions on this level are currently limited by technological constraints. Minimum feature sizes of state-of-the-art semiconductor devices at VLSI levels of integration (10^4 to 10^5 devices/cm^2) are now in the 1 μm range. For the moment, hard limits such as tunneling are unimportant in comparison to limits imposed by the minimum pixel size in pattern-generating equipment, registration between mask levels in aligners, and minimum spacing tolerances required by steps in the fabrication processes (such as thin-film deposition and etching). The upper limit to size for level 1 components is set by the field size of the mask aligners and the defect density at the end of the fabrication process. The larger the substrate, the higher the probability for critical defects degrading or destroying circuit performance—the yield for larger substrates decreases dramatically. These two facts have restricted chip sizes to about 1 cm per side, and prevented systems designers from integrating entire systems on a single chip. Certainly VLSI has made it possible to implement more system functions with fewer chips, but since the system size has simultaneously increased, the dream of fabricating a complete system on-chip is still far from realization.

Wafer-scale integration, a concept in which chiplike subelements are integrated on one wafer with a diameter of 3–6 in. and the overall yield is enhanced using redundant circuitry and error-correcting codes, has been proposed as a solution [14]. Wafer-scale integration has, however, encountered a number of serious problems, which have not yet been resolved. For example, for large systems such as mainframe computers, wafer-scale integration seems impractical because of the high servicing cost—a failure in an individual circuit would require the replacement of large, expensive components, perhaps even the entire system. From this point of view, it may be better to have the system broken into a number of smaller field-replaceable units.

Level 1, then, is currently restricted at both ends by technological limits. Nevertheless, it spans a range of four orders of magnitude, from the semiconductor device to the chip itself. From Figure 32-9, it is clear that level 1 is nearly at the wire limit. This does not mean that evolution must stop, however, but that wire pitch must be scaled down to accommodate the increasing number of interconnection pads forced to reside on chip. A tradeoff should be evident here—scaling the wire pitch brings the level 1 systems closer to the fundamental coupled noise limits (Figure 32-9). Eventually, the two limits will act to pinch off all progress, and the level of integration on chip will saturate. In addition, cooling is a major concern, and new cooling technologies will have to be implemented in order to get close to the fundamental cooling limit. Whether ultimate cooling-limited performance is ever attained at this level will depend on whether the benefit-to-cost ratio for such cooling techniques is sufficient to justify their use.

Next, we consider the properties of level 3 components. Whereas level 1 spanned a range of four order of magnitude, level 3 covers element sizes from about 1 m up to the largest communication system in operation. Level 3 systems tend to be time-of-flight limited, due to the large size of the structures. This limit restricts the functional capability of these systems to certain computational classes. For general-purpose, real-time systems, the cycle time minimum is fixed by the largest system dimension. Special-purpose architectures may be employed, however, that are not subject to this restriction. They may be highly pipelined, dedicated systems, or they may be highly parallel, multiprocessor systems, designed to partition the computational work among a set of parallel pipelines. In the first example, the data path is very narrow, and the cycle time depends on the spacing of the stages in the pipeline. For the second example, the data path is wide, but the processors must have access to memory, and so the cycle time is still restricted by the system size. Because the processors are working in parallel, computational speed is higher than that of a single-processor, general-purpose system, providing the extra processors can be used efficiently and do not require too much additional cycle time for coordinating and partitioning the computation.

If we focus on the lower end of the level 3 range, we realize that the smallest dimension is dictated by the size of one human being. A telephone receiver is designed to match the size of the head, printers are built to provide hardcopy which fits the size of the hand, and computer keyboards accommodate one finger per key. The system as a whole must be accessible for repair and maintenance, unless the entire assembly is designed as a field-replaceable unit. Even if the system is integrated and miniaturized to the point that it is unrepairable, the user must still be provided with an interface with which to communicate with the device; or, if the system requires no input or control, for it to communicate with the outside world. Thus, though the current trend for mainframe computer systems is to reduce the overall size and increase speed as much as possible, ultimately it is the size of the human interface that determines the lower bound for level 3 systems.

As long as level 1 and level 3 do not overlap, there must be an intermediate level to bridge the gap. Level 2 provides the space transformation between levels 1 and 3. In comparison to the levels flanking it, level 2 covers a relatively small range of dimensions—currently less than an order of magnitude. It is neither time-of-flight limited, as is level 3, nor completely wire limited, as are level 1 systems. Problems at the lower level can, however, be transmitted to level 2, where they must be solved. For example, packaging a number of chips that must each be cooled with liquid immersion requires that the package be built with the cooling apparatus installed. Another example is wiring density. It has turned out that higher levels of integration on chip have not alleviated the problem of providing chip interconnections; rather, systems designers have made use of the higher chip functionality to increase the number of chip input/output (I/O) channels— especially by widening data and address busses for microprocessor systems. This tendency was first noticed by Rent at IBM and later analyzed by Landman and Russo, also of IBM [15]. The result of this has been an increase in wire density at level 2. This has generated concerns about noise, wiring space, and cooling on this level.

The current trend for level 2 structures is to integrate as much of the systems function on as few chips as possible—that is, to move the upper bound of level 1 as high as possible. If it is impossible currently to implement the entire system on a single chip, VLSI has at least made it possible to reduce the number of chips dramatically. Of course the same number of interconnections among the subsystems must be made in a smaller space. New technologies must be developed to advance the state-of-the-art in module-level integration, as they have been for chips.

At the other end of the level 2 spectrum stands the human operator. It is the job of the level 2 package to provide a signal path to and from this operator, but it is not necessary that signals not needed by the operator be routed from level 1 up to level 3 and back down again. This type of architecture, while unavoidable in some cases, creates a bottleneck through level 2. Getting signals off the chip adds unnecessary delays and adds to the wiring density. Clearly, at least one data/control path will always be necessary from level 1 to level 3. Such a path should be as narrow as possible, providing only for essential control and I/O functions. The rest of level 2 wiring must be used to interconnect the level 1 subsystems, and efforts should be directed toward reducing this wiring as much as possible.

We now examine the possibilities for improvements on each of the three levels, as well as for the system as a whole. We begin with level 1. Current semiconductor device technologies may be scaled by about two to three more orders of magnitude before fundamental limits are reached. It should be emphasized that practical limits have not been considered here. It is unlikely that more than one order of magnitude of progress can be attained without considerable effort to overcome technological constraints. The first fundamental limits likely to be encountered are wiring space, cooling, and noise. Along the way, new device concepts may also have to be developed, since there is no guarantee that current devices, such as MOS and bipolar, will continue to function properly—classical device and interconnect scaling may break down.

In addition to new classes of devices, containing new materials and possibly new operating principles, several general concepts are currently under consideration for improving the performance of microelectronics. Both pseudo- and true three-dimensional circuitry could be developed to greatly enhance system density and reduce interconnect lengths [16]. True three-dimensional circuitry could be interconnected in space in much the same way that our 15 billion neurons are tightly packed into our brains. Another possibility for improving performance is to cool semiconductor devices to very low temperatures, thereby enhancing their speed. The discovery of high-T_c (high critical temperature) superconductors could possibly be used in this approach. High-T_c superconductors would allow denser interconnects, since dc line resistance would no longer be a problem (especially in power distribution lines). This is true for second-level structures as well as ICs (level 1). Another area of active research, mentioned earlier, is wafer-scale integration. WSI is also a possible avenue of exploration for reducing off-chip delays and improving overall system density and speed. It remains to be seen, however, whether WSI can overcome some of the serious challenges it faces to remain a viable technology alternative. A better way to reduce off-chip delays may be to use optical fibers to interconnect the chips. Perhaps even free-space optical communication could be used to break the I/O bottleneck, allowing for massively parallel I/O communication via high-bandwidth holographically-directed laser light [17].

Finally, new computer architectures may emerge that can more efficiently harness the computational power already technologically available. These new architectures would change the look of every level in the system. For example, neural network architectures, mentioned before, or *bio-systems*, may someday revolutionize the way in which computers are built. *Bio-chips* will consist of devices modeled on or actually constructed from organic systems, coupled together in much the same way as the central nervous system of higher mammals. It is probably safe to say that in order to approach the density and functionality of the human brain, devices that dissipate very little power, a property of neurons and other biological components found in the nervous system, must be developed.

At level 3, evolution can only consist of scaling down the system dimensions, or changing the system geometry to bring all components within the light-sphere of a single clock cycle. The most famous current-day example of such a geometrical alteration is the Cray line of supercomputers. Their cylindrical shape and carefully adjusted lengths of coaxial cable packed inside give away the fact that high-speed systems of this size already must be designed to accommodate time-of-flight limits. To further reduce time-of-flight delays a lower-dielectric-constant material, or free-space communication must be employed. Regardless of how the circuits communicate, properties of systems in this size range are largely determined by geometry.

Since evolution of levels 1 and 3 tends to pinch level 2 in between, the task assigned for these intermediate structures is to follow the trends on the other levels. Level 2 subsystems will therefore move down and to the left, carried by the flanking levels 1 and 3. There is still a long way to go before the potential of this level is exhausted, however. There is a huge gap in the active fraction, that is, the percentage of planar area occupied by active circuitry, between VLSI chips and the densest board configurations. Hybrid, multichip modules, silicon-on-silicon packaging, and WSI should all help to fill that gap. A modular approach to level 2 packaging allows very dense chip configurations. Multichip modules and multimodule packages offer a chance to construct systems with nearly the density of WSI, but with individually tested parts of much higher yield. The challenge is to economically cool such dense modules.

From the above discussion, it is clear that a more unified approach must be taken in the design of the overall systems package. Currently, a systems designer is given a set of requirements, which he uses to produce a paper design for the system. This design is handed off to the electrical designer, and from him to the thermal designer—by this time it is often found that the system cannot be built without major design revisions. An integrated approach recognizes that many critical packaging problems are not independent, and that different systems levels cannot be effectively designed independently from one another.

The most obvious new way of reintegrating the system design is to build it entirely on level 1, except for the human interface to a minimum-size level 3. A simple calculation, using the tunneling limit as a minimum spacing, shows that it should be possible, in principle, to integrate more than the largest current mainframe systems on a 1 cm chip. Unfortunately, there is a long way to go before this ideal is reached, if practical considerations make it possible at all.

An alternate solution consists of modifying the systems architecture: implement the systems function with a number of dedicated subsystems, such that I/O between subsystems, up-down communication between systems levels, and the interface to level 2 is minimized, and lateral communication between subsystems is maximized. Level-specific architectures, for example general-purpose Neumann type at level 1, highly pipelined at level 2, and highly parallel at level 3, could be implemented to take advantage of each level's individual strengths and to remove bottlenecks at each level.

The current trends in electronic systems will require that the design be integrated from top to bottom. Returns from continued scaling of device families will diminish due either to inherent restrictions at that level, or to the fact that we can no longer access efficiently the power integrated into a single chip. The latter case must be avoided, and can be avoided, by concentrating efforts on new packaging concepts, and innovations in the design and technologies of electronic packages.

REFERENCES

1. W. Shockley, "The Path to the Conception of the Junction Transistor," *IEEE Trans. Electron Devices* ED-23, p. 597 (1976).

2. J. S. Kilby, "Invention of the Integrated Circuit," *IEEE Trans. Electron Devices* ED-23, p. 648 (1976).
3. J. Mikkelson, L. Hall, A. Malhotra, S. Seccombe, and M. Wilson, "An NMOS VLSI Process for Fabrication of a 32b CPU Chip," *IEEE International Solid-State Circuits Conference*, p. 106 (February, 1981).
4. S. Suzuki, M. Nakao, T. Takeshima, and M. Yoshida. "256K/1Mb DRAM," *IEEE International Solid-State Circuits Conference*, February, 1981.
5. J. A. Appels, E. Kooi, M. M. Paffen, J. J. H. Schlorje, and W. H. C. G. Verkuylen, "Local Oxidation of Silicon and Its Applications in Semiconductor Technology," *Phillips Research Report* 25, p. 118 (1970).
6. H. C. Jones and D. J. Herrel, "The Characteristics of Chip-to-Chip Signal Propagation in a Package Suitable for Superconducting Circuits," *IBM Journal Res. Dev.* 24, no. 2, pp. 172–177 (March, 1980).
7. C. Mead and L. Conway, *Introduction to VLSI Systems*, Reading, Mass: Addison-Wesley, 1980.
8. A. J. Blodgett, "Thermal Conduction Module: A High-Performance Multilayer Ceramic Package," *IBM Journal Res. Dev.* 26, no. 1, pp. 30–36 (1981).
9. D. B. Tuckerman and R. F. W. Pease, "High-Performance Heat Sinking for VLSI," *IEEE Electron Device Letters* ED-2, no. 5, pp. 126–129 (1981).
10. R. W. Keyes, "The Evolution of Digital Electronics Towards VLSI," *IEEE Journal of Solid-State Circuits*, April 1979, pp. 193–201.
11. R. C. Chu, U.P. Hwang, and R. E. Simons, "Conduction Cooling for an LSI Package: A One-Dimensional Approach," *IBM Journal Res. Dev.* 26, no. 1, pp. 45–54 (1982).
12. A. Feller, H. R. Kaupp, and J. J. Digiacomo, "Crosstalk and Reflections in High-Speed Digital Systems," *Proc. Fall Joint Computer Conference* 1965, pp. 511–519.
13. D. B. Jarvis, "The Effects of Interconnections on High-Speed Logic Circuits," *IEEE Trans. Elec. Computers*, EC-12, pp. 476–487 (Oct. 1963).
14. J. F. Macdonald, E. H. Rogers, K. Rose, and A. J. Steckl, "The Trials of Wafer-Scale Integration," *IEEE Spectrum*, 21(10), pp. 32–39 (Oct. 1984).
15. B. S. Landman and R. L. Russo, "On a Pin versus Block Relationship for Partitions of Logic Graphs," *IEEE Transactions on Computers* C-20, no. 12, pp. 1469–1478 (December, 1971).
16. Y. Akasaka, "Three-Dimensional IC Trends," *IEEE Proceedings*, 74, no. 12, pp. 1703–1714 (December, 1986).
17. W. H. Wu, L. A. Bergman, A. R. Johnston, C. C. Guest, S. C. Esener, M. R. Feldman, and S. H. Lee, "Implementation of Optical Interconnections for VLSI," *IEEE Trans. Electron Devices* ED-34, no. 3, pp. 706–714 (March, 1987).

PROBLEMS

32.1. Calculate the maximum linear dimension of a general-purpose CPU running synchronously from a single clock at
   (a) 10 kHz
   (b) 100 kHz
   (c) 1 MHz
   (d) 100 MHz

32.2. Calculate the minimum subelement spacing for devices fabricated with the following materials used as device insulators. Assume the barrier height is equal to the bandgap and that $P < 1/e$.
   (a) Ge
   (b) Si
   (c) GaAs
   (d) SiO_2

32.3. Estimate the minimum allowable switching energy that a logic device must dissipate during a switching operation under the following conditions:

(a) $T = 300$ K, $P_{noise} \leq 1$
(b) $T = 300$ K, $P_{noise} \leq 0.01$
(c) $T = 300$ K, $P_{noise} \leq 0.0001$
(d) $T = 70$ K, $P_{noise} \leq 0.00001$

P_{noise} is the probability that a noise pulse occurs that has energy equal to the signal energy.

32.4. Which of the following transfer characteristics is restoring? Why?

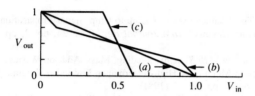

For each of the transfer characteristics (a), (b), and (c), plot the output voltage as a function of N, the number of successive stages with such a transfer characteristic that are traversed by the signal. Assume the input to the first stage is a pulse of amplitude 0.51 V. Plot N along the horizontal axis from 0 to 10.

32.5. Calculate the minimum total delay time predicted by the cooling model under the following conditions:
(a) $m = 1.5$, $E_{sw} = 26$ meV, $Q = 0.01$ W/cm^2
   (thermal noise limited, air cooled)
(b) $m = 1.5$, $E_{sw} = 48$ keV, $Q = 0.01$ W/cm^2
   (restoring logic, air cooled)
(c) $m = 1.5$, $E_{sw} = 48$ keV, $Q = 20$ W/cm^2
   (restoring logic, liquid cooled)
(d) $m = 1.5$, $E_{sw} = 48$ keV, $Q = 1000$ W/cm^2
   (restoring logic, pressurized-liquid cooled with micro–heat exchanger)

*32.6. Generalize the cooling model to three dimensions as follows:
(a) Start with the heat equation:

$$\frac{\partial T}{\partial t} = k\nabla^2 T + P(x, y, z)$$

Assume the machine is a sphere packed with circuits with a density ρ, and that the average thermal conductivity of the machine is k. If the sphere has a radius a, and the power generated by the circuitry is given by $P(r, \theta, \phi, t)$, show that for a spherically symmetric machine,

$$\frac{d^2 T}{dr^2} = \frac{2}{r}\frac{dT}{dr} = -\frac{P}{k}$$

is the correct form of the steady state heat equation, where $T(r)$ is the temperature of the circuitry within the machine.
(b) Find $T(r)$ by first solving the homogeneous equation, and then by finding a solution to the inhomogeneous equation. Add the two solutions and apply the following boundary condition: $T = 0$ at $r = a$. Show that the solution has the form

$$T(r) = \frac{P}{6k}(a^2 - r^2)$$

(c) Show that ΔT, the temperature difference between the center and outer boundary of the sphere, is

$$\frac{P}{6k}a^2$$

(d) Let G be the total number of circuits contained in the sphere. Show that

$$G = \frac{4\pi}{3}R^3\rho = \frac{4\pi}{3}\frac{R^3}{a^3}$$

Solve the above expression for a and substitute for a in the equation found in part (c). Show that

$$\Delta T(center) = \frac{GPr^3}{8\pi ka} = \frac{GP}{8\pi a\lambda}$$

where $\lambda = k/r^3$.

(e) If the total delay is modeled as before, as $t_{delay} = t_{prop} + t_{sw} = ma/c + U/P$, show that

$$t_{delay} = \frac{\beta mG^{2/3}P}{\theta\lambda c} + \frac{U}{P}$$

where $\beta = (2^7 3\pi)^{-1/3} = 0.094$, and

$$\theta = \frac{GP}{8\pi a\lambda}$$

(f) Minimize the total delay with respect to P and show that the minimum delay has the form

$$t_{delay}(min) = 2\left(\frac{\beta mU}{\theta\lambda c}\right)^{1/2}G^{1/3}$$

(g) Find the minimum delay time predicted by this model if $m = 4$, $\theta = 40$ degrees, $\lambda = 0.03$ W/cm · degree, $U = 4 \times 10^{11}$ J, $G = 10^5$, and $c = 5 \times 10^9$ cm/s.

32.7. Calculate the effective electrical length of the following transmission line circuits:
(a) 1.0-m lossless line, driver rise time = 1.0 ms
(b) 1.0-m lossless line, 1.0 μs rise time
(c) As above, 1.0 ns rise time
(d) 1.0-mm lossless line, 1.0 ps rise time

Assume in each case that the signal propagation velocity is equal to the speed of light.

32.8. Derive the partition function used in the wire limit model. Make the following assumptions:
(a) The number of interconnects required in the system is $N/2$.
(b) The fraction of interconnections that require vias is given by the normalized excess wire area.
(c) The number of vias is proportional to the number of signal layers in the system.

32.9. Starting with Equations (32.22) and (32.23), and the definition of γ, derive the two cases (32.26) and (32.27), assuming the fringing fields between the lines can be neglected. Assume $w = s$ for maximum wiring density. Plot the expressions for s

(a) As a function of $V_{coupled}/V_{signal}$ from 0 to 1, $h_1 = h_g = 0.25 \ \mu m$, $x = 1.0$ m, $T_{rise} = 1$ ns.

(b) As a function of line length x from 0 to 10 m, $V_{coupled}/V_{signal} = 0.01$, otherwise as above.

(c) As a function of line dimension $h = h_1 = h_g$, otherwise as above.

Calculate the minimum value of s for the parameter values given in (a), with $V_{coupled}/V_{signal} = 0.01$.

32.10. Local capacitors are often used on PC boards to supply energy for switching drivers. Estimate the maximum time that a 1 μF capacitor could supply 1 mA of current to a switching driver that changes output suddenly from 5 V to 1 V.

32.11. From Figure 32-8, predict the maximum overall speed of a hypothetical system with the following specifications:

(a) State-of-the-art liquid-cooled (nucleate boiling) VHSIC-type package.

(b) 0.1-μm minimum feature size CMOS technology.

(c) 1-meter system size; system is run synchronously from a single clock.

Which is the most important of the above specifications for determining the overall speed of this system?

INDEX

INDEX